Electronic Devices and Circuits

Fourth Edition

David A. Bell

Published by: David A. Bell
                P.O. Box 22003, Twin Lakes Postal Outlet
                Sarnia Ontario Canada N7S 6J4

Canadian Cataloguing in Publication Data

Bell, David A., 1930-
Electronic devices and circuits

4th ed.
Includes index
ISBN 0-9683705-4-3

   1. Semiconductors. 2. Electronic circuits. 3. Electronic
apparatus and appliances. I. Title.

TK7871.B3785 1999   621.3815   C99-900795-5

Earlier editions published by Reston
Publishing Co., Inc. — 1975, 1980, 1986.

10 9 8 7 6 5 4 3 2

ISBN 0-9683705-4-3

Contents

Chapter 11   AC Analysis of FET Circuits   335

Chapter 12   Small Signal Amplifiers   365

Chapter 13   Amplifiers with Negative Feedback   421

Chapter 14 IC Operational Amplifiers and Basic Op-amp Circuits 465

Chapter 15 Operational Amplifier Frequency Response and Compensation 505

Chapter 16 Signal Generators 529

Chapter 17 Linear and Switching Voltage Regulators 563

Chapter 18 Audio Power Amplifiers 605

Chapter 19 Thyristors 669

Chapter 20   Optoelectronic Devices 713

Chapter 21   Miscellaneous Devices 745

Appendices 761

Index 791

X

Preface

This book is intended for use as an electronics technology course text in colleges and universities, and as a reference text for practising professionals.

The objectives of the book are to provide clear explanations of the operation of all important electronics devices generally available today, and to show how each device is used in appropriate circuits. I am convinced that an understanding of devices and circuits is most easily achieved by learning how to design circuits. *Practical circuit design is usually quite simple; much simpler than some methods of circuit analysis.*

After discussing device operation, characteristics, and parameters, typical circuits using the device are explained. Then, circuit design and analysis are treated. Many practical examples are included in the text, using parameters from device manufacturers' data sheets. The circuit design procedure most often involves determining appropriate current and voltage levels, and then applying Ohm's law and the capacitor impedance equation. Most equations are derived, so that the student knows exactly what is going on. Instead of rigorous analysis methods, practical approximations are employed wherever possible.

Conventional current direction is used because it is the direction normally employed by device and integrated circuit manufacturers; also, because every device graphic symbol uses an arrowhead that indicates conventional current direction.

I am always grateful for suggestions that might improve my presentation of the material, or for additional topics that should be treated. Comments concerning this book would be very welcome.

David Bell.

Chapter **1**

Basic Semiconductor and pn-Junction Theory

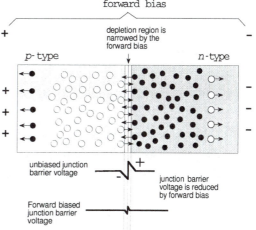

Chapter Contents

Objectives
You will be able to:

1 Describe an atom, and sketch a diagram to show its components.

2 Explain energy levels and energy bands relating to electrons, and define valence band and conduction band.

3 Describe how electric current flow occurs by electron motion and by hole transfer.

4 Explain conventional current direction and direction of electron flow.

5 Sketch a diagram to show the relationship between covalently bonded atoms.

6 Discuss the differences between conductors, insulators, and semiconductors.

7 Explain how n-type and p-type semiconductor materials are created, and discuss the differences between the two types.

8 Sketch a pn-junction, and explain the origin of the junction depletion region.

9 Draw diagrams to show the effects of forward biasing and reverse biasing a pn-junction.

10 Sketch the current/voltage characteristics for forward biased and reverse biased pn-junctions.

11 Discuss temperature effects on conductors, insulators, semiconductors, and pn-junctions.

Introduction

An electronic device controls the movement of electrons. The study of electronic devices requires a basic understanding of the relationship between electrons and the other components of an atom. The movement of electrons within a solid, and the bonding forces between atoms can then be investigated. This leads to a knowledge of the differences between conductors, insulators, and semiconductors, and to an understanding of *p*-type and *n*-type semiconductor material.

Junctions of *p*-type and *n*-type material (*pn*-junctions) are basic to all but a very few semiconductor devices. Forces act upon electrons that are adjacent to a *pn*-junction, and these forces are altered by the presence of an external bias voltage.

1-1   Atomic Theory

The Atom

The atom can be thought of as consisting of a central *nucleus* surrounded by orbiting *electrons* (see Fig. 1-1). Thus, it may be compared to a planet with orbiting satellites. Just as satellites are held in orbit by the attractive force of gravity due to the mass of the planet, so each electron is held in orbit by an *electrostatic force of attraction* between it and the nucleus.

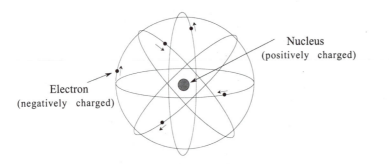

Figure 1-1
The atom consists of a central nucleus surrounded by orbiting electrons. The electrons have a negative charge, and the nucleus contains protons that are positively charged.

Each electron has a negative electrical charge of 1.602×10^{-19} coulombs (*C*), and some particles within the nucleus (*protons*) have a positive charge of the same magnitude. Since opposite charges attract, a force of attraction exists between the oppositely charged electron and nucleus. Compared to the mass of the nucleus, electrons are relatively tiny particles of almost negligible mass. In fact, they can be considered to be little particles of negative electricity having no mass at all.

The nucleus of an atom (Fig. 1-2) is largely a cluster of two types of particles, protons and *neutrons*. Protons have a positive electrical charge, equal in magnitude (but opposite in polarity) to the negative charge on an electron. A neutron has no charge at all. Protons and neutrons each have masses about 1800 times the mass of an electron. For a given atom, the number of protons in

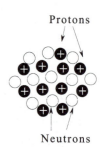

Figure 1-2
The nucleus of an atom is largely a cluster of protons and neutrons.

the nucleus normally equals the number of orbiting electrons.

Since the protons and orbital electrons are equal in number and equal and opposite in charge, they neutralise each other electrically. For this reason, all atoms are normally electrically neutral. If an atom loses an electron, it has lost some negative charge. Thus, it becomes positively charged and is referred to as a *positive ion*, [see Fig 1-3(a)]. Similarly, if an atom gains an additional electron, it becomes negatively charged and is termed a *negative ion*, [Fig. 1-3(b)].

The differences among atoms consist largely of dissimilar numbers and arrangements of the three basic types of particles. However, all electrons are identical, as are all protons and all neutrons. An electron from one atom could replace an electron in any other atom. Different materials are made up of different types of atoms, or differing combinations of several types of atoms.

The number of protons in an atom is referred to as the *atomic number* of the atom. The *atomic weight* is approximately equal to the total number of protons and neutrons in the nucleus of the atom. The atom of the semiconductor material *silicon* has 14 protons and 14 neutrons in its nucleus, as well as 14 orbital electrons. Therefore, the atomic number for silicon is 14, and its atomic weight is approximately 28.

Electron Orbits and Energy Levels

Atoms may be conveniently represented by the two-dimensional diagrams shown in Figs. 1-4. It has been found that electrons can occupy only certain orbital rings or *shells* at fixed distances from the nucleus, and that each shell can contain only a particular number of electrons. The electrons in the outer shell determine the electrical (and chemical) characteristics of each particular type of atom. These electrons are usually referred to as *valence electrons*. An atom may have its outer shell, or *valence shell*, completely filled or only partially filled.

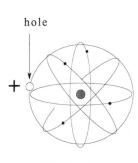

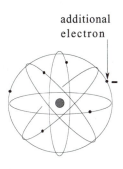

(a) Positive ion

(b) Negative ion

Figure 1-3
Positive and negative ions are created when an atom looses or gains an electron.

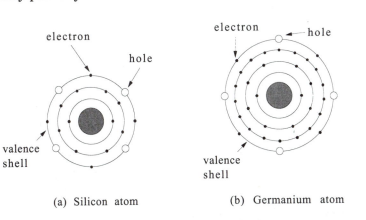

(a) Silicon atom

(b) Germanium atom

Figure 1-4
Two-dimensional representations of silicon and germanium atoms. The outer shells have four electrons and four holes.

The atoms represented in Fig. 1-4 are those of two important semiconductor materials, *silicon (Si)* and *germanium (Ge)*. It is seen that each of these atoms has four electrons in a valence shell that can contain a maximum of eight. Thus, the valence shells have four electrons and four *holes*. A *hole* is defined as an absence of an electron in a shell where one could exist. Even though the valence shells of silicon and germanium have four holes, both types of atoms are electrically neutral because the total number of orbiting (negatively charged) electrons equals the total number of (positively charged) protons in the nucleus.

The closer an electron is to the nucleus, the stronger are the forces that bind it. Each shell has an *energy level* associated with it that represents the amount of energy required to extract an electron from the atom. Since the electrons in the valence shell are farthest from the nucleus, they require the least amount of energy to extract them, [see Fig 1-5(a)]. Conversely, those electrons closest to the nucleus require the greatest energy application to extract them from the atom.

The energy levels of the orbiting electrons are measured in *electron volts (eV)*. An electron volt is defined as the amount of energy required to move one electron through a voltage difference of one volt.

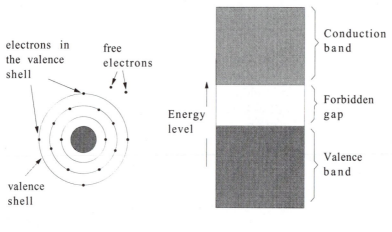

(a) Electrons in the valance shell and free electrons

(b) Energy band diagram

Figure 1-5
The energy level of electrons within a solid is shown by the energy band diagram. Electrons in orbit around a nucleus are in the valence band. Electrons which have broken away from an atom are in the conduction band.

Energy Bands

So far, the discussion has concerned a system of electrons in one isolated atom. The electrons of an isolated atom are acted upon only by the forces within that atom. However, when atoms are brought closer together, as in a solid, the electrons come under the influence of forces from other atoms. Under these circumstances, the energy levels that may be occupied by electrons merge into bands of energy levels. Within any given material there are two distinct energy bands in which electrons may exist; the *valence band* and the *conduction band*. Separating these two bands is an *energy gap*, termed the *forbidden gap*, in which no electrons can normally exist. The valence band, conduction band, and forbidden gap are shown diagrammatically in Fig. 1-5(b).

Electrons within the conduction band have become disconnected from atoms and are drifting around within the material. Conduction band electrons may be easily moved by the application of relatively small amounts of energy. Much larger amounts of energy must be applied to move an electron in the valence band. Electrons in the valence band are usually in orbit around a nucleus. For any given type of material, the forbidden gap may be large, small, or nonexistent. The distinction between conductors, insulators, and semiconductors is largely concerned with the relative widths of the forbidden gap.

It is important to note that *the energy band diagram is simply a graphic representation of the energy levels associated with electrons.* To repeat; those electrons in the valence band are actually in orbit around the nucleus of an atom; those in the conduction band are drifting in the spaces between atoms.

Section 1-1 Review

1-1.1   Define: nucleus, electron, electronic charge, proton, neutron, shell, positive ion, negative ion.

1-1.2   What is meant by atomic number and atomic weight? State the atomic number and atomic weight for silicon.

1-1.3   Define conduction band, valence band, and forbidden gap.

1-2   Conduction in Solids

Electron Motion and Hole Transfer

Conduction occurs in any given material when an applied voltage causes electrons within the material to move in a desired direction. This may be due to one or both of two processes; *electron motion* and *hole transfer*. In *electron motion*, free electrons in the conduction band are moved under the influence of the applied electric field, thus creating an electric current, (see Fig 1-6). Since electrons have a negative charge, they are repelled from the negative terminal of the applied voltage and attracted toward the positive terminal. Hole transfer involves electrons which are still attached to atoms, (those in the valence band).

Atoms   Free electrons

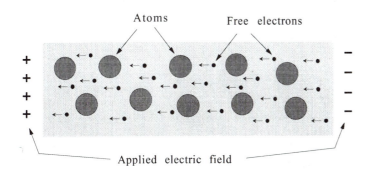

+
+
+
+

−
−
−
−

Applied electric field

Figure 1-6
Free electrons are easily moved by an applied voltage to create an electric current.

If some of the electron positions in the valence shell of an atom are not occupied by electrons, there are holes where electrons could exist. When sufficient energy is applied, an electron may be made to jump from one atom to a hole in another atom. When it jumps, the electron leaves a hole behind it, and thus the hole has moved in a direction opposite to that of the electron.

Figure 1-7 illustrates a situation where there are no free electrons. However, those electrons in orbit around atoms experience a force of attraction to the positive terminal of the applied voltage, and repulsion from the negative terminal. This force can cause an electron to jump from one atom to another, moving toward the positive terminal.

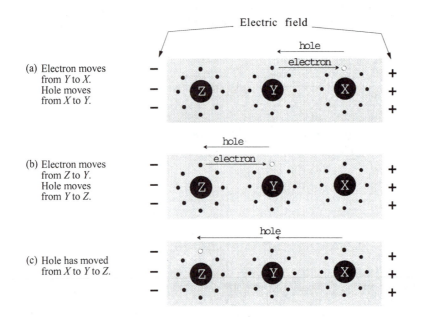

Figure 1-7
Conduction by hole transfer. An electron can be made to jump from one atom to another under the influence of an applied voltage, causing the hole to move in the opposite direction.

Figure 1-7(a) shows an electron jumping from atom Y to atom X (toward the positive terminal of the applied voltage). When this occurs, the hole in the valence shell of atom X is filled, and a hole is left in the valence shell of atom Y, [Fig. 1-7(b)]. If an electron now jumps from atom Z to fill the hole in Y, a hole is left in the valence shell of Z, [Fig. 1-7(c)]. Thus, the hole has moved from atom X to atom Y to atom Z. So, a flow of current (electron motion) has occurred, and this may be said to be due to *hole movement*, or *hole transfer*.

Holes may be thought of as *positively charged particles,* and as such, they move through an electric field in a direction opposite to that of electrons. (Positive particles are attracted toward the negative terminal of an applied voltage.) In the circumstance illustrated in Fig. 1-7 where there are few free electrons, it is more convenient to think in terms of hole movement rather than in terms of electrons jumping from atom to atom.

Since the flow of electric current is constituted by the movement of electrons and holes, electrons and holes are referred to as *charge carriers*. Each time a hole moves, an electron must be supplied with sufficient energy, to enable it to escape from its atom. Free electrons require less application of energy than holes to move them, because they are already disconnected from their atoms. For this reason, electrons have *greater mobility* than holes.

Conventional Current and Electron Flow

In the early days of electrical experimentation it was believed that a positive charge represented an increased amount of electricity, and that a negative charge was a reduced quantity. Thus, it was assumed that current flowed from positive to negative. This is a convention that remains in use today even though current is now known to be a movement of electrons from negative to positive, (see Fig. 1-8).

Current flow from positive to negative is referred to as the **conventional current direction.**

Electron flow from negative to positive is known as the **direction of electron flow.**

It is important to understand both electron flow and conventional current direction. The operation of electronic devices is explained in terms of electron movement. However, *every graphic symbol used to represent an electronic device has an arrowhead which indicates conventional current direction*, (see the *diode* symbol in Fig. 1-9). Consequently, electronic circuits are most easily explained by using current flow from positive to negative.

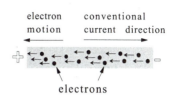

Figure 1-8
Conventional current direction is from positive to negative. Electron flow is from negative to positive.

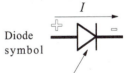

Figure 1-9
All electronic device graphic symbols have arrowheads that indicate conventional current direction.

Section 1-2 Review

1-2.1   Draw sketches to show the process of current flow by electron motion and by hole transfer.

1-2.2   Explain the difference between *conventional current direction* and *direction of electron flow*.

1-3   Conductors, Semiconductors, and Insulators

Bonding Forces Between Atoms

Whether a material is a conductor, semiconductor, or an insulator depends largely upon what happens to the outer-shell electrons when the atoms bond themselves together to form a solid. In the case of copper, the easily detached valence electrons are given up by the atoms. As illustrated in Fig. 1-10, this creates a great mass of free electrons (or electron gas) drifting about in the space between the copper atoms. The electrons are easily moved under the influence of an applied voltage to create a current flow. The bonding force that holds atoms together in a conductor is known as *metallic bonding*.

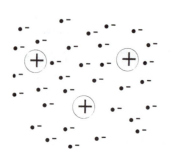

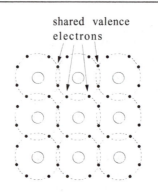

shared valence
electrons

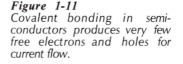

Figure 1-10
In metallic bonding there is a mass of free electrons drifting around in the spaces between atoms.

Figure 1-11
Covalent bonding in semi-conductors produces very few free electrons and holes for current flow.

Semiconductor atoms normally have four outer-shell electrons and four holes, and are so close together that the outer-shell electrons behave as if they were orbiting in the valence shells of two atoms. In this way each valence-shell electron fills one of the holes in the valence shell of a neighbouring atom. This produces an atomic bonding force known as *covalent bonding*. As shown in Fig. 1-11, it would appear that there are no holes and no free electrons drifting about within the semiconductor material. In fact, some of the electrons are so weakly attached to their atoms that they can be made to break away to create a current flow when a voltage is applied.

In some insulating materials the atoms bond together in a similar way to semiconductor atoms *(covalent bonding)*. But the valence shell electrons are so strongly attached to their atoms in an insulator that no charge carriers are available for current flow. In other types of insulating materials, some atoms give up outer-shell electrons which are accepted into the orbit of neighbouring atoms. Because the atoms are *ionized*, this is termed *ionic bonding*, [Fig. 1-12]. Here again, all of the electrons are very strongly attached to the atoms, and the possibility of current flow is virtually zero.

negative ion

positive ion

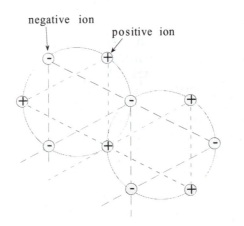

Figure 1-12
Ionic bonding occurs in some insulating materials. There are no holes or free electrons to facilitate current flow.

Energy Bands in Different Materials

The energy band diagrams in Fig. 1-13 show that insulators have a wide forbidden gap, semiconductors have a narrow forbidden gap, and conductors have no forbidden gap at all.

In the case of insulators, there are practically no electrons in the conduction band, and the valence band is filled. Also, the forbidden gap is so wide [Fig. 1-13(a)] that it would require the application of relatively large amounts of energy to cause an electron to cross from the valence band to the conduction band. Therefore, when a voltage is applied to an insulator, conduction cannot normally occur either by electron motion or hole transfer.

For semiconductors at absolute zero of temperature (-273°C) the valence band is usually full and there may be no electrons in the conduction band. As shown in Fig. 1-13(b), the forbidden gap in a semiconductor is very much narrower than that in an insulator, and the application of small amounts of energy can raise electrons from the valence band to the conduction band. Sufficient thermal energy for this purpose is available when the semiconductor is at normal room temperature. If a voltage is applied to the semiconductor, conduction can occur by both electron movement in the conduction band and by hole transfer in the valence band.

In the case of a conductor [Fig. 1-13(c)] there is no forbidden gap, and the valence and conduction energy bands overlap. For this reason, very large numbers of electrons are available for conduction, even at extremely low temperatures.

Typical resistance values for a 1-centimeter-cube sample are: Conductor 10^{-6} Ω, Semiconductor 10 Ω, Insulator 10^{14} Ω.

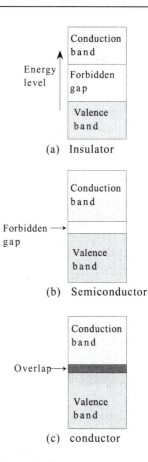

Figure 1-13
Energy band diagrams for insulators, semiconductors, and conductors.

Section 1-3 Review

1-3.1   Briefly describe the type of atomic bonding that occurs in (a) insulators, (b) conductors.

1-3.2   Draw a sketch to show the bonding of atoms in semiconductor material. Briefly explain.

1-3.3   Sketch and explain the energy band diagrams for conductors, insulators, and semiconductor.

1-4   n-Type and p-Type Semiconductor

Doping

Pure semiconductor material is known as *intrinsic* material. Before intrinsic material can be used for device manufacture, *impurity atoms* must be added to improve its conductivity. The process of adding the atoms is termed *doping*. Two different types of doping are possible; *donor doping,* and *acceptor doping.* Donor doping generates free electrons in the conduction band (i.e., electrons that are not tied to an atom). Acceptor doping produces valence band holes, or a shortage of valence electrons in the material. After doping, the semiconductor is known as *extrinsic* material.

n-Type Material

In donor doping, illustrated in Fig. 1-14, impurity atoms which have five electrons and three holes in their valence shells are added to the undoped material. The impurity atoms form covalent bonds with the silicon or germanium atoms. Because the semiconductor atoms have only four electrons and four holes in their valence shells, there is one extra valence-shell electron for each impurity atom added. Each additional electron produced in this way enters the conduction band as a free electron. As shown in Fig. 1-14, there is no hole for the fifth electron from the outer shell of the impurity atom; therefore, this electron becomes a free electron. Because the free electrons have *negative charges,* donor-doped semiconductor is known as *n-type material.*

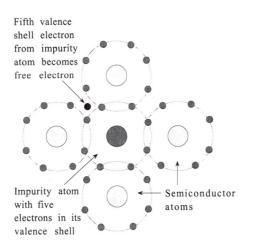

Fifth valence shell electron from impurity atom becomes free electron

Impurity atom with five electrons in its valence shell

Semiconductor atoms

Figure 1-14
In donor doping, impurity atoms with five valence shell electrons and three holes are added to the semiconductor material.

Free electrons in the conduction band are easily moved around under the influence of an electric field. Consequently, conduction occurs largely by electron motion in donor-doped semiconductor material. The doped material remains electrically neutral (it is neither positively charged nor negatively charged), because the total number of electrons (including the free electrons) is still equal to the total number of protons in the atomic nuclei.

The term *donor doping* comes from the fact that an electron is donated to the conduction band by each impurity atom. Typical donor materials are *antimony, phosphorus,* and *arsenic.* Since these atoms have five valence electrons, they are referred to as *pentavalent atoms.*

p-Type Material

The impurity atoms used for acceptor doping (Fig. 1-15) have outer shells containing three electrons and five holes. Suitable atoms with three valence electrons (*trivalent atoms*) are *boron, aluminum,* and *gallium.* These atoms form covalent bonds with the semiconductor atoms, but the bonds lack one electron for a complete outer shell of eight. In Fig. 1-15 the impurity atom illustrated has only three valence electrons; so, a hole exists in its bond with the surrounding atoms. Thus, in acceptor doping, holes are introduced into the valence band, so that conduction may occur by the process of hole transfer.

Since holes can be said to have a *positive charge*, acceptor-doped semi-conductor material is referred to as *p-type material*. As with *n*-type, the material remains electrically neutral, because the total number of orbital electrons in each impurity atom is equal to the total number of protons in its atomic nucleus. Holes can accept a free electron, hence the term *acceptor doping*.

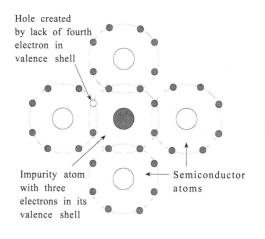

Hole created by lack of fourth electron in valence shell

Impurity atom with three electrons in its valence shell

Semiconductor atoms

Figure 1-15
Acceptor doping requires the use of impurity atoms that have three electrons and five holes in the valence shells. This creates holes in the semiconductor material.

Majority and Minority Charge Carriers
In undoped semiconductor material at room temperature there are a number of free electrons and holes. These are due to thermal energy causing some electrons to break the bonds with their atoms and enter the conduction band. This process creates pairs of holes and electrons, and is appropriately termed *hole-electron pair generation*. The opposite effect, called *recombination*, occurs when an electron *falls into* a hole in the valence band. Because there are many more electrons than holes in *n*-type material, electrons are said to be the *majority charge carriers*, and holes are said to be *minority carriers* in *n*-type material. In *p*-type material, holes are the majority carriers and electrons are minority carriers.

Effects of Heat and Light
When a conductor is heated, the atoms (which are in fixed locations) tend to vibrate, and the vibration impedes the movement of the surrounding electron gas. This causes a reduction in the flow of the electrons that constitute the electric current. The reduced current flow means that the conductor resistance has increased. A conductor has a *positive temperature coefficient (PTC)* of resistance; a resistance which increases with

increasing temperature. This is illustrated in Fig. 1-16(a).

When undoped semiconductor material is at a temperature of absolute zero (-273°C), all electrons are in normal orbit around the atoms, and there are virtually no free electrons in the conduction band and no holes in the valence band. Consequently, at -273°C a semiconductor behaves as an insulator.

When the semiconductor temperature is raised, electrons break away from their atoms and move from the valence band to the conduction band. This produces holes in the valence band and free electrons in the conduction band, allowing conduction to occur by electron movement and by hole transfer. Increasing application of thermal energy generates an increasing number of hole-electron pairs.

As in the case of a conductor, thermal vibration of atoms occurs in a semiconductor. However, there are very few electrons to be impeded in a semiconductor compared to the large quantities in a conductor. The thermal generation of electrons is the dominating factor, and the semiconductor current increases with increasing temperature. This represents a decrease in resistance with increasing temperature; a *negative temperature coefficient (NTC)*, [Fig. 1-16(b)]. Heavily doped semiconductor material is an exception, behaving more like a conductor than a semiconductor.

Just as thermal energy can cause electrons to break their atomic bonds, so hole-electron-pairs can be generated by energy applied to a semiconductor in the form of light. A material which has few free electrons available for conduction when not illuminated is said to have a high *dark resistance.* When the semiconductor material is illuminated, its resistance decreases and may become comparable to that of a conductor.

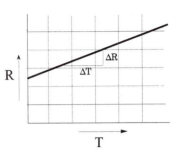

(a) A conductor has a positive temperature coefficient

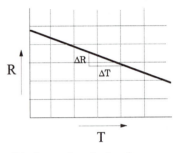

(b) A semiconductor has a negative temperature coefficient

Figure 1-16
The resistances of conductors and semiconductors are differently affected by temperature change.

Section 1-4 Review

1-4.1   Draw a sketch to illustrate donor doping.

1-4.2   Draw a sketch to illustrate acceptor doping.

1-4.3   Define: *n-type material, p-type material, minority charge carriers, majority charge carriers, positive temperature coefficient, negative temperature coefficient, dark resistance.*

1-5   The pn-Junction

Junction of p-Type and n-Type

Two blocks of semiconductor material are represented in Fig. 1-17; one block is *p*-type material, and the other is *n*-type. The small circles in the *p*-type material represent holes, which are the majority charge carriers in *p*-type. The dots in the *n*-type material represent the majority charge carrier free electrons within that material. Normally, the holes are uniformly distributed throughout the volume of the *p*-type semiconductor, and the electrons are uniformly distributed in the *n*-type.

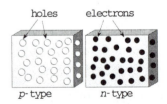

Figure 1-17
P-type and n-type semiconductor materials.

In Figure 1-18, *p*-type and *n*-type semiconductor materials are shown side by side, representing a *pn-junction*. Holes and electrons are close together at the junction, so some free electrons from the *n*-side are attracted across the junction to fill adjacent holes on the *p*-side. They are said to *diffuse* across the junction from a region of high carrier concentration to one of low concentration. The free electrons crossing the junction create negative ions on the *p*-side by giving some atoms one more electron than their total number of protons. The electrons also leave positive ions (atoms with one fewer electron than the number of protons) behind them on the *n*-side.

Barrier Voltage

The *n*-type and *p*-type materials are both electrically neutral before the charge carriers diffuse across the junction. When negative ions are created on the *p*-side, the portion of the *p*-side close to the junction acquires a negative voltage, (see Fig. 1-18). Similarly, the positive ions created on the *n*-side gives the *n*-side a positive voltage close to the junction. The negative voltage on the *p*-side tends to repel additional electrons crossing from the *n*-side. Also, (thinking of the holes as positive particles) the positive voltage on the *n*-side tends to repel any hole movement from the *p*-side. Thus, the initial diffusion of charge carriers creates a *barrier voltage* at the junction, which is negative on the *p*-side and positive on the *n*-side. The transfer of charge carriers and the resultant creation of the barrier voltage occur when the *pn*-junctions are formed during the manufacturing process, (see Chapter 6).

The magnitude of the barrier voltage at a *pn*-junction can be calculated from a knowledge of the doping densities, electronic charge, and junction temperature. Typical barrier voltages at 25°C are 0.3 V for germanium junctions and 0.7 V for silicon.

It has been explained that the barrier voltage at the junction opposes both the flow of electrons from the *n*-side and the flow of holes from the *p*-side. Because electrons are the majority charge carriers in the *n*-type material, and holes are the majority charge carriers in the *p*-type, it is seen that *the barrier voltage opposes the flow of majority carriers* across the *pn*-junction, (see Fig. 1-19). Any free electrons generated by thermal energy on the *p*-side of the junction are attracted across the positive barrier to the *n*-side. Similarly, thermally generated holes on the *n*-side are attracted to the *p*-side through the negative barrier presented to them at the junction. Electrons on the *p*-side and holes on the *n*-side are minority charge carriers. Therefore, *the barrier voltage assists the flow of minority carriers* across the junction, (Fig. 1-19).

Depletion Region

The movement of charge carriers across the junction leaves a layer on each side which is depleted of charge carriers. This is the *depletion region* shown in Fig. 1-20(a). On the *n*-side, the depletion region consists of donor impurity atoms which, having lost the free electron associated with them, have become positively charged.

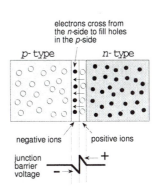

Figure 1-18
At a pn-junction, electrons cross from the n-side to fill holes in a layer of the p-side close to the junction.

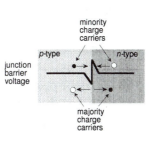

Figure 1-19
The barrier voltage at a pn-junction assists the flow of minority charge carriers and opposes the flow of majority carriers.

The depletion region on the *p*-side is made up of acceptor impurity atoms which have become negatively charged by losing the hole associated with them. (The hole has been filled by an electron.)

On each side of the junction, equal numbers of impurity atoms are involved in the depletion region. If the two blocks of semiconductor material have equal doping densities, the depletion layers on each side have equal widths, [Fig. 1-20(a)]. If the *p*-side is more heavily doped than the *n*-side, as illustrated in Fig. 1-20(b), the depletion region penetrates more deeply into the *n*-side in order to include an equal number of impurity atoms on each side of the junction. Conversely, when the *n*-side is most heavily doped, the depletion region penetrates deepest into the *p*-type material.

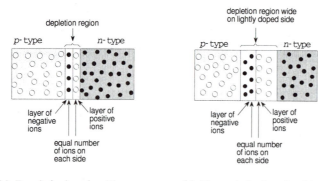

(a) Equal doping densities   (b) Unequal doping densities

Figure 1-20
Charge carrier diffusion across a pn-junction creates a region depleted of charge carriers which penetrates deepest into the most lightly doped side.

Summarizing:

- A region depleted of charge carriers spreads across both sides of a *pn*-junction, penetrating furthest into the lowest doped side.
- The depletion region contains an equal number of ionized atoms on opposite sides of the junction.
- A barrier voltage is created by the charge carrier depletion effect, positive on the *n*-side and negative on the *p*-side.
- The barrier voltage opposes majority charge carrier flow and assists the flow of minority charge carriers across the junction.

Section 1-5 Review

1-5.1   Sketch a *pn*-junction showing the depletion region. Briefly explain how the depletion region is created.

1-5.2   Explain the origin of the barrier voltage at a *pn*-junction. Discuss the effect of the barrier voltage on minority and majority charge carriers.

1-6   Biased Junctions

Reverse Biased Junction

When an external bias voltage is applied to a *pn*-junction, positive to the *n*-side and negative to the *p*-side, electrons from the *n*-side are attracted to the positive terminal, and holes from the *p*-side are attracted to the negative terminal. As shown in Fig. 1-21, holes on the *p*-side of the junction are attracted away from the junction, and electrons are attracted away from the junction on the *n*-side. This causes the depletion region to be widened and the barrier voltage to be increased, as illustrated. With the barrier voltage increase there is no possibility of majority charge carrier current flow across the junction, and the junction is said to be *reverse biased*. Because only a very small reverse current flows, a reverse-biased *pn*-junction can be said to have a high resistance.

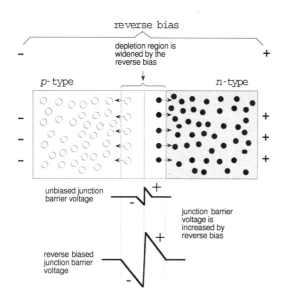

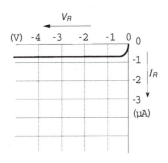

Figure 1-21
A reverse bias applied to a pn-junction (positive on the n-side, negative on the p-side) causes the depletion region to widen, and increases the barrier voltage. Only a very small reverse current flows across the junction.

Although there is no possibility of a majority charge carrier current flowing across a reverse-biased junction, minority carriers generated on each side can still cross the junction. Electrons in the *p*-side are attracted across the junction to the positive voltage on the *n*-side. Holes on the *n*-side may flow across to the negative voltage on the *p*-side. This is shown by the junction *reverse characteristic*, or the graph of reverse current (I_R) versus reverse voltage (V_R), (Fig. 1-22). Only a very small reverse bias voltage is necessary to direct all available minority carriers across the junction, and further increases in bias voltage do not increase the current. This current is referred to as a *reverse saturation current*.

The reverse saturation current is normally a very small quantity, ranging from nanoamps to microamps, depending on the junction area, temperature, and semiconductor material.

Forward Biased Junction

Consider the effect of an external bias voltage applied with the polarity shown in Fig. 1-23; positive on the *p*-side, negative on the *n*-side. The holes on the *p*-side, being positively charged particles, are repelled from the positive terminal and driven toward the junction. Similarly, the electrons on the *n*-side are repelled from the negative terminal toward the junction. The result is that the

Figure 1-22
Current versus voltage characteristic for a reverse biased pn-junction.

depletion region width and the barrier potential are both reduced.

When the applied bias voltage is progressively increased from zero, the barrier voltage gets smaller until it effectively disappears and charge carriers easily flow across the junction. Electrons from the *n*-side are now attracted across to the positive bias terminal on the *p*-side, and holes from the *p*-side flow across to the negative terminal on the n-side, (thinking of holes as positively-charged particles). Thus, a majority carrier current flows, and the junction is said to be *forward biased.*

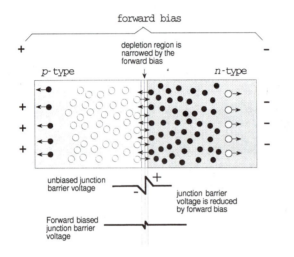

Figure 1-23
Forward biasing a pn-junction (positive on the p-side, negative on the n-side) narrows the depletion region, reduces the barrier voltage, and causes a relatively large current to flow across the junction.

The graph in Fig. 1-24 shows the forward current (I_F) plotted against forward voltage (V_F) for typical germanium and silicon *pn*-junctions. In each case, the graph is known as the *forward characteristic* of the junction. It is seen that very little forward current flows until V_F exceeds the junction barrier voltage (0.3 V for germanium, 0.7 V for silicon). When V_F is increased from zero toward the knee of the characteristic, the barrier voltage is progressively overcome, allowing more majority charge carriers to flow across the junction. Above the knee of the characteristic I_F increases almost linearly with increase in V_F. The level of current that can be made to flow across a forward-biased *pn*-junction largely depends on the junction area.

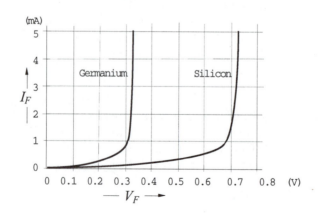

Figure 1-24
pn-junction forward character-istics. Germanium junctions are forward biased at approximately 0.3 V. A silicon junction requires approximately 0.7 V for forward bias.

Junction Temperature Effects

As already discussed, the reverse saturation current (I_R) at a *pn*-junction consists of minority charge carriers. When the temperature of semiconductor material is increased, increasing numbers of electrons break away from their atoms. This generates additional minority charge carriers, causing I_R to increase as the junction temperature rises. It can be shown that the reverse current level approximately doubles with each 10°C increase in temperature, and that for a given junction, there is a definite I_R level for each temperature level, see Fig. 1-25(a).

The junction forward voltage drop (V_F) is also affected by temperature. The horizontal line (at 3 mA) on Fig. 1-25(b) shows that, if I_F is held constant while the junction temperature is changing, the forward voltage decreases with increasing junction temperature. This means that V_F has a negative temperature coefficient. It is found that the temperature coefficient for the forward voltage of a *pn*-junction is approximately -1.8 mV/°C for a silicon junction, and -2.02 mV/°C for germanium.

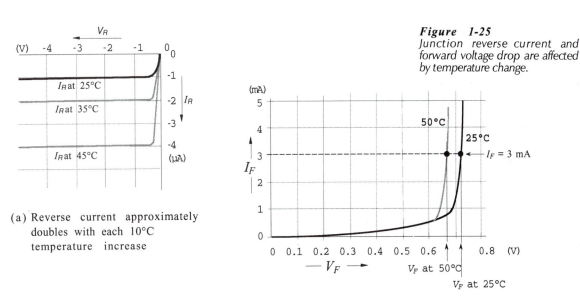

Figure 1-25
Junction reverse current and forward voltage drop are affected by temperature change.

(a) Reverse current approximately doubles with each 10°C temperature increase

(b) Forward voltage drop changes by approximately -2 mV/°C

Chapter-1 Review Questions

Section 1-1

1-1   Describe the atom and draw a two-dimensional diagram to illustrate your description. Compare the atom to a planet with orbiting satellites.

1-2   Define: nucleus, electron, electronic charge, proton, neutron, shell, positive ion, negative ion.

1-3   Sketch two-dimensional diagrams of silicon and germanium atoms. Describe the valence shell of each atom.

1-4   Explain atomic number and atomic weight. State the atomic number and atomic weight for silicon.

1-5   Explain what is meant by energy levels and energy bands. Sketch an energy band diagram, and define conduction band, valence band, and forbidden gap.

Section 1-2

1-6   Draw a sketch to show the process of current flow by electron motion. Briefly explain.

1-7   Draw sketches to show the process of current flow by hole transfer. Which have greater mobility, electrons or holes? Explain why.

1-8   Define *conventional current direction* and *direction of electron flow*. State why each is important.

Section 1-3

1-9   Name the three kinds of bonds that hold atoms together in a solid. What kind of bonding might be found in: (a) conductors, (b) insulators, (c) semiconductors?

1-10   Draw sketches to illustrate *metallic bonding* and *ionic bonding.* Explain what occurs in each case.

1-11   Draw a sketch to illustrate *covalent bonding.* Explain the bonding process.

1-12   Draw energy band diagrams for conductors, insulators, and semiconductors. Explain the reasons for the differences between the diagrams.

Section 1-4

1-13   Define *acceptor doping,* and draw a sketch to illustrate the process. Explain.

1-14   Define *donor doping,* and draw a sketch to illustrate the process. Explain.

1-15   State the names given to acceptor-doped material and donor-doped material. Explain.

1-16   What is meant by *majority charge carriers* and *minority charge carriers* ? Which are majority carriers and why in: (a) donor-doped material, (b) acceptor-doped material?

1-17 Explain what happens to resistance with increasing temperature in the case of: (a) a conductor, (b) a semiconductor, (c) a heavily doped semiconductor. What do you think would happen to the resistance of an insulator with increasing temperature? Why?

1-18 Explain *hole-electron pair generation* and *recombination.*

1-19 Discuss the effects that illumination can produce on semiconductor materials.

1-20 Define: *n-type material, p-type material, majority charge carriers, minority charge carriers, positive temperature coefficient, negative temperature coefficient, dark resistance.*

Section 1-5

1-21 Using illustrations, explain how the *depletion region* and *barrier voltage* are produced at a *pn*-junction. List the characteristics of the depletion region.

1-22 Draw a sketch to show the depletion region and barrier voltage at a *pn*-junction, with unequal doping of each side. Briefly explain.

Section 1-6

1-23 A bias is applied to a *pn*-junction, positive to the *p*-side, negative to the *n*-side. Show, by a series of sketches, the effect of this bias on depletion region width, barrier voltage, minority carriers, and majority carriers. Briefly explain the effect in each case.

1-24 Repeat Question 1-23 for a bias applied negative to the *p*-side and positive to the *n*-side.

1-25 Sketch the voltage-current characteristics for a *pn*-junction: (a) with forward bias, and (b) with reverse bias. Show how temperature change affects the characteristics.

1-26 State typical values of barrier voltage for silicon and germanium junctions. Discuss the resistances of forward-biased and reverse-biased *pn*-junctions.

1-27 State typical *reverse saturation current* levels for *pn*-junctions. Explain the origin of reverse saturation current.

Chapter *2*

Semiconductor Diodes

Chapter   Contents

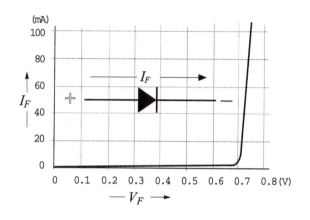

Objectives

You will be able to:

1 *Sketch a diode circuit symbol, identify the terminals, and discuss diode circuit behaviour.*

2 *Explain diode forward and reverse characteristics.*

3 *List important diode parameters, and determine parameter values from the characteristics.*

4 *Sketch approximate diode characteristics and dc equivalent circuits, and apply them to circuit analysis.*

5 *Draw dc load lines on diode characteristics to precisely analyse diode circuits.*

6 *Determine maximum diode power dissipations and forward voltages at various temperatures.*

7 *Sketch diode ac equivalent circuits and calculate switching times.*

8 *Determine diode parameter values from device data sheets.*

9 *Test diodes and plot the characteristics.*

10 *Sketch a Zener diode circuit symbol, identify the terminals, and explain the characteristics.*

11 *Determine Zener diode parameter values from device characteristics and from data sheets*

12 *Analyse basic Zener diode circuits.*

Introduction

The term *diode* identifies a two-electrode, or two-terminal, device. A semiconductor diode is simply a *pn*-junction with a connecting lead on each side. A diode is a one-way device, offering a low resistance when forward biased, and behaving almost as an open switch when reverse biased. An approximately constant voltage drop across occurs across a forward-biased diode, and this simplifies diode circuit analysis. Diode (forward and reverse) characteristics are graphs of corresponding current and voltage levels. For precise circuit analysis, dc load lines are drawn on the diode forward characteristic.

Some diodes are low-current devices for use in switching circuits. High-current diodes are most often used as rectifiers for *ac* to *dc* conversion. Zener diodes are operated in reverse breakdown, because they have a very stable breakdown voltage. There are many other types of diodes designed for specialised applications.

2-1   pn-Junction Diode

As discussed in Sections 1-5 and 1-6, a *pn*-junction has the ability to permit substantial current flow when forward-biased, and to block current when reverse-biased. Thus, it can be used as a switch; *on* when forward-biased, and *off* when biased in reverse. A *pn*-junction provided with copper wire connecting leads becomes an electronic device known as a *diode*, (see Fig. 2-1).

The circuit symbol (or *graphic symbol*) for a diode is an arrowhead and bar, (Fig. 2-2). The arrowhead indicates the conventional direction of current flow when the diode is forward biased, (from the positive terminal through the device to the negative terminal). The *p*-side of the diode is always the positive terminal for forward bias and is termed the *anode.* The *n*-side, called the *cathode,* is the negative terminal when the device is forward biased.

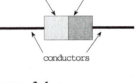

Figure 2-1
A semiconductor diode is a pn-junction with conductors for connecting the device to a circuit.

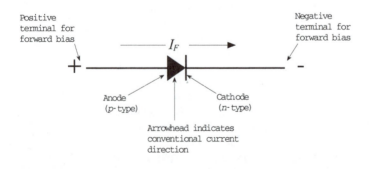

Figure 2-2
Diode circuit symbol. Current flows when the diode is forward biased: + on the anode, - on the cathode.

A *pn*-junction diode can be destroyed by a high level of forward current overheating the device. It can also be destroyed by a large reverse voltage causing the junction to breakdown. The maximum level of forward current and reverse voltage for diodes are specified on the manufacturer's data sheets, (see Section 2-6). In general physically large diodes pass the largest currents and survive the largest reverse voltages. Small diodes are limited to low current levels and low reverse voltages.

Figure 2-3 shows the appearance of low-, medium-, and high-current diodes. The body of the low-current device in Fig. 2-3(a) may be only 0.3 cm long, so the cathode is usually denoted by a coloured band. This type of diode is typically capable of passing a maximum forward current of approximately 100 mA. It can also survive about a 75 V reverse bias without breaking down, and its reverse current at 25°C is usually less than 1 μA.

The medium-current diode shown in Fig. 2-3(b) can typically pass a forward current of about 400 mA and survive over 200 V of reverse bias. The anode and cathode terminals may be indicated by a diode symbol on the side of the device.

Low-current and medium-current diodes are usually mounted by soldering the connecting leads to terminals. Power dissipated in the device is then carried away by air convection and by heat conduction along the connecting leads. High-current diodes, or *power diodes* [Fig. 2-3(c)], generate a lot of heat. So, air convection would be completely inadequate. Such devices are designed for mechanically connecting to a metal heat sink. Power diodes can pass forward currents of many amperes and can survive several hundred volts of reverse bias.

(a) Low-current diode

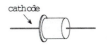

(b) Medium-current diode

(c) High-current diode

Figure 2-3
The size and appearance of a diode depends upon the level of forward current that the device is required to pass.

2-2 *Characteristics and Parameters*

Forward and Reverse Characteristics

The semiconductor diode is essentially a *pn*-junction, and its characteristics are those discussed in Chapter 1. Figures 2-4 and 2-5 show typical forward and reverse characteristics for low-current silicon and germanium diodes. From the silicon diode characteristics in Fig. 2-4, it is seen that the forward current (I_F) remains very low (less than microamps) until the diode forward-bias voltage (V_F) exceeds approximately 0.7 V. Above 0.7 V, I_F increases almost linearly with increase in V_F.

Because the diode reverse current (I_R) is very much smaller than its forward current, the reverse characteristics are plotted with expanded current scales. For a silicon diode, I_R is normally less than 100 nA, and it is almost completely independent of the reverse-bias voltage. As already explained in Chapter 1, I_R is largely a minority charge carrier *reverse saturation current*. A small increase in I_R can occur with increasing reverse-bias voltage, due to some minority charge carriers leaking along the junction surface. For a diode with the characteristics in Fig. 2-4, the reverse current is typically less than 1/10,000 of the lowest normal forward current level. Therefore, I_R is quite negligible when compared to I_F, and a reverse-biased diode may be treated almost as an open switch. This is further investigated in Example 2-1

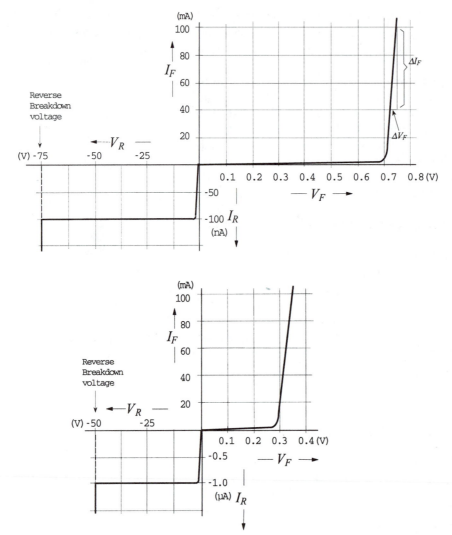

Figure 2-4
Typical forward and reverse characteristics for a silicon diode.

Figure 2-5
Typical forward and reverse characteristics for a germanium diode.

Example 2-1
Calculate the forward and reverse resistances offered by a silicon diode with the characteristics in Fig. 2.4, at $I_F = 100$ mA, and at $V_R = 50$ V.

Solution

at $I_F = 100$ mA, $V_F \approx 0.75$ V

$$R_F = \frac{V_F}{I_F} = \frac{0.75\text{ V}}{100\text{ mA}} \quad \text{[see Fig. 2-6(a)]}$$

$$= 7.5\ \Omega$$

at $V_R = 50$ V, $I_R = 100$ nA

$$R_R = \frac{V_R}{I_R} = \frac{50\text{ V}}{100\text{ nA}} \quad \text{[see Fig. 2-6(b)]}$$

$$= 500\ \text{M}\Omega$$

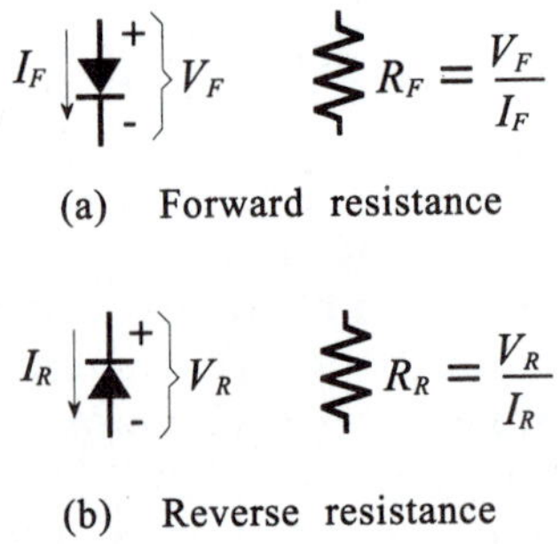

$I_F \quad V_F \qquad R_F = \dfrac{V_F}{I_F}$

(a) Forward resistance

$I_R \quad V_R \qquad R_R = \dfrac{V_R}{I_R}$

(b) Reverse resistance

Figure 2-6
Diode forward and reverse resistance determination.

When the diode reverse voltage (V_R) is sufficiently increased, the device goes into *reverse breakdown*. For the characteristics shown in Fig. 2-4, reverse breakdown occurs at 75 V. Reverse breakdown can destroy a diode unless the current is limited by a suitable series-connected resistor. Reverse breakdown is usefully applied in *Zener diodes* which are discussed in Section 2-9.

The characteristics of a germanium diode are similar to those of a silicon diode, with some important differences, (see Fig. 2-5). The forward voltage drop of a germanium diode is typically 0.3 V, compared to 0.7 V for silicon. For a germanium device, the reverse saturation current at 25°C may be around 1 μA, which is much larger than the reverse current for a silicon diode. Finally, the reverse breakdown voltage for germanium devices is likely to be substantially lower than that for silicon devices.

The lower forward voltage drop for germanium diodes can be a distinct advantage. However, the lower reverse current and higher reverse breakdown voltage of silicon diodes make them preferable to germanium devices for most applications.

Diode Parameters

The diode parameters of greatest interest are:

V_F	*forward voltage drop*
I_R	*reverse saturation current*
V_{BR}	*reverse breakdown voltage*
r_d	*dynamic resistance*
$I_{F(max)}$	*maximum forward current*

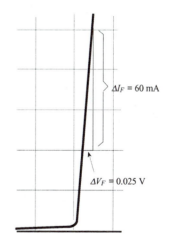

The values of these quantities are normally listed on the diode data sheet provided by device manufacturers (see Section 2-7). Some of the parameters can be determined directly from the diode characteristics. For the silicon diode characteristics in Fig. 2-4, V_F ≈ 0.7 V, I_R = 100 nA, and V_{BR} = 75 V.

The forward resistance calculated in Example 2-1 is a *static quantity*; it is a constant resistance of the diode at a particular constant forward current. The *dynamic resistance* of the diode is the resistance offered to changing levels of forward voltage. The dynamic resistance, also known as the *incremental resistance* or *ac resistance,* is the reciprocal of the slope of the forward characteristics beyond the knee. Referring to Fig. 2-4, and Fig. 2-7,

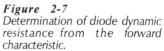

Figure 2-7
Determination of diode dynamic resistance from the forward characteristic.

$$r_d = \frac{\Delta V_F}{\Delta I_F} \qquad \text{(2-1)}$$

The dynamic resistance can also be calculate from the equation,

$$r'_d = \frac{26 \text{ mV}}{I_F} \qquad \text{(2-2)}$$

where I_F is the *dc* forward current at the junction. Thus, for example, the dynamic resistance for a diode passing a 10 mA forward current is, r'_d = 26 mV/10 mA = 26 Ω.

Equation 2-2 shows that the diode dynamic resistance changes with the level of *dc* forward current. This is not shown in Figs. 2-4 and 2-5, so, the characteristics are approximations of the actual device characteristics. It should also be noted that Eq. 2-2 gives the *ac* resistance only for the junction. It does not include the *dc* resistance of the semiconductor material, which might be as large as 2 Ω depending on the design of the device. The resistance derived from the slope of the device characteristic does include the semiconductor *dc* resistance. So, r_d (from the characteristic) should be slightly larger than r'_d calculated from Eq. 2-2.

Example 2-2

Determine the dynamic resistance at a forward current of 70 mA for the diode characteristics given in Fig. 2-4. Also, use Eq. 2-2 to estimate the diode dynamic resistance.

Solution

at I_F = 70 mA,   for ΔI_F = 60 mA, $\Delta V_F \approx$ 0.025 V

Eq. 2-1, $r_d = \dfrac{\Delta V_F}{\Delta I_F} = \dfrac{0.025 \text{ V}}{60 \text{ mA}}$ (see Fig. 2-7)

$= 0.42 \ \Omega$

Eq. 2-2, $r'_d = \dfrac{26 \text{ mV}}{I_F} = \dfrac{26 \text{ mV}}{70 \text{ mA}}$

$= 0.37 \ \Omega$

Practise Problems

2-2.1   Calculate the resistances offered by a diode with the characteristics in Fig. 2.5, when the forward current is 60 mA, and when the reverse voltage is 30 V.

2-2.2   Determine the dynamic resistance at a 50 mA forward current for a diode with the characteristics in Fig. 2-5. Also, use Eq. 2-2 to estimate the diode dynamic resistance.

2-3   Diode Approximations

Ideal Diode Characteristics

As already discussed, a diode is essentially a one-way device, offering a low resistance when forward biased, and a high resistance when biased in reverse. An *ideal diode* (or perfect diode) would have zero forward resistance and zero forward voltage drop. It would also have an infinitely high reverse resistance, which would result in zero reverse current. Figure 2-8(a) shows the current/voltage characteristics of an ideal diode.

Although an ideal diode does not exist, there are many applications where diodes can be assumed to be near-ideal devices.

In circuits with supply voltages much larger than the diode forward voltage drop, V_F can be assumed constant without introducing any serious error. Also, the diode reverse current is normally so much smaller than the forward current that the reverse current can be ignored. These assumptions lead to the near-ideal, or approximate, characteristics for silicon and germanium diodes shown in Fig. 2-8(b) and (c). Example 2-3 investigates a situation where the diode V_F is assumed constant

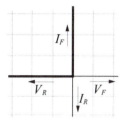

(a) Ideal diode
    characteristics

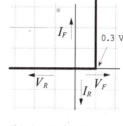

(b) Approximate
    characteristics for
    a germanium diode

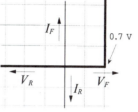

(c) Approximate
    characteristics for
    a silicon diode

Figure 2-8
An ideal diode has $V_F = 0$ and $I_R = 0$. Practical diodes can be treated as near-ideal devices if the forward voltage drop is taken into account.

Example 2-3
A silicon diode is used in the circuit shown in Fig. 2-9. Calculate the diode current.

Solution

$$E = I_F R_1 + V_F$$

or,

$$I_F = \frac{E - V_F}{R_1} = \frac{15 \text{ V} - 0.7 \text{ V}}{4.7 \text{ k}\Omega}$$

$$= 3.04 \text{ mA}$$

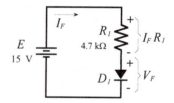

Figure 2-9
Circuit for Example 2-3.

Piecewise Linear Characteristic
When the forward characteristic of a diode is not available, a straight-line approximation, called the *piecewise linear characteristic*, may be employed. To construct the piecewise linear characteristic, V_F is first marked on the horizontal axis, as shown in Fig. 2-10. Then, starting at V_F, a straight line is drawn with a slope equal to the diode dynamic resistance. Example 2-4 demonstrates the process.

Example 2-4
Construct the piecewise-linear characteristic for a silicon diode that has a 0.25 Ω dynamic resistance and a 200 mA maximum forward current.

Solution
Plot point A on the horizontal axis at:

$$V_F = 0.7 \text{ V (see Fig. 2-10)}$$

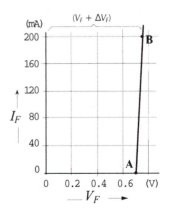

Figure 2-10
Diode piecewise linear characteristic, or straight-line approximation of the diode forward characteristic.

$$\Delta V_F = \Delta I_F \times r_d = 200 \text{ mA} \times 0.25 \text{ }\Omega$$
$$= 0.05 \text{ V}$$

Plot point B (on Fig. 2-10) *at:*
$$I_F = 200 \text{ mA and } V_F = (0.7 \text{ V} + 0.05 \text{ V})$$

Draw the characteristic through points A and B.

(a) Basic dc equivalent circuit

dc Equivalent Circuits

An *equivalent circuit* for a device is a circuit that represents the device behaviour. Usually, the equivalent circuit is made up of a number of components, such as resistors and voltage cells. A diode equivalent circuit may be substituted in place of the device when investigating a circuit containing the diode. Equivalent circuits may also be used as device *models* in computer analysis of circuits.

In Example 2-3 a forward-biased diode is assumed to have a constant forward voltage drop (V_F) and negligible series resistance. In this case the *dc* equivalent circuit is assumed to be a voltage cell with a voltage V_F, [see Fig. 2-11(a)]. This simple *dc* equivalent circuit is quite suitable for a great many diode applications.

A more accurate equivalent circuit includes the diode dynamic resistance (r_d) in series with the voltage cell, as shown in Fig. 2-11(b). This takes account of the small variations in V_F that occur with change in forward current. An ideal diode is also included to show that curent flows only in one direction. The equivalent circuit without r_d assumes that the diode has the approximate characteristics illustrated in Fig. 2-8(b) or (c). With r_d included, the equivalent circuit represents a diode with the type of piecewise linear characteristic in Fig. 2-10. Consequently, the circuit in Fig. 2-11(b) is known as the *piecewise linear equivalent circuit.*

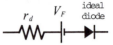

(b) Complete dc equivalent circuit

Figure 2-11
Dc equivalent circuits for a junction diode.

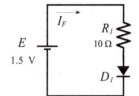

(a) Diode circuit

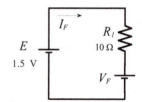

(b) Diode replaced with voltage cell

Example 2-5

Calculate I_F for the diode circuit in Fig. 2-12(a) assuming that the diode has V_F = 0.7 V and r_d = 0. Then, recalculate the current taking r_d = 0.25 Ω.

Solution

Substituting V_F as the diode equivalent circuit, [Fig 2-12(b)]:

$$I_F = \frac{E - V_F}{R_1} = \frac{1.5 \text{ V} - 0.7 \text{ V}}{10 \text{ }\Omega}$$
$$= 80 \text{ mA}$$

Substituting V_F and r_d as the diode equivalent circuit, [Fig 2-12(c)]:

$$I_F = \frac{E - V_F}{R_1 + r_d} = \frac{1.5 \text{ V} - 0.7 \text{ V}}{10 \text{ }\Omega + 0.25 \text{ }\Omega}$$
$$= 78 \text{ mA}$$

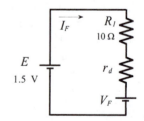

(c) Diode replaced with r_d and V_F

Figure 2-12
Diode circuits for Example 2-5.

Practise Problems

2-3.1   Calculate the new level of diode current in the circuit in Fig. 2-9 when D_1 is replaced with two series-connected silicon diodes.

2-3.2   A germanium diode has a maximum forward current of 100 mA, and a 0.5 Ω dynamic resistance Construct the piecewise linear characteristics for this diode on Fig. 2-10.

2-3.3   Calculate the circuit current when the diode in Problem 2-3.2 is connected in series with a 15 Ω resistor and a 3 V battery which forward biased the diode.

2-4   DC Load Line Analysis

DC Load line

Figure 2-13(a) shows a diode in series with a 100 Ω resistance (R_1) and a supply voltage (E). The polarity of E is such that the diode is forward biased, so a diode forward current (I_F) flows. As already discussed, the circuit current can be determined approximately by assuming a constant diode forward voltage drop (V_F). When the precise levels of the diode current and voltage must be calculated, *graphical analysis* (also termed *dc load line analysis*) is employed.

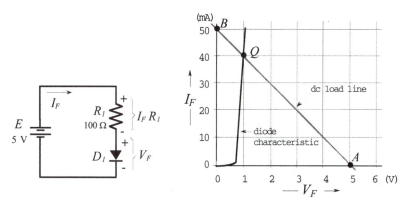

(a) Diode-resistor series circuit

(b) Plotting the dc load line on
the diode characteristics

Figure 2-13
Drawing a dc load line on the diode characteristic.

For graphical analysis, a *dc load line* is drawn on the diode forward characteristics, [Fig. 2-13(b)]. This is a straight line that illustrates all dc conditions that could exist within the circuit. Because the load line is always straight, it can be constructed by plotting any two corresponding current and voltage points and then drawing a straight line through them. To determine two points on the load line, an equation relating voltage, current,and resistance is first derived for the circuit. From Fig. 2-13(a),

$$E = (I_F \, R_1) + V_F \qquad\qquad (2\text{-}3)$$

Any convenient two levels of I_F can be substituted into Eq. 2-3 to calculate corresponding V_F levels, or vice versa. As demonstrated in Example 2-6, it is convenient to calculate V_F with $I_F = 0$, and to determine I_F when $V_F = 0$.

Example 2-6

Draw the dc load line for the circuit in Fig. 2-13(a). The diode forward characteristic is given in Fig. 2-13(b).

Solution

Substitute $I_F = 0$ into Eq. 2-3,

$$E = (I_F R_1) + V_F = 0 + V_F$$

or,
$$V_F = E = 5 \text{ V}$$

Plot point A on the diode characteristic at,

$$I_F = 0, \text{ and } V_F = 5 \text{ V}$$

Now substitute $V_F = 0$ into Eq. 2-3,

$$E = (I_F R_1) + 0$$

giving,
$$I_F = \frac{E}{R_1} = \frac{5 \text{ V}}{100 \text{ }\Omega}$$
$$= 50 \text{ mA}$$

Plot point B on the diode characteristic at,

$$I_F = 50 \text{ mA, and } V_F = 0$$

Draw the dc load line through points A and B.

Q Point

The relationship between the diode forward voltage and current in the circuit in Fig. 2-13(a) is defined by the device characteristic. Consequently, there is only one point on the *dc* load line where the diode voltage and current are compatible with the circuit conditions. That is *point Q*, termed the *quiescent point* or *dc bias point*, where the load line intersects the characteristic. This may be checked by substituting the levels of I_F and V_F at point Q into Eq. 2-3. From the Q point on Fig. 2-13(b), $I_F = 40$ mA and $V_F = 1$ V. Equation 2-3 states that $E = (I_F R_1) + V_F$. Therefore,

$$E = (40 \text{ mA} \times 100 \text{ }\Omega) + 1 \text{ V} = 5 \text{ V}$$

So, with $E = 5$ V and $R_1 = 100$ Ω, the only levels of I_F and V_F that can satisfy Eq. 2-3 on the diode characteristics in Fig. 2-13(b) are $I_F = 40$ mA and $V_F = 1$ V.

Note that, although $V_F = 0$ and $V_F = 5$ V were used when drawing the dc load line in Ex. 2-6, no functioning semiconductor diode would have such forward voltage drops. These are simply convenient theoretical levels for constructing the *dc* load line.

Calculating Load Resistance and Supply Voltage

In a diode series circuit [see Fig. 2-14(a)], resistor R_1 dictates the slope of the *dc* load line, and supply voltage E determines point A on the load line. So, the circuit conditions can be altered by changing either R_1 or E.

When designing a diode circuit, it may be required to use a given supply voltage and set up a specified forward current. In this case, points A and Q are first plotted and the load line is drawn. R_1 is then calculated from the slope of the load line. The problem could also occur in another way. For example, R_1 and the required I_F are known, and the required supply voltage is to be determined. This problem is solved by plotting point Q and drawing the load line with slope $1/R_1$. The supply voltage is then read at point A.

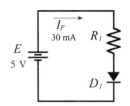

(a) Diode-resistor circuit

Example 2-7

Determine the required load resistance for the circuit in Fig. 2-14(a) using the device characteristics in Fig. 2-14(b).

Solution

From Eq. 2-3, $V_F = E - (I_F R_1)$

Substituting $I_F = 0$, $V_F = E - 0 = 5$ V

Plot point A on the diode characteristic in Fig. 2-14(b) at,

$$I_F = 0, \text{ and } V_F = 5 \text{ V}$$

Now plot point Q in Fig. 2-14(b) at,

$$I_F = 30 \text{ mA}$$

Draw the new dc load line through points A and Q.

From the load line,

$$R_1 = \frac{\Delta V_F}{\Delta I_F} = \frac{5 \text{ V}}{37.5 \text{ mA}}$$

$$= 133 \ \Omega$$

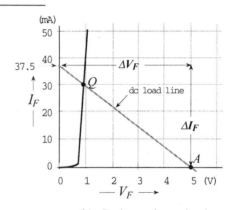

(b) Resistor determination

Figure 2-14
Determination of the series resistance from the slope of the dc load line.

Example 2-8

Determine a new supply voltage for the circuit in Fig. 2-14(a) to give a 50 mA diode forward current when $R_1 = 100 \ \Omega$.

Solution

Plot point Q on the diode characteristic in Fig. 2-15 at,

$$I_F = 50 \text{ mA}$$

and from the characteristic read,

$$V_F = 1.1 \text{ V}$$

From Eq. 2-3, $V_F = E - (I_F R_1)$

When I_F changes from 50 mA to 0,

$$\Delta I_F = 50 \text{ mA (see Fig. 2-15)}$$

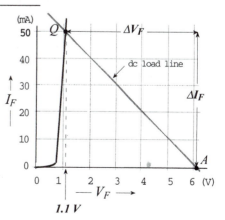

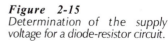

Figure 2-15
Determination of the supply voltage for a diode-resistor circuit.

and, $\Delta V_F = I_F R_1 = 50 \text{ mA} \times 100 \ \Omega$ (see Fig. 2-15)
$$= 5 \text{ V}$$

The new level of E is, $E = V_F + \Delta V_F = 1.1 \text{ V} + 5 \text{ V}$
$$= 6.1 \text{ V}$$

Point A may now be plotted (on Fig. 2-15) at $I_F = 0$, and $E = 6.1 \text{ V}$, and the new dc load line may be drawn through points A and Q.

Practise Problems

2-4.1   A diode with the characteristics in Fig. 2-15 is connected in series with a 3 V supply and a 100 Ω resistor. Draw the dc load line for the circuit, and determine the forward current level.

2-4.2   Calculate a new load resistance for the circuit in Problem 2-4.1 to produce a 40 mA diode forward current.

2-4.3   A diode with the characteristics in Fig. 2-15 is to have a 35 mA forward current when connected in series with a 70 Ω resistor. Determine the required supply voltage.

2-5   Temperature Effects

Diode Power Dissipation

The power dissipation in a diode is simply calculated as the device terminal voltage multiplied by the current level.

$$P = V_F I_F \qquad \qquad \textbf{(2-4)}$$

Device manufacturers specify a maximum power dissipation for each type of diode. If the specified level is exceeded, the device will overheat and it may short-circuit or open-circuit.

The maximum power that may be dissipated in a diode (or any other electronic device) is normally specified for an ambient temperature of 25°C, or sometimes for a 25°C case temperature. When the temperature exceeds the specified level, the device maximum power dissipation must be derated.

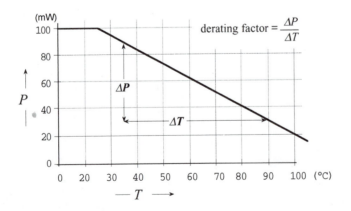

Figure 2-16
A power versus temperature graph shows how a device maximum power dissipation must be derated with increasing temperature.

Figure 2-16 shows the type of power-versus-temperature graph provided on some device data sheets. The maximum power dissipation for any given temperature can simply be read from the graph. Then, the maximum level of forward current is calculated from Eq. 2-4. Instead of a power-versus-temperature graph, the rectifier diode data sheet in Appendix 1-2 has a current-versus-temperature graph. This allows the maximum forward current at a desired temperature to be determined directly.

As an alternative to a power or current graph, a *derating factor* is sometimes listed on a device data sheet. As illustrated in Fig. 2-16, the derating factor defines the slope of the power-versus-temperature graph. It can be used to draw the graph, or employed directly without reference to the graph. The equation for the new maximum power dissipation when the temperature changes involves the specified power at the specified temperature (P_1 at T_1), the temperature change (ΔT), and the derating factor.

$$\boldsymbol{P_2 = (P_1 \text{ at } T_1) - [\Delta T \times (\textit{derating factor})]} \qquad \textbf{(2-5)}$$

Example 2-9
A diode with a 700 mW maximum power dissipation at 25°C has a 5 mW/°C derating factor. If the forward voltage drop remains constant at 0.7 V, calculate the maximum forward current at 25°C and at 65°C temperatures.

Solution
At 25°C:

From Eq. 2-4, $\quad I_F = \dfrac{P}{V_F} = \dfrac{700 \text{ mW}}{0.7 \text{ V}}$

$\qquad\qquad\qquad\quad = 1 \text{ A}$

At 65°C:

Eq. 2-5, $\quad P_2 = (P_1 \text{ at } T_1) - [\Delta T \times (\textit{derating factor})]$

$\qquad\qquad\quad = 700 \text{ mW} - [(65°C - 25°C) \times 5 \text{ mW/°C}]$

$\qquad\qquad\quad = 500 \text{ mW}$

From Eq. 2-4, $\quad I_F = \dfrac{P_2}{V_F} = \dfrac{500 \text{ mW}}{0.7 \text{ V}}$

$\qquad\qquad\qquad = 714 \text{ mA}$

Forward Voltage Drop
There are circumstances in which it is important to know the precise level of a diode forward voltage drop. In these cases, the graphical analysis techniques discussed in Section 2-4 may be appropriate. However, as explained in Section 1-6 and illustrated in Fig. 2-17, the voltage drop across a forward-biased *pn*-junction changes with temperature by approximately -1.8 mV/°C for a silicon device, and by -2.02 mV/°C for germanium. A diode V_F at any temperature can be calculated from a knowledge of V_F at the starting temperature (V_{F1} at T_1), the temperature change (ΔT), and the voltage/temperature coefficient ($\Delta V_F/°C$).

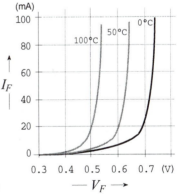

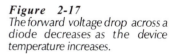

Figure 2-17
The forward voltage drop across a diode decreases as the device temperature increases.

$$V_{F2} = (V_{F1} \text{ at } T_1) + [\Delta T \times (\Delta V_F/°C)] \qquad \textbf{(2-6)}$$

Dynamic Resistance

Equation 2-2 for calculating the dynamic resistance of a forward biased diode is correct only for 25°C junction temperatures. For higher or lower temperatures, the equation must be modified to,

$$r'_d = \frac{26 \text{ mV}}{I_F} \left[\frac{T + 273°C}{298°C} \right] \qquad \textbf{(2-7)}$$

where T is the junction temperature in degrees Celsius.

Example 2-10

A silicon diode with a 0.7 V forward voltage drop at 25°C is to be operated with a constant forward current up to a temperature of 100°C. Calculate the diode V_F at 100°C. Also, determine the junction dynamic resistance at 25°C and at 100 °C if the forward current is 26 mA.

Solution

Eq. 2-6,

$$\begin{aligned} V_{F2} &= (V_{F1} \text{ at } T_1) + [\Delta T \times (\Delta V_F/°C)] \\ &= 0.7 \text{ V} + [(100°C - 25°C) \times (-1.8 \text{ mV/°C})] \\ &= 0.565 \text{ V} \end{aligned}$$

at 25°C,

Eq. 2-7,

$$\begin{aligned} r'_d &= \frac{26 \text{ mV}}{I_F} \left[\frac{T + 273°C}{298°C} \right] \\ &= \frac{26 \text{ mV}}{26 \text{ mA}} \left[\frac{25°C + 273°C}{298°C} \right] \\ &= 1 \ \Omega \end{aligned}$$

at 100°C,

$$\begin{aligned} r'_d &= \frac{26 \text{ mV}}{26 \text{ mA}} \left[\frac{100°C + 273°C}{298°C} \right] \\ &= 1.25 \ \Omega \end{aligned}$$

Practise Problems

2-5.1   A diode with a constant 0.65 V forward drop has the power-versus-temperature graph in Fig. 2-16. Calculate the maximum forward current that may be passed at 25°C and at 80°C.

2-5.2   Calculate the maximum and minimum levels of V_F for a germanium diode with V_F = 0.3 V at 25°C when operated over a temperature range of 10°C to 80°C. Also, determine the junction dynamic resistances at the temperature extremes if I_F is 20 mA.

2-5.3   A diode with a 1 W maximum power dissipation at 25°C has a 4 mW/°C derating factor. Calculate the maximum power that may be dissipated in the diode when its temperature is 80°C.

2-6 AC Equivalent Circuits

Junction Capacitances

The depletion region of a *pn*-junction (see Section 1-5) is a layer depleted of charge carriers situated between two blocks of low resistance material. Because this is the description of a capacitor, the depletion region clearly has a capacitance. The *depletion layer capacitance* (C_{pn}) may be calculated from the equation for a parallel-plate capacitor if the junction dimensions are known. Typically, C_{pn} is 4 pF for a low-current diode.

The depletion layer capacitance is essentially the capacitance of a reverse-biased *pn*-junction, [Fig. 2-18(a)]. Consider the forward-biased junction in Fig. 2-18(b). If the applied voltage is suddenly reversed, forward current I_F ceases immediately leaving some majority charge carriers in the depletion region. These charge carriers must flow back out of the depletion region, which is widened when the junction is reverse biased. The result is that, when a forward-biased junction is suddenly reversed, a reverse current flows which is large initially and slowly decreases to the level of the reverse saturation current. The effect may be likened to the discharging of a capacitor, and so it is represented by a capacitance known as the *diffusion capacitance* (C_d).

It can be shown that C_d is proportional to the forward current I_F. This is to be expected, since the number of charge carriers in the depletion region must be directly proportional to I_F. For a low-current diode with a 10 mA forward current, a typical value of diffusion capacitance is around 1 nF.

(a) A depletion layer capacitance occurs at a reverse-biased diode

(b) A diffusion capacitance is present at a forward-biased diode

Figure 2-18
The capacitance of a diode depends upon the polarity of the applied voltage, and on the current level.

ac Equivalent Circuits (Reverse- and Forward-Biased)

A reverse-biased diode can be simply represented by the (high non-constant) reverse resistance R_R in parallel with the depletion layer capacitance C_{pn}, [Fig. 2-19(a)]. The equivalent circuit for a forward-biased diode consists of the dynamic resistance r_d in series with a voltage cell representing V_F, as discussed in Section 2-3. To allow for the effect of the diffusion capacitance, C_d is included in parallel. This gives the complete equivalent circuit shown in Fig. 2-19(b).

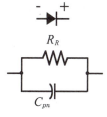

(a) Equivalent circuit for a reverse-biased diode

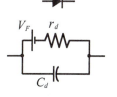

(b) Equivalent circuit for a forward-biase diode

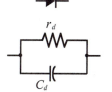

(c) AC equivalent circuit for a forward-biased diode

Figure 2-19
Equivalent circuits for reverse-biased and forward biased diodes.

The equivalent circuit for the forward-biased diode may be modified to form an *ac equivalent circuit*. This applies to diodes that are maintained in a forward-biased condition while subjected to small variations in I_F and V_F. The ac equivalent circuit is created simply by removing the voltage cell representing V_F from the complete equivalent circuit, [see Fig. 2-19(c)].

Reverse Recovery Time

In many applications, diodes must switch rapidly between forward and reverse bias. Most diodes switch very quickly into the forward-biased condition, however, there is a longer *turn-off* time due to the junction diffusion capacitance.

Figure 2-20 illustrates the effect of a voltage pulse on the diode forward current. When the pulse switches from positive to negative, the diode conducts in reverse instead of switching *off* sharply, [Fig. 2-20(a)]. The reverse current (I_R) initially equals the forward current (I_F), then it gradually decreases toward zero. The high level of reverse current occurs because at the instant of reverse bias there are charge carriers crossing the junction depletion region, and these must be removed. (The same effect that produces diffusion capacitance.) The *reverse recovery time* (t_{rr}) is the time required for the current to decrease to the reverse saturation current level. Typical values of t_{rr} for switching diodes range from 4 ns to 50 ns. Figure 2-20(b) shows that, to keep the diode reverse current to a minimum, the *fall time* (t_f) of the applied voltage pulse must be much larger than the diode t_{rr}. Typically,

$$t_{f(min)} = 10\ t_{rr} \qquad\qquad \textbf{(2-8)}$$

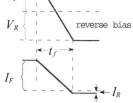

(a) Effect of instantaneous reverse bias

(b) Effect of $t_f \gg t_{rr}$

Figure 2-20
The minimum on/off switching time of a diode is limited by the reverse recovery time.

Example 2-11
Calculate the minimum fall time for voltage pulses applied to *1N915* and *1N917* diodes.

Solution
From the diode data sheet in Appendix 1-1,

$$t_{rr} = 10 \text{ ns for the 1N915, and } t_{rr} = 3 \text{ ns for the 1N917}$$

For the 1N915,
Eq. 2-8,
$$t_{f(min)} = 10\ t_{rr} = 10 \times 10 \text{ ns}$$
$$= 100 \text{ ns}$$

For the 1N917,
Eq. 2-8,
$$t_{f(min)} = 10\ t_{rr} = 10 \times 3 \text{ ns}$$
$$= 30 \text{ ns}$$

Practise Problems

2-6.1   Determine the maximum reverse recovery time for a diode with an applied voltage pulse that has a 0.5 μs fall-time.

2-6.2   Estimate a suitable minimum fall time for switching a diode from *on* to *off* if the diode has $t_{rr} = 15$ ns.

2-7 Diode Specifications

Diode Data Sheets

To select a suitable diode for a particular application, the *data sheets*, or *specifications,* provided by device manufacturers must be consulted. Portions of typical diode data sheets are shown in Fig. 2-21 and in Appendices 1-1 through 1-3.

Most data sheets start off with the device type number at the top of the page, such as *1N914 through 1N917*, or *1N5391 through 1N5399.* The *1 (one)* in the type number signifies a one-junction device; a diode. A short descriptive title follows the type number; for example, *silicon switching diode*, or *silicon rectifier. Mechanical data* is also given, usually in the form of an illustration showing the package shape and dimensions. The *maximum ratings* at 25°C are then listed, (see Fig. 2-21).

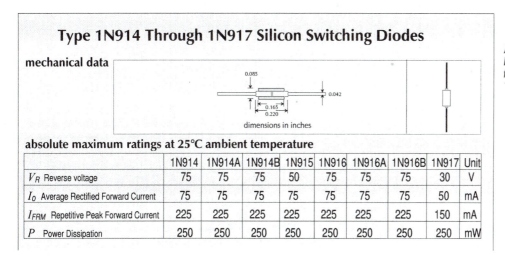

Type 1N914 Through 1N917 Silicon Switching Diodes

mechanical data

dimensions in inches

absolute maximum ratings at 25°C ambient temperature

	1N914	1N914A	1N914B	1N915	1N916	1N916A	1N916B	1N917	Unit
V_R Reverse voltage	75	75	75	50	75	75	75	30	V
I_O Average Rectified Forward Current	75	75	75	75	75	75	75	50	mA
I_{FRM} Repetitive Peak Forward Current	225	225	225	225	225	225	225	150	mA
P Power Dissipation	250	250	250	250	250	250	250	250	mW

Figure 2-21
Portion of data sheet for 1N914 through 1N917 diodes.

The maximum ratings are the maximum voltages, currents, etc., that can be applied without destroying the device. It is very important that these ratings not be exceeded, otherwise failure of the diode is quite likely. *For reliability, the maximum ratings should not even be approached.* If a diode is to survive a 50 V reverse bias, one that has a 75 V peak reverse voltage should be selected. If the diode peak forward current is to be 100 mA, use a device that can handle 150 mA. It also is important to note that the maximum ratings must be adjusted downward for operation at temperatures greater than 25°C, (see Section 2-5).

A list of other electrical characteristics for the device normally follows the maximum ratings. An understanding of all the parameters specified on a data sheet will not be achieved until the data sheets have been consulted frequently. However, some of the most important parameters are considered below:

V_R or V_{RRM} *Peak reverse voltage,* (also termed *peak inverse voltage,* and *DC blocking voltage*). This is the maximum reverse voltage that may be applied across the diode.

I_o or $I_{F(AV)}$ *Steady-state forward current.* The maximum current that may be passed continuously through the diode.

I_{FSM} *Non-repetitive peak surge current.* This current may be passed for a specified time period. The surge current is very much higher than the normal maximum forward current. It is a current that may be allowed flow briefly when a circuit is first switched *on.*

I_{FRM} *Repetitive peak surge current.* Peak current that may be repeated over and over again; for example, during each cycle of a rectified waveform.

V_F Static forward voltage drop. The maximum forward volt drop for a given forward current and device temperature.

P Continuous power dissipation at 25°C. The maximum power that the device can safely dissipate continuously in free air. This rating must be downgraded at higher temperatures (see Section 2-5).

Low-Power Diodes

The data sheet portion in Fig. 2-21 identifies the *1N914* to *1N917* devices as *switching diodes.* The average rectified forward current is listed as 75 mA (except for the *lN917*). Maximum reverse voltage ranges from 30 V to 75 V. Thus, these diodes are intended for relatively low-current, low-voltage applications, in which they may be required to switch rapidly between the *on* and *off* states.

Rectifier Diodes

Appendices 1-2 and 1-3 shows data sheets for low-power rectifiers. It is seen that the *1N4000* range of rectifiers can pass an average forward current of 1 A, and that the *1N5390* range can pass 1.5 A. Both types have maximum reverse voltages ranging from 50 V to 1000 V. Unlike the case of the switching diodes specified in Appendix 1-1, the reverse recovery time is not listed on the rectifier data sheet. Rectifier diodes are generally intended for low-frequency applications (60 Hz to perhaps 400 Hz) in which switching time is not important, (see Sections 3-1 and 3-2).

Example 2-12

Referring to Fig. 2-21, determine the following quantities for a *1N915* diode: peak reverse voltage, steady-stage forward current, peak repetitive forward current, power dissipation.

Solution
For the 1N915,

$$PIV = V_R = 50 \text{ V}$$
$$I_o = 75 \text{ mA}$$
$$I_{FRM} = 225 \text{ mA}$$
$$P = 250 \text{ mW}$$

Practise Problems

2-7.1   Referring to Appendix 1-3, determine the following quantities for a *1N5397* diode: peak reverse voltage, steady-stage forward current, nonrepetitive forward current.

2-7.2   A rectifier has to pass an average current of 600 mA, and survive a repetitive reverse voltage of 75 V. Select a suitable diode from Appendices 1-1 and 1-3.

2-8   Diode Testing

Ohmmeter Tests

Several methods are available for testing diodes. One of the simplest and quickest tests can be made using an ohmmeter to measure the forward and reverse resistance of a diode, [see Fig. 2-22(a)]. The diode should offer a low resistance when forward biased, and a high resistance when reverse biased. A diode is short-circuited when it displays a low resistance for both forward and reverse bias. If a high resistance is measured with both bias polarities, the device is open-circuited. Usually, the actual indicated high resistance is approximately half the selected ohmmeter range.

Obviously, the ohmmeter test will identify the anode and cathode of a diode, if there is doubt about the diode terminals. However, it is very important to note that *when some multifunction instruments are used as ohmmeters, the voltage polarity at the terminals may **not** be the same as the polarity marked on the instrument*. A voltmeter should be used, as illustrated in Fig. 2-22(a), to check the ohmmeter terminal polarity.

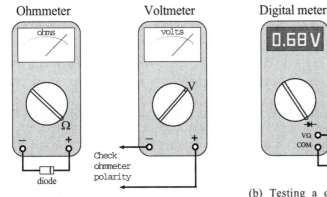
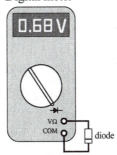

(a) Testing a diode with an ohmmeter

(b) Testing a diode with a digital multimeter

Figure 2-22
Analog or digital multimeters may be used for diode testing.

Use of a Digital Meter

Many portable multifunction digital instruments have a diode testing facility which displays the diode forward voltage when the terminals are connected *positive* to the anode, *negative* to the cathode [Fig. 2-22(b)]. When reverse connected, a functioning diode produces an *OL* display. If an *OL* display occurs with both forward and reverse connections, the diode is open-circuited. When the display is *000*, the device is short-circuited.

Diode Characteristics Plotting

The forward characteristics of a diode can be obtained by use of the circuit illustrated in Fig. 2-23(a). The diode voltage is set at a series of convenient levels, and the corresponding current levels are measured and recorded. The characteristics are then plotted from the table of quantities obtained in this way. The reverse characteristics can be obtained in the same way, except that a very sensitive microammeter is required to measure the diode reverse current, [Fig. 2-23(b)]. The microammeter must be connected directly in series with the diode, as shown; otherwise the voltmeter current may introduce a serious error.

Figure 2-24 shows a method of using an *XY recorder* for drawing the forward characteristics of a diode. The resistor voltage (V_{R1}) is directly proportional to the diode forward current (I_F). So, V_{R1} is applied to the vertical input terminals of the *XY* recorder, as illustrated. The diode forward voltage (V_F) goes to the horizontal input terminals. When the power supply voltage is slowly increased from zero, the diode forward characteristic is traced out by the pen on the *XY* recorder.

If R_1 in Fig. 2-24 is a 1 kΩ resistor, there is a 1 V drop across R_1 for every 1 mA of diode current. Therefore, with the vertical scale of the *XY* recorder set to 1 V/cm, the (vertical) current co-ordinate of the graph is 1 mA/cm. A convenient scale for the (horizontal) voltage co-ordinate is 0.1 V/cm. Diodes can also be investigated by means of a *curve tracer* which displays the device characteristics on oscilloscope-type screen.

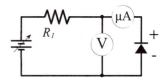

(a) Circuit for obtaining diode forward charactertistics

(b) Circuit for obtaining diode reverse charactertistics

Figure 2-23
Diode characteristics can be plotted from a table of corresponding current and voltage measurements

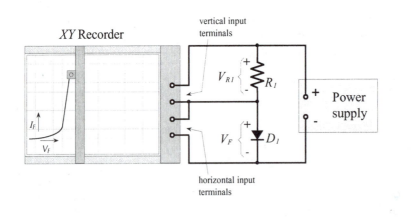

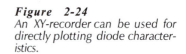

Figure 2-24
An XY-recorder can be used for directly plotting diode character-istics.

Example 2-13

The arrangement shown in Fig. 2-24 is to be used to plot the characteristics of a 1N914 diode. Select resistor R_1, and determine appropriate V/cm scales for the *XY* recorder.

Solution

From Fig. 2-21, $I_o = 75$ mA (for the 1N914)

Select a vertical scale of 5 mA/cm for the characteristic to give,

$$\text{vertical scale length (for } I_F) = \frac{75 \text{ mA}}{5 \text{ mA/cm}}$$
$$= 15 \text{ cm}$$

Select a vertical scale of 1 V/cm for the XY recorder, so that 15 cm represents 15 V, as well as 75 mA.

$$R_1 = \frac{15 \text{ V}}{75 \text{ mA}}$$
$$= 200 \ \Omega$$

and, $P_{R1(max)} = [I_{F(max)}]^2 R_1 = (75 \text{ mA})^2 \times 200 \ \Omega$
$$= 1.1 \text{ W}$$

V_F might be as large as 0.8 V, so select a horizontal scale of 0.1 V/cm to give,

$$\text{horizontal scale length (for } V_F) = \frac{0.8 \text{ V}}{0.1 \text{ V/cm}}$$
$$= 8 \text{ cm}$$

Practise Problems

2-8.1 Plot the forward characteristics for a silicon diode from the following experimental data:

V_F (V)	0.6	0.62	0.64	0.66	0.68	0.7	0.72	0.74	0.76
I_F (mA)	1	1.4	1.8	2.5	10	50	90	130	170

2-8.2 An XY recorder is used, as in Fig. 2-24, to plot silicon diode forward characteristics to approximately fill a 10 cm by 10 cm square. If the maximum forward current is to be 100 mA, select suitable V/cm scales for the recorder, and a suitable resistance for R_1.

2-9 Zener Diodes

Junction Breakdown

When a junction diode is reverse biased, normally only a very small reverse saturation current flows; I_S on the reverse characteristic in Fig. 2-25(a). When the reverse voltage is sufficiently increased, the junction *breaks down* and a large reverse current flows. If the reverse current is limited by means of a suitable series resistor, [R_1 in the circuit in Fig. 2-25(b)], the power dissipation in the junction can be kept to a level that will not destroy the device. In this case, the diode may be operated continuously in reverse breakdown. The reverse current returns to its normal level when the voltage is reduced below the reverse breakdown level.

Diodes designed for operation in reverse breakdown are found to have a breakdown voltage that remains extremely stable over a

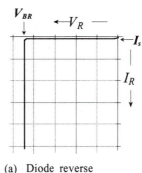

(a) Diode reverse characteristic

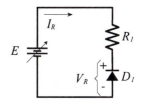

(b) Diode-resistor circuit

Figure 2-25
A diode can be operated in reverse breakdown if the current is limited by means of a series-connected resistor.

wide range of current levels. This property gives the *breakdown diode* many useful applications as a voltage reference source.

There are two mechanisms that cause breakdown in a reverse biased *pn*-junction. With a very narrow depletion region, the electric field strength (volts/width) produced by a reverse bias voltage can be very high. The high intensity electric field causes electrons to break away from their atoms, thus converting the depletion region from an insulating material into a conductor. This is *ionization by electric field,* also called *Zener breakdown,* and it usually occurs with reverse bias voltages less than 5 V.

In cases where the depletion region is too wide for Zener breakdown, the electrons in the reverse saturation current can be given sufficient energy to cause other electrons to break free when they strike atoms within the depletion region. This is termed *ionization by collision.* The electrons released in this way collide with other atoms to produce more free electrons in an *avalanche* effect. *Avalanche breakdown* is normally produced by reverse voltage levels above 5 V. Although *Zener* and *avalanche* are two different types of breakdown, the name *Zener diode* is commonly applied to all breakdown diodes.

Circuit Symbol and Package
The circuit symbol for a Zener diode in Fig. 2-26(a) is the same as that for an ordinary diode, but with the cathode bar approximately in the shape of a letter *Z.* The arrowhead on the symbol still points in the (conventional) direction of forward current when the device is forward biased. As illustrated, for operation in reverse bias, the voltage drop (V_Z) is + on the cathode, - on the anode.

Low-power Zener diodes are available in a variety of packages. For the device package shown in Fig. 2-26(b), the colored band identifies the cathode terminal, as in the case of an ordinary low-current diode. High-current Zener diodes are also available in the type of package that allows for mounting on a heat sink.

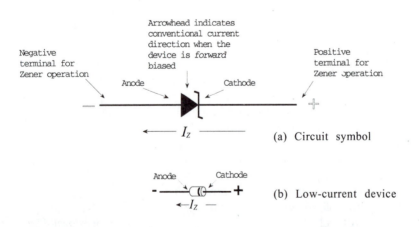

Figure 2-26
Zener diode circuit symbol and low-current zener diode package.

Characteristics and Parameters
The typical characteristics of a Zener diode are shown in detail in Fig. 2-27. Note that the forward characteristic is simply that of an

ordinary forward-biased diode. Some important points on the reverse characteristic are:

V_Z *Zener breakdown voltage*

I_{ZT} *Test current* for measuring V_Z

I_{ZK} Reverse current near the *knee* of the characteristic; the *minimum reverse current* to sustain breakdown

I_{ZM} *Maximum Zener current;* limited by the maximum power dissipation.

The *dynamic impedance* (Z_Z) is another important parameter that may be derived from the characteristics. As illustrated in Fig. 2-27, Z_Z defines how V_Z changes with variations in diode reverse current. When measured at I_{ZT}, the dynamic impedance is designated (Z_{ZT}). The dynamic impedance measured at the knee of the characteristic (Z_{ZK}) is substantially larger than Z_{ZT}.

$$Z_Z = \frac{\Delta V_Z}{\Delta I_Z} \qquad (2\text{-}9)$$

The Zener diode current may be any level between I_{ZK} and I_{ZM}. For greatest voltage stability, the diode is normally operated at the test current. Many low-power Zener diodes have a test current specified as 20 mA, however, some devices have lower test currents.

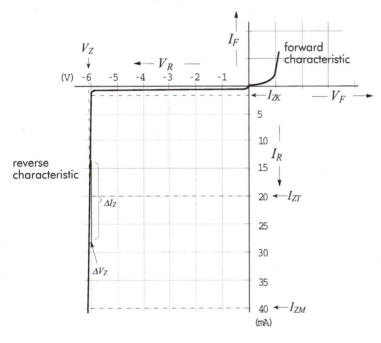

Figure 2-27
Typical characteristics for a Zener diode. The most important parameters are: the Zener voltage V_Z, the knee current I_{ZK}, the test current I_{ZT}, and the maximum current I_{ZM}.

Data Sheet

Portions of a data sheet for low-power Zener diodes with voltages ranging from 3.3 V to 12 V is shown in Fig. 2-28. (A data sheet for 2.4 V to 110 V Zener diodes is presented in Appendix 1-4.) Note in Fig. 2-28 that the V_Z tolerance is ±5% or ±10%. This means, for example, that for a *1N753* with a ±10% tolerance the actual V_Z level is 6.3 V ±10%, or 5.67 V to 6.93 V. The Zener voltage remains stable at whatever it happens to be within this range.

The data sheet also lists the dynamic impedance, reverse leakage current, and the temperature coefficient for the V_Z of each device. The Zener voltages at any temperature can be calculated from,

$$V_{Z2} = (V_{Z1} \text{ at } T_1) + [\Delta T \, \alpha_Z \, V_Z/100)] \qquad (2\text{-}10)$$

Temperature compensated Zener diodes are also available with extremely low temperature coefficients.

Type 1N746 Through 1N759 Silicon Zener Diodes

3.3 V to 12 V (±5% or ±10%)
P_D = 400 mW (derate linearly above 50°C at 3.2 mW/°C)

electrical characteristics at 25°C ambient temperature

Type number	Nominal Zener voltage V_Z (V)	Test current I_Z (mA)	Zener impedance Z_{ZT} (Ω)	leakage current I_R (µA)	Reverse Temperature coefficient α_Z (%/°C)
1N746	3.3	20	28	10	-0.062
1N747	3.6	20	24	10	-0.055
1N753	6.2	20	7	0.1	+0.022
1N755	7.5	20	6	0.1	+0.045
1N757	9.1	20	10	0.1	+0.056
1N759	12.0	20	30	0.1	+0.060

Figure 2-28
Portions of a data sheet for low-power Zener diodes.

Low-power Zener diodes are typically limited to a maximum power dissipation of 400 mW, (P_D in Fig. 2-28). Higher-power devices are available. All of the power dissipations must be derated with temperature increase, exactly as explained in Section 2-5. When the maximum Zener current is not listed on the device data sheet, it may be calculated from the power dissipation equation.

$$P_D = V_Z \times I_{ZM} \qquad (2\text{-}11)$$

Example 2-14
Calculate the maximum current that may be allowed to flow through a *1N755* Zener diode at device temperatures of 50°C and 100 °C.

Solution
From the data sheet in Fig. 2-28, the 1N755 Zener diode has,

V_Z = 7.5 V, P_D = 400 mW at 50°C, and the derating factor = 3.2 mW/°C

At 50°C:
From Eq. 2-11,
$$I_{ZM} = \frac{P_D}{V_Z} = \frac{400 \text{ mW}}{7.5 \text{ V}}$$
$$= 53.3 \text{ mA}$$

At 100°C:
Eq. 2-5,
$$P_2 = (P_1 \text{ at } T_1) - [\Delta T \times (\text{derating factor})]$$
$$= 400 \text{ mW} - [(100°C - 50°C) \times 3.2 \text{ mW/°C}]$$
$$= 240 \text{ mW}$$

From Eq. 2-11, $I_{ZM} = \dfrac{P_2}{V_z} = \dfrac{240 \text{ mW}}{7.5 \text{ V}}$

$$= 32 \text{ mA}$$

Example 2-15

The Zener diode circuit in Fig. 2-29 has $E = 20$ V, $R_1 = 620\,\Omega$, and a *1N755* Zener diode. Calculate the diode current and power dissipation.

Solution

From the data sheet in Fig. 2-28, the 1N755 Zener diode has, $V_Z = 7.5$ V

$$V_{R1} = E - V_Z = 20 \text{ V} - 7.5 \text{ V}$$
$$= 12.5 \text{ V}$$

$$I_Z = I_{R1} = \frac{V_{R1}}{R_1} = \frac{12.5 \text{ V}}{620\,\Omega}$$

$$= 20.16 \text{ mA}$$

$$P_D = V_Z\, I_Z = 7.5 \text{ V} \times 20.16 \text{ mA}$$
$$= 151 \text{ mW}$$

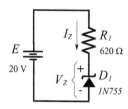

Figure 2-29
Circuit for Example 2-15

Equivalent Circuit

The *dc* equivalent circuit for a Zener diode is simply a voltage cell with a voltage V_Z, as in Fig. 2-30(a). This is the complete equivalent circuit for the device for all dc calculations. For the ac equivalent circuit [Fig. 2-30(b)], the dynamic impedance is included in series with the voltages cell. The *ac* equivalent circuit is used in situations where the Zener current is varied by small amounts. It must be understood that *these equivalent circuits apply only when the Zener diode is maintained in reverse breakdown.* If the device becomes forward biased, then the equivalent circuit for a forward-biased diode must be used.

(a) *dc* eqivalent circuit

(b) *ac* eqivalent circuit

Figure 2-30
Equivalent circuits for a Zener diode.

Example 2-16

A Zener diode with $V_Z = 4.3$ V has $Z_Z = 22\,\Omega$ at $I_Z = 20$ mA. Calculate the upper and lower limits of V_Z when I_Z changes by ± 5 mA.

Solution

$$\Delta V_Z = \pm(\Delta I_Z \times Z_Z) = \pm(5 \text{ mA} \times 22\,\Omega)$$
$$= \pm 110 \text{ mV}$$

$$V_{Z(max)} = V_Z + \Delta V_Z = 4.3 \text{ V} + 110 \text{ mV}$$
$$= 4.41 \text{ V}$$

$$V_{Z(min)} = V_Z - \Delta V_Z = 4.3 \text{ V} - 110 \text{ mV}$$
$$= 4.19 \text{ V}$$

Chapter-2 Review Questions

Section 2-1
2-1 Sketch the symbol for a semiconductor diode, labelling the anode and cathode, and showing the polarity and current direction for forward bias. Also, show the direction of movement of charge carriers when the device is (a) forward biased, (b) reverse biased.

2-2 Draw sketches to show the appearance of low-current and medium-current diodes, and show how the cathode is identified in each case.

Section 2-2
2-3 Sketch typical forward and reverse characteristics for a germanium diode and for a silicon diode. Discuss the characteristics, and compare silicon and germanium diodes.

2-4 For diodes, define: forward voltage drop, maximum forward current, dynamic resistance, reverse saturation current, and reverse breakdown voltage.

2-5 Show how the diode dynamic resistance can be determined from the forward characteristics. Also, write an equation for calculating the dynamic resistance from *dc* forward current.

Section 2-3
2-6 Sketch the characteristics for an ideal diode, and approximate characteristics for practical diodes. Briefly explain each characteristic.

2-7 Draw the *dc* equivalent circuit for a diode and the piecewise linear equivalent circuit. Discuss the application of each.

Section 2-4
2-8 Explain the purpose of a dc load line. Write the equation for drawing a dc load line for a series circuit consisting of a supply voltage (E), a resistor (R_1), and a diode (D_1).

2-9 Define the Q point in a diode circuit, and explain how it is related to the diode characteristics and the dc load line.

Section 2-5
2-10 Sketch and explain a power-versus-temperature graph for a diode. Define the power derating factor for a diode.

2-11 Discuss how temperature change affects diode forward voltage drop.

Section 2-6

2-12 Explain the origins of depletion layer capacitance and diffusion capacitance, and discuss the importance of each.

2-13 Sketch the complete equivalent circuits for forward-biased and reverse-biased diodes. Also, sketch the ac equivalent circuit for a forward-biased diode. Briefly explain each circuit.

2-14 Define reverse recovery time. Sketch waveforms to show the effect of reverse recovery time on a diode switched rapidly from *on* to *off*. Explain each waveform.

Section 2-7

2-15 Discuss the major differences between switching diodes and rectifier diodes.

2-16 Define the following diode quantities: peak reverse voltage, repetitive peak surge current, steady-state forward current.

Section 2-8

2-17 Describe how an ohmmeter may be used for diode testing.

2-18 Sketch circuits for obtaining diode forward and reverse characteristics by measuring corresponding current and voltage levels. Explain.

2-19 Draw a sketch to show how the forward characteristics of a diode may be plotted on an *XY* recorder.

Section 2-9

2-20 Discuss the different types of junction breakdown that can occur in a reverse-biased diode. Sketch the circuit symbol for a Zener diode, and briefly explain its operation.

2-21 Sketch typical characteristics for a Zener diode. Explain the shape of the characteristics, and identify the important points.

2-22 Sketch the equivalent circuit for a Zener diode. Briefly explain.

Chapter-2 Problems

Section 2-2

2-1 Calculate the static forward resistance for the characteristics in Fig. 2-31 at a 200 mA forward current. Also, determine the reverse resistance at a 75 V reverse voltage.

2-2 Determine the dynamic resistance at a 150 mA forward current for a diode with the characteristics Fig. 2-31.

2-3   Calculate the static forward resistance for the characteristics in Fig. 2-32 at a 25 mA forward current. Also, determine the dynamic resistance for the device at I_F = 25 mA, using the characteristic and using Eq. 2-2.

Section 2-3

2-4   Calculate the forward current in a circuit consisting of a germanium diode connected in series with a 9 V battery and a 3.3 kΩ resistor.

2-5   A silicon diode in series with a 2.7 kΩ resistor and a battery is to have I_F = 1.96 mA. Calculate the battery voltage.

2-6   Draw the piecewise linear characteristics for a silicon diode with a 0.6 Ω dynamic resistance and a 75 mA maximum forward current.

2-7   Draw a straight-line approximation of the forward characteristic for a silicon diode that has a 0.5 Ω dynamic resistance and a maximum forward current of 300 mA.

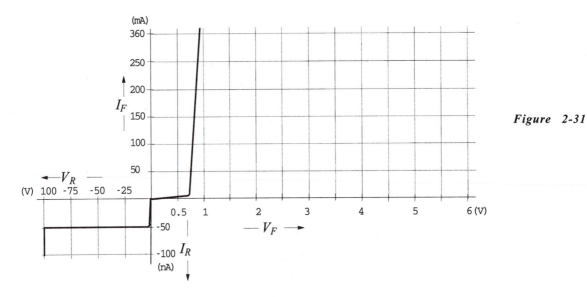

Figure   2-31

Section 2-4

2-8   A diode with the characteristics in Fig. 2-13(b) is to pass a 35 mA current from a 5.5 V supply. Draw the dc load line and calculate the required series resistance value. Determine the new current level when the supply is reduced to 3.5 V.

2-9   A diode with the forward characteristic in Fig. 2-31 is connected in series with a 30 Ω resistance and a 6 V supply. Determine the diode current, and find the new current when the resistance is changed to 20 Ω.

2-10   A diode which has the characteristics shown in Fig. 2-32 is to pass a 20 mA forward current when the supply is 12 V. Determine the value of resistance that must be connected in series with the diode.

2-11   Calculate the new level of supply voltage for the circuit in Problem 2-10 to give a 15 mA current level.

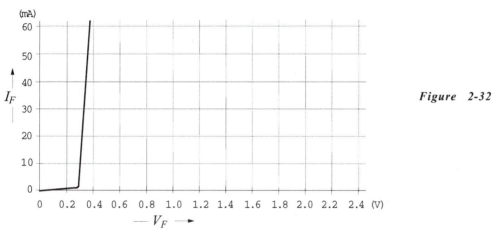

Figure   2-32

Section   2-5

2-12   A diode with a maximum power dissipation of 1000 mW at 25°C is to pass an average forward current of 500 mA. The forward voltage drop for the device is 0.8 V, and the power derating factor is 10 mW/°C. Calculate the maximum temperature at which the diode may be safely operated.

2-13   The diode specified in Problem 2-12 is to be operated at a temperature of 75°C. Calculate the maximum level of average forward current that the diode can safely pass.

2-14   Draw the power dissipation-versus-temperature graph for the diode specified in Problem 2-12.

2-15   A diode with a 0.9 V forward drop has a 1.5 W maximum power dissipation at 25°C. If the device derating factor is 7.5 mW/°C, calculate the maximum forward current level at 25°C and at 75°C. Assume that V_F remains constant.

2-16   Estimate the forward voltage drop at 75°C of the silicon diode in Problem 2-15. Determine the junction dynamic resistances at the temperature extremes if I_F is 20 mA.

2-17   A 5 V supply is applied via a 150 Ω resistor to a silicon diode and a germanium diode which are connected in series. Determine the diode current at 25°C and at 100°C.

Section   2-6

2-18   Calculate the minimum fall time for a voltage pulse applied to a diode with a reverse recovery time of (a) 6 ns, (b) 50 ns.

2-19   A diode has an applied voltage with a 200 ns fall time. Determine the maximum reverse recovery time for the diode.

2-20   Calculate a suitable minimum fall time for a switching waveform applied to a diode which has a 12 ns reverse recovery time.

2-21 Calculate the minimum fall time for a voltage pulse applied to a *1N914* diode.

Section 2-7

2-22 A diode connected in series with a 560 Ω resistor has a supply voltage that alternates between peak levels of +150 V and -150 V. Select a suitable device from the data sheets in Appendices 1-1 and 1-2.

2-23 Referring to Appendix 1-3, determine the peak reverse voltage and average rectified forward current for a *1N5398* diode.

2-24 A rectifier diode has to pass an average current of 55 mA and survive a 40 V peak reverse voltage. Select a suitable device from Appendices 1-1 through 1-3.

Section 2-8

2-25 The circuit shown in Fig. 2-24 is to be used to plot the forward characteristics of a *lN917* diode. Determine the resistance value for R_1, and select appropriate V/cm scales for the horizontal and vertical inputs of the *XY* recorder. The graph size should be approximately 20 cm x 20 cm.

2-26 Plot the forward characteristics of a diode from the following experimental data:

V_F (V)	0.24	0.26	0.28	0.29	0.30	0.31	0.32	0.33	0.34
I_F (mA)	0.05	0.07	0.09	0.5	0.9	20	40	60	80

Section 2-9

2-27 Determine the maximum current that may be used with a *1N757* Zener diode at temperatures of 25°C and 80°C.

2-28 A *1N750* Zener diode is connected in series with an 470 Ω resistor and a 10 V supply voltage. Calculate the diode current and power dissipation.

Practise Problem Answers

2-2.1	5.5 Ω, 30 MΩ	2-6.1	50 ns
2-2.2	0.5 Ω, 0.5 Ω	2-6.2	150 ns
2-3.1	2.89 mA	2-7.1	600 V, 1.5 A, 50 A
2-3.2	(Point *A*: 0 mA, 0.3 V),	2-7.2	*1N5392*
	(Point *B*: 100 mA, 0.35 V)	2-8.2	Vertical 1 V/cm, Horizontal 0.1 V/cm, 100 Ω
2-3.3	174 mA	2-9.1	64.5 mA, 38.7 mA
2-4.1	20 mA	2-9.2	40.4 mW
2-4.2	50 Ω		
2-4.3	3.45 V		
2-5.1	154 mA, 61.5 mA		
2-5.2	267 mV, 421 mV, 1.23 Ω, 1.54 Ω		
2-5.3	780 mV		

Chapter *3*

Diode Applications

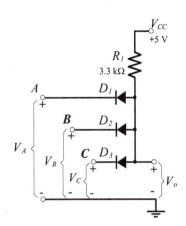

Chapter   Contents

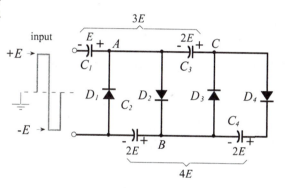

Objectives

You will be able to:

1   Sketch half-wave and full-wave rectifier circuits showing input and output waveforms, and explain the operation of each circuit.

2   Draw and explain circuit diagrams for basic dc power supplies using rectifiers and capacitor filters.

3   Determine capacitor values and diode current and voltage levels for basic dc power supplies.

4   Define line effect, load effect, line regulation, and load regulation for dc power supplies. Calculate these quantities for various power supplies.

5   Sketch Zener diode voltage regulator circuits, and explain their operation.

6   Design and analyze Zener diode regulator circuits.

7   Draw diagrams for series and shunt clipping circuits, clamping circuits, and dc voltage multipliers.

8   Sketch the input and output waveforms, and explain the operation of each of the circuit in objective 7.

9   Design and analyze the types of circuits listed in objective 7.

10   Sketch diode AND and OR gate circuits, and explain their operation.

Introduction

One of the most important application of diodes is rectification; that is, conversion of a sinusoidal *ac* waveform into single-polarity half cycles. The rectified wave is smoothed by the use of capacitors to process it into direct voltage. A Zener diode may be applied to regulate the direct voltage.

Other important diode applications include clipping, clamping, *dc* voltage multiplication, and logic circuits. Diode clipping circuits are used for *clipping-off* an unwanted portion of a waveform. Clamping circuits change the *dc* voltage level of a waveform without affecting the wave shape. DC voltage multipliers are applied to change the level of a *dc* voltage source to a desired higher level. Logic circuits produce a *high* or *low* output voltage, depending upon the voltage levels at several input terminals.

3-1   Half-Wave Rectification

Half-Wave Rectifier Circuit

A basic diode *half-wave rectifier* circuit is shown in Fig. 3-1(a). An alternating input voltage is applied to a single diode connected in series with a load resistor R_L. The diode is forward biased during the positive half-cycles of the input waveform, and reverse biased during the negative half-cycles. Substantial current flows through R_L only during the positive half-cycles of the input. For the duration of the negative half-cycles, the diode behaves almost as an open switch. The output voltage waveform developed across R_L is a series of positive half-cycles of alternating voltage with intervening very small negative voltage levels produced by the diode reverse saturation current.

When the diode is forward biased [Fig. 3-1(a)], the voltage drop across it is V_F, and the output voltage is (input voltage) - V_F. So, the peak output voltage is,

$$V_{po} = V_{pi} - V_F \qquad \text{(3-1)}$$

Note that $V_{pi} = 1.414\, V_i$, where V_i is the *rms* level of the sinusoidal input voltage.

The diode peak forward current is,

$$I_p = \frac{V_{po}}{R_L} \qquad \text{(3-2)}$$

During the negative half-cycle of the input [Fig. 3-1(b)], the reverse-biased diode offers a very high resistance. So, only a very small reverse current (I_R) flows, giving an output voltage,

$$-V_o = -I_R \times R_L \qquad \text{(3-3)}$$

While the diode is reverse biased, the peak voltage of the negative half-cycle of the input is applied to its terminals. Thus, the *peak reverse voltage*, or *peak inverse voltage (PIV)*, applied to the diode is,

$$V_R = PIV = V_{pi} \qquad\qquad (3\text{-}4)$$

Most rectifier circuits use a *reservoir capacitor* at the output terminals to smooth the rectified voltage wave into direct voltage [see Fig. 3-1(c)]. It is important to note that

the presence of the reservoir capacitor substantially changes the rectified voltage waveform and affects the diode current and voltage requirements.

This is discussed in Section 3-3.

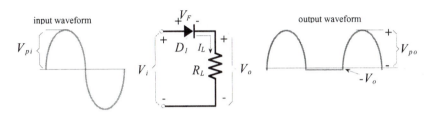

(a) Half-wave rectifier circuit with input and output waveforms

Figure 3-1
Half-wave rectifier circuit. The diode is forward biased during one half-cycle of the applied waveform, and reverse-biased during the other half-cycle.

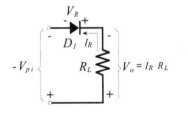

(b) Effect of negative input

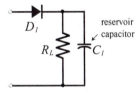

(c) Use of a reservoir capacitor

Example 3-1
A diode with $V_F = 0.7$ V is connected as a half-wave rectifier. The load resistance is 500 Ω, and the *(rms)* ac input is 22 V. Determine the peak output voltage, the peak load current, and the diode peak reverse voltage.

Solution

$$V_{pi} = 1.414\,V_i = 1.414 \times 22\ \text{V} = 31.1\ \text{V}$$

Eq. 3-1,
$$V_{po} = V_{pi} - V_F = 31.1\ \text{V} - 0.7\ \text{V}$$
$$= 30.4\ \text{V}$$

Eq. 3-2,
$$I_P = \frac{V_{po}}{R_L} = \frac{30.4\ \text{V}}{500\ \Omega}$$
$$= 60.8\ \text{mA}$$

Eq. 3-4,
$$PIV = V_{pi} = 31.1\ \text{V}$$

Practise Problems

3-1.1   A half-wave rectifier circuit has a 15 V ac input and a 330 Ω load resistance. Calculate the peak output voltage, peak load current, and the diode maximum reverse voltage.

3-1.2   A half-wave rectifier (as in Fig. 3-1) produces a 40 mA peak load current through a 1.2 kΩ resistor. If the diode is silicon, calculate the rms input voltage and the diode PIV.

3-2   Full-Wave Rectification

Two-Diode Full-Wave Rectifier

The *full-wave rectifier* circuit in Fig. 3-2 uses two diodes, and its input voltage is supplied from a transformer (T_1) with a center-tapped secondary winding. The circuit is essentially a combination of two half-wave rectifier circuits, each supplied from half of the transformer secondary.

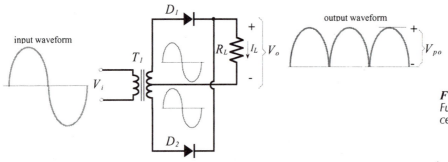

Figure 3-2
Full-wave rectifier circuit using a center-tapped transformer.

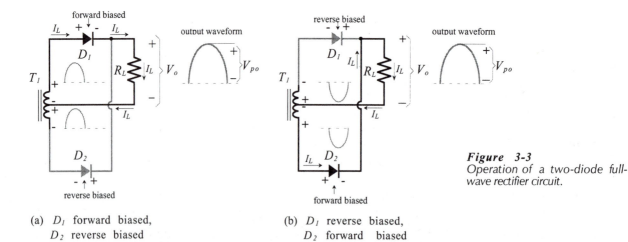

(a) D_1 forward biased,
       D_2 reverse biased

(b) D_1 reverse biased,
       D_2 forward biased

Figure 3-3
Operation of a two-diode full-wave rectifier circuit.

When the transformer output voltage is positive at the top, as illustrated in Fig. 3-3(a), the anode of D_1 is positive, and the center-tap of the transformer is connected to the cathode of D_1 via R_L. Consequently, D_1 is forward biased, and load current (I_L) flows from the top of the transformer secondary through D_1, through R_L from top to bottom, and back to the transformer center-tap. During this time, the polarity of the voltage from the bottom half of the transformer secondary causes diode D_2 to be reverse biased.

For the duration of the negative half-cycle of the transformer output, the polarity of the transformer secondary voltage causes D_1 to be reverse biased and D_2 to be forward biased, [see Fig. 3-3(b)]. I_L flows from the bottom terminal of the transformer secondary through diode D_2, through R_L from top to bottom, and back to the transformer center-tap. The output waveform is the combination of the two half-cycles; that is, a continuous series of positive half-cycles of sinusoidal waveform. This is *full-wave rectification.*

Bridge Rectifier

The center-tapped transformer used in the circuit of Fig. 3-2 is usually more expensive and requires more space than additional diodes. So, a *bridge rectifier* is the circuit most frequently used for full-wave rectification.

The bridge rectifier circuit in Fig. 3-4 is seen to consist of four diodes connected with their arrowhead symbols all pointing toward the positive output terminal of the circuit. Diodes D_1 and D_2 are series-connected, as are D_3 and D_4. The *ac* input terminals are the junction of D_1 and D_2 and the junction of D_3 and D_4. The positive output terminal is at the cathodes of D_1 and D_3, and the negative output is at the anodes of D_2 and D_4.

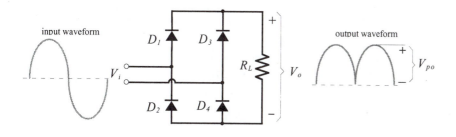

Figure 3-4
Full-wave bridge rectifier circuit.

During the positive half-cycle of input voltage, diodes D_1 and D_4 are in series with R_L, as illustrated in Figs. 3-5(a) and (b). Thus, load current (I_L) flows from the positive input terminal through D_1 to R_L, and then through R_L and D_4 back to the negative input terminal. Note that the direction of the load current through R_L is from top to bottom. During this time, the positive input terminal is applied to the cathode of D_2, and the negative output is at D_2 anode, [see Fig. 3-5(a)]. So, D_2 is reverse biased during the positive half-cycle of the input. Similarly, D_3 has the negative input at its anode and the positive output at its cathode during the positive input half-cycle, causing D_3 to be reverse biased.

Figures 3-5(c) and (d) show that diodes D_2 and D_3 are forward biased during the negative half-cycle of the input waveform, while D_1 and D_4 are reverse biased. Although the circuit input terminal polarity is reversed, I_L again flows through R_L from top to bottom, via D_3 and D_2.

It is seen that during both half-cycles of the input, the output terminal polarity is always positive at the top of R_L, negative at the bottom. Both positive and negative half-cycles of the input are passed to the output. The negative half-cycles are inverted, so that the output is a continuous series of positive half-cycles of sinusoidal voltage.

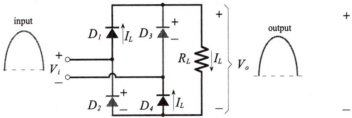

(a) Effect of positive half-cycle of input   (b) D_1 and D_4 forward biased

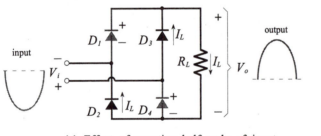

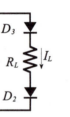

Figure 3-5
Operation of the bridge rectifier circuit.

(c) Effect of negative half-cycle of input   (d) D_3 and D_2 forward biased

The bridge rectifier has two forward-biased diodes in series with the supply voltage and the load. Because each diode has a forward voltage drop (V_F), the peak output voltage is,

$$V_{po} = V_{pi} - 2V_F \qquad\qquad \textbf{(3-5)}$$

As in the case of the half-wave rectifier,

a reservoir capacitor substantially changes the full-wave rectified output voltage waveform and affects the diode current and voltage requirements.

See the discussions in Sections 3-3 and 3-4.

Example 3-2
Determine the peak output voltage and current for the bridge rectifier circuit in Fig. 3-4 when V_i = 30 V, R_L = 300 Ω, and the diodes have V_F = 0.7 V.

Solution

$$V_{pi} = 1.414\,V_i = 1.414 \times 30 \text{ V}$$
$$= 42.42 \text{ V}$$

Eq. 3-5, $V_{po} = V_{pi} - 2 V_F = 42.42 \text{ V} - (2 \times 0.7 \text{ V})$
     $= 41 \text{ V}$

Eq. 3-2, $I_p = \dfrac{V_{po}}{R_L} = \dfrac{41 \text{ V}}{300 \text{ } \Omega}$
     $= 137 \text{ mA}$

Figure 3-6 shows two common methods of drawing a bridge rectifier circuit. Although they both look more complex than the circuit in Fig. 3-4, comparison shows that they are exactly the same circuit. The cathodes of D_1 and D_3 in all three circuits are connected to the positive output terminal, and the anodes of D_2 and D_4 are connected to the negative output terminal. Also, the *ac* input is applied to the junction of D_1 and D_2 and the junction of D_3 and D_4.

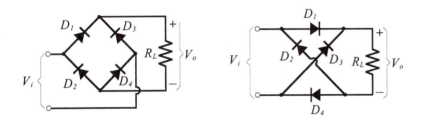

Figure 3-6
Two more ways of showing bridge rectifier circuits. These are exactly the same as the circuit in Fig. 3-4.

Practise Problems
3-2.1   Determine the peak load voltage, peak current, and the power dissipation in a 470 load resistor connected to a bridge rectifier circuit that has a 24 V ac input. The rectifier diodes are germanium.

3-2.2   A bridge rectifier with silicon diodes and a 680 Ω load resistor has an 18 V peak output. Calculate the power dissipated in the load resistor and the *rms* input voltage.

3-3   Half-Wave Rectifier DC Power Supply

Capacitor Filter Circuit
When a sinusoidal alternating voltage is rectified, the resultant waveform is a series of positive (or negative) half-cycles of the input waveform; it is not direct voltage. To convert to direct voltage (*dc*), a *smoothing circuit* or *filter* must be employed.

Figure 3-7(a) shows a half-wave rectifier circuit with a single capacitor filter (C_1) and a load resistor (R_L). The capacitor, termed a *reservoir capacitor*, is charged almost to the peak level of the circuit input voltage when the diode is forward biased. This occurs at V_{pi}, as shown in Fig. 3-7(b), giving a peak capacitor voltage,

$$\mathbf{V_C = V_{pi} - V_F} \qquad\qquad \textbf{(3-6)}$$

When the instantaneous level of input (at the diode anode) falls below V_{pi} the diode becomes reverse biased, because the capacitor voltage (V_C) (at the diode cathode) remains close to $(V_{pi} - V_F)$, [see Fig. 3-7(c)]. With the diode reverse biased, the capacitor begins to discharge through the load resistor (R_L). So, V_C falls slowly, as shown by the capacitor voltage waveform in Fig. 3-7(a).

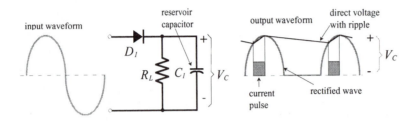

(a) Half-wave rectifier circuit with a reservoir capacitor

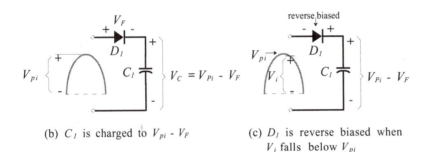

(b) C_1 is charged to $V_{pi} - V_F$

(c) D_1 is reverse biased when V_i falls below V_{pi}

Figure 3-7
A reservoir capacitor smoothes the output from a rectifier circuit by charging up to the peak output voltage and retaining most of its charge between voltage peaks.

The diode remains reverse biased through the remainder of the input positive half-cycle, the negative half-cycle, and the first part of the positive half-cycle again until the instantaneous level of V_i becomes greater than V_C once more. At this point current flows through the diode to recharge the capacitor, causing the capacitor voltage to return to $(V_{pi} - V_F)$. The charge and discharge of the capacitor causes the small increase and decrease in the capacitor voltage, which is also the circuit output voltage. It is seen that the circuit output is a direct voltage with a small *ripple voltage* waveform superimposed, [Fig. 3-7(a)].

Capacitance Calculation

The amplitude of the ripple voltage is affected by the load current, the reservoir capacitor value, and the capacitor discharge time. The discharge time depends upon the frequency of the ripple waveform, which is the same as the *ac* input frequency in the case of a half-wave rectifier. With a constant load current, the ripple amplitude is inversely proportional to the capacitance; the largest capacitance produces the smallest ripple.

The capacitance for the reservoir capacitor can be calculated from the load current, the acceptable ripple amplitude, and the capacitor discharge time. Consider the circuit output voltage waveform illustrated in Fig. 3-8(a). The waveform quantities are:

E_{ave}	average *dc* output voltage
$E_{o(max)}$	maximum output voltage level
$E_{o(min)}$	minimum output voltage
V_r	ripple voltage peak-to-peak amplitude
T	time period of the *ac* input waveform
t_1	capacitor discharge time
t_2	capacitor charge time
θ_1	phase angle of the input wave from zero to $E_{o(min)}$
θ_2	phase angle of the input wave from $E_{o(min)}$ to $E_{o(max)}$

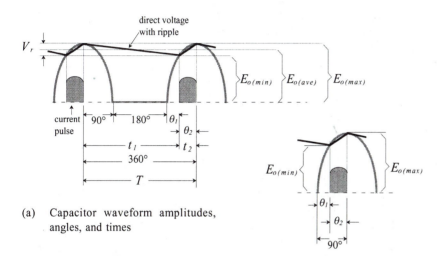

(a)   Capacitor waveform amplitudes, angles, and times

(b)   Relationship between $E_{o(min)}$ and $E_{o(max)}$

Figure 3-8
The capacitance value for a reservoir capacitor can be calculated from a knowledge of the load current, ripple voltage, and input frequency.

Figure 3-8(b) shows that, because the input wave is sinusoidal,

$$E_{o(min)} = E_{o(max)} \sin \theta_1$$

which gives,

$$\theta_1 = \sin^{-1} \frac{E_{o(min)}}{E_{o(max)}} \qquad \text{(3-7)}$$

Also from Fig. 3-8(b),

$$\theta_2 = 90° - \theta_1 \qquad \text{(3-8)}$$

The time period is,

$$T = \frac{1}{f}$$

where *f* is the frequency of the *ac* input waveform.

During *T*, the input waveform goes through a 360° phase angle, which gives the time per degree as,

$$t/degree = \frac{T}{360°}$$

so,

$$t_2 = \frac{\theta_2\, T}{360°} \qquad \textbf{(3-9)}$$

and,

$$t_1 = T - t_2 \qquad \textbf{(3-10)}$$

Taking the current as a constant quantity,

$$C = \frac{I_L\, t_1}{V_r} \qquad \textbf{(3-11)}$$

Capacitor Selection

When a capacitance value is calculated, an appropriate capacitor has to be selected from a manufacturer's list of available standard values. With a reservoir capacitor, the calculated capacitance is always the minimum value required to give a specified maximum ripple voltage amplitude. If a larger-than-calculated capacitance is used, the ripple voltage will be lower than the specified maximum. So, a larger standard value capacitor is always selected in the case of a reservoir capacitor.

The standard-value capacitors listed in Appendix 2-2 are typically available with ±20% tolerance. In the case of capacitors greater than 10 µF, the tolerance is often listed as -10% +50%. This means that a 100 µF capacitor might have a capacitance as low as 90 µF, or as high as 150 µF. In most circuit situations, a minimum capacitance value is calculated, and a larger value is quite acceptable.

The voltage that a capacitor will be subjected to must be taken into consideration. The maximum voltage that may be safely applied to a capacitor is stated in terms of its *dc working voltage*. The *dc* working voltages can be quite small for large-value capacitors. For example, some 10 µF capacitors have 6.3 V working voltages. The capacitor dielectric may break down if the specified voltage is exceeded.

Capacitor Polarity

It is very important that polarized capacitors be correctly connected. Sometimes polarized capacitors **explode** when they are incorrectly connected, and this could have **tragic consequences** for the eyes of an experimenter. The positive terminal is represented by the straight bar on the component graphic symbol, or identified by the plus sign on the alternative symbol, (see Fig. 3-9). This should be connected to the most positive point in the circuit where the capacitor is to be installed. Non-polarized capacitors should be used in situations where the voltage polarity might be reversed.

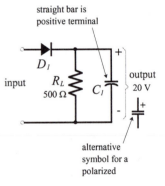

Figure 3-9
It is very important that polarized capacitors be connected with the correct polarity.

Example 3-3

The half-wave rectifier *dc* power supply in Fig. 3-9 is to supply 20 V to a 500 Ω load. The peak-to-peak ripple voltage is not to exceed 10% of the average output voltage, and the *ac* input frequency is 60 Hz. Calculate the required reservoir capacitor value.

Solution

$$V_r = 10\% \text{ of } E_{o(ave)} = 10\% \text{ of } 20 \text{ V}$$
$$= 2 \text{ V}$$

$$E_{o(min)} = E_{o(ave)} - 0.5 \, V_r = 20 \text{ V} - (0.5 \times 2 \text{ V})$$
$$= 19 \text{ V}$$

$$E_{o(max)} = E_{o(ave)} + 0.5 \, V_r = 20 \text{ V} + (0.5 \times 2 \text{ V})$$
$$= 21 \text{ V}$$

Eq. 3-7,
$$\theta_1 = sin^{-1} \frac{E_{o(min)}}{E_{o(max)}} = sin^{-1} \frac{19 \text{ V}}{21 \text{ V}}$$
$$\approx 65°$$

Eq. 3-8,
$$\theta_2 = 90° - \theta_1 = 90° - 65°$$
$$= 25°$$

$$T = \frac{1}{f} = \frac{1}{60 \text{ Hz}}$$
$$= 16.7 \text{ ms}$$

Eq. 3-9,
$$t_2 = \frac{\theta_2 \, T}{360°} = \frac{25° \times 16.7 \text{ ms}}{360°}$$
$$= 1.16 \text{ ms}$$

Eq. 3-10,
$$t_1 = T - t_2 = 16.7 \text{ ms} - 1.16 \text{ ms}$$
$$= 15.54 \text{ ms}$$

$$I_L = \frac{E_{o(ave)}}{R_L} = \frac{20 \text{ V}}{500 \, \Omega}$$
$$= 40 \text{ mA}$$

Eq. 3-11,
$$C = \frac{I_L \, t_1}{V_r} = \frac{40 \text{ mA} \times 15.54 \text{ ms}}{2 \text{ V}}$$
$$= 310 \, \mu F \text{ (use 330 } \mu F \text{ - see Appendix 2-2)}$$

Approximate Calculations

The capacitance calculation in Example 3-3 assumes that the load current is a constant quantity. This is an approximation, because the load current changes by a small amount as the output voltage increases and decreases. Normally, the load current change is so small that it has no significant effect on the calculation.

Another approximation that can be made to simplify the capacitance calculation is to take the discharge time (t_1) as equal to the input waveform time period (T), [see Fig. 3-8(a)].

$$t_1 \approx T \qquad\qquad\qquad \textbf{(3-12)}$$

Equation 3-12 assumes that the capacitor charging time (t_2) is so much smaller than t_1 that it can be neglected. This is a reasonable assumption where the ripple voltage is small. Also, use of Eq. 3-12 gives a larger capacitance value than the more precise calculation, and this is acceptable because a larger-than-calculated standard value capacitor is normally selected.

Example 3-4

Assuming that $t_2 \ll t_1$, recalculate the required reservoir capacitor value for the circuit in Example 3-3

Solution

$$V_r = 10\% \text{ of } E_{o(ave)} = 10\% \text{ of } 20\text{ V}$$
$$= 2\text{ V}$$

$$T = \frac{1}{f} = \frac{1}{60\text{ Hz}}$$
$$= 16.7\text{ ms}$$

Eq. 3-12, $\qquad t_1 \approx T = 16.7\text{ ms}$

$$I_L = \frac{E_{o(ave)}}{R_L} = \frac{20\text{ V}}{500\ \Omega}$$
$$= 40\text{ mA}$$

Eq. 3-11, $\qquad C = \frac{I_L\, t_1}{V_r} = \frac{40\text{ mA} \times 16.7\text{ ms}}{2\text{ V}}$

$$= 334\ \mu F$$

Diode Specification

Rectifier diodes must be specified in terms of the currents and voltages that they are subjected to. The calculated levels are normally minimum quantities, and the selected diodes must be able to survive higher levels.

Consider Fig. 3-10 which illustrates the situation when the *ac* input wave is at its negative peak voltage ($-V_p$). The capacitor has already been charged up to approximately the positive peak level of the input ($+V_p$). Consequently, the diode has $-V_p$ at its anode and $+V_p$ at its cathode, so the diode peak reverse voltage is,

$$V_R \approx 2\ V_p \qquad\qquad \textbf{(3-13)}$$

The *average forward rectified current* ($I_{F(av)}$) that the diode must pass is equal to the *dc* output current.

$$I_{F(av)} = I_L \qquad\qquad \textbf{(3-14)}$$

The diode in a half-wave rectifier circuit with a reservoir capacitor does not conduct continuously, but repeatedly passes pulses of current to recharge the capacitor each time the diode becomes forward biased. This is illustrated in Fig. 3-8 and again in Fig. 3-11. The current pulse is known as the *repetitive surge current*, and is designated (I_{FRM}). The average input current to the rectifier circuit

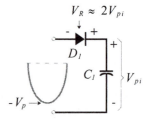

Figure 3-10
Diode reverse voltage in a half-wave rectifier.

must equal the average load current (I_L), so I_{FRM} averaged over time period T equals I_L, (see Fig. 3-11).

$$I_L = \frac{I_{FRM}\ t_2}{(t_1 + t_2)}$$

giving,

$$I_{FRM} = \frac{I_L\,(t_1 + t_2)}{t_2} \qquad \textbf{(3-15)}$$

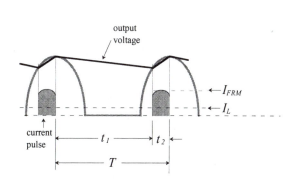

Figure 3-11
Diode average (load) current (I_L) and repetitive surge current (I_{FRM}) in a half-wave rectifier.

Figure 3-12 shows a half-wave rectifier circuit with a resistor (R_S) connected in series with the diode. This is a low-resistance component known as a *surge limiting resistor.* As its name suggests, the purpose of R_S is to limit the level of any surge current that might pass through the diode. The highest surge current occurs when the *ac* supply is first switched on to the rectifier circuit. Before switch-on, the reservoir capacitor normally contains no charge, so it behaves as a short-circuit at the instant of switch-on. If switch-on occurs when the *ac* input is at its peak level, the surge current is,

$$I_{F(surge)} = \frac{V_p}{R_S}$$

For a diode with a specified maximum *non-repetitive surge current* (I_{FSM}), the surge limiting resistor is calculated as,

$$R_S = \frac{V_p}{I_{FSM}} \qquad \textbf{(3-16)}$$

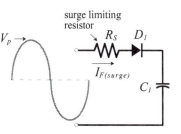

Figure 3-12
Surge limiting resistor in a half-wave rectifier.

Example 3-5
Specify the diode for the half-wave rectifier circuit in Example 3-3. Select a suitable device from Appendix 1, and calculate a suitable surge limiting resistance.

Solution

$$V_p = E_{o(max)} + V_F = 21\ \text{V} + 0.7\ \text{V}$$
$$= 21.7\ \text{V}$$

Eq. 3-13, $V_R \approx 2\,V_p = 2 \times 21.7\,V$
$$= 43.4\,V$$

Eq. 3-14, $I_{F(av)} = I_L = 40\,mA$

Eq. 3-15, $I_{FRM} = \dfrac{I_L(t_1 + t_2)}{t_2} = \dfrac{40\,mA \times 16.7\,ms}{1.16\,ms}$
$$= 576\,mA$$

In Appendix 1-2, the *1N4001* has $V_R = 50\,V$, and $I_{F(av)} = I_o = 1\,A$, and $I_{FRM} = I_{Fm(rep)} = 10\,A$. So, its specification is better than required for this application. Any of the *1N4002* through *1N4007* diodes could be used, but, those with higher reverse voltage are more expensive than the *1N4001*.

For the *1N4001*, $I_{F(surge)} = 30\,A$

Eq. 3-16, $R_S = \dfrac{V_p}{I_{FSM}} = \dfrac{21.7\,V}{30\,A}$
$$\approx 0.7\,\Omega$$

Practise Problems

3-3.1   A *dc* power supply consisting of a half-wave rectifier and a reservoir capacitor has to supply 120 V to a 200 Ω load. Determine the capacitance required if the maximum ripple voltage is to be ±5% of the average output, and the input frequency is 60 Hz.

3-3.2   For Problem 3-3.1, recalculate the capacitance approximately by assuming that the capacitor discharge time is very much larger than the charging time.

3-3.3   For the circuit in Problem 3-3.1, specify the diode, select a suitable device from the available data sheets, and determine the required surge limiting resistance.

3-4   Full-Wave Rectifier DC Power Supply

Like half-wave rectifiers, full-wave rectifiers require filter circuits to convert the output waveform to direct voltage. Figure 3-13 shows a full-wave rectifier circuit with a reservior capacitor and a surge limiting resistor. These components operate exactly as explained for the half-wave rectifier circuit, with a few important exceptions.

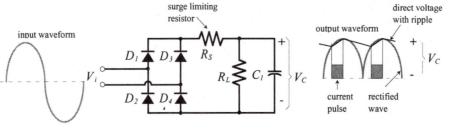

Figure 3-13
Bridge rectifier circuit with a reservior capacitor and a surge limiting resistor

The capacitor-smoothed full-wave rectifier waveforms are shown in detail in Fig. 3-14. Eqation 3-7, derived for the half-wave rectifier circuit, still applies for determining the angle θ_1.

Eq. 3-7, $\theta_1 = \sin^{-1} \dfrac{E_{o(min)}}{E_{o(max)}}$

Also, θ_2 can be determined from Eq.3-8,

Eq.3-8, $\theta_2 = 90° - \theta_1$

and Eq. 3-9 can be used for time t_2,

Eq. 3-9, $t_2 = \dfrac{\theta_2 \, T}{360°}$

Comparing Fig. 3-14 and Fig. 3-8, it is seen that the capacitor discharge time (t_1) for the half-wave circuit is approximately equal to the waveform time period (T), whereas for the full-wave circuit t_1 approximately equals $T/2$. More precisely,

$$t_1 = (T/2) - t_2 \qquad\qquad\qquad \textbf{(3-17)}$$

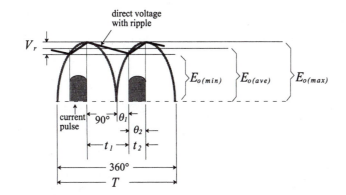

Figure 3-14
Ripple voltage, waveform angles, and time periods for a full-wave rectifier with capacitor smoothing.

Using the correct value of t_1, Eq. 3-11 can be used for calculating the reservior capacitance for a full-wave rectifier circuit.

Eq. 3-11, $C = \dfrac{I_L \, t_1}{V_r}$

Similarly, (using the correct value of t_1 in Fig. 3-14), the repetitive current (I_{FRM}) can be determined from Eq. 3-15,

Eq. 3-15, $I_{FRM} = \dfrac{I_L \, (t_1 + t_2)}{t_2}$

The average forward current passed by the bridge rectifier circuit is equal to the load current. But, each pair of diodes supplies current for no more than a half cycle of the input wave. The other pair conducts during the other half cycle. So, as illustrated in Fig. 3-15, the diode average forward current is half of the load current.

$$I_{F(av)} = \frac{I_L}{2}$$

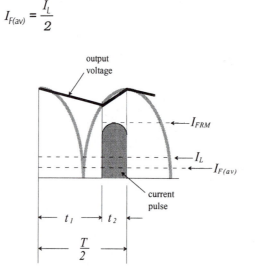

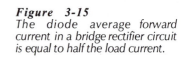

Figure 3-15
The diode average forward current in a bridge rectifier circuit is equal to half the load current.

Another difference between the half-wave and full-wave rectifier power supply circuits concerns the reverse voltage applied to the diodes. Consider Fig. 3-16. When the instantaneous input voltage is $+V_p$, as illustrated, V_p is applied across forward-biased diode D_1 in series with reverse-biased diode D_3. Therefore, the reverse voltage across D_3 is,

$$V_R = V_p - V_F$$

or, $$V_R \approx V_p \qquad\qquad \textbf{(3-18)}$$

Examination of the circuit shows that Eq. 3-18 applies to each diode when reverse biased.

As in the case of the half-wave circuit, the capacitor calculation can be simplified by assuming that time t_2 is very much smaller than t_1. This gives the approximation that the capacitor discharge time is equal to half the input waveform time period.

$$t_1 \approx \frac{T}{2} \qquad\qquad \textbf{(3-19)}$$

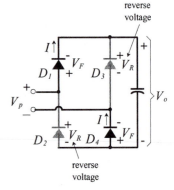

Figure 3-16
Diode reverse voltage (V_R) in a bridge rectifier circuit.

Example 3-6

The full-wave rectifier *dc* power supply in Fig. 3-17 is to supply 20 V to a 500 Ω load. The peak-to-peak ripple voltage is not to exceed 10% of the average output voltage, and the *ac* input frequency is 60 Hz. Calculate the required reservoir capacitor value. (Note that the specifications for this example are similar to those for the half-wave circuit in Example 3-3.)

Solution

$$V_r = 10\% \text{ of } E_{o(ave)} = 10\% \text{ of } 20 \text{ V}$$
$$= 2 \text{ V}$$

$$E_{o(min)} = E_{o(ave)} - 0.5\,V_r = 20\,V - (0.5 \times 2\,V)$$
$$= 19\,V$$

$$E_{o(max)} = E_{o(ave)} + 0.5\,V_r = 20\,V + (0.5 \times 2\,V)$$
$$= 21\,V$$

Eq. 3-7, $\quad \theta_1 = sin^{-1}\dfrac{E_{o(min)}}{E_{o(max)}} = sin^{-1}\dfrac{19\,V}{21\,V}$

$$\approx 65°$$

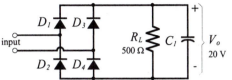

Figure 3-17
Circuit for Example 3-6.

Eq. 3-8, $\quad \theta_2 = 90° - \theta_1 = 90° - 65°$
$$= 25°$$

$$T = \dfrac{1}{f} = \dfrac{1}{60\,Hz}$$
$$= 16.7\,ms$$

Eq. 3-9, $\quad t_2 = \dfrac{\theta_2\,T}{360°} = \dfrac{25° \times 16.7\,ms}{360°}$
$$= 1.16\,ms$$

Eq. 3-17, $\quad t_1 = (T/2) - t_2 = (16.7\,ms/2) - 1.16\,ms$
$$= 7.19\,ms$$

$$I_L = \dfrac{E_{o(ave)}}{R_L} = \dfrac{20\,V}{500\,\Omega}$$
$$= 40\,mA$$

Eq. 3-11, $\quad C = \dfrac{I_L\,t_1}{V_r} = \dfrac{40\,mA \times 7.19\,ms}{2\,V}$

$$= 143.8\,\mu F \text{ (use 150 } \mu F \text{ - see Appendix 2-2)}$$

Example 3-7

Assuming that t_2 is very much smaller than t_1, recalculate the required reservoir capacitor value for the circuit in Example 3-5

Solution

$$V_r = 10\% \text{ of } E_{o(ave)} = 10\% \text{ of } 20\,V$$
$$= 2\,V$$

$$T = \dfrac{1}{f} = \dfrac{1}{60\,Hz}$$
$$= 16.7\,ms$$

Eq. 3-19, $\quad t_1 \approx \dfrac{T}{2} = \dfrac{16.7\,ms}{2}$
$$= 8.35\,ms$$

$$I_L = \dfrac{E_{o(ave)}}{R_L} = \dfrac{20\,V}{500\,\Omega}$$
$$= 40\,mA$$

Eq. 3-11, $C = \dfrac{I_L\, t_1}{V_r} = \dfrac{40\ mA \times 8.36\ ms}{2\ V}$

$= 167\ \mu F$

Example 3-8

Specify the diode for the bridge rectifier circuit in Example 3-6. Select a suitable device and calculate the surge limiting resistance.

Solution

$$V_p = E_{o(max)} + (2\ V_F) = 21\ V + (2 \times 0.7\ V)$$
$$= 22.4\ V$$

Eq. 3-18, $V_R \approx V_p = 22.4\ V$

Eq. 3-14, $I_{F(av)} = I_L = 40\ mA$

Eq. 3-15, $I_{FRM} = \dfrac{I_L(t_1 + t_2)}{t_2} = \dfrac{40\ mA \times 8.35\ ms}{1.16\ ms}$

$= 288\ mA$

From Appendix 1-2, the *1N4001* is suitable.

For the *1N4001*, $I_{F(surge)} = 30\ A$

Eq. 3-16, $R_S = \dfrac{V_p}{I_{FSM}} = \dfrac{22.4\ V}{30\ A}$

$\approx 0.75\ \Omega$

Practise Problems

3-4.1   A dc power supply consisting of a bridge rectifier and a reservoir capacitor has to supply 15 V to a 150 Ω load. Determine the capacitance required if the maximum ripple voltage is to be ±5% of the average output, and the input frequency is 60 Hz.

3-4.2   For Problem 3-4.1, recalculate the capacitance approximately by assuming that the capacitor discharge time is very much larger than the charging time.

3-4.3   For the circuit in Problem 3-4.1, specify the diode, select a suitable device from the available data sheets, and determine the required surge limiting resistance.

3-5 Power Supply Performance

Source Effect

The *ac* supply to the input of a transformer in a *dc* power supply does not always remain constant. A ±10% variation in the *ac source voltage* (V_s) (also termed the *line voltage*) is not unusual. When the source voltage varies, there is some variation in the *dc* output voltage from a power supply, (see Fig. 3-18). This output voltage change (ΔE_o) due to a change in the input is termed the *source effect*. If the output varies by 100 mV when the source voltage changes by ±10%, the *source effect* is 100 mV. An alternative way of stating this output change is to express ΔE_o as a percentage of the *dc* output voltage (E_o). In this case, the term *line regulation* is used.

$$\textbf{Source effect} = \Delta E_o \textbf{ for a 10\% change in } V_S \qquad \textbf{(3-20)}$$

$$\textbf{Line regulation} = \frac{(\Delta E_o \textbf{ for a 10\% change in } V_S) \times 100\%}{E_o} \qquad \textbf{(3-21)}$$

Load Effect

Power supply output voltage is also affected by changes in load current (I_L). The output voltage decreases when I_L is increased, and rises when I_L is reduced. The *load effect* defines how the output voltage changes when the load current is increased from zero to its specified maximum level ($I_{L(max)}$). If the load current change (ΔI_L) produces a voltage change (ΔE_o) of 100 mV, the *load effect* is 100 mV. As for the source effect, the load effect can also be expressed as a percentage of the output voltage. This is termed the *load regulation*.

$$\textbf{Load effect} = \Delta E_o \textbf{ for } \Delta I_{L(max)} \qquad \textbf{(3-22)}$$

$$\textbf{Load regulation} = \frac{(\Delta E_o \textbf{ for } \Delta I_{L(max)}) \times 100\%}{E_o} \qquad \textbf{(3-23)}$$

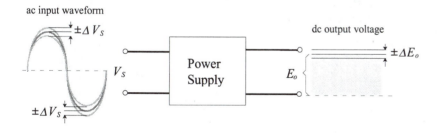

ac input waveform

dc output voltage

V_S

Power
Supply

E_o

Figure 3-18
Power supply source effect is the output voltage change that occurs when the input voltage changes by ±10%.

Example 3-9

The output voltage of a *dc* power supply changes from 20 V to 19.7 V when the load is increased from zero to maximum. The voltage also increases to 20.2 V when the *ac* supply increases by 10%. Calculate the load and source effects, and the load and line regulations.

Solution

Eq. 3-22,   Load effect $= \Delta E_o$ for $\Delta I_{L(max)} = 20\text{ V} - 19.7\text{ V}$
$$= 300\text{ mV}$$

Eq. 3-23,   Load regulation $= \dfrac{(\Delta E_o \text{ for } \Delta I_{L(max)}) \times 100\%}{E_o} = \dfrac{300\text{ mV} \times 100\%}{20\text{ V}}$

= 1.5%

Eq. 3-20,   *Source effect* $= \Delta E_o$ *for 10% change in* $V_S = 20.2\ V - 20\ V$

= 200 mV

Eq. 3-21,   *Line regulation* $= \dfrac{(\Delta E_o \text{ for 10\% change in } V_S) \times 100\%}{E_o}$

$= \dfrac{200\ mV \times 100\%}{20\ V}$

= 1%

Practise   Problems

3-5.1   A dc power supply output drops from 15 V to 14.95 V when the ac source voltage falls by 10%. The output also falls from 15 V to 14.9 V when the load current is increased from zero to maximum. Calculate the source effect, load effect, line regulation, and load regulation.

3-6   Zener Diode Voltage Regulators

Regulator Circuit With No Load

The most important application of Zener diodes (discussed in Section 2-9) is *dc* voltage regulator circuits. These can be the simple regulator circuit shown in Fig. 3-19, or the more complex regulators discussed in Chapters 17. The circuit in Fig. 3-19 is usually employed as a voltage reference source that supplies only a very low current (much lower than I_Z) to the output. Resistor R_1 in Fig. 3-19 limits the Zener diode current to the desired level.

$$I_Z = \frac{E_S - V_Z}{R_1} \tag{3-24}$$

The Zener current may be just greater than the diode knee current (I_{ZK}). However, for the most stable reference voltage, I_Z should be selected as I_{ZT} (the specified test current). Example 3-10 demonstrates the circuit design procedure.

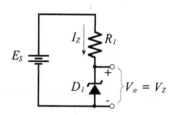

Figure 3-19
Zener diode reference voltage source, or voltage regulator.

Example 3-10

A 9 V reference source is to be designed using a series-connected Zener diode and resistor connected to a 30 V supply. Select suitable components, and calculate the circuit current when the supply voltage drops to 27 V.

Solution

From the Zener diode data sheet in Appendix 1-4, the most suitable device is a *1N757* which has $V_Z = 9.1$ V, and $I_{ZT} = 20$ mA.

With $E_S = 30$ V,

From Eq. 3-24,

$$R_1 = \frac{E_S - V_Z}{I_Z} = \frac{30 \text{ V} - 9.1 \text{ V}}{20 \text{ mA}}$$

$$= 1.05 \text{ k}\Omega \text{ (use 1 k}\Omega — \text{see Appendix 2-1)}$$

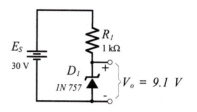

P_D for R_1,

$$P_{R1} = I_1{}^2 R_1 = (20 \text{ mA})^2 \times 1 \text{ k}\Omega$$
$$= 0.4 \text{ W}$$

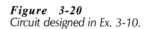

When $E_S = 27$ V,

$$I_Z = \frac{E_S - V_Z}{R_1} = \frac{27 \text{ V} - 9.1 \text{ V}}{1 \text{ k}\Omega}$$

$$= 17.9 \text{ mA}$$

Figure 3-20
Circuit designed in Ex. 3-10.

Loaded Regulator

When a Zener diode regulator is required to supply a load current (I_L), as shown in Fig. 3-21, the total supply current (flowing through resistor R_1) is the sum of I_L and I_Z. Care must be taken to ensure that the minimum Zener diode current is large enough to keep the diode in reverse breakdown. Typically, $I_{Z(min)} = 5$ mA for a Zener diode with $I_{ZT} = 20$ mA. The circuit current equation is,

$$I_Z + I_L = \frac{E_S - V_Z}{R_1} \qquad \text{(3-25)}$$

In some cases, the load current in the type of circuit shown in Fig. 3-21 may be reduced to zero. Because the voltage drop across R_1 remains constant, the supply current remains constant at,

$$I_{R1} = I_Z + I_L$$

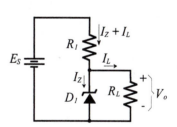

All of this current flows through the Zener diode when R_L is disconnected. The circuit design must ensure that the total current does not exceed the maximum Zener diode current.

Example 3-11 demonstrates the design process for a loaded voltage reference source that uses a low-power Zener diode. It should be noted that high-power Zener diodes are available for higher current levels.

Figure 3-21
Zener diode voltage regulator circuit supplying a load current. The diode must be able to pass a current of ($I_L + I_Z$).

Example 3-11

Design a 6 V dc reference source to operate from a 16 V supply. The circuit is to use a low-power Zener diode (as in Appendix 1-4) and is to produce maximum possible load current. Calculate the maximum load current that can be produced by the circuit.

Solution

From the Zener diode data sheet in Appendix 1-4, the *1N753* has $V_Z = 6.2$ V, and $P_D = 400$ mW.

from Eq. 2-11,

$$I_{ZM} = \frac{P_D}{V_Z} = \frac{400 \text{ mW}}{6.2 \text{ V}}$$

$$= 64.5 \text{ mA}$$

$$I_{L(max)} + I_{Z(min)} = I_{ZM} = 64.5 \text{ mA}$$

from Eq. 3-24, $R_1 = \dfrac{E_S - V_Z}{I_{ZM}} = \dfrac{16 \text{ V} - 6.2 \text{ V}}{64.5 \text{ mA}}$

$$= 152 \; \Omega \text{ (use 150 } \Omega \text{ standard value)}$$

$$P_{R1} = I_1^2 R_1 = (64.5 \text{ mA})^2 \times 150 \; \Omega$$

$$= 0.62 \text{ W}$$

select $I_{Z(min)} = 5 \text{ mA}$

$$I_{L(max)} = I_{ZM} - I_{Z(min)} = 64.5 \text{ mA} - 5 \text{ mA}$$

$$= 59.5 \text{ mA}$$

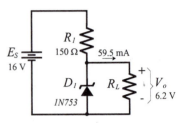

Figure   3-22
Circuit designed in Ex. 3-11.

Regulator Performance

The performance of a Zener diode voltage regulator may be expressed in terms of the source and load effects, and the line and load regulations, exactly as discussed in Section 3-5. Equations 3-20 through 3-23 may be applied. If there is an input ripple voltage, the output ripple will be severely attenuated. The *ripple rejection ratio* is the ratio of the output to input ripple amplitudes.

To assess the performance of a Zener diode voltage regulator, the *ac* equivalent circuit is first drawn by replacing the diode with its dynamic impedance (Z_Z), as shown in Fig. 3-23(a). The complete *ac* equivalent circuit is seen to be a simple voltage divider. When the input voltage changes by ΔE_S, the output voltage change is,

$$\Delta V_o = \frac{\Delta E_S \times Z_Z}{R_1 + Z_Z} \qquad \textbf{(3-26)}$$

Equation 3-26 assumes that there is no load connected to the regulator output. When a load is present, R_L appears in parallel with Z_Z in the *ac* equivalent circuit, [see Fig. 3-23(b)]. The equation for the output voltage change now becomes,

$$\Delta V_o = \frac{\Delta E_S \times (Z_Z \| R_L)}{R_1 + (Z_Z \| R_L)} \qquad \textbf{(3-27)}$$

The regulator source effect can be determined from Eq. 3-26 or Eq. 3-27, as appropriate. The equations can also be used for calculating ripple rejection ratio. The input ripple amplitude (V_{ri}) and the output ripple (V_{ro}) are substituted in place of the input and output voltages in Eq. 3-26 and Eq. 3-27. Thus, Eq. 3-26 can be modified to give a ripple rejection ratio equation,

$$\frac{V_{ro}}{V_{ri}} = \frac{Z_Z}{R_1 + Z_Z} \qquad \textbf{(3-28)}$$

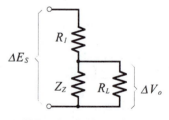

(a) Regulator without a load

(b) Regulator with a load

Figure   3-23
AC equivalent circuits for Zener diode voltage regulators.

and, for a loaded regulator, Eq. 3-27 gives,

$$\frac{V_{ro}}{V_{ri}} = \frac{Z_Z||R_L}{R_1 + (Z_Z||R_L)} \qquad (3\text{-}29)$$

To determine the load effect of the Zener diode voltage regulator, the circuit output resistance has to be calculated. The regulator Thevenin equivalent circuit in Fig. 3-24 shows that, assuming a zero source resistance, the circuit output resistance is,

$$R_o = Z_Z||R_1 \qquad (3\text{-}30)$$

When the load current changes by ΔI_L, the output voltage change is,

$$\Delta V_o = \Delta I_L (Z_Z||R_1) \qquad (3\text{-}31)$$

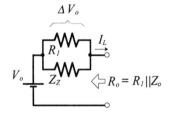

Figure 3-24
Thevenin equivalent circuit for a Zener diode regulator.

Example 3-12
Calculate the line regulation, load regulation, and ripple rejection ratio for the voltage regulator in Example 3-11.

Solution
From Appendix 1-4, the *1N753* has $Z_Z = 7\ \Omega$.

Source effect,

$$\Delta E_S = 10\% \text{ of } E_S = 10\% \text{ of } 16\text{ V}$$
$$= 1.6\text{ V}$$

$$R_L = \frac{V_o}{I_L} = \frac{6.2\text{ V}}{59.5\text{ mA}}$$
$$= 104\ \Omega$$

Eq. 3-27, $\Delta V_o = \dfrac{\Delta E_S \times (Z_Z|| R_L)}{R_1 + (Z_Z|| R_L)} = \dfrac{1.6\text{ V} \times (7\ \Omega||104\ \Omega)}{150\ \Omega + (7\ \Omega||104\ \Omega)}$

$$= 67\text{ mV}$$

Eq. 3-21,   *Line regulation* $= \dfrac{(\Delta E_o \text{ for } 10\% \text{ change in } E_S) \times 100\%}{E_o}$

$$= \frac{67\text{ mV} \times 100\%}{6.2\text{ V}}$$

$$= 1.08\%$$

Load effect,
Eq. 3-31, $\Delta V_o = \Delta I_L (Z_Z||R_1) = 59.5\text{ mA} \times (7\ \Omega||150\ \Omega)$

$$= 398\text{ mV}$$

Eq. 3-23,   *Load regulation* $= \dfrac{(\Delta E_o \text{ for } \Delta I_{L(max)}) \times 100\%}{E_o} = \dfrac{398\text{ mV} \times 100\%}{6.2\text{ V}}$

$$= 6.4\%$$

Ripple rejection,

Eq. 3-30,

$$\frac{V_{ro}}{V_{ri}} = \frac{Z_Z \| R_L}{R_1 + (Z_Z \| R_L)} = \frac{7\ \Omega \| 104\ \Omega}{150\ \Omega + (7\ \Omega \| 104\ \Omega)}$$

$$= 4.19 \times 10^{-2}$$

Practise Problems

3-6.1   Design a 12 V *dc* reference source (consisting of a Zener diode and series-connected resistor) to operate from a 25 V supply. Determine the effect on the diode current when the supply drops to 22 V.

3-6.2   An 8 V *dc* reference source is to be designed to produce the maximum possible output current from a low-power Zener diode. The supply voltage is 20 V. Design the circuit and determine the maximum load current.

3-6.3   Calculate the line effect, load effect, and ripple rejection for the circuit designed in Problem 3-6.2.

3-7 Series Clipping Circuits

Series Clipper

The function of a *clipper* (or *limiter*) is to clip off an unwanted portion of a waveform. This is sometimes necessary to protect a device or circuit that might be destroyed by a large amplitude (negative or positive) signal.

A half-wave rectifier can be described as a clipper, because it passes only the positive (or negative) half-cycle of an alternating waveform and clips off the other half-cycle. In fact, a diode *series clipper* is simply a half-wave rectifier circuit.

Figure 3-25(a) shows a *negative series clipper circuit* with a square wave input symmetrical above and below ground level. While the input is positive, D_1 is forward biased and the positive half-cycle is passed to the output.

$$V_o = E - V_F \qquad \textbf{(3-32)}$$

During the negative half-cycle of the input, the diode is reverse biased. Consequently, the output remains at zero and the negative half-cycle is effectively clipped off.

The *zero level* output from a series clipper circuit is not exactly zero. The reverse saturation current (I_R) of the diode produces a voltage drop across resistor R_1,

$$V_o = -I_R R_1 \qquad \textbf{(3-33)}$$

This is almost always so small that it can be neglected.

If the diode in Fig. 3-25(a) is reconnected with reversed polarity, as shown in Fig. 3-25(b), the *positive* half-cycles are clipped off, and the circuit becomes a *positive series clipper*. The input waveforms to a clipper may be square, or sinusoidal, or any other shape.

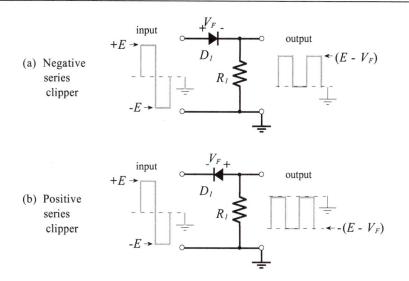

(a) Negative series clipper

(b) Positive series clipper

Figure 3-25
Diode series clipping circuits clip off unwanted portions of input waveforms. These are essentially half-wave rectifier circuits.

The output terminals of series clippers are usually connected to circuits that have a high input resistance, so (current-limiting) resistor R_1 is selected to pass an acceptable minimum current through the diode. This current must be sufficient to operated the diode beyond the *knee* of its forward characteristic, (see Fig. 2-4). Typically, a current of 1 mA is appropriate. The resistor value is calculated from the output voltage and the selected current level. The remaining part of the design process is specifying the diode.

Example 3-13

The negative series clipper in Fig. 3-26 has a ±9 V input, and zero load current. Determine a suitable resistance for R_1, and specify the diode forward current and reverse voltage. Use a silicon diode.

Solution

$$V_o = E - V_F = 9\text{ V} - 0.7\text{ V}$$
$$= 8.3\text{ V}$$

select $I_F = 1\text{ mA}$

$$R_1 = \frac{V_o}{I_F} = \frac{8.3\text{ V}}{1\text{ mA}}$$

$$= 8.3\text{ mA (use 8.2 k}\Omega\text{ standard value)}$$

Diode specification,

$$V_R = E = 9\text{ V}$$
$$I_F = 1\text{ mA}$$

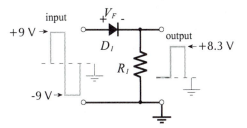

Figure 3-26
Series clipper circuit for Example 3-13.

Current Level Selection

In most electronic circuits it is best to select the smallest possible level of current. One reason for this is to keep the total supply current to a minimum. High supply currents require larger, more expensive, power supplies than low current levels. Where a battery supply is used, batteries last longer with low supply current levels. Another reason to keep circuit current levels low whenever possible, is to minimise component power dissipations. Low power dissipation allows the smallest components to be used, and avoids circuit heating problems.

Series Noise Clipper

Digital signal waveforms sometimes have unwanted lower-level *noise* voltages. The noise can be removed by a *series noise clipping circuit*, which simply consists of two inverse-parallel connected diodes, as illustrated in Fig. 3-27. When the noise amplitude is much smaller than the diode forward voltage drop (V_F), and the signal amplitude is larger than V_F, the signals are passed and the noise is blocked. The signal peak output voltage is ($E - V_F$).

A *dead zone* of $\pm V_F$ exists around ground level in the output waveform, indicating that inputs must exceed $\pm V_F$ to pass to the output. For noise voltage amplitudes that approach $\pm V_F$, two diodes may be connected in series to give a larger dead zone.

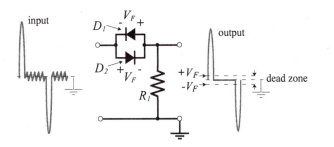

Figure 3-27
Series noise clipping circuit. Noise amplitudes lower than $\pm V_F$ cannot pass to the output.

Practise Problems

3-7.1   A positive series clipping circuit, as in Fig. 3-25(b), has a ± 7 V input, and zero load current. Calculate a suitable resistance for R_1, and specify the diode.

3-7.2   A ± 6 V square wave is applied via a series clipping circuit to a device that cannot survive $+6$ V. The device input current is $100\,\mu A$ when its input voltage is negative. Select a suitable clipping circuit, determine the required resistance, and specify the diode.

3-8   Shunt Clipping Circuits

Shunt Clipper

A *positive shunt clipper circuit* is illustrated in Fig. 3-28(a). Here the diode is connected in *shunt* (or parallel) with the output terminals. When the input is negative, the diode is reverse biased and only a small voltage drop occurs across R_1, due to load current I_L. This means that the circuit output voltage (V_o) is approximately equal to the negative input peak (-E). When the input is +E, D_1 is forward biased, and the output voltage equals the diode voltage drop ($+V_F$). Thus, the positive half of the waveform is clipped off. As illustrated, the upper and lower levels of the output of a positive shunt clipper are approximately $+V_F$ and $-E$.

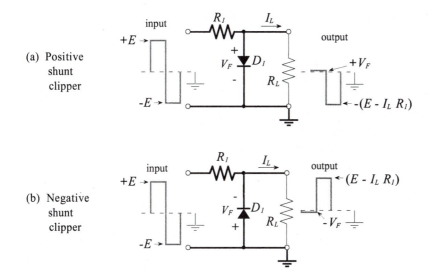

(a) Positive
shunt
clipper

(b) Negative
shunt
clipper

Figure 3-28
Shunt clipping circuits. The output voltage cannot exceed the diode voltage drop (V_F) when the input voltage forward biases the diode.

A *negative shunt clipper* circuit is exactly the same as a positive shunt clipper with the diode polarity reversed, [see Fig. 3-28(b)]. The negative half-cycle of the waveform is clipped off.

The load current on a shunt clipper produces a voltage drop ($I_L R_1$) across the resistor, which might be insignificant where the load current is very low.

$$V_o = E - (I_L \, R_1) \qquad\qquad (3\text{-}34)$$

As in the case of series clipping circuits, shunt clippers may be used with square, sinusoidal, or other input waveforms.

Example 3-14
The negative shunt clipper in Fig. 3-29 has a ±5 V input, and is to produce a $+4.5$ V minimum output when the load current is 2 mA. Determine a suitable resistance for R_1, and specify the diode forward current and reverse voltage.

Solution
When the diode is reverse biased,
Eq. 3-34, $\qquad V_o = E - (I_L R_1)$

or, $\qquad R_1 = \dfrac{E - V_o}{I_L} = \dfrac{5\text{ V} - 4.5\text{ V}}{2\text{ mA}}$

$\qquad\qquad\quad = 250\ \Omega \quad$ (use 220 Ω)

Diode reverse voltage,

$\qquad V_R = E = 5$ V

When the diode is forward biased,

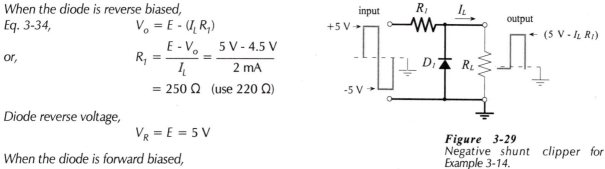

Figure 3-29
Negative shunt clipper for Example 3-14.

$$I_F = \frac{E - V_F}{R_1} = \frac{5\text{ V} - 0.7\text{ V}}{220\ \Omega}$$

$$= 19.5\text{ mA}$$

Shunt Noise Clipper

The *shunt noise clipper* shown in Fig. 3-30 removes noise riding on the peaks of an input waveform. The signal amplitude must be greater than the diode forward voltage drop. The output waveform is clipped at $\pm V_F$, so that the noise is not passed to the output. This type of clipper is used with pulse signals, where the pulse amplitude is not important. In this case, information is contained in the pulse width, or simply in its presence or absence.

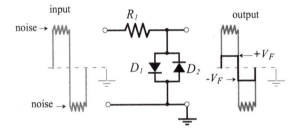

Figure 3-30
Shunt noise clipper circuit. Input amplitudes greater than $\pm V_F$ are clipped off.

Biased Shunt Clipper

Figure 3-31 shows a clipping circuit that uses two diodes which have different bias voltages. The cathode of D_1 is connected to a +2 V bias (V_{B1}), and the anode of D_2 has a -2 V bias (V_{B2}). While the input waveform amplitude is less than $\pm(V_B + V_F)$, neither diode is forward biased, and the input is simply passed to the output. When the positive input is greater than ($V_{B1} + V_F$), D_1 becomes forward biased, and the output cannot exceed this voltage. Similarly, when the negative input goes below ($-V_{B2} - V_F$), D_2 is forward biased, and the output is limited to $-(V_{B2} + V_F)$.

$$V_o = \pm(V_B + V_F) \tag{3-35}$$

For obvious reasons, the circuit is termed a *biased shunt clipper*. Biased shunt clippers are used to protect circuits or devices from (positive/negative) input voltages that must not exceed specified levels.

The voltage across resistor R_1 is ($E - V_o$), and the resistor current is the sum of the load current and the diode forward current ($I_L + I_F$). As in other diode circuits, a minimum level of I_F is selected, and the resistor value is calculated as,

$$R_1 = \frac{E - V_o}{I_L + I_F} \tag{3-36}$$

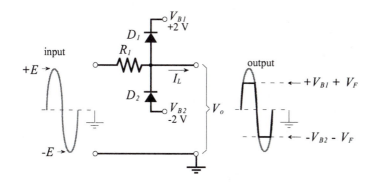

Figure 3-31
Biased shunt clipper circuit. Inputs amplitudes greater than $\pm(V_B + V_F)$ are clipped off.

Example 3-15

The biased shunt clipper in Fig. 3-31 has a ±9 V input, and its output is to be limited to ±2.7 V. Determine a suitable resistance for R_1 if the clipper output current is to be ±1 mA.

Solution

Eq. 3-35,
$$V_o = \pm(V_B + V_F)$$

giving,
$$V_B = \pm(V_o - V_F) = \pm(2.7\text{ V} - 0.7\text{ V})$$
$$= \pm2\text{ V}$$

Select the diode forward current as,

$$I_F = 1\text{ mA}$$

Eq 3-36,
$$R_1 = \frac{E - V_o}{I_L + I_F} = \frac{9\text{ V} - 2.7\text{ V}}{1\text{ mA} + 1\text{ mA}}$$
$$= 3.15\text{ k}\Omega \quad (\text{use } 2.7\text{ k}\Omega \text{ to give } I_F > 1\text{ mA})$$

Zener Diode Shunt Clipper

A *Zener diode shunt clipper* produces the same kind of result as a biased shunt clipper without the need for bias voltages. The clipper circuit in Fig. 3-32 has two back-to-back series-connected Zener diodes. When the input voltage is positive and has sufficient amplitude, D_1 is forward bias and D_2 is biased into reverse breakdown. At this time, the output voltage is limited to $(V_F + V_{Z2})$. A negative input voltage produces a maximum negative output of $-(V_F + V_{Z1})$. With equal-voltage Zener diodes, the maximum output voltage is,

$$V_o = \pm(V_F + V_Z) \tag{3-37}$$

The resistor voltage is $(E - V_o)$, and the resistor current is $(I_L + I_Z)$. A minimum level of I_Z (greater than the device knee current) is selected, and the resistor value is calculated as,

$$R_1 = \frac{E - V_o}{I_L + I_Z} \tag{3-38}$$

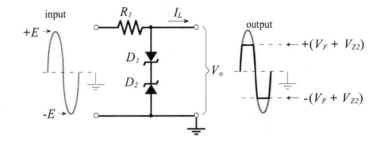

Figure 3-32
Zener diode shunt clipper circuit. Inputs amplitudes greater than $\pm(V_F + V_Z)$ are clipped off.

Example 3-16

A Zener diode shunt clipper, as in Fig. 3-32, is to be connected between a ±20 V square wave signal and a circuit which cannot accept inputs greater than ±5 V. Select suitable Zener diodes, and determine R_1. The clipper output current is to be ±1 mA.

Solution

From Eq. 3-37,
$$V_Z = V_o - V_F = 5 \text{ V} - 0.7 \text{ V}$$
$$= 4.3 \text{ V}$$

From Appendix 1-4, the 1N749 has $V_Z = 4.3$ V. Use 1N749 Zener diodes.

Select,
$$I_{Z(min)} = 5 \text{ mA}$$

Eq. 3-38,
$$R_1 = \frac{E - V_o}{I_L + I_Z} = \frac{20 \text{ V} - 5 \text{ V}}{1 \text{ mA} + 5 \text{ mA}}$$

$$= 1.86 \text{ k}\Omega \quad \text{(use 1.8 k}\Omega \text{ to give } I_Z > 5 \text{ mA)}$$

Practise Problems

3-8.1   A positive shunt clipper, as in Fig. 3-28(a), has a ±7 V input. The negative output voltage is to be 6.5 V when the load current is 1.5 mA. Calculate the required resistance for R_1, and specify the diode forward current and reverse voltage.

3-8.2   A ±8 V square wave is applied to a device that cannot accept inputs greater than ± 3.5 V. The device input current is ± 600 μA. Design a suitable biased shunt clipping circuit.

3-8.3   A Zener diode shunt clipping circuit is to be used to clip a ±18 V square wave off at approximately ±7 V. The output current is to be ±1.2 mA. Design the circuit.

3-9   Clamping Circuits

Negative and Positive Voltage Clamping Circuits

A *clamping circuit*, also known as a *dc restorer*, changes the *dc* voltage level of a waveform, but does not affect its shape. Consider the clamping circuit shown in Fig. 3-33. When the square wave input is positive, diode D_1 is forward biased, and the output voltage equals the diode forward voltage drop (V_F).

$$V_o = V_F \qquad \text{(3-39)}$$

During the positive half-cycle of the input, the voltage on the right side of the capacitor is $+V_F$, while that on the left side is $+E$. Thus, C_1 is charged with the polarity shown to a voltage,

$$V_C = (E - V_F) \qquad \text{(3-40)}$$

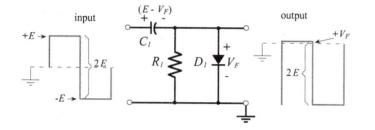

Figure 3-33
A negative voltage clamping circuit passes the complete input waveform to the output, but clamps the positive peak of the output close to ground level.

When the input goes negative, the diode is reverse biased and has no further effect on the capacitor voltage. Also, R_1 has a very high resistance, so that it cannot discharge C_1 significantly during the negative portion of the input waveform. While the input is negative, the output voltage is the sum of the input and capacitor voltages. Since the polarity of the capacitor voltage is the same as the (negative) input, the output is,

$$V_o = -(E + V_c)$$
$$= -[E + (E - V_F)]$$

or, $$V_o = -[2E - V_F] \qquad \text{(3-41)}$$

The peak-to-peak output voltage ($V_{o(pp)}$) is the difference between the positive output peak (V_F) and negative output peak $-(2E - V_F)$.

$$V_{o(pp)} = V_F - [-(2E - V_F)]$$

or, $$V_{o(pp)} = 2E \qquad \text{(3-42)}$$

It is seen that the peak-to-peak amplitude of the output waveform from the clamping circuit is exactly the same as the peak-to-peak input. Instead of the waveform being symmetrical above and below ground, however, the output positive peak is clamped at a level of $+V_F$.

The circuit in Fig. 3-33 is known as a *negative voltage clamping circuit*, because the output waveform is (almost) completely negative. Reversing the polarity of the diode and capacitor in Fig. 3-33 creates the *positive voltage clamping circuit* shown in Fig. 3-34. As illustrated, the output waveform is clamped to keep it (almost) completely positive.

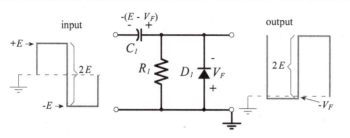

Figure 3-34
A positive voltage clamping circuit passes the complete input waveform to the output, but clamps the negative peak of the output close to ground level.

Output Slope

Resistor R_1 in Figs. 3-33 and 3-34 is sometimes termed a *bleeder resistor*. Its function is to gradually discharge the capacitor over several cycles of input waveform, so that it can be charged to a new voltage level if the input changes. However, R_1 partially discharges the capacitor during the time that D_1 is reverse biased, and this produces a *slope*, or *tilt* (ΔV_c) on the output waveform, as illustrated in Fig. 3-35. The tilt voltage can be calculated from a knowledge of the waveform frequency and the circuit component values. The constant-current equation for charging or discharging a capacitor may be applied,

$$\Delta V_c = \frac{I_c \times t}{C_1} \tag{3-43}$$

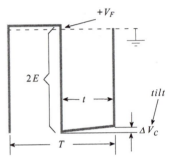

Figure 3-35
The output voltage from a clamping circuit has a slope (ΔV_c) produced by capacitor discharge.

Example 3-17

The diode clamping circuit in Fig. 3-36, has ±10 V, 1 kHz square wave input. Calculate the tilt on the output waveform.

Solution

While the diode is reverse biased,

$$V_{o(PP)} = 2E = 2 \times 10 \text{ V}$$
$$= 20 \text{ V}$$

and,
$$I_c \approx \frac{V_{o(pp)}}{R_1} = \frac{20 \text{ V}}{56 \text{ k}\Omega}$$
$$= 367 \text{ } \mu A$$

$$t = \frac{T}{2} = \frac{1}{2f} = \frac{1}{2 \times 1 \text{ kHz}}$$
$$= 500 \text{ } \mu s$$

Eq. 3-43,
$$\Delta V_C = \frac{I_c \times t}{C_1} = \frac{367 \text{ } \mu A \times 500 \text{ } \mu s}{1 \text{ } \mu F}$$
$$= 184 \text{ mV}$$

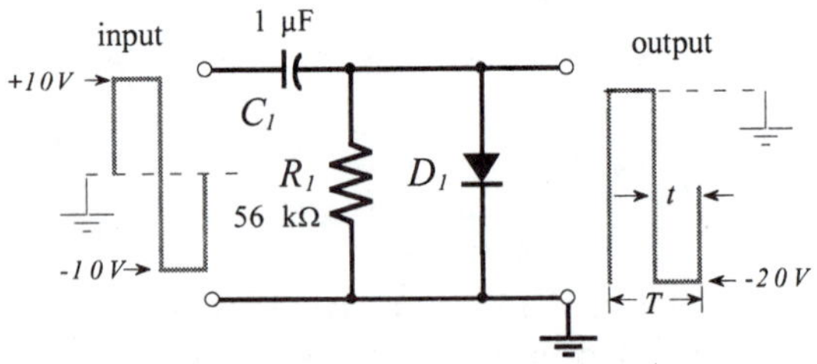

Figure 3-36
Clamping circuit for Ex. 3-17.

Component Determination

When designing a clamping circuit, capacitor C_1 is selected so that it becomes completely charged in approximately five cycles of the input waveform. As already discussed (and illustrated in Fig. 3-37), diode D_1 is forward biased when C_1 is being charged. Resistor R_1 is not part of the C_1 charging circuit, because D_1 bypasses R_1. As illustrated, C_1 charge is controlled by the signal source resistance (R_S). A capacitor is completely charged in approximately five time constants of the circuit. In the case of a resistor-capacitor circuit, the time constant is RC. So, the charging time is,

$$5\,RC = 5 \times [input\ pulse\ width\ (PW)]$$
or,
$$RC = PW$$

This gives, $$\mathbf{C_1\,R_S = PW} \qquad (3\text{-}44)$$

When the pulse width and source resistance of the input waveform is known, C_1 can be determined from Eq. 3-44.

So long as the slope on the output waveform is very small, the capacitor discharge current is essentially a constant quantity.

$$I_C = \frac{V_o}{R_1}$$

or, $$I_C = \frac{2E}{R_1} \qquad (3\text{-}45)$$

When the output waveform slope is specified, I_C can be calculated from Eq. 3-43, and a suitable resistance for R_1 is then determined by substituting I_C into Eq. 3-45.

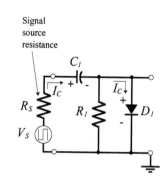

Signal source resistance

Figure 3-37
Capacitor C_1 is charged via signal source resistance R_S.

Example 3-18
The negative voltage clamping circuit in Fig. 3-38, has a ± 8 V, 500 Hz square wave input with a 600 Ω signal source resistance. The output waveform is to have a maximum slope of 1%. Determine suitable values for R_1 and C_1.

Solution

$$t = \frac{T}{2} = \frac{1}{2f} = \frac{1}{2 \times 500\ \text{Hz}}$$
$$= 1\ \text{ms}$$

$$PW = t = 1\ \text{ms}$$

From Eq. 3-44, $$C_1 = \frac{PW}{R_S} = \frac{1\ \text{ms}}{600\ \Omega}$$
$$= 1.7\ \mu\text{F (use } 1.8\ \mu\text{F)}$$

$$V_{o(PP)} = 2E = 2 \times 8\ \text{V}$$
$$= 16\ \text{V}$$

$$\Delta V_C = 1\% \text{ of } V_{o(PP)} = 1\% \text{ of } 16\ \text{V}$$
$$= 0.16\ \text{V}$$

From Eq. 3-43, $$I_c = \frac{\Delta V_C\,C_1}{t} = \frac{0.16\ \text{V} \times 1.8\ \mu\text{F}}{1\ \text{ms}}$$
$$= 288\ \mu\text{A}$$

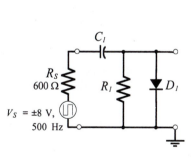

Figure 3-38
Clamping circuit for Ex. 3-18.

From Eq. 3-45, $R_1 = \dfrac{2E}{I_C} = \dfrac{16\,\text{V}}{288\,\mu\text{A}}$

$\qquad\qquad\qquad = 55.5\,\text{k}\Omega$ (use 56 kΩ)

Biased Clamping Circuit

The *biased clamping circuit* shown in Fig. 3-39 has a bias voltage (V_B) at the cathode of diode D_1. In this case, the output voltage cannot exceed $(V_{B1} + V_F)$, as illustrated. The capacitor charges to $(E - V_{B1} - V_F)$ when the input is positive. When the input goes negative, the output voltage is $-(E + V_C)$, giving,

$$V_o = -[E + (E - V_{B1} - V_F)]$$
$$= -2E + V_{B1} + V_F$$

The peak-to-peak output voltage is the difference between the positive and negative output peak levels.

$$V_o = (V_{B1} + V_F) - [-2E + V_{B1} + V_F]$$
$$= 2E$$

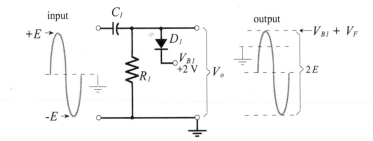

Figure 3-39
A biased clamping circuit clamps one peak of the output voltage at a selected level.

It is seen that, as for other clamping circuits, the peak-to-peak output waveform from the biased clamper is the same as the input. The positive level of the output is clamped to $(V_B + V_F)$.

Virtually any bias voltage level (positive or negative) can be used with a biased clamping circuit. The diode polarity determines whether the positive or the negative peak of the output is clamped. Care must be taken to ensure that the capacitor is connected with the correct polarity. Biased clamping circuits are designed in exactly the same way as unbiased clamping circuits.

Example 3-19

Determine the upper and lower levels of the output voltage for the circuits shown in Fig. 3-40 when the input voltage is ±6 V. Also, calculate the capacitor voltage in each case.

Solution

For Fig. 3-40(a),
D_1 is forward biased when the input goes negative.

When the input is -E, $V_C = V_{B1} - V_F - E = 3\ V - 0.7\ V - (-6\ V)$
$$= 8.3\ V\ (+\text{ on the right})$$

$$V_o = V_{B1} - V_F = 3\ V - 0.7\ V$$
$$= 2.3\ V\ (\text{low level of output})$$

When the input is +E, $V_o = E + V_C = 6\ V + 8.3\ V$
$$= 14.3\ V\ (\text{high level of output})$$

For Fig. 3-40(b),
D_1 is forward biased when the input goes positive.

When the input is +E, $V_C = E - V_{B1} - V_F = 6\ V - 3\ V - 0.7\ V$
$$= 2.3\ V\ (+\text{ on the left})$$

$$V_o = V_{B1} + V_F = 3\ V + 0.7\ V$$
$$= 3.7\ V\ (\text{high level of output})$$

When the input is -E, $V_o = E + V_C = -6\ V - 2.3\ V$
$$= -8.3\ V\ (\text{low level of output})$$

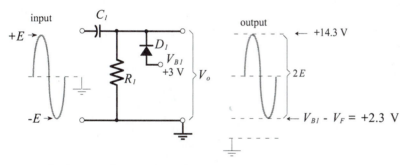

(a) Output voltage cannot go below $V_{B1} - V_F$

Figure 3-40
Biased clamping circuits use positive or negative bias voltages to clamp the level of one of the peaks of the output waveform.

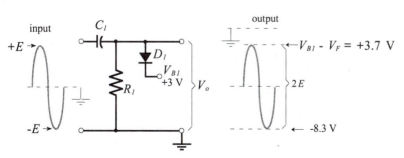

(b) Output voltage cannot go above $V_{B1} - V_F$

Zener Diode Clamping Circuit

Zener diode clamping circuits operate just like biased voltage clampers, with the Zener diode voltage (V_Z) substituting for the bias voltage. The circuit in Fig. 3-41 has an ordinary diode (D_1) connected in series with a Zener diode (D_2). When the input voltage is at its positive peak, the output voltage is,

$$V_o = V_Z + V_F$$

and C_1 charges (+ on the left) to,

$$V_C = E - (V_Z + V_F)$$

When the input goes to its negative peak, the output voltage is,

$$V_o = -(E + V_C)$$
$$V_o = -[E + (E - (V_Z + V_F))]$$
$$= -(2E - V_Z - V_F)$$

The peak-to-peak output remains equal to the peak-to-peak input voltage (2E). If the polarity of the diodes (and the capacitor) are reversed, the negative output peak is clamped at $-(V_Z + V_F)$.

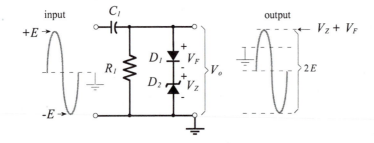

Figure 3-41
Zener diode clamping circuits clamp the output voltage at ($V_Z + V_F$). The positive peak or the negative peak can be clamped above or below ground

---

Practise Problems

3-9.1   A ±8 V, 300 Hz square wave is applied to the clamping circuit in Fig. 3-34. If C_1 = 1.5 µF and R_1 = 33 kΩ, determine the tilt voltage on the output waveform.

3-9-2   A positive voltage clamping circuit, as in Fig. 3-34, is to have an output waveform with a maximum of 0.5% tilt. The signal is a ±7 V, 1 kHz square wave with a 500 Ω source resistance. Calculate suitable values for R_1 and C_1.

3-9.3   A clamping circuit, as in Fig. 3-41, uses a *1N754* Zener diode. Determine the upper and lower levels of the output voltage if the input is a ±12 V square wave. Also, calculate the capacitor voltage.

3-10   DC Voltage Multipliers

Voltage Doubler

A *voltage doubling circuit* produces an output voltage which is approximately double the peak voltage of the input waveform. Consideration of the voltage doubling circuit in Fig. 3-42 shows that it is simply a combination of two diode-capacitor clamping circuits without the discharge resistors. In fact, the circuit operation is similar to that of clamping circuits.

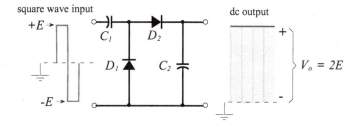

Figure 3-42
A voltage doubling circuit produces a dc output voltage which is approximately double the peak input voltage.

When the input voltage is negative, as shown in Fig. 3-43(a), diode D_1 is forward biased, and C_1 charges to $(E - V_{F1})$ with the polarity illustrated. D_2 is reverse biased during the negative half-cycle of the input, so the charge on C_2 is not affected at this time.

Figure 3-43(b) shows what occurs during the input positive half-cycle. D_1 is now reverse biased, and D_2 is forward biased. The voltage applied to D_2 and C_2 is the sum of the input voltage and the voltage on C_1. So, as illustrated, capacitor C_2 is charged to,

$$V_{C2} = E + V_{C1} - V_{F2}$$
$$= E + (E - V_{F1}) - V_{F2}$$
$$= 2(E - V_F)$$

so, $$V_o = 2(E - V_F) \qquad\qquad \textbf{(3-46)}$$

It is seen that, when the diode voltage drop is much smaller than the input voltage, the output is approximately double the peak input amplitude. The polarity of the output voltage can be reversed by reversing the polarity of the diodes and capacitors.

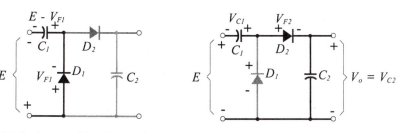

(a) C_1 charges while E is negative   (b) C_2 charges while E is positive

Figure 3-43
Capacitor C_1 charges to $(E - V_{F1})$ during the negative half-cycle of the input. Then, $+[E + (E - V_{F1}) - V_{F2}]$ is passed to C_2 during the input positive half-cycle.

The output terminals of the voltage doubler are the terminals of capacitor C_2. The load current partially discharges the capacitors, producing an output voltage drop in the same way that tilt is created on a clamping circuit output. The repeated charge and discharge of C_1 and C_2 results in a ripple waveform on the output.

A sinusoidal input waveform to a voltage doubler produces exactly the same type of output ripple that occurs with a half-wave rectifier, [Fig. 3-44(a)]. However, the input most often used is a *dc* voltage source which has been *chopped,* or converted into a square waveform, [Fig. 3-44(b)].

Capacitor C_2 supplies the load current (I_l) while diode D_2 is reverse biased, [Fig. 3-45(a)]. The discharge of C_2 accounts for half the output ripple voltage amplitude, and the discharge of C_1 produces the other half of the ripple amplitude. Equation 3-43 can be modified for calculating the capacitance of C_2.

$$C_2 = \frac{I_L t}{\Delta V_C} \qquad (3\text{-}47)$$

While D_2 is forward biased [Fig. 3-45(b)], capacitor C_1 supplies I_L and the recharging current to C_2. The recharging current must equal I_L to maintain the full charge on C_2, so C_1 supplies $2I_L$. Applying Eq. 3-43 to calculate C_1, it is found that,

$$C_1 = 2C_2 \qquad (3\text{-}48)$$

(a) Sine wave input

(b) Square wave input

Figure 3-44
A ripple waveform occurs at the output of a voltage doubler.

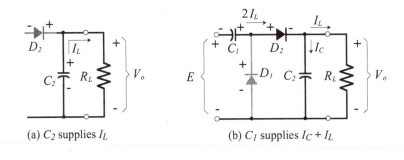

(a) C_2 supplies I_L

(b) C_1 supplies $I_C + I_L$

Figure 3-45
Capacitor C_2 supplies I_L while diode D_2 is reverse biased, and C_1 supplies $2I_L$ when D_2 is forward biased.

Example 3-20

Determine C_1 and C_2 for the voltage doubling circuit in Fig. 3-46 to produce a 1% maximum output ripple. The input is a ±12 V, 5 kHz square wave.

Solution

Eq. 3-46,
$$V_o = 2(E - V_F) = 2(12\text{ V} - 0.7\text{ V})$$
$$= 22.6\text{ V}$$

$$I_L = \frac{V_o}{R_L} = \frac{22.6\text{ V}}{47\text{ k}\Omega}$$
$$= 481\ \mu A$$

Capacitor discharge time, $t = \dfrac{T}{2} = \dfrac{1}{2f} = \dfrac{1}{2 \times 5\text{ kHz}}$
$$= 100\ \mu s$$

For 1% output ripple, allow 0.5% due to discharge of C_2 and 0.5% due to discharge of C_1.
$$\Delta V_C = 0.5\% \text{ of } V_{o(PP)} = 0.5\% \text{ of } 22.6\text{ V}$$
$$= 113\text{ mV}$$

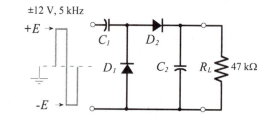

Figure 3-46
Diode voltage doubling circuit for Ex. 3-20.

Eq. 3-47,
$$C_2 = \frac{I_L\, t}{\Delta V_C} = \frac{481\ \mu A \times 100\ \mu s}{113\ mV}$$

$$= 0.43\ \mu F \text{ (use } 0.47\ \mu F\text{)}$$

Eq. 3-48,
$$C_1 = 2\, C_2 = 2 \times 0.47\ \mu F$$

$$= 0.94\ \mu F \text{ (use } 1\ \mu F\text{)}$$

Multistage Voltage Multipliers

A four-stage *dc* voltage multiplier is shown in Fig. 3-47. Comparing to the voltage doubler in Fig. 3-42, it is seen that this circuit consists of two cascade-connected voltage doubling circuits. To simplify the explanation of the circuit operation, assume ideal diodes with $V_F = 0$.

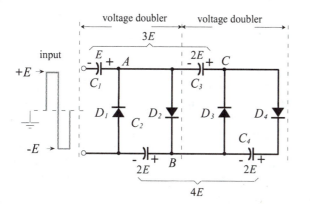

Figure 3-47
Four stage dc voltage multiplying circuit. Neglecting the diode voltage drops, C_1 is charged to E volts, and all other capacitors are charged to approximately 2E. The final output is 4E volts.

- When $V_i = -E$; D_1 is forward biased, and C_1 charges via D_1 to E.
- When $V_i = +E$; point A is at $+2E$, D_1 is reverse biased, D_2 is forward biased, C_2 charges via D_2 to 2E.
- When $V_i = -E$; point A is close to ground level, point B is at 2E (because of V_{C2}), D_2 is reverse biased, D_3 is forward biased, and C_3 charges via D_3 to 2E volts.
- When $V_i = +E$; point A is at $+2E$, point B is at $+2E$, and point C is at $+4E$, D_4 is forward biased, and C_4 charges via D_4 to 2E volts.

The resulting output voltage, taken across C_2 and C_4 is 4E volts, as illustrated. Additional stages may be added to the circuit to produce higher levels of *dc* output voltage. Figure 3-48 shows another way that *dc* voltage multiplier circuit diagrams are often drawn. Examination of the circuit shows that it is exactly the same as Fig. 3-47.

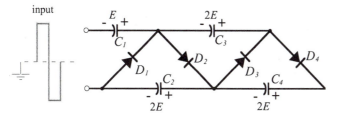

Figure 3-48
Another way of drawing the circuit of a four-stage dc voltage multiplier circuit.

3-11   Diode Logic Circuits

AND Gate

A logic circuit produces an output voltage which is either *high* or *low*, depending upon the levels of several input voltages. The two basic logic circuits are the *AND* gate and the *OR* gate.

Figure 3-49 shows the circuit diagram of a diode *AND* gate. It is seen that the diode anodes are connected to a 5 V supply (V_{CC}) via resistor R_1. The circuit has a single output terminal at the diode anodes, and three input terminals at the device cathodes. (An *AND* gate could have any number of inputs from 2 up to perhaps 50.)

If one (or more) of the input terminals is grounded, current flows from the supply through R_1 and through the forward-biased diode to ground. In this case, the output voltage is just V_F above ground, (0.7 V for silicon). The output is said to be at a *low* level. When input levels of 5 V are applied to all three input terminals, none of the diodes is forward biased, no resistor current flows, and no significant voltage drop occurs across R_1. Thus, the output voltage is equal to V_S, and it is referred to as a *high* output level.

> An *AND* gate produces a high output only when input A is high, **AND** input B is high, **AND** input C is high.

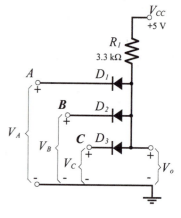

Figure 3-49
Circuit of a three-input diode *AND* gate. A high output occurs only when high inputs are present at input A, **and** input B, **and** input C.

Example 3-21

For the *AND* gate circuit in Fig. 3-49, determine each diode forward current level, (a) when all three inputs are *low*, (b) when only input A is *high*, (c) when inputs A and B are both *high*, and C is *low*.

Solution

(a)
$$I_{R1} = \frac{V_{CC} - V_F}{R_1} = \frac{5\ V - 0.7\ V}{3.3\ k\Omega}$$

$$= 1.3\ mA$$

For each diode,

$$I_F = \frac{I_{R1}}{3} = \frac{1.3\ mA}{3}$$

$$= 433\ \mu A$$

(b)
$$I_{R1} = 1.3\ mA \text{ (as above)}$$

$$I_{F2} = I_{F3} = \frac{I_{R1}}{2} = \frac{1.3 \text{ mA}}{2}$$

$$= 650 \,\mu A$$

(c) $I_{F3} = I_{R1} = 1.3 \text{ mA}$

OR Gate

The circuit diagram of an *OR* gate with three input terminals is shown in Fig. 3-50. Like the *AND* gate, the *OR* gate could have two or more inputs. It is obvious that the output voltage (V_o) of the *OR* gate in Fig. 3-50 is *low* when all three inputs are *low*. Now suppose that a +5 V input is applied to terminal A, while terminals B and C remain grounded. Diode D_1 becomes forward biased, and its cathode voltage is $V_o = (5 \text{ V} - V_F)$. The gate output voltage is *high*, and diodes D_2 and D_3 are reverse biased with $+V_o$ at the cathodes and ground at the anodes. The output will be *high* if a *high* input voltage is applied to any one (or more) of the inputs.

> *An OR gate produces a high output when input A is high,*
> **OR** *or when input B is high,* **OR** *when input C is high.*

Diode *AND* and *OR* gates can be designed and constructed using discrete components. However, *integrated-circuit (IC)* packages are available containing many diodes already fabricated in the form of one or more gates.

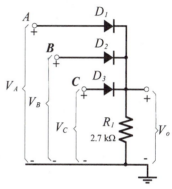

Figure 3-50
*Circuit of a three-input OR gate. A high output occurs when high inputs are present at input A, **or** at input B, **or** at input C.*

Practise Problems

3-11.1   The diodes in the *AND* gate in Fig. 3-49 are each to have a 2 mA maximum forward current. Calculate the required resistance for R_1.

3-11.2   Calculate the maximum and minimum levels of diode forward current for the *OR* gate in Fig. 3-50 when $V_i = 4.5$ V, and $R_1 = 2.7$ kΩ.

Chapter-3 Review Questions

Section 3-1

3-1   Sketch a half-wave rectifier circuit showing input and output waveforms, and explain the circuit operation.

Section 3-2

3-2   Sketch a two-diode full-wave rectifier circuit showing input and output waveforms. Explain the circuit operation.

3-3   Draw the circuit diagram for a bridge rectifier, together with its input and output waveforms. Carefully explain the operation of the bridge rectifier circuit, identifying the forward-biased and reverse-biased diodes during each half-wave of the input waveform.

Section 3-3

3-4   Draw the circuit diagram for a *dc* power supply that uses a half-wave rectifier and a capacitor filter circuit. Sketch the output waveform, and explain the circuit operation.

3-5   For a rectifier circuit with capacitor filtering, explain: ripple voltage, repetitive surge current, non-repetitive surge current, diode peak reverse voltage.

Section 3-4

3-6   Draw the circuit diagram for a *dc* power supply that uses a bridge rectifier and a capacitor filter circuit. Sketch the output waveform, and explain the circuit operation.

3-7   Compare half-wave and full-wave rectifier circuits with capacitor filtering, referring to capacitor size for a given ripple amplitude, and regarding diode specification.

Section 3-5

3-8   Define power supply source effect and load effect, and write equations for each. Also, write equations for line regulation and load regulation.

Section 3-6

3-9   Sketch the circuit diagram for a Zener diode voltage regulator. Briefly explain the circuit operation, and discuss the effects of load current.

3-10   Sketch the *ac* equivalent circuit for a Zener diode voltage regulator. Write equations for source effect and load effect.

Section 3-7

3-11   Draw circuit diagrams for negative and positive series clipping circuits. Show input and output waveforms and explain the operation of each circuit.

3-12   Sketch the circuit of a diode series noise clipper circuit, and explain its application.

Section 3-8

3-13   Draw circuit diagrams for negative and positive shunt clipping circuits. Show input and output waveforms, and explain the operation of each circuit.

3-14   Sketch a shunt noise clipping circuit, and explain its application.

3-15   Draw a biased shunt clipping circuit. Sketch input and output waveforms, and explain the circuit operation.

3-16   A biased shunt clipper has ±9 V bias voltages. Sketch the output waveform produced by a ±12 V square wave input.

3-17   Draw a diagram for a Zener diode shunt clipping circuit, and explain its application.

Section 3-9

3-18 Explain the difference between clipping circuits and clamping circuits. A positive voltage clamping circuit and a positive voltage clipping circuit each have a ±12 V square wave input. Sketch the output waveform from each circuit.

3-19 Draw circuit diagrams for negative and positive voltage clamping circuits. Show input and output waveforms, and explain the operation of each circuit.

Section 3-10

3-20 Draw a voltage doubling circuit. Sketch input and output waveforms, and explain the circuit operation.

3-21 Draw the circuit diagram for a four-stage voltage multiplier circuit. Briefly explain its operation, and identify the output terminals and the output voltage level relative to the input. Show how two additional diode-capacitor circuits may be added, and estimate the new output-to-input voltage ratio.

Section 3-11

3-22 Sketch the circuit diagram for a diode AND gate with five input terminals. Briefly explain the circuit operation.

3-23 Sketch the circuit diagram for a diode OR gate with five input terminals. Briefly explain the circuit operation.

Chapter-3 Problems

Section 3-1

3-1 A half-wave rectifier circuit has a 25 V (*rms*) sinusoidal *ac* input and a 600 Ω load resistance. Calculate the peak output voltage, peak load current, and diode peak reverse voltage. Assume $V_F = 0.7$ V.

3-2 A half-wave rectifier circuit produces a 55 mA peak current in an 820 Ω load resistor. Calculate the *rms ac* input voltage, and the diode peak reverse voltage if $V_F = 0.7$ V.

3-3 Calculate the power dissipated in a 560 Ω load resistor connected to the output of a half wave rectifier circuit. The *ac* input is 25 V (*rms*), and the diode voltage drop is 0.7 V.

Section 3-2

3-4 A 470 Ω load resistor is connected at the output of a bridge rectifier circuit that has a 15 V (*rms*) input. Calculate the peak output voltage, peak load current, and load power dissipation, if the rectifiers have 0.3 V forward voltage drop.

3-5 A bridge rectifier circuit has a 25 V (*rms*) sinusoidal *ac* input and a 600 Ω load resistance. Calculate the peak output voltage, peak load current, diode power dissipation, and diode peak reverse voltage. Assume $V_F = 0.7$ V.

3-6   A two-diode full-wave rectifier circuit (as in Fig. 3-2) dissipates 640 mW in a 400 Ω load resistor. If the diodes are silicon, calculate the peak output voltage from each half of the transformer secondary.

Section 3-3

3-7   A dc power supply consisting of a half-wave rectifier and a reservoir capacitor is required to supply 24 V with a 200 mA maximum load current. The output ripple is not to exceed ±500 mV, and the ac input frequency is 60 Hz. Determine a suitable capacitance for the reservoir capacitor.

3-8   Recalculate the capacitance approximately for the circuit in Problem 3-7, assuming that the capacitor discharge time is much greater than the charge time.

3-9   Specify the diodes required for the half-wave rectifier circuit in Problem 3-7. Select a suitable device from Appendix 1, and calculate the required surge limiting resistance.

3-10   Calculate the reservoir capacitance for a half-wave rectifier dc power supply which produces a 15 V output with a 300 mA maximum load current. The peak-to-peak output ripple voltage is to be a maximum of ±10% of V_o, and the ac input frequency is 60 Hz.

3-11   Specify the diodes required for the power supply in Problem 3-10. Select a suitable device from Appendix 1, and calculate the required surge limiting resistance.

3-12   Recalculate the capacitance for the power supply in Problem 3-10 for a 400 Hz ac supply.

Section 3-4

3-13   A dc power supply consisting of a bridge rectifier circuit and a reservoir capacitor is to supply 24 V with a 200 mA maximum load current. The output ripple is not to exceed ±500 mV, and the ac input frequency is 60 Hz. Determine a suitable capacitance for the reservoir capacitor.

3-14   Recalculate the capacitance for the circuit in Problem 3-13, using the approximation that the capacitor discharge time is much greater than the charge time.

3-15   Specify the diodes required for the rectifier circuit in Problem 3-13. Select a suitable device from Appendix 1, and calculate the required surge limiting resistance.

3-16   Calculate the reservoir capacitance for a dc power supply which produces an 18 V output with a 300 mA maximum load current. A bridge rectifier circuit is used and the ac input frequency is 60 Hz. The peak-to-peak output ripple voltage is to be a maximum of ±10% of V_o.

3-17   Specify the diodes required for the power supply in Problem 3-16. Select a suitable device from Appendix 1, and calculate the required surge limiting resistance.

3-18 Recalculate the capacitance for the power supply in Problem 3-16 for a 400 Hz *ac* supply.

Section 3-5

3-19 A *dc* power supply output voltage changes from 12 V to 11.6 V when the input drops by 10%, and from 12 V to 11.5 V when the load current increases from zero to maximum. Determine the source and load effects, and the line and load regulations.

3-20 A *dc* power supply with a 20 V output has a 3% line regulation and a 5% load regulation. Calculate the source effect and the load effect.

3-21 Calculate the line and load regulation for the circuit designed in Problem 3-7.

3-22 Calculate the line and load regulation for the circuit designed in Problem 3-16.

Section 3-6

3-23 A series-connected Zener diode and resistor is to be used as a 10 V reference source with a load current less than 1 mA. The available supply voltage is 25 V. Select suitable components, and calculate the effect of a ±10% supply voltage change on the diode current.

3-24 A 5 V Zener diode voltage source is to be designed to produce maximum possible output current from a low-power Zener diode. Design the circuit to operate from an 22 V supply, and calculate the maximum load current.

3-25 Determine the source effect, load effect, and ripple rejection for the circuit designed in Problem 3-24.

3-26 Recalculate the maximum output current for the circuit designed in Problem 3-24 when the ambient temperature is raised to 100°C. The diode derating factor is 3.2 mW/°C for temperatures above 50°C.

3-27 A *1N751* Zener diode is connected in series with a 330 Ω resistor to a 25 V supply. Determine the maximum and minimum level of the Zener diode voltage, and calculate the minimum resistance that may be connected in parallel with the diode.

3-28 Calculate the source effect, load effect, and ripple rejection for the circuit in Problem 3-27.

Section 3-7

3-29 A diode-resistor negative series clipping circuit [as in Fig. 3-25(a)] has a ±12 V input, and zero load current. Determine a suitable resistor value and specify the diode.

3-30 A device is to be protected from the negative half-cycle of a ±8

V square wave input. The device input current is 50 μA when the input is positive. Select a suitable clipping circuit, determine an appropriate resistor, and specify the diode.

3-31 Specify the diodes for the series noise clipper circuit in Fig. 3-27, if R_1 = 270 Ω, and the input peaks are ±7 V.

Section 3-8

3-32 A positive shunt clipping circuit [as in Fig. 3-28(a)] with a ±10 V input is to produce a -9 V output when the load current is 500 μA. Determine a suitable resistor value and specify the diode.

3-33 A device is to be protected from the negative half-cycle of a ±6 V square wave input. The device current is 750 μA when its input is +4 V. Select a suitable shunt clipping circuit, determine an appropriate resistor, and specify the diode.

3-34 Specify the diodes for the shunt noise clipper circuit in Fig. 3-30, if R_1 = 470 Ω, and the input peaks are ±5 V.

3-35 A ±12 V square wave is applied to a circuit that cannot accept inputs in excess of ±4 V. The circuit input current is ±100 μA. Design a suitable biased shunt clipping circuit.

3-36 Design a Zener diode shunt clipping circuit to satisfy the requirements in Problem 3-35.

3-37 A Zener diode shunt clipping circuit with a ±15 V input is to have a maximum output of ±9 V with an output current of ±750 μA. Design the circuit.

3-38 A biased shunt clipper circuit is to produce a maximum output of ±6 V to a 10 kΩ load. The input voltage peaks can be as high as ±11 V. Design the circuit.

3-39 Design a Zener diode shunt clipping circuit to satisfy the requirements in Problem 3-38.

3-40 Sketch the output waveforms from each of the circuits in Fig. 3-51 when a ±8 V square wave is applied at the input.

Section 3-9

3-41 A positive voltage clamping circuit (as in Fig. 3-34) has C_1 = 2 μF and R_1 = 27 kΩ. Determine the tilt on the output when a ±5 V, 700 Hz square wave input is applied.

3-42 A negative voltage clamping circuit has a ±9 V, 1 kHz square wave input with a 500 Ω source resistance. The slope on the output waveform is not to exceed 0.5%. Determine suitable resistance and capacitance values.

3-43 A biased clamping circuit (as in Fig. 3-39) has a ±12 V, 330 Hz sine wave input with a 600 Ω source resistance. The positive output peak is to be clamped at 5 V, and the slope between peaks is not to exceed 1%. Determine the required bias voltage and suitable resistance and capacitance values.

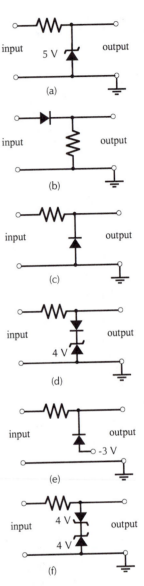

Figure 3-51
Circuits for Problem 3-40.

3-44 Design a Zener diode clamping circuit that will perform as specified in Problem 3-43.

3-45 A Zener diode clamping circuit, as in Fig. 3-41, uses a germanium diode and a *1N746* Zener diode. The input voltage is a ±8 V, 1 kHz square wave. Determine the output voltage levels, and calculate the capacitor voltage.

3-46 Sketch the output waveforms from each of the circuits in Fig. 3-52 when a ±15 V square wave is applied at the input.

3-47 A ±20 V, 700 Hz square wave with a 500 Ω source resistance is to be clamped to a maximum level of approximately 12 V. Design a suitable Zener diode circuit that will produce a maximum output tilt of 3%.

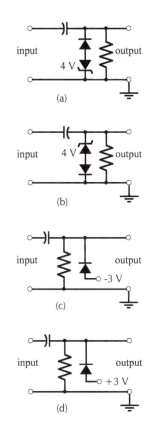

(a)

(b)

(c)

(d)

Figure 3-52
Circuits for Problem 3-46.

Section 3-10

3-48 A voltage doubling circuit is to be designed to produce an output of 25 V to a 56 kΩ resistor. Determine the required amplitude for a 1 kHz square wave input, and calculate suitable capacitor values if the output ripple is to be a maximum of 0.5 V.

3-49 A voltage doubling circuit with a ±10 V, 2 kHz square wave input has a 33 kΩ load resistance. Determine suitable capacitance values to give a 1% maximum output ripple.

3-50 Refer to the voltage doubling circuit designed for Problem 3-48. Determine the effect of (a) changing R_L to 33 kΩ, (b) changing the input frequency to 1.5 kHz.

Section 3-11

3-51 A three-input diode *AND* gate, as in Fig. 3-49, is to have a minimum diode current of 1 mA. If the circuit supply voltage is 9 V, determine a suitable resistance for R_1.

3-52 A five-input diode *OR* gate used diodes which require a 500 µA minimum forward current. Determine a suitable resistance value if the input voltages are +5 V.

Practise Problem Answers

3-1.1	20.51 V, 62.2 mA, 21.2 V
3-1.2	33.95 V, 48.7 V
3-2.1	32.54 V, 69.2 mA, 1.12 V
3-2.2	238 mW, 13.7 V
3-3.1	1000 µF
3-3.2	835 µF
3-3.3	253.4 V, 600 mA, 8.6 A, 1N4004, 4.2 Ω
3-4.1	500 µF
3-4.2	557 µF
3-4.3	1N4001, 0.57 Ω
3-5.1	50 mV, 100 mV, 0.33%, 0.66%
3-6.1	1N756, 680 Ω, 14.7 mA

3-6.2	1N756, (220Ω + 22 Ω), 43.8 mA
3-6.3	61.4 mV, 336 mV, 3.07×10^{-2}
3-7.1	5.6 kΩ, 7 V, 1 mA
3-7.2	4.7 kΩ, 6 V, 1 mA
3-8.1	330 Ω, 7 V, 19 mA
3-8.2	±2.8 V, 2.7 kΩ
3-8.3	1N753, 1.8 kΩ
3-9.1	550 mV
3-9.2	1 μF, 100 kΩ
3-9.3	7.5 V, -16.5 V, 4.5 V
3-10.1	1.6%, 0.65%
3-10.2	±10.7 V, 3.9 μF, 8 μF
3-11.1	680 Ω
3-11.2	1.41 mA, 469 μA

Chapter *4*

Bipolar Junction Transistors

Chapter Contents

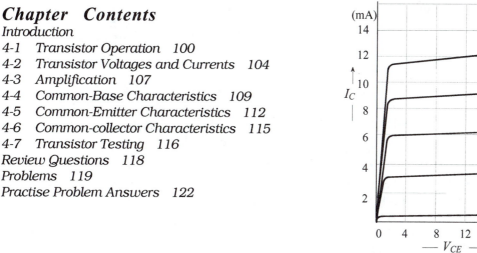

Objectives

You will be able to:

1   *Draw block representations of npn and pnp transistors showing depletion regions and barrier voltages. Explain the operation of each device in terms of junction bias voltages and charge carrier movement.*

2   *Sketch the circuit symbols for npn and pnp transistors. Identify current directions and junction bias polarities, and state typical terminal voltage levels.*

3   *Identify and describe the various current components in a transistor. Define α and β. Write equations relating the base, emitter, and collector currents, and calculate transistor current levels.*

4   *Describe how a transistor can be used for current amplification and voltage amplification. Calculate current and voltage gains.*

5   *Sketch typical transistor common-base, common-emitter, and common-collector characteristics. Draw circuits to show how each type of characteristic is obtained.*

6   *Discuss various methods of transistor testing.*

Introduction

A *bipolar junction transistor* (*BJT*) has three layers of semiconductor material. These are arranged either in *npn* (n-type—p-type—n-type) sequence, or in *pnp* sequence, and each of the three layers has a terminal. A small current at the central region terminal controls the much larger total current flow through the device. This means that the transistor can be used for current amplification. As will be explained, it can also perform voltage amplification.

Because the transistor has two *pn*-junctions, its operation can be understood by applying *pn*-junction theory. The currents that flow in a transistor are similar to those that flow across a single *pn*-junction, and the transistor equivalent circuit is essentially two *pn*-junction equivalent circuits. There are three sets of current-voltage characteristics for defining the performance of a transistor.

4-1   Transistor Operation

pnp and npn Transistors

A junction transistor is simply a sandwich of one type of semiconductor material (*p*-type or *n*-type) between two layers of the opposite type. A block representation of a layer of *p*-type material between two layers of *n*-type is shown in Fig. 4-1(a). This is described as an *npn* transistor. Figure 4-1(b) shows a *pnp* transistor, consisting of a layer of *n*-type material between two layers of *p*-type. For reasons which will be understood later, the centre layer is called the *base*, one of the outer layers is termed the *emitter*, and the other outer layer is referred to as the *collector*. The emitter, base, and collector are provided with terminals which are appropriately labelled *E, B,* and *C.* Two *pn*-junctions exist within each transistor; the *collector-base junction* and the *emitter-base junction*. Each of these junctions has the characteristics discussed in Chapters 1 and 2.

Circuit symbols for *pnp* and *npn* transistors are shown in Fig. 4-2. The arrowhead on each symbol identifies the transistor emitter terminal, and indicates the conventional direction of current flow. For an *npn* transistor, the arrowhead points from the *p*-type base to the *n*-type emitter. For a *pnp* device, the arrowhead points from the *p*-type emitter to the *n*-type base. Thus, the arrowhead always points from *p* to *n.*

Two transistor packages are illustrated in Fig. 4-3; a low-power transistor, and a high power device. Low-power transistors typically pass currents of 1 mA to 20 mA. Current levels for high-power transistors range from 100 mA to several amps. The high-power transistor is designed for mechanical connection to a heat sink in order to prevent device overheating. Semiconductor device packages are further investigated in Chapter 7.

npn Transistor Operation

Figure 4-4 illustrates the depletion regions and barrier voltages at

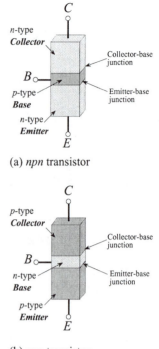

(a) *npn* transistor

(b) *pnp* transistor

Figure 4-1
Block representation of npn and pnp transistors.

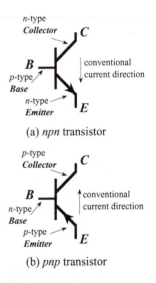

(a) *npn* transistor

(b) *pnp* transistor

Figure 4-2
Transistor circuit symbols. The arrowhead identifies the emitter terminal, and always points in the conventional current direction.

the junctions of an unbiased *npn* transistor. These were originally explained in Chapter 1. Although it is not shown in the illustration, the centre layer of the transistor is very much narrower than the two outer layers. Also, the outer layers are much more heavily doped than the centre layer, causing the depletion regions to penetrate deep into the base, as illustrated. Because of this penetration, the distance between the two depletion regions is very short (within the base). Note that the junction barrier voltages are positive on the emitter and collector, and negative on the base of the *npn* device.

Consider Fig. 4-5 which shows an *npn* transistor with external bias voltages. For normal operation, the base-emitter (*BE*) junction is forward biased and the collector-base (*CB*) junction is reverse biased. Note the external bias voltage polarities. The forward bias at the *BE* junction reduces the barrier voltage and causes electrons to flow from the *n*-type emitter to the *p*-type base. The electrons are *emitted* into the base region; hence the name *emitter*. Holes also flow from the *p*-type base to the *n*-type emitter, but because the base is much more lightly doped than the collector, almost all of the current flow across the *BE* junction consists of electrons entering the base from the emitter. Thus, electrons are the *majority charge carriers* in an *npn* device.

The reverse bias at the *CB* junction causes the *CB* depletion region to penetrate deeper into the base than when the junction is unbiased, (see Fig. 4 5). The electrons crossing from the emitter to the base arrive quite close to the large negative-positive electric field (or barrier voltage) at the *CB* depletion region. Because electrons have a negative charge, they are drawn across the *CB* junction by this bias voltage. They are said to be *collected*.

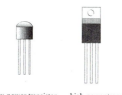

low-power transistor   high-power transistor

Figure 4-3
Low-power and high-power transistors.

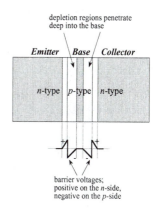

Figure 4-4
Unbiased npn transistor depletion regions and barrier voltages.

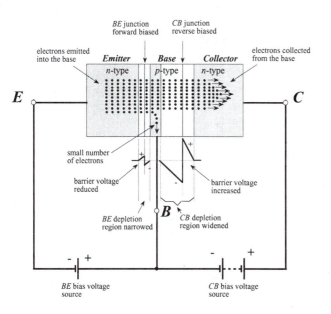

Figure 4-5
For transistor operation, the base-emitter junction is forward biased, and the collector-base junction is reverse biased. Almost all the charge carriers emitted into the base flow to the collector.

Some of the charge carriers entering the base from the emitter do not reach the collector, but flow out via the base connection, as illustrated in Fig. 4-5. However, the path from the *BE* junction to the *CB* depletion region is much shorter than that to the base terminal. So, only a very small percentage of the total charge carriers flow out of the base terminal. Also, because the base region is very lightly doped, there are few holes in the base to recombine with electrons from the emitter. The result is that about 98% of the charge carriers from the emitter are drawn across the *CB* junction to flow via the collector terminal and the voltage sources back to the emitter.

Another way of looking at the effect of the reverse-biased *CB* junction is from the point of view of minority and majority charge carriers. It has already been shown (in Section 1-6) that a reverse-biased junction opposes the flow of majority charge carriers and assists the flow of minority carriers. Majority carriers are, of course, holes from the *p*-side of a junction and electrons from the *n*-side. Minority carriers are holes from the *n*-side and electron from the *p*-side, (see Fig. 4-6). In the case of an *npn* transistor, the charge carriers arriving at the *CB* junction are electrons (from the emitter) travelling through the *p*-type base. Consequently, to the *CB* junction they appear a minority charge carriers, and the reverse bias assists them to cross the junction.

Because the *BE* junction is forward biased, it has the characteristics of a forward-biased diode. Substantial current will not flow until the forward bias is about 0.7 V for a silicon device, or about 0.3 V for germanium. Reducing the level of the *BE* bias voltage reduces the *pn*-junction forward bias, and thus reduces the current that flows from the emitter through the base to the collector. Increasing the *BE* bias voltage increases this current. Decreasing the *BE* bias voltage to zero, or reversing it, cuts the current *off* completely. Thus, variation of the small forward-bias voltage on the *BE* junction controls the emitter and collector currents, and the *BE* controlling voltage source has to supply only the small base current. This is illustrated in Fig. 4-7.

pnp Transistor Operation

In an unbiased *pnp* transistor, the barrier voltages are positive on the base and negative on the emitter and collector, (see Fig. 4-8). As in the case of the *npn* device, the collector and emitter are heavily-doped, so that the *BE* and *CB* depletion regions penetrate deep into the lightly-doped base.

A *pnp* transistor behaves exactly the same as an *npn* device, with the exception that the majority charge carriers are holes. As illustrated in Fig. 4-9, the *BE* junction is forward biased by an external voltage source, and the *CB* junction is reverse biased. Holes are emitted from the *p*-type emitter across the forward-biased *BE* junction into the base. In the lightly doped *n*-type base, the holes find few electrons to absorb. Some of the holes flow out via the base terminal, but most are drawn across to the collector by the positive-negative electric field at the reverse-biased *CB*

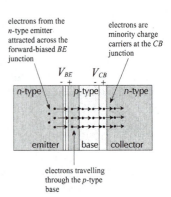

Figure 4-6
Electrons crossing from the emitter into the base become minority charge carriers at the collector-base junction.

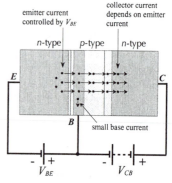

Figure 4-7
The emitter and collector currents are controlled by the base-emitter voltage.

junction. Variation of the forward bias voltage at the *BE* junction controls the small base current and the much larger collector and emitter currents.

Bipolar Devices

Although one type of charge carrier is in the majority in a *pnp* or *npn* transistor, two types of charge carrier (holes and electrons) are involved in the current flow. Consequently, these devices are termed *bipolar junction transistors (BJT)*. This is to distinguish them from *field-effect transistors (FET)*, (see Chapter 9), which are termed *unipolar* devices, because they use only one type of charge carrier.

Summary

A transistor is a sandwich of *npn* or *pnp* semiconductor materials. The outer layers are called the emitter and the collector, and the centre layer is termed the base. Two *pn*-junctions are formed, with depletion regions and barrier voltages at each junction. The barrier voltages are negative on the *p*-side and positive on the *n*-side.

The base-emitter junction is forward biased, so that charge carriers are emitted into the base. The collector-base junction is reverse biased, and its depletion region penetrates deep into the base. The base section is made as narrow as possible so that charge carriers can easily move across from emitter to collector. The base is also lightly doped, so that few charge carriers are available to recombine with the majority charge carriers from the emitter. Most charge carriers from the emitter flow to the collector; few flow out through the base terminal. Variation of the base-emitter junction bias voltage alters the base, emitter, and collector currents.

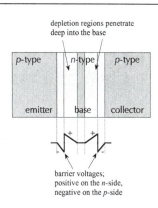

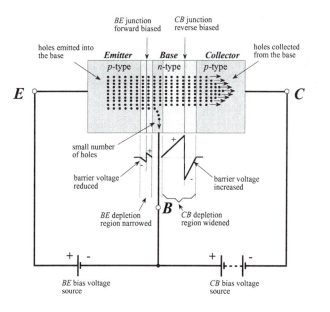

Figure 4-8
Unbiased pnp transistor depletion regions and barrier voltages.

Figure 4-9
In a pnp transistor, the forward-biased base-emitter junction causes charge carriers (holes) to be emitted into the base. Most of these flow across the reverse-biased collector-base junction.

4-2   Transistor Voltages and Currents

Terminal Voltages

The terminal voltage polarities for an *npn* transistor are shown in Fig. 4-10(a). As well as conventional current direction, the direction of the arrowhead indicates the transistor bias polarities. For an *npn* transistor, the base is biased positive with respect to the emitter, and the arrowhead points from the (positive) base to the (negative) emitter. The collector is then biased to a higher positive voltage than the base.

Figure 4-10(b) shows that the voltage sources are usually connected to the transistor via resistors. The base bias voltage (V_B) is connected via resistor R_B, and the collector supply (V_{CC}) is connected via R_C. The negative terminals of the two voltage sources are connected at the transistor emitter terminal. V_{CC} is always much larger than V_B, and this ensures that the *CB* junction remains reverse biased; positive on the collector (*n*-side), and negative on the base (*p*-side).

Typical transistor base-emitter voltages are similar to diode forward voltages; 0.7 V for a silicon transistor, and 0.3 V for a germanium device. Typical collector voltages might be anything from 3 V to 20 V for most types of transistors, although in many the collector voltage may be greater than 20 V.

For a *pnp* device [Fig. 4-11(a)] the base is biased negative with respect to the emitter. The arrowhead points from the (positive) emitter to the (negative) base, and the collector is made more negative than the base. Figure 4-11(b) shows the voltage sources connected via resistors, and the source positive terminals connected at the emitter. With V_{CC} larger than V_B, the (*p*-type) collector is more negative than the (*n*-type) base, keeping the *CB* junction reverse biased.

All transistors (*npn* and *pnp*) are normally operated with the *CB* junction reverse biased and the *BE* junction forward biased. In the case of a switching transistor, (a transistor that is not operated as an amplifier, but is either switched *on* or *off*), the *CB* junction may become forward biased by approximately 0.5 V. Also, in transistor switching circuits (and some other circuits) the *BE* junction can become reverse biased.

Transistor Currents

The various current components that flow within a transistor are illustrated in Fig. 4-12. The current flowing into the emitter terminal is referred to as the *emitter current* and is identified as I_E. For the *pnp* device shown, I_E can be thought of as a flow of holes

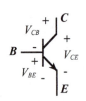

(a) *npn* terminal voltage polarities

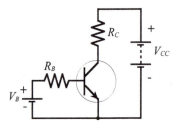

(b) Voltage source connections

Figure 4-10
An npn transistor must have its base biased positive with respect to its emitter, and its collector more positive than the base.

(a) *pnp* terminal voltage polarities

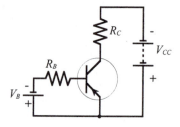

(b) Voltage source connections

Figure 4-11
An pnp transistor must have its base biased negative with respect to its emitter, and its collector more negative than the base.

from the emitter to the base. Note that the indicated I_E direction is the conventional current direction from positive to negative. *Base current I_B* and *collector current I_C* are also shown as conventional current direction. Both I_C, and I_B flow *out* of the transistor. while I_E flows *into* the transistor, (see Fig. 4-12 and 4-13). Therefore,

$$I_E = I_C + I_B \qquad (4\text{-}1)$$

As already discussed, almost all of I_E crosses to the collector, and only a small portion flows out of the base terminal. Typically 96% to 99.5% of I_E flows across the collector-base junction to become the collector current. As shown in Fig. 4-12,

$$I_C = \alpha_{dc}\, I_E \qquad (4\text{-}2)$$

α_{dc} *(alpha dc)* is the *emitter-to-collector current gain*, or the ratio of collector current to emitter current. From Eq. 4-2, $\alpha_{dc} = I_C/I_E$. Numerically, α_{dc} is typically 0.96 to 0.995. So, the collector current is almost equal to the emitter current, and in many circuit situations I_C is assumed equal to I_E. For reasons that will be explained later, α_{dc} is termed *the common-base dc current gain.*

Because the *CB* junction is reverse biased, a very small *reverse saturation current* (I_{CBO}) flows across the junction, (see Fig. 4-12). I_{CBO} is named the *collector-to-base leakage current*, and it is normally so small that it can be neglected.

Substituting for I_E from Eq. 4-1 into Eq. 4-2,

$$I_C = \alpha_{dc}\, (I_C + I_B) \qquad (4\text{-}3)$$

which gives, $\quad I_C = \dfrac{\alpha_{dc}\, I_B}{1 - \alpha_{dc}} \qquad (4\text{-}4)$

Equation 4-4 can be rewritten as,

$$I_C = \beta_{dc}\, I_B \qquad (4\text{-}5)$$

where, $\quad \beta_{dc} = \dfrac{\alpha_{dc}}{1 - \alpha_{dc}} \qquad (4\text{-}6)$

β_{dc} *(beta dc)* is the *base-to-collector current gain*, or the ratio of collector current to base current, (see Fig. 4-13). From Eq. 4-5, $\beta_{dc} = I_C/I_B$. Typically, β_{dc} ranges from 25 to 300. As will be explained, β_{dc} is also termed the *common-emitter dc current gain.* Instead of β_{dc}, another symbol for common-emitter *dc* current gain is h_{FE}. This originates from the *h-parameter* circuit discussed in Section 6-3. *h_{FE} is the symbol used on transistor data sheets.*

Although Equations 4-1 through 4-6 were derived for a *pnp* device, they apply equally to *npn* transistors. One difference in the case of an *npn* transistor is that I_B and I_C are assumed to flow *into* the device (conventional current direction), and I_E is taken as flowing *out*, (see Fig, 4-14). As explained in Section 4-1, electrons are the majority charge carriers in an *npn* transistor, and they move in a direction opposite to conventional current direction.

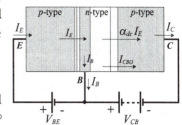

Figure 4-12
Currents in a pnp transistor. I_C approximately equals I_E, and I_B is very much smaller than I_E.

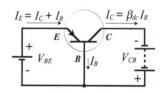

Figure 4-13
Currents in a pnp transistor. I_E is the sum of I_C and I_B. $I_C = \beta_{dc}I_B$, where β_{dc} is typically 25 to 300.

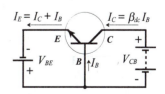

Figure 4-14
Currents in an npn transistor. I_E is the sum of I_C and I_B. $I_C = \beta_{dc}I_B$.

Typical collector and emitter currents for low-power transistors range from 1 mA to 25 mA, and base currents are usually less than 100 µA. High power transistors handle currents of several amps.

Example 4-1

Calculate I_C and I_E for a transistor that has $\alpha_{dc} = 0.98$ and $I_B = 100\,\mu A$. Also, determine the value of β_{dc} for the transistor.

Solution

Eq. 4-4,
$$I_C = \frac{\alpha_{dc} I_B}{1 - \alpha_{dc}} = \frac{0.98 \times 100\,\mu A}{1 - 0.98}$$
$$= 4.9 \text{ mA}$$

from Eq. 4-2,
$$I_E = \frac{I_C}{\alpha_{dc}} = \frac{4.9 \text{ mA}}{0.98}$$
$$= 5 \text{ mA}$$

Eq. 4-6,
$$\beta_{dc} = \frac{\alpha_{dc}}{1 - \alpha_{dc}} = \frac{0.98}{1 - 0.98}$$
$$= 49$$

Example 4-2

Calculate α_{dc} and β_{dc} for the transistor in Fig. 4-15 if I_C is measured as 1 mA, and I_B is 25 µA. Also, determine the new base current to give $I_C = 5$ mA.

Solution

Eq. 4-5,
$$\beta_{dc} = \frac{I_C}{I_B} = \frac{1 \text{ mA}}{25\,\mu A}$$
$$= 40$$

Eq. 4-1,
$$I_E = I_C + I_B = 1 \text{ mA} + 25\,\mu A$$
$$= 1.025\,\mu A$$

from Eq. 4-2,
$$\alpha_{dc} = \frac{I_C}{I_E} = \frac{1 \text{ mA}}{1.025 \text{ mA}}$$
$$= 0.976$$

Eq. 4-5,
$$I_B = \frac{I_C}{\beta_{dc}} = \frac{5 \text{ mA}}{40}$$
$$= 125\,\mu A$$

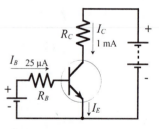

Figure 4-15
Circuit for Example 4-2.

Practise Problems

4-2.1   Determine α_{dc} and I_B for a transistor that has $I_C = 2.5$ mA and $I_E = 2.55$ mA. Also, calculate β_{dc} for the transistor.

4-2.2   A transistor has measured currents of $I_C = 3$ mA and $I_E = 3.03$ mA. Calculate the new current levels when the transistor is replaced with a device that has $\beta_{dc} = 75$. Assume that I_B remains constant.

4-3 Amplification

Current Amplification

The equations derived in Section 4-2 demonstrate that a transistor can be used for current amplification. A small change in the base current (ΔI_B) produces a large change in collector current (ΔI_C) and a large emitter current change (ΔI_E), [see Fig. 4-16(a) and (b)]. Rewriting Eq. 4-5, the current gain from the base to collector can be stated in terms of current level changes.

$$\beta_{dc} = \frac{\Delta I_C}{\Delta I_B}$$

The increasing and decreasing levels of input and output currents may be defined as alternating quantities. In this case, small (lower-case) letters are used for the subscripts. Thus, I_b is an *ac* base current, I_c is an *ac* collector current, and I_e is an *ac* emitter current. The alternating current gain from base to collector may now be stated as,

$$\beta_{ac} = \frac{I_c}{I_b} \qquad (4\text{-}7)$$

As in the case of *dc* current gain, two parameter symbols are available for common-emitter *ac* current gain; β_{ac} and h_{fe}. Either symbol may be used, but, once again, h_{fe} *is the symbol employed on transistor data sheets.*

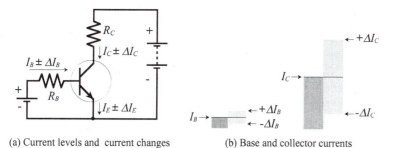

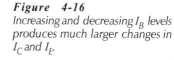

(a) Current levels and current changes   (b) Base and collector currents

Figure 4-16
Increasing and decreasing I_B levels produces much larger changes in I_C and I_E.

Voltage Amplification

Refer to the circuit in Fig. 4-17(a), and assume that the transistor (Q_1) has $\beta_{dc} = 50$. Note that the 0.7 V *dc* voltage source (V_B) forward biases the transistor base-emitter junction. An *ac* signal source (v_i) in series with V_B provides a ±20 mV input voltage. The transistor collector is connected to a 20 V *dc* voltage source (V_{CC}) via the 12 $k\Omega$ collector resistor (R_1).

If Q_1 has the I_B/V_{BE} characteristic shown in Fig. 4-17(b), the 0.7 V level of V_B produces a 20 μA base current. This gives,

$$I_C = \beta_{dc} I_B = 50 \times 20 \ \mu A$$
$$= 1 \ mA$$

The *dc* level of the transistor collector voltage can now be

calculated as,

$$V_C = V_{CC} - (I_C R_1) = 20\text{ V} - (1\text{ mA} \times 12\text{ k}\Omega)$$
$$= 8\text{ V}$$

The V_C and I_C levels are shown on Fig. 4-17(a).

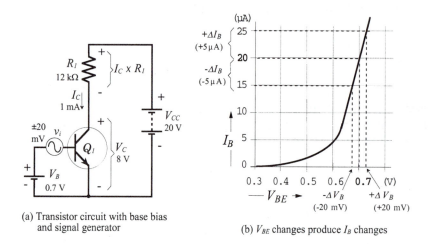

Figure 4-17
Increasing and decreasing V_B levels ($\pm\Delta V_B$) produces I_B changes ($\pm\Delta I_B$). This results in I_C changes and (output) collector voltage variations.

(a) Transistor circuit with base bias and signal generator

(b) V_{BE} changes produce I_B changes

While the *ac* input voltage (v_i) is zero, the transistor collector voltage remains at 8 V. When v_i causes a base voltage variation (ΔV_B) of ±20 mV, the base current changes by ±5 µA, as shown in Fig. 4-17(b). The I_B change produces a collector current change.

$$\Delta I_C = \beta_{dc}\,\Delta I_B = 50 \times (\pm 5\text{ µA})$$
$$= \pm 250\text{ µA}$$

Figure 4-18(a) shows that ΔI_C causes a change in the voltage drop across R_1, and thus produces a variation in the transistor collector voltage.

$$\Delta V_C = \Delta I_C\,R_1 = \pm 250\text{ µA} \times 12\text{ k}\Omega$$
$$= \pm 3\text{ V}$$

The circuit *ac* input is the base voltage change (ΔV_B), and the *ac* output is the collector voltage change (ΔV_C). Because the output is greater than the input, the circuit has a voltage gain; it is a *voltage amplifier*. The *voltage gain* (A_v) is the ratio of the output voltage to the input voltage.

$$A_v = \frac{\Delta V_C}{\Delta V_B} = \frac{\pm 3\text{ V}}{\pm 20\text{ mV}}$$
$$= 150$$

As already discussed for current changes, the increasing and decreasing levels of voltage can be referred to as *ac* quantities. The *ac* signal voltage (v_i) produces the *ac* base current (I_b), and this generates the *ac* collector current (I_c) which produces the *ac* voltage change across R_1, [see Fig. 4-18(b)]. The equation for *ac* voltage gain is,

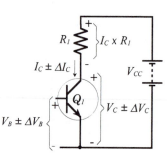

(a) ΔI_C changes V_{R1} and V_C

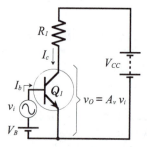

(b) An *ac* input (v_i) produces an *ac* output (v_o)

Figure 4-18
I_C changes produce changes in V_C. An ac input (v_i) produces an ac output ($v_o = A_v\,v_i$).

$$A_v = \frac{v_o}{v_i} \qquad\qquad \textbf{(4-8)}$$

Substituting the *ac* quantities for Fig. 4-17(a) into Eq. 4-8 gives A_v = 150 once again.

Example 4-3
Determine the *dc* collector voltage for the circuit in Fig. 4-19(a), if the transistor has the I_B/V_{BE} characteristics shown in Fig. 4-19(b), and $\beta_{dc} = \beta_{ac} = 80$. Also, calculate the circuit voltage gain when v_i is ±50 mV.

Solution
From Fig. 4-19(b),

$I_B = 15\,\mu A$ for $V_B = 0.7$ V

$\begin{aligned}I_C &= \beta_{dc}\,I_B = 80 \times 15\,\mu A\\ &= 1.2 \text{ mA}\end{aligned}$

$\begin{aligned}V_C &= V_{CC} - (I_C\,R_1)\\ &= 18 \text{ V} - (1.2 \text{ mA} \times 10 \text{ k}\Omega)\\ &= 6 \text{ V}\end{aligned}$

From Fig. 4-18(b),

$I_b = \pm3\,\mu A$ for $v_i = \pm50$ mV

$\begin{aligned}I_c &= \beta_{ac}I_b = 80 \times \pm3\,\mu A\\ &= \pm240\,\mu A\end{aligned}$

$\begin{aligned}v_o &= I_c\,R_1 = \pm240\,\mu A \times 10 \text{ k}\Omega\\ &= \pm2.4 \text{ V}\end{aligned}$

Eq. 4-9,
$A_v = \frac{V_o}{v_i} = \frac{\pm2.4 \text{ V}}{\pm50 \text{ mV}}$

$= 48$

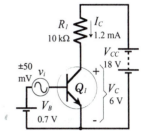

(a) Transistor circuit

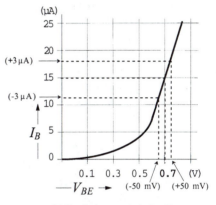

(b) V_{BE} I_B characteristic for Q_1.

Figure 4-19
Circuit and characteristics for Example 4-3.

Practise Problems
4-3.1 A circuit similar to Fig. 4-17 has $V_{CC} = 15$ V, $R_1 = 5.6$ kΩ, $V_C = 7$ V, and $I_B = 20\,\mu A$. Calculate the *dc* current gain.

4-3.2 A ±50 mV *ac* input is applied in series with the transistor base in Problem 4-3.1. Assuming that the transistor has the input characteristics in Fig. 4-17(b), calculate the circuit *ac* voltage gain.

4-4 Common-Base Characteristics

Common-Base Connection
To investigate the characteristics of a diode (a two-terminal device), several levels of forward or reverse bias voltage are applied and the resulting current levels are measured. The characteristics of the device are then derived by plotting the graph of current versus

voltage. Because a transistor is a three-terminal device, there are three possible connection arrangements (*configurations*) for investigating its characteristics. From each of these configurations, three sets of characteristics may be derived.

Figure 4-20 shows a *pnp* transistor with its base terminal common to both the input (*EB*) voltage and the output (*CB*) voltage. The transistor is said to be connected in *common-base* configuration. Voltmeters and ammeters are included to measure the input and output voltages and currents.

Input Characteristics

To determine the *input characteristics*, the output (*CB*) voltage is maintained constant, and the input (*EB*) voltage is set at several convenient levels. For each level of input voltage, the input current I_E is recorded. I_E is then plotted versus V_{EB} to give the common-base input characteristics shown in Fig. 4-21.

Because the *EB* junction is forward biased, the common-base input characteristics are essentially those of a forward biased *pn*-junction. Figure 4-21 also shows that for a given level of input voltage, more input current flows when higher levels of *CB* voltage are used. This is because larger *CB* (reverse bias) voltages cause the depletion region at the *CB* junction to penetrate deeper into the base of the transistor, thus shortening the distance and reducing the resistance between the *EB* and *CB* depletion regions.

Output Characteristics

The emitter current I_E is held constant at each of several fixed levels. For each fixed level of I_E, the output voltage V_{CB} is adjusted in convenient steps, and the corresponding levels of collector current I_C are recorded. In this way, a table of values is obtained from which a family of output characteristics may be plotted. In Fig. 4-22 the corresponding I_C and V_{CB} levels obtained when I_E was held constant at 1 mA are plotted, and the resultant characteristic is identified as I_E = 1 mA. Similarly, other characteristics are plotted for I_E levels of 2 mA, 3 mA, 4 mA, and 5 mA.

The common-base output characteristics in Fig. 4-22 show that for each fixed level of I_E, I_C is almost equal to I_E and appears to remain constant when V_{CB} is increased. In fact, there is a very small increase in I_C with increasing V_{CB}. This is because the increase in collector-to-base bias voltage (V_{CB}) expands the *CB* depletion region, and thus shortens the distance between the two depletion regions. With I_E held constant, the increase in I_C is so small that it is noticeable only for large variations in V_{CB}.

As illustrated in Fig. 4-22, when V_{CB} is reduced to zero, I_C still flows. Even when the externally applied bias voltage is zero, there is still a barrier voltage existing at the *CB* junction, and this assists the flow of I_C. The charge carriers which constitute I_C are minority carriers as they cross the *CB* junction. Thus, the reverse-bias voltage V_{CB} and the (unbiased) *CB* barrier voltage assist their movement across the junction, (see Fig. 4-12). To stop the flow of charge carriers, the *CB* junction has to be forward biased.

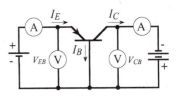

Figure 4-20
Circuit for determining transistor common-base characteristics.

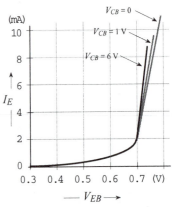

Figure 4-21
The common-base input characteristics for a transistor are the characteristics of a forward-biased pn-junction.

Consequently, as shown in Fig. 4-22, I_C is reduced to zero only when V_{CB} is increased positively.

The region of the graph for the forward-biased CB junction is known as the *saturation region*, (see Fig. 4-22). The region in which the junction is *reverse biased* is named the *active region*, and this is the normal operating region for the transistor.

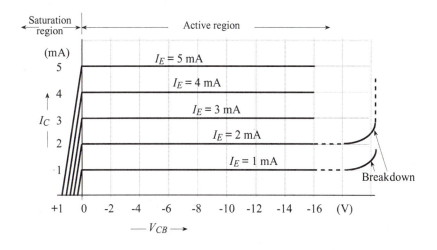

Figure 4-22
The common-base output characteristics (or collector characteristics) for a transistor show that I_C remains substantially constant for each level of I_E.

If the reverse-bias voltage on the CB junction is allowed to exceed the maximum safe limit specified by the manufacturer, device breakdown may occur. Breakdown, illustrated by the dashed lines in Fig. 4-22, can be caused by the same effects that make diodes breakdown, (see Section 2-9). Breakdown can also result from the CB depletion region penetrating into the base until it makes contact with the EB depletion region, (Fig. 4-23). This condition is known as *punch-through*, or *reach-through*, and very large currents can flow when it occurs, possibly destroying the device. The extension of the depletion region is produced by the increase in V_{CB}. So, it is very important to maintain V_{CB} below the maximum safe limit specified by the device manufacturer. Typical maximum V_{CB} levels range from 25 V to 80 V.

Current Gain Characteristics

The *current gain characteristics* (also termed the *forward transfer characteristics*) are a graph of output current (I_C) versus input current (I_E). They are obtained experimentally by use of the circuit in Fig. 4-20. V_{CB} is held constant at a convenient level, and I_C is measured for various levels of I_E. I_C is then plotted versus I_E, and the resultant graph is identified by the V_{CB} level, (see Fig. 4-24).

The CB current gain characteristics can be derived from the CB output characteristics, as shown in Fig. 4-25. A vertical line is drawn through a selected V_{CB} value, and corresponding levels of I_E and I_C are read along the line. The I_C levels are then plotted versus I_E, and the characteristic is labelled with the V_{CB} used. Because almost all of I_E flows out of the collector terminal as I_C, V_{CB} has only a small effect on the current gain characteristics.

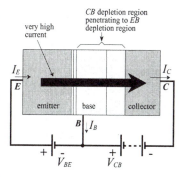

Figure 4-23
Excessive levels of V_{CB} can cause the depletion regions to make contact, resulting in destructive current levels.

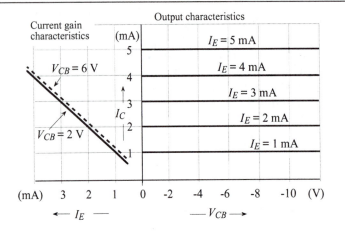

Figure 4-24
The common-base current gain characteristics for a transistor are graphs of output current I_C versus input current I_E.

Example 4-4

Derive the current gain characteristic for $V_{CB} = 2$ V from the common-base output characteristics in Fig. 4-25.

Solution

Draw a vertical line at $V_{CB} = 2$ V. Where the line intersects the characteristics at points A and B, read;

$$I_C \approx 1 \text{ mA for } I_E = 1 \text{ mA}$$

and, $I_C \approx 4$ mA for $I_E = 4$ mA

Plot points C and D for the output characteristic at the corresponding levels of I_C and I_E.

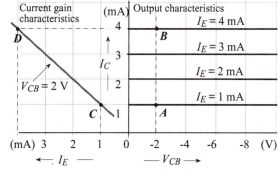

Figure 4-25
Derivation of common-base current gain characteristics from the output characteristics.

Practise Problems
4-4.1   Plot two lines of common-base output characteristics from the following measured quantities:
for $I_E = 1.5$ mA: ($I_C = 1.40$ mA when $V_{CB} = 0$ V)
          ($I_C = 1.45$ mA when $V_{CB} = 10$ V)
for $I_E = 3$ mA: ($I_C = 2.80$ mA when $V_{CB} = 0$ V)
          ($I_C = 2.95$ mA when $V_{CB} = 10$ V)
Derive the current gain characteristics for $V_{CB} = 5$ V from the transistor output characteristics

4-5 Common-Emitter Characteristics

Common-Emitter Circuit

Figure 4-26 shows a circuit for determining transistor *common-emitter* characteristics. The input voltage is applied between the *B* and *E* terminals, and the output is taken at the *C* and *E* terminals. The emitter terminal is common to both input and output. Voltage and current levels are measured as shown.

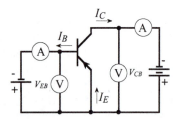

Figure 4-26
Circuit for determining transistor common emitter characteristics.

Input Characteristics

To prepare a table of values for constructing the *common-emitter input characteristics*, V_{CE} is held constant, V_{BE} is set at convenient levels, and the corresponding I_B levels are recorded. I_B is then plotted versus V_{BE}, as shown in Fig. 4-27. It is seen that the common-emitter input characteristics (like the common-base input characteristics) are those of a forward-biased *pn*-junction. It should be remembered that I_B is only a small portion of the total current (I_E) that flows across the forward-biased *BE* junction.

Figure 4-27 also shows that, for a given level of V_{BE}, I_B is reduced when higher V_{CE} levels are employed. This is because higher V_{CE} produces greater depletion region penetration into the base, reducing the distance between the *CB* and *EB* depletion regions. Consequently, more of the charge carriers from the emitter flow across the *CB* junction, and fewer flow out via the base terminal.

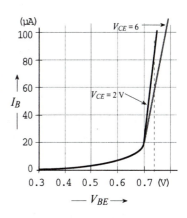

Figure 4-27
The common-emitter input characteristics for a transistor are those of a forward-biased pn-junction.

Output Characteristics

To prepare a table of values for plotting the *common-emitter output characteristics*, I_B is maintained constant at several convenient levels. V_{CE} is adjusted in steps at each I_B level, and the I_C level is recorded at each V_{CE} step. For each I_B level, I_C is plotted versus V_{CE} to give a family of characteristics as shown in Fig. 4-28.

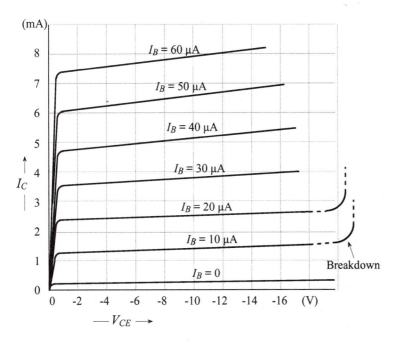

Figure 4-28
The common-emitter output characteristics (or collector characteristics) for a transistor are a plot of I_C versus V_{CE} for various I_B levels.

Because I_E is not held constant (as it is for the common-base output characteristics) the shortening of the distance between the depletion regions (when V_{CB} is increased) draws more charge carriers from the emitter to the collector. Thus, I_C increases to some extent with increasing V_{CE} although I_B is held constant. So, the slopes of the common-emitter output characteristics are much more pronounced than those of the common-base characteristics.

On Fig. 4-28 note that I_C reduces to zero when V_{CE} becomes zero. This is because the horizontal axis voltage is V_{CE}, which equals $(V_{CB} + V_{BE})$, (see Fig. 4-29). At the *knee* of the characteristic, the *CB* junction voltage (V_{CB}) has been reduced to zero, (because $V_{CE} = V_{BE}$). Further reduction in V_{CE} causes the *CB* junction to be forward biased. The forward bias repels the minority charge carriers, thus reducing I_C to zero. The dashed lines in Fig. 4-28 show that, if V_{CE} exceeds a maximum safe level, I_C increases rapidly and the device may be destroyed. As for common-base configuration, this is due to *punch-through*.

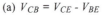

(a) $V_{CB} = V_{CE} - V_{BE}$

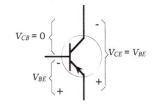

(b) $V_{CB} = 0$ when $V_{CE} = V_{BE}$

Figure 4-29
In a common-emitter circuit, V_{CB} can be reduced to zero, and can be reversed, when V_{CE} is reduced.

Example 4-5

Determine the I_B and I_C levels for a device with the characteristics in Figs. 4-27 and 4-28, when V_{BE} is 0.7 V and V_{CE} is 6 V. Also, calculate the β_{dc} value.

Solution

From Fig. 4-27, at ($V_{BE} = 0.7$ V and $V_{CE} = 6$ V): $I_B \approx 20\,\mu A$

From Fig. 4-28, at ($V_{CE} = 6$ V and $I_B = 20\,\mu A$): $I_C = 2.5$ mA

$$\beta_{dc} = \frac{I_C}{I_B} = \frac{2.5\ \text{mA}}{20\ \mu A}$$

$$= 125$$

Current Gain Characteristics

The *common-emitter current gain characteristics* are (output current) I_C plotted versus (input current) I_B for various fixed levels of V_{CE}. Like the common-base current gain characteristics, they can be obtained experimentally, or derived from the output characteristics. To experimentally obtain the table of I_C and I_B values, V_{CE} is held at a selected level, and the base current I_B is adjusted in steps. The corresponding I_C level is recorded at each step of I_B.

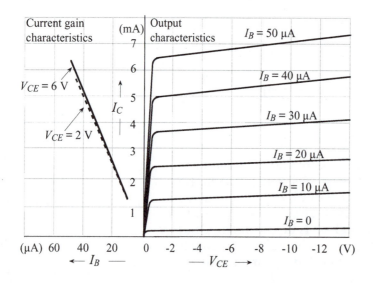

Figure 4-30
The common-emitter current-gain characteristics for a transistor are a plot of I_C versus I_B.

Practise Problems

4-5.1   Plot two lines of common-emitter transistor output characteristics from the following quantities:

for $I_B = 25\,\mu A$: ($I_C = 1.5$ mA when $V_{CE} = 1$ V)
($I_C = 1.75$ mA when $V_{CE} = 12$ V)

for $I_B = 50\,\mu A$: ($I_C = 3$ mA when $V_{CE} = 1$ V)
($I_C = 3.5$ mA when $V_{CE} = 12$ V)

Derive the current gain characteristics for $V_{CE} = 12$ V from the device output characteristics.

4-6   Common-Collector   Characteristics

In the circuit arrangement of Fig. 4-31, the collector terminal is common to both input *CB* voltage and output *CE* voltage. Using this circuit, the *common-collector input, output,* and *current gain characteristics* can be determined. The output and current gain characteristics are shown in Fig. 4-32. The common-collector output characteristics are I_E plotted versus V_{CE} for several fixed values of I_B. The common-collector current gain characteristics are I_E plotted versus I_B for several fixed values of V_{CE}.

It will be recalled that the common-emitter output characteristics are I_E plotted against V_{CE}, and that the common-emitter current gain characteristics are I_E plotted against I_B. Because I_C is approximately equal to I_E, the common-collector output and current gain characteristics are practically identical to those of the common-emitter circuit.

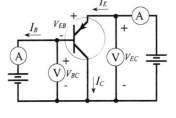

Figure 4-31
Transistor circuit used for determining common-collector characteristics.

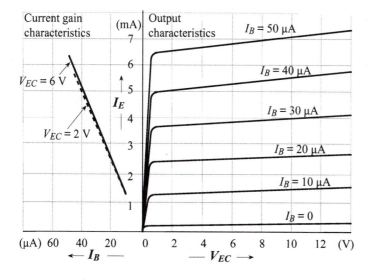

Figure 4-32
The common-collector output and current-gain characteristics are identical to the common-emitter characteristics.

The common-collector input characteristics shown in Fig. 4-33 are quite different from either common-base or common-emitter input characteristics. The difference is due to the fact that the input voltage (V_{BC}) is largely determined by the V_{EC} level.

Referring to Fig. 4-31,

$$V_{EC} = V_{EB} + V_{BC}$$

or,

$$V_{EB} = V_{EC} - V_{BC}$$

Increasing the level of (input voltage) V_{BC} with V_{EC} held constant, reduces the level of V_{EB}, and thus reduces I_B.

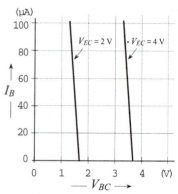

Figure 4-33

The common-collector input characteristics are quite different from input characteristics for common-base and common-emitter connections.

Practise Problems

4-6.1   Referring to the common-collector input characteristics in Fig. 4-33, determine the level of V_{BC} when $I_B = 60\ \mu A$ and $V_{EC} = 2$ V, also when $I_B = 60\ \mu A$ and $V_{EC} = 4$ V.

4-7   Transistor Testing

In-Circuit Testing

A quick test to check if a transistor is operational can be performed while the device is still connected in a circuit. Consider Fig. 4-34, which shows a voltmeter connected to measure the transistor collector voltage (V_C). The V_C measurement is noted, then the base and emitter terminals are temporarily short-circuited, as illustrated. This should turn the transistor *off*, and V_C should jump to approximately the circuit supply voltage (V_{CC}). When the shorting wire is removed, V_C should return to its previous level. If the change in the V_C level does not occur, the transistor is not operational. This may be due to a faulty transistor, or to some other problem in the circuit.

Ohmmeter Tests

An ohmmeter may be used for checking the transistor emitter-base and collector-base junctions. As in the case of a diode, the measured resistance of a forward-biased junction depends upon the ohmmeter range. With an analog ohmmeter, a good forward-biased *pn*-junction typically indicates half-scale, (see Fig. 4-35). A good reverse-biased junction gives a very high resistance measurement. Recall from Section 2-8 that the terminal polarity of some multimeters is reversed when used as an ohmmeter. This should be checked with a voltmeter. The measured resistance between the collector and emitter terminals of a good transistor should be very high regardless of the ohmmeter terminal polarity.

A digital ohmmeter may show a resistance of a few megohms when measuring a good forward-biased junction, and an open-circuit indication (*OL*) for a reverse-biased junction, (Fig. 4-36). It should also indicate open-circuit when measuring between the collector and emitter terminals. The diode testing terminals on a digital multimeter may also be used for testing transistor *CB* and *EB* junctions. Some digital multifunction instruments have testing facilities for measuring transistor h_{FE} values.

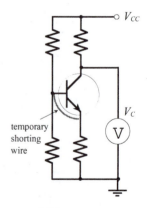

Figure 4-34

Short-circuiting the base and emitter terminals should cause V_C to jump to V_{CC} if the transistor is operating.

Characteristic Plotting

A transistor may be rapidly checked by use of a *curve tracer*, (see Fig. 4-37). Two transistor sockets are normally provided to display the characteristics of two devices. A switch permits selection of either socket. The characteristics of a known good transistor should first be displayed. Then, the switch should be operated to display the characteristics of the transistor under investigation.

Transistor characteristics may be plotted by obtaining a table of corresponding levels of current and voltage, as already discussed. A more convenient method of plotting characteristics using an *XY recorder* is illustrated in Fig. 4-38. One power supply is connected to produce base current (I_B) via resistor R_1, and the other provides I_C flow through R_2. The collector and emitter terminals are connected to the horizontal input terminals of the *XY* recorder, and the voltage drop across resistor R_2 provides the vertical input. The base current is set at a several convenient levels, and (at each I_B level) V_{CC} is slowly increased from zero to cause the pen to trace one line of the transistor output characteristics.

A convenient selection for the horizontal and vertical scales of the *XY* recorder is 1 V/cm. If $R_2 = 1$ kΩ, each 1 V drop across R_2 represents 1 mA of collector current. So, the I_C (vertical) scale on the characteristics is 1 mA/cm, and the V_{CE} scale is 1 V/cm.

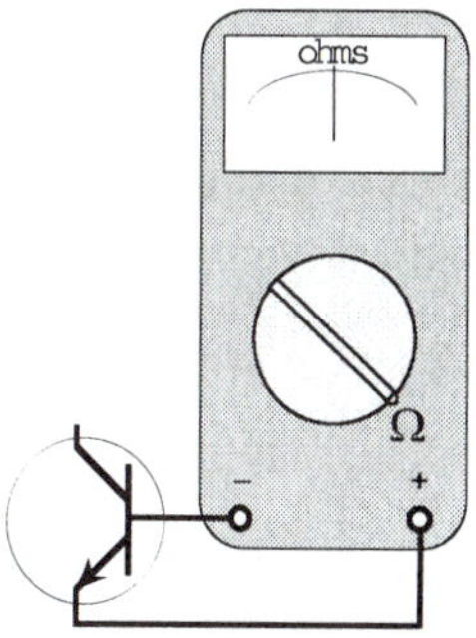

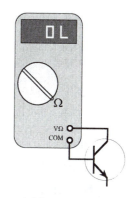

Figure 4-35
Use of an analog ohmmeter for checking the junctions of a transistor.

Example 4-6

The circuit and *XY* recorder in Fig. 4-38 are to be used to plot transistor common-emitter output characteristics to $V_{CE(max)} = 10$ V and $I_{C(max)} = 20$ mA. Select V_{CC} and R_2, and determine appropriate V/cm scales for the *XY* recorder. The plotted characteristics are to be approximately 20 cm x 20 cm.

Solution

For $I_{C(max)}$ corresponding to 20 cm,

$$vertical\ scale\ = \frac{I_{C(max)}}{scale\ length} = \frac{20\ mA}{20\ cm}$$

$$= 1\ mA/cm$$

Select a vertical scale of 1 V/cm, so that 20 cm represents 20 V and 20 mA.

$$R_2 = \frac{20\ V}{20\ mA}$$

$$= 1\ k\Omega$$

$$V_{CC} = V_{CE(max)} + (I_{C(max)} \times R_2)$$
$$= 10\ V + (20\ mA \times 1\ k\Omega)$$
$$= 30\ V$$

$$horizontal\ scale\ = \frac{V_{CE(max)}}{scale\ length} = \frac{10\ V}{20\ cm}$$

$$= 0.5\ V/cm$$

Figure 4-36
Transistor testing by use of a digital ohmmeter.

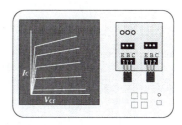

Figure 4-37
A curve tracer can be use to compare the characteristics of a transistor of unknown quality with the characteristics of a known good device.

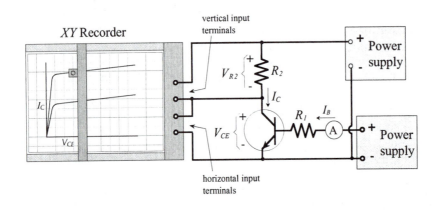

Figure 4-38
Transistor characteristic plotting on an XY recorder. V_{CE} produces horizontal deflection, and V_{R2} causes vertical deflection.

Practise Problems

4-7.1   A transistor with $V_{CE(max)} = 40$ V and $I_C = 100$ mA is to have its common-emitter output characteristics plotted on an XY-recorder as in Fig. 4-38. Select an appropriate supply voltage and collector resistance value, and determine suitable V/cm scales for the XY-recorder to give characteristics approximately 25 cm x 25 cm.

Chapter-4   Review   Questions

Section   4-1

4-1   Draw a block diagrams of an unbiased *npn* transistor. Identify each part of the device, and show the depletion regions and barrier voltages. Briefly explain.

4-2   Repeat Question 4-1 for a *pnp* transistor.

4-3   Sketch a correctly-biased *npn* transistor. Show the effect of the bias voltages on the depletion regions and barrier voltages, and show the movement of charge carriers through the transistor. Explain the transistor operation.

4-4   Repeat Question 4-3 for a correctly-biased *pnp* transistor.

4-5   Draw circuit symbols for *npn* and *pnp* transistors. Identify current directions and bias voltage polarities.

Section   4-2

4-6   Draw a sketch to show the various current components in a transistor. Briefly explain the origin of each current. Write an equations relating I_E, I_B, and I_C.

4-7   Write equations for I_C in terms of I_E, and for I_C in terms of I_B. Define α_{dc} and β_{dc}, and state typical values for each.

Section   4-3

4-8   Sketch a circuit diagram to show how a transistor can be used for amplification of direct current changes and alternating signal currents. Briefly explain.

4-9   Draw a circuit diagram to show how a transistor can amplify direct voltage changes and alternating signal voltages. Briefly explain.

Section 4-4

4-10   Explain transistor *common-base* configuration, and sketch a circuit for determining common-base characteristics.

4-11   Sketch typical transistor *common-base* input characteristics and output characteristics. Explain the shape of each set of characteristics.

4-12   Explain transistor *punch-through,* and how it occurs.

4-13   Sketch typical transistor *common-base* current-gain characteristics. Explain the shape of the characteristics.

Section 4-5

4-14   Explain transistor *common-emitter* configuration, and draw a circuit for determining common-emitter characteristics.

4-15   Sketch typical transistor *common-emitter* input and output characteristics. Explain the shape of the characteristics.

4-16   Sketch typical transistor *common-emitter* current-gain characteristics. Explain the shape of the characteristics.

Section 4-6

4-17   Explain transistor *common-collector* configuration, and draw a circuit for determining common-collector characteristics.

4-18   Sketch typical transistor *common-collector* output and current-gain characteristics, and explain the shape of the characteristics.

4-19   Sketch typical transistor *common-collector* input characteristics. Discuss the shape of the characteristics.

Section 4-7

4-20   Describe how a transistor may be quickly tested while still connected in a circuit.

4-21   Discuss how a transistor may be tested by means of an ohmmeter, and by use of a digital multimeter.

4-22   Sketch a diagram to show how an *XY* recorder may be used for plotting transistor characteristics.

Chapter-4 Problems

Section 4-2

4-1   Calculate the values of I_C and I_E for a transistor with α_{dc} = 0.97 and I_B = 50 μA. Also, determine β_{dc} for the device.

4-2   A transistor has measured current levels of I_C = 5.25 mA and

I_B = 100 µA. Calculate I_E, α_{dc}, and β_{dc}, and determine the new I_B level to give I_C = 15 mA.

4-3   Calculate the collector and emitter current levels for a transistor with α_{dc} = 0.99 and I_B = 20 µA.

4-4   The following current measurements were made on a transistor: I_C = 12.42 mA, I_B = 200 µA. Determine I_E, α_{dc}, and β_{dc}. Also, determine the new I_C level when I_B is 150 µA.

4-5   A transistor with measured current levels of I_C = 16 mA and I_E = 16.04 mA is replaced with another transistor that has β_{dc} = 25. Calculate the new level of I_C and I_E, assuming that the base current remains constant.

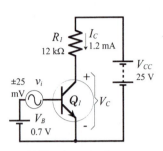

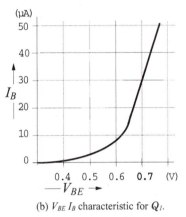

(a) Transistor circuit

Section 4-3

4-6   The transistor in the circuit in Fig. 4-39(a) has the input characteristics in Fig. 4-39(b). If β_{dc} = 50, calculate the collector voltage when V_{BE} = 0.7 V.

4-7   A ±25 mV *ac* input is applied in series with the base of the transistor in Problem 4-6. Calculate the circuit voltage gain.

4-8   Determine the new V_{CE} level for the circuit in Fig. 4-39(a) if the transistor V_{BE} is adjusted to 0.75 V.

4-9   Calculate the new voltage gain for the circuit in Fig. 4-39(a) when R_1 is changed to 6.8 kΩ.

4-10   When the transistor in the circuit of Fig. 4-39(a) is changed, V_C is measured as 9 V. Assuming that I_B remains constant at 30 µA, calculate the current gain for the new transistor.

(b) V_{BE} I_B characteristic for Q_1.

Figure 4-39

Section 4-4

4-11   Derive the *common-base* current gain characteristics for V_{CB} = -5 V from the output characteristics in Fig. 4-40.

4-12   Refer to the *common-base* output characteristics in Fig. 4-22. Derive the current gain characteristics for V_{CB} = -8 V.

4-13   From the *common-base* input characteristics in Fig. 4-21, estimate the level of I_E when V_{EB} = 0.72 V and V_{CB} = 6 V.

Section 4-5

4-14   Derive the current gain characteristics for V_{CE} = 9 V from the *common-emitter* output characteristics in Fig. 4-41.

4-15   From the *common-emitter* input characteristics in Fig. 4-27, estimate the level of I_B when V_{BE} = 0.72 V and V_{CE} = 2 V.

4-16   Refer to the *common-emitter* output characteristics in Fig. 4-28. Derive the current gain characteristics for V_{CE} = 14 V.

4-17   Determine the level of I_C for a transistor with the *common-emitter* output characteristics in Fig. 4-28 when I_B = 60 µA and when I_B = 30 µA, with V_{CE} = 5 V in each case. Also calculate β_{dc}.

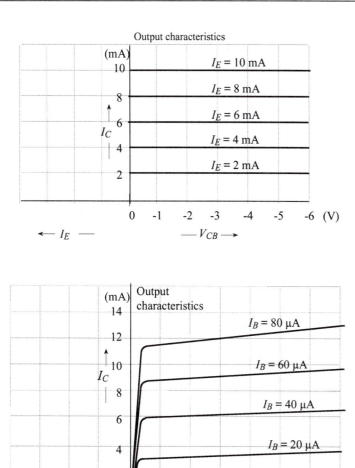

Figure 4-40

Figure 4-41

Section 4-6

4-18 Refer to the *common-collector* output characteristics in Fig. 4-32. Derive the current gain characteristics for V_{EC} = 10 V.

4-19 Calculate the input voltage (V_{BC}) for the *common-collector* circuit in Fig. 4-31 when V_{EC} = 10 V and V_{EB} = 0.7 V.

4-20 Determine the level of V_{BC} from the *common-collector* input characteristics in Fig. 4-33 when I_B = 20 μA and V_{EC} = 4 V.

Section 4-7

4-21 An *XY* recorder is to be used to plot transistor *common-emitter* output characteristics as in Fig. 4-38. The transistor current and voltage levels are to be $V_{CE(max)}$ = 20 V and $I_{C(max)}$ = 50 mA. Select V_{CC} and resistor R_2, and determine appropriate V/cm scales for the *XY* recorder. The plotted characteristics size is to be approximately 25 cm x 25 cm.

4-22 Sketch a circuit diagram for using an *XY* recorder to determine the *common-emitter* input characteristics for a transistor. Select suitable circuit components and V/cm scales to plot 20 cm x 20 cm characteristics with maximum levels of V_{BE} = 1 V and I_B = 100 μA.

Practise Problem Answers

4-2.1 0.98, 50 μA, 50
4-2.2 2.25 mA, 2.28 mA
4-3.1 71.5
4-3.2 80
4-6.1 1.5 V, 3.5 V
4-7.1 1 V/cm, 250 Ω, 65 V, 1.5 V/cm

Chapter 5

BJT Biasing

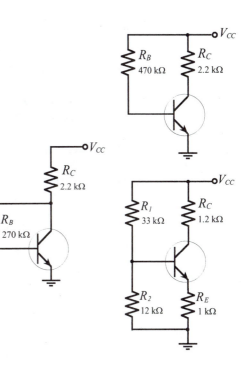

Chapter   Contents

Objectives

You will be able to:

1   *Draw dc load lines for transistor circuits, identify circuit Q-points, and estimate the maximum circuit output voltage variations.*

2   *Sketch base bias, collector-to-base bias, and voltage divider bias circuits, and explain the operation of each circuit.*

3   *Analyze various types of BJT bias circuits to determine circuit voltage and current levels.*

4   *Trouble-shoot non-operational BJT bias circuits.*

5   *Design base bias, collector-to-base bias, and voltage divider bias circuits, and select appropriate standard value components.*

6   *Sketch several unusual types of BJT bias circuit, and explain the operation of each circuit.*

7   *Analyze and design the bias circuit types referred to in item 6.*

8   *Discuss the thermal stability of BJT bias circuits, and determine the effects of I_{CBO} and V_{BE} changes with temperature.*

9   *Sketch bias circuits for direct-coupled and capacitor-coupled switching transistors, and explain the operation of each circuit.*

10   *Analyze and design switching transistor bias circuits.*

Introduction

Transistors used in amplifier circuits must be biased into an *on* state with constant (direct) levels of collector, base, and emitter current, and constant terminal voltages. The levels of I_C and V_{CE} define the transistor *dc operating point,* or *quiescent point.* The circuit that provides this state is known as a *bias circuit.* Ideally, the current and voltage levels in a bias circuit should remain absolutely constant. In practical circuits these quantities are affected by the transistor current gain (h_{FE}) and by temperature changes. The best bias circuits have the greatest stability; they hold the currents and voltages substantially constant regardless of the h_{FE} and temperature variations. The simplest bias circuits are the least costly because they use the smallest number of components, but they are not as stable as more complex circuits.

5-1 DC Load Line and Bias Point

DC Load Line

The *dc load line* for a transistor circuit is a straight line drawn on the transistor output characteristics. For a common-emitter (*CE*) circuit, the load line is a graph of collector current (I_C) versus collector-emitter voltage (V_{CE}), for a given value of collector resistance (R_C) and a given supply voltage (V_{CC}). The load line shows all corresponding levels of I_C and V_{CE} that can exist in a particular circuit.

Consider the common-emitter circuit in Fig. 5-1. Note that the polarity of the transistor terminal voltages are such that the base-emitter junction is forward biased and the collector-base junction is reverse biased. These are the normal bias polarities for the transistor junctions. The *dc* load line for the circuit in Fig. 5-1 is drawn on the device common-emitter characteristics in Fig. 5-2.

From Fig. 5-1, the collector-emitter voltage is,

V_{CE} = (supply voltage) - (voltage drop across R_C)

or, $$V_{CE} = V_{CC} - I_C R_C \qquad (5\text{-}1)$$

If the base-emitter voltage (V_{BE}) is zero, the transistor is not conducting and I_C = 0. Substituting the V_{CC} and R_C values from Fig. 5-1 into Eq. 5-1,

$$V_{CE} = 20 \text{ V} - (0 \times 10 \text{ k}\Omega)$$
$$= 20 \text{ V}$$

Plot *point A* on the common-emitter characteristics in Fig. 5-2 at I_C = 0 and V_{CE} = 20 V. This is one point on the dc load line.

Now assume a collector current of 2 mA, and calculate the corresponding collector-emitter voltage level.

$$V_{CE} = 20 \text{ V} - (2 \text{ mA} \times 10 \text{ k}\Omega)$$
$$= 0 \text{ V}$$

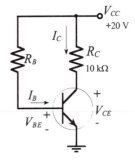

Figure 5-1
Transistor circuit with collector resistor R_C.

Plot *point B* on Fig. 5-2 at $V_{CE} = 0$ and $I_C = 2$ mA. The straight line drawn through *point A* and *point B* is the *dc load line* for $R_C = 10$ kΩ and $V_{CC} = 20$ V. If either of these two quantities is changed, a new load line must be drawn.

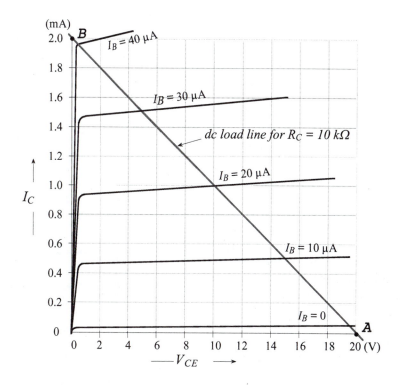

Figure 5-2
DC load line drawn upon transistor common-emitter output characteristics.

As already stated, the *dc* load line represents all corresponding I_C and V_{CE} levels that can exist in the circuit, as represented by Eq. 5-1. For example, a point plotted at $V_{CE} = 16$ V and $I_C = 1.5$ mA on Fig. 5-2 does not appear on the load line. This combination of voltage and current cannot exist in this particular circuit, (Fig. 5-1). Knowing any one of I_B, I_C, or V_{CE}, it is easy to determine the other two from a *dc* load line drawn on the device characteristics.

It is not always necessary to have the device characteristics in order to draw the *dc* load line. A simple graph of I_C versus V_{CE} can be used, as demonstrated in Example 5-1.

Example 5-1
Draw the new *dc* load line for the circuit in Fig. 5-1 when $R_C = 12$ kΩ.

Solution
Prepare a graph for I_C versus V_{CE}, (Fig. 5-3).

when $I_C = 0$,
Eq. 5-1,
$$V_{CE} = V_{CC} - I_C R_C = 20 \text{ V} - 0$$
$$= 20 \text{ V}$$

Plot *point A* on Fig. 5-3 at,

$$I_C = 0 \text{ and } V_{CE} = 20 \text{ V}$$

when $V_{CE} = 0$,
from Eq. 5-1,

$$0 = V_{CC} - I_C R_C$$

giving,

$$I_C = \frac{V_{CC}}{R_C} = \frac{20 \text{ V}}{12 \text{ k}\Omega}$$

$$\approx 1.7 \text{ mA}$$

Plot *point B* on Fig. 5-3 at,

$$I_C = 1.7 \text{ mA and } V_{CE} = 0 \text{ V}.$$

Draw the *dc* load line through *points A* and *B*.

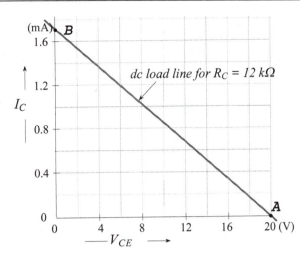

Figure 5-3
DC load line for Example 5-1.

DC Bias Point (Q-Point)

The *dc bias point*, or *quiescent point (Q-point)* (also known as the *dc operating point*), identifies the transistor collector current and collector-emitter voltage when there is no input signal at the base terminal. Thus, it defines the *dc* conditions in the circuit. When a signal is applied to the transistor base, I_B varies according to the instantaneous amplitude of the signal. This causes I_C to vary, and consequently produces a variation in V_{CE}.

Consider the circuit in Fig. 5-4, and the 10 kΩ load line drawn for the circuit in Fig. 5-5. Assume that the bias conditions are as identified by the *Q-point* on the load line,

$$I_B = 20 \text{ μA, } I_C = 1 \text{ mA, and } V_{CE} = 10 \text{ V}$$

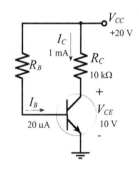

Figure 5-4
Transistor circuit with a bias point (Q-point) at $V_{CE} = 10$ V and $I_C = 1$ mA.

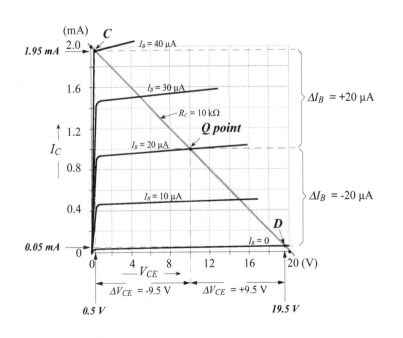

Figure 5-5
DC load line for transistor with a bias point (Q-point) at $V_{CE} = 10$ V and $I_C = 1$ mA. The transistor may be biased to any point on the dc load line.

When I_B is increased from 20 μA to 40 μA, I_C becomes approximately 1.95 mA and V_{CE} becomes 0.5 V, as illustrated at point C on the load line. The V_{CE} change from the Q-point is,

$$\Delta V_{CE} = 10\ V - 0.5\ V$$
$$= 9.5\ V$$

So, increasing I_B by 20 μA (from 20 μA to 40 μA) caused V_{CE} to decrease by 9.5 V, (from 10 V to 0.5 V).

Now look at the effect of decreasing the base current. When I_B is reduced from 20 μA to zero, I_C goes down to approximately 0.05 mA, and V_{CE} goes up to 19.5 V (point D on the load line in Fig. 5-5). So, the V_{CE} change is,

$$\Delta V_{CE} = 19.5 - 10\ V$$
$$= 9.5\ V$$

Decreasing I_B by 20 μA (From 20 μA to zero) caused V_{CE} to increase by 9.5 V (from 10 V to 19.5 V). It is seen that with the Q-point at I_C = 1 mA and V_{CE} = 10 V, an I_B variation of ±20 μA produces a collector voltage swing of ΔV_{CE} = ±9.5 V.

The base current does not have to be varied by the maximum amounts discussed above; it can be increased and decreased by smaller amounts. For example, a base current change of ±10 μA (from the Q-point on Fig. 5-5) would produce a collector current change of ±0.5 mA, and a collector-emitter voltage change of ±5 V.

The maximum possible transistor collector-emitter voltage swing for a given circuit can be determined without using the transistor characteristics. For convenience, it may be assumed that I_C can be driven to zero at one extreme and to V_{CC}/R_C at the other extreme, [see Fig. 5-6]. This changes the collector-emitter voltage from V_{CE} = V_{CC} to V_{CE} = 0, as illustrated in Fig. 5-7. Thus, with the Q-point at the center of the load line, the maximum possible collector voltage swing is seen to be approximately ±$V_{CC}/2$

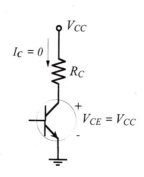

(a) $V_{CE} \approx V_{CC}$ when $I_C = 0$

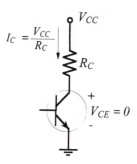

(b) $V_{CE} \approx 0$ when $I_C = V_{CC}/R_C$

Figure 5-6
The collector-emitter voltage of a transistor can range from approximately zero to V_{CC}.

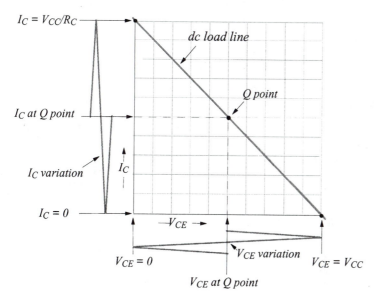

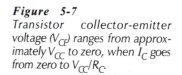

Figure 5-7
Transistor collector-emitter voltage (V_{CE}) ranges from approximately V_{CC} to zero, when I_C goes from zero to V_{CC}/R_C.

Q-point Selection

Suppose that, instead of being biased half way along the load line, the transistor is biased at I_C = 0.5 mA, and V_{CE} = 15 V, as shown in Fig. 5-8(a). Increasing the collector current to 2 mA reduces V_{CE} to zero, giving ΔV_{CE} = -15 V. Reducing I_C to zero increases V_{CE} to V_{CC}, producing ΔV_{CE} = +5 V

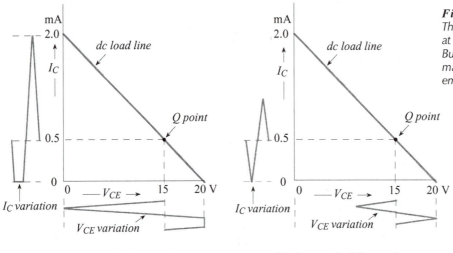

(a) Assymmetrical V_{CE} swing

(b) Symmetrical V_{CE} swing

Figure 5-8
The Q-point does not have to be at the centre of the dc load line. But, its position determines the maximum symmetrical collector-emitter voltage swing.

When used as an amplifier, the transistor output (collector-emitter) voltage must swing up and down by equal amounts; that is, the output voltage swing must be symmetrical above and below the bias point. So, the asymmetrical V_{CE} swing of -15 V +5 V illustrated in Fig. 5-8(a) is unsuitable. If I_C is driven up and down by ±0.5 mA [see Fig. 5-8(b)], a symmetrical output voltage swing of ±5 V is obtained. This is the maximum symmetrical output voltage swing that can be achieved with the bias point shown in Fig. 5-8.

In many cases circuits are designed to have the *Q-point* at the center of the load line (as in Figs. 5-5 and 5-7) to give the largest possible symmetrical output voltage swing. This is especially true for some large signal amplifiers, (see Chapter 18). Small signal amplifiers (discussed Chapter 12) usually require an output voltage swing not greater than ±1 V. So,

> **transistors in amplifiers do NOT all have to be biased at the center of the dc load line.**

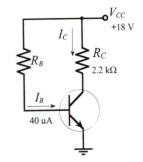

Example 5-2

The transistor circuit in Fig. 5-9 has the collector characteristics shown in Fig. 5-10. Determine the circuit *Q-point* and estimate the maximum symmetrical output voltage swing. Note that V_{CC} = 18 V, R_C = 2.2 kΩ, and I_B = 40 μA.

Figure 5-9
Circuit for Example 5-2.

Solution

Eq. 5-1, $V_{CE} = V_{CC} - I_C R_C$

when $I_C = 0$, $V_{CE} = V_{CC} - 0$

$= 18\ V$

Plot *point A* on Fig. 5-10 at,

$I_C = 0$ and $V_{CE} = 18\ V$

when $V_{CE} = 0$, $0 = V_{CC} - I_C R_C$

or, $I_C = \dfrac{V_{CC}}{R_C} = \dfrac{18\ V}{2.2\ k\Omega}$

$\approx 8.2\ mA$

Plot *point B* on Fig. 5-10 at,

$I_C = 8.2\ mA$ and $V_{CE} = 0\ V$,

Draw the *dc load line* through *points A* and *B*. The *Q-point* is at the intersection of the load line and the $I_B = 40\ \mu A$ characteristic.

The *dc* bias conditions are,

$I_C \approx 4.1\ mA$ and $V_{CE} \approx 9\ V$

The maximum symmetrical output voltage swing is,

$\Delta V_{CE} \approx \pm 9\ V$

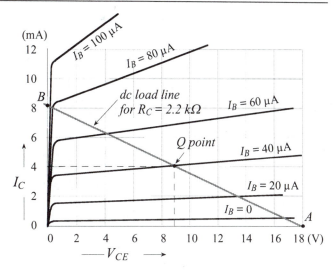

Figure 5-10
Transistor Common-emitter output characteristics and dc load line for Example 5-2.

Effect of Emitter Resistor

Figure 5-11(a) shows a circuit that has a resistor (R_E) in series with the transistor emitter terminal, and the supply voltage connected directly to the collector terminal. In this case R_E is the *dc* load, and Eq. 5-1 is rewritten as,

$$V_{CE} = V_{CC} - I_E R_E$$

The *dc* load line is drawn exactly as discussed, with I_E taken as equal to I_C for convenience.

In Fig. 5-11(b) collector and emitter resistors R_C and R_E are both present, and the total *dc* load in series with the transistor is ($R_C + R_E$). For drawing the dc load line, Eq. 5-1 is modified to,

$$V_{CE} = V_{CC} - I_C (R_C + R_E)$$

Note that the voltage drop across the emitter resistor is actually ($I_E R_E$), but again for convenience I_E is taken as equal to I_C.

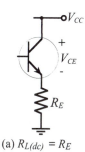

(a) $R_{L(dc)} = R_E$

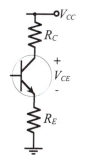

(b) $R_{L(dc)} = (R_C + R_E)$

Figure 5-11
The transistor dc load is the sum of the resistors in series with the collector and emitter terminals.

5-2   Base Bias

Circuit Operation and Analysis

The transistor bias arrangement shown in Fig. 5-12 is known as *base bias* and also as *fixed current bias*. The base current is a constant quantity determined by supply voltage V_{CC} and base resistor R_B. Because V_{CC} and R_B are constant quantities, I_B remains fixed at a particular level. Unlike some other bias circuits, the base current in a base bias circuit is not affected by the transistor current gain.

From Fig. 5-12, the voltage drop across R_B is $(V_{CC} - V_{BE})$, and the base current is,

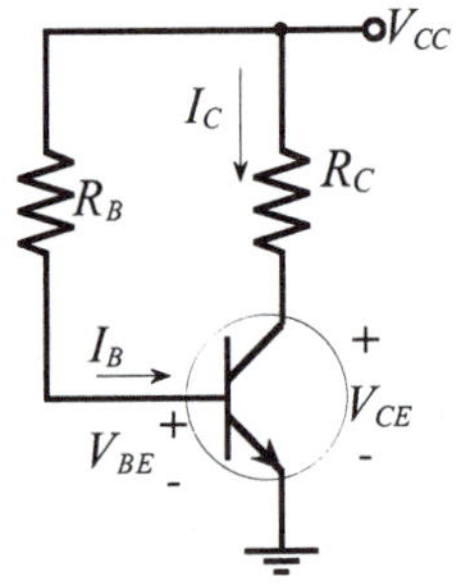

$$I_B = \frac{V_{CC} - V_{BE}}{R_B} \tag{5-2}$$

In Eq. 5-2 the base-emitter voltage (V_{BE}) is taken as 0.7 V for a silicon transistor, and as 0.3 V for a germanium device.

The transistor collector current is calculated as,

$$I_C = h_{FE} I_B \tag{5-3}$$

The collector current is now used with Eq. 5-1 $(V_{CE} = V_{CC} - I_C R_C)$ to calculate the collector-emitter voltage. Thus, when the supply voltage and component values are known, a base bias circuit is easily analysed to determine the circuit current and voltage levels.

Figure 5-12
Base bias circuit, also known as fixed current bias. The base current remains constant at,
$I_B = (V_{CC} - V_{BE})/R_B.$

Example 5-3

The base bias circuit in Fig. 5-13 has R_B = 470 kΩ, R_C = 2.2 kΩ, V_{CC} = 18 V, and the transistor has h_{FE} = 100. Determine I_B, I_C, and V_{CE}.

Solution

Eq. 5-2, $I_B = \dfrac{V_{CC} - V_{BE}}{R_B} = \dfrac{18\text{ V} - 0.7\text{ V}}{470\text{ k}\Omega}$

$= 36.8\ \mu A$

Eq. 5-3, $\qquad I_C = h_{FE} I_B = 100 \times 36.8\,\mu A$

$\qquad\qquad\qquad = 3.68\text{ mA}$

Eq. 5-1, $\qquad V_{CE} = V_{CC} - I_C R_C = 18\text{ V} - (3.68\text{ mA} \times 2.2\text{ k}\Omega)$

$\qquad\qquad\qquad = 9.9\text{ V}$

The circuit conditions are illustrated in Fig. 5-13.

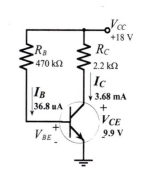

Figure 5-13
Circuit for Example 5-3.

Effect of $h_{FE(max)}$ and $h_{FE(min)}$

When the transistor *dc* current gain is known, it is quite easy to determine the circuit bias conditions exactly as in Example 5-3. However, in practise the precise current gain of each transistor is normally not known. The transistor is usually identified by its type number, and then the maximum and minimum values of current gain can be obtained from the manufacturer's data sheet. In circuit analysis it is sometimes convenient to use a typical h_{FE} value. However, as demonstrated in Example 5-4, $h_{FE(max)}$ and $h_{FE(min)}$ must be used to calculate the range of possible levels of I_C and V_{CE}.

Example 5-4

Calculate the maximum and minimum levels of I_C and V_{CE} for the base bias circuit in Ex. 5-3 when $h_{FE(min)} = 50$ and $h_{FE(max)} = 200$.

Solution

The base current in this circuit is unaffected by h_{FE}, and so (as in Ex. 5-3),

$$I_B = \frac{V_{CC} - V_{BE}}{R_B} = \frac{18\text{ V} - 0.7\text{ V}}{470\text{ k}\Omega}$$

$$= 36.8\,\mu A$$

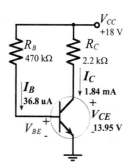

(a) Conditions for $h_{FE(min)}$

for $h_{FE(min)}$, $\qquad I_C = h_{FE(min)} I_B = 50 \times 36.8\,\mu A$

$\qquad\qquad\qquad = 1.84\text{ mA}$

and, $\qquad V_{CE} = V_{CC} - I_C R_C = 18\text{ V} - (1.84\text{ mA} \times 2.2\text{ k}\Omega)$

$\qquad\qquad\qquad = 13.95\text{ V}$ [See Fig. 5-14(a)]

for $h_{FE(max)}$, $\qquad I_C = h_{FE(max)} I_B = 200 \times 36.8\,\mu A$

$\qquad\qquad\qquad = 7.36\text{ mA}$

and, $\qquad V_{CE} = V_{CC} - I_C R_C = 18\text{ V} - (7.36\text{ mA} \times 2.2\text{ k}\Omega)$

$\qquad\qquad\qquad = 1.8\text{ V}$ [See Fig. 5-14(b)]

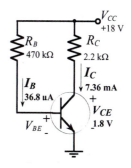

(b) Conditions for $h_{FE(max)}$

Note that the typical h_{FE} value of 100 used in Ex. 5-3 gives $I_C = 3.68$ mA and $V_{CE} = 9.9$ V. But, applying the $h_{FE(min)}$ and $h_{FE(max)}$ values in Example 4-4 results in an I_C range of 1.84 mA to 7.36 mA) and V_{CE} ranging from 1.8 V to 13.95 V. The different I_C and

Figure 5-14
Circuit current and voltage levels for $h_{FE(min)}$ and $h_{FE(max)}$ as determined in Example 5-4.

V_{CE} levels determined in Examples 5-3 and 5-4 are illustrated by the three *Q points* in Fig. 5-15. Transistors of a given type number always have a wide range of h_{FE} values (the h_{FE} *spread*), so

> $h_{FE(max)}$ *and* $h_{FE(min)}$ *should always be used for practical circuit analysis.*

The base bias circuit is rarely employed because of the uncertainty of the *Q point*. More predictable results can be obtained with other types of bias circuit.

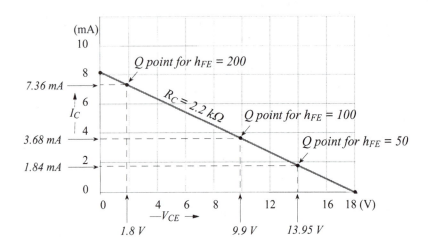

Figure 5-15
The transistor h_{FE} value has a major effect on the Q point for a base bias circuit.

Base Bias Using a pnp Transistor

All of the base bias circuits discussed so far use *npn* transistors. Figure 5-16 shows circuits that use *pnp* transistors. In Fig. 5-16(a) note that the voltage polarities and current directions are reversed compared to *npn* transistor base bias circuits. Because it is normal to show the supply voltage with the positive terminal uppermost, the circuit of Fig. 5-16(a) is redrawn in Fig. 5-16(b). This is the way that the circuit would usually be shown.

Equations 5-1, 5-2, and 5-3 can be applied for analysing *pnp* transistor base bias circuit exactly as for *npn* transistor circuits.

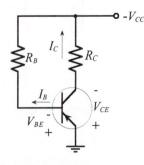

(a) Base bias circuit using an *pnp* transistor

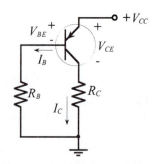

(b) Usual way to show a *pnp* transistor circuit; with the + terminal of the supply uppermost

Figure 5-16
Base bias circuits using pnp transistors.

5-3   Collector-to-Base Bias

Circuit Operation and Analysis

The *collector-to-base bias* circuit shown in Fig. 5-17(a) has the base resistor (R_B) connected between the transistor collector and base terminals. As will be demonstrated, this circuit has significantly improved bias stability for h_{FE} changes compared to base bias.

Refer to Fig. 5-17(b) and note that the voltage across R_B is dependent on V_{CE},

$$V_{CE} = V_{BE} + I_B R_B \qquad\qquad \textbf{(5-4)}$$

giving, $\qquad I_B = \dfrac{V_{CE} - V_{BE}}{R_B}$

Also, V_{CE} is dependent on the level of I_C and I_B,

$$V_{CE} = V_{CC} - R_C(I_C + I_B) \qquad\qquad \textbf{(5-5)}$$

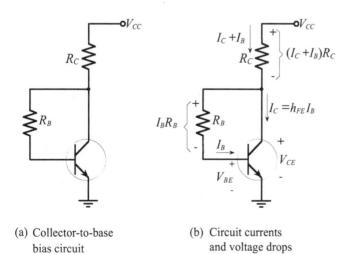

(a) Collector-to-base
      bias circuit

(b) Circuit currents
      and voltage drops

Figure 5-17
Collector-to-base bias circuits. Any change in V_{CE} changes I_B. The I_B change causes I_C to change, and this tends to return V_{CE} toward its original level.

If I_C increases above the design level, there is an increased voltage drop across R_C, resulting in a reduction in V_{CE}. The reduced V_{CE} level causes I_B to be lower than its design level, and because I_C = $h_{FE} I_B$, the collector current is also reduced. Thus, an increase in I_C produces a feedback effect that tends to return I_C toward its original level. Similarly, a reduction in I_C produces an increase in V_{CE} which increases I_B, thus tending to increase I_C back to its original level.

Analysis of this circuit is a little more complicated than base bias analysis. To simplify the process, an equation is first derived for the base current. Substituting for V_{CE} from Eq. 5-4 into Eq. 5-5,

$$V_{BE} + I_B R_B = V_{CC} - R_C(I_C + I_B)$$

Substituting $I_C = h_{FE} I_B$ (from Eq. 5-3) into the above equation,

$$V_{BE} + I_B R_B = V_{CC} - R_C (h_{FE} I_B + I_B)$$

This gives, $$I_B = \frac{V_{CC} - V_{BE}}{R_B + R_C(h_{FE} + 1)}$$ **(5-6)**

Example 5-5

A collector-to-base bias circuit as in Fig. 5-18 has $R_B = 270$ kΩ, $R_C = 2.2$ kΩ, $V_{CC} = 18$ V, and a transistor with $h_{FE} = 100$. Analyse the circuit to determine I_B, I_C, and V_{CE}.

Solution

Eq. 5-6, $I_B = \dfrac{V_{CC} - V_{BE}}{R_B + R_C(h_{FE} + 1)} = \dfrac{18\text{ V} - 0.7\text{ V}}{270\text{ k}\Omega + 2.2\text{ k}\Omega(100 + 1)}$

$= 35.1\ \mu A$

Eq. 5-3, $I_C = h_{FE} I_B = 100 \times 35.1\ \mu A$

$= 3.51$ mA

Eq. 5-5, $V_{CE} = V_{CC} - R_C(I_C + I_B)$

$= 18$ V $- 2.2$ k$\Omega(3.51$ mA $+ 35.1\ \mu A)$

$= 10.2$ V

The circuit conditions are illustrated in Fig. 5-18.

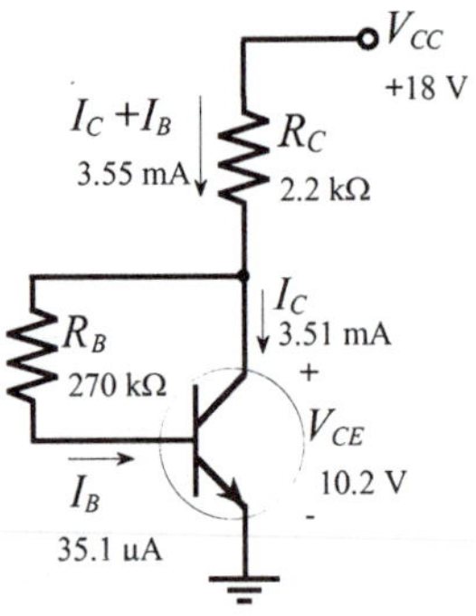

Figure 5-18
Collector-to-base bias circuit for Example 5-5.

Effect of $h_{FE(max)}$ and $h_{FE(min)}$

As discussed for the base bias circuit, transistors of a given type number have a wide range of h_{FE} values, (h_{FE} spread). This affects the current and voltage levels in all bias circuits. In the collector-to-base bias circuit, the feedback from the collector to the base reduces the effects of h_{FE} spread. Thus, as demonstrated in Ex. 5-6, collector-to-base bias has greater stability than base bias for a given range of h_{FE} values.

It is important to note that, unlike the situation in a base bias circuit, *the base current in a collector-to-base bias circuit does NOT remain constant* when the transistor h_{FE} value is changed. This is also demonstrated in Example 5-6.

Example 5-6

Calculate the maximum and minimum levels of I_C, and V_{CE} for the bias circuit in Ex. 5-5 when $h_{FE(min)} = 50$ and $h_{FE(max)} = 200$.

Solution

for $h_{FE(min)}$,

Eq. 5-6,
$$I_B = \frac{V_{CC} - V_{BE}}{R_B + R_C(h_{FE} + 1)} = \frac{18\ V - 0.7\ V}{270\ k\Omega + 2.2\ k\Omega(50 + 1)}$$
$$= 42.3\ \mu A$$

Eq. 5-3,
$$I_C = h_{FE}\,I_B = 50 \times 42.3\ \mu A$$
$$= 2.26\ mA$$

Eq. 5-5,
$$V_{CE} = V_{CC} - R_C(I_C + I_B)$$
$$= 18\ V - 2.2\ k\Omega\ (2.26\ mA + 42.3\ \mu A)$$
$$= 12.9\ V\ [\text{see Fig. 5-19(a)}]$$

for $h_{FE(max)}$,

Eq. 5-6,
$$I_B = \frac{V_{CC} - V_{BE}}{R_B + R_C(h_{FE} + 1)} = \frac{18\ V - 0.7\ V}{270\ k\Omega + 2.2\ k\Omega(200 + 1)}$$
$$= 24.3\ \mu A$$

Eq. 5-3,
$$I_C = h_{FE}\,I_B = 200 \times 24.3\ \mu A$$
$$= 4.86\ mA$$

Eq. 5-5,
$$V_{CE} = V_{CC} - R_C(I_C + I_B)$$
$$= 18\ V - 2.2\ k\Omega(4.86\ mA + 24.3\ \mu A)$$
$$= 7.25\ V\ [\text{see Fig. 5-19(b)}]$$

The circuit Q points are shown in Fig. 5-20.

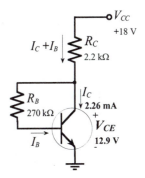

(a) Conditions for $h_{FE(min)}$

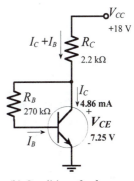

(b) Conditions for $h_{FE(max)}$

Figure 5-19
Circuit conditions for $h_{FE(min)}$ and $h_{FE(max)}$ in the collector-to-base bias circuit analyzed in Ex. 5-6.

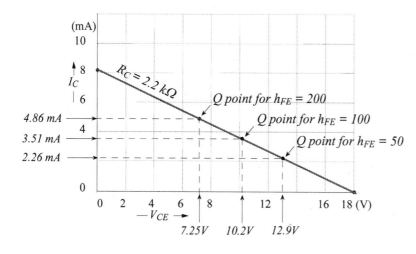

Figure 5-20
Effect of $h_{FE(min)}$ and $h_{FE(max)}$ on the Q point for the collector-to-base bias circuit analyzed in Example 5-6.

As illustrated by the Q points in Fig. 5-20, the typical h_{FE} value used in Ex. 5-5 gives $I_C = 3.51$ mA and $V_{CE} = 10.2$ V. Also, applying the $h_{FE(min)}$ and $h_{FE(max)}$ values produces an I_C range of 2.26 mA to 4.86 mA and V_{CE} levels ranging from 7.25 V to 12.9 V.

Recall that the base bias circuit analysed in Examples 5-3 and 5-4 has the same V_{CC}, R_C and h_{FE} values as used in the collector-to-base bias circuit in Examples 5-5 and 5-6. Comparing the collector-to-base bias circuit Q points in Fig. 5-20 to the base bias circuit Q points Fig. 5-15, it is seen that the Q *points* for the collector-to-base bias circuit are much closer than those for the base bias circuit. The collector-to-base bias circuit clearly has greater stability against h_{FE} spread than the base bias circuit.

Collector-to-base Bias Using a pnp Transistor

A collector-to-base bias circuit using a *pnp* transistor is illustrated in Fig. 5-21. Note that the voltage polarities and current directions are reversed compared to *npn* transistor collector-to-base bias circuit. This circuit can be analysed in exactly the same way as the *npn* transistor circuit.

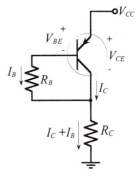

Practise Problems

5-3.1   The collector-to-base bias circuit in Example 5-5 uses a transistor with $h_{FE(min)} = 75$ and $h_{FE(max)} = 250$. Determine the new maximum and minimum levels for I_C and V_{CE}.

5-3.2   A collector-to-base bias circuit has $V_{CC} = 24$ V, $R_B = 180$ kΩ, $R_C = 3.3$ kΩ, and V_{CE} is measured as 10 V. Calculate the transistor h_{FE} value, and determine the V_{CE} level when a new transistor is substituted with $h_{FE} = 120$.

Figure 5-21
Collector-to-base bias circuit using a pnp transistor.

5-4   Voltage Divider Bias

Circuit Operation

Voltage divider bias, also known as *emitter current bias,* is the most stable of the three basic transistor bias circuits. A voltage divider bias circuit is shown in Fig. 5-22(a), and the current and voltage conditions throughout the circuit are illustrated in Fig. 5-22(b). It is seen that, as well as the collector resistor (R_C), there is an emitter resistor (R_E) connected in series with the transistor. As discussed in Section 5-1, the total *dc* load in series with the transistor is ($R_C + R_E$), and this total resistance must be used when drawing the *dc* load line for the circuit. Resistors R_1 and R_2 constitute a voltage divider that divides the supply voltage to produce the base bias voltage (V_B).

Voltage divider bias circuits are normally designed to have the voltage divider current (I_2) very much larger than the transistor base current (I_B). In this circumstance, V_B is largely unaffected by I_B, so V_B can be assumed to remain constant.

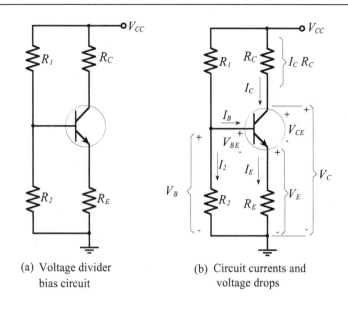

(a) Voltage divider
bias circuit

(b) Circuit currents and
voltage drops

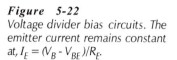

Figure   5-22
Voltage divider bias circuits. The
emitter current remains constant
at, $I_E = (V_B - V_{BE})/R_E$.

Referring to Fig. 5-22(b),

$$V_B = \frac{V_{CC} \times R_2}{R_1 + R_2} \tag{5-7}$$

With V_B constant, the voltage across the emitter resistor is also a
constant quantity,

$$V_E = V_B - V_{BE} \tag{5-8}$$

This means that the emitter current is constant,

$$I_E = \frac{V_B - V_{BE}}{R_E} \tag{5-9}$$

The collector current is approximately equal to the emitter current,
so I_C is held at a constant level.

Again referring to Fig. 5-22(b), the transistor collector voltage is,

$$V_C = V_{CC} - (I_C R_C) \tag{5-10}$$

The collector-emitter voltage is,

$$V_{CE} = V_C - V_E \tag{5-11}$$

V_{CE} can also be determined as,

$$V_{CE} \approx V_{CC} - I_C (R_C + R_E) \tag{5-12}$$

Clearly, with I_C and I_E constant, the transistor collector-emitter
voltage remains at a constant level.

It should be noted that the transistor h_{FE} value is not involved in
any of the above equations.

Approximate Circuit Analysis

If the transistor base current is assumed to be much smaller than the voltage divider current, as discussed above, the circuit currents and voltages can be readily determined by the use of Equations 5-7 through 5-12. Example 5-7 demonstrates the process.

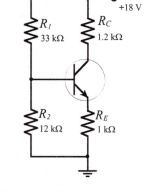

Example 5-7

Analyse the voltage divider bias in circuit in Fig. 5-23 to determine the emitter voltage, collector voltage, and collector-emitter voltage.

Solution

Eq. 5-7,
$$V_B = \frac{V_{CC} \times R_2}{R_1 + R_2} = \frac{18 \text{ V} \times 12 \text{ k}\Omega}{33 \text{ k}\Omega + 12 \text{ k}\Omega}$$
$$= 4.8 \text{ V}$$

Figure 5-23
Voltage divider bias circuit for Example 5-7.

Eq. 5-8,
$$V_E = V_B - V_{BE} = 4.8 \text{ V} - 0.7 \text{ V}$$
$$= 4.1 \text{ V}$$

Eq. 5-9,
$$I_E = \frac{V_B - V_{BE}}{R_E} = \frac{4.8 \text{ V} - 0.7 \text{ V}}{1 \text{ k}\Omega}$$
$$= 4.1 \text{ mA}$$

$$I_C \approx I_E = 4.1 \text{ mA}$$

Eq. 5-10,
$$V_C = V_{CC} - (I_C R_C) = 18 \text{ V} - (4.1 \text{ mA} \times 1.2 \text{ k}\Omega)$$
$$= 13.1 \text{ V}$$

Eq. 5-11,
$$V_{CE} = V_C - V_E = 13.1 \text{ V} - 4.1 \text{ V}$$
$$= 9 \text{ V}$$

The circuit conditions are illustrated in Fig. 5-24.

Precise Circuit Analysis

To precisely analyse a voltage divider bias circuit, the voltage divider must be replaced with its *Thevenin equivalent circuit* (V_T in series with R_T) as illustrated in Fig. 5-25. From Figure 5-25(a),

$$V_T = \frac{V_{CC} \times R_2}{R_1 + R_2} \qquad \textbf{(5-13)}$$

and R_T is calculated as R_1 in parallel with R_2,

$$R_T = R_1 \| R_2 \qquad \textbf{(5-14)}$$

Referring to Fig. 5-25(b), an equation may be written for the voltage drops around the base-emitter circuit;

$$V_T = I_B R_T + V_{BE} + R_E(I_B + I_C)$$

Substituting $I_C = h_{FE} I_B$,

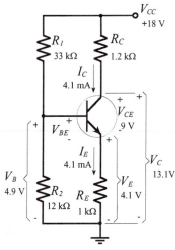

Figure 5-24
Voltage and current conditions for the circuit in Example 5-7.

$$V_T = I_B R_T + V_{BE} + R_E I_B(1 + h_{FE})$$

giving, $$I_B = \frac{V_T - V_{BE}}{R_T + R_E(1 + h_{FE})}$$ (5-15)

Once I_B has been determined, I_C can be calculated using the appropriate h_{FE} value, and the transistor terminal voltages can then be calculated. The effects of the maximum and minimum h_{FE} values can also be investigated.

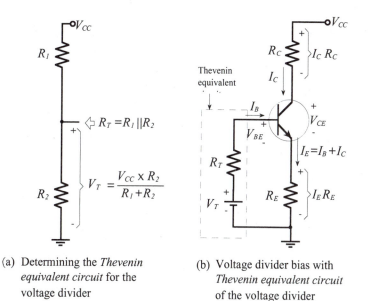

(a) Determining the *Thevenin equivalent circuit* for the voltage divider

(b) Voltage divider bias with *Thevenin equivalent circuit* of the voltage divider

Figure 5-25
For precise analysis of a voltage divider bias circuit, the voltage divider is replaced with its Thevenin equivalent circuit.

Example 5-8

Accurately analyse the voltage divider bias circuit in Example 5-7 to determine I_C, V_E, V_C, and V_{CE} when the transistor $h_{FE} = 100$.

Solution

Eq. 5-13, $$V_T = \frac{V_{CC} \times R_2}{R_1 + R_2} = \frac{18\,V \times 12\,k\Omega}{33\,k\Omega + 12\,k\Omega}$$

$$= 4.8\,V$$

Eq. 5-14, $$R_T = R_1 \| R_2 = 33\,k\Omega \| 12\,k\Omega$$

$$= 8.8\,k\Omega$$

Eq. 5-15, $$I_B = \frac{V_T - V_{BE}}{R_T + R_E(1 + h_{FE})} = \frac{4.8\,V - 0.7\,V}{8.8\,k\Omega + 1\,k\Omega(1 + 100)}$$

$$= 37.3\,\mu A$$

$$I_C = h_{FE}\,I_B = 100 \times 37.3\,\mu A$$

$$= 3.73\,mA$$

$$I_E = I_B + I_C = 37.3\,\mu A + 3.73\text{ mA}$$
$$= 3.77\text{ mA}$$

$$V_E = I_E R_E = 3.77\text{ mA} \times 1\text{ k}\Omega$$
$$= 3.77\text{ V}$$

Eq. 5-10,
$$V_C = V_{CC} - (I_C R_C) = 18\text{ V} - (3.73\text{ mA} \times 1.2\text{ k}\Omega)$$
$$= 13.52\text{ V}$$

Eq. 5-11,
$$V_{CE} = V_C - V_E = 13.52\text{ V} - 3.77\text{ V}$$
$$= 9.75\text{ V} \text{ (See Fig. 5-26)}$$

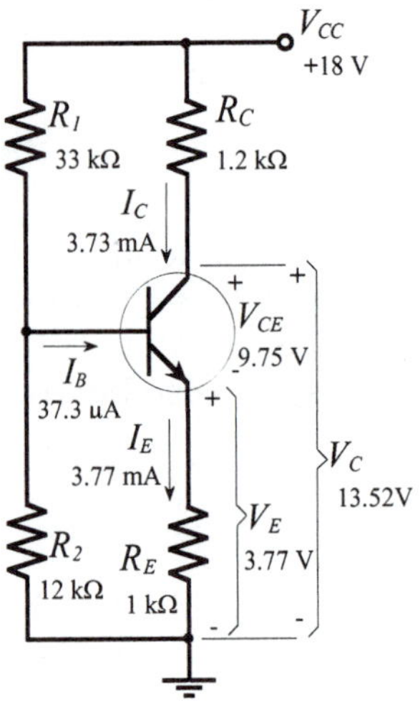

Figure 5-26
Voltage and current levels for the for the voltage divider bias circuit in Example 5-8.

Example 5-9

Accurately analyse the voltage divider bias circuit in Examples 5-7 and 5-8 for the conditions of $h_{FE(min)} = 50$.

Solution

From Ex. 5-8, $\quad V_T = 4.8\text{ V}$ and $R_T = 8.8\text{ k}\Omega$

Eq. 5-15,
$$I_B = \frac{V_T - V_{BE}}{R_T + R_E(1 + h_{FE})} = \frac{4.8\text{ V} - 0.7\text{ V}}{8.8\text{ k}\Omega + 1\text{ k}\Omega(1 + 50)}$$
$$= 68.6\,\mu A$$

$$I_C = h_{FE} I_B = 50 \times 68.6\,\mu A$$
$$= 3.43\text{ mA}$$

$$I_E = I_B + I_C = 68.6\,\mu A + 3.43\text{ mA}$$
$$= 3.5\text{ mA}$$

$$V_E = I_E R_E = 3.5\text{ mA} \times 1\text{ k}\Omega$$
$$= 3.5\text{ V}$$

Eq. 5-10,
$$V_C = V_{CC} - (I_C R_C) = 18\text{ V} - (3.43\text{ mA} \times 1.2\text{ k}\Omega)$$
$$= 13.9\text{ V}$$

Eq. 5-11,
$$V_{CE} = V_C - V_E = 13.9\text{ V} - 3.5\text{ V}$$
$$= 10.4\text{ V} \text{ [see Fig. 5-27(a)]}$$

Example 5-10

Accurately analyse the voltage divider bias circuit in Examples 5-7, 5-8 and 5-9 for the condition of $h_{FE(max)} = 200$.

Solution

From Ex. 5-8, $\quad V_T = 4.8\text{ V}$ and $R_T = 8.8\text{ k}\Omega$

Eq. 5-15,
$$I_B = \frac{V_T - V_{BE}}{R_T + R_E(1 + h_{FE})} = \frac{4.8\text{ V} - 0.7\text{ V}}{8.8\text{ k}\Omega + 1\text{ k}\Omega(1 + 200)}$$
$$= 19.5\,\mu A$$

$$I_C = h_{FE} I_B = 200 \times 19.5 \ \mu A$$
$$= 3.9 \text{ mA}$$

$$I_E = I_B + I_C = 19.5 \ \mu A + 3.9 \text{ mA}$$
$$= 3.92 \text{ mA}$$

$$V_E = I_E R_E = 3.92 \text{ mA} \times 1 \text{ k}\Omega$$
$$= 3.92 \text{ V}$$

Eq. 5-10, $$V_C = V_{CC} - (I_C R_C)$$
$$= 18 \text{ V} - (3.9 \text{ mA} \times 1.2 \text{ k}\Omega)$$
$$= 13.3 \text{ V}$$

Eq. 5-11, $$V_{CE} = V_C - V_E = 13.3 \text{ V} - 3.92 \text{ V}$$
$$= 9.4 \text{ V} \ [\text{see Fig. 5-27(b)}]$$

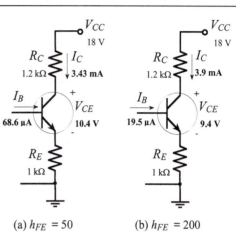

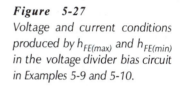

(a) $h_{FE} = 50$ (b) $h_{FE} = 200$

Figure 5-27
Voltage and current conditions produced by $h_{FE(max)}$ and $h_{FE(min)}$ in the voltage divider bias circuit in Examples 5-9 and 5-10.

The maximum and minimum h_{FE} values used in Example 5-9 gives $I_C = (3.43 \text{ mA to } 3.91 \text{ mA})$ and $V_{CE} = (9.4 \text{ V to } 10.4 \text{ V})$, while the typical h_{FE} value of 100 used in Example 5-8 gives $I_C = 3.73 \text{ mA}$ and $V_{CE} = 9.75 \text{ V}$. The I_C and V_{CE} levels determined in Examples 5-8 and 5-9 are shown in Fig. 5-27, and are illustrated by the *Q points* in Fig. 5-28. Note that the analysis using the highest h_{FE} value gives results that are closest to those from the approximate analysis in Example 5-7. This is because the approximate analysis assumes that I_B is very much smaller than the voltage divider current, and the highest h_{FE} gives the lowest I_B level.

For most practical purposes, precise analysis of voltage divider bias circuits is unnecessary. The approximate analysis method gives quite satisfactory results.

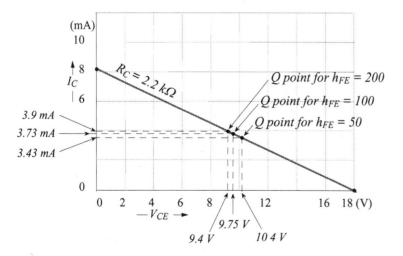

Figure 5-28
Effect of three different h_{FE} values on the Q point for the voltage divider bias circuit analyzed in Examples 5-8, 5-9, and 5-10.

Voltage Divider Bias Using a pnp Transistor

A voltage divider bias circuit using a *pnp* transistor is shown in Fig. 5-29. Note that the positions of the collector and emitter resistors are reversed compared to the *npn* transistor circuit. Also, note that the base voltage (V_B) in Fig. 5-29 is the voltage drop across resistor R_1, *not* that across R_2. As in the case of other circuits using *pnp* transistors, the current directions and voltage polarities are the reverse of those in *npn* transistor circuits. Apart from these differences, a *pnp* transistor voltage divider bias circuit is analysed in exactly the same way as an *npn* transistor circuit.

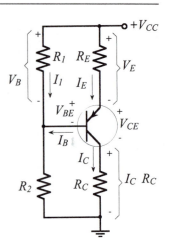

Figure 5-29
Voltage divider bias using a pnp transistor.

Practise Problems

5-4.1   A voltage divider bias circuit (as in Fig. 5-23) has V_{CC} = 24 V, R_1 = 180 kΩ, R_2 = 56 kΩ, R_E = 4.7 kΩ, R_C = 8.2 kΩ. Calculate the approximate levels of I_C, V_E, V_C, and V_{CE}.

5-4.2   The circuit in Practise Problem 5-4.1 uses a transistor with $h_{FE(min)}$ = 75 and $h_{FE(max)}$ = 250. Accurately analyse the circuit to determine the maximum and minimum levels of V_{CE}.

5-4.3   The voltage divider bias circuit in Fig. 5-29 has V_{CC} = 20 V, R_1 = 33 kΩ, R_2 = 100 kΩ, R_E = 3.9 kΩ, R_C = 6.8 kΩ. Calculate the approximate levels of V_E and V_C.

5-5   Comparison of Basic Bias Circuits

When comparing the performance of the three basic bias circuits, it must be recalled that transistor manufacturers specify maximum and minimum h_{FE} values for each transistor type number at various levels of collector current. Normally, the current gain of each individual transistor is not known, so that (as already stated) *the specified maximum and minimum values of h_{FE} must be used when analysing (or designing) a transistor bias circuit.* It is completely impractical to use some kind of average h_{FE} value.

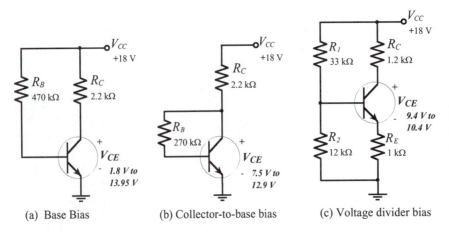

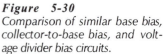

(a) Base Bias   (b) Collector-to-base bias   (c) Voltage divider bias

Figure 5-30
Comparison of similar base bias, collector-to-base bias, and voltage divider bias circuits.

Consider the three basic bias circuits reproduced in Fig. 5-30. Each circuit uses an 18 V supply and has a 2.2 kΩ (total) load resistance. The circuits were analysed in Examples 5-4, 5-6, 5-9, and 5-10 to determine the I_C and V_{CE} levels for $h_{FE(min)}$ = 50 and $h_{FE(max)}$ = 200. The calculated V_{CE} levels are indicated on each circuit and listed in Table 5-1. The Q point ranges are also illustrated in Fig. 5-31.

Table 5-1 V_{CE} maximum and minimum levels for similar bias circuits.

	Base bias	Collector-to-base bias	Voltage divider bias
$V_{CE(min)}$	1.8 V	7.25 V	9.4 V
$V_{CE(max)}$	13.75 V	12.9 V	10.4 V

To put these results in the most practical terms, suppose that a base bias circuit is employed as part of a complex electronics system, and that its V_{CE} level is measured as 10 V. If the transistor fails and has to be replaced with another (same type number) device with an h_{FE} range of 50 to 200, the new V_{CE} level could be anywhere from 1.8 V to 13.95 V. This would normally be unacceptable. A collector-to-base bias circuit in the same circumstance would have a V_{CE} ranging from 7.25 V to 12.9 V when the transistor is replaced. For a voltage divider bias circuit in this situation, the measurable V_{CE} range with a new transistor would be 9.4 V to 10.4 V.

Clearly, collector-to-base bias gives more predictable bias conditions than base bias. Voltage divider bias gives the most predictable bias conditions, because it has the greatest stability against h_{FE} spread.

Because of its excellent stability, voltage divider bias is almost always preferable. Collector-to-base bias, or some variation of it, is used in many circuits. Its one major advantage is that the feedback from the collector prevents the transistor from going into *saturation* (see Section 5-10). Base bias is most often used in switching circuits. This is also explained in Section 5-10.

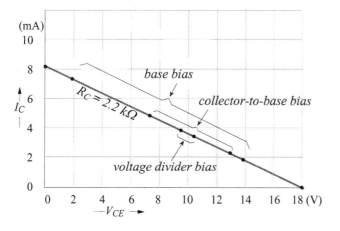

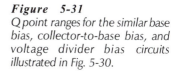

Figure 5-31
Q point ranges for the similar base bias, collector-to-base bias, and voltage divider bias circuits illustrated in Fig. 5-30.

5-6   *Trouble-Shooting BJT Bias Circuits*

Voltage Measurement

When a bias circuit is constructed in a laboratory situation, the supply voltage (V_{CC}) and the voltage levels at the transistor terminals $(V_C, V_B,$ and $V_E)$ should be measured with respect to the ground or negative supply terminal, as illustrated in Fig. 5-32. When the measured voltages are not as expected, the circuit must be further investigated to locate the fault.

Common Errors

The following is a list of errors that commonly occur with both experienced and inexperienced individuals:

- Power supply not switched *on.*
- Power supply current limiter control incorrectly set.
- Cables incorrectly connected to the power supply.
- Volt-Ohm-Milliammeter *(VOM)* function incorrectly selected.
- Wrong *VOM* terminals used.
- Incorrect component connections.
- Incorrect resistor values.
- Resistors in the wrong places.

Obviously, it is important to correctly connect the cables to the power supply, and to switch the power supply *on.* However, with many things to think about, people often neglect basic items. Similarly, it is just too easy to connect the *VOM* cables to the wrong terminals, or to select the wrong function. With plug-in-type breadboards, the sockets are so small and close together that incorrect connections often occur. These items should all be checked before looking for other circuit faults. Care should also be taken when selecting resistors. For example, a 100 Ω resistor used instead of a 100 kΩ resistor can have serious consequences.

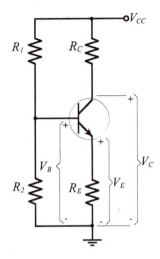

Figure 5-32
When testing a bias circuit, all voltage levels should first be measured with respect to the ground or negative supply terminal.

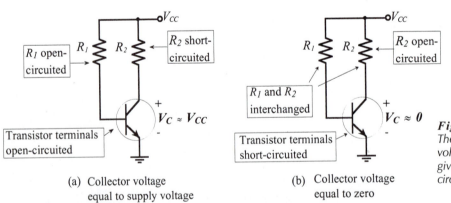

(a) Collector voltage equal to supply voltage

(b) Collector voltage equal to zero

Figure 5-33
The measured transistor terminal voltage on a base bias circuit can give an indication of possible circuit faults.

Base Bias

Figure 5-33 illustrates typical error sources in a base biased circuit. Normal voltage levels in a base bias circuit are $V_{BE} \approx 0.7$ V, and V_C anywhere in the range of 2 V to $(V_{CC} - 2$ V). If $V_C = V_{CC}$, as in Fig. 5-33(a), R_1 or one of its connecting leads might be open-circuited, or the transistor may have an open-circuited junction. Alternatively,

R_2 might be shorted. When $V_C \approx 0$, as in Fig. 5-33(b), R_1 and R_2 could be in the wrong places; that is, the high-value and low-value resistors might be interchanged. Other possibilities are that R_2 is open-circuited, or the transistor *CE* terminals are short-circuited.

Collector-to-Base Bias

Some reasons for incorrect voltage levels in a collector-to-base bias circuit are illustrated in Fig. 5-34. These are similar to those discussed for the base bias circuit. When V_C equals V_{CC}, either resistor R_1 or one of the transistor terminals is likely to be open-circuited, [Fig 5-34(a)]. If V_C is in the range of 0 V to 0.7 V, R_1 and R_2 might be interchanged, the device terminals might be short-circuited, or R_2 might be open-circuited, [Fig 5-34(b)].

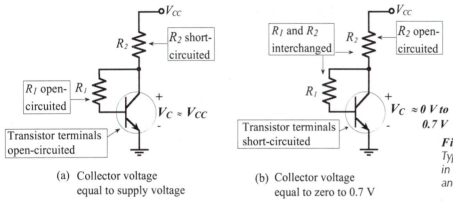

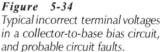

(a) Collector voltage equal to supply voltage

(b) Collector voltage equal to zero to 0.7 V

Figure   5-34
Typical incorrect terminal voltages in a collector-to-base bias circuit, and probable circuit faults.

Voltage Divider Bias

Figure 5-35 shows unsuitable measured voltages and probable errors in a voltage divider bias circuit. If V_C equals V_{CC}, either R_1 or R_4 might be open-circuited. Alternatively, R_2 or R_3 might be short-circuited, [Fig. 5-35(a)], or the transistor terminals might be open-circuited. When V_C approximately equals V_E, R_2 might be open-circuited, or R_1 and R_2 might be interchanged, [Fig. 5-35(b)]. Alternatively, the transistor terminals might be short-circuited.

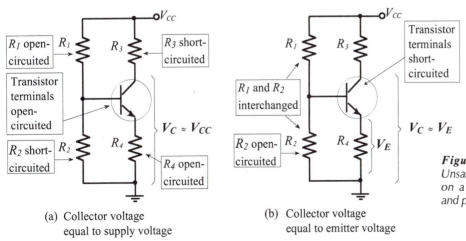

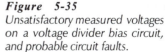

(a) Collector voltage equal to supply voltage

(b) Collector voltage equal to emitter voltage

Figure   5-35
Unsatisfactory measured voltages on a voltage divider bias circuit, and probable circuit faults.

5-7 Bias Circuit Design

Bias circuit design can be amazingly simple. Usually, it is just a matter of determining the required voltage across each resistor and the appropriate current levels. Then, the resistor values are calculated by application of Ohm's law.

Designs usually begin with specification of the supply voltage and the required levels of I_C and V_{CE}. The resistor values are calculated to meet these requirements, and standard value resistors are selected. (See Appendix 2-1.) Usually, resistors with a tolerance of ±10% are used wherever possible. These are less expensive than ±5% and ±1% components.

Base Bias Circuit Design

A base bias circuit is very easily designed by application of Equations 5-1 and 5-2. This is illustrated in Fig. 5-36 and demonstrated in Example 5-11.

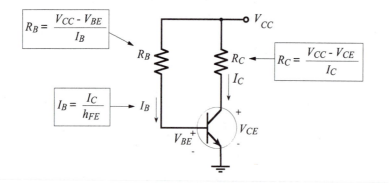

Figure 5-36
Base bias circuit design.

Example 5-11

Design a base bias circuit (as in Fig. 5-36) to have V_{CE} = 5 V and I_C = 5 mA. The supply voltage is 15 V, and the transistor has h_{FE} =100.

Solution

From Eq. 5-1,

$$R_C = \frac{V_{CC} - V_{CE}}{I_C} = \frac{15\ V - 5\ V}{5\ mA}$$

$$= 2\ k\Omega\ \text{(use 1.8 k}\Omega\text{ or 2.2 k}\Omega\text{ standard value)}$$

$$I_B = \frac{I_C}{h_{FE}} = \frac{5\ mA}{100}$$

$$= 50\ \mu A$$

From Eq. 5-2,

$$R_B = \frac{V_{CC} - V_{BE}}{I_B} = \frac{15\ V - 0.7\ V}{50\ \mu A}$$

$$= 286\ k\Omega\ \text{(use 270 k}\Omega\text{ or 330 k}\Omega\text{ standard value)}$$

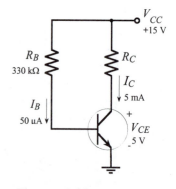

Figure 5-37
Base bias circuit designed in Example 5-11.

When selecting the standard value resistors, a decision must be made whether to select the next smaller resistance value or the next larger value. In general, it is best to select the resistance value that tends to increase the transistor collector-emitter voltage, thus keeping V_{CE} from approaching zero. In Example 5-11, selecting R_C = 1.8 kΩ (instead of 2.2 kΩ) gives the smallest voltage drop across R_C, and results in a larger V_{CE} than the (5 V) design value. Selecting R_B = 330 kΩ (instead of 270 kΩ) gives the lowest level of I_B, and consequently keeps I_C at a minimum, to produce the largest level of V_{CE}.

After design, the circuit should be analysed using the selected standard-value components and the maximum and minimum h_{FE} values for the transistor, (see Section 5-2).

Collector-to-Base Bias Circuit Design

The design procedure for a collector-to-base bias circuit is similar to that for base bias, with the exception that the voltage and current levels are different for calculating R_B and R_C. In collector-to-base bias, the voltage across R_B is $(V_C - V_{BE})$ and the current through R_C is $(I_B + I_C)$. The design equations are shown in Fig. 5-38, and the design procedure is demonstrated in Example 5-12.

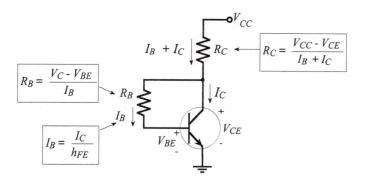

Figure 5-38
Collector-to-base bias circuit design.

Example 5-12

Design a collector-to-base bias circuit (as in Fig. 5-38) to have V_{CE} = 5 V and I_C = 5 mA when the supply voltage is 15 V and the transistor h_{FE} is 100.

Solution

$$I_B = \frac{I_C}{h_{FE}} = \frac{5\ mA}{100}$$

$$= 50\ \mu A$$

From Eq. 5-5,

$$R_C = \frac{V_{CC} - V_{CE}}{I_C + I_B} = \frac{15\ V - 5\ V}{5\ mA + 50\ \mu A}$$

$$= 1.98\ k\Omega\ \text{(use 1.8 kΩ or 2.2 kΩ standard value)}$$

From Eq. 5- 4,

$$R_B = \frac{V_{CE} - V_{BE}}{I_B} = \frac{5\text{ V} - 0.7\text{ V}}{50\,\mu A}$$

$$= 86\text{ k}\Omega \text{ (use 82 k}\Omega \text{ or 100 k}\Omega \text{ standard value)}$$

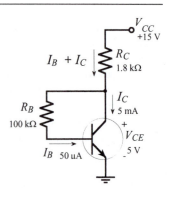

Figure 5-39
*Collector-to-base bias   circuit
designed in Example 5-12.*

Once again decisions must be made (in Example 5-12) about selecting the next larger or the next smaller standard value resistors. As in the case of the base bias circuit in Example 5-11, selection of the smaller value for R_C and the larger value for R_B tends to produce a larger V_{CE} than the specified level. The design should be analysed using the selected standard-value components and the transistor $h_{FE(max)}$ and $h_{FE(min)}$ values.

Voltage Divider Bias Circuit Design

When designing a voltage divider bias circuit the voltage divider current (I_2 in Fig. 5-40) should be selected much larger than the transistor base current (I_B). This makes the base voltage (V_B) a stable quantity largely unaffected by the transistor h_{FE} value. However, a high level of I_2 results in small resistance values for R_1 and R_2, and (as explained in Chapter 12) this gives the circuit an undesirable low input impedance.

A rule-of-thumb approach to selection of I_2 is to use a voltage divider current approximately equal to one-tenth of the transistor collector current.

$$I_2 = \frac{I_C}{10} \tag{5-16}$$

As can be easily demonstrated, this gives reasonably large values for R_1 and R_2 while still keeping I_2 much larger than I_B.

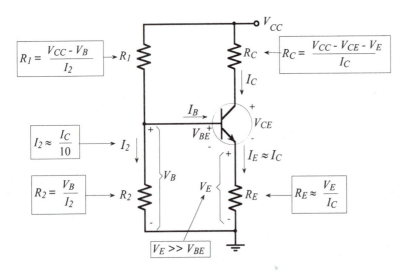

Figure 5-40
Voltage divider bias circuit design.

If V_E is not specified, it should be selected much larger than the transistor V_{BE},

$$V_E >> V_{BE}$$

This is because V_{BE} can vary from transistor to transistor, and it can also change with temperature increase or decrease. Making V_E very much larger than V_{BE} minimises the effect of V_{BE} changes on the circuit bias conditions. Typically, as another rule-of-thumb, V_E is selected as 5 V regardless of the supply voltage. When V_{CC} is low (for example, 9 V), V_E can be as low as 3 V.

Figure 5-40 shows the equations used for calculating each resistor value, and Example 5-13 demonstrates the design procedure.

Example 5-13

Design the voltage divider bias circuit in Fig. 5-41 to have $V_{CE} = V_E = 5$ V and $I_C = 5$ mA when the supply voltage is 15 V. Assume the transistor h_{FE} is 100.

Solution

$$R_E = \frac{V_E}{I_E} \approx \frac{V_E}{I_C} = \frac{5\ V}{5\ mA}$$

$$= 1\ k\Omega \text{ (standard value)}$$

$$R_C = \frac{V_{CC} - V_{CE} - V_E}{I_C} = \frac{15\ V - 5\ V - 5\ V}{5\ mA}$$

$$= 1\ k\Omega \text{ (standard value)}$$

Eq. 5-16, $$I_2 = \frac{I_C}{10} = \frac{5\ mA}{10}$$

$$= 500\ \mu A$$

From Eq. 5-8, $V_B = V_E + V_{BE} = 5\ V + 0.7\ V$

$$= 5.7\ V$$

$$R_2 = \frac{V_B}{I_2} = \frac{5.7\ V}{500\ \mu A}$$

$$= 11.4\ k\Omega \text{ (use 12 k}\Omega \text{ standard value)}$$

$$R_1 = \frac{V_{CC} - V_B}{I_2} = \frac{15\ V - 5.7\ V}{500\ \mu A}$$

$$= 18.6\ k\Omega \text{ (use 18 k}\Omega \text{ standard value)}$$

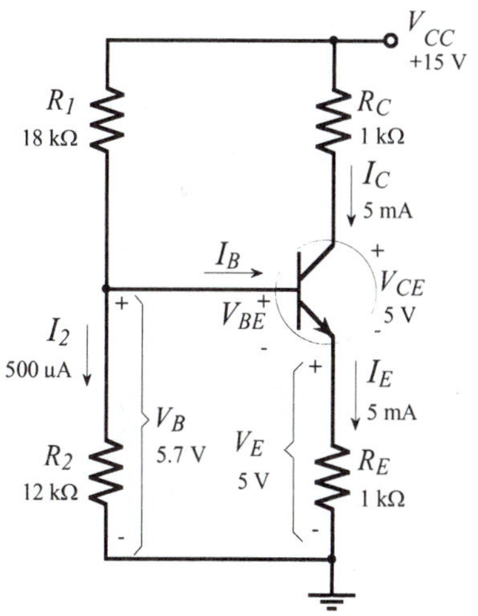

Figure 5-41
Voltage divider bias circuit designed in Example 5-13.

The calculated resistor values in Example 5-13 are conveniently close to standard values. When this is not the case, some thought must be put into deciding whether to select higher or lower standard value. As explained for the other bias circuits, it is best to select the resistance value that tend to increase the transistor collector-emitter voltage, in order to keep V_{CE} from approaching zero. As with all bias circuit designs, the designed circuit should be analysed using the standard value components and the transistor $h_{FE(max)}$ and $h_{FE(min)}$ values.

Designing with Standard Resistor Values

In all circuit designs a suitable standard resistor should normally be selected when each resistor value is calculated, instead of first completing the design. Then, the new resistor voltage drop or current level should be determined before calculating the next component value. This is demonstrated in Example 5-14.

Example 5-14

Design the voltage divider bias circuit in Fig. 5-42 to operate from a 12 V supply. The bias conditions are to be $V_{CE} = 3$ V, $V_E = 5$ V and $I_C = 1$ mA.

Solution

$$R_4 = \frac{V_E}{I_E} \approx \frac{V_E}{I_C} = \frac{5 \text{ V}}{1 \text{ mA}}$$

$$= 5 \text{ k}\Omega \text{ (use a 4.7 k}\Omega \text{ standard value)}$$

With $I_C = 1$ mA and $R_4 = 4.7$ kΩ, V_E becomes,

$$V_E = I_C R_4 = 1 \text{ mA} \times 4.7 \text{ k}\Omega$$

$$= 4.7 \text{ V}$$

and,

$$V_C = V_E + V_{CE} = 4.7 \text{ V} + 3 \text{ V}$$

$$= 7.7 \text{ V}$$

$$V_{R3} = V_{CC} - V_C = 12 \text{ V} - 7.7 \text{ V}$$

$$= 4.3 \text{ V}$$

$$R_3 = \frac{V_{R3}}{I_C} = \frac{4.3 \text{ V}}{1 \text{ mA}}$$

$$= 4.3 \text{ k}\Omega \text{ (use a 3.9 k}\Omega \text{ standard value to}$$
$$\text{reduce } V_{R3} \text{ and increase } V_{CE}.)$$

$$V_B = V_E + V_{BE} = 4.7 \text{ V} + 0.7 \text{ V}$$

$$= 5.4 \text{ V}$$

Eq. 5-16,

$$I_2 = \frac{I_C}{10} = \frac{1 \text{ mA}}{10}$$

$$= 100 \text{ } \mu\text{A}$$

$$R_2 = \frac{V_B}{I_2} = \frac{5.4 \text{ V}}{100 \text{ } \mu\text{A}}$$

$$= 54 \text{ k}\Omega \text{ (use 56 k}\Omega \text{ standard value)}$$

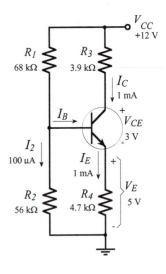

Figure 5-42
Voltage divider bias circuit for Example 5-14.

With R_2 = 56 kΩ and V_B = 5.4 V, I_2 becomes,

$$I_2 = \frac{V_B}{R_2} = \frac{5.4\ V}{56\ k\Omega}$$

$$= 96.4\ \mu A$$

$$R_1 = \frac{V_{CC} - V_B}{I_2} = \frac{12\ V - 5.4\ V}{96.4\ \mu A}$$

$$= 68.5\ k\Omega\ \ (\text{use } 68\ k\Omega \text{ standard value})$$

Practise Problems

5-7.1 A base bias circuit is to be designed to have V_{CE} = 4 V and I_C = 3 mA. The supply voltage is 12 V and the transistor h_{FE} is 125.

5-7.2 Design a collector-to-base bias circuit to have V_{CE} = 4 V and I_C = 2 mA. The supply voltage is 18 V and the transistor h_{FE} is 90.

5-7.3 Design a voltage divider bias circuit to have V_{CE} = V_E = 6 V and I_C = 1.5 mA. The supply voltage is 24 V and the transistor h_{FE} is 80.

5-7.4 Design a voltage divider bias circuit to have V_{CE} = 7 V, V_E = 6 V and I_C = 2 mA. The supply voltage is 20 V.

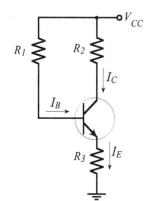

Figure 5-43
A base bias circuit with an emitter resistor has essentially the same stability as a similar base bias circuit.

5-8 More Bias Circuits

Base Bias with Emitter Resistor

The bias circuit shown in Fig. 5-43 is the usual base bias arrangement with the addition of an emitter resistor. Analysis reveals that this circuit has essentially the same stability characteristics as a similar base bias circuit. The only advantage of the emitter resistor in this case is that it gives the circuit a higher *input resistance* (see Section 6-5). An emitter resistor could also be employed with collector-to-base bias, and this would also produce a higher input resistance without significantly altering the circuit stability.

Voltage Divider and Collector-to-Base Combination

Figure 5-44 shows a voltage divider bias circuit with resistor R_1 connected to the transistor collector instead of to the supply. Thus, the circuit combines collector-to-base bias with voltage divider bias. An analysis of the circuit shows that this combination produces even greater bias stability than voltage divider bias alone.

The design procedure for this circuit is similar to voltage divider bias design, except that for calculating the resistances of R_1 and R_3, the voltage across R_1 is ($V_C - V_B$) instead of ($V_{CC} - V_B$), and the current through R_3 is ($I_C + I_2$).

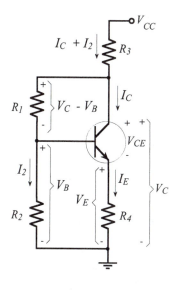

Figure 5-44
A combination of voltage divider bias and collector-to-base bias has excellent stability.

Example 5-15

Design the bias circuit in Fig. 5-44 to operate from an 18 V supply. The bias conditions are to be $V_{CE} = 9$ V, $V_E = 4$ V and $I_C = 4$ mA.

Solution

$$R_4 = \frac{V_E}{I_E} \approx \frac{V_E}{I_C} = \frac{4\ V}{4\ mA}$$

$$= 1\ k\Omega\ \ (standard\ value)$$

$$V_B = V_E + V_{BE} = 4\ V + 0.7\ V$$

$$= 4.7\ V$$

Eq. 5-16,
$$I_2 = \frac{I_C}{10} = \frac{4\ mA}{10}$$

$$= 400\ \mu A$$

$$R_2 = \frac{V_B}{I_2} = \frac{4.7\ V}{400\ \mu A}$$

$$= 11.8\ k\Omega\ \ (use\ 12\ k\Omega\ standard\ value)$$

With $R_2 = 12$ kΩ, I_2 becomes,

$$I_2 = \frac{V_B}{R_2} = \frac{4.7\ V}{12\ k\Omega}$$

$$= 392\ \mu A$$

$$R_1 = \frac{V_C - V_B}{I_2} = \frac{(9\ V + 4\ V) - 4.7\ V}{392\ \mu A}$$

$$= 21.2\ k\Omega\ \ (use\ 22\ k\Omega\ standard\ value)$$

$$V_{R3} = V_{CC} - V_C = 18\ V - (9\ V + 4\ V)$$

$$= 5\ V$$

$$R_3 = \frac{V_{R3}}{I_C + I_2} = \frac{5\ V}{4\ mA + 392\ \mu A}$$

$$= 1.14\ k\Omega\ \ (use\ a\ 1.2\ k\Omega\ standard\ value)$$

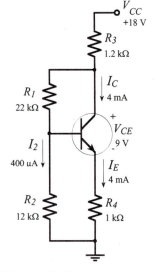

Figure 5-45
Circuit designed in Example 5-15.

Emitter Current Bias

The *emitter current bias* circuit in Fig. 5-46(a) uses a *plus and minus* voltage supply (+V_{CC} and -V_{EE}) and has the transistor base grounded via resistor R_1. This is similar to voltage divider bias, in fact, as illustrated in Fig. 5-46(b), a voltage divider (R_1 and R_2) could be used to provide V_B instead of grounding the base via R_1.

So long as there is very little voltage drop across base resistor R_1, in Fig. 5-46(a), this circuit (like voltage divider bias), has excellent bias stability. Normally, R_1 is selected to have a voltage drop much smaller than the transistor base-emitter voltage:

$$I_{B(max)} R_1 << V_{BE}$$

A reasonable rule-of-thumb to use is,

$$I_{B(max)} R_1 \approx \frac{V_{BE}}{10} \qquad \text{(5-17)}$$

This mean that the transistor base voltage can be treated as ground level, and the voltage at the emitter terminal is always V_{BE} below ground, [see Fig. 5-46(a)]. So, the voltage across R_E is a constant quantity:

$$V_E = V_{EE} - V_{BE}$$

The transistor collector voltage is,

$$V_C = V_{CC} - V_{RC}$$

and, as illustrated, the collector-emitter voltage is,

$$V_{CE} = V_C + V_{BE}$$

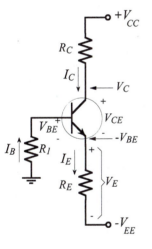

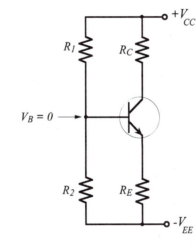

Figure 5-46
Emitter current bias using a plus-minus supply voltage is similar to voltage divider bias.

(a) Emitter current bias circuit

(b) Voltage divider equivalent of emitter current bias circuit

Example 5-16

Design the bias circuit in Fig. 5-47(a) to operate from a ±9 V supply. The bias conditions are to be $V_C = 5$ V and $I_C = 1$ mA, and the transistor has $h_{FE} = 70$

Solution

$$V_E = V_{EE} - V_{BE} = 9\text{ V} - 0.7\text{ V}$$
$$= 8.3\text{ V}$$

$$R_3 = \frac{V_E}{I_E} \approx \frac{V_E}{I_C} = \frac{8.3\ V}{1\ mA}$$

$$= 8.3\ k\Omega \text{ (use 8.2 k}\Omega \text{ standard value)}$$

New level of I_E,

$$I_E = \frac{V_E}{R_3} = \frac{8.3\ V}{8.2\ k\Omega}$$

$$= 1.01\ mA$$

$$V_{R2} = V_{CC} - V_C = 9\ V - 5\ V$$

$$= 4\ V$$

$$R_2 = \frac{V_{R2}}{I_C} \approx \frac{V_{R2}}{I_E} = \frac{4\ V}{1.01\ mA}$$

$$= 3.96\ k\Omega \text{ (use 3.9 k}\Omega \text{ standard value)}$$

$$I_B = \frac{I_C}{h_{FE}} = \frac{1\ mA}{70}$$

$$= 14.3\ \mu A$$

$$V_{R1} \approx \frac{V_{BE}}{10} = \frac{0.7\ V}{10}$$

$$= 70\ mV$$

$$R_1 = \frac{V_{R1}}{I_B} \approx \frac{70\ mV}{14.3\ \mu A}$$

$$= 4.9\ k\Omega \text{ (use 4.7 k}\Omega \text{ standard value)}$$

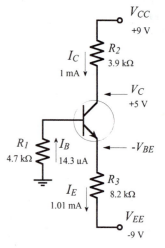

Figure 5-47
Emitter current bias circuit designed in Example 5-16.

The circuit in Fig. 5-47 can be quickly analysed almost by just looking at it. The transistor emitter terminal is 0.7 V below ground, and the voltage across R_E is,

$$V_E = 9\ V - 0.7\ V = 8.3\ V$$

This gives,

$$I_E \approx \frac{V_E}{R_E} = \frac{8.3\ V}{8.2\ k\Omega}$$

$$\approx 1\ mA$$

and,

$$V_C = V_{CC} - V_{RC} = 9\ V - (1\ mA \times 4.7\ k\Omega)$$

$$= 4.3\ V$$

Note that, as in the case of voltage divider bias, the transistor h_{FE} value is not used in the approximate circuit analysis. For accurate analysis, the same approach as used for voltage divider bias must be employed; the voltage source at the transistor base must be replaced with its Thevenin equivalent circuit.

Practise Problems

5-8.1 Design the bias circuit in Fig. 5-43 to have $V_{CE} = 4$ V, $V_E = 5$ V and $I_C = 1.5$ mA. The supply voltage is 18 V and the transistor h_{FE} is 70.

5-8.2 A bias circuit that combines voltage divider and cathode-to-base bias (as in Fig. 5-44) is to have $V_{CE} = 5$ V and $I_C = 2$ mA. Design the circuit using $V_{CC} = 15$ V.

5-8.3 Design an emitter current bias circuit [as in Fig. 5-46 (a)] to have $V_C = 6$ V and $I_C = 3$ mA. The supply voltage is ±10 V and the transistor h_{FE} is 120.

5-9 Thermal Stability of Bias Circuits

V_{BE} and I_{CBO} Variations

Many transistor circuits are required to operate over a wide temperature range. So, another aspect of bias circuit stability is *thermal stability*, or how stable I_C and V_{CE} remain when the circuit temperature changes.

Measures to deal with the effects of h_{FE} variations have already been discussed. These apply whether the different h_{FE} values are due to temperature changes or to h_{FE} differences from one transistor to another.

The base-emitter voltage (V_{BE}) and the collector-base reverse saturation current (I_{CBO}) are the two temperature-sensitive quantities that largely determine the thermal stability of a transistor circuit, [see Fig. 5-48(a)]. The base-emitter and collector-base *pn*-junctions have the temperature characteristics discussed in Section 1-6. For a silicon transistor, V_{BE} changes by approximately -1.8 mV/°C, and I_{CBO} approximately doubles for every 10°C rise in temperature. These effects are illustrated by the characteristics in Fig. 5-48(b) and (c).

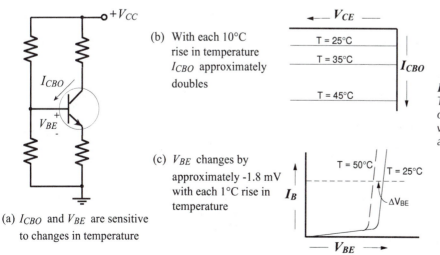

(b) With each 10°C rise in temperature I_{CBO} approximately doubles

(c) V_{BE} changes by approximately -1.8 mV with each 1°C rise in temperature

(a) I_{CBO} and V_{BE} are sensitive to changes in temperature

Figure 5-48
Transistor collector-base leakage current (I_{CBO}) and base-emitter voltage (V_{BE}) can be seriously affected by temperature change.

An increase in I_{CBO} causes I_C to be larger, and the I_C increase raises the collector-base junction temperature. This, in turn, results in a further increase in I_{CBO}. The effect is cumulative, so that the end result might be a substantial collector current increase. This could produce a significant shift in the circuit Q-point, or in the worst case, I_C *might keep on increasing until the transistor collector-base junction overheats and burns out.* This effect is known as *thermal runaway.* Measures taken to avoid thermal runaway are similar to those required for good bias stability against h_{FE} spread.

Changes in V_{BE} may also produce significant changes in I_C and consequently in the circuit Q-point. However, because of the possibility of thermal runaway, I_{CBO} changes are the most important. The thermal stability of a bias circuit is assessed by calculating a stability factor.

Stability   Factor

The *stability factor* (S) of a circuit is the ratio of the change in collector current to the change in collector-base leakage current.

$$S = \frac{\Delta I_C}{\Delta I_{CBO}}$$

or, $$\Delta I_C = S \times \Delta I_{CBO} \qquad \text{(5-18)}$$

The value of S depends on the circuit configuration and on the resistor values. The minimum values of S is 1. This means that if I_{CBO} increases by 1 µA I_C will increase by 1 µA. If a circuit has an S of 50, then $\Delta I_C = 50 \times \Delta I_{CBO}$. A stability factor of 50 (or larger) is considered poor, while a factor of 10 or less is considered good.

An equation for the stability factor of a bias circuit can be derived by writing an equation for the circuit I_C and investigating the effect of I_{CBO} change. The stability factors for the three basic bias circuit types (reproduced in Fig. 5-49) can be shown to be:

For base bias,

$$S = 1 + h_{FE} \qquad \text{(5-19)}$$

For collector-to-base bias,

$$S = \frac{1 + h_{FE}}{1 + h_{FE} R_C/(R_C + R_B)} \qquad \text{(5-20)}$$

For voltage divider bias,

$$S = \frac{1 + h_{FE}}{1 + h_{FE} R_E/(R_E + R_1 \| R_2)} \qquad \text{(5-21)}$$

The change in I_{CBO} over a given temperature range can be calculated by recalling that I_{CBO} doubles for every 10°C increase in temperature. The temperature change (ΔT) is divided by 10 to give the number of 10°C changes (n). If the starting level of collector-base leakage current is $I_{CBO(1)}$, the new level is,

$$I_{CBO(2)} = I_{CBO(1)} \times 2^n \qquad\qquad\qquad \text{(5-22)}$$

The I_{CBO} change and the circuit stability factor can be used to determine the change in I_C, (Eq. 5-18). Then the resulting V_{CE} change can be investigated.

$$\boxed{S = 1 + h_{FE}}$$

$$\boxed{S = \dfrac{1 + h_{FE}}{1 + h_{FE}\, R_C/(R_C + R_B)}}$$

$$\boxed{S = \dfrac{1 + h_{FE}}{1 + h_{FE}\, R_E/(R_E + R_1 \| R_2)}}$$

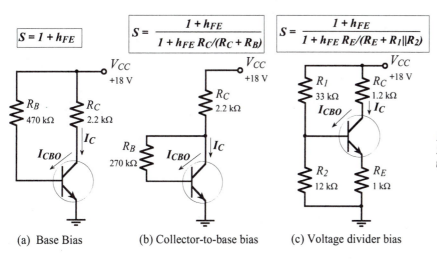

(a) Base Bias   (b) Collector-to-base bias   (c) Voltage divider bias

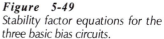

Figure 5-49
Stability factor equations for the three basic bias circuits.

Example 5-17
Calculate the stability factors for the three bias circuits shown in Fig. 5-49, if each circuit uses a transistor with $h_{FE} = 100$. Note that these are the bias circuits analysed in Examples 5-4, 5-6, and 5-8.

Solution
For base bias,

Eq. 5-19, $S = 1 + h_{FE} = 1 + 100$

$\qquad\qquad\quad = 101$

For collector-to-base bias,

Eq. 5-20, $S = \dfrac{1 + h_{FE}}{1 + h_{FE}\, R_C/(R_C + R_B)}$

$\qquad\qquad = \dfrac{1 + 100}{1 + [100 \times 2.2\ \text{k}\Omega/(2.2\ \text{k}\Omega + 270\ \text{k}\Omega)]}$

$\qquad\qquad \approx 56$

For voltage divider bias,

Eq. 5-21, $S = \dfrac{1 + h_{FE}}{1 + h_{FE}\, R_E/(R_E + R_1 \| R_2)}$

$\qquad\qquad = \dfrac{1 + 100}{1 + [100 \times 1\ \text{k}\Omega/(1\ \text{k}\Omega + 33\ \text{k}\Omega \| 12\ \text{k}\Omega)]}$

$\qquad\qquad = 8.2$

Example 5-18

Determine the I_C change produced in each bias circuit referred to in Example 5-17 when the circuit temperature increases from 25°C to 105°C, and I_{CBO} = 15 nA at 25°C. The circuits are reproduced in Fig. 5-50.

Solution

$$\Delta T = 105°C - 25°C$$
$$= 80°C$$

or, $\quad \Delta T$ in 10°C steps $= \dfrac{80°C}{10°C}$

$$= 8$$

Eq. 5-22, $\quad I_{CBO(2)} = I_{CBO(1)} \times 2^n = 15 \text{ nA} \times 2^8$

$$= 3.84 \,\mu A$$

$$\Delta I_{CBO} = I_{CBO(2)} - I_{CBO(1)} = 3.84 \,\mu A - 15 \text{ nA}$$
$$\approx 3.83 \,\mu A$$

For base bias,
Eq. 5-18, $\quad \Delta I_C = S \times \Delta I_{CBO} = 101 \times 3.83 \,\mu A$

$$\approx 386 \,\mu A$$

For collector-to-base bias,
Eq. 5-18, $\quad \Delta I_C = S \times \Delta I_{CBO} = 56 \times 3.83 \,\mu A$

$$= 214 \,\mu A$$

For voltage divider bias,
Eq. 5-18, $\quad \Delta I_C = S \times \Delta I_{CBO} = 8.2 \times 3.83 \,\mu A$

$$= 31.4 \,\mu A$$

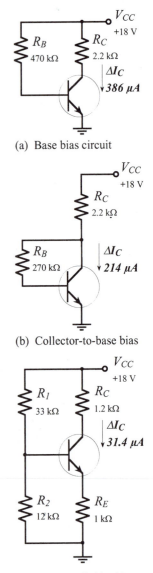

(a) Base bias circuit

(b) Collector-to-base bias

(c) Voltage divider bias

Figure 5-50
Effect of temperature change from 25°C to 105°C on bias circuits, as determined in Example 5-18.

The collector current changes produced by the increase in I_{CBO} (as calculated in Examples 5-17 and 5-18) are small compared to the effects of transistor h_{FE} spread. However, many transistors have higher levels of I_{CBO} (at 25°C) than the 15 nA used in Example 5-18, (some have lower levels). It is important to note that the bias circuit that offers the best thermal stability is also the one that gives the best stability against h_{FE} spread — voltage divider bias.

Effect of V_{BE} Changes

Consider the voltage divider bias circuits in Fig. 5-51(a) and (b), which each have V_{CC} = 12 V and I_C = 1 mA. From Eq. 5-9,

$$I_C \approx I_E = \frac{V_B - V_{BE}}{R_E}$$

Assuming that V_B remains substantially constant, an equation for I_C change with V_{BE} change can be written:

$$\Delta I_C \approx \Delta I_E = \frac{\Delta V_{BE}}{R_E} \qquad\qquad \textbf{(5-23)}$$

As discussed in Section 5-7, the emitter resistor voltage should typically be selected as 5 V, ($V_E \gg V_{BE}$). This is to ensure that I_C is not significantly affected by changes in V_{BE}. Thus, the circuit in Fig. 5-51(a) (with $V_E \approx 5$ V) has greater stability against V_{BE} changes than the one in Fig. 5-51(b) which was designed for $V_E = V_{CC}/10$, ($V_E = 1.2$ V). This is demonstrated in Example 5-19.

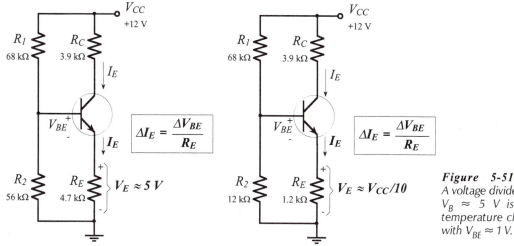

(a) Circuit designed for $V_E \approx 5$ V   (b) Circuit designed for $V_E \approx V_{CC}/10$

Figure 5-51
A voltage divider bias circuit with $V_B \approx 5$ V is less affected by temperature changes than one with $V_{BE} \approx 1$ V.

Example 5-19

Determine the I_C change produced in each of the two circuits in Fig. 5-51 by the effect of V_{BE} changes over a temperature range of 25°C to 125°C.

Solution

$$\Delta T = 125°C - 25°C$$
$$= 100°C$$

$$\Delta V_{BE} = \Delta T \times (1.8 \text{ mV/°C}) = 100°C \times 1.8 \text{ mV}$$
$$= 180 \text{ mV}$$

For Fig. 5-51 (a),

$$\Delta I_C \approx \Delta I_E = \frac{\Delta V_{BE}}{R_E} = \frac{180 \text{ mV}}{4.7 \text{ k}\Omega}$$
$$= 38 \, \mu A$$

For Fig. 5-51 (b),

$$\Delta I_C \approx \Delta I_E = \frac{\Delta V_{BE}}{R_E} = \frac{180 \text{ mV}}{1.2 \text{ k}\Omega}$$
$$= 150 \, \mu A$$

Diode Compensation

The use of a diode to compensate for V_{BE} changes is illustrated in Fig. 5-52. In this case,

$$V_B = V_{R2} + V_{D1}$$

and, $$I_C \approx I_E = \frac{V_{R2} + V_{D1} - V_{BE}}{R_E} \qquad (5\text{-}24)$$

When V_{BE} changes by ΔV_{BE}, the diode voltage changes by an approximately equal amount (ΔV_{D1}). ΔV_{BE} and ΔV_{D1} tend to cancel each other, leaving I_C largely constant at,

$$I_C \approx I_E = \frac{V_{R2}}{R_E}$$

Base bias and collector-to-base bias are less affected by V_{BE} changes than voltage divider bias. This can easily be demonstrated by considering the equations for the base current in each case.

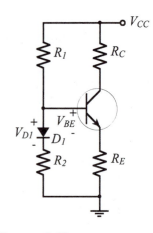

Figure 5-52
A diode can be used to compensate for V_{BE} changes in a voltage divider bias circuit.

Practise Problems

5-9.1   Calculate the stability factor for the bias circuits designed in Practise Problems 5-7.1, 5-7.2, and 5-7.3.

5-9.2   The circuit temperature of each of the bias circuits in Practise Problem 5-9.1 increases from 25°C to 125°C. Calculate the collector current change in each circuit if the transistor I_{CBO} = 10 nA at 25°C.

5-9.3   Calculate the I_C change produced by ΔV_{BE} in the circuit designed in Example 5-14 when the temperature changes from -35°C to +100°C.

5-10   Biasing Transistor Switching Circuits

Direct-Coupled Switching Circuit

When a transistor is used as a switch, it is either biased *off* to I_C = 0, or biased *on* to its maximum collector current level. Figure 5-53 illustrates the two conditions. Note that the circuit in Fig. 5-53 is termed a *direct-coupled* switching circuit, because the signal source is directly connected to the circuit. In Fig. 5-53(a) the negative polarity of the base input voltage (V_s) biases the transistor (Q_1) *off*. In this case, the only current flowing is the collector base leakage current (I_{CBO}) which is normally so small that it can be neglected. The transistor collector-emitter voltage is,

$$V_{CE} = V_{CC} - (I_C R_C)$$

With Q_1 *off* [Fig. 5-53(a)],

$$V_{CE} \approx V_{CC}$$

In Fig. 5-53(b) V_S is positive, and it biases Q_1 *on* to the maximum possible I_C level. The collector current is limited only by the

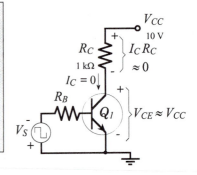

(a) *OFF*-biased transistor

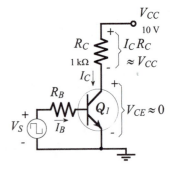

(b) *ON*-biased transistor

Figure 5-53
Direct-coupled transistor switching circuit. When V_B is zero or negative, $I_C = 0$ and $V_{CE} \approx V_{CC}$. When V_B is positive, I_B flows driving the transistor into saturation.

collector supply voltage (V_{CC}) and the collector resistor (R_C). So,

$$I_C R_C \approx V_{CC} \qquad \text{(5-25)}$$

and, $$V_{CE} = V_{CC} - (I_C R_C) \approx 0$$

Now consider Fig. 5-54, which shows the output characteristics and *dc* load line for the switching circuit in Fig. 5-53. The load line is drawn by the usual process of plotting point A at ($I_C = 0$ and $V_{CE} = V_{CC}$), and point B at ($V_{CE} = 0$ and $I_C = V_{CC}/R_C$), [see Section 5-1].

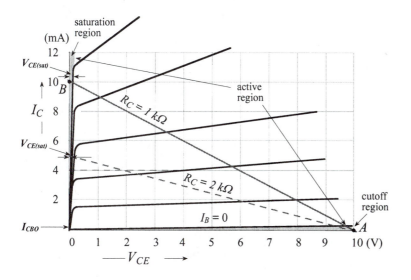

Figure 5-54
DC load line for a transistor switching circuit. When ON, the transistor is operating in the saturation region. When OFF, it is in the cutoff region.

The 1 kΩ load line shows that, when $I_B = 0$, I_C is close to zero at,

$$I_C = I_{CBO}$$

At this point the transistor is said to be *cut off*. The region of the characteristics below $I_B = 0$ is termed the *cutoff region*. When I_C is at its maximum level, the transistor is said to be *saturated*, and the collector-emitter voltage is the *saturation voltage* ($V_{CE(sat)}$).

$$V_{CE} = V_{CE(sat)}$$

The region of the transistor characteristics at $V_{CE(sat)}$ is termed the *saturation region*. The region between saturation and cutoff is the *active region*, which is where a transistor is normally biased for amplification. From the load line, it is seen that $V_{CE(sat)}$ is dependent upon the I_C level. For the 2 kΩ load line shown dashed in Fig. 5-54, $V_{CE(sat)}$ is smaller than for $R_C = 1$ kΩ.

Returning to Fig. 5-53(b), it is seen that I_B is a constant quantity,

$$I_B = \frac{V_S - V_{BE}}{R_B} \qquad \text{(5-26)}$$

Also, the level of I_C depends upon I_B,

$$I_C = h_{FE} \times I_B$$

The collector current can be determined from Eq. 5-25 (which assumes that $V_{CE(sat)} = 0$), and the base current can be calculated from Eq. 5-26. Then, the minimum required current gain for the transistor is,

$$h_{FE(1)} = \frac{I_C}{I_B} \qquad\qquad (5\text{-}27)$$

If the transistor h_{FE} value is less than the calculated $h_{FE(1)}$ value, I_C will be lower than the level required for transistor saturation. If the actual h_{FE} of the transistor is greater than the calculated $h_{FE(1)}$, I_C tends to be greater than the current level required to saturate the transistor. However, I_C cannot exceed V_{CC}/R_C. Consequently, an h_{FE} value larger than $h_{FE(1)}$ will adjust down to $h_{FE(1)}$. An h_{FE} value lower than $h_{FE(1)}$ cannot adjust up. So, to ensure that the transistor in a switching circuit saturates, it must have an h_{FE} value equal to or greater than the calculated $h_{FE(1)}$ for the circuit.

$V_{CE(sat)}$ is typically 0.2 V for a low-current silicon transistor, while V_{BE} is typically 0.7 V. Consider the circuit in Fig. 5-53(b) once again. If $V_{BE} = 0.7$ V and $V_{CE(sat)} = 0.2$ V, the transistor base is 0.5 V more positive than the collector,(see Fig. 5-55). This means that the collector-base junction, which is usually reverse biased, is forward biased when the transistor is in saturation. With the collector-base junction forward biased, fewer charge carriers from the emitter are drawn across to the collector, and the device current gain is lower than normal.

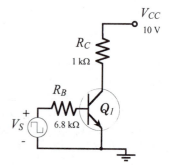

Figure 5-55
When a transistor is in saturation, the CB junction is forward biased.

Example 5-20
Calculate the minimum h_{FE} for the transistor in the circuit in Fig. 5-56 when $V_{CC} = 10$ V, $R_C = 1$ kΩ, $R_B = 6.8$ kΩ, and $V_S = 5$ V. Also, determine the transistor V_{CE} level when $h_{FE} = 10$.

Solution
$h_{FE(1)}$ calculation
From Eq. 5-25,

$$I_C \approx \frac{V_{CC}}{R_C} = \frac{10\text{ V}}{1\text{ k}\Omega}$$

$$= 10\text{ mA}$$

Eq. 5-26,

$$I_B = \frac{V_S - V_{BE}}{R_B} = \frac{5\text{ V} - 0.7\text{ V}}{6.8\text{ k}\Omega}$$

$$= 632\ \mu\text{A}$$

Eq. 5-27,

$$h_{FE(1)} = \frac{I_C}{I_B} = \frac{10\text{ mA}}{632\ \mu\text{A}}$$

$$= 15.8$$

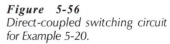

Figure 5-56
Direct-coupled switching circuit for Example 5-20.

When $h_{FE} = 10,$

$$I_C = h_{FE} I_B = 10 \times 632 \, \mu A$$
$$= 6.32 \text{ mA}$$
$$V_{CE} = V_{CC} - (I_C R_C) = 10 \text{ V} - (6.32 \text{ mA} \times 1 \text{ k}\Omega)$$
$$= 3.68 \text{ V}$$

Capacitor-Coupled Switching Circuit

The base bias circuit in Fig. 5-57 is similar to circuits of that type that have already been considered, with the exception that the transistor is biased into saturation. Although too unpredictable for biasing amplifier circuits, base bias is quite satisfactory for switching circuits. The transistor in Fig. 5-57 is in a normally-*on* state with $V_{CE} = V_{CE(sat)}$. The *capacitor-coupled* (pulse waveform) input turns the device *off*, giving $V_{CE} = V_{CC}$.

The transistor base current is,

$$I_B = \frac{V_{CC} - V_{BE}}{R_B} \qquad \text{(5-28)}$$

The collector current can be determined From Eq. 5-25, and then the minimum required h_{FE} value can be calculated from Eq. 5-27.

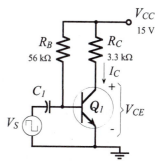

Figure 5-57
Capacitor-coupled switching circuit with the transistor biased ON into saturation.

Example 5-21

A new transistor is to be substituted in place of Q_1 in the circuit in Fig. 5-57. Calculate the minimum required h_{FE} for the transistor.

Solution

From Eq. 5-25,

$$I_C \approx \frac{V_{CC}}{R_C} = \frac{15 \text{ V}}{3.3 \text{ k}\Omega}$$
$$= 4.55 \text{ mA}$$

Eq. 5-28,

$$I_B = \frac{V_{CC} - V_{BE}}{R_B} = \frac{15 \text{ V} - 0.7 \text{ V}}{56 \text{ k}\Omega}$$
$$= 255 \, \mu A$$

Eq. 5-27,

$$h_{FE(1)} = \frac{I_C}{I_B} = \frac{4.55 \text{ mA}}{255 \, \mu A}$$
$$= 17.8$$

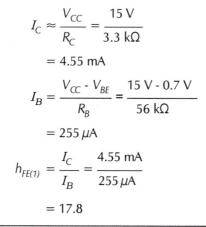

Another type of capacitor-coupled switching circuit is illustrated in Fig. 5-58. In this case, resistor R_B keeps the transistor base-emitter voltage at zero, to ensure that the device is in a normally-*off* state. The capacitor-coupled input voltage biases the transistor *on* into saturation.

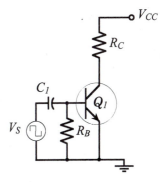

Figure 5-58
Capacitor-coupled switching circuit with the transistor biased OFF.

Switching Circuit Design

The resistance of R_C for any one of the switching circuits discussed can be calculated by using the specified V_{CC} and I_C levels with Eq. 5-25. The transistor $h_{FE(min)}$ value can be used with I_C to determine the minimum I_B level required for transistor saturation. However, concern about the actual transistor current gain can be avoided by using an h_{FE} value of 10. It is very unlikely that the transistor will have an $h_{FE(min)}$ less than 10, so a circuit designed in this way will work with virtually any low-current transistor.

The resistance of R_B for the direct-coupled circuit in Fig. 5-53 is calculated from Eq. 5-26. For the normally-*on* capacitor-coupled circuit in Fig. 5-57, Eq. 5-28 is used for determining the R_B value.

A different approach must be taken for calculating a suitable resistance for R_B in the normally-*off* capacitor-coupled circuit in Fig. 5-58. The current that flows through R_B when the transistor is *off*, and the allowable voltage drop across R_B must be considered, (see Fig. 5-59). The resistor current is the collector-base leakage current (I_{CBO}), and the maximum voltage drop produced by I_{CBO} must be much smaller than the normal transistor V_{BE} level when the device is *on*. I_{CBO} at maximum transistor temperature is unlikely to exceed 5 μA, and a V_{RB} of 0.1 V will normally keep the transistor biased *off*. So, a maximum resistance value for R_B is,

$$R_B = \frac{0.1\text{ V}}{5\text{ μA}} = 20\text{ k}\Omega$$

Typically, an R_B value of 22 kΩ or lower is suitable for keeping the transistor biased *off*.

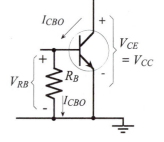

Figure 5-59
A voltage drop is produced across R_B by the collector-base leakage current (I_{CBO}).

Example 5-22

The circuit in Fig. 5-60 uses $V_{CC} = 12$ V, the collector current is to be approximately 1.5 mA, and $V_S = +5$ V. Determine suitable resistances for R_B and R_C.

Solution

From Eq. 5-25,

$$R_C \approx \frac{V_{CC}}{I_C} = \frac{12\text{ V}}{1.5\text{ mA}}$$

$$= 8\text{ k}\Omega \text{ (use 8.2 k}\Omega \text{ standard value)}$$

From Eq. 5-27,

$$I_B = \frac{I_C}{h_{FE(min)}} = \frac{1.5\text{ mA}}{10}$$

$$= 150\text{ μA}$$

From Eq. 5-26,

$$R_B = \frac{V_S - V_{BE}}{I_B} = \frac{5\text{ V} - 0.7\text{ V}}{150\text{ μA}}$$

$$= 28.6\text{ k}\Omega \text{ (Use 27 k}\Omega \text{ to ensure that } I_B \text{ is}$$
$$\text{larger than required for saturation)}$$

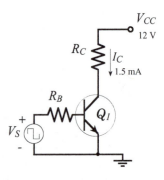

Figure 5-60
Direct-coupled switching circuit for Example 5-22.

Example 5-23

Determine suitable resistor values for the capacitor-coupled switching circuit in Fig. 5-61.

Solution

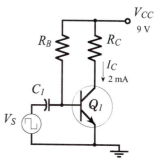

From Eq. 5-25,

$$R_C \approx \frac{V_{CC}}{I_C} = \frac{9 \text{ V}}{2 \text{ mA}}$$

$$= 4.5 \text{ k}\Omega \text{ (use 4.7 k}\Omega \text{ standard value)}$$

From Eq. 5-27,

$$I_B = \frac{I_C}{h_{FE(min)}} = \frac{2 \text{ mA}}{10}$$

$$= 200 \, \mu A$$

From Eq. 5-28,

$$R_B = \frac{V_{CC} - V_{BE}}{I_B} = \frac{9 \text{ V} - 0.7 \text{ V}}{200 \, \mu A}$$

$$= 41.5 \text{ k}\Omega \text{ (Use 39 k}\Omega\text{)}$$

Figure 5-61
Capacitor-coupled   switching
circuit for Example 5-23.

Practise   Problems

5-10.1   A direct-coupled transistor switching circuit as in Fig. 5-56 has $V_{CC} = $ 15 V, $V_S = $ 9 V, $R_C = $ 3.3 kΩ,and $R_B = $ 22 kΩ. Calculate the minimum h_{FE} for the transistor.

5-10.2   A normally-*on* capacitor-coupled switching circuit (as in Fig. 5-57) uses a transistor with $h_{FE(min)} = $ 25. The supply voltage is 12 V and R_C = 4.7 kΩ. Determine a suitable resistance for R_B.

5-10.3   A direct-coupled transistor switching circuit has $V_{CC} = $ 5 V and $V_S = $ 3 V. Calculate suitable resistances for R_C and R_B to give $I_C = $ 2.5 mA.

Chapter-5 Review Questions

Section 5-1

5-1   Identify the components that constitute the *dc* load in a *BJT* bias circuit, Explain the procedure for drawing the *dc* load line on the transistor *CE* output characteristics.

5-2   Explain the selection of a *Q*-point for a transistor bias circuit, and discuss the limitations on the output voltage swing.

Section 5-2

5-3   Sketch a base bias circuit, (a) using an *npn* transistor, (b) using a *pnp* transistor. In each case show the polarity of V_{CC}, V_{BE}, and V_{CE}. Also, show the direction of I_C and I_B.

5-4   Explain the operation of the base bias circuits drawn for Question 5-3, and write equations for I_B, I_C, and V_{CE}.

Section 5-3

5-5   Sketch a collector-to-base bias circuit using an *npn* transistor. Show the polarity of V_{CC}, V_{BE}, and V_{CE}. Also, identify the direction of I_C and I_B.

5-6   Explain the operation of the collector-to-base bias circuit drawn for Question 5-5. Write equations for I_B, I_C, and V_{CE}.

5-7   Sketch a collector-to-base bias circuit that uses a *pnp* transistor. Show all voltage polarities and current directions, and discuss the circuit operation.

Section 5-4

5-8   Sketch a voltage divider bias circuit using an *npn* transistor. Show all voltage polarities and current directions.

5-9   Explain the operation of the voltage divider bias circuit drawn for Question 5-8, and write approximate equations for V_B, I_E, I_C, and V_{CE}.

5-10   Explain the procedure for precise analysis of a voltage divider bias circuit.

5-11   Sketch a voltage divider bias circuit using an *pnp* transistor. Show all voltage polarities and current directions, and discuss the circuit operation.

Section 5-5

5-12   Compare base bias, collector-to-base bias, and voltage divider bias with regard to stability of the transistor collector voltage with spread in h_{FE} value. Discuss the advantages and disadvantages of the three types of bias circuit.

Section 5-6

5-13   List common errors involved in constructing and testing a transistor bias circuit.

5-14   Suggest possible errors in a base bias circuit, (a) when $V_C \approx V_{CC}$, (b) when $V_C \approx 0$.

5-15   Suggest possible errors in a collector-to-base bias circuit, (a) when $V_C \approx V_{CC}$, (b) when $V_C \approx 0$.

5-16   Suggest possible errors in a voltage divider bias circuit, (a) when $V_C \approx V_{CC}$, (b) when $V_C \approx V_E$.

Section 5-7

5-17   Write the equations for calculating R_B and R_C for a base bias circuit.

5-18   Write equations for calculating R_B and R_C for a collector-to-base bias circuit.

5-19   For a voltage divider bias circuit, write equations for calculating R_E and R_C, and the voltage divider resistors.

Section 5-8

5-20 Sketch a base bias circuit that uses an emitter resistor with an *npn* transistor. Briefly discuss the circuit operation.

5-21 Sketch an *npn* transistor bias circuit that uses a combination of voltage divider bias and collector-to-base bias. Show all voltage polarities and current directions, and explain the circuit operation.

5-22 Repeat Question 5-21 for a circuit that uses a *pnp* transistor.

5-23 Sketch an emitter current bias circuit using a plus and minus supply. Explain the circuit operation and compare it to voltage divider bias.

Section 5-9

5-24 Discuss the thermal stability of transistor bias circuit with regard to I_{CBO} and V_{BE}. State approximations for the variation in V_{BE} and I_{CBO} with temperature changes.

5-25 Define the stability factor (*S*) for a transistor bias circuit. Compare the three basic bias circuits with regard to thermal stability.

5-26 Show how a voltage divider bias circuit may be compensated for V_{BE} changes with temperature. Derive an equation for I_C.

Section 5-10

5-27 Draw a direct-coupled transistor switching circuit that may be turned *on* or *off* by application of an input voltage. Explain the circuit operation.

5-28 Sketch the typical output characteristics and *dc* load line for a transistor used as a switch. Identify the regions of the characteristics and briefly explain.

5-29 Sketch the circuit of a capacitor-coupled transistor switch biased in, (a) a normally-*on* condition, (b) a normally-*off* condition. Explain the operation of each circuit.

Chapter-5 Problems

Section 5-1

5-1 Plot the *dc* load line for the transistor circuit in Fig. 5-62 on the output characteristics shown in Fig. 5-63, (a) with V_{CC} = 15 V and R_C = 7.5 kΩ, (b) with V_{CC} = 12 V and R_C = 8 kΩ. Specify the *Q*-point in each case if I_B = 20 µA.

5-2 A circuit with the configuration and characteristics in Figs. 5-62 and 5-63 has V_{CC} = 10 V, R_C = 4.7 kΩ, and I_B = 10 µA. Draw the *dc* load line, and specify the circuit *Q*-point.

5-3 Determine the maximum symmetrical output voltage swing for each of the two circuits in Problem 5-1.

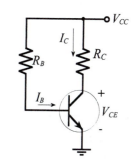

Figure 5-62
Circuit for Problem 5-1.

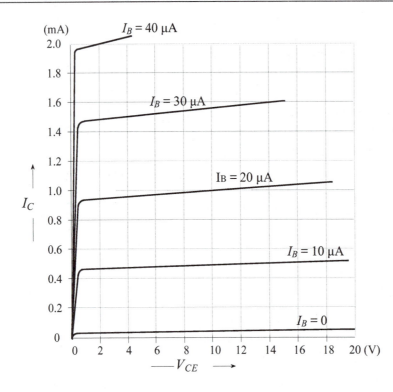

Figure 5-63
Transistor characteristics for Problem 5-1.

5-4   Determine the maximum symmetrical output voltage swing for the circuit in Problem 5-2.

5-5   Using the transistor characteristics in Fig. 5-63, draw the *dc* load line for the circuit in Fig. 5-64. Determine I_C for the circuit when $I_B = 10$ μA, and calculate V_{RC}, V_{RE}, and V_{CE}.

5-6   A circuit with the characteristics in Fig. 5-10 has $V_{CC} = 7.5$ V, $R_C = 1.2$ kΩ, and $I_B = 40$ μA. Draw the *dc* load line, specify the circuit Q-point, and determine the upper and lower limits of V_{CE}.

Section 5-2

5-7   A base bias circuit (as in Fig. 5-62) has $V_{CC} = 15$ V, $R_C = 1.8$ kΩ, and $R_B = 120$ kΩ. Assuming $V_{BE} = 0.7$ V and $h_{FE} = 50$, determine I_C and V_{CE}.

5-8   Determine the maximum and minimum I_C and V_{CE} levels for the circuit in Problem 5-7 when the transistor used has $h_{FE(max)} = 60$ and $h_{FE(min)} = 20$.

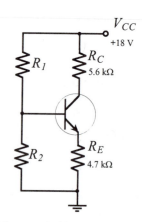

Figure 5-64
Circuit for Problem 5-5.

5-9   A base bias circuit has $V_{CC} = 20$ V, $R_C = 5.6$ kΩ, $R_B = 270$ kΩ, and $V_{CE} = 10$ V. Determine the transistor h_{FE} value, and calculate the new V_{CE} level when a transistor with $h_{FE} = 40$ is substituted.

5-10   A base bias circuit with $V_{CC} = 12$ V, $R_C = 3.9$ kΩ, and $R_B = 330$ kΩ has a measured V_{CE} of 3 V. Determine the transistor h_{FE} value.

Section 5-3

5-11   A collector-to-base bias circuit (as in Fig. 5-65) has V_{CC} = 15 V, R_C = 1.8 kΩ, R_B = 39 kΩ, and h_{FE} = 50. Determine the I_C and V_{CE} levels.

5-12   Calculate the maximum and minimum levels of I_C and V_{CE} for the circuit in Problem 5-11 when the transistor used has $h_{FE(max)}$ = 60 and $h_{FE(min)}$ = 20.

5-13   A collector-to-base bias circuit has V_{CC} = 15 V, R_C = 5.6 kΩ, R_B = 82 kΩ, and V_{CE} = 5 V. Determine the transistor h_{FE} value, and calculate the new V_{CE} level when a transistor with h_{FE} = 50 is substituted.

5-14   Determine the new I_C and V_{CE} levels for the circuit in Problem 5-11 when the supply voltage is changed to 20 V.

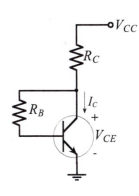

Figure 5-65
Circuit for Problem 5-11.

Section 5-4

5-15   The voltage divider bias circuit in Fig. 5-66, has V_{CC} = 15 V, R_1 = 6.8 kΩ, R_2 = 3.3 kΩ, R_3 = 900 Ω, R_4 = 900 Ω, and h_{FE} = 50. Analyze the circuit approximately to determine the levels of I_C and V_{CE}.

5-16   Precisely analyze the circuit in Problem 5-15 to determine the maximum and minimum levels of I_C and V_{CE} when $h_{FE(max)}$ = 60 and $h_{FE(min)}$ = 20.

5-17   A voltage divider bias circuit with a 25 V supply has R_C = 4.7 kΩ, R_E = 3.3 kΩ, R_1 = 33 kΩ, R_2 = 12 kΩ, and h_{FE} = 50. Use the approximate analysis method to calculate the V_{CE} level.

5-18   Precisely analyze the circuit in Problem 5-17 to determine the levels of I_C and V_{CE} when the transistor has h_{FE} = 60.

5-19   A transistor circuit using voltage divider bias has the following components: R_C = 2.2 kΩ, R_E = 3.3 kΩ, R_1 = 6.8 kΩ, R_2 = 4.7 kΩ. The supply voltage is V_{CC} = 15 V. Analyze the circuit approximately to determine V_{CE}.

5-20   A voltage divider bias circuit has V_{CC} = 15 V, R_C = 2.7 kΩ, R_E = 2.2 kΩ, R_1 = 22 kΩ, R_2 = 12 kΩ. Calculate the V_{CE} level.

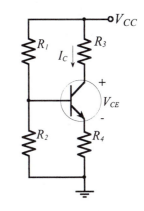

Figure 5-66
Circuit for Problem 5-15.

Section 5-5

5-21   Using the transistor characteristics in Fig. 5-10, plot the *dc* load line and maximum and minimum *Q*-points for the circuits in Problems 5-8, 5-12, 5-16.

Section 5-6

5-22   Analyze the circuit in Problem 5-7 to investigate the effect of interchanging the base and collector resistors.

5-23   Analyze the circuit in Problem 5-11 to investigate the effect of interchanging the base and collector resistors.

5-24   Analyze the circuit in Problem 5-15 to investigate the effect of interchanging the voltage divider resistors.

Section 5-7

5-25   A base bias circuit with V_{CC} = 18 V uses a transistor with the characteristics in Fig. 5-63. The circuit is to have V_{CE} = 9 V and I_C = 2 mA. Plot the Q-point, draw the dc load line, and determine the required resistance for R_C.

5-26   A base bias circuit with a 12 V supply uses a transistor with h_{FE} = 70. Design the circuit to have I_C = 2 mA and V_{CE} = 9 V.

5-27   Analyze the circuit designed for Problem 5-26 to determine the maximum and minimum levels of V_{CE} when the transistor type used has $h_{FE(min)}$ = 30 and $h_{FE(max)}$ = 200.

5-28   A base bias circuit has V_{CC} = 20 V, R_C = 6.8 kΩ, and the transistor h_{FE} = 120. Calculate the required base resistance value to give V_{CE} = 5 V.

5-29   Design a collector-to-base bias circuit to have I_C = 3.5 mA and V_{CE} = 12 V. The supply voltage is V_{CC} = 20 V, and the transistor h_{FE} = 80.

5-30   Analyze the circuit designed for Problem 5-29 to determine the maximum and minimum levels of V_{CE} when the transistor type used has $h_{FE(min)}$ = 20 and $h_{FE(max)}$ = 100.

5-31   A collector-to-base bias circuit is to have I_C = 3 mA and V_{CE} = 10 V. The supply voltage is V_{CC} = 25 V, and the transistor h_{FE} = 80. Calculate suitable resistor values.

5-32   A collector-to-base bias circuit has V_{CC} = 30 V, R_C = 8.2 kΩ, and the transistor h_{FE} = 100. Calculate the required base resistance value to give V_{CE} = 7 V.

5-33   Design a voltage divider bias circuit to have V_{CE} = 10 V and I_C = 1 mA. The supply is V_{CC} = 30 V.

5-34   Precisely analyze the circuit designed for Problem 5-33 to determine the maximum and minimum levels of V_{CE} when the transistor type used has $h_{FE(min)}$ = 40 and $h_{FE(max)}$ = 80.

5-35   A voltage divider bias circuit with V_{CC} = 20 V and R_C = 6 kΩ uses a transistor with h_{FE} = 80. Calculate suitable resistor values to give V_{CE} = 8 V.

5-36   Design a voltage divider bias circuit using a 2N3906 transistor. The supply voltage is 22 V, the collector current is to be 10 mA, and the collector-emitter voltage is to be 4 V.

5-37   Precisely analyze the circuit in Problem 5-35 to determine the levels of V_{CE} and I_C.

5-38   Design a voltage divider bias circuit to have V_{CE} = 4 V, V_E = 5 V, and I_C = 1.3 mA. the supply is V_{CC} = 18 V.

Section 5-8

5-39   Design an emitter current bias circuit (as in Fig. 5-67) to have V_{CE} = 6 V and I_C = 10 mA. The supply is V_{CC} = ±18 V and the transistor has h_{FE} = 100.

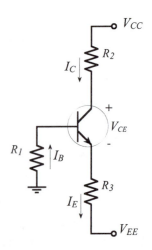

Figure 5-67
Emitter current bias circuit for Problem 5-39.

5-40 Precisely analyze the circuit in Problem 5-39 to determine the levels of V_{CE} and I_C.

5-41 An emitter current bias circuit has a 10 kΩ emitter resistor and V_{CC} = ± 20 V. Calculate suitable base and collector resistor values to give V_{CE} = 8 V when h_{FE} = 75.

5-42 Determine suitable resistor values for the circuit in Fig. 5-68.

5-43 A circuit employing collector-to-base bias and voltage divider bias as in Fig. 5-68 uses a 30 V supply. Design the circuit to have V_{CE} = 10 V and I_C = 10 mA.

Section 5-9

5-44 Determine the stability factors for the circuit in Problems 5-7, 5-11, and 5-15.

5-45 Calculate the I_C changes produced in each of the circuits referred to in Problem 5-44 when the circuit temperature changes from 25°C to 75°C. The transistor collector-base leakage current is specified as 25 nA at 25°C.

5-46 The voltage divider bias circuit in Problems 5-33 and 5-34 is to have a maximum I_C change of 15 μA due to change in I_{CBO} when the circuit temperature increases from 25°C to 35°C. Determine the maximum acceptable level of I_{CBO} for the transistor at 25°C.

5-47 Determine the I_C change produced in the circuits designed in Problems 5-33, 5-35, and 5-39 by the effect of transistor V_{BE} changes over a temperature range of 25°C to 75°C.

5-48 A voltage divider bias circuit is to have a maximum I_C change of 50 μA due to change in V_{BE} when the circuit temperature increases from 25°C to 75°C. Determine the minimum acceptable emitter resistance value.

Section 5-10

5-49 Calculate $h_{FE(min)}$ for the transistor in the direct-coupled switching circuit in Fig. 5-69.

5-50 The normally-*on* capacitor-coupled switching circuit in Fig. 5-70 uses a transistor with $h_{FE(min)}$ = 40. Determine a suitable resistance for R_B.

5-51 A direct-coupled transistor switching circuit (as in Fig. 5-69) with V_{CC} = 9 V and V_S = 6 V is to have I_C = 5 mA. Calculate suitable resistances for R_C and R_B.

5-52 A direct-coupled transistor switching circuit has V_{CC} = 9 V and V_S = 1.6 V. The resistor values are R_C = 2.7 kΩ and R_B = 4.7 kΩ. Calculate the minimum transistor h_{FE} value for saturation. Will the transistor be saturated if R_C is changed to 1 kΩ, and h_{FE} = 40 ?

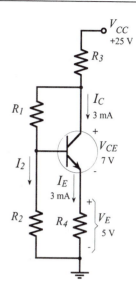

Figure 5-68
Circuit for Problem 5-42.

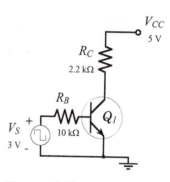

Figure 5-69
Circuit for Problem 5-49.

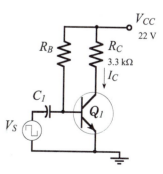

Figure 5-70
Circuit for Problem 5-50.

5-53 The normally-*off* capacitor-coupled switching circuit in Fig. 5-71 uses a transistor with $h_{FE(min)} = 40$. Determine suitable resistances for R_B and R_C.

5-54 A normally-*on* capacitor-coupled switching circuit has $V_{CC} = 12$ V and $R_C = 3.9$ kΩ. Determine a suitable resistance for R_B.

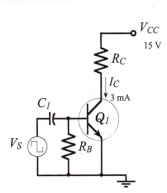

Figure 5-71
Circuit for Problem 5-53.

Practise Problem Answers

5-1.1   Q-point (40 μA, 4 V, 3.6 mA)
5-1.2   Q-point (10 μA, 0.5 mA, 9 V), $\Delta V_o = \pm 6$ V
5-1.3   Q-point (g0 μA, 6.25 mA, 4.5 V), $\Delta V_o \approx \pm 4.5$ V
5-2.1   (9.2 mA, 2.76 mA), (11.93 V, 0 V)
5-2.2   71, 4.3 V
5-3.1   (5.26 mA, 2.97 mA), (11.38 V, 6.38 V)
5-3.2   81, 7.93 V
5-4.1   1.06 mA, 15.3 V, 10.3 V
5-4.2   11.8 V, 10.7 V
5-4.3   4.26 V, 12.7 V
5-7.1   470 kΩ, 2.7 kΩ
5-7.2   150 kΩ, 6.8 kΩ
5-7.3   100 kΩ, 39 kΩ, 8.2 kΩ, 3.9 kΩ
5-7.4   (56 kΩ + 5.6 kΩ), 27 kΩ, 3.9 kΩ, 2.7 kΩ
5-8.1   (560 kΩ + 15 kΩ), 5.6 kΩ, 3.3 kΩ
5-8.2   22 kΩ, 27 kΩ, 2.7 kΩ, 2.2 kΩ
5-8.3   2.7 kΩ, 1.2 kΩ, 3.3 kΩ
5-9.1   126, 18.6, 7.5
5-9.2   1.29 mA, 190 μA, 76.5 μA
5-9.3   51.7 μA
5-10.1 12.1
5-10.2 100 kΩ
5-10.3 2.2 kΩ, 8.2 kΩ

Chapter **6**

AC Analysis of BJT Circuits

Chapter Contents

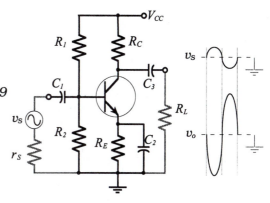

Objectives

You will be able to:

1   Explain the need for coupling and bypass capacitors in transistor circuits, and draw ac equivalents for circuits containing capacitors.

2   Draw ac load lines for basic transistor circuits.

3   Sketch and explain transistor r-parameter and h-parameter models.

4   Show how h-parameters can be derived from device characteristics.

5   Convert between CE, CC, and CB h-parameters, and between h-parameters and r-parameters.

6   Sketch h-parameter circuits for various CE, CC, and CB transistor circuits.

7   Analyze various CE, CC, and CB transistor circuits to determine input resistance, output resistance, voltage gain, current gain, and power gain.

8   Compare the performance of CE, CC, and CB circuits.

Introduction

A transistor circuit used as an *ac* amplifier normally has the signal source capacitor-coupled to the circuit input terminal. A load resistor at the output is also usually capacitor-coupled to the circuit. Capacitor-coupling ensures that the source and load do not affect the *dc* bias conditions in the circuit. An *ac* load line can be drawn on the transistor output characteristics to study the *ac* performance of a circuit. All transistor circuits, however complicated, are based on one of the three basic circuits configurations — common-emitter, common-collector, or common-base. The three basic circuits have different input impedances, output impedances, and voltage gains, and these are most easily determined by the use of *h*-parameters. Multistage transistor circuit can also be analyzed by application of *h*-parameters.

6-1   Coupling and Bypassing Capacitors

Coupling Capacitors

To use a transistor circuit to amplify or otherwise process an *ac* signal, the signal source must be connected to the circuit input. If the source is directly connected to the input, as illustrated in Fig. 6-1(a), the circuit bias conditions will be altered. Figure 6-1(b) shows that the signal source resistance (r_s) is in parallel with voltage divider resistor R_2 when the signal is directly connected to the circuit. Thus, the bias voltage is altered to,

$$V_B = \frac{V_{CC} \times (r_s \| R_2)}{R_1 + (r_s \| R_2)}$$

instead of, $$V_B = \frac{V_{CC} \times R_2}{R_1 + R_2}$$

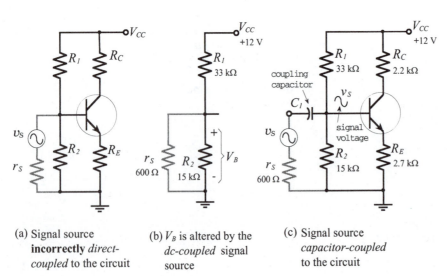

(a) Signal source
incorrectly *direct-coupled* to the circuit

(b) V_B is altered by the *dc-coupled* signal source

(c) Signal source *capacitor-coupled* to the circuit

Figure 6-1
AC input signals must be capacitor-coupled to the input terminal of a transistor circuit. Direct coupling would alter the circuit bias conditions.

Figure 6-1(c) shows the use of capacitor C_1 to couple the signal source to the circuit input. Because C_1 is an open-circuit to direct currents, r_S does not affect the level of V_B. Capacitor C_1 behaves as a short-circuit for the *ac* signals, so that the signal voltage (v_s) appears at the transistor base, as illustrated. In this case, the signal is said to be *ac coupled* to the circuit input, and C_1 is referred to as a *coupling-capacitor*. Input coupling capacitors are normally used with all types of bias circuits, otherwise the circuit bias conditions will be altered.

A coupling capacitor is usually required at the output of a transistor circuit (as well as at the input) to couple to a load resistor, or to another amplification stage. Figures 6-2(a) and (b) show the effect of directly coupling a load (R_L) to the circuit output. The supply voltage at the transistor collector terminal is reduced from V_{CC} to,

$$V = \frac{V_{CC} \times R_L}{R_C + R_L}$$

and the collector resistance becomes,

$$R = R_C \| R_L$$

This has the effect of altering the circuit *dc* load line and *Q*-point.

The use of an output coupling capacitor (C_2) is illustrated in Fig. 6-2(c). Like the input coupling capacitor, C_2 offers a *dc* open-circuit and behaves as an *ac* short-circuit. Thus, it passes the output waveform to the load without affecting the circuit bias conditions.

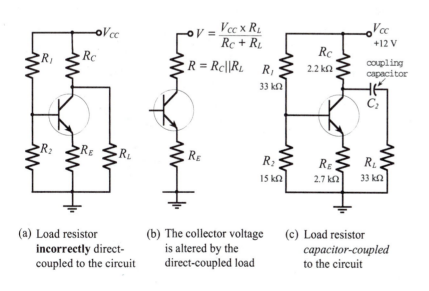

(a) Load resistor **incorrectly** direct-coupled to the circuit

(b) The collector voltage is altered by the direct-coupled load

(c) Load resistor *capacitor-coupled* to the circuit

Figure 6-2
Output loads must be capacitor-coupled to the output terminal of a transistor circuit. Direct coupling will alter the circuit bias conditions.

Example   6-1

Calculate the base bias voltage for the circuit in Fig. 6-1(c) when no signal source is present, and when the signal source is directly connected as in Fig. 6-1(a).

Solution

With no signal source,

$$V_B = \frac{V_{CC} \times R_2}{R_1 + R_2} = \frac{12\text{ V} \times 15\text{ k}\Omega}{33\text{ k}\Omega + 15\text{ k}\Omega}$$

$$= 3.75\text{ V}$$

With the signal source directly connected,

$$V_B = \frac{V_{CC} \times (r_S \| R_2)}{R_1 + (r_S \| R_2)} = \frac{12\text{ V} \times (600\ \Omega \| 15\text{ k}\Omega)}{33\text{ k}\Omega + (600\ \Omega \| 15\text{ k}\Omega)}$$

$$= 0.21\text{ V}$$

AC Degeneration

In the discussion on collector-to-base bias, it was explained that an increase in I_B would produce a decrease in V_{CE}, and this would tend to cancel the original increase in I_B. This, of course, is an effect that produces good bias stability, however, this same reaction occurs when an *ac* signal is applied to the circuit input for amplification. The voltage change at the transistor collector (produced by the *ac* input) is fed back to the base where it tends to partially cancel the signal. The effect is termed *ac degeneration*, and it can result in a very low voltage gain.

Figure 6-3(a) shows how *ac* degeneration is eliminated in collector-to-base bias circuits. R_B is replaced with two approximately equal resistors (R_{B1} and R_{B2}), which add up to R_B. A *bypass capacitor* (C_B) is connected from the junction of R_{B1} and R_{B2} to the ground terminal, as illustrated. Capacitor C_B offers a short-circuit to *ac* signals, so that there is no feedback from the transistor collector to the base. Figure 6-3(b) shows that, when C_B is replaced with a short-circuit, R_{B1} and R_{B2} appear in parallel with the circuit input and output respectively. Because they are large-value resistors, they do not affect the circuit voltage gain.

Emitter Bypassing

Voltage divider bias circuits also suffer *ac* degeneration because of the presence of the emitter resistor (R_E). In this case, the *ac* signal voltage is developed across R_E and the transistor *BE* junction in series, instead of just across the *BE* junction alone. The problem is eliminated by the *emitter bypass* capacitor (C_E) illustrated in Fig. 6-4, which provides an *ac* short-circuit across R_E.

Capacitor polarity

Bypass capacitors are usually polarized (electrolytic-type), and coupling capacitors can also be polarized, so it is very important that capacitors are correctly connected. As discussed in Section 3-3, *incorrectly connected polarized capacitors can* **explode!** Also, electrolytic capacitors tend to have a high leakage current when incorrectly connected, so even if they do not explode, *they can affect circuit bias conditions.* The positive terminal must be connected to the more positive of the two circuit points where the capacitor is installed.

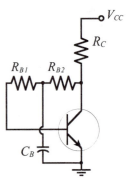

(a) Bypass capacitor used with collector-to-base bias

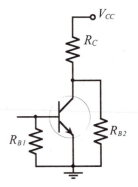

(b) The bypass capacitor is an *ac* short-circuit

Figure 6-3
A collector-to-base bias circuit must use two resistors for R_B with the junction of the two ac grounded via a capacitor. This is necessary to eliminate ac degeneration.

The capacitor positive terminal is represented by the straight line on the capacitor symbol, or is identified by the plus sign on the alternative capacitor symbol, (see Fig 6-4). The transistor emitter terminal in Fig. 6-4 is more positive than ground level, as illustrated. Consequently, the positive terminal of the emitter bypass capacitor is connected at the transistor emitter, and the negative terminal goes to ground.

Figure 6-5 shows a circuit with correctly-connected coupling and bypass capacitors. The *dc* voltage level at the right side of C_1 is +0.7 V, and the left side is grounded via the signal source. So, the polarity is *plus* on the right, *minus* on the left. The voltage at the junction of resistors R_{B1} and R_{B2} is +3.2 V, consequently, the positive terminal of C_2 must be connected at that point, and C_2 negative terminal is grounded. Output coupling capacitor C_3 has its left side at +5.7 V, and the right side is grounded via R_L, requiring the capacitor to be connected as illustrated.

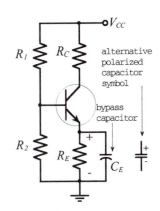

Figure 6-4
A resistor connected in series with the transistor emitter must be ac short-circuited by means of a bypass capacitor to ensure maximum amplification.

Practise Problems

6-1.1 Calculate the transistor collector voltage for the circuit in Fig. 6-2(c) with C_2 present, and with R_L directly connected.

6-2 AC Load Lines

AC Equivalent Circuits

Capacitors behave as short-circuits to *ac* signals, so in the *ac equivalent circuit* for a transistor circuit all capacitors must be replaced with short-circuits. Power supplies also behave as *ac* short-circuits, because the *dc* supply voltage is not affected by *ac* signals. Also, all power supplies have large-value capacitors at the output terminals, and these will offer short-circuits to *ac* signals. Substituting short-circuits in place of the power supply and all capacitors in the circuit in Fig. 6-6(a) gives the *ac* equivalent circuit in Fig. 6-6(b). If R_L is present, as shown, it appears in parallel with R_C in the *ac* equivalent circuit.

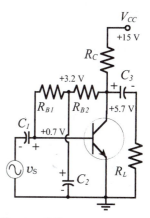

Figure 6-5
Polarized capacitors must be correctly connected, to avoid high leakage current and the possibility of capacitor explosion.

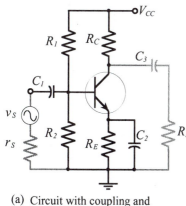

(a) Circuit with coupling and bypass capacitors

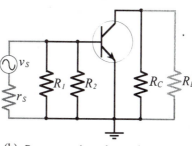

(b) Power supply and capacitors behave as *ac* short-circuits

Figure 6-6
Voltage divider bias circuit with coupling and bypass capacitors, and the ac equivalent circuit

Once the *ac* equivalent circuit is drawn, the circuit *ac* performance can be investigated by drawing an *ac* load line, and by substituting a transistor model in place of the device.

AC Load Lines

The *dc* load for the circuit in Fig. 6-6(a) is $(R_C + R_E)$, consequently, the *dc* load line is drawn (as discussed in Section 5-1) for a total resistance of $(R_C + R_E)$. Because the emitter resistor is capacitor bypassed in Fig. 6-6(a), resistor R_E is not part of the circuit *ac* load. If external load R_L were not present, the circuit *ac* load would simply be R_C. With R_L capacitor-coupled to the circuit output, the *ac* load is $R_C \| R_L$. An *ac load line* may now be drawn to represent the circuit *ac* performance.

When there is no input signal, the transistor voltage and current conditions are exactly as indicated by the *Q*-point on the *dc* load line. An *ac* signal causes the transistor voltage and current levels to vary above and below the *Q*-point. Therefore, the *Q*-point is common to both the *ac* and *dc* load lines. Starting from the *Q*-point, another point is found on the *ac* load line by taking a convenient collector current change (usually $\Delta I_C = I_{CQ}$) and calculating the corresponding collector-emitter voltage change (ΔV_{CE}), (see Fig. 6-8). As demonstrated in Ex. 6-2, *point C* is plotted by measuring ΔI_C and ΔV_{CE} from the *Q*-point. The *ac* load line is then drawn through points *C* and *Q*, as illustrated.

Example 6-2

Draw the *dc* and *ac* load lines for the transistor circuit in Fig. 6-7, using the transistor common-emitter characteristics in Fig. 6-8.

Solution

dc load line,

$$R_{L(dc)} = R_C + R_E = 2.2\ k\Omega + 2.7\ k\Omega$$
$$= 4.9\ k\Omega$$

The equation for V_{CE} is,

$$V_{CE} = V_{CC} - I_C (R_C + R_E)$$

When $I_C = 0$, $V_{CE} = V_{CC} = 20\ V$

Plot point A at, $I_C = 0$ and $V_{CE} = 20\ V$

When $V_{CE} = 0$, $I_C = \dfrac{V_{CC}}{R_C + R_E} = \dfrac{20\ V}{4.9\ k\Omega}$

$$= 4.08\ mA$$

Plot point B at, $I_C = 4.08\ mA$ and $V_{CE} = 0$
Draw the *dc* load line through points A and B.

$$V_B = \dfrac{V_{CC} \times R_2}{R_1 + R_2} = \dfrac{20\ V \times 8.2\ k\Omega}{18\ k\Omega + 8.2\ k\Omega}$$

$$= 6.3\ V$$

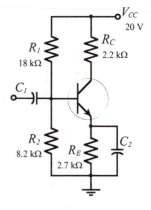

Figure 6-7
Voltage divider bias circuit with a bypassed emitter resistor. The dc load is $(R_C + R_E)$, and the ac load is R_C when there is no capacitor-coupled load.

$$V_E = V_B - V_{BE} = 6.3 \text{ V} - 0.7 \text{ V}$$
$$= 5.6 \text{ V}$$

$$I_C \approx I_E = \frac{V_E}{R_E} = \frac{5.6 \text{ V}}{2.7 \text{ k}\Omega}$$
$$= 2.07 \text{ mA}$$

Mark the Q-point on the dc load line at
$$I_C = 2.07 \text{ mA}.$$

ac load line,
when there is no external R_L,
$$R_{L(ac)} = R_C = 2.2 \text{ k}\Omega$$

When I_C changes by $\Delta I_C = 2.07$ mA,

$$\Delta V_{CE} = \Delta I_C \times R_C$$
$$= 2.07 \text{ mA} \times 2.2 \text{ k}\Omega$$
$$= 4.55 \text{ V}$$

Plot point C at, $\Delta I_C = 2.07$ mA and $\Delta V_{CE} = 4.55$ V from the Q point.
Draw the ac load line through points C and Q.

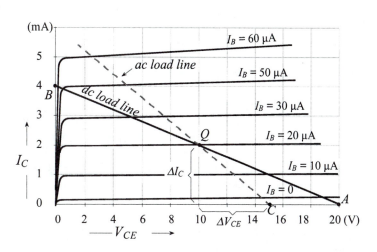

Figure 6-8
The ac load line for a transistor circuit is drawn through the Q-point.

The Output Voltage Swing

In Section 5-1 it is explained that the maximum symmetrical output voltage swing ($\pm V_{o(max)}$) from a common-emitter circuit depends upon the Q-point position on the dc load line. In the case of a circuit with a bypassed emitter resistor or a capacitor-coupled load, the maximum symmetrical output swing depends upon the Q-point position on the ac load line. For the ac load line shown in Fig. 6-10,

$$V_{o(max)} \approx \pm 4.5 \text{ V}$$

Practise Problems

6-2.1   Draw the new ac load line for the transistor circuit in Ex. 6-2, when a 5.6 kΩ external load is capacitor-coupled to the circuit output. Also, determine the new maximum possible symmetrical output voltage swing.

6-3   Transistor Models and Parameters

T-Equivalent Circuit

Because a transistor consists of two *pn*-junctions with a common centre block, it should be possible to use two *pn*-junction *ac* equivalent circuits as the transistor model. Figure 6-9 shows the *ac* equivalent circuit for a transistor connected in common-base configuration. Resistor r_e represents the *BE* junction resistance, r_c represents the *CB* junction resistance, and r_b represents the resistance of the base region which is common to both junctions. Junction capacitances C_{BE} and C_{BC} are also included.

If the transistor equivalent circuit is simply left as a combination of resistances and capacitances, it could not account for the fact that most of the emitter current flows out of the collector terminal as collector current. To represent this, a current generator is included in parallel with r_c and C_{BC}. The current generator is given the value αI_e, where $\alpha = I_c/I_e$.

The complete circuit is known as the *T-equivalent circuit*, or the *r-parameter equivalent circuit*. The equivalent circuit can be rearranged in common-emitter or common-collector configuration.

The currents in Fig. 6-9 are designated I_b, I_c, and I_e (instead of I_B, I_C, and I_E) to indicate that they are *ac* quantities rather than *dc*. The circuit parameters r_c, r_b, r_e, and α, are also *ac* quantities.

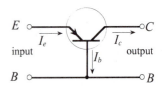

(a) *CB* connected transistor

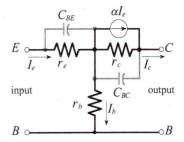

(b) *CB* equivalent circuit

Figure 6-9
The T-equivalent circuit, or r-parameter circuit for a transistor is essentially a combination of two pn-junction equivalent circuits.

r-Parameters

Referring to Fig. 6-9, r_e represents the *ac* resistance of the forward-biased *BE* junction, so it has a low resistance value, (typically 25 Ω). The resistance of the reverse-biased *CB* junction (r_c) is high, (typically 100 kΩ to 1 MΩ). The base region resistance (r_b) depends upon the doping density of the base material. Usually, r_b ranges from 100 Ω to 300 Ω.

C_{BE} is the capacitance of a forward-biased *pn*-junction, and C_{BC} is that of a reverse-biased junction, (see Section 2-6). At medium and low frequencies the junction capacitances may be neglected. Instead of the current generator (αI_e) in parallel with r_c, a voltage generator ($\alpha I_e r_c$) may be used in series with r_c. Two transistor *r*-parameter models are shown in Fig. 6-10.

Determination of r_e

Because r_e is the *ac* resistance of the *BJT* forward-biased base-emitter junction, it can be determined from a plot of I_E versus V_{BE}. As illustrated in Fig. 6-11,

$$r_e = \frac{\Delta V_{BE}}{\Delta I_E} \qquad (6\text{-}1)$$

This is similar to the determination of the dynamic resistance (r_d) for a forward-biased diode, (see Section 2-2). Also like the case of a diode, the *ac* resistance for the transistor *BE* junction can be calculated in terms of the current crossing the junction.

$$r'_e = \frac{26 \text{ mV}}{I_E} \qquad (6\text{-}2)$$

As in the case of r'_d for a diode, r'_e does not include the resistance of the device semiconductor material. Consequently, r'_e is slightly smaller than the actual measured value of r_e for a given transistor.

Equation 6-2 applies only to transistors at a temperature of 25°C. For determination of r'_e at higher or lower temperatures, the equation must be modified.

$$r'_e = \frac{26 \text{ mV}}{I_E} \left[\frac{T + 273°C}{298°C} \right] \qquad (6\text{-}3)$$

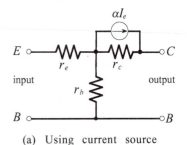

(a) Using current source

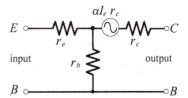

(b) Using voltage source

Figure 6-10
Transistor low-frequency r-parameter ac equivalent circuits.

h-Parameters

It has been shown that transistor circuits can be represented by an r-parameter or T-equivalent circuit. In circuits involving more than a single transistor, analysis by r-parameters can be virtually impossible. The *hybrid parameters*, or *h-parameters* are much more convenient for circuit analysis. These are used only for *ac* circuit analysis, although *dc* current gain factors are also expressed as h-parameters. Transistor h-parameter models simplify transistor circuit analysis by separating the input and output stages of a circuit to be analyzed.

In Fig. 6-12 a common-emitter *h-parameter equivalent circuit* is compared with a common-emitter r-parameter circuit. In each case an external collector resistor (R_C) is included, as well as a signal source voltage (v_s) and source resistance (r_s). Note that the output current generator in the r-parameter circuit has a value of αI_e, which equals βI_b.

The input to the h-parameter circuit is represented as an input resistance (h_{ie}) in series with a voltage source $(h_{re}v_{ce})$. Looking at the r-parameter circuit, it is seen that a change in output current I_c causes a voltage variation across r_e. This means that a voltage is *fed back* from the output to the input. In the h-parameter circuit this feedback voltage is represented as the portion h_{re} of the output voltage v_{ce}. The parameter h_{re} is appropriately termed the *reverse voltage transfer ratio*.

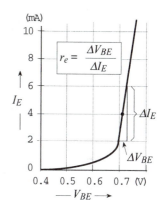

Figure 6-11
Determination of r_e from the transistor base-emitter junction forward characteristic.

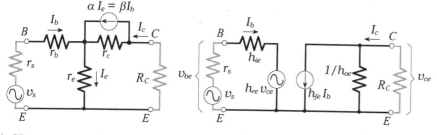

(a) CE r-parameter equivalent circuit   (b) CE h-parameter equivalent circuit

Figure 6-12
Comparison of common-emitter r-parameter and h-parameter equivalent circuits.

The output of the h-parameter circuit is represented as an output resistance $(1/h_{oe})$ in parallel with a current generator $(h_{fe}I_b)$, where I_b is the (input) base current. So, $h_{fe}I_b$ is produced by the input current I_b, and it divides between the device output resistance $1/h_{oe}$ and the collector resistor R_C. I_c is the current passed to R_C. This compares with the r-parameter equivalent circuit, where some of the generator current (βI_b) flows through r_c. The current generator parameter (h_{fe}) is termed the *forward transfer current ratio*. The *output conductance* is h_{oe}, so that $1/h_{oe}$ is a resistance.

r_π Equivalent Circuit

An approximate h-parameter model for a transistor CE circuit is shown in Fig. 6-13(a). In this case, the feedback generator $[h_{re}v_{ce}$ in Fig. 6-12(b)] is omitted. The effect of $h_{re}v_{ce}$ is normally so small that it can be neglected for most practical purposes.

The approximate *h*-parameter circuit is reproduced in Fig. 6-13(b) with the components labelled as *r*-parameters; $r_\pi = h_{ie}$, $\beta I_b = h_{fe}I_b$, and $r_c = 1/h_{oe}$. This circuit, known as a *hybrid-π model*, is sometimes used instead of the *h*-parameter circuit.

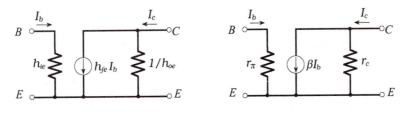

(a) *CE h*-parameter transistor model   (b) *CE r_π* transistor model

Figure 6-13
Transistor h-parameter and r_π equivalent circuits.

Definition of h-Parameters

The **e** in the subscript of h_{ie} identifies the parameter as a *common-emitter* quantity, and the **i** signifies that it is an *input* resistance. Common-base and common-collector input resistances are designated h_{ib} and h_{ic}, respectively.

As an *ac* input resistance, h_{ie} can be defined as the *ac* input voltage divided by the *ac* input current.

$$h_{ie} = \frac{V_{be}}{I_b}$$

This is usually stated as,

$$\boldsymbol{h_{ie}} = \frac{V_{be}}{I_b}\bigg|_{V_{CE}} \qquad \textbf{(6-4)}$$

This means that the collector-emitter voltage (V_{CE}) must remain constant when h_{ie} is measured.

The input resistance can also be defined in terms of changes in *dc* levels;

$$\boldsymbol{h_{ie}} = \frac{\Delta V_{BE}}{\Delta I_B}\bigg|_{V_{CE}} \qquad \textbf{(6-5)}$$

Equation 6-5 can be used for determining h_{ie} from the transistor common-emitter input characteristics. As illustrated in Fig. 6-14, ΔV_{BE} and ΔI_B are measured at one point on the characteristics, and h_{ie} is calculated.

The reverse transfer ratio h_{re} can also be defined in terms of *ac* quantities, or as the ratio of *dc* current changes. In both cases, the input current (I_B) must be held constant.

$$\boldsymbol{h_{re}} = \frac{\Delta V_{BE}}{\Delta V_{CE}}\bigg|_{I_B} \qquad \textbf{(6-6)}$$

The forward current transfer ratio h_{fe} can similarly be defined in terms of *ac* quantities, or as the ratio of *dc* current changes. In both cases, the output voltage (V_{CE}) must be held constant.

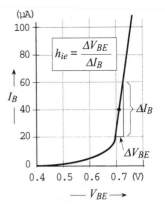

Figure 6-14
Derivation of h_{ie} from the CE input characteristics.

$$h_{fe} = \frac{I_c}{I_b}\bigg|_{V_{CE}} \qquad \text{(6-7)}$$

or,

$$h_{fe} = \frac{\Delta I_C}{\Delta I_B}\bigg|_{V_{CE}} \qquad \text{(6-8)}$$

Equation 6-8 can be used to determine h_{fe} from the CE current gain characteristics. Figure 6-15 shows the measurement of ΔI_C and ΔI_B at one point on the characteristics for the h_{fe} calculation.

The output conductance, h_{fe}, is the ratio of ac collector current to ac collector-emitter voltage, and its value can be determined from the common emitter output characteristics, (see Fig. 6-16).

$$h_{oe} = \frac{I_c}{V_{ce}}\bigg|_{I_B} \qquad \text{(6-9)}$$

or,

$$h_{oe} = \frac{\Delta I_C}{\Delta V_{CE}}\bigg|_{I_B} \qquad \text{(6-10)}$$

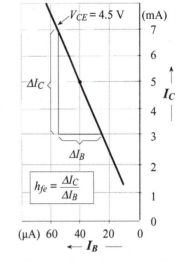

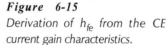

Figure   6-15
Derivation of h_{fe} from the CE current gain characteristics.

Example   6-3
Determine h_{fe} and h_{oe} for $V_{CE} = 4.5\ V$ and $I_B = 40\ \mu A$ from the transistor characteristics in Figs. 6-15 and 6-16.

Solution
From the current gain characteristics, at $V_{CE} = 4.5\ V$ and $I_B = 40\ \mu A$,

$$\Delta I_C = 4\ mA \text{ and } \Delta I_B \approx 30\ \mu A$$

Eq. 6-8,

$$h_{fe} = \frac{\Delta I_C}{\Delta I_B} = \frac{4\ mA}{30\ \mu A}$$

$$= 133$$

From the output characteristics, at $V_{CE} = 4.5\ V$ and $I_B = 40\ \mu A$,

$$\Delta I_C \approx 400\ \mu A \text{ and } \Delta V_{CE} = 6\ V$$

Eq. 6-9,

$$h_{oe} = \frac{\Delta I_C}{\Delta V_{CE}} = \frac{0.2\ mA}{6\ V}$$

$$= 33.3\ \mu S \ \text{(micro Siemens)}$$

$$1/h_{oe} = \frac{1}{33.3\ \mu S}$$

$$= 30\ k\Omega$$

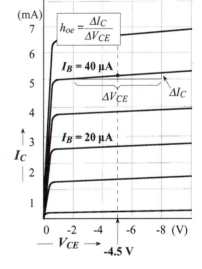

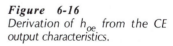

Figure   6-16
Derivation of h_{oe} from the CE output characteristics.

Common-Base   and   Common-Collector   h-Parameters
The common-base and common-collector h-parameters are defined similarly to common-emitter h-parameters. They may also be

derived from the *CB* and *CC* characteristics. Common-base parameters are identified as h_{ib}, h_{fb}, etc., and common-collector parameters are designated h_{ic}, h_{fc}, and so on. Common-emitter, common-base, and common-collector *h*-parameter equivalent circuits are further investigated in Sections 6-4 through 6-9.

Parameters Relationships

Device manufacturers do not list the values of all parameters on transistor data sheets. Usually, only the *CE* *h*-parameters are stated. However, *CB* and *CC* *h*-parameters can be determined from the *CE* *h*-parameters. *r*-parameters can also be calculated from the *CE* *h*-parameters. Table 6-1 shows the parameter relationships.

Table 6-1   Conversion from *CE* *h*-parameters to *CB* and *CC* *h*-parameters, and to *r*-parameters.

CE to CB h-parameters	CE to CC h-parameters	CE h-parameters to r-parameters
$h_{ib} \approx \dfrac{h_{ie}}{1+h_{fe}}$	$h_{ic} = h_{ie}$	$\alpha \approx \dfrac{h_{fe}}{1+h_{fe}}$
$h_{rb} \approx \dfrac{h_{ie}h_{oe}}{1+h_{fe}} - h_{re}$	$h_{rc} = 1 - h_{re}$	$r_c = \dfrac{1+h_{fe}}{h_{oe}}$
$h_{fb} \approx \dfrac{h_{fe}}{1+h_{fe}}$	$h_{fc} = 1+h_{fe}$	$r_e = \dfrac{h_{ie}}{1+h_{fe}} \approx r'_e$
$h_{ob} \approx \dfrac{h_{oe}}{1+h_{fe}}$	$h_{oc} = h_{oe}$	$r_b = h_{ie} - \dfrac{h_{re}(1+h_{fe})}{h_{oe}}$
		$r_\pi = h_{ie}$
		$\beta = h_{fe}$

Example 6-4

Calculate h_{fc}, h_{ob}, and α from the parameters determined in Example 6-3.

Solution

from Table 6-1, $h_{fc} = 1 + h_{fe} = 1 + 133$

$= 134$

$h_{ob} \approx \dfrac{h_{oe}}{1+h_{fe}} \approx \dfrac{33.3\,\mu S}{1+133}$

$= 249\ mS$

$\alpha \approx \dfrac{h_{fe}}{1+h_{fe}} \approx \dfrac{133}{1+133}$

$= 0.993$

Example 6-5

A transistor in a circuit has its current levels measured as; $I_B = 20\,\mu A$, and $I_C = 1$ mA. Estimate the CE input resistance. Also, determine r_π, and β.

Solution

Eq. 6-2,
$$r'_e = \frac{26\ mV}{I_E} \approx \frac{26\ mV}{1\ mA}$$
$$= 26\ \Omega$$

$$h_{fe} \approx \frac{I_C}{I_B} \approx \frac{1\ mA}{20\,\mu A}$$
$$= 50$$

$$h_{ie} = (1 + h_{fe})\,r'_e = (1 + 50) \times 26\ \Omega$$
$$\approx 1.33\ k\Omega$$

$$r_\pi = h_{ie} = 1.33\ k\Omega$$

$$\beta = h_{fe} = 50$$

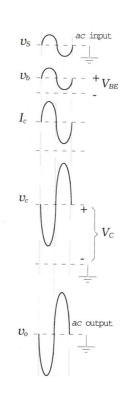

Figure 6-17
Transistor common-emitter amplifier with coupling and bypass capacitors.

Practise Problems

6-3.1   Determine h_{fe} and h_{oe} at $V_{CE} = 6$ V and $I_B = 20\ \mu A$ from the common-emitter output and current gain characteristics in Fig. 4-30.

6-3.2   Calculate α and r_c from the h-parameter values determined in Problem 6-3.1.

6-3.3   Calculate h_{ie} for a transistor at a temperature of 50°C when the emitter current is 1.3 mA and $h_{fe} = 80$.

6-4   Common-Emitter Circuit Analysis

Common-Emitter Circuit

Consider the transistor amplifier circuit shown in Fig. 6-17. When the capacitors are regarded as *ac* short-circuits, it is seen that the circuit input terminals are the transistor base and emitter, and the output terminals are the collector and the emitter. So, the emitter terminal is *common* to both input and output, and the circuit configuration is termed *common-emitter* (CE).

The current and voltage waveforms for the *CE* circuit in Fig. 6-17 are illustrated in Fig. 6-18. It is seen that there is a 180° *phase shift* between the input and output waveforms. This can be understood by considering the effect of a positive-going input signal. When v_s increases in a positive direction, it increases the transistor base-emitter voltage (V_{BE}). The increase in V_{BE} raises the level of I_C, thereby increasing the voltage drop across R_C, and thus reducing the level of the collector voltage (V_C). The changing level of V_C is capacitor-coupled to the circuit output to produce the *ac* output voltage (v_o). As v_s increases in a positive direction, v_o goes

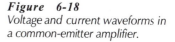

Figure 6-18
Voltage and current waveforms in a common-emitter amplifier.

in a negative direction, as illustrated. Similarly, when v_s changes in a negative direction, the resultant decrease in V_{BE} reduces the I_C level, thereby reducing V_{RC}, and producing a positive-going output.

The circuit in Fig. 6-17 has an *input impedance* (Z_i), and an *output impedance* (Z_o). These can cause voltage division of the circuit input and output voltages, as illustrated in Fig. 6-19. So, for most transistor circuits, Z_i and Z_o are important parameters. The circuit voltage amplification (A_v), or *voltage gain,* depends on the transistor parameters and on resistors R_C and R_L.

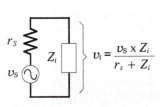

$$v_i = \frac{v_s \times Z_i}{r_s + Z_i}$$

(a) Division of signal voltage

h-Parameter Equivalent Circuit

The first step in *ac* analysis of a transistor circuit is to draw the *ac* equivalent circuit, by substituting short-circuits in place of the power supply and capacitors. When this is done for the circuit in Fig. 6-17, it gives the *ac* equivalent circuit shown in Fig. 6-20(a) [reproduced from Fig. 6-6(b)]. The *h*-parameter circuit is now drawn simply by replacing the transistor in the *ac* equivalent circuit with its *h*-parameter model, [from Fig. 6-12(b)]. This gives the CE *h*-parameter equivalent circuit in Fig. 6-20(b). Note that the feedback voltage generator $(h_{re}v_c)$ in Fig. 6-12(b) is omitted in Fig. 6-20(b). As discussed, the effect of $h_{re}v_c$ in a CE circuit is unimportant for most practical purposes.

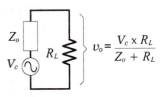

$$v_o = \frac{V_c \times R_L}{Z_o + R_L}$$

(b) Division of output voltage

Figure 6-19
Circuit input and output voltages are divided by the effects of Z_i and Z_o.

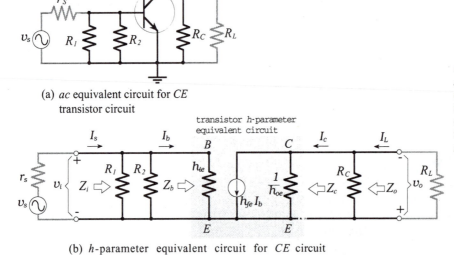

(a) *ac* equivalent circuit for CE transistor circuit

(b) *h*-parameter equivalent circuit for CE circuit

Figure 6-20
A common-emitter h-parameter circuit is drawn by first replacing the power supply and all capacitors with short-circuits, then the h-parameter model is substituted for the transistor.

The current directions and voltage polarities in Fig. 6-20(b) are those that occur when the instantaneous level of the input voltage is moving in a positive direction.

Input Impedance

The input section of the *h*-parameter circuit in Fig. 6-20(b) reproduced in Fig. 6-21 shows that the input impedance at the transistor terminals is,

$$Z_b \approx h_{ie} \qquad\qquad\qquad \text{(6-11)}$$

A typical value of h_{ie} for a low-current transistor is 1.5 kΩ. If h_{ie} is not known, it can be estimated by first calculating r'_e from Eq. 6-2 or Eq. 6-3, as appropriate. Then $r'_e = h_{ib}$, and $h_{ie} = (1 + h_{fe})h_{ib}$.

Z_b is the input impedance at the device base terminal. At the circuit input terminals, resistors R_1 and R_2 are seen to be in parallel with Z_b. So, the circuit input impedance is,

$$Z_i = R_1\|R_2\|Z_b \qquad (6\text{-}12)$$

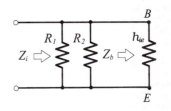

Figure 6-21
The transistor input impedance in a CE circuit is Z_b and the circuit input impedance is $Z_b\|R_1\|R_2$.

Output Impedance

'Looking into' the output of the *CE* h-parameter circuit reproduced in Fig. 6-22, the output impedance at the transistor collector is,

$$Z_c \approx 1/h_{oe} \qquad (6\text{-}13)$$

At the circuit output terminals, resistor R_C is in parallel with Z_c. So, the circuit output impedance is,

$$Z_o = R_C\|(1/h_{oe}) \qquad (6\text{-}14)$$

Because $1/h_{oe}$ is typically 50 kΩ, and R_C is usually much less than 50 kΩ, the circuit output impedance is approximately equal to R_C. Using this information, it is possible to tell the output impedance of a *CE* circuit simply by reading the resistance of R_C.

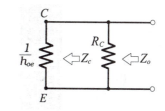

Figure 6-22
The output impedance for a CE circuit is $R_C\|(1/h_{oe})$.

Voltage Gain

The circuit voltage gain is given by the equation,

$$A_v = \frac{v_o}{v_i}$$

Figure 6-23(a) and (b) reproduced from Fig. 6-20 shows that,

$$v_i = I_b\,h_{ie}, \text{ and } v_o = -I_C\,(R_C\|R_L\|1/h_{oe}),$$

so, $$A_v = \frac{-I_C\,(R_C\|R_L\|1/h_{oe})}{I_b\,h_{ie}}$$

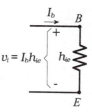

$v_i = I_b h_{ie}$

(a) Input voltage

I_C/I_b can be replaced with h_{fe}. Also, because $1/h_{oe}$ can be omitted from the equation because it is usually much larger than $(R_C\|R_L)$. The normally used voltage gain equation now becomes,

$$A_v = \frac{-h_{fe}\,(R_C\|R_L)}{h_{ie}} \qquad (6\text{-}15)$$

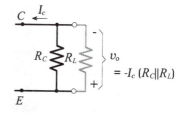

$v_o = -I_c\,(R_C\|R_L)$

(b) output voltage

The minus sign in Eq. 6-15 indicates that v_o is 180° out of phase with v_i. (When v_i increases, v_o decreases, and vice versa.) Knowing the appropriate h-parameters as well as R_C and R_L, the voltage gain of a *CE* circuit can be quickly estimated. Using values of $R_C = 4.7$ kΩ, $h_{fe} = 70$, $h_{ie} = 1.5$ kΩ, and assuming that $R_L \gg R_C$, a typical *CE* voltage gain is -220.

Figure 6-23
Derivation of equations for v_i and v_o in a CE circuit.

Substituting from Table 6-1, the *CE* voltage gain equation can be rewritten as,

$$A_v \approx \frac{-(R_C \| R_L)}{h_{ib}} \tag{6-16}$$

h_{ib} can be determined approximately as r'_e from the emitter current level. So, Eq. 6-16 can be useful in cases where the device parameters are not known.

Current Gain

The transistor current gain is,

$$h_{fe} = \frac{I_C}{I_b} \tag{6-17}$$

This is the device current gain, *not* the circuit current gain. Examination of the circuit in Fig. 6-20 shows that the signal current (I_s) is divided between h_{ie} and the bias resistors [$R_B = (R_C \| R_L)$]. Also, that the output current from the transistor collector (I_c) is divided between R_C and R_L. The overall circuit current gain can be shown to be,

$$A_i = \frac{h_{fe} R_C R_B}{(R_C + R_L)(R_C + h_{ie})} \tag{6-18}$$

*In virtually all practical applications, the current gain of a transistor amplifier is normally **not** an important quantity.*

Power Gain

The power gain of an amplifier can be shown to be,

$$A_p = A_v \times A_i \tag{6-19}$$

Like current gain, the power gain of a transistor amplifier is largely unimportant for practical applications.

Summary of Typical CE Circuit Performance

Device input impedance	$Z_b \approx h_{ie}$
Circuit input impedance	$Z_i = R_1 \| R_2 \| Z_b$
Device output impedance	$Z_c \approx 1/h_{oe}$
Circuit output impedance	$Z_o \approx R_C \|(1/h_{oe}) \approx R_C$
Circuit voltage gain	$A_v = \frac{-h_{fe}(R_C \| R_L)}{h_{ie}} \approx \frac{-(R_C \| R_L)}{h_{ib}}$

The common-emitter circuit has good voltage gain, with 180° phase shift, medium input impedance, and relatively high output impedance. As a voltage amplifier, the *CE* circuit is by far the most often used of the three basic circuit configurations.

Example 6-6

The transistor parameters in the *CE* circuit in Fig. 6-24 are h_{ie} = 2.1 kΩ, h_{fe} = 75, and h_{oe} = 1 μS. Calculate the circuit input impedance, output impedance, and voltage gain. This circuit is designed in Example 5-14.

Solution

From Eq. 6-11 and Eq. 6-12,

$$Z_i = R_1 \| R_2 \| h_{ie} = 68\ k\Omega \| 56\ k\Omega \| 2.1\ k\Omega$$

$$= 1.97\ k\Omega$$

Eq. 6-14, $\quad Z_o = R_C \| (1/h_{oe}) = 3.9\ k\Omega \| (1/1\ \mu S)$

$$\approx 3.9\ k\Omega$$

Eq. 6-15, $\quad A_v = \dfrac{-h_{fe}\ (R_C \| R_L)}{h_{ie}} = \dfrac{-75 \times (3.9\ k\Omega \| 82\ k\Omega)}{2.1\ k\Omega}$

$$= -133$$

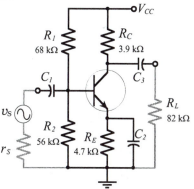

Figure 6-24
CE circuit for Example 6-6. Designed in Example 5-14.

Example 6-7

A *CE* circuit (as in Fig. 6-24) has I_C = 1.5 mA, R_C = 4.7 kΩ, and R_L = 56 kΩ. Estimate the value of r'_e, and calculate the circuit voltage gain.

Solution

Eq. 6-2, $\quad r'_e = \dfrac{26\ mV}{I_C} \approx \dfrac{26\ mV}{1.5\ mA}$

$$= 17.3\ \Omega$$

Eq. 6-16, $\quad A_v \approx \dfrac{-(R_C \| R_L)}{h_{ib}} = \dfrac{-(4.7\ k\Omega \| 56\ k\Omega)}{17.3\ \Omega}$

$$= -250$$

Practise Problems

6-4.1 Calculate Z_i, Z_o, and A_v for a *CE* circuit (as in Fig. 6-24) with the following quantities: R_1 = 18 kΩ, R_2 = 8.2 kΩ, R_C = 5.6 kΩ, R_E = 2.7 kΩ, R_L = 68 kΩ, h_{ie} = 1 kΩ, h_{fe} = 100, h_{oe} = 1.67 μS.

6-4.2 A *CE* circuit has a 0.8 mA emitter current, a 6.8 kΩ collector resistor, and a 68 kΩ capacitor-coupled load. Calculate the circuit voltage gain at 25°C and at 80°C.

6-5 CE Circuit with Unbypassed Emitter Resistor

h-parameter Equivalent Circuit

When an unbypassed emitter resistor (R_E) is present in a *CE* circuit, as in Fig. 6-25(a), it is also present in the *ac* equivalent circuit, [Fig. 6-25(b)]. R_E must also be shown in the *h*-parameter circuit between the transistor emitter terminal and the circuit common input-output terminal, (Fig. 6-26). As with the previous *h*-parameter circuit, the current directions and voltage polarities shown are those that occur when the instantaneous input voltage is positive-going. The presence of R_E without a bypass capacitor significantly affects the circuit input impedance and voltage gain.

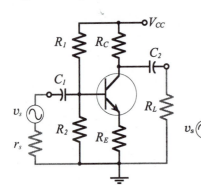

(a) Circuit with unbypassed R_E (b) *ac* equivalent circuit

Figure 6-25
CE circuit with unbypassed emitter resistor, and ac equivalent circuit.

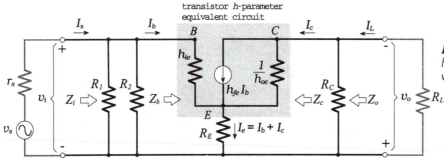

Figure 6-26
h-parameter circuit for CE circuit with unbypassed emitter resistor.

Input Impedance

An equation for the input impedance at the transistor base can be determined from v_i and I_b. From Fig. 6-26,

$$v_i = I_b h_{ie} + I_e R_E$$
$$= I_b h_{ie} + R_E(I_b + I_c)$$
$$= I_b h_{ie} + R_E(I_b + h_{fe} I_b)$$
$$= I_b[h_{ie} + R_E(1 + h_{fe})]$$

so,
$$Z_b = \frac{v_i}{I_b} = \frac{I_b[h_{ie} + R_E(1 + h_{fe})]}{I_b}$$

or,
$$\mathbf{Z_b = h_{ie} + R_E(1 + h_{fe})} \tag{6-20}$$

Equation 6-20 gives the input impedance at the transistor base terminal. The circuit input impedance is again given by Eq. 6-12,

$$\mathbf{Z_i = R_1 \| R_2 \| Z_b}$$

Examination of Eq. 6-20 shows that it is possible to very quickly estimate the input impedance at the transistor base in a *CE* circuit with an unbypassed emitter resistor. For example, a circuit with R_E = 1 kΩ and h_{fe} = 100, has $Z_b \approx$ 100 kΩ.

Output Impedance

The output impedance at the transistor collector can be shown to be substantially increased (above $1/h_{oe}$) by the presence of the unbypassed emitter resistor. Consequently, the circuit output impedance should be taken as,

$$Z_o \approx R_C$$

Voltage Gain

From the derivation of the input impedance equation,

$$v_i = I_b[h_{ie} + R_E(1 + h_{fe})]$$

and,

$$v_o = -I_c(R_C\|R_L)$$

so,

$$A_v = \frac{v_o}{v_i} = \frac{-I_c(R_C\|R_L)}{I_b[h_{ie} + R_E(1 + h_{fe})]}$$

giving,

$$A_v = \frac{-h_{fe}(R_C\|R_L)}{h_{ie} + R_E(1 + h_{fe})} \qquad (6\text{-}21)$$

Usually,

$$R_E(1 + h_{fe}) \gg h_{ie}$$

so,

$$A_v \approx \frac{-(R_C\|R_L)}{R_E} \qquad (6\text{-}22)$$

The voltage gain of a *CE* circuit with an unbypassed emitter resistor can be quickly estimated using Eq. 6-22. For the circuit in Fig. 6-25(a), with R_C = 4.7 kΩ, R_E = 1 kΩ, and $R_L \gg R_C$, $A_v \approx$ -4.7.

Performance Summary for CE Circuit with Unbypassed R_E

Device input impedance	$Z_b = h_{ie} + R_E(1 + h_{fe}) \approx h_{fe}R_E$
Circuit input impedance	$Z_i \approx R_B\|Z_b$
	$= R_1\|R_2\|Z_b$
Circuit output impedance	$Z_o \approx R_C$
Circuit voltage gain	$A_v = \dfrac{-h_{fe}(R_C\|R_L)}{h_{ie} + R_E(1 + h_{fe})}$
Circuit voltage gain	$A_v \approx \dfrac{-(R_C\|R_L)}{R_E}$

The most significant feature of the performance of a *CE* circuit with an unbypassed emitter resistor is that its voltage gain is much lower than it would be normally. Its input impedance is also much higher than Z_i for a *CE* circuit that has R_E bypassed.

Example 6-8

The *CE* circuit in Fig. 6-27 is the same as in Fig. 6-24, but with bypass capacitor C_2 omitted. The transistor parameters are $h_{ie} = 2.1 \text{ k}\Omega$, $h_{fe} = 75$, and $h_{oe} = 1 \text{ }\mu\text{S}$, (as in Example 6-6). Calculate the circuit input and output impedance and voltage gain.

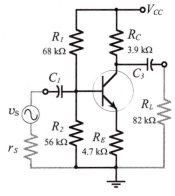

Solution

From Eq. 6-20, $Z_b = h_{ie} + R_E(1 + h_{fe}) = 2.1 \text{ k}\Omega + 4.7 \text{ k}\Omega(1 + 75)$

$$= 359 \text{ k}\Omega$$

From Eq. 6-12, $Z_i = R_1\|R_2\|Z_b = 68 \text{ k}\Omega\|56 \text{ k}\Omega\|359 \text{ k}\Omega$

$$= 28.3 \text{ k}\Omega$$

$$Z_o \approx R_C = 3.9 \text{ k}\Omega$$

Eq. 6-21, $A_v = \dfrac{-h_{fe}\,(R_C\|R_L)}{h_{ie} + R_E(1 + h_{fe})} = \dfrac{-75 \times (3.9 \text{ k}\Omega\|82 \text{ k}\Omega)}{2.1 \text{ k}\Omega + 4.7 \text{ k}\Omega(1 + 75)}$

$$= -0.78$$

Figure 6-27
CE circuit for Example 6-8.

Practise Problems

6-5.1   Calculate Z_i, Z_o, and A_v for a *CE* circuit with an unbypassed emitter resistor (as in Fig. 6-27) with the following component values and parameters: $R_1 = 18 \text{ k}\Omega$, $R_2 = 8.2 \text{ k}\Omega$, $R_C = 5.6 \text{ k}\Omega$, $R_E = 2.7 \text{ k}\Omega$, $R_L = 68 \text{ k}\Omega$, $h_{ie} = 1 \text{ k}\Omega$, $h_{fe} = 100$, $h_{oe} = 1.66 \text{ }\mu\text{S}$.

6-6   Common-Collector Circuit Analysis

Common-Collector Circuit

In the *common-collector* (CC) circuit shown in Fig. 6-28 the external load (R_L) is capacitor-coupled to the transistor emitter terminal. The circuit uses voltage divider bias to derive the transistor base voltage (V_B) from the supply. The transistor collector terminal is directly connected to V_{CC}; no collector resistor is used. The circuit output voltage is developed across the emitter resistor (R_E), and there is no bypass capacitor.

To understand the operation of a *CC* circuit, note that V_B is a constant quantity, and that $V_E = V_B - V_{BE}$. When a signal is applied via C_1 to the transistor base, V_B increases and decreases as the signal goes positive and negative. If the signal voltage (v_s) increases to +0.5 V, V_B is increased by 0.5 V. Also, V_E increases by 0.5 V, because V_{BE} remains substantially constant, and $V_E = V_B - V_{BE}$. The change in V_E is coupled via C_2 to give an *ac* output voltage ($v_o = 0.5$ V), (see the waveforms in Fig. 6-29). Similarly, when v_s decreases to -0.5 V, both V_B and V_E decrease by 0.5 V, giving $v_o = -0.5$ V.

It is seen that the *ac* output voltage from a *CC* circuit is essentially the same as the input voltage; there is no voltage gain or phase shift. Thus, the *CC* circuit can be said to have a voltage

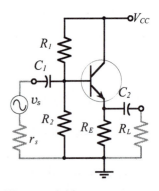

Figure 6-28
Transistor common-collector amplifier circuit, also called an emitter follower.

gain of 1. (Actually, the output voltage can be shown to be slightly smaller than the input because of a very small change in V_{BE}.)

The fact that the *CC* output voltage follows the changes in signal voltage gives the circuit its other name: *emitter follower*.

h-Parameter Equivalent Circuit

As for a *CE* circuit, the power supply and capacitors must be replaced with short-circuits to study the *CC ac* performance. This gives the *CC ac* equivalent circuit in Fig. 6-30(a). The input terminals of the *ac* equivalent circuit are seen to be the transistor base and collector, and the output terminals are the emitter and collector. Because the collector terminal is common to both input and output, the circuit configuration is named *common-collector*.

The *CC h*-parameter circuit is now drawn by substituting the transistor *h*-parameter model into the *ac* equivalent circuit, to give the circuit in Fig. 6-30(b). The current directions and voltage polarities indicated in Fig. 6-30(b) are, once again, those that are produced by a positive-going signal voltage. It should be noted that $h_{rc} = 1$ for a *CC* circuit; all of v_o is fed back to the input. So, unlike the case of a *CE* circuit, the feedback generator cannot be omitted in the equivalent circuit of a *CE* amplifier.

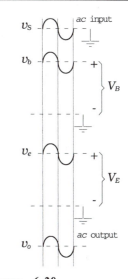

Figure 6-29
Voltage waveforms in a common-collector amplifier circuit.

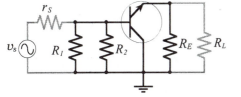

(a) *ac* equivalent circuit for *C C* transistor circuit

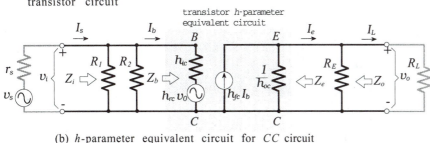

(b) *h*-parameter equivalent circuit for *CC* circuit

Figure 6-30
Common-collector amplifier ac equivalent circuit and h-parameter circuit.

Input Impedance

The input impedance for the *CC* circuit is determined by first writing an equation for the input voltage. Referring to Fig. 6-30 and Fig. 6-31,

$$v_i = I_b h_{ic} + h_{rc} v_o$$
$$= I_b h_{ic} + I_e(R_E \| R_L)$$
$$= I_b h_{ic} + h_{fc} I_b(R_E \| R_L)$$
$$= I_b [h_{ic} + h_{fc}(R_E \| R_L)]$$

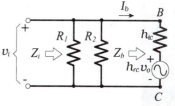

Figure 6-31
Input impedance determination for a CC circuit.

and,

$$Z_b = \frac{v_i}{I_b} = \frac{I_b[h_{ic} + h_{fc}(R_E\|R_L)]}{I_b}$$

or,

$$Z_b = h_{ic} + h_{fc}(R_E\|R_L) \qquad \textbf{(6-23)}$$

Equation 6-23 is similar to the equation for the transistor input impedance in a *CE* circuit with an unbypassed emitter resistor (Eq. 6-20), except that R_L is now in parallel with R_E. The circuit input impedance is again given by Eq. 6-12,

$$Z_i = R_1\|R_1\|Z_b$$

Using Eq. 6-23, the input impedance at the transistor base in a *CC* circuit can be quickly estimated. A circuit with $R_E = 1$ kΩ and $h_{fe} = 100$, has $Z_b \approx 100$ kΩ if $R_L \gg R_E$.

Output Impedance

As already discussed, the *ac* voltage at the output of a *CC* circuit is all fed back to the input. This fact is used in determining the output impedance (Z_e) at the emitter terminal. The signal voltage is assumed to be zero, and v_o is used to calculate I_e. With $v_s = 0$, I_b is produced by the fed back voltage $(h_{rc} v_o = v_o)$, [see Fig. 6-32(a)].

$$I_b = \frac{v_o}{h_{ic} + (R_1\|R_2\|r_s)}$$

and,

$$I_e = h_{fc} I_b = \frac{h_{fc}\, v_o}{h_{ic} + (R_1\|R_2\|r_s)}$$

$$Z_e = \frac{v_o}{I_e} = \frac{v_o[h_{ic} + (R_1\|R_2\|r_s)]}{h_{fc}\, v_o} \qquad \text{[see Fig. 6-32(b)]}$$

so,

$$Z_e = \frac{h_{ic} + (R_1\|R_2\|r_s)}{h_{fc}} \qquad \textbf{(6-24)}$$

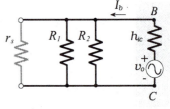

(a) I_b determination with $v_s = 0$

(b) Z_e and Z_o determination

Figure 6-32
Determination of output impedance for a CC circuit.

Note that the output impedance at the emitter terminal is

$$\frac{h_{ic} + (Z \text{ in series with the base terminal})}{h_{fc}}$$

It is interesting to compare this to the base input impedance (Eq. 6-23), which is,

$$h_{ic} + h_{fc}\,(Z \text{ in series with the emitter terminal})$$

Equation 6-24 gives the device output impedance. (Actually, $1/h_{oc}$ should be included, but it has a negligible effect on Z_e.) The circuit output impedance also involves R_E, [see Fig. 6-32(b)].

$$Z_o = Z_e\|R_E \qquad \textbf{(6-25)}$$

R_E is usually much larger than Z_e, so that $Z_o \approx Z_e$.

Voltage Gain

As already explained, the *ac* output voltage from a *CC* circuit is almost exactly equal to the input voltage. However, the precise equation for A_v is easily derived by considering the ac output and input voltages as derived from Fig, 6-30(b).

$$v_i = I_b[h_{ic} + h_{fc}(R_E\|R_L)]$$

and, $$v_o = I_e(R_E\|R_L)$$

so, $$A_v = \frac{v_o}{v_i} = \frac{I_e(R_E\|R_L)}{I_b[h_{ic} + h_{fc}(R_E\|R_L)]}$$

this reduces to,

$$A_v = \frac{(R_E\|R_L)}{h_{ib} + (R_E\|R_L)} \qquad \text{(6-26)}$$

Because $r'_e \approx h_{ib}$, Eq. 6-26 can be used in situations where the device parameters are unknown. Usually, $R_E\|R_L$ is so much larger than h_{ib} that the CC circuit voltage gain is simply taken as,

$$A_v \approx 1 \qquad \text{(6-27)}$$

Summary of CC Circuit Performance

Device input impedance $Z_b = h_{ic} + h_{fc}(R_E\|R_L)$

Circuit input impedance $Z_i \approx R_B\|Z_b = R_1\|R_2\|Z_b$

Device output impedance $Z_e = \dfrac{h_{ic} + (R_1\|R_2\|r_s)}{h_{fc}}$

Circuit output impedance $Z_o = Z_e\|R_E$

Circuit voltage gain $A_v = \dfrac{(R_E\|R_L)}{h_{ib} + (R_E\|R_L)} \approx 1$

A CC circuit has a voltage gain of 1, no phase shift between input and output, high input impedance, and low output impedance. Because of its high Z_i, low Z_o, and unity gain, the CC circuit is normally used as a *buffer amplifier*, placed between a high impedance signal source and a low impedance load.

Example 6-9
The transistor parameters for the CC circuit in Fig. 6-33 are h_{ie} = 2.1 kΩ and h_{fe} = 75. (a) Calculate the circuit input and output impedance with R_L not connected. (b) Calculate Z_i and A_v with R_L connected.

Solution
(a) R_L not connected

From Table 6-1, $h_{ic} = h_{ie}$ = 2.1 kΩ

and, $h_{fc} = 1 + h_{fe}$ = 76

Eq. 6-23, $Z_b = h_{ic} + h_{fc}(R_E\|R_L)$ = 2.1 kΩ + 76(4.7 kΩ)

                  = 359.3 kΩ

Eq. 6-12, $Z_i = R_1\|R_2\|Z_b = 10\ k\Omega\|10\ k\Omega\|359.3\ k\Omega$
$= 4.93\ k\Omega$

Eq. 6-24, $Z_e = \dfrac{h_{ic} + (R_1\|R_2\|r_s)}{h_{fc}} = \dfrac{2.1\ k\Omega + (10\ k\Omega\|10\ k\Omega\|1\ k\Omega)}{76}$
$= 38.6\ \Omega$

Eq. 6-25, $Z_o = Z_e\|R_E = 38.6\ \Omega\|4.7\ k\Omega$
$= 38.3\ \Omega$

(b) R_L connected

Eq. 6-23, $Z_b = h_{ic} + h_{fc}(R_E\|R_L) = 2.1\ k\Omega + 76(4.7\ k\Omega\|12\ k\Omega)$
$= 258.8\ k\Omega$

Eq. 6-12, $Z_i = R_1\|R_2\|Z_b = 10\ k\Omega\|10\ k\Omega\|258.8\ k\Omega$
$= 4.91\ k\Omega$

From Table 6-1, $h_{ib} = \dfrac{h_{ie}}{1 + h_{fe}} = \dfrac{2.1\ k\Omega}{1 + 75}$
$= 27.6\ \Omega$

Eq. 6-26, $A_v = \dfrac{(R_E\|R_L)}{h_{ib} + (R_E\|R_L)} = \dfrac{(4.7\ k\Omega\|12\ k\Omega)}{27.6\ \Omega + (4.7\ k\Omega\|12\ k\Omega)}$
$= 0.99$

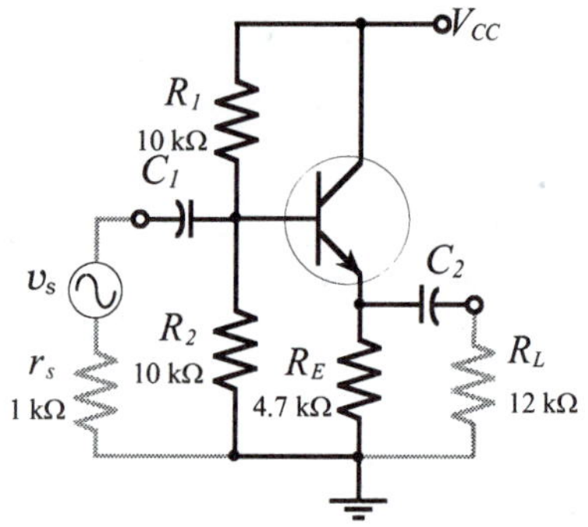

Figure 6-33
CC circuit for Example 6-9.

Practise Problems
6-6.1   Calculate Z_i and Z_o, for a CC circuit (as in Fig. 6-33) with the following component values and parameters: $R_1 = 56\ k\Omega$, $R_2 = 39$ $k\Omega$, $R_E = 5.6\ k\Omega$, $R_L = 22\ k\Omega$, $r_s = 400\ \Omega$, $h_{ie} = 1\ k\Omega$, $h_{fe} = 100$.

6-7 Common-Base Circuit Analysis

Common-Base Circuit

The common-base (CB) circuit in Fig. 6-34 is very similar to a CE circuit, except that the input signal is applied to the transistor emitter terminal (via C_2), instead of the base. Also, there is no bypass capacitor across the R_E, but the base terminal is ac grounded via capacitor C_1. Because the base is ac grounded, all of the signal voltage is developed across the transistor base-emitter junction. As for a CE circuit, the output voltage is developed across the collector resistor (R_C).

A positive-going signal voltage at the input of a CB circuit pushes the transistor emitter in a positive direction while the base voltage remains fixed. Thus, a positive-going signal reduced the base-emitter voltage, (see the waveforms in Fig. 6-35). The reduction in V_{BE} reduces the transistor collector current, and the I_C reduction reduces V_{RC}, and consequently, causes the transistor collector voltage to increase. The rise in V_C is effectively a rise in the circuit output voltage (v_o). It is seen that a positive-going input voltage produced a positive-going output. It can also be demonstrated that a negative-going input produces a negative-going output. There is no phase shift from input to output in a *CB* circuit.

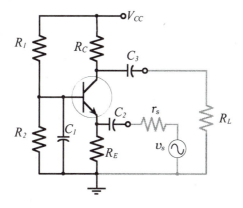

Figure 6-34
In a common-base circuit, the signal is capacitor-coupled to the transistor emitter terminal, and the output is taken from the collector.

h-parameter Equivalent Circuit

The *CB ac* equivalent circuit is drawn, as always, by replacing the supply voltage and capacitors with short circuits. This gives the circuit in Fig. 6-36(a), which shows that the transistor base terminal (grounded via capacitor C_1) is common to both input and output. Hence the name *common base*.

The *CB h*-parameter circuit is now drawn by substituting the transistor *h*-parameter model into the *ac* equivalent circuit, giving the circuit in Fig. 6-36(b). Once again, the current directions and voltage polarities indicated in the *h*-parameter circuit are those produced by a positive-going signal voltage. Note that the feedback voltage generator $(h_{rb} \, v_o)$ is not included in the *CB h*-parameter circuit. This is because the feedback voltage effect is so small that it can be neglected when deriving practical approximate equations for the circuit performance. This corresponds with the *CE h*-parameter circuit, but *not* with the *CC h*-parameter circuit, where the feedback voltage is very important.

Input Impedance

Referring to Fig. 6-36(b), the input impedance to the transistor emitter is simply h_{ib}.

$$Z_e = h_{ib} \qquad\qquad \textbf{(6-28)}$$

Typically h_{ib} is around 21 Ω for a low-current transistor. (Also, $h_{ib} \approx r'_e$, see Section 6-3.) Resistor R_E is in parallel with Z_e, so the circuit input impedance is,

$$Z_i = Z_e \| R_E \qquad\qquad \textbf{(6-29)}$$

R_E is usually much larger than Z_e, so $Z_i \approx Z_e$.

Output Impedance

'Looking into' the collector-base terminals of the *CB h*-parameter circuit in Fig. 6-36(b), a very large resistance $(1/h_{ob})$ is *seen*.

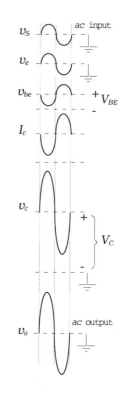

Figure 6-35
Voltage and current waveforms for a common-base amplifier circuit.

$$Z_c = 1/h_{ob} \tag{6-30}$$

Once again, Z_c is the device output impedance, and the circuit output impedance is R_C in parallel with Z_c.

$$Z_o = Z_c \| R_c \tag{6-31}$$

Because Z_c is normally much larger than R_C, $Z_o \approx R_C$.

As in the case of a *CE* circuit, the output impedance of a *CB* circuit can be determined by just reading the resistance of R_C.

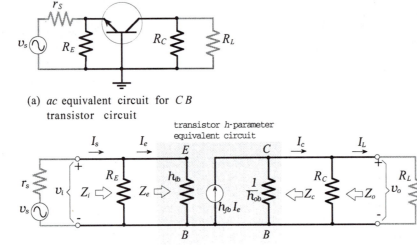

(a) *ac* equivalent circuit for *C B* transistor circuit

(b) *h*-parameter equivalent circuit for *CB* circuit

Figure 6-36
Common-base amplifier, ac equivalent circuit, and h-parameter circuit.

Voltage Gain

The input voltage equation for the circuit in Fig. 6-36(b) is,

$$v_i = I_e h_{ib}$$

and the output voltage is,

$$v_o = I_c (R_C \| R_L)$$
$$= h_{fb} I_e (R_C \| R_L)$$

so,
$$A_v = \frac{v_o}{v_i} = \frac{h_{fb} I_e (R_C \| R_L)}{I_e h_{ib}}$$

and,
$$A_v = \frac{h_{fb} (R_C \| R_L)}{h_{ib}} \tag{6-32}$$

Effect of Unbypassed Base Resistors

If capacitor C_1 is not present in the *CB* circuit in Fig. 6-34, the transistor base is *not ac* short-circuited to ground. So, a resistance ($R_B = R_1 \| R_2$) must be shown in the *ac* equivalent circuit [Fig. 6-37(a)], and in the *h*-parameter circuit [Fig. 6-37(b)]. The presence of the unbypassed base resistors can substantially affect the transistor input impedance and the circuit voltage gain. Analysis of the *h*-parameter circuit shows that,

$$Z_e = h_{ib} + R_B(1 - h_{fb}) \qquad \text{(6-33)}$$

and, $\qquad A_v = \dfrac{h_{fb}(R_C\|R_L)}{h_{ib} + R_B(1 - h_{fb})} \qquad \text{(6-34)}$

Comparing Eq. 6-33 to Eq. 6-28 (Z_e with C_1 present), it is seen that $R_B(1 - h_{fb})$ is added to h_{ib} to give Z_e without C_1 present, (Eq. 6-33). Eq. 6-34 is similar to Eq. 6-32 (A_v with C_1 present), except that $R_B(1 - h_{fb})$ is again added to h_{ib} to give the CB circuit voltage gain without C_1, (Eq. 6-34). Typically $h_{fb} = 0.99$ and $(1 - h_{fb}) = 0.01$.

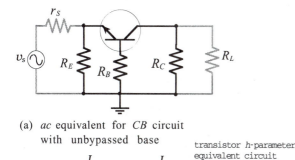

(a) *ac* equivalent for *CB* circuit with unbypassed base

Figure 6-37
AC equivalent circuit and h-parameter circuit for a CB amplifier with an unbypassed base resistor.

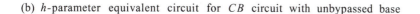

(b) *h*-parameter equivalent circuit for *CB* circuit with unbypassed base

Summary of CB Circuit Performance

With the base bypassed to ground:

Device input impedance	$Z_e = h_{ib}$
Circuit input impedance	$Z_i = Z_e\|R_E$
Device output impedance	$Z_c = \dfrac{1}{h_{ob}}$
Circuit output impedance	$Z_o \approx R_c$
Circuit voltage gain	$A_v = \dfrac{h_{fb}(R_C\|R_L)}{h_{ib}}$

With the base unbypassed:

Device input impedance	$Z_e = h_{ib} + R_B(1 - h_{fb})$
Circuit voltage gain	$A_v = \dfrac{h_{fb}(R_C\|R_L)}{h_{ib} + R_B(1 - h_{fb})}$

A *CB* circuit has good voltage gain and relatively high output impedance, like a *CE* circuit. Unlike a *CE* circuit, a *CB* circuit has a very low input impedance, and this makes it unsuitable for most voltage amplifier applications. It is normally employed only as a high-frequency amplifier. The transistor base terminal in a *CB* circuit should always be bypassed to ground. When this is not done, the input impedance is increased and the voltage amplification is substantially reduced.

Example 6-10

The transistor parameters in the *CB* circuit in Fig. 6-38 are $h_{ie} = 2.1$ kΩ and $h_{fe} = 75$. Calculate the circuit input and output impedances and voltage gain.

Solution

From Table 6-1, $h_{ib} = \dfrac{h_{ie}}{1 + h_{fe}} = \dfrac{2.1 \text{ k}\Omega}{1 + 75}$

$\qquad\qquad = 27.6 \ \Omega$

and, $h_{fb} = \dfrac{h_{fe}}{1 + h_{fe}} = \dfrac{75}{1 + 75}$

$\qquad\qquad = 0.987$

Eq. 6-29, $Z_i = h_{ib}\|R_E = 27.6 \ \Omega\|4.7 \text{ k}\Omega$

$\qquad\qquad = 27.4 \ \Omega$

Eq. 6-31, $Z_o = (1/h_{ob})\|R_C \approx R_C$

$\qquad\qquad = 3.9 \text{ k}\Omega$

Eq. 6-32, $A_v = \dfrac{h_{fb}(R_C\|R_L)}{h_{ib}} = \dfrac{0.987 \ (3.9 \text{ k}\Omega\|82 \text{ k}\Omega)}{27.6 \ \Omega}$

$\qquad\qquad = 133$

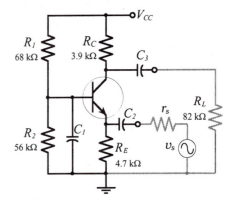

Figure 6-38
CB circuit for Example 6-10.

Example 6-11

Calculate the input impedance and voltage gain for the *CB* circuit in Ex. 6-10 when capacitor C_1 is disconnected.

Solution

Eq. 6-33, $Z_e = h_{ib} + R_B(1 - h_{fb}) = 27.6 \ \Omega + (68 \text{ k}\Omega\|56 \text{ k}\Omega)(1 - 0.987)$

$\qquad\qquad = 426.8 \ \Omega$

Eq. 6-29, $Z_i = Z_e\|R_E = 426.8 \ \Omega\|4.7 \text{ k}\Omega$

$\qquad\qquad = 391 \ \Omega$

Eq. 6-34, $A_v = \dfrac{h_{fb}(R_C\|R_L)}{h_{ib} + R_B(1 - h_{fb})} = \dfrac{0.987 \ (3.9 \text{ k}\Omega\|82 \text{ k}\Omega)}{27.6 \ \Omega + (68 \text{ k}\Omega\|56 \text{ k}\Omega)(1 - 0.987)}$

$\qquad\qquad = 8.6$

6-8   Comparison of CE, CC, and CB Circuits

Table 6-2 compares Z_i, Z_o, and A_v for CE, CC, and CB circuits. As already discussed, the CE circuit has high voltage gain, medium input impedance, high output impedance, and a 180° phase shift from input to output. The CC circuit has high input impedance, low output impedance, a voltage gain of 1, and no phase shift. The CB circuit offers low input impedance, high output impedance, high voltage gain, and no phase shift.

Table 6-2   Comparison of common-emitter, common-collector, and common-base circuits.

Circuit configuration	Z_i	Z_o	A_v	Phase shift
CE	medium	$\approx R_C$	high	180°
CC	high	low	≈ 1	0
CB	low	$\approx R_C$	high	0

Device manufacturers normally only list the CE h-parameters on a transistor data sheet. Although these can be converted to CC and CB parameters, it is convenient to use CE parameters for all three types of circuits. Table 6-3 gives the circuit impedance equations in terms of CE h-parameters. In the case of input and output impedances, it is helpful to think in terms of the terminal being *looked into*.

Impedance at the Transistor Base

Consideration of each type of circuit shows that the input impedance (Z_i) depends upon which transistor terminal is involved. In both the CE and CC circuits, the input signal is applied to the transistor base terminal. So, Z_i is the impedance 'looking into' the base. Figure 6-39 shows that, for a CE circuit with an unbypassed emitter resistor,

$$Z_b = h_{ie} + R_E(1 + h_{fe})$$

When R_E is bypassed, the $R_E(1 + h_{fe})$ portion can be treated as zero, so that,

$$Z_b = h_{ie}$$

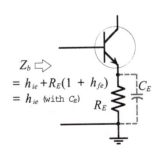

$Z_b \Rightarrow$

$= h_{ie} + R_E(1 + h_{fe})$

$= h_{ie}$ (with C_E)

Figure 6-39
Impedance at the transistor base terminal.

The *CC* equations for Z_b are almost identical to the *CE* equations, except that they use h_{ic} and h_{fc}, which are essentially equal to h_{ie} and h_{fe}, (see Table 6-3). A rough approximation for the base input impedance of any transistor circuit with an unbypassed emitter resistor is,

$$Z_b \approx h_{fe}(\text{impedance in series with the emitter})$$

The circuit input impedance for both *CE* and *CC* configurations is Z_b in parallel with the bias resistors,

$$Z_i = Z_b \| R_1 \| R_2$$

Table 6-3	Common-emitter, common-collector, and common-base circuit equations using *CE* h-parameters.		
Circuit configuration	Z_b	Z_e	Z_c
CE	h_{ie}	—	$1/h_{oe}$
CE with unbypassed R_E	$h_{ie} + R_E(1 + h_{fe})$	—	$1/h_{oe}$
CC	$h_{ie} + (R_E\|R_L)(1 + h_{fe})$	$\dfrac{h_{ie} + R_B}{1 + h_{fe}}$	—
CB	—	$\dfrac{h_{ie}}{1 + h_{fe}}$	$\dfrac{1 + h_{fe}}{h_{oe}}$
CB with unbypassed R_B	—	$\dfrac{h_{ie} + R_B}{1 + h_{fe}}$	$\dfrac{1 + h_{fe}}{h_{oe}}$

Impedance at the Transistor Emitter

The transistor emitter is the output terminal for a *CC* circuit and the input terminal for a *CB* circuit. So, the device impedance in each case is the impedance 'looking into' the transistor emitter terminal (Z_e). Although the Z_e equations for the two circuits look different, they can be shown to give exactly the same result in similar situations. For a *CC* circuit or a *CB* circuit with an unbypassed base (using *CE* parameters),

$$Z_e = \frac{h_{ie} + R_B}{1 + h_{fe}}$$

where R_B is the impedance 'looking back' from the transistor base.

As shown in Fig. 6-40, R_B includes the signal source resistance (r_s) for a *CC* circuit. A *CB* circuit normally has its base bypassed as shown, so that the impedance at the emitter is,

$$Z_e = \frac{h_{ie}}{1 + h_{fe}}$$

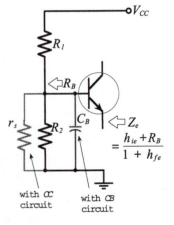

Figure 6-40
Impedance at the transistor emitter terminal.

For all circuit arrangements, the impedance at the transistor emitter terminal is,

$$Z_e = \frac{Impedance\ in\ series\ with\ the\ base}{1 + h_{fe}}$$

Impedance at the Transistor Collector

The output for *CE* and *CB* circuits is produced at the transistor collector terminal. So, the impedance 'looking into' the collector is the device output impedance. This is normally a very large quantity at low and medium signal frequencies. As illustrated in Fig. 6-41, the circuit output impedance at the collector is essentially,

$$Z_o \approx R_C$$

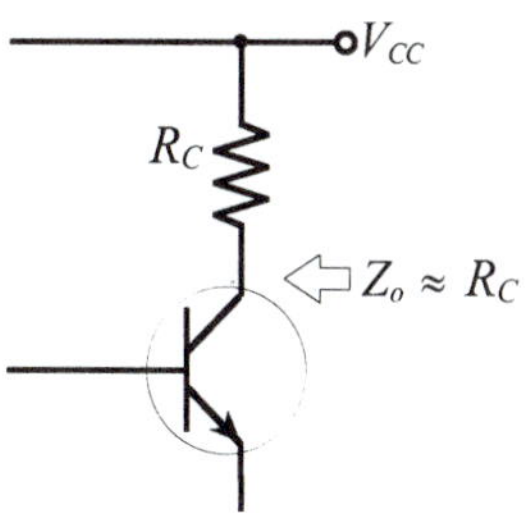

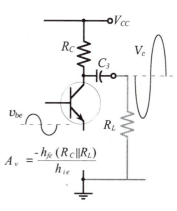

Figure 6-41
Impedance at the transistor collector terminal.

Voltage gain

In the case of a circuit with an unbypassed emitter resistor, the *ac* voltage at the emitter follows the *ac* input at the transistor base. So, a *CC* circuit (an emitter follower) has a voltage gain of 1.

With both *CE* and *CB* circuits, the *ac* input (v_i) is developed across the base-emitter junction, and the *ac* output (v_o) is produced at the transistor collector terminal, (see Fig 6-42). Thus, the magnitude of the voltage gain is the same for *CB* and *CE* circuits with similar component values and transistor parameters. The *CE* voltage gain equation can be used for the *CB* circuit, with the omission of the minus sign that indicates *CE* phase inversion.

Although the voltages gains are equal for similar *CB* and *CE* circuits, the low input impedance of the *CB* circuit can substantially attenuate the signal voltage, and result in a low amplitude output. This is demonstrated in Example 6-12.

$$A_v = \frac{-h_{fe}(R_C \| R_L)}{h_{ie}}$$

Figure 6-42
CE and CB voltage gain.

Example 6-12

A 50 mV signal with a 600 Ω source resistance is applied to the circuit in Fig. 6-43. Calculate v_o for: (a) *CE* circuit operation with v_s at the transistor base, and R_E bypassed. (b) *CB* circuit operation with v_s at the emitter, and the base resistors bypassed. The transistor parameters are $h_{ie} = 1.5\ k\Omega$ and $h_{fe} = 100$.

Solution

(a) *CE* circuit

$$A_v = \frac{h_{fe}(R_C \| R_L)}{h_{ie}} = \frac{100\ (5.6\ k\Omega \| 33\ k\Omega)}{1.5\ k\Omega}$$

$$= 319$$

$$Z_b = h_{ie} = 1\ k\Omega$$

$$Z_i = Z_b \| R_1 \| R_2 = 1\ k\Omega \| 100\ k\Omega \| 47\ k\Omega$$

$$= 970\ \Omega$$

$$v_i = \frac{v_s \times Z_i}{r_s + Z_i} = \frac{50\ mV \times 970\ \Omega}{600\ \Omega + 970\ \Omega}$$

$$= 30.9\ mV$$

$$v_o = A_v \times v_i = 319 \times 30.9 \text{ mV}$$
$$= 9.9 \text{ V}$$

(b) CB circuit

$$A_v = \frac{h_{fe}(R_C \| R_L)}{h_{ie}} = 319$$

from Table 6-3, $Z_e = \dfrac{h_{ie}}{1 + h_{fe}} = \dfrac{1.5 \text{ k}\Omega}{1 + 100}$

$$= 14.85 \text{ }\Omega$$

$$Z_i = Z_e \| R_E = 14.85 \text{ }\Omega \| 5.6 \text{ k}\Omega$$
$$= 14.81 \text{ }\Omega$$

$$v_i = \frac{v_s \times Z_i}{r_s + Z_i} = \frac{50 \text{ mV} \times 14.81 \text{ }\Omega}{600 \text{ }\Omega + 14.81 \text{ }\Omega}$$
$$= 1.2 \text{ mV}$$

$$v_o = A_v \times v_i = 319 \times 1.2 \text{ mV}$$
$$= 383 \text{ mV}$$

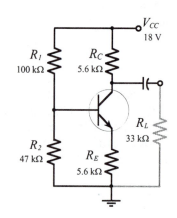

Figure 6-43
Circuit for Example 6-12.

Practise Problems

6-8.1 Using the appropriate CE equations from Table 6-3, calculate Z_i, Z_o, and A_v for a CB circuit (as in Fig. 6-38) with the following component values and parameters: $R_1 = 56 \text{ k}\Omega$, $R_2 = 39 \text{ k}\Omega$, $R_C = 5.6 \text{ k}\Omega$, $R_E = 3.3 \text{ k}\Omega$, $R_L = 47 \text{ k}\Omega$, $h_{ie} = 1 \text{ k}\Omega$, $h_{fe} = 100$.

6-8.2 Determine v_o for the circuit in Problem 6-8.1 when a 30 mV ac signal with $r_s = 400 \text{ }\Omega$ is capacitor-coupled to the transistor emitter.

Chapter-6 Review Questions

Section 6-1

6-1 Sketch the circuit of a CE amplifier with a capacitor-coupled signal source and capacitor-coupled load. Explain the need for coupling capacitors, and discuss the correct polarity for connecting the capacitors.

6-2 Explain what is meant by *ac* degeneration in a transistor circuit. Show how *ac* degeneration is eliminated in voltage divider bias and collector-to-base bias circuits.

Section 6-2

6-3 Sketch the *ac* equivalent circuit for a *CE* circuit with voltage divider bias, a bypassed emitter resistor, a capacitor-coupled signal source, and a capacitor-coupled load. Briefly explain.

6-4 Sketch the *ac* equivalent circuit for a *CE* circuit with collector-to-base bias [as in Fig. 6-3(a)], a capacitor-coupled signal source, and a capacitor-coupled load. Briefly explain.

6-5 Discuss the purpose of an *ac* load line for a transistor circuit. State how *ac* load lines differ from *dc* load lines for circuits with bypassed emitter resistors, and for circuits with capacitor-coupled loads.

Section 6-3

6-6 Sketch the *T*-equivalent (*r*-parameter) circuit for a transistor connected in *CB* configuration. Identify each component of the circuit and discuss its origin. Also, sketch a simplified form of the transistor low-frequency equivalent circuit.

6-7 Sketch the *h*-parameter equivalent circuit for a transistor connected in *CE* configuration. Identify each component of the circuit and discuss its origin.

6-8 Define h_{ie}, h_{fe}, h_{oe}, and h_{FE}, and show how each parameter may be derived from the transistor characteristics. Also, state typical *h*-parameter values for a low-current transistor.

6-9 Sketch an approximate *h*-parameter *CE* transistor model that may be used for most purposes. Briefly explain.

6-10 Draw a *CE* transistor hybrid-π model. Briefly explain.

6-11 Define r'_e, and write equations for calculating r'_e at 25°C, and at other temperatures.

Section 6-4

6-12 Briefly explain the operation of the *CE* circuit described in Question 6-3.

6-13 Draw an *h*-parameter equivalent circuit for the *CE* circuit described in Question 6-3. Identify all components.

6-14 Write equations for Z_i, Z_o, and A_v for the *h*-parameter circuit in Question 6-13.

6-15 Draw an *h*-parameter equivalent circuit for the *CE* circuit described in Question 6-4. Identify all components.

6-16 Write equations for Z_i, Z_o, and A_v for the *h*-parameter circuit in Question 6-15.

6-17 Sketch a base-biased *pnp* transistor *CE* circuit. Include a capacitor-coupled signal source and load resistor. Draw the *h*-parameter equivalent circuit identifying all components.

6-18 Write equations for Z_i, Z_o, and A_v for the *h*-parameter circuit in Question 6-17.

Section 6-5

6-19 Referring to the *CE* circuit described in Question 6-3, explain the effect of removing the bypass capacitor from R_E.

6-20 Draw an *h*-parameter equivalent circuit for the *CE* circuit in Question 6-19.

6-21   Write equations for Z_i, Z_o, and A_v for the h-parameter circuit in Question 6-20.

6-22   Sketch a collector-to-base biased *pnp* transistor *CE* circuit with two center-bypassed base resistors and an unbypassed emitter resistor. Include a capacitor-coupled signal source and load resistor. Draw the h-parameter equivalent circuit identifying all components.

6-23   Write equations for Z_i, Z_o, and A_v for the h-parameter circuit in Question 6-22.

Section 6-6

6-24   Sketch a practical *npn* transistor *CC* circuit that uses base bias. Include a capacitor-coupled signal source and load resistor. Explain the circuit operation.

6-25   Draw the h-parameter equivalent circuit for the circuit in Question 6-24. Identify all components.

6-26   Write equations for Z_i, Z_o, and A_v for the h-parameter circuit in Question 6-25.

6-27   Sketch the h-parameter equivalent for a *CC* circuit with voltage divider bias, a capacitor-coupled signal source, and a capacitor-coupled load.

6-28   Write the Z_i and Z_o equations for the h-parameter circuit in Question 6-27.

6-29   Sketch a base-biased *pnp* transistor *CC* circuit. Include a capacitor-coupled signal source and load resistor.

6-30   Draw the h-parameter equivalent circuit for the circuit in Question 6-29.

Section 6-7

6-31   Sketch a practical *npn* transistor *CB* circuit that uses collector-to-base bias. Include a capacitor-coupled signal source and load resistor. Explain the circuit operation.

6-32   Draw the h-parameter equivalent circuit for the circuit in Question 6-31. Identify all components.

6-33   Write equations for Z_i, Z_o, and A_v for the *CB* h-parameter circuit in Question 6-32.

6-34   Sketch a practical *pnp* transistor *CB* circuit that uses voltage divider bias. Include a capacitor-coupled signal source and load resistor.

6-35   Sketch the h-parameter equivalent for a *CB* circuit with voltage divider bias, a capacitor-coupled signal source, and a capacitor-coupled load.

6-36   Write the Z_i, Z_o, and A_v equations for the *CB* h-parameter circuit in Question 6-35.

Section 6-8

6-37 Compare the performance of *CE*, *CC*, and *CB* circuits, and discuss typical applications of each type of circuit.

6-38 Write equations for the impedances 'seen looking into' the base, emitter, and collector terminals of a transistor in a circuit. Briefly explain each equation.

Chapter-6 Problems

Section 6-1

6-1 Calculate V_{CE} for the circuit in Fig. 6-43 when the load is capacitor-coupled, as shown, and when the load is directly connected to the collector terminal.

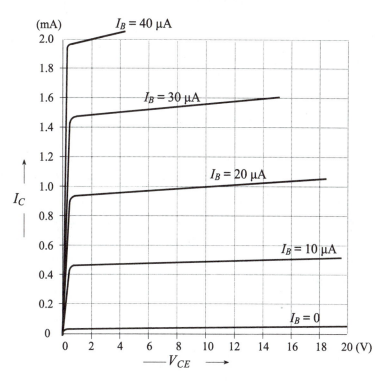

Figure 6-44

6-2 Determine V_B for the circuit in Fig. 6-43 when a 600 Ω signal source is capacitor-coupled to the input, and when the signal source is directly connected.

Section 6-2

6-3 The transistor characteristics for the circuit in Fig. 6-43 are shown in Fig. 6-44. Draw the *dc* and *ac* load lines for the circuit when it is operated in *CE* configuration with R_E bypassed. Also, determine the maximum possible symmetrical output voltage swing.

6-4   Draw the new *ac* load line for the circuit in Problem 6-3 when R_L is changed to 12 kΩ.

6-5   Using the device characteristics in Fig. 6-44, draw the *dc* and *ac* load lines for the *CE* circuit shown in Fig. 6-45.

6-6   The *CE* circuit in Fig. 6-46 uses collector-to-base bias with a bypass capacitor. Draw the *dc* and *ac* load lines for the circuit using the device characteristics in Fig. 6-44. The transistor has h_{FE} = 100.

Section 6-3

6-7   Referring to the *CE* input and output characteristics in Figs. 4-27 and 4-28, determine h_{ie}, h_{oe}, and h_{fe} for V_{CE} = 6 V, and I_B = 40 μA.

6-8   A transistor in a *CE* circuit has a constant I_B level of 40 μA. When V_{CE} = 5 V, I_C is measured as 4.9 mA, and when V_{CE} is adjusted to 10 V, I_C becomes 5 mA. Calculate h_{oe} and h_{FE}.

6-9   A transistor has V_{CE} maintained constant at 7.5 V, and I_C measured at several levels of I_B as follows:

I_B	25	50	75	100	(μA)
I_C	3.06	6.12	9.2	12.3	(mA)

Determine the h_{fe} value for the transistor.

6-10   The transistor in Problem 6-9 has V_{CE} maintained constant at 7.5 V, and I_B measured at several levels of V_{BE} as follows:

V_{BE}	0.68	0.72	0.76	0.8	(V)
I_B	20	40	60	80	(μA)

Determine the h_{ie} value for the transistor.

6-11   The transistor in Problems 6-9 and 6-10 has I_B maintained constant at 40 μA, and corresponding levels of I_C and V_{CE} measured as follows:

V_{CE}	2	4	6	8	(V)
I_C	4.7	4.8	4.9	5	(mA)

Determine the h_{oe} value for the transistor.

6-12   Calculate α and r_c for the transistor referred to in Problems 6-9 through 6-11.

6-13   Determine the value of r'_e for a transistor with a 2 mA emitter current if its junction temperature is 80°C.

6-14   The collector and base currents of a transistor are measured as 1.28 mA and 20 μA, respectively. Calculate r'_e, h_{ie}, and r_π for the transistor.

Section 6-4

6-15   The *CE* circuit in Fig. 6-47 has the following transistor parameters: h_{ie} = 1 kΩ, h_{fe} = 85, h_{oe} = 2 μS. Calculate Z_i, Z_o, and A_v.

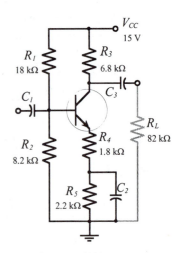

Figure   6-45

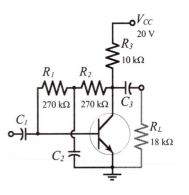

Figure   6-46

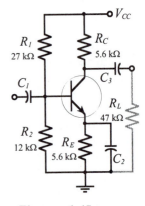

Figure   6-47

6-16 Recalculate Z_i, Z_o, and A_v for the circuit in Problem 6-15 when the transistor is replaced with one having $h_{ie} = 1.4$ kΩ and $h_{fe} = 55$.

6-17 The base biased *CE* circuit in Fig. 6-48 has the following transistor parameters: $h_{ie} = 1.3$ kΩ, $h_{fe} = 40$, $h_{oe} = 1.5$ μS. Calculate Z_i, Z_o, and A_v.

6-18 The collector-to-base biased *CE* circuit in Fig. 6.49 has the following transistor parameters: $h_{ie} = 1.3$ kΩ, $h_{fe} = 40$, $h_{oe} = 1.5$ μS. Calculate Z_i, Z_o, and A_v.

6-19 Recalculate Z_i, Z_o, and A_v for the circuit in Problem 6-17 when the transistor is replaced with one having $h_{ie} = 1.5$ kΩ, $h_{oe} = 1$ μS, and $h_{fe} = 100$.

6-20 The transistor in the circuit in Problem 6-18 is replaced with one having $h_{ie} = 1.5$ kΩ, $h_{oe} = 1$ μS, and $h_{fe} = 100$. Recalculate Z_i, Z_o, and A_v.

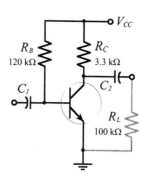

Figure 6-48

Section 6-5

6-21 For the circuit in Problem 6-15, recalculate Z_i and A_v when the bypass capacitor is removed from R_E.

6-22 The circuit in Problem 6-15 has R_E replaced with two series-connected resistors, $R_{E1} = 1.5$ kΩ, and $R_{E2} = 3.9$ kΩ. Calculate Z_i, and A_v when only R_{E1} is capacitor-bypassed, and when only R_{E2} is capacitor-bypassed.

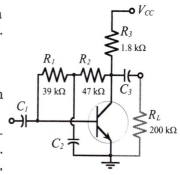

Figure 6-49

Section 6-6

6-23 The transistor parameters for the *CC* circuit in Fig. 6-50 are: $h_{ie} = 1$ kΩ, $h_{fe} = 85$, $h_{oe} = 2$ μS. Calculate Z_i and Z_o.

6-24 Recalculate Z_i and Z_o for Problem 6-23 when the transistor is replaced with one having $h_{ie} = 1.4$ kΩ and $h_{fe} = 55$,

6-25 The *CC* circuit in Fig. 6-51 has the following transistor parameters: $h_{ie} = 1.3$ kΩ and $h_{fe} = 40$. Calculate Z_i and Z_o.

6-26 A *CC* circuit using voltage divider bias (as in Fig. 6-50) has the following component values: $R_1 = 33$ kΩ, $R_2 = 47$ kΩ, $R_E = 15$ kΩ, $R_L = 47$ kΩ. The transistor parameters are: $h_{ie} = 1.2$ kΩ and $h_{fe} = 120$. Calculate Z_i and Z_o.

Section 6-7

6-27 The *CB* circuit in Fig. 6-52 has a transistor with the following parameters: $h_{ie} = 1$ kΩ, $h_{fe} = 85$, $h_{oe} = 2$ μS. Calculate Z_i, Z_o, and A_v.

6-28 Recalculate Z_i and A_v for the *CB* circuit in Problem 6-27 when the transistor is replaced with one having $h_{ie} = 1.4$ kΩ and $h_{fe} = 55$.

6-29 Calculate Z_i, Z_o, and A_v for the *CB* circuit in Fig. 6-53. The transistor parameters are: $h_{ie} = 1.3$ kΩ, $h_{fe} = 40$, $h_{oe} = 1.5$ μS.

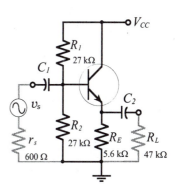

Figure 6-50

6-30   Recalculate Z_i, Z_o, and A_v for the circuit in Problem 6-29 when the transistor is replaced with one having h_{ie} = 1.5 kΩ, h_{oe} = 1 μS, and h_{fe} = 100.

6-31   A *CB* circuit (as in Fig. 6-52) has the following component values and transistor parameters: R_1 = 33 kΩ, R_2 = 22 kΩ, R_C = 3.3 kΩ, R_E = 3.9 kΩ, R_L = 39 kΩ, h_{ie} = 1.6 kΩ, h_{fe} = 200, h_{oe} = 1.5 μS. Calculate Z_i, Z_o, and A_v.

Section 6-8

6-32   The transistor in the *CE* circuit in Problem 6-15 is replaced with a *2N3904* transistor (see Appendix 1-5). Calculate the maximum and minimum values of Z_i, Z_o, and A_v.

6-33   The *CC* circuit in Problem 6-23 is reconstructed to use a *2N3903* transistor. Calculate the maximum and minimum values of Z_i and Z_o.

6-34   The transistor in the *CB* circuit in Problem 6-31 is replaced with a *2N3903* transistor. Calculate the maximum and minimum values of Z_i and Z_o.

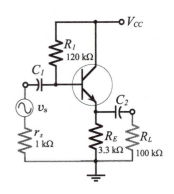

Figure 6-51

Practise Problem Answers

6-1.1   9.5 V, 8.9 V
6-2.1   point C at 2.07 mA and 3.3 V from Q
6-3.1   133, 20 μS
6-3.2   0.993, 6.7 MΩ
6-3.3   1.76 kΩ
6-4.1   849 Ω, 5.6 kΩ, -517
6-4.2   190, 161
6-5.1   5.52 kΩ, 5.6 kΩ, -1.87
6-6.1   21.9 kΩ, 14 Ω
6-7.1   9.87 Ω, 5.6 kΩ, 500
6-7.2   224 Ω, 20.6
6-8.1   9.87 Ω, 5.6 kΩ, 500
6-8.2   361 mV

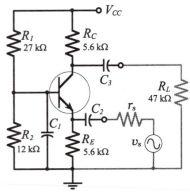

Figure 6-52

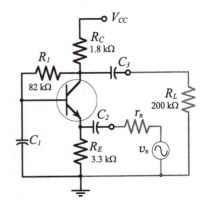

Figure 6-53

Chapter *7*

Semiconductor Device and IC Fabrication

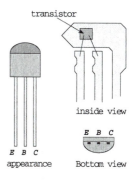

transistor

inside view

E B C

appearance   Bottom view

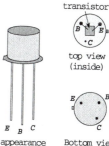

transistor

top view
(inside)

E B C

appearance   Bottom view

Chapter Contents

Objectives

You will be able to:

1 Describe the process of preparing semiconductor material for device manufacture, and explain diffusion and epitaxial growth.

2 Draw sketches to show the fabrication of alloy and diffused diodes.

3 Explain the manufacturing requirements for transistors to produce satisfactory current gain, frequency response, power dissipation, etc.

4 Describe the various transistor fabrication methods and their performance characteristics.

5 Discuss manufacturing methods for the production of integrated circuits.

6 Draw sketches to show the construction of monolithic IC components, and discuss problems that occur with the use of IC components.

7 Sketch various diode, transistor, and IC packages, and show the terminal identification systems.

Introduction

The method employed to manufacture a semiconductor device determines its electrical characteristics and, thus, dictates its applications. For example, low-current fast-switching diodes and transistors must be fabricated differently from high-power devices. The process that produces a transistor with high current gain may result in undesirable large junction capacitances. Also, a technique that gives a small junction capacitance might produce devices with low breakdown voltages. An *integrated circuit (IC)* consists of many transistors, diodes, and other components in one package. Usually, all of the components are fabricated on one small silicon chip.

7-1   Processing of Semiconductor Materials

Preparation of Silicon and Germanium

Silicon is one of the commonest elements on earth. It occurs as silicon dioxide (SiO_2) and as silicates; mixtures of silicon and other materials. Germanium is derived from zinc or copper ores. When converted to bulk metal, silicon and germanium contain large quantities of impurities. Both metals must be carefully refined before they can be used for semiconductor device manufacture.

Semiconductor material is normally *polycrystalline* after it is refined. This means that it is made up of many individual formations of atoms with no overall fixed pattern. For use in semiconductors, the metal must be converted into *single-crystal* material; that is, all of its atoms must be arranged into a single pattern.

In its final condition for device manufacture, silicon and germanium are in the form of single-crystal bars about 2.5 cm in diameter and perhaps 30 cm long. The bars are sliced into disc-shaped wafers about 0.4 cm thick, and the wafers are polished to a mirror surface. Several thousand devices are usually fabricated on the surface of each wafer, then the wafers are scribed and cut like glass, (see Fig. 7-1).

(a) Semiconductor bar sliced into wafers

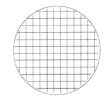

(b) Scribed wafer

Figure 7-1
Bars of single-crystal semiconductor material are sliced into wafers for device manufacture.

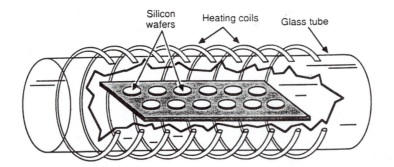

Silicon wafers   Heating coils   Glass tube

Figure 7-2
Semiconductor wafers are heated in an atmosphere containing n-type or p-type impurities which diffuse into the surface of the wafers.

Diffusion

For processing, semiconductor wafers contained in an enclosure are heated by means of *radio frequency* (RF) heating coils. This is illustrated in Fig. 7-2. When wafers of *n*-type material are raised to a high temperature in an atmosphere containing *p*-type impurity atoms some of the impurities *diffuse* into each wafer. This converts the outer layer of the *n*-type material into *p*-type, (Fig 7-3). The diffusion process can be continued by further heating the wafers in an atmosphere containing *n*-type impurity atoms, so that *npn* layers are produced, as illustrated. Because the diffusion process is very slow (about 2.5 μm per hour), very narrow diffused regions can be created by careful timing.

Epitaxial Growth

The *epitaxial* process is very similar to the diffusion process, except that silicon or germanium atoms are contained in the gas surrounding the semiconductor wafers. The semiconductor atoms in the gas *grow* (accumulate) on the wafer in the form of a thin layer, (Fig. 7-4). This layer is single-crystal material and it may be *p*-type or *n*-type, according to the impurity content of the gas. The epitaxial layer may be doped by the diffusion process.

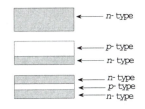

Figure 7-3
The diffusion process converts n-type semiconductor into p-type, and vice versa.

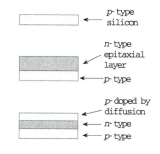

Figure 7-4
Epitaxial growth adds layers of n-type or p-type material on the surface of a wafer.

Section 7-1 Review
7-1.1   Briefly explain polycrystalline material, single-crystal material, diffusion, and epitaxial growth.

7-2 Diode Fabrication and Packaging

Alloy and Diffused Diodes

Two commonly used techniques for diode manufacture are the *alloy* method and the *diffusion* method. To construct an alloy diode, a *pn*-junction is formed by melting a tiny pellet of aluminum (or some other *p*-type impurity) on the surface of an *n*-type crystal. Alternatively, an *n*-type impurity may be melted on the surface of a *p*-type crystal. The process is illustrated in Fig. 7-5(a).

Figure 7-5(b) shows the diffusion technique for diode construction. When an *n*-type semiconductor is heated in a chamber containing an acceptor impurity in vapour form, some of the acceptor atoms are diffused (or absorbed) into the *n*-type crystal. This produces a *p*-region in the *n*-type material and thus creates a *pn*-junction. The size of the *p*-region can be precisely defined by uncovering only part of the *n*-type material during the diffusion process, (the rest of the surface has a thin coating of silicon dioxide). Metal contacts are electroplated on to each region for connecting leads.

The diffusion technique lends itself to the simultaneous fabrication of many hundreds of diodes on one small disc of semiconductor material. This process is also used in the production of transistors and integrated circuits.

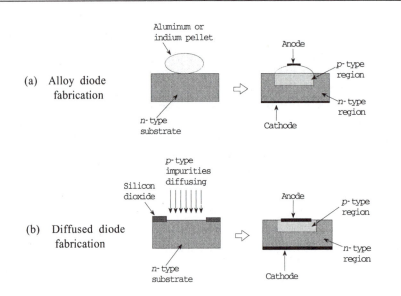

(a) Alloy diode fabrication

(b) Diffused diode fabrication

Figure 7-5
Diode fabrication methods. For the alloy diode, a pellet containing p-type impurities is melted into n-type material. A diffused diode is created by heating a wafer of n-type material in an atmosphere containing p-type impurities.

Diode Packaging

As discussed in Section 2-1, semiconductor diodes vary widely in size, depending on their application. Figure 7-6 shows typical single diode packages for low, medium, and high current devices. In Fig. 7-7 multiple diode packages are illustrated. Two diodes in a surface mount package (see Section 7-7) are shown in Fig. 7-7(a), and two in a power-transistor-type package are illustrated in Fig. 7-7(b). Note from the circuits that the cathode terminals are internally commoned in each case.

A rectifier assembly is shown in Fig. 7-7(c). This is four diodes connected to function as a bridge rectifier and contained in one package. One side of the package has a metal plate for mounting on a heat sink (see Section 8-8). The two output terminals are identified as **+** and **-**, and the other two terminals are for the *ac* input.

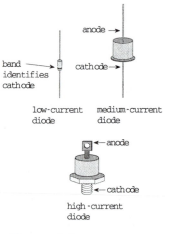

Figure 7-6
Individual diodes vary in size according to the current they are required to pass.

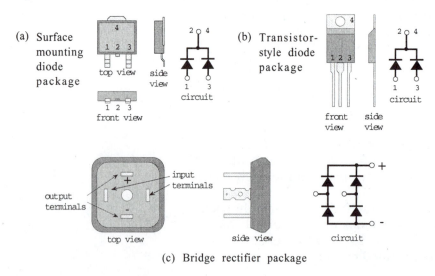

(a) Surface mounting diode package

(b) Transistor-style diode package

(c) Bridge rectifier package

Figure 7-7
Some multiple diode packages

7-3   Transistor Construction and Performance

Current Gain

Good current gain requires that most charge carriers from the emitter pass rapidly to the collector. So, there should be little outflow of charge carriers via the base terminal, and there should be few carrier recombinations within the base. These conditions dictate a very narrow, lightly doped, base, (see Fig. 7-8).

High Power

High-power transistors must have large emitter-base surfaces to provide the required quantity of charge carriers. Large collector-base surfaces are also required to dissipate power without overheating the collector-base junction, (Fig. 7-8).

Frequency Response

For the highest possible frequency response, the transistor base region must be very narrow to ensure a short transit time of charge carriers from emitter to collector, (see Fig. 7-9). Input capacitance must also be kept to a minimum, and this requires a small area emitter-base junction as well as a highly-resistive (lightly doped) base region.

Because power transistors require large junction surfaces, and high-frequency performance demands small junction areas, there is a conflict in high-frequency power transistors. To keep the junction area to a minimum and still provide adequate charge-carrier emission, the emitter-base junction is usually in the form of a long thin zig-zag strip.

Switching Transistors

Fast switching demands the same low junction capacitance required for good high-frequency performance. A switching transistor should have a low *saturation voltage* and a short *storage time,* (see Section 8-5). For low saturation voltages, the collector region must have low resistivity, (Fig. 7-9). Short storage time requires fast recombination of charge carriers left in the depletion region at the collector-base junction. Fast recombination is assisted by additional doping of the collector with gold atoms.

Breakdown and Punch-Through

The transistor collector-base junction is normally reverse biased, so the maximum collector voltage is limited by the junction reverse breakdown voltage. To achieve high breakdown voltages, either the collector or the base must be very lightly doped. If the base is more lightly doped than the collector, the collector-base depletion region penetrates deeply into the base. This will cause transistor breakdown by *punch-through* when it links with the emitter-base

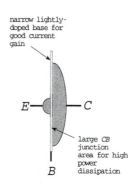

Figure 7-8
High current gain requires a narrow lightly-doped transistor base, and high power dissipation demands a large CB junction area.

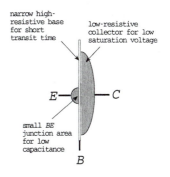

Figure 7-9
A small BE junction area and narrow base are required for high frequency performance and fast switching. Low saturation voltage is promoted by a low-resistive collector region.

depletion region. To avoid punch through, the collector region close to the base is usually more lightly doped than the base. The collector-base depletion region then spreads into the collector, rather than into the base.

Section 7-3 Review

7-3.1 Briefly discuss the requirements for constructing transistors with (a) good current gain, (b) high frequency response.

7-3.2 Explain the design requirements for transistors with (a) high power dissipation, (b) high breakdown voltage.

7-3.3 List the requirements for switching transistor construction.

7-4 Transistor Fabrication

Alloy Transistors

For manufacture of *alloy transistors*, single-crystal *n*-type wafers are scribed into many small sections, or *dice*, each of which forms the substrate for one transistor. A small pellet of *p*-type material is melted on one surface of each section until it partially penetrates and forms an alloy with the substrate, [Fig. 7-10(a)], thus creating a *pn*-junction. The process is repeated on the other side of the wafer to constitute a *pnp* transistor.

One of the junctions has a large area and one has a small area, as illustrated in Fig. 7-10(b). The small-area junction becomes the emitter-base junction, and the larger one becomes the collector-base junction. One reason for this is that the large area junction most easily collects all of the charge carriers emitted from the small-area junction. Another more important reason is that most of the power is dissipated at the transistor collector-base junction.

Suppose a silicon transistor has $I_C = 10$ mA and $V_{CE} = 10$ V, (see Fig. 7-11). The total power dissipate in the transistor is,

$$P = V_{CE} \times I_C = 10 \text{ V} \times 10 \text{ mA}$$
$$= 100 \text{ mW}$$

The base-emitter voltage is $V_{BE} = 0.7$ V, and the collector-base voltage is,

$$V_{CB} = V_{CE} - V_{BE} = 10 \text{ V} - 0.7 \text{ V}$$
$$= 9.3 \text{ V}$$

The power dissipated at the base-emitter junction is,

$$V_{BE} \times I_C = 0.7 \text{ V} \times 10 \text{ mA}$$
$$= 7 \text{ mW}$$

and, the power dissipated at the collector-base junction is,

$$V_{CB} \times I_C = 9.3 \text{ V} \times 10 \text{ mA}$$
$$= 93 \text{ mW}$$

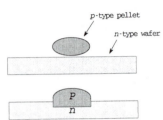

(a) *p*-type pellet melted on an *n*-type wafer

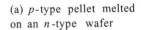

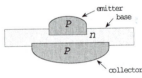

(b) Alloy transistor

Figure 7-10
A pnp alloy transistor is constructed by melting p-type pellets on the surface of an n-type wafer.

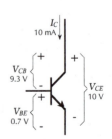

Figure 7-11
Most of the power dissipation in a transistor occurs at the collector-base junction.

Microalloy Transistors

Because very narrow base widths are difficult to obtain with alloy transistors, they are not suitable for high frequency applications. To improve the high-frequency performance, holes are first etched into the wafer from each side, leaving a very thin layer, (Fig 7-12). By a plating process, surfaces of impurity material are formed on each side of the thin n-type portion. Heat is then applied to alloy the impurities into the base region. This process produces very thin base regions and good high-frequency performance. The device is termed a *microalloy* transistor.

Microalloy-Diffused Transistors

In microalloy transistors, the collector-base depletion region penetrates deeply into the very thin base. Thus, a major disadvantage is that punch-through can occur at very low collector voltages. *Microalloy-diffused* transistors use a substrate that is initially undoped. After the holes are etched into each side of the wafer to produce the thin base region, the base is doped by diffusion from the collector side. The diffusion can be carefully controlled so that the base region is heavily doped at the collector side, with the doping becoming progressively less until the material is almost intrinsic (undoped) at the emitter. With this kind of doping, the collector-base depletion region penetrates only a short distance into the base, so that high punch-through voltages are achieved.

Diffused Mesa

In the production of *mesa transistors*, several thousand transistors are simultaneously formed on the wafer, by the diffusion process. As shown in Fig. 7-13, the main body of the wafer becomes the n-type collectors, the diffused p-regions become the bases, and the final n-regions are the emitters. Metal films are deposited on the base and emitter surfaces for contacts.

The transistors could be separated by the usual process of scribing lines on the surface of the wafer and breaking it into individual units. However, this produces very rough edges that result in relatively high leakage currents between collector and base. So, before cutting the wafer, the transistors are isolated by etching away the unwanted portions of the diffused area to form troughs between devices. As illustrated, the base and emitter regions now project above the main wafer which forms the collector region. This is the *mesa* structure. The narrow base widths that can be achieved by the diffusion process make the mesa transistor useful at very high frequencies.

Epitaxial Mesa

One of the disadvantages of the process just described is that, because the collector region is highly resistive, diffused mesa transistors have a high saturation voltage, (see Section 5-10). Such devices are unsuitable for saturated switching applications. This same characteristic (high collector resistance) is desirable for

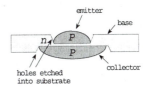

Figure 7-12
A very narrow base region gives microalloy transistors good high-frequency performance.

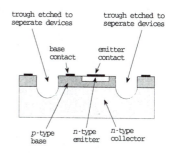

Figure 7-13
Mesa transistors are created by etching troughs between individual devices to give low collector-base leakage currents.

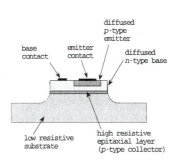

Figure 7-14
Epitaxial mesa transistor. The epitaxial process combined with the mesa structure gives high breakdown voltages and low saturation levels.

high punch-through voltage. One method of achieving both high punch-through voltage and low saturation level uses the epitaxial process, to produce an *epitaxial mesa* transistor.

Starting with a low-resistive (highly doped) wafer, a thin, highly resistive epitaxial layer is grown. This layer becomes the collector, and the base and emitter are diffused as already discussed. The arrangement is illustrated in Fig. 7-14. The punch-through voltage is high because the collector-base depletion region spreads deepest into the lightly doped collector. Saturation voltage is low, because the collector region is very narrow and the main body of the wafer through which collector current must flow has a very low resistance.

Diffused Planar Transistor

In all the previously described transistors the collector-base junction is exposed (within the transistor package), and substantial charge carrier leakage can occur at the junction surface. In the *diffused planar* transistor illustrated in Fig. 7-15, the collector-base junction is covered with a layer of silicon dioxide. This construction gives a very low collector-base leakage current (I_{CBO}); typically 0.1 nA.

Annular Transistor

A problem which occurs particularly with *pnp* planar transistors is the *induced channel*. This results when a relatively high voltage is applied to the silicon dioxide surface, (at one of the terminals). Consider the *pnp* structure shown in Fig. 7-16(a). If the surface of the silicon dioxide has a positive voltage, minority charge carriers within the lightly doped *p*-type substrate are attracted by the positive potential. The minority charge carriers concentrate at the upper edge of the substrate and form an *n*-type channel from the base to the edge of the device. This becomes an extension of the *n*-type base region and results in charge carrier leakage at the exposed edge of the collector-base junction.

The problem arises because the *p*-type substrate is highly resistive. If it were heavily doped with *p*-type charge carriers, the concentration of *n*-type carriers would be absorbed (the electrons would be swallowed by holes). The introduction of a heavily doped *p*-type ring around the base, as in Fig. 7-16(b), interrupts the induced channel and isolates the collector-base junction from the device surface. The *annular transistor*, therefore, is a high-voltage device with the low collector-base leakage of the planar transistor.

Section 7-4 Review

7-4.1   Briefly explain the construction of alloy, microalloy, and microalloy-diffused transistors.

7-4.2   Describe diffused mesa and epitaxial mesa transistor construction.

7-4.3   Discuss the construction of diffused planar and annular transistors.

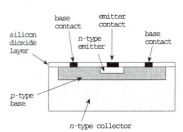

Figure 7-15
A diffused planar transistor has the collector-base junction covered, so that I_{CBO} is reduced to very low levels.

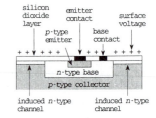

(a) *n*-type channel induced by terminal voltage

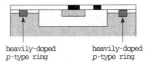

(b) Heavily doped *p*-type ring interrupts the induced channel

Figure 7-16
Annular transistors have a heavily-doped p-type ring included to interrupt unwanted n-type channels that are induced by high voltages.

7-5 Integrated Circuits

IC Types

An *integrated circuit* (*IC*) consists of several interconnected transistors, resistors, etc., all contained in one small package with external connecting terminals. The circuit may be entirely self-contained, requiring only input and output connections and a supply voltage. Alternatively, a few external components may have to be connected to make the circuit operate.

Integrated circuits may be classified in terms of their function as *analog* or *digital*. Analog *ICs* (also termed *linear ICs*) may be amplifiers, voltage regulators, etc. Digital *ICs* (logic gates and counting circuits) contain transistors which are normally either in a switched-*off* or switched-*on* state.

The techniques used in manufacturing integrated circuits provide another method of classification. The major *IC* manufacturing technique is termed *monolithic*. Other fabrication techniques are *thin-film*, *thick-film*, and *hybrid*.

Monolithic Integrated Circuits

In a *monolithic* integrated circuit all components are fabricated by the diffusion process on a single chip of silicon. Interconnections between components are provided on the surface of the structure, and external connecting wires are taken out to terminals, as illustrated in Fig. 7-17. The vast majority of integrated circuits are monolithic because it is the most economical process for mass production.

Thin Film

Thin-film integrated circuits are constructed by depositing films of conducting material on the surface of a glass or ceramic base. By controlling the width and thickness of the films, and by using different materials selected for their resistivity, resistors and conductors are fabricated. Capacitors are produced by sandwiching a film of insulating oxide between two conducting films. Inductors are made by depositing metal film in a spiral formation. Transistors and diodes are usually tiny discrete components connected into the circuit.

Thick Film

Thick-film integrated circuits are sometimes referred to as *printed thin-film* circuits. In this process, silk-screen printing techniques are employed to create the desired circuit pattern on a ceramic substrate. The inks used are pastes which have conductive, resistive, or dielectric properties. After printing, the circuits are high-temperature-fired in a furnace to fuse the films to the substrate. Thick-film passive components are fabricated in the same way as those in thin-film circuits. As with thin-film circuits, active components are added as separate devices. A portion of a thick-film circuit is shown in Fig. 7-18.

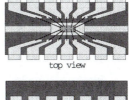

(a) Can-type *IC* enclosure

top view

side view

(b) Plastic *IC* enclosure

Figure 7-17
Monolithic integrated circuits consist of a single-chip circuit contained in a metal or plastic enclosure with connecting pins.

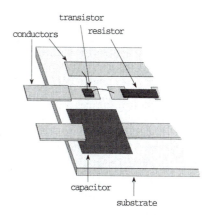

Figure 7-18
Enlarged portion of a thick film integrated circuit.

Integrated circuits produced by thin- or thick-film techniques usually have better component tolerances and give better high-frequency performance than monolithic integrated circuits.

Hybrid

Figure 7-19 illustrates the structure of a *hybrid* or *multichip* integrated circuit. As the name implies, the circuit is constructed by interconnecting a number of individual chips. The active components are diffused transistors or diodes. The passive components may be groups of diffused resistors or capacitors on a single chip, or they may be thin-film components. Connections between chips are provided by fine wire or metal film.

Like thin-film and thick-film *ICs*, multichip circuits usually have better performance than monolithic circuits. Although the process is too expensive for mass production, multichip techniques are quite economical for small quantities and are frequently used as prototypes for monolithic integrated circuits.

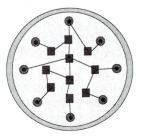

Figure 7-19
Uncovered top view of hybrid integrated circuit. Individual components are interconnected to form a single circuit.

Section 7-5 Review
7-5.1   Briefly define monolithic, thin-film, thick-film, and hybrid integrated circuits.

7-6 Integrated Circuit Components

Transistors and Diodes

The epitaxial planar diffusion process described in Section 7-4 is normally employed for the manufacture of *IC* transistors and diodes. Collector, base, and emitter regions are diffused into a silicon substrate and surface terminals are provided for connection, as illustrated in Fig. 7-20(a).

In discrete transistors the substrate is normally used as a collector. If this were done with transistors in a monolithic integrated circuit, all transistors fabricated on one substrate would have their collectors connected together. For this reason, separate collector regions must be diffused into the substrate.

Even though separate collector regions are formed, they are not completely isolated from the substrate. Figure 7-20(a) shows that an unwanted (*parasitic*) *pn*-junction is formed by the substrate and the transistor collector region. If the circuit is to function correctly, these junctions must never become forward biased. Thus, in the case of a *p*-type substrate, the substrate must always be kept negative with respect to the transistor collectors, [Fig. 7-20(b)]. This requires that the substrate be connected to the most negative terminal of the circuit supply.

The parasitic junctions can affect the circuit performance even when they are reverse biased. The junction reverse leakage current can be a serious problem in circuits that operate at very low current levels. The capacitance of the reverse-biased junction can affect the circuit high-frequency performance, and the junction

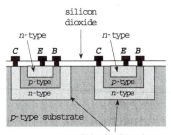

(a) Cross-section of two adjacent *IC* transistors

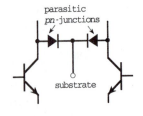

(b) Circuit diagram showing parasitic junctions

Figure 7-20
Integrated circuit components have unwanted (parasitic) interconnecting pn-junctions that must be kept reverse biased.

breakdown voltage imposes limits on the usable level of supply voltage. All these factors can be minimized by using highly resistive (lightly doped) material for the substrate.

Integrated circuit diodes are fabricated by diffusion exactly like transistors. Two of the regions are used to form one *pn*-junction. Alternatively, the collector region of a transistor may be connected directly to the base base (Fig. 7-21), so that the device behaves as a diode while operating similarly to a saturated transistor.

Resistors

The resistivity of semiconductor material is a function of doping density, so resistors can be produced by doping strips of material and providing terminals, (Fig. 7-22). The range of resistor values that may be produced by the diffusion process varies from ohms to hundreds of kilohms. The typical tolerance, however, may be no better than ±5%, and may even be as high as ±20%. However, if all resistors are diffused at the same time, then the *tolerance ratio* can be good. For example, several resistors with the same nominal value may all have a ±20% tolerance, but have actual resistance values within a few percent of each other.

Another method of producing resistors for integrated circuits uses the thin-film technique. In this process a metal film is deposited on a glass or silicon dioxide surface. The thickness, width, and length of the film are regulated to give a desired resistance value. Since diffused resistors can be processed while diffusing transistors, the diffusion technique is the least expensive and the most frequently used.

Capacitors

Because all *pn*-junctions have capacitance, capacitors may be produced by fabricating suitable junctions. As in the case of other diffused components, parasitic junctions are unavoidable. Both the parasitic and the main junction must be kept reverse biased to avoid direct current flow. The depletion region width and junction capacitance vary with changes in reverse bias, so for capacitance value stability, a *dc* reverse bias greater than maximum signal voltage levels must be maintained across the junction.

Integrated circuit capacitors may also be fabricated by utilizing the silicon dioxide surface layer as a dielectric. A heavily-doped *n*-region is diffused to form one plate of the capacitor. The other plate is created by depositing a film of aluminum on the silicon dioxide, (see Fig. 7-23). Voltages of any polarity may be employed with this type of capacitor, and the breakdown voltage is very much larger than that for diffused capacitors. The junction areas available for creation of *IC* capacitors are very small indeed, so that normally only picofarad capacitance values are possible.

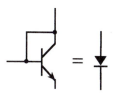

Figure 7-21
A transistor can be made to behave as a diode by connecting the collector and base.

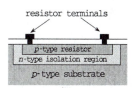

Figure 7-22
A monolithic IC resistor is created by doping a strip of semiconductor material to give the desired resistance.

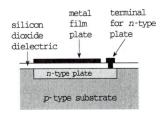

Figure 7-23
IC capacitor using silicon dioxide as the dielectric. The lower plate is heavily-doped n-type semiconductor, and the upper plate is a metal film.

Section 7-6 Review

7-6.1   Discuss the fabrication of integrated circuit transistors and diodes.

7-6.2   Explain how *IC* resistors and capacitors are constructed.

7-7 Transistor and IC Packaging

Discrete Transistor Packaging

Many low-power transistors are encapsuled in resin with protruding metal connecting leads, as illustrated in Fig. 7-24. This is known as a *TO-92* package. Note the emitter, base, and collector terminal connections. These are in the sequence *E,B,C*, left to right, looking at the bottom of the transistor with the flat side uppermost. This kind of package is economical, but it has a limited operating temperature range.

Figure 7-25 shows another method of low-power transistor packaging where the device is hermetically scaled in a metal can. Depending on the size of the can, this is identified as a *TO-5* to *TO-18* package. The transistor is first mounted with its collector in (mechanically and electrically) contact with a heat-conducting metal plate. Wires (insulated from the plate) pass through for the emitter and base connections, and the covering metal can is welded to the plate. Looking at the bottom of the transistor, and moving clockwise from the tab, the terminals are identified as *E,B,C*. The metal can enclosure gives a greater temperature range and greater power dissipation than resin encapsulation.

For high-power transistor packaging, a sealed can (*TO-3*) is often used, [Fig. 7-26(a)]. In this case the heat-conducting plate is much larger than in the *TO-5* package, and it is designed for mounting directly on a heat sink. Connecting pins are provided for the base and the emitter, and the collector connection is made by means of the metal plate. Note the terminal identifications, again looking at the bottom of the device.

Figure 7-26(b) shows a plastic-encapsuled power transistor with a metal tab for fastening to a heat sink. Once again, note the transistor terminal identifications, and note that the the metal tab is connected to the collector. Several other package types are available for low-power and high-power devices.

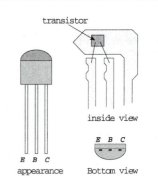

Figure 7-24
Low-power resin-encapsuled transistor; TO-92 package.

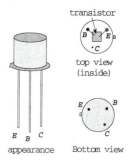

Figure 7-25
Low-power metal can-enclosed transistor; TO-18 package.

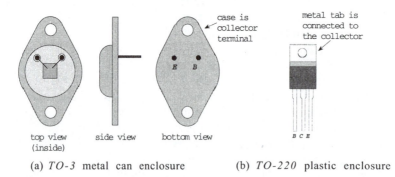

(a) *TO-3* metal can enclosure   (b) *TO-220* plastic enclosure

Figure 7-26
Metal can-type and plastic-encapsuled high power transistors.

IC Packages

Like semiconductor devices, integrated circuits must be packaged to provide mechanical protection and terminals for electrical connection. Several standard packages in general use are illustrated in Fig. 7-27.

The metal-can-type of container affords electromagnetic shielding for the *IC* chip which cannot be obtained with the plastic packages, [Fig. 7-27(a)]. Note that the terminals are numbered clockwise from the tab, looking at the bottom of the can.

Plastic *dual-in-line (DIL)* packages [Fig. 7-27(b)] are more economical than metal cans, and are widely used for industrial and consumer applications. The terminal numbering system again follows the bottom-view clockwise rule, starting from a notch in the plastic. A dimple, or other marker, is sometimes used instead of a notch. This is located on top of the package close to pin 1. *DIL*-type packages are much more convenient for circuit board use than cans, because the lead arrangement and flat package allows greater circuit densities. The surface mounting package shown in Fig. 7-27(c) is most economical for printed circuit board use, because the terminals are soldered directly on to the board without the need for hole drilling.

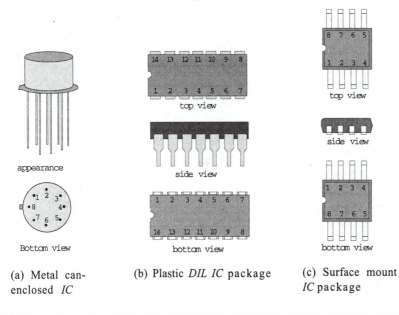

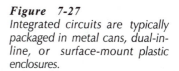

Figure 7-27
Integrated circuits are typically packaged in metal cans, dual-in-line, or surface-mount plastic enclosures.

(a) Metal can-enclosed *IC*

(b) Plastic *DIL IC* package

(c) Surface mount *IC* package

Section 7-7 Review

7-7.1   Sketch the terminal arrangements for *TO-92*, *TO-5*, and *TO-3* transistor packages, and identify the emitter, base, and collector terminals on each package.

7-7.2   Sketch metal can and *DIL* integrated circuit packages, and show the terminal numbering systems.

Chapter-7 Review Questions

Section 7-1

7-1   Describe the process of preparing semiconductor material for device manufacture.

7-2   Explain diffusion and epitaxial growth, and discuss the application of each process to transistor manufacture.

Section 7-2
7-3   Draw sketches to show the construction of alloy diodes and diffused diodes. Briefly explain in each case.

Section 7-3
7-4   Explain the various transistor manufacturing requirements for maximum performance with respect to (a) current gain, (b) power dissipation, (c) frequency response, (d) switching response, (e) breakdown voltage.

Section 7-4
7-5   Describe the microalloy and microalloy diffusion techniques for transistor manufacture. Explain the advantages and disadvantages of transistors made by these techniques.

7-6   Using sketches, explain the diffused mesa and epitaxial mesa transistors. Discuss the advantages and disadvantages of mesa transistors.

7-7   Explain the manufacturing process for diffused-planar and annular transistors, and discuss the advantages and disadvantages of each.

7-8   Show that most of the power dissipation in a transistor occurs at the collector-base junction.

Section 7-5
7-9   Briefly explain the thin-film and thick-film methods of integrated circuit manufacture, and discuss their advantages and disadvantages.

7-10 Using illustrations, explain the fabrication process for monolithic integrated circuits. Discuss the advantages and disadvantages of monolithic *ICs*.

7-11 Describe hybrid integrated circuits.

Section 7-6
7-12 Sketch the construction and circuit diagram for two monolithic integrated circuit transistors showing the parasitic components. State any precautions necessary in the use of the circuit.

7-13 Briefly explain how diodes, resistors, and capacitors are fabricated in monolithic integrated circuits.

Section 7-7
7-14 Sketch *TO-92*, *TO-18*, and *TO-3* transistor packages, and identify the emitter, base, and collector terminals.

7-15 Sketch metal can and *DIL* integrated circuit packages, and show the terminal numbering system.

Chapter *8*

BJT Specifications and Performance

Chapter Contents

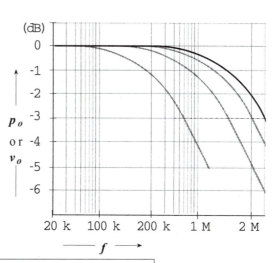

Objectives

You will be able to:

1   Identify important parameters on transistor data sheets for small-signal, high-power, high-frequency, and switching transistors, and determine the parameter values.

2   Calculate circuit power gains and power output changes in decibels.

3   Define Miller effect, and determine the input capacitance for transistor circuits using the data sheet specifications.

4   Explain the effects of input and stray capacitances on BJT circuit frequency response, and calculate cutoff frequencies for CE, CB, and CC circuits.

5   Determine transistor switching times from the data sheet, and calculate device turn-on and turn-off times.

6   Define noise figure and noise factor, and estimate circuit output noise levels originating from resistors and transistors.

7   Determine transistor maximum power dissipations at any temperature using power derating factors and derating graphs.

8   Sketch a thermal resistance circuit for a transistor with a heat sink. Determine the required thermal resistance for transistor heat sinks, and select suitable heat sinks for given circuit conditions.

Introduction

The performance characteristics and parameters for each type of transistor are specified on a data sheet published by the device manufacturer. The specifications must be correctly interpreted to achieve optimum transistor performance, and to avoid breakdown. Transistors must be selected to have a specified maximum collector-emitter voltage greater than the voltage applied to the *CE* terminals. The specified maximum collector current must be greater than the highest I_C level that may flow in the circuit. The maximum power that may be dissipated in a device is normally listed for a case temperature of 25°C. This value must be derated for operation at higher temperatures, and heat sinks must be employed where appropriate. Transistor cutoff frequency, switching times, noise figure, and input capacitance are all items that can be very important in various applications.

8-1   Transistor Data Sheets

To select a transistor for a particular application, the data sheets provided by device manufacturers must be consulted. Typical transistor data sheets are shown in Appendix 1-5 through 1-9, and portions are reproduced in this chapter.

Most data sheets start off with the device type number at the top of the page, a descriptive title, and a list of major applications for the device. This information is usually followed by mechanical data in the form of an illustration showing the package shape and dimensions, as well as indicating which leads are collector, base, and emitter, (see Fig. 8-1).

2N3903, 2N3904 (Silicon) NPN Transistors

NPN silicon annular transistors, designed for general purpose switching and amplifier applications.

E B C

Maximum ratings at $T_A = 25°C$ unless otherwise noted.

Characteristic	Symbol	Rating	Unit
Collector-emitter voltage	V_{CEO}	40	V_{dc}
Emitter-base voltage	V_{EB}	6	V_{dc}
Collector Current	I_C	200	mA

Figure 8-1
Portion of 2N3903-2N3904 transistor data sheet listing maximum terminal voltages and collector current.

The absolute maximum ratings of the transistor at a temperature of 25°C are listed next. These are the maximum voltages, currents, and power dissipation that the device can survive without breaking down. It is very important that these ratings never be exceeded; otherwise, failure of the device is quite possible. *For reliability, the maximum ratings should not even be approached.* The maximum transistor ratings must also be adjusted downward for operation at temperatures greater than 25 °C.

Following the absolute maximum ratings, the data sheet normally shows a complete list of electrical characteristics for the device. Again, these are specified at 25°C, and allowances are necessary for temperature variations.

A complete understanding of all the quantities specified on a data sheet will not be achieved until circuit analysis and design is studied. Some of the most important quantities are considered below. It is important to note that the ratings for a given transistor are stated for specified circuit conditions. The ratings are no longer valid if these conditions change.

V_{CBO} — *Collector-base voltage* — maximum *dc* voltage for reverse-biased collector-base junction.

V_{CEO} — *Collector-emitter voltage* — maximum collector-emitter *dc* voltage with base open-circuited.

V_{EBO} — *Emitter-base voltage* — maximum emitter-base reverse-bias *dc* voltage.

I_C — *Collector current* — maximum *dc* collector current.

I_{CBO} or I_{CO} — *Collector cutoff current* — *dc* collector current with the collector-base junction reverse biased and the emitter open-circuited.

$V_{CE(sat)}$ — *Collector-emitter saturation voltage* — collector-emitter voltage with the device in saturation.

h_{FE} — *Static forward current transfer ratio* — common emitter ratio of *dc* collector current and base current.

NF — *Noise figure* — ratio of total noise output to total noise input expressed as a decibel (*dB*) ratio (see Sections 8-2). Defines the amount of noise added by the device, (see Sections 8-5).

f_{hfe} or $f_{\alpha e}$ — *Common emitter cutoff frequency* — common emitter operating frequency at which the device current gain falls to 0.707 of its normal (mid-frequency) value.

f_{hfb} or $f_{\alpha b}$ — *Common base cutoff frequency* — as above, for common base.

From the section of the *2N3903-2N3904* data sheet reproduced in Fig. 8-1, the maximum collector-emitter voltage is specified as V_{CEO} = 40 V. Obviously, this device should not be used in any circuit with a supply exceeding 40 V; otherwise the transistor might break down. Preferably, the circuit supply voltage should always be less than the specified maximum V_{CEO}.

The maximum V_{EB} is listed as 6 V. This is the highest voltage that may be applied *in reverse* across the base-emitter junction of the transistor. If a greater voltage is used, the base-emitter junction of the transistor is likely to break down. The maximum (reverse) V_{EB} for most transistors is 5 V. In circumstances where the applied voltage might exceed this level, a clipping circuit may be used, (see Sections 3-7 and 3-8). Alternatively, a diode can be connected in series with the transistor emitter to increase the breakdown voltage, as shown in Fig. 8-2. The normal forward-biased base-emitter voltage for a transistor is, of course, approximately 0.7 V for a silicon transistor and approximately 0.3 V for a germanium transistor.

Figure 8-3 shows sections of a transistor data sheet which specify the *dc current gain* h_{FE}, and the *small signal* (or *ac*) *current gain* h_{fe}. For a *2N3904* device with I_C = 10 mA, h_{FE} is listed as 100 minimum and 300 maximum, (see the solid underlined quantities). If the transistor collector current is to be around 1 mA, then the minimum h_{FE} is 70, (dashed underlined). The maximum h_{FE} for I_C = 1 mA is not specified. It is unlikely that $h_{FE(max)}$ would exceed 300 for this condition. However, because the manufacturer does not list a maximum value, it remains uncertain. This situation illustrates the importance of reliable bias circuits that make I_C and V_{CE} largely independent of h_{FE}, (See Ch. 5).

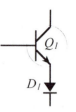

Figure 8-2
A diode connected in series with a transistor emitter protects the BE junction against excessive reverse voltage.

2N3903, 2N3904

Characteristic		Symbol	Min	Max	Unit
DC current gain		h_{FE}			—
(I_C = 0.1 mA dc, V_{CE} = 1 V dc)	2N3903		20	—	
	2N3904		40	—	
(I_C = 1.0 mA dc, V_{CE} = 1 V dc)	2N3903		35	—	
	2N3904		70	—	
(I_C = 10 mA dc, V_{CE} = 1 V dc)	2N3903		50	150	
	2N3904		100	300	
(I_C = 50 mA dc, V_{CE} = 1 V dc)	2N3903		30	—	
	2N3904		60	—	
(I_C = 100 mA dc, V_{CE} = 1 V dc)	2N3903		15	—	
	2N3904		30	—	
Small Signal Current Gain		h_{fe}			—
(I_C = 1 mA dc, V_{CE} = 10 V, f = 1 kHz)					
	2N3903		50	200	
	2N3904		100	400	

Figure 8-3
Portion of 2N3903-2N3904 transistor data sheet listing dc current gain and small-signal ac current gain.

The *small-signal current gain* (h_{fe}) for a *2N3904* is listed in Fig. 8-3 as 100 minimum and 400 maximum. Once again, this parameter is specified for a particular set of bias conditions. The signal frequency at which h_{fe} was measured is specified as 1 kHz. Section 8-2 discusses the fact that h_{fe} decreases at high frequencies.

Example 8-1

Referring to the *2N3905* specification in Appendix 1-6, determine the h_{FE} range for I_C = 10 mA, and the h_{fe} range for f = 1 kHz. Also determine the emitter-base breakdown voltage.

Solution

at I_C = 10 mA, h_{FE} = 50 to 150

at f = 1 kHz, h_{fe} = 50 to 200

$$V_{(BR)EBO} = 5 \text{ V}$$

Practise Problems

8-1.1   From the data sheet in Appendix 1-7, determine the following quantities for a *2N3251* transistor: $h_{FE(min)}$ at I_C = 1 mA, $h_{fe(min)}$ at f = 1 kHz, $V_{CE(max)}$, $V_{EB(max)}$, noise figure.

8-2   Decibels and Half-Power Points

Power Measurement in Decibels

The *power gain* (A_p) of an amplifier may be expressed in terms of the log of the ration of output power (p_o) to input power (p_i). This is illustrated in Fig. 8-4.

$$A_p = \log_{10} \left[\frac{p_o}{p_i} \right] \text{ bels}$$

or, $$\mathbf{A_p = 10 \log_{10} \left[\frac{p_o}{p_i} \right] \textbf{ decibels (dB)}} \qquad (8\text{-}1)$$

Thus, the *decibel* is the unit of *power gain*, or *power level change.*

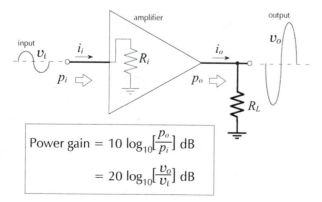

Figure 8-4
Amplifier power gain is measured in decibels (dB). Output power level changes are also measured in decibels.

Power gain = $10 \log_{10} \left[\frac{p_o}{p_i} \right]$ dB

= $20 \log_{10} \left[\frac{v_o}{v_i} \right]$ dB

The power dissipated in a resistance is (v^2/R).

So,
$$A_p = 10 \log_{10} \left[\frac{(v_o)^2/R_L}{(v_i)^2/R_i} \right] dB$$

$$= 10 \log_{10} \left[\frac{v_o}{v_i} \right]^2 dB \qquad (when \ R_i = R_L)$$

or,
$$A_p = 20 \log_{10} \left[\frac{v_o}{v_i} \right] dB \qquad (8\text{-}2)$$

Also, $P_o = i_o^2 R_L$ and $P_i = i_i^2 R_i$

So,
$$A_p = 20 \log_{10} \left[\frac{i_o}{i_i} \right] dB \qquad (8\text{-}3)$$

It is important to note that *the above equations are correct only when the load resistance connected to the circuit output is equal to the circuit input resistance.* However, it is common practice in the electronics industry to apply these equations regardless of the relationship between input and load resistances!

Amplifier *output power level changes* (ΔP_o) are also measured in decibels, and in this case the equations become,

$$\Delta P_o = 10 \log_{10} \left[\frac{p_2}{p_1} \right] dB \qquad (8\text{-}4)$$

where p_1 is the initial output power level, and p_2 is the new level.

Also,
$$\Delta P_o = 20 \log_{10} \left[\frac{v_2}{v_1} \right] dB \qquad (8\text{-}5)$$

and,
$$\Delta P_o = 20 \log_{10} \left[\frac{i_2}{i_1} \right] dB \qquad (8\text{-}6)$$

where v_1 and i_1 are the initial output voltage and current levels, and v_2 and i_2 are the new levels. Equations 8-4 through 8-6 are correct so long as the amplifier load resistance remains constant. Examples 8-2 and 8-3 demonstrate that the output power of an amplifier is reduced by 3 dB when the measured power falls to half its normal level, or when the measured voltage falls to 0.707 of its normal level.

Example 8-2

The output power from an amplifier is 50 mW when the signal frequency is 5 kHz. The power output falls to 25 mW when the frequency is increased to 20 kHz. Calculate the output power change in decibels.

Solution

Eq. 8-4,
$$\Delta P_o = 10 \log_{10} \left[\frac{p_2}{p_1} \right] = 10 \log_{10} \left[\frac{25 \ mW}{50 \ mW} \right]$$
$$= -3 \ dB$$

Half-Power Points

Figure 8-5 shows a typical graph of amplifier output voltage or power plotted versus frequency. Note that the frequency scale is logarithmic. The output normally remains constant over a middle range of frequencies and falls off at low and high frequencies, due to effects explained in Sections 8-3 and 8-4. The gain over this middle range is termed the *mid-frequency gain.* The low frequency and high frequency at which the gain falls by 3 dB are designated f_1 and f_2, respectively. This range is normally considered the useful range of operating frequency for the amplifier, and the frequency difference $(f_2 - f_1)$ is termed the *amplifier bandwidth* (BW).

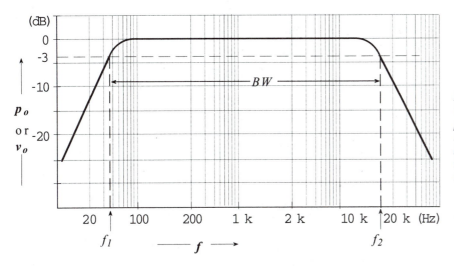

Figure 8-5
The output voltage and output power of an amplifier normally remains constant over a middle band of signal frequencies, and falls-off at high and low frequencies.

Frequencies f_1 and f_2 are termed the *half-power points,* or the *3 dB points.* This is because, as shown in Example 8-2, the power output is -3 *dB* from its normal level when p_2 is half p_1. When the amplifier output is expressed as a voltage on the graph of frequency response, the *3 dB* points $(f_1$ and $f_2)$ occur when v_2 is 0.707 v_1. This relationship is demonstrated in Example 8-3.

Example 8-3
The output voltage of an amplifier is measured as 1 V at 5 kHz, and as 0.707 V at 20 kHz. Calculate the output power change.

Solution
Eq. 8-5,
$$\Delta P_o = 20 \log_{10} \left[\frac{V_2}{V_1} \right] = 20 \log_{10} \left[\frac{0.707 \text{ V}}{1 \text{ V}} \right]$$
$$= -3 \ dB$$

Practise Problems
8-2.1   The output voltage of an amplifier is 2 V when the signal frequency is 3 kHz, and 0.5 V when the signal frequency is increased to 50 kHz. Calculate the decibel change in output power.

8-3   BJT Cutoff Frequency and Capacitances

Device Cutoff Frequency

All transistors have junction capacitances, (see Sections 2-6 and 6-3). The junction capacitances and the transit time of charge carriers through the semiconductor material limit the high frequency performance of the device. This limitation is expressed as a *cutoff frequency* (f_α), which is the frequency at which the transistor current gain falls to 0.707 of its gain at low and medium frequencies. The cutoff frequency can be expressed in two ways, the *common emitter cutoff frequency* $f_{\alpha e}$, and the *common base cutoff frequency* $(f_{\alpha b})$. $f_{\alpha e}$ is the frequency at which the common emitter current gain (h_{fe}), falls to 0.707 × (mid-frequency h_{fe}). $f_{\alpha b}$ is the frequency at which the common base current gain (h_{fb}) falls to 0.707 × (mid-frequency h_{fb}). It can be shown that,

$$f_{\alpha b} = h_{fe} f_{\alpha e} \qquad (8\text{-}7)$$

Equation 8-7 shows that $f_{\alpha b}$ can be identified as the *current-gain-bandwidth product*, or *total bandwidth* (f_T). This is the designation used on many transistor data sheets. The data sheet portion in Fig. 8-6 shows that f_T is 250 MHz for a *2N3903* and 300 MHz for a *2N3904*.

Characteristic	Symbol	Min	Max	Unit
Small-Signal Characteristics				
Current-gain—Bandwidth Product	f_T			MHz
$(I_C$ = 10 mA dc, V_{CE} = 20 V, f = 100 MHz)				
2N3903		250	—	
2N3904		300	—	

Figure 8-6
Specification of transistor cutoff frequency on the 2N3903, 2N3904 data sheet.

Junction Capacitances

As already discussed, all *BJTs* have junction capacitances. These are identified as the base-collector capacitance (C_{bc}) and the base-emitter capacitance (C_{be}), (see Fig. 8-7). The device data sheet portion in Fig. 8-8 shows that the capacitances are usually listed as an *output capacitance* (C_{obo}), which is the same as C_{bc}, and an *input capacitance* (C_{ibo}) which corresponds to C_{be}. It should be noted that in both cases, the capacitance values are specified for reverse-biased junctions with zero current levels. C_{obo} varies when V_{CB} is altered, however, the changes are usually a maximum of approximately ±3 pF. C_{ibo} changes substantially from the specified capacitance when the base-emitter junction is forward biased, (as it always is for linear *BJT* operation).

When an emitter current flows across the *BE* junction, there is a diffusion capacitance (see Section 2-6) at the junction. This is directly proportional to the current level. It can be shown that,

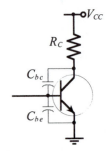

Figure 8-7
All transistors have junction capacitances.

$$C_{be} \approx \frac{6.1\, I_E}{f_T} \qquad\qquad \text{(8-8)}$$

Characteristic	Symbol	Min	Max	Unit
Output Capacitance (V_{CB} = 5 Vdc, I_E = 0, f = 1.0 MHz)	C_{obo}	—	4.0	pF
Input Capacitance (V_{BE} = 0.5 Vdc, I_C = 0, f = 1.0 MHz)	C_{ibo}	—	8.0	pF

Figure 8-8
Junction capacitance specifications for 2N3903 and 2N3904 transistors.

Miller Effect

Figure 8-9 shows an input signal $(+\Delta V_i)$ applied to the base of a transistor connected in *CE* configuration. If the circuit voltage amplification is $-A_v$, then the collector voltage change is,

$$\Delta V_o = -A_v \times \Delta V_i$$

Note that, because of the phase reversal between input and output, the collector voltage is *reduced* by $(A_v\, \Delta V_i)$ when the base voltage is *increased* by ΔV_i. The increase in V_B and decrease in V_C results in a total collector-base voltage reduction of,

$$\Delta V_{CB} = \Delta V_i + A_v \Delta V_i$$
$$= \Delta V_i (1 + A_v)$$

This voltage change appears across the collector-base capacitance.

Using the equation $Q = C \times \Delta V$, it is found that the charge supplied to the input of the circuit is,

$$Q = C_{bc} \times \Delta V_i (1 + A_v)$$

or,
$$Q = (1 + A_v) C_{bc} \times \Delta V_i$$

Therefore, "*looking into*" the base, the collector-base capacitance appears to be $(1 + A_v)C_{bc}$. So, the capacitance is *amplified* by a factor of $(1 + A_v)$. This is known as the *Miller effect*.

The total input capacitance (C_{in}) to the transistor is $(1 + A_v)C_{bc}$ in parallel with the base-emitter capacitance (C_{be});

$$C_{in} = C_{be} + (1 + A_v)C_{bc} \qquad\qquad \text{(8-9)}$$

It should be noted that the Miller effect occurs only with amplifiers that have a 180° phase shift between input and output, (an inverting amplifier). Consequently, it occurs with *CE* circuits, but not with *CB* and emitter follower and circuits.

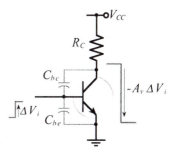

(a) ΔV_i and $-A_v \Delta V_i$ change V_{CB}

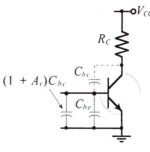

(b) $C_{in} = C_{be} + (1 + A_v)C_{bc}$

Figure 8-9
Because of the Miller effect, the input capacitance for a common-emitter amplifier is
$C_{in} = C_{be} + (1 + A_v)C_{bc}.$

Example 8-4

The transistor in the circuit in Fig. 8-10 has I_C = 1 mA, h_{fe} = 50, h_{ie} = 1.3 kΩ, f_T = 250 MHz, and C_{bc} = 5 pF. Calculate the input capacitance when the circuit is operated as a *CE* amplifier with R_E bypassed.

Solution

From Eq. 6-12,
$$|A_v| = \frac{h_{fe}\,(R_C\|R_L)}{h_{ie}} = \frac{50 \times (8.2\ k\Omega\|100\ k\Omega)}{1.3\ k\Omega}$$
$$= 291$$

Eq. 8-8,
$$C_{be} \approx \frac{6.1\,I_E}{f_T} \approx \frac{6.1 \times 1\ mA}{250\ MHz}$$
$$= 24.4\ pF$$

Eq. 8-9,
$$C_{in} = C_{be} + (1 + A_v)C_{bc} = 24.4\ pF + [(1 + 291) \times 5\ pF]$$
$$= 1.48\ nF$$

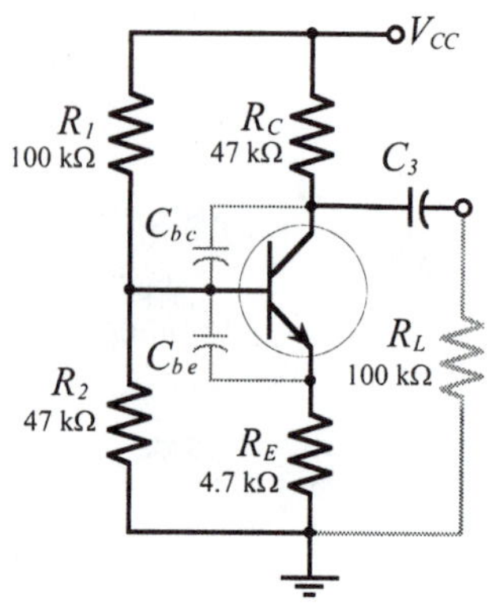

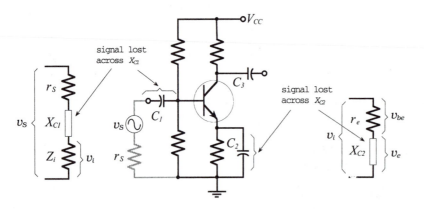

Figure 8-10
The upper cutoff frequency for a transistor circuit can be limited by the input capacitance.

Practise Problems

8-3.1   A CE circuit with a voltage gain of 100 has $I_C = 0.75$ mA, $C_{bc} = 3$ pF, and $f_T = 300$ MHz. Calculate C_{in} when a 45 pF capacitor is connected across the collector-base terminals.

8-4   BJT Circuit Frequency Response

Coupling and Bypass Capacitor Effects

Consider the typical transistor amplifier frequency response illustrated in Fig. 8-5. As discussed, the amplifier voltage gain is constant over a middle range of signal frequencies, and it falls at the low and high ends of the frequency range.

Figure 8-11
The low-frequency fall off in voltage gain in a transistor amplifier is due to signal loss across coupling and bypass capacitors.

The gain fall-off at low signal frequencies is due to the effect of coupling and bypass capacitors. Recall that the reactance of a capacitor is $X_C = 1/(2\pi fC)$. At medium and high frequencies, the factor f makes X_C very small, so that all coupling and bypass capacitors behave as *ac* short circuits. At low frequencies, X_C is large enough to divide the voltages across the capacitors and series resistances, (see Fig. 8-11). As the signal frequency gets lower, the capacitive reactance increases, more of the signal is lost across the capacitors, and the circuit gain continues to fall. Coupling and bypass capacitors are further investigated in Section 12-1.

Input Capacitance Effect on CE and CB Circuits

The input capacitance of an amplifier (discussed in Section 8-3) reduces the circuit gain by 3 dB when the capacitive impedance equals the resistance in parallel with the input, (see Fig. 8-12). That is when,

$$X_{Ci} = Z_i || r_s \qquad \text{(8-10)}$$

So, all circuits have an *input capacitance limited upper cutoff frequency* $(f_{2(i)})$. As already explained, the input capacitance of a *CE* circuit is amplified by the Miller effect, but Miller effect does not occur in a *CB* circuit. Consequently, a *CB* circuit operates to a much higher signal frequency than a similar *CE* circuit.

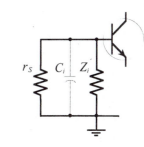

Figure 8-12
The input resistance (Z_i) and the signal source resistance (r_s) are in parallel with the circuit input capacitance (C_{in}).

Example 8-5

Calculate the input capacitance limited upper cutoff frequency for the circuit in Fig. 8-13 (reproduced from Fig. 8-10); (a) when the circuit is used in *CE* configuration with R_E bypassed, (b) when operating as a *CB* circuit with the base bypassed to ground. As in Ex. 8-4, $C_{bc} = 5$ pF and $C_{be} = 24.4$ pF. Also, $h_{fe} = 50$, $h_{ie} = 1.3$ kΩ, $h_{ib} = 24.5$ Ω, and $r_s = 600$ Ω.

Solution

(a) Common emitter circuit,

$$Z_i = R_1 || R_2 || h_{ie} = 100 \text{ kΩ} || 47 \text{ kΩ} || 1.3 \text{ kΩ}$$
$$= 1.25 \text{ kΩ}$$

$$r_s || Z_i = 600 \text{ Ω} || 1.25 \text{ kΩ}$$
$$= 405 \text{ Ω}$$

From Ex. 8-4, $C_{in} = 1.48$ nF

From Eq. 8-10, $f_{2(i)} = \dfrac{1}{2\pi C_{in}(r_s||Z_i)} = \dfrac{1}{2\pi \times 1.48 \text{ nF} \times 405 \text{ Ω}}$

$$= 266 \text{ kHz}$$

(b) Common base circuit,
$$Z_i = R_E || h_{ib} = 4.7 \text{ kΩ} || 24.5 \text{ Ω}$$
$$= 24.4 \text{ Ω}$$

$$r_s || Z_i = 600 \text{ Ω} || 24.4 \text{ Ω}$$
$$= 23.4 \text{ Ω}$$

$$C_{in} = C_{be} + C_{bc} = 24.4 \text{ pF} + 5 \text{ pF}$$
$$= 29.4 \text{ pF}$$

from Eq. 8-10, $f_{2(i)} = \dfrac{1}{2\pi C_{in}(r_s||Z_i)} = \dfrac{1}{2\pi \times 29.4 \text{ pF} \times 23.4 \text{ Ω}}$

$$= 231 \text{ MHz}$$

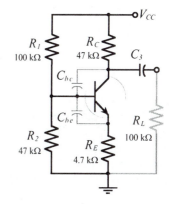

Figure 8-13
Circuit that may be connected to function in either CE or CB configuration.

Input Capacitance Effects on Emitter Follower

When a transistor is used as an emitter follower its *BE* junction voltage is not significantly altered by the *ac* input signal, because virtually all of v_i appears at the emitter as v_o, (see Fig. 8-14). Also, there is no Miller effect to amplify C_{bc}, so the input capacitance is $C_{ibo} \| C_{obo}$. The input capacitance limited cutoff frequency for an emitter follower is very much higher than that for a *CE* circuit.

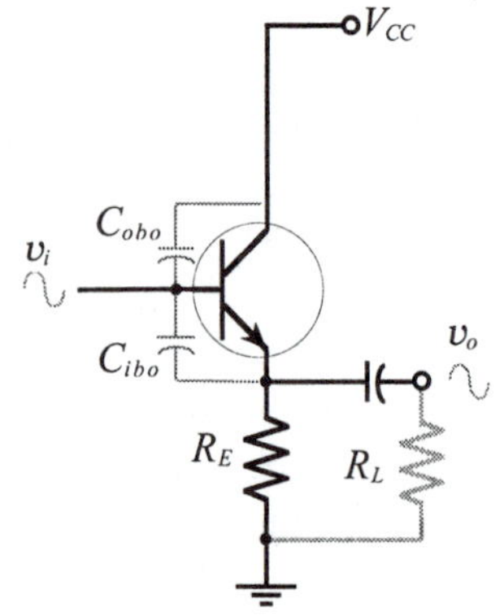

Stray Capacitance

Figure 8-15 illustrates the fact that *stray capacitances* (C_{si} and C_{so}) exists in all transistor circuits. This is capacitance between connecting wires and ground, and normally it is extremely small. The stray capacitance at the device base is usually much smaller than the input capacitance at the base, so that it can normally be neglected. When this is not the case, the stray capacitance must be included with the input capacitance for *CE* circuit cutoff frequency calculations.

At the circuit output, the impedance of the stray capacitance (C_{so}) is very high at low and medium signal frequencies, so that it has no effect on the gain. At high frequencies, the stray capacitive impedance becomes small enough to shunt away some of the input and output currents, and thus it reduces the circuit gain. Referring to Fig. 8-15, it is seen that circuit *ac* load consists of the output stray capacitance (C_{so}) in parallel with R_C and R_L. Rewriting the voltage gain equation for *CE* and *CB* circuits,

$$|A_v| = \frac{h_{fe}(R_C \| R_L \| X_{cso})}{h_{ie}}$$

If the transistor cutoff frequency has not caused the gain to fall off at a lower frequency, then the gain falls by 3 dB when $X_{cso} = R_C \| R_L$. This is an *output capacitance limited cutoff frequency* ($f_{2(o)}$).

If the voltage gain falls by 3 dB at f_{ae} due to the transistor, and by 3 dB at the same frequency due to the stray capacitance, then the gain is down by 6 dB. Consequently, the amplifier upper cutoff frequency is lower than f_{ae}. This is illustrated in Fig. 8-16.

As discussed, there is an additional 3 dB attenuation when $X_{cso} = R_C \| R_L$. If $X_{cso} = 2(R_C \| R_L)$ at f_{ae}, the additional attenuation can be shown to be 1 dB, (see Fig. 8-16). With $X_{cso} = 5(R_C \| R_L)$ at f_{ae}, there is only 0.2 dB additional attenuation, as illustrated.

In some circumstances it is desirable to set the upper cutoff frequency of a circuit at a frequency well below the transistor cutoff frequency, (see Fig. 8-16). This is done simply by connecting a capacitor from the collector terminal to ground exactly as C_{so} is shown in Fig. 8-15. The capacitance value is calculated at the desired cutoff frequency to give,

$$\mathbf{X_c = R_C \| R_L} \qquad\qquad \textbf{(8-11)}$$

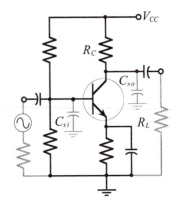

Figure 8-14
There is no Miller effect input capacitance amplification with an emitter follower circuit.

Figure 8-15
Stray capacitance can cause transistor amplifier gain reduction at high frequency.

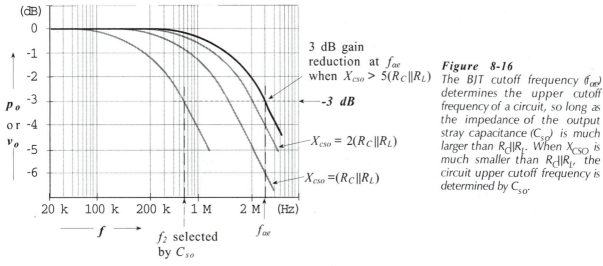

3 dB gain reduction at $f_{\alpha e}$ when $X_{cso} > 5(R_C \| R_L)$

-3 dB

$X_{cso} = 2(R_C \| R_L)$

$X_{cso} = (R_C \| R_L)$

f

f_2 selected by C_{so}

$f_{\alpha e}$

Figure 8-16
The BJT cutoff frequency ($f_{\alpha e}$) determines the upper cutoff frequency of a circuit, so long as the impedance of the output stray capacitance (C_{so}) is much larger than $R_C \| R_L$. When X_{CSO} is much smaller than $R_C \| R_L$, the circuit upper cutoff frequency is determined by C_{so}.

Example 8-6

A transistor with f_T = 50 MHz and h_{fe} = 50 is employed in the *CE* amplifier in Fig. 8-17. Determine the upper *3 dB* frequency for the device. Also, calculate the required capacitance for C_4 to give a 60 kHz upper cutoff frequency. Assume that $R_L >> R_C$.

Solution

From Eq. 8-7,

$$f_{\alpha e} = \frac{f_T}{h_{fe}} = \frac{50\ \text{MHz}}{50}$$

$$= 1\ \text{MHz}$$

From Eq. 8-11,

$$C_4 = \frac{1}{2\pi f_{2(o)} R_C} = \frac{1}{2 \times \pi \times 60\ \text{kHz} \times 10\ \text{k}\Omega}$$

$$= 265\ \text{pF}$$

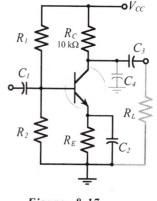

Figure 8-17
Circuit for Ex. 8-7.

Practise Problems

8-4.1   A *CE* amplifier circuit has R_C = 5.6 kΩ, and R_L = 56 kΩ. The transistor used has h_{fe} = 60, h_{ie} = 1.5 kΩ, C_{be} = 192 pF and C_{bc} = 6 pF. The signal source resistance is 1 kΩ, and the voltage divider bias resistors are R_1 = 82 kΩ and R_2= 39 kΩ. Determine the input capacitance limited upper cutoff frequency.

8-4.2   A transistor in a single-stage *CE* amplifier has h_{fe} = 75 and f_T = 12 MHz. The load resistance is $R_C \| R_L$ = 20 kΩ. Determine the upper *3 dB* frequency (f_2) for the circuit. Also, calculate the output stray capacitance that will produce an additional 3 dB attenuation at f_2.

8-5   *Transistor Switching Times*

For transistor switching circuits (see Section 5-10), the switching speed of the device can be an important quantity. Consider the circuit in Fig. 8-18(a). When the base input current is applied, the transistor does not switch *on* immediately. Like frequency response, the switching time is affected by junction capacitance and the transit time of electrons across the junctions. The time between the application of the input pulse and commencement of collector current flow is termed the *delay time* (t_d), [see Fig. 8-18(b). Even when the transistor begins to switch *on*, a finite time elapses before I_C reaches its maximum level. This is known as the *rise time* (t_r). The rise time is specified as the time required for I_C to go from 10% to 90% of its maximum level. As illustrated, the *turn-on time* (t_{on}), is the sum of t_d and t_r.

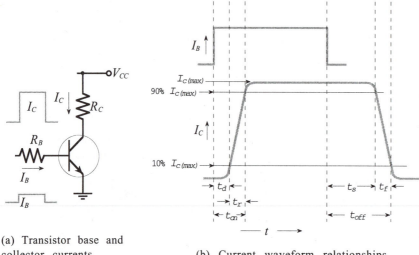

(a) Transistor base and collector currents

(b) Current waveform relationships

Figure 8-18
The turn-on time for a switching transistor is the sum of the delay time (t_d) and the rise time (t_r). The transistor turn-off time is the sum of the storage time (t_s) and the fall time (t_f).

When the input current is switched *off*, I_C does not go to zero until after a *turn-off time* t_{off}, made up of a *storage time* (t_s), and a *fall time* (t_f), as illustrated. The fall time is specified as the time required for I_C to go from 90% to 10% of its maximum level. The storage time is the result of charge carriers being trapped in the depletion region when a junction polarity is reversed. When a transistor is in a saturated *on* condition (see Section 5-10), both the collector-base and emitter-base junctions are forward biased. At switch-*off*, both junctions are reverse biased, and before I_C begins to fall, the stored charge carriers must be withdrawn or made to recombine with opposite-type charge carriers.

For a fast-switching transistor, t_{on} and t_{off} must be of the order of nanoseconds. The *2N3904* transistor data sheet portion in Fig. 8-19 specifies the following switching times: t_d = 35 ns, t_r = 35 ns, t_s = 200 ns, and t_f = 50 ns.

Characteristic		Symbol	Min	Max	Unit
Switching Characteristics (2N3904)					
Delay time	V_{CC} = 3 V, V_{BE} = 0.5 V,	t_d	—	35	ns
Rise time	I_C = 10 mA, I_{B1} = 1 mA	t_r	—	35	ns
Storage time	V_{CC} = 3 V, I_C = 10 mA,	t_s	—	200	ns
Fall time	I_{B1} = I_{B2} = 1 mA	t_f	—	50	ns

Figure 8-19
Transistor switching times, as specified on a data sheet.

Example 8-7

The circuit shown in Fig. 8-18(a) uses a *2N3904* transistor and has an input pulse with a 5 μs pulse width (*PW*). Determine the time from commencement of I_C until it turns *off*.

Solution

I_C commences at t_d (at 10% of $I_{C(max)}$) after the start of the input pulse, and ceases at (t_s + t_f) (at 10% of $I_{C(max)}$) after the end of the input pulse.

$$t = PW - t_d + t_s + t_f = 5\ \mu s \text{ - } 35\ ns + 200\ ns + 50\ ns$$
$$= 5.215\ \mu s$$

Practise Problems

8-5.1　Calculate the turn-*on* time and the turn-*off* time for a transistor with: t_d = 10 ns, t_r = 12 ns, t_s = 15 ns, t_f = 12 ns. Also, determine the time from commencement of a 100 ns input pulse to the end of the transistor *on* time.

8-6　Transistor Circuit Noise

Unwanted signals at the output of an electronics system are termed *noise*. The noise amplitude may be large enough to severely distort, or completely swamp, the wanted signals. Consequently, the noise level dictates the minimum signal amplitude that can be handled. Noise originates as atmospheric noise from outside the system and as circuit noise generated within resistors and devices.

Consider a conductive material at room temperature, (Fig. 8-20). The motion of free electrons drifting around within the material constitutes a flow of many tiny random electric currents. These currents cause minute voltage drops, which appear across the ends (or terminals) of the material. Because the number of free electrons available and the random motion of the electrons are both increased as temperature rises, the generated voltage amplitude is proportional to temperature. This unwanted, randomly varying voltage is termed *thermal noise*.

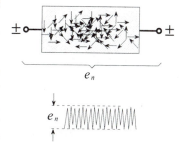

Figure 8-20
Noise voltages are generated within a resistor by random movements of electrons.

Thermal noise is generated within resistors, and when the resistors are at the input stage of an amplifier, the noise is amplified and produced as an output. Noise from other resistors is not amplified as much as that from the resistors right at the input; consequently, only the input stage resistors need be considered in noise calculations. Noise is also generated within a transistor, and like resistors, the input stage transistor is the most important because its noise is amplified more than that from any other stage.

Because thermal noise is an alternating quantity, its rms output level from any amplifier is dependent upon the bandwidth of the amplifier. It can be shown that the rms noise voltage generated in a resistance is,

$$e_n = \sqrt{4\,kTBR} \qquad \text{(8-12)}$$

where, k = Boltzmann's constant = 1.374×10^{-23} J/K
         (i.e., joules per degree Kelvin)
T = absolute temperature (Kelvin)
R = resistance in ohms
B = circuit bandwidth

In the circuit shown in Fig. 8-21(a) R_1 and R_2 are bias resistors; e_s is a signal voltage with source resistance r_s. The total noise-generating resistance (R_G) in parallel with the amplifier input terminal is,

$$R_G = r_s \| R_2 \| R_2 \qquad \text{(8-13)}$$

In the noise equivalent circuit [Fig. 8-21(b)], e_n is the noise voltage generated by R_G. If the device input resistance is R_i, then the noise voltage is divided,

$$e_{ni} = e_n \times \frac{R_i}{R_i + R_G} \qquad \text{(8-14)}$$

If the amplifier voltage gain is A_v, the output noise due to R_G is,

$$e_{no} = A_v\, e_{ni} \qquad \text{(8-15)}$$

With a circuit collector resistance R_C and load resistance R_L, the noise output power produced by R_G is,

$$P_{nG} = \frac{e_{no}^2}{R_C \| R_L} \qquad \text{(8-16)}$$

To specify the amount of noise generated by a transistor, manufacturers usually quote a *noise figure* (*NF*). To arrive at this figure, the transistor noise output is measured under specified bias conditions and with a specified source resistance, temperature, and noise bandwidth. The noise figure defines the amount of noise added by the transistor to the noise generated by the specified resistance (R_G) at the input. Recall that R_G is the combined bias and signal source resistances, as seen from the amplifier input.

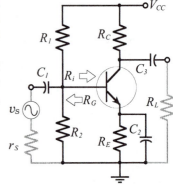

(a) *CE* amplifier circuit

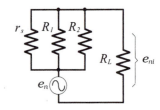

(b) Equivalent noise input circuit

Figure 8-21
Thermal noise at the input of an amplifier depends upon the equivalent resistance of the bias and signal source resistances.

The data sheet portion in Fig. 8-22 specifies *spot noise figures* for a *2N4104* transistor. This means that the noise has been measured for a bandwidth of 1 Hz. The bias conditions are listed because the transistor noise can be affected by V_{CE} and I_C. Note that the specified I_C levels are very low (5 μA to 30 μA), because transistor noise increases with increasing current levels.

Noise Characteristics (2N4104)

Parameter	Test Conditions	Min	Max
NF spot noise figure	V_{CE} = 5 V, I_C = 30 μA, R_G = 10 kΩ, f = 10 Hz		15 dB
	V_{CE} = 5 V, I_C = 30 μA, R_G = 10 kΩ, f = 100 Hz		4 dB
	V_{CE} = 5 V, I_C = 5 μA, R_G = 50 kΩ, f = 1 kHz		1 dB
	V_{CE} = 5 V, I_C = 5 μA, R_G = 50 kΩ, f = 10 kHz		1 d B

Figure 8-22
Data sheet specification of noise characteristics for a 2N4104 transistor.

The *noise factor* (F) is the total circuit noise power output divided by noise output power from the source resistor.

$$F = \frac{P_{no}}{P_{nG}} \qquad \text{(8-17)}$$

The *noise figure* (NF) is the decibel value of F,

$$NF = 10 \log_{10} F \qquad \text{(8-18)}$$

Ideally, a transistor would add no noise to the circuit, and its noise figure would be zero. Obviously, the smallest possible transistor noise figure is the most desirable.

If the circuit in which the transistor is employed does not have the value of source resistance and the bias conditions specified, then the specified noise figure does not apply. In this case the noise figures can still be used to compare transistors, but for accurate estimations of noise, a new measurement of the noise figure must be made. From Eq. 8-17, the total noise output power due to R_G and the input transistor is,

$$P_{no} = F \times P_{nG} \qquad \text{(8-19)}$$

Example 8-8
Calculate the noise output voltage for the amplifier in Fig. 8-23 if the transistor is completely noiseless. The circuit voltage gain is 600, the base input resistance is 3 kΩ, and the cutoff frequencies are f_1 = 100 Hz and f_2 = 40 kHz. The circuit temperature is 25°C.

Solution

Eq. 8-13, $R_G = r_s \| R_2 \| R_2 = 30 \text{ k}\Omega \| 30 \text{ k}\Omega \| 30 \text{ k}\Omega$

$= 10 \text{ k}\Omega$

$T = 25°C = (273 + 25) \text{ K}$

$= 298 \text{ K (degrees Kelvin)}$

$B = f_2 - f_1 = 40 \text{ kHz} - 100 \text{ Hz}$

$= 39.9 \text{ kHz}$

Eq. 8-12, $e_n = \sqrt{4 \, k \, T \, B \, R}$

$= \sqrt{4 \times 1.37 \times 10^{-23} \times 298 \times 39.9 \text{ kHz} \times 10 \text{ k}\Omega}$

$= 2.6 \, \mu V$

Eq. 8-14, $e_{ni} = e_n \times \dfrac{R_i}{R_i + R_G} = 2.6 \, \mu V \times \dfrac{3 \text{ k}\Omega}{3 \text{ k}\Omega + 10 \text{ k}\Omega}$

$= 0.59 \, \mu V$

Eq. 8-15, $e_{no} = A_v e_{ni} = 600 \times 0.59 \, \mu V$

$= 354 \, \mu V$

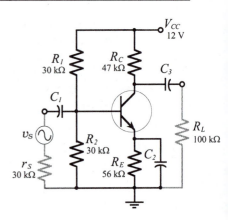

Figure 8-23
Circuit for Examples 8-8 and 8-9.

Example 8-9

The circuit in Ex. 8-8 uses a *2N4104* transistor with bias conditions of $I_C = 30$ μA and $V_{CE} = 5$ V. Calculate the total noise output voltage for the amplifier.

Solution

From Fig. 8-22, $NF = 4 \text{ dB}$

From Eq. 8-18, $F = \text{antilog} \left[\dfrac{NF}{10} \right] = \text{antilog} \left[\dfrac{4 \text{ dB}}{10} \right]$

$= 2.51$

Eq. 8-19, $P_{no} = F \times P_{nG}$

so, $\dfrac{v_n^2}{R_C \| R_L} = F \times \dfrac{e_{no}^2}{R_C \| R_L}$

giving, $v_n = \sqrt{F} \times e_{no} = \sqrt{2.51} \times 354 \, \mu V$

$= 560 \, \mu V$

Practise Problems

8-6.1   A transistor circuit has $A_v = 800$, $B = 20 \text{ kHz}$, $T = 40°C$, and $R_i = 2 \text{ k}\Omega$. Determine the noise output voltage produced by bias resistors with an equivalent resistance of 15 kΩ.

8-6.2   Calculate the total noise output voltage for the circuit in Problem 8-6.1 if the transistor has a noise figure of 6 dB.

8-7   *Transistor Power Dissipation*

Maximum Power Dissipation

Consider the portion of the data sheet for the *2N3903* and *2N3904* transistor reproduced in Fig. 8-24. The *total device dissipation* (P_D) is specified as 625 mW at a maximum ambient temperature (T_A) of 25°C. This means that $(V_{CE} \times I_C)$ must not exceed 625 mW when the surrounding air temperature is 25°C. Figure 8-1 lists the maximum I_C and V_{CE} for the *2N3903* and *2N3904* as 40 V and 200 mA, respectively. If these two quantities exist simultaneously at the transistor, the power dissipation is,

$$P_D = V_{CE} \times I_C = 40\text{ V} \times 200\text{ mA}$$
$$= 8\text{ W}$$

which is much greater than the 625 mW specified maximum.

If the maximum V_{CE} is used with a *2N3904* transistor, then the maximum collector current must not exceed,

$$I_C = \frac{P_{D(max)}}{V_{CE}} = \frac{625\text{ mW}}{40\text{ V}}$$
$$= 15.6\text{ mA}$$

When the transistor has to pass the maximum I_C, the V_{CE} should be limited to,

$$V_{CE} = \frac{P_{D(max)}}{I_C} = \frac{625\text{ mW}}{200\text{ mA}}$$
$$= 3.1\text{ V}$$

If the air temperature is greater than 25°C, the transistor maximum power dissipation must be derated. From Fig. 8-24, the *power derating factor* for *2N3903-2N3904* transistors is 5 mW/°C. This may be used to calculate the device maximum power dissipation at any air temperature.

$$\boldsymbol{P_{D(T2)} = P_{D(25°C)} - D(T_2 - 25°C)} \qquad\qquad \textbf{(8-20)}$$

where *D* is the power dissipation derating factor.

2N3903, 2N3904			
Total Device Dissipation @ $T_A = 25°C$	P_D	625	mW
Derate above 25°C		5	mW/°C

Figure 8-24
Power dissipation specification for low-power transistors.

Example 8-10

Calculate the maximum I_C level that may be used with a *2N3904* transistor at 55°C when V_{CE} is 10 V.

Solution

Eq. 8-20,

$$P_{D(T2)} = P_{D(25°C)} - D(T_2 - 25°C)$$

$$= 625 \text{ mW} - (5 \text{ mW/°C})(55°C - 25°C)$$

$$= 475 \text{ mW}$$

$$I_C = \frac{P_D}{V_{CE}} = \frac{475 \text{ mW}}{10 \text{ V}}$$

$$= 47.5 \text{ mA}$$

The data sheet portion in Fig. 8-25 shows the voltage, current and power ratings for a *2N3055* high-power transistor. The maximum dissipation is specified as 115 W at a *case temperature* (T_C) of 25°C. Note that for a power transistor the *case* temperature is specified, instead of the air temperature. The derating factor is 0.667 W/°C.

2N3055 Silicon NPN transistor

Characteristic	Symbol	Value	Unit
Collector-Emitter Voltage	V_{CEO}	60	Vdc
Collector-Emitter Voltage	V_{CER}	70	Vdc
Collector-Base Voltage	V_{CB}	100	Vdc
Emitter - Base Voltage	V_{EB}	7	Vdc
Collector Current — Continuous	I_C	15	Adc
Base Current	I_B	7	Adc
Total Power Dissipation @ T_C = 25°C	P_D	115	Watts
Derate above 25°C		0.667	W/°C

Figure 8-25
Data sheet portion for 2N3055 power transistor.

The maximum power dissipation at any temperature may be calculated for the *2N3055* by use of Eq. 8-20. Alternatively, a *power derating graph* can be constructed for reading the device maximum power dissipation at any temperature. This is demonstrated in Ex. 8-11.

Example 8-11

Draw a power derating graph for a *2N3055* transistor, and determine the device maximum power dissipation at 100°C.

Solution

Prepare a horizontal temperature scale to 200°C, and a vertical power dissipation scale to 115 W, as in Fig. 8-26.

Plot point A at P_D = 115 W and T = 25 °C.

When $P_D = 0$, $T = 25°C + \dfrac{P_D}{D} = 25°C + \dfrac{15 \text{ W}}{0.667 \text{ W/°C}}$

$= 197°C$

Plot point B at $P_D = 0$ and $T = 197°C$.
Draw the power derating graph through points A and B.

From the graph, at $T = 100°C$,
            $P_D = 65$ W

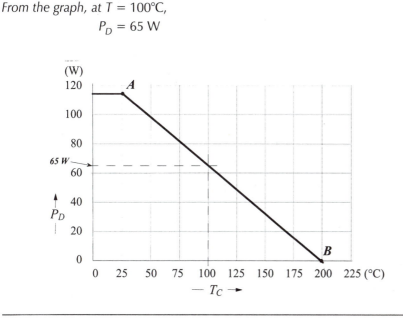

Figure 8-26
Power derating graph for 2N3055 transistor.

Maximum Power Dissipation Curve

When the maximum power that may be dissipated in a transistor is determined, the maximum I_C level may be calculated for any given V_{CE}, or vice versa. The corresponding voltage and current levels may be more easily determined by drawing a *maximum power dissipation curve* on the transistor output characteristics. To draw this curve, the greatest power that may be dissipated at the operating temperature is first calculated. Then, using convenient collector-emitter voltage levels, the corresponding collector current levels are calculated for the maximum power dissipation. Using these current and voltage levels the curve is plotted on the device characteristics. Example 8-12 demonstrates the process.

The transistor voltage and current conditions must at all times be maintained in the portion of the characteristics below the maximum power-dissipation curve. This means, for example, that all point on load lines must be below the curve.

Example 8-12

Draw a maximum power dissipation curve for $P_D = 80$ W on the I_C/V_{CE} characteristics in Fig. 8-27.

Solution

when $V_{CE} = 60$ V, $I_C = \dfrac{P_D}{V_{CE}} = \dfrac{80 \text{ W}}{60 \text{ V}}$

$= 1.3$ A

Plot point 1 on the characteristics at:

$V_{CE} = 60$ V and $I_C = 1.3$ A

when $V_{CE} = 40$ V, $I_C = \dfrac{80 \text{ W}}{40 \text{ V}} = 2$ A   (point 2)

when $V_{CE} = 20$ V, $I_C = \dfrac{80 \text{ W}}{20 \text{ V}} = 4$ A   (point 3)

when $V_{CE} = 10$ V, $I_C = \dfrac{80 \text{ W})}{10 \text{ V}} = 8$ A   (point 4

Draw the maximum power dissipation curve through the points on the graph.

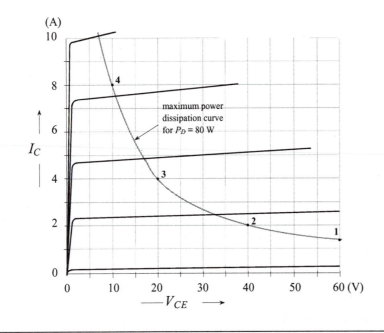

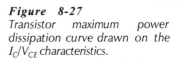

Figure 8-27
Transistor maximum power dissipation curve drawn on the I_C/V_{CE} characteristics.

Practise Problems

8-7.1   Draw a power derating graph for a transistor with $P_{D(25°C)} = 310$ mW and $D = 2.81$ mV/°C. Read the device maximum power dissipation at 80°C from the graph.

8-7.2   Using $V_{CE(max)} = 40$ V, determine four suitable points for drawing the maximum power dissipation curve for the transistor in Practise Problem 8-7.1 when $T_C = 75°C$.

8-8 Heat Sinking

When power is dissipated in a transistor, the heat generated must flow from the collector-base junction to the case, and then to the surrounding atmosphere. When only a very small amount of power is involved, as in a small-signal transistor, the surface area of the transistor case is normally large enough to allow all of the heat to escape. For the large power dissipations that can occur in high-power transistors, the transistor surface area is not large enough. *Heat sinks* must be used to increase the area in contact with the atmosphere. For small transistors, the clip-on *star-type* heat sinks illustrated in Fig. 8-28(a) may be used. For higher-power transistors, sheet-metal and aluminum-extrusion heat sinks are available, as illustrated in Fig. 8-28(b) and (c).

(a) Clip-on type heat sink with
    TO-18 transistor enclosure

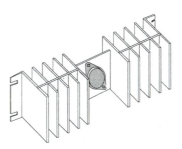

(c) Aluminum extrusion heat
    sink with *TO-3* transistor

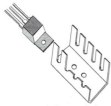

(b) Sheet-metal heat sink with
    TO-220 transistor

Figure 8-28
Heat sinks are used to conduct heat from a transistor case, to keep the device from overheating when power is dissipated.

Figure 8-29(a) shows the cross section of a transistor fastened to a heat sink. The heat generated at the collector-base junction must flow from the junction to the transistor case, then from the case to the heat sink, and finally from the heat sink to the surrounding air. In many cases, a mica gasket is inserted between the transistor case and the heat sink for electrical insulation, [see Fig. 8-29(a)]. Each portion of the path that the heat must pass through has a *thermal resistance*. These are:

θ_{JC} — *junction-to-case* thermal resistance
θ_{CS} — *case-to-sink* thermal resistance
θ_{SA} — *sink-to-air* thermal resistance

Figure 8-29(b) shows the *thermal equivalent circuit* for the transistor and heat sink. This consists of the three thermal resistances connected in series. The size of heat sink required for a transistor with a given power dissipation may be determined from the thermal equivalent circuit.

The temperature difference between the transistor collector-base junction and the air surrounding the heat sink (T_J - T_A) causes the dissipated power (Q) to flow through each of the thermal resistances in turn. The thermal resistance series circuit is analogous to an electrical series resistive circuit. The power flow in the thermal circuit is similar to the current flow in the electrical circuit. Also, the temperature drop across each thermal resistance is analogous to the voltage drop across each electrical resistance, [see Fig. 8-29(b)]. Ohm's Law may be applied to a thermal series circuit exactly as in the case of an electrical series circuit.

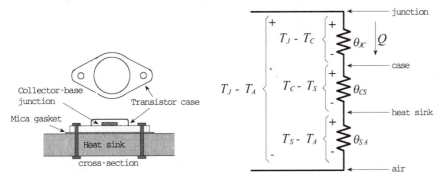

(a) Transistor and heat sink   (b) Thermal resistance circuit

Figure 8-29
The thermal resistances between the collector-base junction of a transistor and the air surrounding the transistor heat sink constitute a series thermal circuit. The thermal equivalent of Ohm's Law may be applied for circuit analysis.

For an electrical series resistive circuit, $I = E/R$, or,

or, $$\text{current flow} = \frac{\text{Voltage difference}}{\text{total resistance}}$$

For a series thermal resistance circuit,

$$\text{power flow} = \frac{\text{temperature difference}}{\text{total thermal resistance}}$$

or, $$Q = \frac{T_J - T_A}{\theta_{JC} + \theta_{CS} + \theta_{SA}}$$ **(8-21)**

When the temperatures are in degrees Celsius (°C), and the thermal resistances are in *degrees Celsius per watt* (°C/W), Q is the power dissipated in watts (W).

Thermal Characteristic (2N3055)

Characteristic		Max	Unit
Junction-to-case thermal resistance θ_{jc}		1.52	°C/W

Figure 8-30
Portion of 2N3055 data sheet showing the transistor junction-to-case thermal resistance.

The value of θ_{JC} depends upon the transistor case style, and it is usually specified on the data sheet. See the *2N3055* data sheet portion in Fig. 8-30. θ_{CS} is dependent on the transistor case and on the mechanical contact between the case and the heat sink. The contact may be *dry*, have a layer of heat-conducting compound, or have an insulating gasket. Table 8-1 shows typical θ_{CS} values for three case styles and three contact conditions. θ_{SA} is determined by the size and style of the heat sink, (see Table 8-2 and Appendix 1-10).

Table 8-1   *Typical case-to-sink thermal resistances*
for various mechanical connections

transistor case	θ_{CS} (°C/W) metal-to-metal		θ_{CS} (°C/W) with mica insulator and compound
	dry	with compound	
TO-220	1.2	0.7	1.0
TO-3	0.6	0.4	0.5
TO-66	1.5	0.5	2.3

Equation 8-21 may be used to calculate θ_{SA} when all other quantities are known. The smallest heat sink with a thermal resistance equal to or lower than the calculated value is then selected. Alternatively, when a given heat sink is used, Eq. 8-20 may be applied to calculate the transistor junction temperature.

Table 8-2   *Typical sink-to-air thermal resistances*
for various heat sinks

Heat sink style	Transistor case	Wakefield heat sink type	Natural convection θ_{SA}	θ_{SA}
Star-type clip-on	TO-18	201	65°C/W	65°C @ 1 W
Sheet metal with fins	TO-220	289	25°C/W	50°C @ 2 W
Aluminum extrusion	TO-3	401	2.7°C/W	80°C @ 30 W
" "	"	403	1.8°C/W	55°C @ 30 W
" "	"	421	1.2°C/W	58°C @ 50 W
" "	"	423	0.94°C/W	47°C @ 50 W
" "	"	441	0.59°C/W	47°C @ 80 W
" "	"	621	5°C/W	75°C @ 15 W

Another method used to specify heat sink thermal resistances is shown in Table 8-2. For the *Wakefield 621* heat sink, θ_{SA} is listed as *75°C at 15 W* for natural convection. This means that the maximum temperature difference between the case and the surrounding air will be 75°C when the heat sink is dissipating 15 W. This can be redefined as,

$$\theta_{SA} = \frac{75°C}{15\ W} = 0.5°C/W$$

Example 8-13

A transistor with $V_{CE} = 20$ V and $I_C = 1$ A has a 1°C/W junction-to-case thermal resistance. The *TO-3* case is fastened directly to the heat sink, and conducting compound is used. Calculate the thermal resistance for a heat sink that will keep the maximum junction temperature at 90°C when the ambient temperature is 25°C. Select a suitable heat sink from Appendix 1-10.

Solution

$$Q = V_{CE}\ I_C = 20\ V \times 1\ A$$
$$= 20\ W$$

From Table 8-1, $\theta_{CS} = 0.4°C/W$ (*TO-3* case, direct contact with compound)

From Eq. 8-21,

$$\theta_{SA} = \frac{T_J - T_A}{Q} - (\theta_{JC} + \theta_{CS})$$

$$= \frac{90°C - 25°C}{20\ W} - (1°C/W + 0.4°C/W)$$

$$= 1.85°C/W$$

From Table 8-2 (and Appendix 1-10), the *403*, *421*, *423*, and *441* all have θ_{SA} less than 1.85°C/W for natural convection. Select the smaller and least expensive of the three — the *403*.

Practise Problems

8-8.1   A *2N6121* transistor (see Appendix 1-9) is fastened to a heat sink with the use of a mica insulator and compound. Calculate the thermal resistance of a suitable heat sink if the device is to dissipate 10 W, and the maximum junction temperature is to be 100 °C. The ambient temperature is 25°C.

8-8.2   A *TO-3* style transistor with $T_{J(max)} = 140$ °C and $\theta_{JC} = 0.6°C/W$ is fastened dry to a heat sink with $\theta_{CA} = 1.3$ °C/W. Determine the maximum power that may be dissipated when $T_A = 40°C$.

Chapter-8 Review Questions

Section 8-1

8-1   Define the following *BJT* quantities listed on a device data sheet: V_{CBO}, V_{CEO}, V_{EBO}, $V_{CE(sat)}$.

8-2   Define the following *BJT* quantities: I_C, I_{CBO}, I_{CO}, h_{FE}, h_{fe}.

Section 8-2

8-3   Write the equations for determining power gains and power level changes in decibels; (a) when two power levels are measured, (b) when two voltage levels are measured.

8-4   Sketch a typical frequency response graph for an amplifier, and identify the upper and lower cutoff frequencies and the bandwidth. Briefly explain.

Section 8-3

8-5   Define f_{ae}, f_{ab}, and f_T for a *BJT*. Briefly explain.

8-6   Explain *CB* and *CE* junction capacitances, and identify the quantities that determine junction capacitance values.

8-7   Describe Miller effect, and derive an equation for the input capacitance of an inverting amplifier.

Section 8-4

8-8   Identify the quantities that cause the gain of an amplifier to fall of at low frequencies. Briefly explain.

8-9   Discuss the quantities that cause the gain of an amplifier to fall off at high frequencies.

Section 8-5

8-10   Sketch the waveforms of input and output currents for a switching transistor. Show the various switching times involved and explain the origin of each.

Section 8-6

8-11   Explain thermal noise, and discuss the effect of resistor noise at the input of a transistor. Also, define noise figure and noise factor for a transistor.

Section 8-7

8-12   Sketch a typical power-temperature derating graph for a transistor. Briefly explain.

8-13   Describe how a transistor maximum power dissipation graph may be drawn on the I_C/V_{CE} characteristics.

Section 8-8

8-14   Sketch the cross section of a power transistor and heat sink. Identify the thermal resistances in the power dissipation path.

8-15   Draw the thermal resistance equivalent circuit for a transistor and heat sink, and write the equation relating power dissipation, temperature difference, and thermal resistance.

Chapter-8 Problems

Section 8-1

8-1   From Appendix 1-9, determine the following quantities for a *2N6125* transistor: maximum V_{CE}, maximum collector current, maximum V_{BE}, $h_{FE(max)}$ and $h_{FE(min)}$ at $I_C = 1.5$ A.

8-2   From the *2N3055* data sheet in Appendix 1-8, determine the following quantities: maximum collector-base voltage, maximum collector current, maximum emitter-base voltage, h_{FE} minimum and h_{FE} maximum at $I_C = 4$ A.

Section 8-2

8-3   The output power from an amplifier is 100 mW when the signal frequency is 1 kHz. When the signal frequency is increased to 25 kHz, the output power falls to 75 mW. Calculate the decibel change in output power.

8-4   Calculate the power gain for an amplifier with equal input and load resistances when $v_i = 100$ mV and $v_o = 3$ V.

8-5   The output voltage of an amplifier is 2 V when the signal frequency is 1 kHz. Calculate the new level of output voltage when it has fallen by 4 dB.

Section 8-3

8-6   The input capacitance of a *CE* circuit is measured as 800 pF. The circuit has $R_C = 7$ kΩ, and the device parameters are: $h_{fe} = 60$, $h_{ie} = 1.5$ kΩ. If the base-emitter capacitance is 15 pF, calculate the collector-base capacitance.

8-7   A transistor with $h_{fe} = 100$, $h_{ie} = 2.2$ kΩ, $C_{cb} = 3$ pF, $I_C = 1.2$ mA, and $f_T = 4$ MHz is connected as a *CE* amplifier with $R_C \| R_L = 6.8$ kΩ. Calculate the amplifier input capacitance.

8-8   Calculate the new value of C_{in} for the amplifier in Problem 8-7 when a 100 pF capacitor is connected (a) between emitter and base, (b) between collector and base.

8-9   A common emitter amplifier has $A_v = 50$, $C_{in} = 414$ pF, and $C_{bc} = 8$ pF. Determine the base-emitter capacitance.

Section 8-4

8-10   The circuit in Problem 8-7 uses voltage divider bias with $R_1 = 56$ kΩ and $R_2 = 15$ kΩ. The signal source resistance is 3.3 kΩ. Calculate the input capacitance limited upper cutoff frequency for the circuit.

8-11   Repeat Problem 8-10 for the circuit connected to function as a *CB* amplifier.

8-12   A voltage divider bias circuit (as in Fig. 8-23) has the following components: $R_1 = 68$ kΩ, $R_2 = 47$ kΩ, $R_C = 5.6$ kΩ, $R_E = 4.7$ kΩ. The supply voltage is 12 V, and the transistor has: $f_T = 35$ MHz, $C_{bc} = 3.5$ pF. Determine the input capacitance limited upper cutoff frequency when the circuit is connected as a *CE* amplifier with R_E bypassed. The signal source resistance is 1.5 kΩ.

8-13 Repeat Problem 8-12 for the circuit employed as a *CB* amplifier with the transistor base bypassed to ground.

8-14 The circuit in Problem 8-12 is modified to function as an emitter follower with R_C shorted and the output taken from the emitter. Determine the input capacitance limited upper cutoff frequency for the circuit.

8-15 A transistor employed in an amplifier has h_{fe} = 75 and $f_{\alpha b}$ = 12 MHz. The collector load is $R_C \| R_L$ = 20 kΩ, and there is 100 pF stray capacitance at the output. Determine the stray capacitance limited upper 3 dB frequency.

8-16 A transistor amplifier with $R_C \| R_L$ = 15 kΩ has a 75 kHz upper 3 dB frequency. Assuming that f_2 is determined by the stray capacitance at the transistor collector terminal, calculate the value of the stray capacitance.

Section 8-5

8-17 Determine the turn-on and turn-off times for a switching circuit with *2N3251* transistor, (see Appendix 1-7).

8-18 If the circuit in Problem 8-17 has a 2 kHz square wave input at the transistor base, calculate the time from the collector current first reaches 90% of its maximum level until it falls to 10% of maximum.

8-19 A switching circuit is to be turned *on* and *off* by a 1 MHz square wave input. The output rise and fall times are not to exceed 10% of the transistor *on* time. Determine the maximum value for t_r and t_f.

Section 8-6

8-20 An amplifier with a *2N4104* input transistor (see Fig. 8-22) has 3 dB points at 2 kHz and 10 kHz, respectively. The transistor bias conditions are V_{CE} = 5 V and I_C = 5 µA, and the amplifier voltage gain is 40. Calculate the noise output voltage at 25°C if R_G = 50 kΩ and R_i = 10 kΩ.

8-21 A transistor amplifier with A_v = 100, B = 15 kHz, and R_i = 12 kΩ has input bias resistors equivalent to R_G = 33 kΩ. If the maximum noise voltage at the output is not to exceed 100 µV, determine the largest tolerable noise figure for the input transistor.

8-22 An amplifier with B = 10 kHz, R_G = 5 kΩ, and R_i = 3.3 kΩ has a maximum output noise of 150 µV. Assuming that the transistor is noiseless, determine the amplifier voltage gain.

Section 8-7

8-23 Calculate the maximum ambient temperature for a transistor with a 200 mW power dissipation if $P_{D(25°C)}$ = 310 mW, and D = 2.81 mW/°C

8-24   The device in Problem 8-23 is to be operated at 80°C maximum ambient, and is to have a 5 mA collector current. Determine the maximum V_{CE} level that may be used.

8-25   Referring to the data sheet in Appendix 1-9, draw a power derating graph for a *2N6121* transistor, and determine the maximum power dissipation at a case temperature of 75°C.

8-26   Calculate the maximum case temperature for a *2N3251* transistor when its power dissipation is 0.25 W.

8-27   A 2N3055 transistor is to be operated at a maximum case temperature of 125°C. Construct a suitable I_C/V_{CE} graph, and draw the maximum power dissipation curve for the device at this temperature.

8-28   For the P_D = 80 W maximum power dissipation curve in Fig. 8-27, draw the dc load line for the smallest possible value of R_C when the circuit supply voltage is 50 V. Determine the value of $R_{C(min)}$.

Section 8-8

8-29   A *2N3055* transistor has V_{CE} = 10 V and I_C = 500 mA. The case temperature is not to exceed 30°C when the ambient is 25°C. The device is fastened to a heat sink using a mica gasket and compound. Calculate the maximum sink-to-air thermal resistance for a suitable heat sink.

8-30   A *2N3055* transistor dissipating 9 W is directly fastened to a *NC403* heat sink using compound, (see Appendix 1-10). If the ambient temperature is 25°C, calculate the transistor case temperature and junction temperature.

8-31   A *2N4900* transistor has a specified maximum power dissipation of 25 W at a case temperature of 25°C, and a derating factor of 0.143 W/°C. If the transistor is to dissipate 20 W, determine its maximum case temperature and calculate the maximum sink-to-air thermal resistance for a suitable heat sink. Assume that the *TO-66* case is directly connected to the heat sink using compound.

8-32   A *TO-220* case style transistor with V_{CE} = 30 V and I_C = 500 mA is fastened to a heat sink using a mica gasket and compound. Calculate the maximum sink-to-air thermal resistance for a suitable heat sink if the transistor maximum junction temperature is 150°C and the junction-to-case thermal resistance is 2.7°C/W.

Practise Problem Answers

8-1.1   90, 100, 40 V, 5 V, 6 dB
8-2.1   -12 dB
8-3.1   4.86 nF
8-4.1   191 kHz

8-4.2 49.7 pF
8-5.1 22 ns, 27 ns, 127 ns
8-6.1 2.14 mV
8-6.2 4.28 mV
8-7.1 (310 mW, 25°C), (0, 135°C)
8-7.2 (40 V, 4.24 mA), (30 V, 5.65 mA),
       (20 V, 8.48 mA), (10 V, 16.95 mA)
8-8.1 3.38°C/W
8-8.2 40 W

Chapter *9*

Field Effect Transistors

Chapter Contents

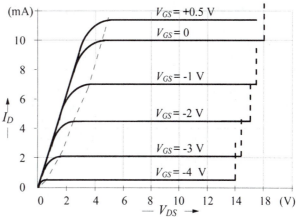

Objectives

You will be able to:

1   Explain the operation of n-channel and p-channel junction field effect transistors (JFETs).

2   Draw typical JFET characteristics. Identify the regions of the characteristics and all important current and voltage levels.

3   For JFETs, define: saturation current, pinch-off voltage, forward transfer admittance, output admittance, and drain-source ON resistance.

4   Determine JFET parameter values from manufacturer's data sheets.

5   From the data sheet information, draw the maximum and minimum transfer characteristics for any given JFET type number.

6   Solve problems involving JFET characteristics and parameters.

7   Show how a JFET can be used for voltage amplification.

8   Explain the operation of enhancement-mode and depletion-enhancement-mode MOSFETs.

9   Draw typical drain and transfer characteristics for MOSFETs, and discuss the differences between MOSFETs and JFETs.

10   Solve problems involving MOSFET characteristics and parameters.

11   Explain the operation of VMOSFETs, sketch typical characteristics, and discuss the advantages of VMOS.

12   Sketch circuit symbols for JFETs, MOSFETs, and VFETs. Identify all terminals, current directions, and voltage polarities.

Introduction

A *field effect transistor (FET)* is a voltage-operated device that can be used in amplifiers and switching circuits, similarly to a bipolar transistor. Unlike a bipolar transistor, a *FET* requires virtually no input current. This gives it an extremely high input resistance, which is its most important advantage over a bipolar transistor. There are two major categories of field effect transistors; *junction FETs* and *MOSFETS*. These are further subdivided into *p*-channel and *n*-channel devices.

9-1   Junction Field Effect Transistors

n-Channel JFET

The operating principle of an *n-channel junction field effect transistor (JFET)* is illustrated by the block representation in Fig. 9-1(a). A piece of *n*-type semiconductor material, referred to as the *channel*, is sandwiched between two smaller pieces of *p*-type (the *gates*). The ends of the channel are designated the *drain (D)* and the *source (S)*, and the two pieces of *p*-type material are connected together and their terminal is named the *gate (G)*.

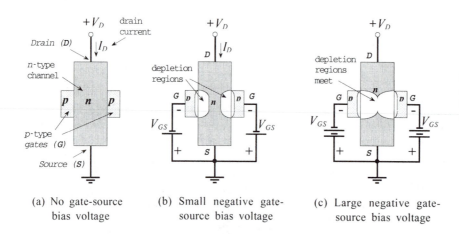

(a) No gate-source bias voltage

(b) Small negative gate-source bias voltage

(c) Large negative gate-source bias voltage

Figure 9-1
An n-channel JFET consists of an n-type channel with p-type gate regions on each side.

With the gate left unconnected, and a drain-source voltage (V_D) applied (positive at the drain, negative at the source), a *drain current* (I_D) flows, as shown in Fig. 9-1(a). When a gate-source voltage (V_{GS}) is applied with the gate negative with respect to the source [Fig. 9-1(b)], the gate-channel *pn*-junctions are reverse biased. The channel is more lightly doped than the gate material, so the depletion regions penetrate deep into the channel. Because the depletion regions are regions depleted of charge carriers, they behave as insulators. The result is that the channel is narrowed, its resistance is increased, and I_D is reduced. When the negative gate-source bias voltage is further increased, the depletion regions meet as the centre of the channel [Fig. 9-1(c)], and I_D is cut off.

An *ac* signal applied to the gate causes the reverse gate-source voltage to increase as the instantaneous level of the signal goes negative, and to decrease when the signal is positive-going. This causes the gate-channel depletion regions to successively widen and decrease. When the signal goes negative, the depletions widen, the channel resistance is increased, and the drain current decreases. As the signal goes positive, the depletion regions recede, the channel resistance is reduced, and the drain current increases. It is seen that the *FET* gate-source voltage controls the drain current. The gate-channel *pn*-junctions are maintained in reverse bias, so the gate current is normally extremely low; much lower than the base current for a bipolar transistor.

The name, *field effect transistor,* comes from the fact that the depletion regions in the channel are produced by the electric field at the reverse-biased gate-channel junctions. The term *unipolar device* is sometimes applied to a *FET*, because, unlike a bipolar transistor, the current consists of only one type of charge carrier; electrons in the case of an *n*-channel device.

Circuit symbols for an *n*-channel *JFET* are shown in Fig. 9-2. As in the case of all semiconductor device symbols, the arrowhead points from *p*-type to *n*-type. For an *n*-channel device, the arrowhead points from the *p*-type gate to the *n*-type channel. This is the direction of conventional current flow if the junctions become forward-biased. Some device manufacturers use the symbol in Fig. 9-2(a) with the gate directly opposite the source terminal. Others show the gate centralised between the drain and source [Fig. 9-2(b)]. This can sometimes make circuit diagrams confusing unless the drain and source terminals are clearly identified. The symbol in Fig. 9-2(c) is used where the terminals of the two gate regions are provided with separate connecting leads. In this case, the device is referred to as a *tetrode-connected FET.*

p-Channel *JFET*

In a *p*-channel *JFET*, shown in block form in Fig. 9-3(a), the channel is *p*-type semiconductor, and the gates are *n*-type. The drain-source voltage (V_D) is applied negative to the drain, positive to the source, as illustrated, and the drain current flows (in the conventional direction) from source to drain. To reverse-bias the gate-channel junctions, the *n*-type gate regions must be made positive with respect to the *p*-type channel. So, the bias voltage is applied positive on the gate terminals, negative on the source.

A positive-going signal at the gate terminal of a *p*-channel *JFET* increases in the gate-channel junction reverse bias, causing the depletions regions to penetrate further into the channel. This increases the channel resistance and decreases the drain current. Conversely, a negative-going signal narrows the depletion regions, reduces the channel resistance, and increases the drain current.

Circuit symbols for a *p*-channel *JFET* are shown in Fig. 9-3(b). The arrowheads again point from the *p*-type material to the *n*-type; in this case, from the *p*-type channel to the *n*-type gate.

(a) Circuit symbol for *n*-channel *JFET*

(b) Alternative circuit symbol

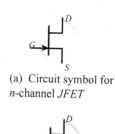

(c) Tetrode-connected *JFET*

Figure 9-2
Circuit symbols for an n-channel JFET. As with all semiconductor symbols, the arrowheads point from the p-type to the n-type material.

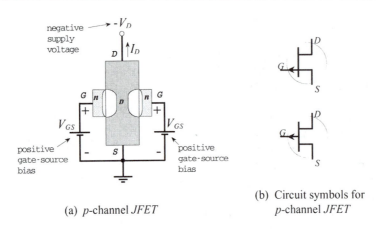

(a) *p*-channel *JFET*

(b) Circuit symbols for *p*-channel *JFET*

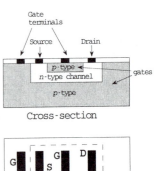

Cross-section

Top view

Figure 9-3
A p-channel JFET operates in exactly the same way as an n-channel device, except that the current directions and voltage polarities are reversed.

Figure 9-4
Cross-section and top view of n-channel JFET. With this type of construction, the drain and source are interchangable.

JFET Fabrication and Packaging

Junction field-effect transistors are normally manufactured by the diffusion process, (see Chapter 7). This type of construction is illustrated in Fig. 9-4. Starting with a *p*-type substrate, an *n*-channel is diffused, then, *p*-type impurities are diffused into the channel. Finally, metal terminal connections are deposited through holes in the silicon dioxide surface, as illustrated.

The *n*-type region is the *FET* channel, and the two *p*-type regions constitute the gates. With this symmetrical construction, the drain and source terminals are interchangeable. Other fabrication techniques produce device geometry that is not symmetrical. In these cases, interchanging the drain and source terminals would radically affect the device performance characteristics.

Figure 9-5 shows several *FET* packages which are similar to *BJT* enclosures. Note the device terminal identifications in each case.

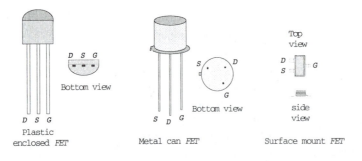

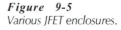

Bottom view

Plastic enclosed FET

Metal can FET

Surface mount FET

Figure 9-5
Various JFET enclosures.

Section 9-1 Review

9-1.1   Sketch a block representation of an *n*-channel *JFET*. Indicate voltages and current direction, and explain the device operation.

9-1.2   Repeat Question 9-1.1 for a *p*-channel *JFET*.

9-1.3   Sketch circuit symbols for *n*-channel and *p*-channel *JFETs*. Identify the terminals and briefly explain.

9-2   JFET Characteristics

Depletion regions

An *n*-channel *JFET* block representation is shown in some detail in Fig. 9-6. With a drain-source voltage applied as illustrated, I_D flows in the direction shown producing voltage drops along the channel. Consider the voltage drops from the source terminal (S) to points *A*, *B*, and *C* within the channel. Point *A* is positive with respect to the source; alternatively, it can be stated that *S* is negative with respect to *A*. Because the gate blocks are connected to *S*, the gates are negative with respect to point *A* by a voltage V_A. This causes the depletion regions to penetrate into the channel by an amount proportional to V_A.

The voltage drop between point *B* and the source is V_B, which is less than V_A. Consequently, at point *B* on the channel the gates are at $-V_B$ with respect to the channel, and the depletion region penetration is less than at point *A*. From point *C* to the source terminal, the voltage drop (V_C) is less than V_B. Thus, the gate-channel reverse bias (at point *C*) is V_C volts, and the depletion region penetration is less than at point *A* or *B*. The differing voltage drops along the channel, and the resulting variation in gate-channel reverse bias, accounts for the shape of the depletion region penetration of the channel.

Drain Characteristics with $V_G = 0$

Figure 9-7 shows a circuit for determining the drain current versus drain-source voltage characteristic for an *n*-channel *JFET* with $V_{GS} = 0$. V_{DS} is increased in convenient steps from zero, and I_D is measured at each V_{DS} level. This produces a table of I_D/V_{DS} values for plotting the characteristic shown in Fig. 9-8.

Referring to the characteristic, it is seen that when $V_{DS} = 0$, $I_D = 0$. There is no channel voltage drop, so the voltage between the gate and all points on the channel is zero, and there is no depletion region penetration. When V_{DS} is increased by a small amount (less than 1 V), a small drain current flows producing some voltage drop along the channel. This results in some depletion penetration of the channel (as explained for Fig. 9-6), but it is so small that it has no significant effect on the channel resistance. With further small increases in V_{DS} the drain current increase is approximately linear, and the channel behaves as an almost constant-value resistance, (see Fig. 9-8).

The channel continues to behave as a fixed-value resistance until the voltage drops along it become large enough to produce considerable depletion region penetration. At this stage the channel resistance begins to be affected by the depletion regions. Further increases in V_{DS} now produce smaller I_D increases, as shown by the curved part of the characteristic. The increased I_D levels, in turn, cause more depletion region penetration and greater channel resistance. Eventually, a *saturation level* of I_D is reached where further V_{DS} increase seems to have no effect on I_D.

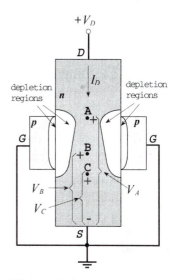

Figure 9-6
Drain current in an n-channel JFET causes voltage drops along the channel which reverse bias the gate channel junctions. This produces different levels of depletion region penetration into the channel.

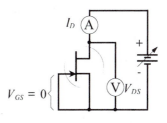

Figure 9-7
Circuit for obtaining the I_D/V_{DS} characteristic for an n-channel junction field effect transistor with $V_{GS} = 0$.

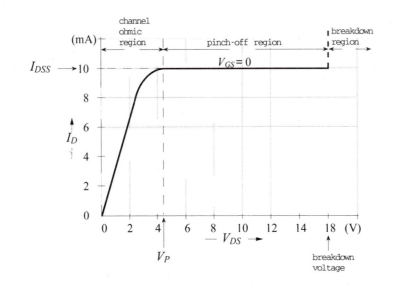

Figure 9-8
I_D/V_{DS} *characteristic for an n-channel JFET with* $V_{GS} = 0$.

At the point on the characteristic where I_D levels off, the drain current is referred to as the *drain-source saturation current* (I_{DSS}), (10 mA in Fig. 9-8). The shape of the depletion regions in the channel at the I_{DSS} level is such that they appear to *pinch off* the channel, (see Fig. 9-6). So, the drain source voltage at this point is termed the *pinch-off voltage* (V_p), (4.5 V in Fig. 9-8.) The region of the characteristic where I_D is constant is called the *pinch-off region*, as illustrated. The channel mostly behaves like a resistance between $V_{DS} = 0$ and $V_{DS} = V_P$, so this part of the characteristic is referred to as the *channel ohmic region.*

If V_{DS} is continuously increased (in the pinch-off region) a voltage is reached at which the (reverse-biased) gate-channel junctions break down, (see Fig. 9-8). When this occurs, I_D increases rapidly, and the device might be destroyed. The pinch-off region of the characteristic is the normal operating region for the *FET*.

Drain Characteristics with External Bias
A circuit for obtaining the I_D/V_{DS} characteristics for an *n*-channel JFET when an external gate-source bias (V_{GS}) is applied is shown in Fig. 9-9. In this case, V_{GS} is set to a convenient (negative) level (such as -1 V). V_{DS} is increased in steps, and the corresponding level of I_D is noted at each V_{DS} step. The I_D/V_{DS} characteristic for $V_{GS} = -1$ V is then plotted as illustrated in Fig. 9-10.

When a -1 V external gate-source bias voltage is applied, the gate-channel junctions are reverse biased even when $I_D = 0$. So, when $V_{DS} = 0$ the depletion regions are already penetrating to some depth into the channel. Because of this, a smaller voltage drop along the channel (smaller than when $V_{GS} = 0$) will increase the depletion regions to the point at which they produce channel pinch-off. Consequently, with $V_{GS} = -1$ V the pinch-off voltage is reached at a lower I_D level than when $V_{GS} = 0$. The $V_{GS} = -1$ V characteristic in Fig. 9-10 has $V_P = 3.5$ V.

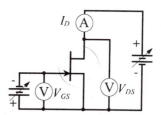

Figure 9-9
Circuit for obtaining the I_D/V_{DS} *characteristic for an n-channel junction field effect transistor with various gate-source bias voltages.*

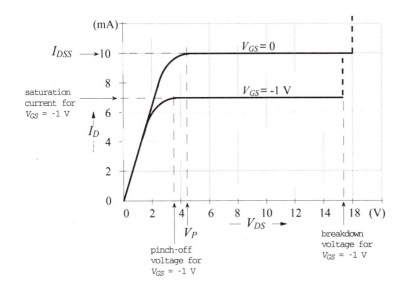

Figure 9-10
I_D/V_{DS} characteristics for $V_{GS} = 0$ and for $V_{GS} = -1$ V for an n-channel JFET.

A family of *drain characteristics* can be obtained by using several levels of negative gate-source bias voltage, (see Fig. 9-11). If a positive V_{GS} is used, a higher level of I_D can be produced, as shown by the characteristic for $V_{GS} = +0.5$ V. However, V_{GS} is normally kept negative to avoid the possibility of forward biasing the gate-channel junctions.

The dashed line on the characteristics in Fig. 9-11 is drawn through the points at which I_D saturates for each level of gate-source bias voltage. When $V_{GS} = 0$, I_D saturates at I_{DSS}, and the characteristic shows $V_p = 4.5$ V. When a -1 V external bias is applied, the gate-channel junctions still require -4.5 V to achieve pinch-off. This means that a 3.5 V drop is now required along the channel instead of 4.5 V, and the lower voltage drop is achieved with a lower level of I_D. Similarly, when $V_{GS} = -2$ V and - 3 V, pinch-off is achieved with 2.5 V and 1.5 V, respectively, along the channel. The 2.5 V and 1.5 V drops are, of course, obtained with further reduced I_D levels.

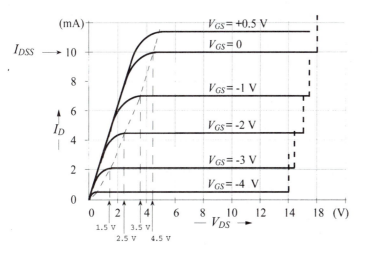

Figure 9-11
Family of I_D/V_{DS} characteristics for an n-channel JFET with various levels of V_{GS}.

Suppose a -4.5 V gate-source bias is applied to a device with the characteristics shown in Fig. 9-11. This is a V_{GS} level equal to the pinch-off voltage V_P. Without any additional channel voltage drop produced by I_D, the depletion regions penetrate so deep into the channel that they meet in the middle, completely cutting I_D *off*. So, a gate-source bias equal to the pinch-off voltage reduces I_D to zero. The bias voltage required to do this is termed the *gate cutoff voltage* ($V_{GS(off)}$), and, as explained, $V_{GS(off)} = V_P$. Note on Fig. 9-11 that the drain-source voltage at which breakdown occurs is reduced as the negative gate-source bias voltage is increased. This is because $-V_{GS}$ adds to the reverse bias at the junctions.

Example 9-1

Plot the I_D/V_{DS} characteristic for a *JFET* from the following table of values obtained with $V_{GS} = 0$. Determine I_{DSS} and V_P from the characteristics.

V_{DS} (V)	0	1	2	2.5	3	3.5	3.75	4	6	9
I_D (mA)	0	3	6	7	7.5	7.8	8	8	8	8

Solution

On Fig. 9-12, plot *point 1* at $V_{DS} = 0$ and $I_D = 0$
Plot *point 2* at $V_{DS} = 1$ V and $I_D = 3$ mA
and so on through $V_{DS} = 9$ V and $I_D = 8$ mA
Draw the drain characteristic for $V_{GS} = 0$ through *pints 1* through *10*.

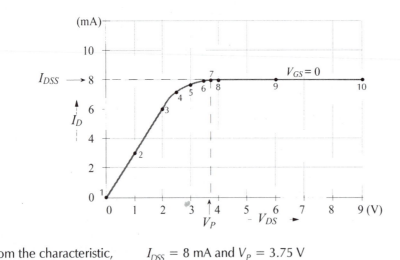

Figure 9-12
FET characteristics for Ex. 9-1.

From the characteristic, $I_{DSS} = 8$ mA and $V_P = 3.75$ V

Transfer Characteristics

The *transfer characteristics* for an *n*-channel *JFET* are a plot of I_D versus V_{GS}. The gate-source voltage of a *FET* controls the level of the drain current, so, the transfer characteristic shows how I_D is controlled by V_{GS}. As illustrated in Fig. 9-13(a), the transfer characteristic extends from $I_D = I_{DSS}$ at $V_{GS} = 0$, to $I_D = 0$ at $V_{GS} = V_P$.

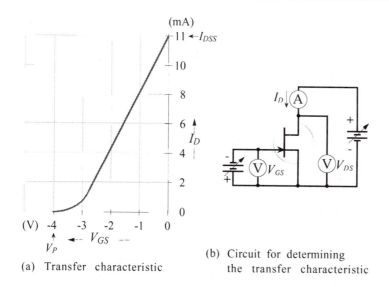

(a) Transfer characteristic

(b) Circuit for determining
the transfer characteristic

Figure 9-13
The transfer characteristics for a FET are a plot of I_D versus V_{GS}.

Figure 9-13(b) shows a circuit for experimentally determining a table of quantities for plotting the transfer characteristic of a given *FET*. The drain-source voltage is maintained constant, V_{GS} is adjusted in convenient step, and the corresponding levels of V_{GS} and I_D are recorded. The characteristic shows that, as $-V_{GS}$ is increased, I_D is progressively reduced from I_{DSS} at $V_{GS} = 0$, to $I_D = 0$ at $V_{GS} = -V_P$.

The transfer characteristic for a *FET* can be derived from the drain characteristics. A line is drawn vertically on the drain characteristics to represent a constant V_{DS} level. The corresponding I_D and V_{GS} values along this line are noted and then used to plot the transfer characteristic. The process is demonstrated in Example 9-2.

Example 9-2
Derive the transfer characteristic for $V_{DS} = 8$ V from the *FET* drain characteristics in Fig. 9-11.

Solution
On Fig. 9-11 draw a vertical line at $V_{DS} = 8$ V.
From the intersection of the line and the characteristics
read the following quantities:

V_{GS} (V)	0	-1	-2	-3	-4
I_D (mA)	10	7	4.5	2.2	0.5

On Fig. 9-14, plot *point 1* at $V_{GS} = 0$ and $I_D = 10$ mA.
Plot *point 2* at $V_{GS} = -1$ V and $I_D = 7$ mA,
and so on through $V_{GS} = -4$ V and $I_D = 0.5$ mA.
Draw the transfer characteristic through the pints.

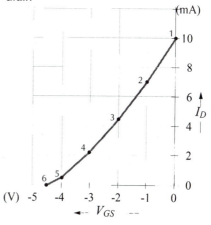

Figure 9-14
Transfer characteristics for Ex. 9-2.

p-channel JFET Characteristics

Figure 9-15 shows a circuit for obtaining the characteristics of a p-channel JFET. Note the direction of the arrowhead on the FET symbol, and the drain current direction. Also, note the supply voltage polarity, and the polarity of the gate-source bias voltage. The drain terminal is negative with respect to the source, and the gate terminal is positive with respect to the source. To obtain a table of quantities to plot a drain characteristic, V_{GS} is maintained constant at the desired (positive) level, $-V_{DS}$ is increased in steps from zero, and the I_D levels are noted at each step.

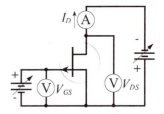

Figure 9-15
Circuit for determining the characteristics of a p-channel JFET.

Typical p-channel JFET drain characteristics and transfer characteristics are shown in Fig. 9-16. It is seen that these are similar to the characteristics for an n-channel JFET, except for the voltage polarities. In Fig. 9-16, when $V_{GS} = 0$, $I_{DSS} = 15$ mA, and progressively more positive levels of V_{GS} reduce the level of I_D toward cutoff at $V_P = +6$ V. Using $V_{GS} = -0.5$ V produces a higher level of I_D than when $V_{GS} = 0$. As in the case of the n-channel JFET, forward bias at the gate-channel junctions should be avoided, consequently, negative V_{GS} levels are normally not used with a p-channel JFET.

The transfer characteristic for a p-channel device can be obtained experimentally, or can be derived from the drain characteristics, just as for an n-channel FET.

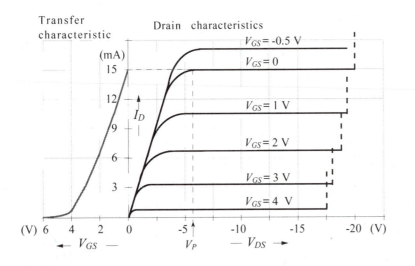

Figure 9-16
Transfer and drain characteristics for a p-channel JFET.

Practise Problems

9-2.1   The following table of I_D/V_{DS} values for a FET was obtained with $V_{GS} = 0$. Plot the drain characteristic and determine I_{DSS} and V_P.

V_{DS} (V)	0	1	2	2.5	3	4	6	8	10	12
I_D (mA)	0	2	4.5	5.3	5.5	5.5	5.5	5.5	5.5	5.5

9-3   JFET Data Sheet and Parameters

Maximum Ratings

Typical *FET* data sheets are shown in Appendix 1-11 and 1-12 and a portion of a *FET* data sheet is reproduced in Fig. 9-17. As with other device data sheets, a device type number and brief description is usually given at the start. Maximum ratings follow, and then the electrical characteristics are stated for specific bias conditions. From Fig. 9-17, the maximum drain-source voltage (V_{DS}) for the *2N5457* through *2N5459* devices is 25 V, and the maximum drain-gate voltage (V_{DG}) is also 25 V. This means, for example, that if a -5 V gate-source bias is used, the drain source voltage should not exceed,

$$V_{DS} = V_{DG(max)} - V_{GS} = 25\ V - 5\ V$$
$$= 20\ V$$

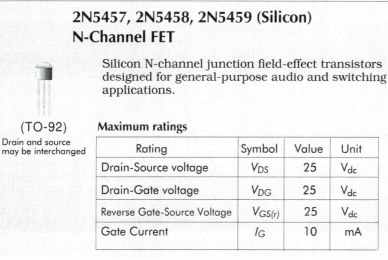

2N5457, 2N5458, 2N5459 (Silicon)
N-Channel FET

Silicon N-channel junction field-effect transistors designed for general-purpose audio and switching applications.

(TO-92)

Drain and source may be interchanged

Maximum ratings

Rating	Symbol	Value	Unit
Drain-Source voltage	V_{DS}	25	V_{dc}
Drain-Gate voltage	V_{DG}	25	V_{dc}
Reverse Gate-Source Voltage	$V_{GS(r)}$	25	V_{dc}
Gate Current	I_G	10	mA

Figure 9-17
Portion of data sheet for n-channel JFET.

Note on Fig. 9-17 that the maximum reverse gate-source voltage ($V_{G(r)}$) is specified as 25 V. This is considerably greater than the (typically 5 V) maximum base-emitter reverse voltage for a *BJT*.

No maximum drain current is specified in Fig. 9-17, but this can be calculated from the maximum power dissipation and the V_{DS} level. The specified gate current (I_G) is the maximum gate current if the gate-channel junctions become forward biased.

Saturation Current and Pinch-off Voltage

The *drain-source saturation current* (I_{DSS}) and the *pinch-off voltage* (V_P or $V_{GS(off)}$) have already been discussed in Section 9-1. Values for these are listed in the *JFET* data sheet portion showing the *Off Characteristics* and *ON Characteristics* in Fig. 9-18. It is seen that $V_{GS(off)}$ for a *2N5457* (underlined) ranges from a minimum of 0.5 V to a maximum of 6 V. Also, the I_{DSS} level for a *2N5457* (dashed underline) is a minimum of 1 mA and a maximum of 5 mA.

The *FET* transfer characteristic approximately follows the equation,

$$I_D = I_{DSS}\left[1 - \frac{V_{GS}}{V_P}\right]^2 \qquad \textbf{(9-1)}$$

When I_{DSS} and V_P are known, a table of corresponding values of I_D and V_{GS} can be determined from Eq. 9-1. These may be used to construct the *FET* transfer characteristic. Because of the wide range of specified values for I_{DSS} and V_P, the transfer characteristic can differ substantially from one device to another one having the same type number. This creates a problem in *FET* bias circuits (see Chapter 10).

2N5457, 2N5458, 2N5459
Off Characteristics

Characteristic	Symbol	Min	Typ	Max	Unit
Gate-Source Cutoff Voltage (V_{DS} = 15 V dc, I_D = 10 nA dc)	$V_{GS(off)}$				V dc
2N5457		0.5	—	6.0	
2N5458		1.0	—	7.0	
2N5459		2.0	—	8.0	

ON Characteristics

Zero gate voltage drain current (V_{DS} = 15 V dc, V_{GS} = 0)	I_{DSS}				mA dc
2N5457		1.0	3.0	5.0	
2N5458		2.0	6.0	9.0	
2N5459		4.0	9.0	16.0	

Figure 9-18
Pinch-off voltage and drain-source saturation current specifications for n-channel JFETs.

Example 9-3
Using the information provided on the data sheet (Fig. 9-18) construct the minimum and maximum transfer characteristics for a *2N5459 JFET*.

Solution

From Fig. 9-18; $V_P = V_{GS(off)}$ = 2 V (min), 8 V (max)
and, I_{DSS} = 4 mA (min), 16 mA (max)

To construct the minimum transfer characteristic, substitute $V_{P(min)}$ and $I_{DSS(min)}$ into Eq. 9-1, together with convenient levels of V_{GS}.

Eq. 9-1, $I_D = I_{DSS}\left[1 - \dfrac{V_{GS}}{V_P}\right]^2$

For V_{GS} = 0, I_D = 4 mA[1 - (0/2 V)]^2 = 4 mA

Plot point 1 for the minimum transfer characteristic on Fig. 9-19 at V_{GS} = 0 and I_D = 4 mA.

For V_{GS} = 0.5 V, I_D = 2.25 mA   *point 2*
For V_{GS} = 1 V, I_D = 1 mA   *point 3*
For V_{GS} = 1.5 V, I_D = 0.25 mA   *point 4*
For V_{GS} = 2 V, I_D = 0   *point 5*

The minimum transfer characteristic is now drawn
through points 1 through 5.

For the maximum transfer characteristic, the above process
is repeated using V_P = 8 V and I_{DSS} = 16 mA.

For V_{GS} = 0, I_D = 16 mA[1 - (0/8 V)]^2 = 16 mA

Plot point 6 for the maximum transfer characteristic
on Fig. 9-19 at V_{GS} = 0 and I_D = 16 mA.

For V_{GS} = 2 V, I_D = 9 mA   *point 7*
For V_{GS} = 4 V, I_D = 4 mA   *point 8*
For V_{GS} = 6 V, I_D = 1 mA   *point 9*
For V_{GS} = 8 V, I_D = 0   *point 10*

Draw the maximum transfer characteristic through
points 6 through 10.

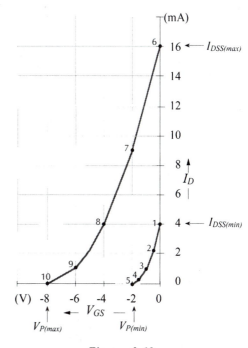

Figure 9-19
Maximum and minimum transfer
characteristic for a 2N5459 JFET
constructed for Ex. 9-3.

Forward Transfer Admittance

The *forward transfer admittance (Y_{fs})* [also known as the *transconductance (g_m or g_{FS})*] for a *FET* defines how the drain current is controlled by the gate-source voltage.

$$Y_{fs} = \frac{variation\ in\ I_D}{variation\ in\ V_{GS}}\ (When\ V_{DS}\ remains\ constant)$$

$$\mathbf{Y_{fs} = \frac{\Delta I_D}{\Delta V_{GS}}\bigg|_{V_{DS}}} \qquad \textbf{(9-2)}$$

The units used for Y_{fs} are *microSiemens (μS)*, which can be restated as *microamps per volt ($\mu A/V$)*. *MilliSiemens (mS)*, or *milliamps/volt (mA/V)* might also be used. For a *FET* with Y_{fs} = 2 mA/V, I_D changes by 2 mA when V_{GS} is altered by 1 V. In the portion of a *FET* data sheet in Fig. 9-20, the Y_{fs} units are *μmhos*. The *mho* (ohm written backwards) is another (older) name for the *Siemen*, the unit of conductance. For the *2N5457* (underlined) the value of Y_{fs} is specified as a minimum of 1000 μmhos, and a maximum of 5000 μmhos. An inverted ohm symbol is also sometimes used instead of the Siemens symbol.

Because Y_{fs} defines the relationship between I_D and V_{GS}, it can be determined from the slope of the *FET* transfer characteristic. This is demonstrated in Ex. 9-4, and illustrated in Fig. 9-21.

2N5457, 2N5458, 2N5459
Dynamic Characteristics

Characteristic	Symbol	Min	Typ	Max	Unit
Forward Transfer Admittance (V_{DS} = 15 V dc, V_{GS} = 0, f = 1 kHz)	Y_{fs}				μmhos
2N5457		1000	3000	5000	
2N5458		1500	4000	5500	
2N5459		2000	4500	6000	

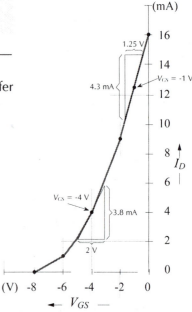

Figure 9-20
JFET forward transfer admittance specification.

Example 9-4

Determine Y_{fs} at V_{GS} = -1 V and at V_{GS} = -4 V from the *2N5459 FET* transfer characteristic in Fig. 9-21.

Solution

At V_{GS} = -1 V,

Eq. 9-2,
$$Y_{fs} = \frac{\Delta I_D}{\Delta V_{GS}} = \frac{4.3 \text{ mA}}{1.25 \text{ V}}$$
$$= 3.4 \text{ mA/V} = 3400 \text{ } \mu S$$

At V_{GS} = -4 V,
$$Y_{fs} = \frac{\Delta I_D}{\Delta V_{GS}} = \frac{3.8 \text{ mA}}{2 \text{ V}}$$
$$= 1.9 \text{ mA/V} = 1900 \text{ } \mu S$$

Figure 9-21
The value of the forward transfer admittance for a FET can be determined from the slope of the transfer characteristic.

The transfer characteristic for a *FET* is defined by Eq. 9-1, and Y_{fs} is determined from the slope of the transfer characteristic. So, an equation for Y_{fs} can be derived by differentiating Eq. 9-1.

$$Y_{fs} = \frac{2 \, I_{DSS}}{V_P} \left[1 - \frac{V_{GS}}{V_P} \right] \qquad (9\text{-}3)$$

Equation 9-3 may be used to calculate the value of Y_{fs} for any V_{GS} level. As in Eq. 9-1, the negative sign for V_{GS} is already included, so that only the numerical value should be entered.

Example 9-5

Using Eq. 9-3, determine Y_{fs} at V_{GS} = -1 V and at V_{GS} = -4 V for a *2N5458* JFET with a maximum transfer characteristic, (see Fig. 9-19).

Solution

From the data sheet (Fig. 9-18), the 2N5458 maximum transfer characteristic is defined by;

$$V_{P(max)} = V_{GS(off)(max)} = 7 \text{ V}$$

and,
$$I_{DSS(max)} = 9 \text{ mA}$$

At $V_{GS} = -1$ V,

Eq. 9-3,
$$Y_{fs} = \frac{2\,I_{DSS}}{V_P}\left[1 - \frac{V_{GS}}{V_P}\right] = \frac{2 \times 9 \text{ mA}}{7 \text{ V}}\left[1 - \frac{1 \text{ V}}{7 \text{ V}}\right]$$

$$= 2200\,\mu S$$

At $V_{GS} = -4$ V,

Eq. 9-3,
$$Y_{fs} = \frac{2\,I_{DSS}}{V_P}\left[1 - \frac{V_{GS}}{V_P}\right] = \frac{2 \times 9 \text{ mA}}{7 \text{ V}}\left[1 - \frac{4 \text{ V}}{7 \text{ V}}\right]$$

$$= 1100\,\mu S$$

Output Admittance

The *drain resistance* (r_d) of a *FET* is the *ac* resistance between drain and source terminals when the device is operating in the pinch-off region of its drain characteristics. It is also the slope of the drain characteristics in the pinch-off region, (see Fig. 9-22). The drain characteristics are almost flat, so, r_d is not easily determined from the characteristic. Because r_d is usually the output resistance of the *FET* at the drain terminal, it may also be expressed as an *output admittance* $(Y_{os} = 1/r_d)$.

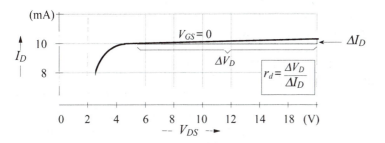

Figure 9-22
The ac drain resistance of a FET (r_d) is the slop of the drain characteristic. The output admittance (Y_{fs}) is the inverse of r_d.

The output admittance is defined as,

$$Y_{os} = \frac{\text{variation in } I_D}{\text{variation in } V_{DS}} \text{ (When } V_{GS} \text{ remains constant)}$$

$$\boldsymbol{Y_{os} = \frac{\Delta I_D}{\Delta V_{DS}}\bigg|_{V_{GS}}} \qquad (9\text{-}4)$$

From Appendix 1-11, Y_{os} for the *2N5457-5459 JFETs* is 10 μS typically, and 50 μS maximum. This corresponds to a drain resistance of 100 kΩ typically, and 20 kΩ minimum.

Drain-Source ON Resistance

The *drain-source ON resistance* $(r_{DS(on)})$ (also designated $R_{D(on)}$), is a *dc* quantity, not to be confused with the *ac* drain-source resistance (r_{ds}). $r_{DS(on)}$ is the resistance of the *FET* channel when the depletion regions are absent; when the device is biased *ON* in the channel ohmic region of the drain characteristics, (see Fig. 9-23). In this condition, the voltage drop along the channel from drain to source is ($I_D \times r_{DS(on)}$). This is the *drain-source ON voltage* ($V_{DS(on)}$), which is similar to the $V_{CE(sat)}$ of a *BJT*.

The drain-source *ON* resistance might typically be 60 Ω, or lower, (see Appendix 1-12). It can be an important quantity for *FETs* used in switching circuits. $V_{DS(on)}$ can be much smaller than $V_{CE(sat)}$, making *FET* gates superior to *BJT* gates for some applications.

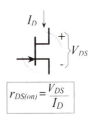

Figure 9-23
The drain-source ON resistance ($r_{DS(on)}$) of a FET is the channel resistance.

Gate Cutoff Current and Input Resistance

The gate-channel junctions in a *JFET* are *pn*-junctions, and because they are normally reverse biased, a minority charge carrier current flows. This is the *gate-source cutoff current* (I_{GSS}), or *gate reverse current*, [see Fig. 9-24(a)]. For a *2N5457 FET*, I_{GSS} is specified as 1 nA at 25°C, and 200 nA at 100°C, (see Appendix 1-11).

The *gate input resistance* (R_{GS}) is the resistance of the reverse-biased gate-channel junctions [Fig. 9-24(b)], and it is inversely proportional to I_{GSS}. Typical values of R_{GS} for a *JFET* are 10^9 Ω at 25°C and 10^7 Ω at 100°C.

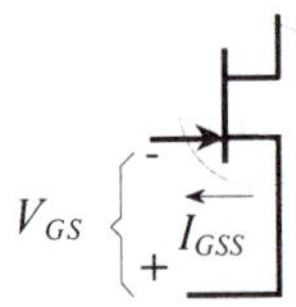

(a) Gate reverse current

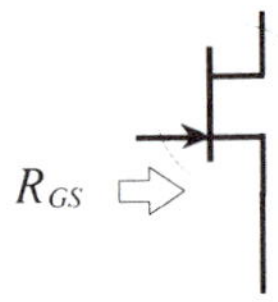

(b) Gate input resistance

Breakdown Voltage

The are several ways used for expressing *JFET* breakdown voltage. BV_{DGO} is the *drain-gate breakdown voltage* with the source terminal open-circuited. BV_{GSS} is the *gate-source breakdown voltage* with the drain terminal shorted to the source. Both are a specification of the voltage at which the gate-channel junctions might break down. Appendix 1-11 shows a BV_{GSS} of 25 V for a *2N5457*, (this is identified as $V_{(BR)GSS}$ on the data sheet).

Figure 9-24
Gate cutoff current and input resistance.

Noise Figure

A *FET* usually has a much lower level of thermal noise than a *BJT*. This is because there are very few charge carriers crossing junctions in a *FET*. As in the case of a *BJT*, the *FET noise figure* (*NF*) is specified as a *spot noise figure* at a particular frequency and bias conditions, and for a given resistance at the input. The figure varies if any of these conditions are altered. Noise calculations for a *FET* circuit are performed exactly as for a *BJT* circuit.

Capacitances

Terminal capacitances for *FETs* may be specified as *gate-drain capacitance* (C_{gd}), *gate-source capacitance* (C_{gs}), and *drain-source capacitance* (C_{ds}). The input capacitance is sometimes expressed as the *common-source input capacitance* (C_{iss} or C_{gss}). This is the gate-source capacitance measured with the drain terminal shorted to the source, (Fig. 9-25). In this case, a *reverse transfer capacitance* (C_{rss}) is also specified; C_{rss} being another term for C_{gd}. These quantities are very important for *FET* high-frequency and switching circuits. From the *2N5457* specification in Appendix 1-11, the maximum capacitance values are C_{iss} = 7 pF, and C_{rss} = 3 pF.

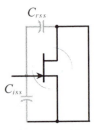

Figure 9-25
FET common-source input capacitances and reverse transfer capacitance.

9-4   FET Amplification

Consider the *n*-channel *JFET* circuit in Fig. 9-26. Note that drain-source terminals are provided with a dc supply (V_{DD}), connected via the drain resistor (R_1). The gate-source junctions are reverse-biased by the gate voltage (V_G). An *ac* signal generator (with voltage v_i) is connected in series with the gate terminal. As already discussed, field effect transistors are voltage-operated devices. The drain current is controlled by the gate-source voltage. A change in gate-source voltage (ΔV_{GS}) produces a change in drain current (ΔI_D), and this causes a variation in the voltage drop across the resistor connected in series with the drain terminal.

Suppose that, for the circuit in Fig. 9-26, V_{DD} = 20 V, R_1 = 6 kΩ, and V_G is adjusted to give I_D = 2 mA. Also, assume that the *FET* has a forward transfer admittance of Y_{fs} = 4000 µS. The dc level of the drain voltage is,

$$V_D = V_{DD} - (I_D R_1) = 20 \text{ V} - (2 \text{ mA} \times 6 \text{ kΩ})$$
$$= 8 \text{ V}$$

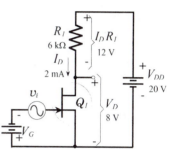

Figure 9-26
FET voltage amplifier circuit. An ac input (v_i) produces a variation in I_D which varies the voltage drop across resistor R_1. This results in an (amplified) ac output voltage.

Now assume that the instantaneous level of v_i is +50 mV. This produces an increase,

$$\Delta I_D = Y_{fs} \times v_i = 4000 \text{ µS} \times 50 \text{ mV}$$
$$= 0.2 \text{ mA}$$

The new level of drain voltage is,

$$V_D = V_{DD} - R_1 (I_D + \Delta I_D) = 20 \text{ V} - 6 \text{ kΩ}(2 \text{ mA} + 0.2 \text{ mA})$$
$$= 6.8 \text{ V}$$

So, V_D changed from 8 V to 6.8 V when the ac input increased from zero to 50 mV.

or, $\Delta V_D = 6.8$ V - 8 V
              $= -1.2$ V

When the instantaneous level of v_i is -50 mV,

$$\Delta I_D = Y_{fs} \times v_i = 4000 \ \mu S \times (-50 \ mV)$$
$$= -0.2 \ mA$$

This gives a new drain voltage,

$$V_D = V_{DD} - R_1 (I_D + \Delta I_D) = 20 \ V - 6 \ k\Omega(2 \ mA - 0.2 \ mA)$$
$$= 9.2 \ V$$

So, $$\Delta V_D = 9.2 \ V - 8 \ V$$
$$= 1.2 \ V$$

It is seen that an ac input of $v_i = \pm 50$ mV produces an output voltage change at the *FET* drain terminal of $\Delta V_D = \pm 1.2$ V. This can be stated as an ac output voltage of $v_o = \pm 1.2$ V. The circuit voltage amplification is,

$$A_v = \frac{v_o}{v_i} = \frac{1.2 \ V}{50 \ mV}$$
$$= 24$$

Example 9-6
A *2N5457 JFET* is used in the amplifier circuit in Fig. 9-27. Calculate the maximum and minimum output voltage produced by a ± 100 mV ac input. Also, calculate the circuit voltage gain in each case.

Solution
From Appendix 1-11,
$$Y_{fs(max)} = 5000 \ \mu S \text{ and } Y_{fs(min)} = 1000 \ \mu S$$

For $Y_{fs(max)}$, $$\Delta I_D = Y_{fs(max)} \times v_i = 5000 \ \mu S \times (\pm 100 \ mV)$$
$$= \pm 0.5 \ mA$$

and, $$v_o = \Delta I_D \times R_1 = \pm 0.5 \ mA \times 5.6 \ k\Omega$$
$$= \pm 2.8 \ V$$

$$A_v = \frac{v_o}{v_i} = \frac{\pm 2.8 \ V}{\pm 100 \ mV}$$

$$= 28$$

For $Y_{fs(min)}$, $$\Delta I_D = Y_{fs(max)} \times v_i = 1000 \ \mu S \times (\pm 100 \ mV)$$
$$= \pm 0.1 \ mA$$

and, $$v_o = \Delta I_D \times R_1 = \pm 0.1 \ mA \times 5.6 \ k\Omega$$
$$= \pm 0.56 \ V$$

$$A_v = \frac{v_o}{v_i} = \frac{\pm 0.56 \ V}{\pm 100 \ mV}$$

$$= 5.6$$

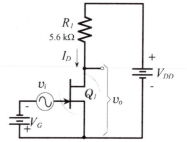

Figure 9-27
FET amplifier circuit for Ex. 9-6.

Practise Problems

9-4.1   A *FET* amplifier circuit, as in Fig. 9-27 , has V_{DD} = 15 V, R_1 = 4.7 kΩ, and v_i = ±30 mV. If V_G is adjusted to give I_D = 1.5 mA, calculate V_D. Also, calculate the typical voltage gain for the circuit if Q_1 is a 2N5459 with Y_{fs} = 4500 μS.

9-5   MOSFETs

Enhancement MOSFET

Figure 9-28 shows the construction of a *metal oxide semiconductor FET (MOSFET)*, also known as an *insulated gate FET*. Starting with a high-resistive *p*-type substrate, two blocks of heavily-doped *n*-type material are diffused into the substrate, and then the surface is coated with a layer of silicon dioxide. Holes are cut through the silicon dioxide to make contact with the *n*-type blocks. Metal is deposited through the holes for source and drain terminals, as illustrated, and a metal plate is deposited on the surface area between drain and source. As will be explained, this plate functions as a gate.

Consider the situation illustrated in Fig. 9-29(a). The drain terminal of the *MOSFET* is positive with respect to the source, and the gate is open-circuited. The two *n*-type blocks and the *p*-type substrate form back-to-back *pn*-junctions connected by the resistance of the *p*-type material, as illustrated. The *pn*-junction close to the drain terminal is reverse biased, so only a very small (reverse leakage) current flows from *D* to *S*.

Now assume that the source terminal is connected to the substrate, and that a positive gate voltage is applied, as shown in Fig. 9-29(b). Negative (minority) charge carriers within the substrate are attracted to the (positive) plate that constitutes the gate. These charge carriers (electrons) cannot cross the silicon dioxide to the gate, so they accumulate close to the surface of the substrate, as shown. The minority charge carriers constitute an *n*-type channel between drain and source, and as the gate-source voltage is made more positive, more electrons are attracted into the channel causing the channel resistance to decrease. A drain current flows along the channel between the *D* and *S* terminals, and because the channel resistance is controlled by the gate-source voltage (V_{GS}), the drain current is also controlled by the level of V_{GS}. The channel conductivity is said to be *enhanced* by the positive gate-source voltage, so the device is known as an *enhancement-mode MOSFET, (E-MOSFET)*.

Typical drain and transfer characteristics for an enhancement mode *n*-channel *MOSFET* are shown in Fig. 9-30. Note on both characteristics that the drain current increases as the positive gate-source bias voltage is increased. Because the gate of the *MOSFET* is insulated from the channel, there is no gate-source leakage current, and the device has an extremely high (gate) input resistance; typically 10^{15} Ω, or greater. Forward transfer admittance

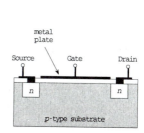

Figure 9-28
Metal oxide semiconductor FET (MOSFET) construction.

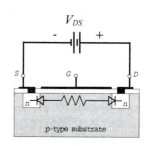

(a) V_{DS} applied and gate open-circuited

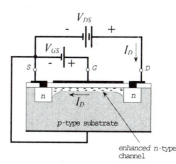

(b) Effect of positive V_{GS}

Figure 9-29
Effect of +V_{DS} on the MOSFET with the gate terminal open-circuited, and with +V_{GS} applied to the gate.

values for this type of (low-power) *MOSFET* typically range from 1 mS (1 mA/V) to a maximum of perhaps 6 mS, which is similar to *JFET* Y_{fs} values.

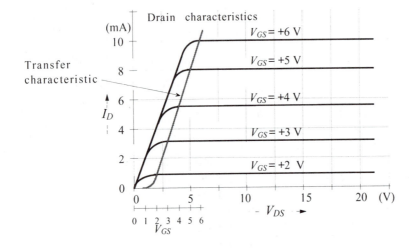

Figure 9-30
Typical drain and transfer characteristics for an n-channel enhancement-mode MOSFET.

Two graphic symbols for the *n*-channel *E-MOSFET* are shown in Fig. 9-31. One symbol shows the source and substrate connected internally, while the other has a separate substrate terminal. The line representing the device channel is broken into three sections, to indicate that the channel does not exist until an appropriate gate voltage is applied. The fact that the device has an insulated gate is illustrated on the symbols by the gate not making direct contact with the channel. The arrowhead points from the *p*-type substrate to the *n*-type channel.

A *p*-channel *E-MOSFET* is constructed by starting with an *n*-type substrate and diffusing *p*-type drain and source blocks, as illustrated in Fig. 9-32(a). The device characteristics are similar to those in Fig. 9-30, except that all voltage polarities and current directions are reversed. The drain-source voltage is negative, and a negative gate-source voltage is required to create the *p*-type channel. The arrowhead in the circuit symbols are also reversed, [see Fig. 9-32(b)].

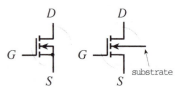
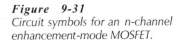

Figure 9-31
Circuit symbols for an n-channel enhancement-mode MOSFET.

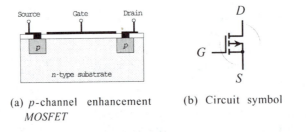

(a) *p*-channel enhancement MOSFET

(b) Circuit symbol

Figure 9-32
Construction and circuit symbols for a p-channel enhancement-mode MOSFET.

Depletion-Enhancement MOSFET

The device cross-section shown in Fig. 9-33(a) is similar to that for an enhancement-mode *MOSFET*, except that a lightly-doped *n*-type channel is included between the drain and source blocks. When a positive drain-source voltage (V_{DS}) is applied, a drain current (I_D) flows even when the gate-source voltage (V_{GS}) is zero. If a negative V_{GS} is applied, as shown in Fig. 9-33(b), some of the negative charge carriers are repelled from the gate and driven out of the n-type channel. This creates a depletion region in the channel, as illustrated, causing an increase in channel resistance and a decrease in drain current. The effect is similar to that in an n-channel *JFET*. Because of the channel depletion regions, the device can be termed a *depletion-mode MOSFET*.

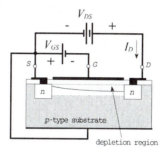

Figure 9-33
n-channel depletion-enhancement MOSFET.

(a) I_D flows even when $V_{GS} = 0$ (b) I_D decreases when V_{GS} is negative, and increases when V_{GS} is positive

Now consider what occurs when a positive gate-source voltage is applied. Additional n-type charge carriers are attracted from the substrate into the channel, decreasing its resistance and increasing the drain current. So, the depletion-mode *MOSFET* can also be operated as an enhancement-mode device. Thus, these devices are referred to as *depletion-enhancement MOS*, or *DE-MOS*.

Typical drain and transfer characteristics for a *DE-MOSFET* are shown in Fig. 9-34. The device operates in the depletion-mode when V_{GS} is negative, and in the enhancement-mode when positive levels of V_{GS} are used.

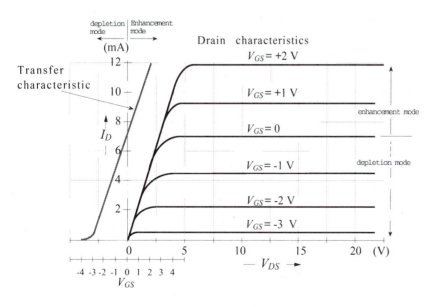

Figure 9-34
Drain and transfer characteristics for an n-channel depletion-enhancement MOSFET.

The circuit symbols used for *DE-MOSFETs* are similar to those already discussed for *E-MOSFETs*, except that the line representing the channel is made solid to show that a channel is present when $V_{GS} = 0$, (see Fig. 9-35).

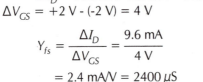

Figure 9-35
Circuit symbols for an n-channel depletion-enhancement MOSFET.

Example 9-7

From the *DE-MOSFET* characteristics in Fig. 9-34, determine the forward transfer admittance at $V_{GS} = 0$.

Solution

At $V_{GS} = -2$ V, $I_D = 2.2$ mA
and at $V_{GS} = +2$ V, $I_D = 11.8$ mA

$$\Delta I_D = 11.8 \text{ mA} - 2.2 \text{ mA} = 9.6 \text{ mA}$$

and, $\Delta V_{GS} = +2 \text{ V} - (-2 \text{ V}) = 4 \text{ V}$

Eq. 9-2, $$Y_{fs} = \frac{\Delta I_D}{\Delta V_{GS}} = \frac{9.6 \text{ mA}}{4 \text{ V}}$$

$$= 2.4 \text{ mA/V} = 2400 \ \mu S$$

VMOS

A disadvantage of the *MOSFET* types already discussed is that the minimum channel length depends upon the dimensions of the photographic masks used in the manufacturing process. Shorter channel lengths can be produced by changing the geometry of the *MOSFET* to create a vertical channel (instead of horizontal). The channel length is then easily determined by the diffusion process. The shorter channel results in lower resistance, greater power dissipation, higher frequency response, and larger forward transfer admittance values than are possible with other *MOSFETs*.

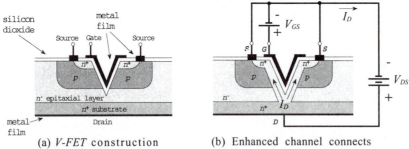

(a) *V-FET* construction

(b) Enhanced channel connects the drain and source

Figure 9-36
VMOSFET construction and operation. The vertical channel gives an improved frequency response, lower channel resistance, and greater power dissipation than other FETs.

Figure 9-36(a) shows a device referred to as a *VMOSFET* because of its *V*-shaped configuration, and because it uses a vertical channel between drain and source. The *V*-cut penetrates from the surface of the device through n^+, p, and n^- layers almost to the n^+ substrate. The n^+ layers are low-resistive, and the n^- is a high-resistive region. The silicon dioxide layer covers both the horizontal surface and the *V*-cut surface. The gate is a metal film deposited on the silicon dioxide surface in the *V*-cut. The drain terminal is at the bottom of the n^+ substrate, and the source connection is made to the top n^+ region and to the *p* region.

The *VFET* operates in the enhancement-mode; no channel exists until a positive gate source voltage is applied. An *n*-type channel is created as shown in Fig. 9-36(b) when V_{GS} is made positive, and a current flows vertically from drain to source.

Some versions of *VMOS* use a vertical channel, but do not use the *V*-cut. *MOTOROLA* currently manufactures devices referred to as *TMOS*®, because the drain current flow tends to be *T*-shaped. *INTERNATIONAL RECTIFIER* use the term *HEXFET*®, because of the hexagonal configuration employed in the device construction.

Typical *VFET* drain and transfer characteristics are shown in Fig. 9-37. It is seen that they are similar to the enhancement-mode device characteristics in Fig. 9-30, although much higher drain current levels are usually involved with a *VFET* than with other *MOSFETs*. The characteristics shown are for a device capable of dissipating 40 W.

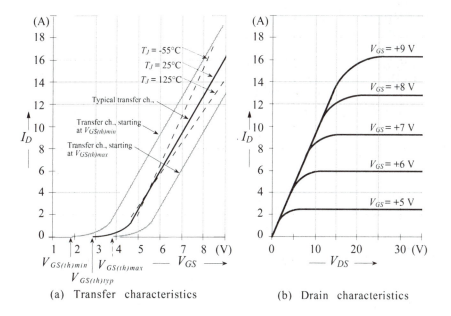

(a) Transfer characteristics   (b) Drain characteristics

Figure 9-37
Typical transfer and drain characteristics for a VMOSFET.

The *VFET* transfer characteristic is seen to be approximately linear over most of its length, and it is curved (non-linear) only at the low current levels. For the typical characteristic shown, I_D commences at the typical gate threshold voltage ($V_{GS(th)typ}$), and the slope of the linear portion of the graph is set by the typical g_{FS} value (same as Y_{FS}). It should be noted that for a device with a given type number, I_D might commence at $V_{GS(th)min}$ or $V_{GS(th)max}$, and that the slope of the characteristic depends upon the actual g_{FS} for the device. Temperature changes can also have a serious effect on the characteristics, as shown by the dashed lines.

The graphic symbols used for *VMOS* devices are the same as for other enhancement-mode *MOSFETs*. However, a *parasitic pn-junction* is present between the *p* and *n⁻* layers in a *VFET*, and this must not be allowed to become forward biased. So, the diode is usually shown on the device symbol, (see Fig. 9-38).

VMOS devices can be described as high-voltage, *E-MOSFETs*. They are very suitable for high-frequency or fast switching applications. Normally, they have much larger forward transfer admittance (g_{FS}) values and lower drain-source *ON* resistances than other *FETs*. Figure 9-39 lists some performance data for two representative (low-power and high-power) *VMOSFETs*. Note that the 80 mS minimum g_{FS} for the *2N7002* is substantially larger than the 4 mS typical for other *FET* types. The *IRF520* has a typical g_{FS} of 2.9 S, which is greater again than that for the *2N7002*. *FET* voltage amplification is directly proportional to the forward transfer admittance, so *VMOS* circuits can have much greater voltage gains than circuits using other *FETs*.

MOSFET specifications normally list *gate threshold voltage* ($V_{GS(th)}$) values. The drain current remains at zero with gate-source voltage levels below $V_{GS(th)}$, and commences to flow when V_{GS} is increased above $V_{GS(th)}$. So, $V_{GS(th)}$ is similar to the pinch-off voltage for a *JFET*. Like the *JFET* pinch-off voltage, $V_{GS(th)}$ has maximum and minimum values for each device type number.

For a *JFET*, I_{DSS} is the drain current at $V_{GS} = 0$. The *E-MOSFET* parameter identified as I_{DSS} is still the drain current at zero gate-source voltage. However, because the device is *off* at $V_{GS} = 0$, I_{DSS} is a drain-source leakage current with a level measured in microamps. For *E-MOSFET* devices, I_D is usually specified as $I_{D(on)}$ at $V_{GS} = 10$ V.

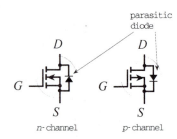

Figure 9-38
Circuit symbols for VMOSFETs.

2N7002

$V_{DS(max)}$	$I_{D(max)}$	$P_{D(max)}$	$r_{d(on)}$	g_{FS}	$V_{GS(th)}$
60 V	115 mA	200 mW	7.5 Ω	80 mS (min)	1 V (min) 2.5 V (max)

IRF520

$V_{DS(max)}$	$I_{D(max)}$	$P_{D(max)}$	$r_{d(on)}$	g_{FS}	$V_{GS(th)}$
100 V	8 A	40 W	0.3 Ω	1.5 S (min) 2.9 S (typ)	2 V (min) 4 V (max)

Figure 9-39
Performance data for low-power and high-power VMOSFETs.

E-MOSFETs are often applied in situations where they are biased *on* to a low drain current level, and then I_D is increased to a desired higher level by the application of a signal voltage. The precise levels of the required gate-source bias and signal voltages can only be determined from the transfer characteristic for each individual device. However, for a given I_D level beyond the curved portion of the characteristic, a rough approximation for the required V_{GS} can be calculated from the maximum $V_{GS(th)}$ and the typical g_{FS} values.

$$V_{GS} \approx V_{GS(th)max} + \frac{I_D}{g_{FS(typ)}} \qquad \text{(9-5)}$$

Example 9-8

Using the information in Fig. 9-39, calculculate the approximate gate-source voltage required to produce a 7 A drain current in an *IRF520*. Also,determine the drain-source *ON* voltage and the device power dissipation at $I_D = 7$ A.

Solution

Eq. 9-5,
$$V_{GS} \approx V_{GS(th)(max)} + \frac{I_D}{g_{FS(typ)}} = 4 \text{ V} + \frac{7 \text{ A}}{2.9 \text{ S}}$$
$$= 6.4 \text{ V}$$

$$V_{DS(on)} = I_D \times r_{D(on)} = 7 \text{ A} \times 0.3 \text{ }\Omega$$
$$= 2.1 \text{ V}$$

$$P_D = I_D \times V_{DS(on)} = 7 \text{ A} \times 2.1 \text{ V}$$
$$= 14.7 \text{ W}$$

Handling MOSFETs

Figure 9-40 shows the kind of label normally found on packages containing *MOSFETs*. This indicates that the devices can be very easily destroyed by *electrostatic discharge (ESD)*; that is, discharge of static electricity that accumulates on individuals or objects. Usually, the very thin layer of silicon dioxide at the gate breaks down when an excessive gate-source voltage is applied. Special precautions are necessary to protect *MOSFETs* from *ESD*:

1. Always store *MOSFETs* in closed conductive containers. Usually, they are packaged with the terminals inserted into conducting foam rubber.
2. Use a work station with a grounded anti-static bench-top pad and a grounded anti-static floor pad.
3. Wear anti-static clothing and grounded wrist bands.
4. Use a soldering iron with a grounded tip.
5. Avoid touching the device terminals.

Some *MOSFETs* have built-in protection against static electricity. Back-to-back series-connected Zener diodes are located between the gate and source terminals, as illustrated in Fig. 9-41. When the gate-source voltage is high enough to reverse bias one of the diodes into breakdown, the other diode is forward biased. Thus, the gate-source voltage cannot exceed $\pm(V_Z + V_F)$.

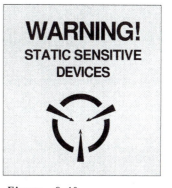

Figure 9-40
Packaging symbol used for indicating that the enclosed devices can be destroyed by static electricity.

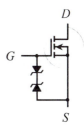

Figure 9-41
MOSFET with built-in Zener diodes for protection against static electricity.

Section 9-5 Review

9-5.1   Using illustrations, explain the operation of *n-channel E-MOS*.

9-5.2   Sketch circuit symbols for *n-channel* and *p-channel E-MOS* and *DE-MOS* field effect transistors.

9-5.3   Sketch typical transfer characteristics for *DE-MOS* devices.

9-5.4   Briefly explain the operation of *VMOS*.

Practise Problems

9-5.1   Determine the forward transfer admittance at I_D = 8 mA for the E-MOS device with the characteristics in Fig. 9-30.

9-5.2   Calculate the approximate gate-source voltage required to produce a 7.5 A drain current in an *E-MOSFET* that has maximum $V_{GS(th)}$ = 2.5 V and typical g_{FS} = 1.5 S.

Chapter-9 Review Questions

Section 9-1

9-1   Sketch a block representation for an *n*-channel *JFET*, showing bias voltages, depletion regions, and current directions. Identify the device terminals and explain its operation. Also explain the effect of increasing levels of negative gate-source voltage.

9-2   Repeat question 9-1 for a *p*-channel *JFET*, explaining the effect of increasing levels of positive gate-source voltage.

9-3   Sketch circuit symbols for *n*-channel and *p*-channel *JFETs*. Identify the terminals in each case, show supply and bias voltage polarities, and indicate the drain current directions.

9-4   Sketch the cross-section of a diffused *JFET*. Identify all parts of the device and briefly explain.

Section 9-2

9-5   Draw a circuit diagram for obtaining the drain and transfer characteristics for an *n*-channel *JFET*. Briefly explain how a table of values should be determined for each characteristic.

9-6   Sketch a typical drain characteristic for V_{GS} = 0 for an *n*-channel *JFET*. Explain the shape of the characteristic, identify the regions, and indicate the important current and voltage levels.

9-7   Draw a typical drain characteristics family for an *n*-channel *JFET*. Identify the V_{GS} for each characteristic. Briefly explain.

9-8   Draw a typical transfer characteristic for an *n*-channel *JFET*. Identify the drain-source saturation current and pinch-off voltage.

9-9   Sketch typical drain and transfer characteristics for a *p*-channel *JFET*. Show typical current and voltage scales and gate-source voltage levels.

Section 9-3

9-10   Define the saturation current and pinch-off voltage for a *JFET*, and state typical values for each quantity.

9-11   Define forward transfer admittance, transconductance, output admittance, and drain resistance for a *JFET*. State typical values for each quantity.

9-12 Define drain-source *ON* resistance, drain-source *ON* voltage and gate cutoff current for a *JFET*. State typical values.

Section 9-4

9-13 Sketch a basic voltage amplifier circuit using a *JFET*, and explain its operation.

Section 9-5

9-14 Draw a cross-section diagram for an enhancement-mode *MOSFET*. Identify all parts of the device and explain its operation.

9-15 Sketch typical drain and transfer characteristics for an *n*-channel *E-MOSFET*. Show typical current and voltage scales and gate-source voltage levels.

9-16 Sketch graphic symbols for *n*-channel and *p*-channel *E-MOSFETs*. Identify the terminals in each case, show the supply and bias voltage polarities, and indicate the drain and source current directions.

9-17 Draw a diagram of the cross-section of a depletion-enhancement-mode *MOSFET*. Identify all parts of the device and explain its operation.

9-18 Sketch typical drain and transfer characteristics for an *n*-channel *DE-MOSFET*. Show typical current and voltage scales and gate-source voltage levels.

9-19 Sketch graphic symbols for *n*-channel and *p*-channel *DE-MOSFETs*. Identify the terminals in each case, show the supply and bias voltage polarities, and indicate the drain and source current directions.

9-20 Draw a diagram of the cross-section of a *V-MOSFET*. Identify all parts of the device, explain its operation, and discuss its advantages compared to other *MOSFETs*.

9-21 Sketch typical drain and transfer characteristics for an *n*-channel *V-MOSFET*. Briefly explain the characteristics.

9-22 Sketch graphic symbols for *n*-channel and *p*-channel *V-MOSFETs*. Identify the terminals in each case, show the supply and bias voltage polarities, and indicate the drain and source current directions.

Chapter 9 Problems

Section 9-2

9-1 Plot the I_D/V_{DS} characteristic for an *n*-channel *JFET* from the following table of current and voltage levels. Determine I_{DSS} and V_P from the characteristics.

$V_{GS} = 0$	V_{DS} (V)	0	1.25	2	2.5	5	10	15	25
	I_D (mA)	0	2	3	4	4.8	4.9	4.95	5

$V_{GS} = -1$ V	V_{DS} (V)	0	1	2	2.5	5	10	15	25
	I_D (mA)	0	1	2	2.3	2.4	2.5	2.6	2.7

$V_{GS} = -2$ V	V_{DS} (V)	0	1	1.5	5	10	15	25
	I_D (mA)	0	0.5	0.8	0.9	0.95	0.99	1

9-2   Derive the device transfer characteristic from the drain characteristics plotted for Problem 9-1.

9-3   Plot the I_D/V_{DS} characteristic for a p-channel JFET from the following table of current and voltage levels. Determine I_{DSS} and V_P from the characteristics.

$V_{GS} = 0$	$-V_{DS}$ (V)	0	1	2.2	3	5	7	9.5	15	25
	I_D (mA)	0	2	4	6	8	8.6	9	9.2	9.3

$V_{GS} = +1$ V	$-V_{DS}$ (V)	0	1	2.5	3	5	6	10	25
	I_D (mA)	0	2	4	5	5.8	6	6.2	6.3

$V_{GS} = +2$ V	$-V_{DS}$ (V)	0	1.2	2.4	3	5	6	10	25
	I_D (mA)	0	2	3	3.5	3.9	4	4.2	4.3

9-4   Derive the device transfer characteristic from the drain characteristics plotted for Problem 9-3.

Section 9-3

9-5   Determine the value of Y_{fs} at V_{GS} = -2 V from the transfer characteristics constructed for Problem 9-2.

9-6   Determine the value of Y_{fs} at V_{GS} = 1 V from the transfer characteristics constructed for Problem 9-4.

9-7   From the data sheet information in Appendix 1-12, determine the following quantities for a 2N4857 JFET: maximum drain-source voltage, maximum power dissipation, maximum pinch-off voltage, maximum drain-source saturation current, drain-source ON resistance.

9-8   Using the data sheet information in Appendix 1-11, construct the maximum transfer characteristics for a 2N5457 JFET.

9-9   Determine the value of Y_{fs} at V_{GS} = -4 V and at V_{GS} = -2 V from the transfer characteristic constructed for Problem 9-8.

9-10   Using the data sheet information in Appendix 1-12, construct the maximum transfer characteristics for a 2N4858 JFET.

9-11   Determine the value of Y_{fs} at V_{GS} = -2 V and at V_{GS} = -1 V from the characteristic constructed for Problem 9-10.

9-12 Using the maximum values of I_{DSS} and $V_{GS(off)}$ for a *2N5457 JFET*, calculate Y_{fs} at V_{GS} = -4 V, and at V_{GS} = -2 V.

9-13 Using the maximum values of I_{DSS} and $V_{GS(off)}$ for a *2N4858 JFET*, calculate Y_{fs} at V_{GS} = -2 V, and at V_{GS} = -1 V.

9-14 Calculate Y_{os} and r_d from the following measurements made on a *JFET*: [I_D = 9 mA when V_{GS} = 0 and V_{DS} = 5 V], [I_D = 9.25 mA when V_{GS} = 0 and V_{DS} = 15 V].

9-15 Referring to the specification in Appendix 1-11, determine the following quantities for a *2N5457 JFET*: maximum drain-source voltage, maximum power dissipation, maximum pinch-off voltage, maximum drain-source saturation current.

Section 9-4

9-16 The circuit in Fig. 9-42 uses a *2N5459 JFET* with the maximum transfer characteristics shown in Fig. 9-19. Calculate V_{DS} when V_{GS} = -1 V and when V_{GS} = -2 V. Also, calculate the circuit voltage gain.

9-17 The *2N5459* in the circuit in Fig. 9-42 is replaced with a *2N5457*. Calculate the maximum and minimum variations in V_{DS} produced by a ±0.1 V change at the gate. Also, calculate the circuit voltage gain.

9-18 Determine a suitable *JFET* forward transfer admittance for the amplifier circuit in Fig. 9-42 to have a minimum voltage gain of 15, when R_1 is changed to 2.2 kΩ.

9-19 A *JFET* amplifier circuit, as in Fig. 9-42, is to have a minimum voltage gain of 5. If the *FET* used is a *2N5458*, determine a suitable resistance for R_1.

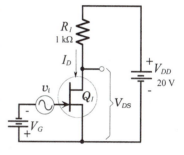

Figure 9-42

Section 9-5

9-20 Plot the transfer characteristics for a *DE-MOSFET* from the following table of measured current and voltage levels. From the characteristics, determine the value of Y_{fs} at V_{GS} = 0.

V_{GS} (V)	-3	-2	-1	0	+1	+2	+3
I_D (mA)	0.5	2.7	4.8	7	9	11.3	13.4

9-21 A *VMOSFET* has $r_{d(on)}$ = 0.5 Ω, g_{FS} = 2.5 S, and $V_{GS(th)}$ = 2 V. Calculate the approximate level of V_{GS} required to produce a 5 A drain current. Also, calculate $V_{DS(on)}$ and the device power dissipation when I_D = 5 A.

9-22 From the *2N7002* data sheet information in Fig. 9-39, determine $V_{DS(on)}$ and the power dissipation when I_D = 50 mA.

9-23 Calculate the approximate gate-source voltage, drain-source *ON* voltage, and power dissipation in a *IRF520* with a 5 A drain current. Refer to the partial specification in Fig. 9-39.

Practise Problem Answers

9-2.1   5.5 mA, 3 V
9-3.1   1127 μS
9-3.2   3160 μS, 1975 μS
9-3.3   -6 V, 100 mA, 40 Ω
9-4.1   7.95 V, 21.15
9-5.1   5 mS
9-5.2   7.5 V

Chapter *10*

FET Biasing

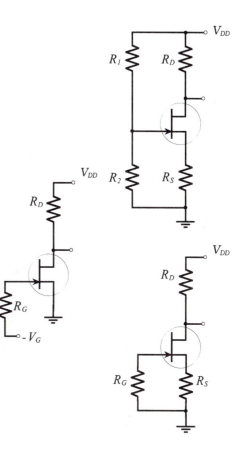

Chapter Contents

Introduction

Objectives

You will be able to:

1 Draw dc load lines for FET circuits, identify circuit Q-points, and estimate the maximum circuit output voltage variations.

2 Sketch gate bias, self bias, and voltage divider bias circuits, and explain the operation of each circuit.

3 Analyze various types of FET bias circuits using the device maximum and minimum transfer characteristics to determine circuit voltage and current levels.

4 Trouble-shoot non-operational FET bias circuits.

5 Design gate bias, self bias, and voltage divider bias circuits, and select appropriate standard value components.

6 Sketch constant current and other unusual bias circuit types, and explain the operation of each circuit.

7 Analyze and design each of the bias circuit types referred to in item 6.

8 Construct a universal transfer characteristic, and use it in analysis and design of FET bias circuits.

9 Sketch various bias circuits for MOSFET devices and explain the operation of each circuit.

10 Analyze and design MOSFET bias circuits.

11 Analyze and design bias circuits used with FET switches.

Introduction

Like bipolar transistor circuits, field effect transistors used in amplifier (or other) circuits must be biased into an *on* state with constant current and terminal voltage levels. The widely differing maximum and minimum transfer characteristics for a FET can result in a wide range of drain current levels. This is similar to the situation caused by $h_{FE(max)}$ and $h_{FE(min)}$ in a *BJT* bias circuit. For reasonable upper and lower drain current limits in a *FET* circuit, source resistors and voltage divider networks must be employed. A graphical approach is most convenient for analysis or design of *FET* bias circuits.

10-1   DC Load Line and Bias Point

DC Load Line

The *dc load line* for a *FET* circuit is drawn on the device output characteristics (or drain characteristics) in exactly the same way as for a *BJT* circuit, (see Section 5-1). Refer to the *n*-channel *FET* circuit in Fig. 10-1. The source terminal is grounded, the gate is biased via resistor R_G to a negative voltage (-V_G), and the drain terminal is supplied from +V_{DD} via resistor R_D. The *dc* load line for this circuit is a graph of drain current (I_D) versus drain-source voltage (V_{DS}), for a given value of drain resistance and a given supply voltage.

From Fig. 10-1, the drain-source voltage is,

$$V_{DS} = \text{(supply voltage) - (voltage drop across } R_D)$$

or,
$$\boldsymbol{V_{DS} = V_{DD} - I_D R_D} \tag{10-1}$$

Substituting V_{DD}, R_D, and convenient levels of I_D into Eq. 10-1, produces corresponding V_{DS} and I_D values. These values are then plotted on the device characteristics, and the *dc* load line is drawn through them.

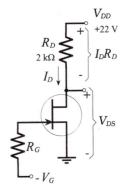

Figure   10-1
FET circuit with dc load resistor (R$_D$), supply voltage (+V$_{DD}$), and gate bias (-V$_G$).

Example   10-1

Draw the *dc* load line for the circuit in Fig. 10-1 on the *FET* characteristics in Fig. 10-2.

Solution

when $I_D = 0$,
Eq. 10-1, $V_{DS} = V_{DD} - I_D R_D = 22\,\text{V} - 0$
                          $= 22\,\text{V}$

Plot *point A* on Fig. 10-2 at,
$$I_D = 0 \text{ and } V_{DS} = 22\,\text{V}$$

when $V_{DS} = 0$,
from Eq. 10-1, $0 = V_{DD} - I_D R_D$

giving,
$$I_D = \frac{V_{DD}}{R_D} = \frac{22\,V}{2\,k\Omega}$$
$$= 11\,mA$$

Plot *point B* on Fig. 10-2 at,

$$I_D = 11\,mA \text{ and } V_{DS} = 0\,V$$

Draw the *dc* load line through *points A* and *B*.

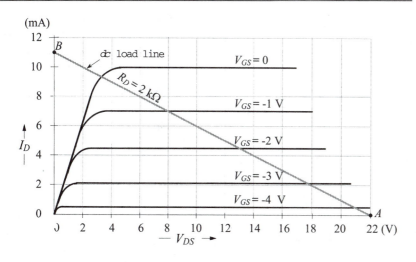

(mA)

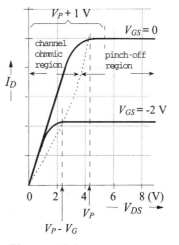

Figure 10-2
DC load line drawn upon FET drain characteristics.

As already explained, the *dc* load line represents all corresponding I_D and V_{DS} levels that can exist in a *FET* circuit, as represented by Eq. 10-1. A point plotted at $V_{DS} = 16\,V$ and $I_D = 10$ mA on Fig. 10-2 does not appear on the load line, and so this combination of voltage and current cannot exist in this particular circuit (Fig. 10-1). The straight line drawn in Ex. 10-1 is the *dc* load line for $R_D = 2\,k\Omega$ and $V_{DD} = 22\,V$. If either of these two quantities is changed, a new load line must be drawn.

Bias Point (Q-Point)
Just as in the case of a *BJT* circuit, the *dc bias point*, or *quiescent point (Q-point)*, identifies the device current and terminal voltages when there is no *ac* input signal. When a signal is applied to the gate, I_D varies according to the instantaneous amplitude of the signal, producing a variation in V_{DS}.

For a *FET* amplifier circuit, V_{DS} must remain in the pinch-off region of the characteristics. This means that it must not be allowed to fall below the the device pinch-off voltage (V_P in Fig. 10-3), to avoid going into the channel ohmic region. Therefore, in a *FET* bias circuit the drain-source voltage should always be a minimum of ($V_P + 1\,V$), as illustrated.

$$\mathbf{V_{DS(min)} = V_P + 1\,V} \qquad \textbf{(10-2)}$$

Figure 10-3
In a FET amplifier circuit the drain-source voltage must be a minimum of $V_P + 1\,V$.

Where an external bias voltage (V_G) is included (as in Fig. 10-1), the pinch-off voltage for that bias level on the device characteristics is ($V_P - V_G$), (see Fig. 10-3). Consequently, the minimum drain-source voltage may be reduced to,

$$V_{DS(min)} = V_P - V_G + 1 \text{ V} \qquad\qquad \textbf{(10-3)}$$

The bias point for a *BJT* can be selected half-way along the load line, to give the maximum possible symmetrical output voltage variation, (see Section 5-1). To achieve the same result with a *FET*, the bias point must be half way between $V_{DS} = V_{DD}$ and $V_{DS} = V_P$. This is illustrated at Q_1 in Fig. 10-4. When maximum V_{DS} variations are not required, the *FET* bias point can be selected at any convenient point on the *dc* load line (Q_2 in Fig. 10-4), just as in the case of a *BJT* amplifier.

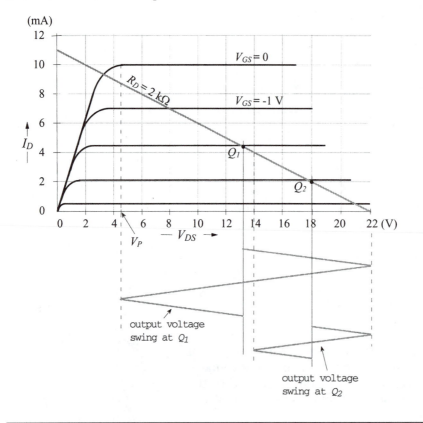

Figure 10-4
The maximum output voltage swing from a FET amplifier circuit depends upon the Q point.

Example 10-2
Estimate the maximum symmetrical output voltage swing for the 2 kΩ load line in Fig. 10-4 when the *FET* is biased to give a Q-point at $V_{DS} = 11$ V, and when it is biased to give $V_{DS} = 20$ V.

Solution
For a Q-point at $V_{DS} = 11$ V:
when I_D is reduced to zero, $\qquad \Delta V_{DS} = 22 \text{ V} - 11 \text{ V}$
$$= 11 \text{ V}$$

when I_D is increased to 9 mA, $\qquad \Delta V_{DS} = 11\ V - 4\ V$
$$= 7\ V$$

So, the maximum symmetrical output voltage swing is,

$$V_{o(max)} = \pm 7\ V$$

For a Q-point at $V_{DS} = 20\ V$:
when I_D is reduced to zero, $\qquad \Delta V_{DS} = 22\ V - 20\ V$
$$= 2\ V$$

when I_D is increased to 9 mA, $\qquad \Delta V_{DS} = 20\ V - 4\ V$
$$= 16\ V$$

So, the maximum symmetrical output voltage swing is,

$$V_{o(max)} = \pm 2\ V$$

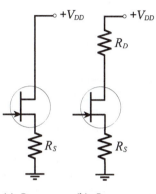

(a) $R_{L(dc)}$ $\qquad$ (b) $R_{L(dc)}$
$\quad = R_S$ $\qquad\qquad = R_D + R_S$

Figure 10-5
When a source resistor (R_S) is included in a circuit, R_S is the dc load or part of the total load.

Effect of Source Resistor

Figure 10-5(a) shows a circuit that has a resistor (R_S) in series with the *FET* source terminal, and the supply voltage connected directly to the drain. In this case R_S is the *dc* load, and Eq. 10-1 is rewritten as,

$$V_{DS} = V_{DD} - I_D\,R_S$$

The dc load line is drawn exactly as discussed.

In Fig. 10-5(b) drain and source resistors R_D and R_S are both present, and the total *dc* load in series with the *FET* is $(R_D + R_S)$. For drawing the dc load line, Eq. 10-1 is modified to,

$$V_{DS} = V_{DD} - I_D\,(R_D + R_S)$$

Practise Problems

10-1.1 A 1 kΩ source resistor (R_S) is added to the circuit of Fig. 10-1, as in Fig. 10-5(b). Draw the *dc* load line on the characteristic in Fig. 10-2, and determine V_{DS} when $V_{GS} = -2$ V.

10-1.2 Determine the maximum symmetrical output voltage swing for the circuit in Problem 10-1.1 when the Q-point is at $V_{DS} = 17$ V.

10-2 Gate Bias

Circuit Operation

Consider the *gate bias* circuit shown in Fig. 10-6. The *FET* gate terminal is connected via resistor R_G to a bias voltage V_G. If the gate is directly connected to the bias source (instead of using R_G), any *ac* signal applied to the gate would be short-circuited to V_G.

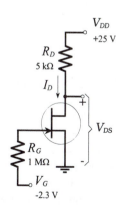

Figure 10-6
FET circuit with gate bias. The gate-source voltage is, $V_{GS} = -V_G$.

There is no gate current to produce a voltage drop across R_G, so the gate-source voltage remains constant at V_G. Depending on the device transfer characteristics, V_G sets the level of I_D in the *FET*. The I_D level can be readily determined, if the transfer characteristic is available for the particular *FET* used in the circuit. If the device is replaced, I_D is determined by the transfer characteristic of the new device. Like the base current in a *BJT* base bias circuit, the *FET* gate voltage does not change when the device characteristics change.

Circuit Analysis

As discussed in Section 9-3, each *FET* of a given type number has maximum and minimum transfer characteristics, and these must be employed to determine the maximum and minimum possible levels of I_D in the circuit. A gate bias circuit is most easily analyzed by drawing a *bias line* vertically on the transfer characteristics from $V_{GS} = -V_G$. The levels of $I_{D(min)}$ and $I_{D(max)}$ are read from the intersections of the bias line with the transfer characteristics, and the maximum and minimum drain-source voltages can then be calculated using these I_D levels.

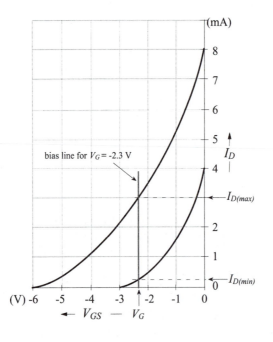

Figure 10-7
The possible drain current levels in a gate bias circuit are determined simply by drawing a bias line vertically on the transfer characteristics at $V_{GS} = V_G$.

Example 10-3

The gate bias circuit in Fig. 10-6 uses a *FET* with the transfer characteristics shown in Fig. 10-7. Determine the maximum and minimum levels of I_D and V_{DS} for the circuit.

Solution

Draw the bias line on the transfer characteristics at,

$$V_{GS} = V_G = -2.3 \text{ V}$$

From the bias line, $I_{D(min)} = 0.25$ mA, and $I_{D(max)} = 3$ mA

$$V_{DS(min)} = V_{DD} - (I_{D(max)} R_D) = 25 \text{ V} - (3 \text{ mA} \times 5 \text{ k}\Omega)$$
$$= 10 \text{ V}$$

$$V_{DS(max)} = V_{DD} - (I_{D(min)} R_D) = 25 \text{ V} - (0.25 \text{ mA} \times 5 \text{ k}\Omega)$$
$$= 23.75 \text{ V}$$

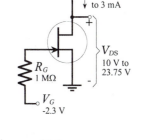

The maximum and minimum I_D and V_{DS} levels determined in Ex. 10-3 are illustrated by the circuit current and voltage levels in Fig. 10-8, and by the *Q-points* plotted in Fig. 10-9. As already discussed, JFETs with a given type number have a wide range of transfer characteristics, so

FET maximum and minimum transfer characteristics should always be used for practical circuit analysis.

The gate bias technique produces a wide range of I_D and V_{DS} levels for any *FET* with a particular type number. It is possible to adjust V_G to set V_{DS} to a desired voltage. However, this is normally done only in an experimental situation. Slightly more complicated bias circuit must be employed for more predictable voltage and current conditions.

Figure 10-8
Maximum and minimum voltage and current levels for the gate bias circuit in Ex. 10-3.

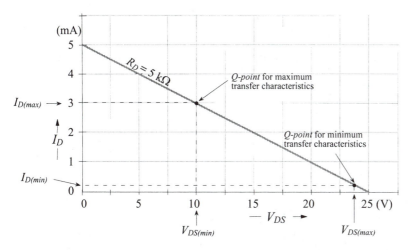

Figure 10-9
Q-points produced by the maximum and minimum transfer characteristics with the gate bias circuit in Ex. 10-3.

Analysis of a gate bias circuit can be performed without drawing the *FET* transfer characteristics. Equation 9-1 is used for plotting the transfer characteristics, so the equation can be applied directly to determine the drain current levels.

Example 10-4
Use Eq. 9-1 to determine $I_{D(max)}$ and $I_{D(min)}$ for the bias circuit in Ex. 10-3.

Solution
From Fig. 10-7, $I_{DSS(max)} = 8$ mA, and $V_{P(max)} = 6$ V

Eq. 9-1, $I_{D(max)} = I_{DSS(max)} \left[1 - \dfrac{V_{GS}}{V_{P(max)}}\right]^2 = 8 \text{ mA} \left[1 - \dfrac{2.3 \text{ V}}{6 \text{ V}}\right]^2$

$= 3.04 \text{ mA}$

From Fig. 10-7, $I_{DSS(min)} = 4 \text{ mA}$, and $V_{P(min)} = 3 \text{ V}$

Eq. 9-1, $I_{D(min)} = I_{DSS(min)} \left[1 - \dfrac{V_{GS}}{V_{P(min)}}\right]^2 = 4 \text{ mA} \left[1 - \dfrac{2.3 \text{ V}}{3 \text{ V}}\right]^2$

$= 0.22 \text{ mA}$

Gate Bias for a p-channel JFET

A gate bias circuit using a *p*-channel *JFET* is shown in Fig. 10-10(a). This is similar to *n*-channel *JFET* gate bias circuits except that V_{DD} is a negative voltage and V_G is a positive quantity. Figure 10-10(b) shows the same *p*-channel *JFET* circuit with a positive supply voltage. In this case, the *FET* source terminal is connected to V_{DD}, and R_D is connected to ground. To reverse bias the gate-channel junctions, the gate terminal must be positive with respect to the source. So, gate bias voltage V_G must be positive with respect to V_{DD}, as illustrated. The *p*-channel gate bias circuit can be analyzed in exactly the same way as an *n*-channel circuit.

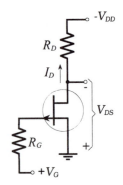

(a) *p*-channel *FET* circuit using -V_{DD}

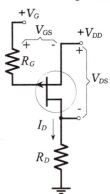

(b) *p*-channel *FET* circuit using +V_{DD}

Figure   10-10
JFET gate bias circuits using p-channel devices.

Practise Problem

10-2.1   A gate bias circuit has $V_{DD} = 20$ V, $R_D = 2.7$ kΩ, $R_G = 1$ MΩ, and $V_G = -1.5$ V. If the *FET* has the characteristics shown in Fig. 10-7, determine the maximum and minimum V_{DS} levels for the circuit.

10-2.2   The *FET* in the circuit in Problem 10-2.1 is changed to one that has $I_{DSS} = 6$ mA and $V_P = 4.5$ V. Calculate the circuit V_{DS}.

10-3   Self Bias

Circuit Operation

In a *self bias JFET* circuit, gate-source bias is provided by the voltage drop across a resistor in series with the device source terminal. Consider the *n*-channel *JFET* self-bias circuit illustrated in Fig. 10-11. The voltage drop across resistor R_S is,

$$V_S = I_D R_S$$

If $I_D = 1$ mA and $R_S = 1$ kΩ, then $V_S = 1$ V. This means that the source terminal in Fig. 10-11 is 1 V positive with respect to ground. Alternatively, ground can be said to be 1 V negative with respect to the source terminal. Because the *FET* gate terminal is grounded via R_G (and there is no voltage drop across R_G), the gate is 1 V negative with respect to the source. That is, the gate-source bias voltage is $V_{GS} = -1$ V. So, for a self-bias circuit,

$$\boldsymbol{V_{GS} = -I_D\, R_S} \qquad\qquad \textbf{(10-4)}$$

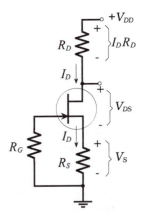

Figure   10-11
JFET self bias circuit using an n-channel device. The gate-source bias voltage is $V_{GS} = -I_D R_S$.

The fact that I_D determines V_{GS}, and that V_{GS} sets the I_D level, means that there is a feedback effect tending to control I_D. Thus, if I_D increases when the device is changed, the increased voltage across R_S results in an increased (negative) gate-source voltage that tends to reduce I_D back toward its original level. Similarly, a fall in I_D produces a reduced V_{GS} which tends to increase I_D toward its original level.

Circuit Analysis
As for all *FET* bias circuits, a graphical technique is most convenient for analysis of a self-bias circuit. In this case, the bias line drawn on the transfer characteristics is quite different from that for the gate bias circuit. The value of R_S and convenient I_D levels are inserted into Eq. 10-4 to determine corresponding V_{GS} levels. The I_D and V_{GS} values are plotted on the transfer characteristics, and the bias line is drawn through them, (see Fig 10-12). The intersection points of the bias line and the transfer characteristics gives the maximum and minimum I_D levels.

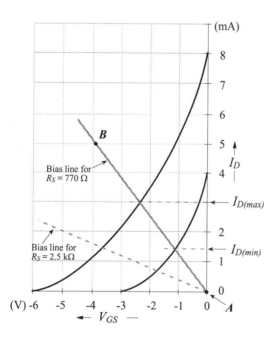

Figure 10-12
The bias line for a self bias circuit is drawn from $I_D = 0$ and $V_{GS} = 0$ at a slope determined by R_S.

Summing the voltage drops across R_D, R_S, and the *FET* drain-source terminals in Fig. 10-11 gives,

$$V_{DD} = I_D R_D + V_{DS} + I_D R_S \qquad \text{(10-5)}$$

Equation 10-5 may be used to calculate the maximum and minimum levels of V_{DS} when $I_{D(max)}$ and $I_{D(min)}$ have been determined.

Example 10-5

Analyze the self-bias circuit in Fig. 10-13 to determine $I_{D(max)}$, $I_{D(min)}$, V_{DSmax}, and V_{DSmin}. The transfer characteristics for the *FET* are given in Fig. 10-12.

Solution

when $I_D = 0$,

Eq. 10-4, $V_{GS} = -I_D R_S = 0$

Plot *point A* on Fig. 10-12 at,

$$I_D = 0 \text{ and } V_{GS} = 0$$

when $I_D = 5$ mA,

Eq. 10-4, $V_{GS} = -I_D R_S = -(5 \text{ mA} \times 770 \text{ } \Omega)$
            $= -3.85$ V

Plot *point B* on Fig. 10-12 at,

$$I_D = 5 \text{ mA and } V_{GS} = -3.85 \text{ V}$$

Draw the bias line for $R_S = 770 \text{ } \Omega$ through *points A* and *B*.

Where the bias line intersects the transfer characteristics, read

$$I_{D(max)} = 3 \text{ mA and } I_{D(min)} = 1.4 \text{ mA}$$

From Eq. 10-5, $V_{DS(min)} = V_{DD} - I_{D(max)}(R_D + R_S)$
               $= 25 \text{ V} - 3 \text{ mA}(4.23 \text{ k}\Omega + 770 \text{ } \Omega)$
               $= 10$ V

and, $V_{DS(max)} = V_{DD} - I_{D(min)}(R_D + R_S)$
               $= 25 \text{ V} - 1.4 \text{ mA}(4.23 \text{ k}\Omega + 770 \text{ } \Omega)$
               $= 18$ V

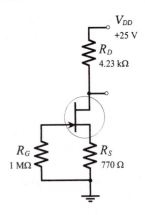

Figure 10-13
FET self bias circuit for Ex. 10-5.

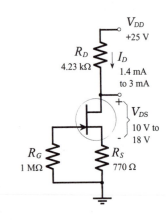

Figure 10-14
Maximum and minimum voltage and current levels for the self bias circuit in Ex. 10-5.

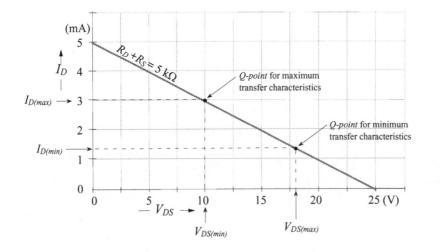

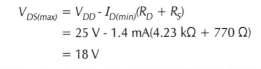

Figure 10-15
Q-points produced by the maximum and minimum transfer characteristics with the self bias circuit in Ex. 10-5.

The maximum and minimum I_D and V_{DS} levels determined in Ex. 10-5 are shown in the circuit in Fig. 10-14, and are illustrated by the *Q-points* plotted in Fig. 10-15. Consideration of Ex. 10-5 shows that closer levels of $I_{D(max)}$ and $I_{D(min)}$ could be obtained by using a larger resistance for R_S. However, as shown by the (dashed) bias line for $R_S = 2.5$ kΩ in Fig. 10-12, I_D is substantially reduced when larger values of R_S are used.

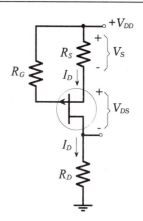

Figure 10-16
JFET self bias circuit using a p-channel device.

Self-Bias for a p-channel JFET

A self-biased circuit using a *p*-channel *JFET* is shown in Fig. 10-16. The gate-source bias voltage is once again provided by the voltage drop across source resistor R_S. In this case, the source is negative with respect to the supply voltage $(+V_{DD})$. Because the gate is connected to $+V_{DD}$ via R_G, the gate is is positive with respect to the source terminal. The positive level of V_{GS} provides the reverse bias at the gate-channel junctions of the *p*-channel *FET*. A *p*-channel self-biased circuit can be analyzed in exactly the same way as an *n*-channel circuit.

Practise Problems

10-3.1   A self bias circuit has $V_{DD} = 22$ V, $R_D = 3.3$ kΩ, $R_G = 1$ MΩ, and $R_S = 3.3$ kΩ. Using the *FET* characteristics in Fig. 10-12, analyze the circuit to determine the maximum and minimum V_{DS} levels.

10-4 Voltage Divider Bias

Circuit Operation

For the self-bias circuit, it was seen that increasing the resistance of R_S brings $I_{D(max)}$ and $I_{D(min)}$ closer together, but that increased R_S values result in lower I_D levels. As will be demonstrated, *voltage divider bias* allows R_S to be increased without making I_D very small.

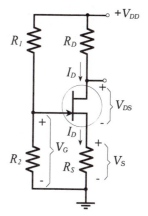

Figure 10-17
JFET voltage divider bias circuit. The gate-source voltage is,
$$V_{GS} = V_G - V_S.$$

Figure 10-17 shows that a voltage divider bias circuit combines the use of a source resistor (R_S) with a gate bias voltage (V_G). The gate bias voltage is derived from the supply voltage by means of the voltage divider $(R_1$ and $R_2)$, and the source voltage (V_S) depends upon R_S and I_D. The gate terminal and the source terminal are both positive with respect to ground, and the gate-source voltage (V_{GS}) is the difference between V_G and V_S. Because V_S is larger than V_G, the *FET* source terminal is at a higher level than the gate. So, the gate is negative with respect to the source, and the gate-channel junctions are reverse biased.

As in the case of a self-biased circuit, the voltage drop across R_S increases if I_D increases. This produces an increase in -V_{GS}, which drives I_D back toward its original level. A decrease in I_D causes a reduction in -V_{GS} which tends to increase I_D.

Circuit Analysis

Once again, the graphical technique of drawing a bias line on the *FET* transfer characteristics is most convenient for analyzing a voltage divider bias circuit. An equation relating V_{GS} and I_D must be derived, so that corresponding values of each quantity may be calculated for plotting the bias line.

As discussed,

$$V_{GS} = V_G - I_D R_S \qquad \textbf{(10-6)}$$

Because there is no gate current,

$$V_G = \frac{V_{DD} \times R_2}{R_1 + R_2} \qquad \textbf{(10-7)}$$

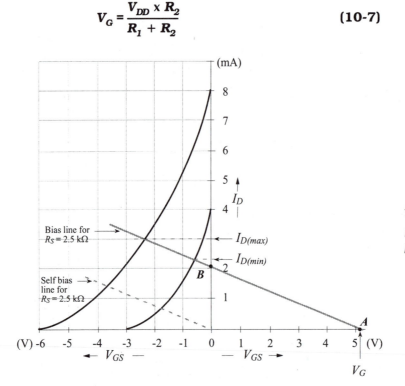

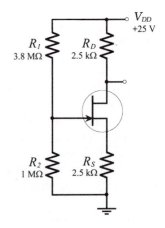

Figure 10-18
The bias line for a voltage divider bias circuit is drawn from the point where $I_D = 0$ and $V_{GS} = +V_G$ at a slope determined by R_S.

Referring to Eq. 10-6, it is seen that

when $I_D = 0$, $\qquad\qquad V_{GS} = V_G$

This is a positive quantity, so the bias line is drawn from $+V_G$, as shown in Fig. 10-18. It is seen that this gives higher drain current levels than those obtained when the bias line is drawn from $I_D = 0$ and $V_{GS} = 0$, (dashed line in Fig. 10-18). The analysis of a voltage divider bias circuit is demonstrated in Ex. 10-6.

Example 10-6
Analyze the bias circuit in Fig. 10-19 to determine $I_{D(max)}$, $I_{D(min)}$, V_{DSmax}, and V_{DSmin}. Figure 10-18 shows the transfer characteristics for the FET.

Solution
Eq. 10-7, $\qquad V_G = \dfrac{V_{DD} \times R_2}{R_1 + R_2} = \dfrac{25\ V \times 1\ M\Omega}{3.8\ M\Omega + 1\ M\Omega}$

$\qquad\qquad\qquad = 5.2\ V$

Figure 10-19
FET voltage divider bias circuit for Ex. 10-6.

when $I_D = 0$,

Eq. 10-6, $V_{GS} = V_G - I_D R_S = 5.2 \text{ V} - 0$

$= 5.2 \text{ V}$

Plot *point A* on Fig. 10-18 at,

$I_D = 0$ and $V_{GS} = +5.2 \text{ V}$

when $V_{GS} = 0$,

from Eq. 10-6, $I_D = \dfrac{V_G}{R_S} = \dfrac{5.2 \text{ V}}{2.5 \text{ k}\Omega}$

$= 2.08 \text{ mA}$

Plot *point B* on Fig. 10-18 at,

$I_D = 2.08 \text{ mA}$ and $V_{GS} = 0$

Draw the bias line for $R_S = 2.5 \text{ k}\Omega$ through *points A* and *B*.
Where the bias line intersects the transfer characteristics, read

$I_{D(max)} = 3 \text{ mA}$ and $I_{D(min)} = 2.3 \text{ mA}$

From Eq. 10-5, $V_{DS(min)} = V_{DD} - I_{D(max)}(R_D + R_S)$

$= 25 \text{ V} - 3 \text{ mA}(2.5 \text{ k}\Omega + 2.5 \text{ k}\Omega)$

$= 10 \text{ V}$

and, $V_{DS(max)} = V_{DD} - I_{D(min)}(R_D + R_S)$

$= 25 \text{ V} - 2.3 \text{ mA}(2.5 \text{ k}\Omega + 2.5 \text{ k}\Omega)$

$= 13.5 \text{ V}$

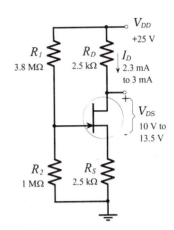

Figure 10-20
*Maximum and minimum voltage
and current levels for the voltage
divider bias circuit in Ex. 10-6.*

The maximum and minimum I_D and V_{DS} levels determined for the
voltage divider bias circuit in Ex. 10-6 are shown on the circuit
diagram in Fig. 10-20, and are illustrated by the Q-points plotted
in Fig. 10-21.

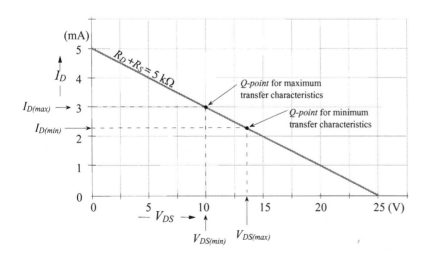

Figure 10-21
*Q-points produced by the
maximum and minimum transfer
characteristics with the voltage
divider bias circuit in Ex. 10-6.*

Voltage Divider Bias for a p-channel JFET

A voltage divider bias circuit using a *p*-channel *JFET* is shown in Fig. 10-22. As for the *n*-channel circuit, the gate-source bias voltage is the difference between V_G and V_S. In this case, these are both negative with respect to the positive supply voltage (V_{DD}). Because V_S is larger than V_G, the gate terminal is positive with respect to the source, and this reverse biases the gate-channel junctions. A *p*-channel voltage divider bias circuit is analyzed in exactly the same way as an *n*-channel circuit.

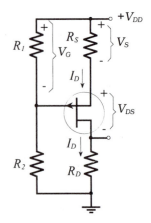

Figure 10-22
JFET voltage divider bias circuit using a p-channel device.

Practise Problems

10-4.1 A voltage divider bias circuit has V_{DD} = 30 V, $R_D = R_S$ = 4.7 kΩ, R_1 = 1 MΩ, and R_2 = 150 kΩ. Using the *FET* characteristics in Fig. 10-18, determine the circuit maximum and minimum V_{DS} levels.

10-5   Comparison of Basic JFET Bias Circuits

The three basic *FET* bias circuits (gate bias, self-bias, and voltage divider bias) are similar in performance to the three basic *BJT* bias circuits, (base bias, collector-to-base bias, and voltage divider bias). Comparing the performance of *FET* bias circuits, it should be recalled that each *FET* type number has maximum and minimum transfer characteristics. The particular transfer characteristic for each individual transistor is normally not known. Therefore, as already stated, *the specified maximum and minimum transfer characteristics must be used when analysing (or designing) a FET bias circuit.*

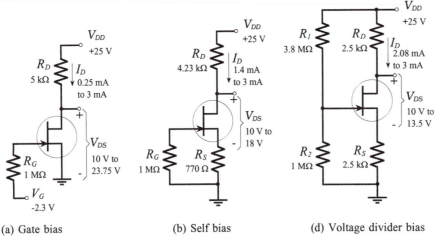

(a) Gate bias   (b) Self bias   (d) Voltage divider bias

Figure 10-23
Comparison of similar JFET gate bias, self bias, and voltage divider bias circuits.

Consider the basic *JFET* bias circuits reproduced in Fig. 10-23. Each circuit uses a 25 V supply and a 5 kΩ (total) load resistance, and each uses a *FET* with the same maximum and minimum transfer characteristics. The circuits are analysed in Examples 10-3, 10-5, and 10-6 to determine the maximum and minimum I_D and

V_{DS} levels. The calculated quantities are indicated on each circuit in Fig. 10-23, and the Q point ranges are illustrated in Fig. 10-24.

If the *FET* in the gate bias circuit fails and has to be replaced with another (same type number) device, the new V_{DS} level could be anywhere from 10 V to 23.75 V. The self-bias circuit in the same circumstance (with the same *FET* type number) would have a V_{DS} ranging from 10 V to 18 V when the *FET* is replaced. For the voltage divider bias circuit in this situation, the measurable V_{DS} range with a new device would be 10 V to 13.5 V. Clearly, self-bias gives more predictable bias conditions than gate bias, and voltage divider bias gives the most predictable bias conditions. Because of its better performance, voltage divider bias is usually preferred.

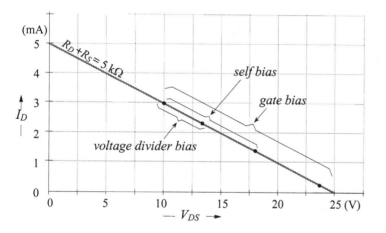

Figure 10-24
Q-point ranges for the gate bias, self bias, and voltage divider bias circuits shown in Fig. 10-23.

Note that, for all three basic bias circuits in Fig. 10-23, the minimum I_D and V_{DS} levels are 3 mA and 10 V, respectively. In Section 10-7, it is shown how *FET* bias circuits may be designed for a given minimum V_{DS}.

Section 10-5 Review
10-5.1 Briefly discuss the three basic *FET* bias circuits, comparing the effect of the bias lines on the maximum and minimum I_D levels.

10-6   *Trouble-Shooting JFET Bias Circuits*

Voltage Measurement
Trouble-shooting procedures for *FET* bias circuits are similar to those for *BJT* bias circuits. The major difference is that there is only one junction in the *FET* (the gate-channel junction) that might become short-circuited or open-circuited. To determine if the circuit is functioning correctly, the *FET* terminal voltages should all be measured with respect to ground, as illustrated in Fig. 10-25. This is exactly as done for a *BJT* circuit, and, here again, if the voltage levels are not satisfactory, the circuit must be further investigated to locate the fault.

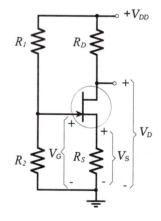

Figure 10-25
When testing a FET bias circuit, all voltage levels should first be measured with respect to the ground or negative supply terminal.

Common Errors

Before proceeding further, the following list of common errors (reproduced from Section 5-6) should be considered. These apply equally to *FET* and *BJT* circuits.

- Power supply not switched *on*.
- Power supply current limiter control incorrectly set.
- Cables incorrectly connected to the power supply.
- Volt-Ohm-Milliammeter (*VOM*) function incorrectly selected.
- Wrong *VOM* terminals used.
- Incorrect component connections.
- Incorrect resistor values.
- Resistors in the wrong places.

Gate Bias

Figure 10-26 illustrates typical error sources in a gate bias circuit. Normally, the gate should be negative with respect to ground, and the drain voltage should be anywhere in the range of perhaps 3 V to $(V_{DD} - 2\text{ V})$. If $V_D \approx V_{DD}$, as in Fig. 10-26(a), the *FET* terminals might be open-circuited, or R_2 might be shorted. When $V_C \approx 0$, as in Fig. 10-26(b), R_1 and R_2 might be interchanged (the high-value and low-value resistors in the wrong places). Other possibilities are that R_2 is open-circuited, or the *FET* terminals are short-circuited.

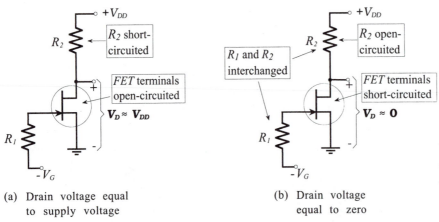

(a) Drain voltage equal to supply voltage   (b) Drain voltage equal to zero

Figure 10-26
The measured FET terminal voltages in a gate bias circuit can indicate possible circuit errors.

Self Bias

Possible reasons for incorrect voltage levels in a self-bias circuit are illustrated in Fig. 10-27. These are similar to those discussed for gate bias. If V_D is approximately equal to V_{DD} [Fig 10-27(a)], resistor R_3 or one of the transistor terminals could to be open-circuited, or R_1 and R_3 might be interchanged. Alternatively, R_2 might be short circuited. If V_D equals V_S [Fig 10-27(b)], the *FET* transistor terminals might be short-circuited, or R_1 and R_2 might be in the wrong places (interchanged).

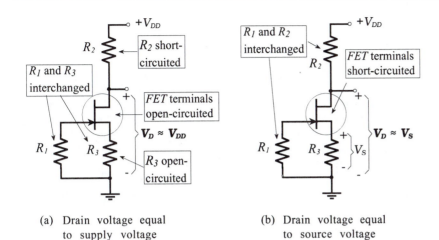

(a) Drain voltage equal to supply voltage

(b) Drain voltage equal to source voltage

Figure 10-27
Typical incorrect terminal voltages in a FET self bias circuit, and probable circuit faults.

Voltage Divider Bias

Unsatisfactory voltage measurements and probable errors in a voltage divider bias circuit are shown in Fig. 10-28. If V_D equals V_{DD} [Fig. 10-28(a)], R_1 or R_4 might be open-circuited. Alternatively, R_2 or R_3 might be short-circuited, or the *FET* terminals might be open-circuited. When V_D approximately equals V_S [Fig. 10-28(b)], R_2 might be open-circuited, or R_1 and R_2 might be interchanged. Another possibility is that the *FET* terminals are short-circuited.

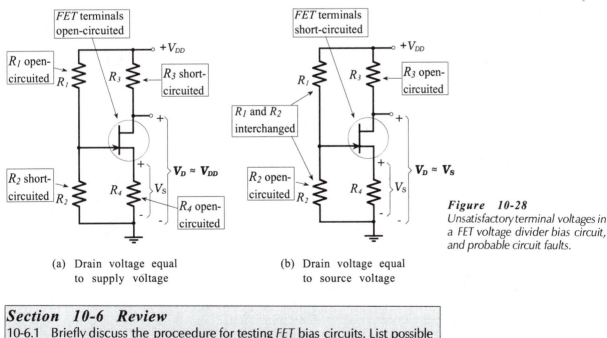

(a) Drain voltage equal to supply voltage

(b) Drain voltage equal to source voltage

Figure 10-28
Unsatisfactory terminal voltages in a FET voltage divider bias circuit, and probable circuit faults.

Section 10-6 Review

10-6.1 Briefly discuss the proceedure for testing *FET* bias circuits. List possible faults for each bias circuit type (a) when $V_{DS} \approx 0$, (b) when $V_D \approx V_{DD}$.

10-7   JFET   Bias   Circuit   Design

Design   Approach

Design of *FET* bias circuits is just as simple as design of *BJT* bias circuits. One major difference is that *FET* circuit design normally uses a graphical approach involving the drawing of a bias line on the device transfer characteristics, as in the case of *FET* circuit analysis. Normally the type number of the *FET* to be used in the circuit is known, and the maximum and minimum transfer characteristics are available or can be plotted. A maximum drain current ($I_{D(max)}$) is selected at a point on the maximum transfer characteristic (see Fig. 10-29), and the bias line is drawn through this point. As will be explained, this procedure applies to all types of bias circuits. The selected level of $I_{D(max)}$ is used to calculate the values of resistors connected to the drain and source terminals.

Instead of using the maximum transfer characteristics, suppose that the minimum characteristics, or some characteristic between the maximum and minimum, is used in the circuit design. Now, if the *FET* happens to have the maximum transfer characteristic for the device type number, I_D will be larger than the design value. Consequently, the resistor voltage drops ($I_D R_D$ and $I_D R_S$) will be greater than intended (see Fig. 10-30). In this case, V_{DS} will be smaller than its design level, the *FET* might be forced into the channel ohmic region of its characteristics, and the circuit is unlikely to function correctly. Therefore,

> **always   use   the   device   maximum   transfer characteristic   when   designing   a   FET   bias   circuit**.

As already explained, a *FET* has a very high input resistance, so high-value bias resistors can be used at the gate terminal. However, there are disadvantages to using extremely high resistance values. A charge accumulated at the gate can take a long time to *leak off* through a very high resistance. In this case, the gate voltage might not be a stable quantity, and consequently, the drain current could be unpredictable. High-value bias resistors also generate unwanted thermal noise (see Section 8-6), and they readily pick-up stray radio-frequency signals. For these reasons, 1 MΩ is normally a reasonable upper limit for bias resistors.

In all situations where resistor values are calculated, the closest standard-value components are selected, (see Appendix 2-1). Normally, ±10% tolerance resistors are always used, and the more expensive ±5% and ±1% components are employed only where greater precision is required.

Gate   Bias   Circuit   Design

Design of a gate bias circuit is very simple. As explained, the desired maximum I_D level is marked on the maximum transfer characteristics for the *FET*. The bias line is drawn vertically through this point, and the required gate bias voltage ($-V_G$) is indicated where the bias line intersects the V_{GS} scale. Alternatively, V_{GS} can be calculated by rewriting Eq. 9-1,

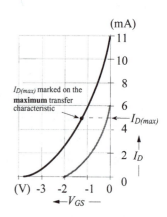

Figure   10-29
The first step in FET bias circuit design is selection of I_D on the maximum transfer characteristic.

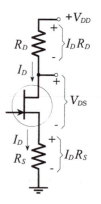

Figure   10-30
*When the maximum transfer characteristic is **not** used in FET circuit design calculations, the I_D level can be greater than the design value, and the FET can be forced into the channel ohmic region of its characteristic.*

$$V_{GS} = V_P\left[1 - \sqrt{(I_D/I_{DSS})}\right] \qquad\qquad \textbf{(10-8)}$$

Because the circuit design uses $I_{D(max)}$, the voltage drop across R_D is a maximum, and consequently, the drain-source voltage (V_{DS}) is a minimum. So, the drain resistor is calculated as,

$$R_D = \frac{V_{DD} - V_{DS(min)}}{I_{D(max)}} \qquad\qquad \textbf{(10-9)}$$

As discussed, the bias resistor (R_G) is usually selected as 1 MΩ, although a lower resistance value may be used.

Figure 10-31 shows the equations involved in designing a gate bias circuit, and Ex. 10-7 demonstrates the design procedure.

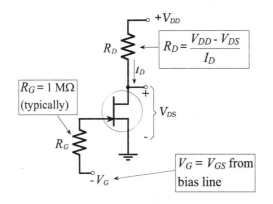

Figure 10-31
FET gate bias circuit design.

Example 10-7
Design a gate bias circuit (as in Fig. 10-31) to have $I_{D(max)} = 3$ mA and $V_{DS(min)} = 10$ V. The *FET* to be used has the transfer characteristics in Fig. 10-32, and the supply voltage is $V_{DD} = 25$ V.

Solution
On the maximum transfer characteristic, mark

$$I_{D(max)} = 3 \text{ mA (point } X \text{ on Fig. 10-32)}$$

Draw the bias line vertically through point X.

Read the required bias voltage where the bias line intersects the V_{GS} scale,

$$V_{GS} = -2.3 \text{ V}$$

Alternatively,

Eq. 10-8,
$$\begin{aligned} V_{GS} &= V_P\left[1 - \sqrt{(I_D/I_{DSS})}\right] \\ &= -6 \text{ V}\left[1 - \sqrt{(3 \text{ mA}/8 \text{ mA})}\right] \\ &= -2.3 \text{ V} \end{aligned}$$

Select $\qquad R_G = 1$ MΩ (see text)

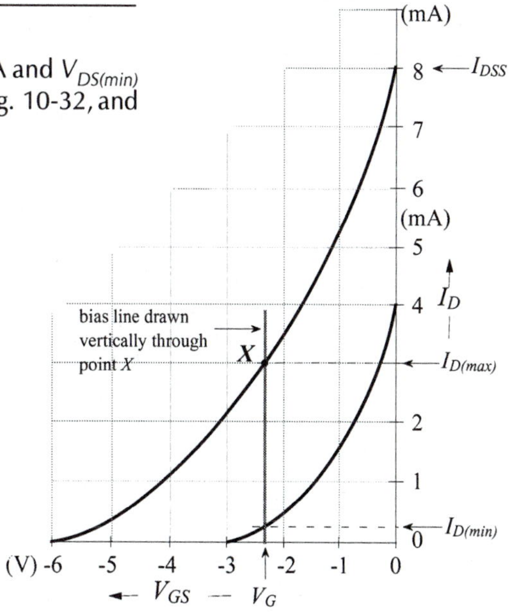

Figure 10-32
FET transfer characteristic and bias line for Ex. 10-7.

Eq. 10-9,

$$R_D = \frac{V_{DD} - V_{DS(min)}}{I_{D(max)}} = \frac{25\ V - 10\ V}{3\ mA}$$

$$= 5\ k\Omega\ \ (\text{use } 4.7\ k\Omega \text{ standard value})$$

From Fig. 10-32, note that, $I_{D(min)} = 0.25$ mA

The calculated resistor values are shown in Fig. 10-33, and the circuit is analyzed in Ex. 10-3.

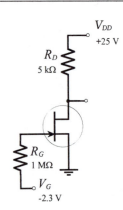

Figure 10-33
FET gate bias circuit designed in Ex. 10-7.

Self Bias Circuit Design

As for gate bias, the procedure for designing a self bias circuit (Fig. 10-34) commences with $I_{D(max)}$ being marked on the *FET* maximum transfer characteristic (Fig. 10-35). The bias line is then drawn through this point and the point where $I_{DS} = 0$ and $V_{GS} = 0$. The value of the source resistor (R_S) is determined from the slope of the bias line,

$$\mathbf{R_S} = \frac{\Delta \mathbf{V_{GS}}}{\Delta \mathbf{I_D}} \tag{10-10}$$

The total resistance in series with the *FET* drain-source terminals (Fig. 10-34) is ($R_D + R_S$). So, Eq. 10-9 can be modified to calculate this quantity,

$$\mathbf{R_D + R_S} = \frac{\mathbf{V_{DD} - V_{DS(min)}}}{\mathbf{I_{D(max)}}} \tag{10-11}$$

With R_S already known, R_D is easily determined. Gate resistor R_G is normally selected as 1 MΩ, or lower, as discussed.

Figure 10-34 shows the equations for designing a self bias circuit, and Ex. 10-8 demonstrates the design procedure.

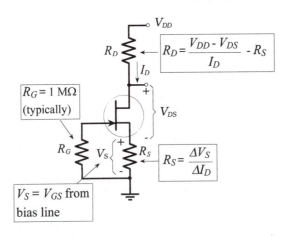

Figure 10-34
FET self bias circuit design.

Example 10-8

Design a self bias circuit (as in Fig. 10-34) to have $I_{D(max)} = 3$ mA and $V_{DS(min)} = 10$ V. The *FET* transfer characteristics are shown in Fig. 10-35, and the supply voltage is $V_{DD} = 25$ V.

Solution

On maximum the transfer characteristics, mark

$$I_{D(max)} = 3 \text{ mA (point } X \text{ on Fig. 10-35)}$$

Mark point Y at $I_D = 0$ and $V_{DS} = 0$

Draw the bias line through points X and Y.

From the bias line: $\Delta V_{GS} = 2.3$ V, and $\Delta I_D = 3$ mA

Eq. 10-10, $\qquad R_S = \dfrac{V_{GS}}{I_D} = \dfrac{2.3 \text{ V}}{3 \text{ mA}}$

$$\approx 770 \text{ }\Omega \text{ (use 750 }\Omega\text{, }\pm 5\%$$
$$\text{standard value)}$$

Eq. 10-11, $\quad R_D + R_S = \dfrac{V_{DD} - V_{DS(min)}}{I_{D(max)}} = \dfrac{25 \text{ V} - 10 \text{ V}}{3 \text{ mA}}$

$$= 5 \text{ k}\Omega$$

$$R_D = (R_D + R_S) - R_S = 5 \text{ k}\Omega - 770 \text{ }\Omega$$

$$= 4.23 \text{ k}\Omega \text{ (use 3.9 k}\Omega \text{ standard value)}$$

Select $\qquad R_G = 1 \text{ M}\Omega \quad$ (see text)

From Fig. 10-35, note that $\quad I_{D(min)} = 1.4 \text{ mA}$

The calculated resistor values are shown in Fig. 10-36, and the circuit is analyzed in Ex. 10-5.

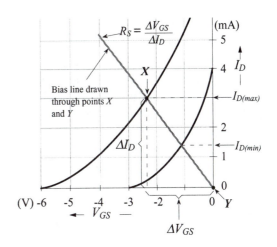

Figure 10-35
FET transfer characteristic and bias line for Ex. 10-8.

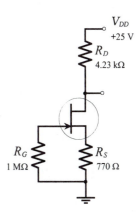

Figure 10-36
FET self bias circuit designed in Ex. 10-8.

Voltage Divider Bias Circuit Design

As always, circuit design commences with the specification of V_{DD}, $I_{D(max)}$, and $V_{DS(min)}$. The maximum drain current is plotted on the device maximum transfer characteristic, then either the resistance of R_S or the level of V_G must be determined before the bias line can be drawn for a voltage divider bias circuit.

With $V_{DS(min)}$ specified, the remaining portion of the supply voltage ($V_{DD} - V_{DS(min)}$) is split between V_S and V_{RD}. It is shown in Chapter 11 that the voltage gain of a *FET* amplifier stage is directly proportional to R_D. Maximum gain is achieved by selecting R_D as large as possible, and for a given level of I_D this requires the greatest possible voltage V_{RD}. However, the best bias stability is achieved by selecting R_S as large as possible, and this dictates maximum V_S. A reasonable compromise between these two conflicting requirements (maximum V_{RD} and maximum V_S) is to simply make the two voltage drops (and the two resistors) equal. Equation 10-11 is used to calculate ($R_D + R_S$), and this is divided to give two equal resistors.

Once R_S is determined, the bias line can be drawn through $I_{D(max)}$ with the slope of the line set by R_S. This requires that the V_{GS} scale be extended positively, as shown in Fig. 10-37.

From Eq. 10-10,

$$\Delta V_{GS} = \Delta I_D \times R_S \qquad \textbf{(10-12)}$$

The gate bias voltage (V_G) is read from the intersection of the bias line and the (extended) V_{GS} scale (Fig. 10-37). R_2 is normally the smallest resistor in the voltage divider. So, R_2 is usually selected as 1 MΩ, for the reasons already discussed for R_G in the gate bias and self bias circuits. R_1 is then calculated from R_2 and the resistor voltage drops. From Eq. 10-7,

$$R_1 = \frac{(V_{DD} - V_G) \times R_2}{V_G} \qquad \textbf{(10-13)}$$

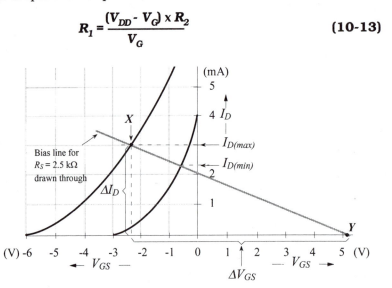

Figure 10-37
FET transfer characteristic and bias line for voltage divider bias circuit design.

Another approach that can be used to determine R_S is to specify a minimum drain current, and mark $I_{D(min)}$ on the minimum drain characteristic (after marking $I_{D(max)}$). The bias line is then drawn through the $I_{D(max)}$ and $I_{D(min)}$ points. The slope of the bias line and its intersection with the V_{GS} scale dictates R_S and V_G, respectively.

The design equations for a voltage divider bias circuit are shown on Fig. 10-38, and Ex. 10-9 demonstrates the design procedure.

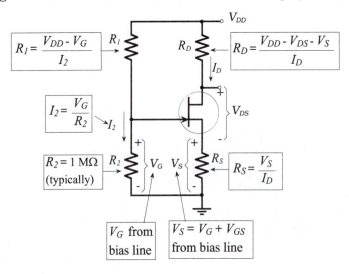

Figure 10-38
FET voltage divider bias circuit design.

Example 10-9

Design a voltage divider bias circuit (as in Fig. 10-38) to have $I_{D(max)}$ = 3 mA and $V_{DS(min)}$ = 10 V. Use the *FET* transfer characteristics in Fig. 10-37, and a supply voltage of V_{DD} = 25 V.

Solution

Eq. 10-11,
$$R_D + R_S = \frac{V_{DD} - V_{DS(min)}}{I_{D(max)}} = \frac{25\,V - 10\,V}{3\,mA}$$
$$= 5\,k\Omega$$

Select,
$$R_D = R_S = \frac{R_D + R_S}{2} = \frac{5\,k\Omega}{2}$$
$$= 2.5\,k\Omega\ \ (\text{use } 2.2\,k\Omega \text{ standard value})$$

On the maximum transfer characteristics, mark
$$I_{D(max)} = 3\,mA\ (\text{point } X \text{ on Fig. 10-37})$$

For ΔI_D = 3 mA,
Eq. 10-12,
$$\Delta V_{GS} = \Delta I_D \times R_S = 3\,mA \times 2.5\,k\Omega$$
$$= 7.5\,V$$

Mark point Y at ΔI_D= 3 mA and ΔV_{DS} = 7.5 V from point X.

Draw the bias line through points X and Y.

From the bias line: V_G = 5.2 V

Alternatively,
Eq. 10-8,
$$V_{GS} = V_P\left[1 - \sqrt{(I_D/I_{DSS})}\right] = -6\,V\left[1 - \sqrt{(3\,mA/8\,mA)}\right]$$
$$= -2.3\,V$$

$$V_S = I_D \times R_S = 3\,mA \times 2.5\,k\Omega$$
$$= 7.5\,V$$

$$V_G = V_S - V_{GS} = 7.5\,V - 2.3\,V$$
$$= 5.2\,V$$

Select R_2 = 1 MΩ (see text)

Eq. 10-13,
$$R_1 = \frac{(V_{DD} - V_G) \times R_2}{V_G} = \frac{(25\,V - 5.2\,V) \times 1\,M\Omega}{5.2\,V}$$
$$= 3.8\,M\Omega\ \ (\text{use } 3.9\,M\Omega \text{ standard value})$$

From Fig. 10-37, note that $I_{D(min)}$= 2.3 mA

The calculated resistor values are shown in Fig. 10-39, and the circuit is analyzed in Ex. 10-6.

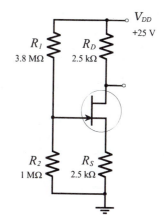

Figure 10-39
FET voltage divider bias circuit designed in Ex. 10-9.

Designing with Standard-Value Resistors

Each time a resistor value is calculated the nearest (higher or lower) standard value should be selected, as shown in the preceding examples. Sometimes it will be appropriate to select a resistance higher than the calculated value, and sometimes a lower value should be used. For example, if a resistance is calculated using a minimum current level, it is usually best to select a lower resistance value so that the current is higher than the specified minimum. Regardless of how the standard values are selected, the new current level or voltage drop produced by the standard value component should be calculated before proceeding with the design. This approach is employed in design examples in the next section.

Practise Problems

10-7.1 A *JFET* gate bias circuit using $V_{DD} = 30$ V is to have $I_{D(max)} = 4$ mA and $V_{DS(min)} = 11$ V. Design the circuit using the transfer characteristics in Fig. 10-32.

10-7.2 Design a *JFET* self bias circuit to have $I_{D(max)} = 2$ mA and $V_{DS(min)} = 8$ V. The supply voltage is $V_{DD} = 20$ V, and the *FET* transfer characteristics are shown in Fig. 10-35.

10-7.3 Design a *JFET* voltage divider bias circuit to have $I_{D(max)} = 1.5$ mA and $V_{DS(min)} = 10$ V. The supply voltage is $V_{DD} = 27$ V, and the *FET* transfer characteristics are shown in Fig. 10-37.

10-8 More JFET Bias Circuits

Use of Plus/Minus Supplies

When plus/minus supply voltages are to be used with a *FET* bias circuit, the gate terminal is usually grounded via R_G, as illustrated in Fig. 10-40(a). In this case, the circuit is essentially a voltage divider bias circuit with the gate bias voltage equal to the level of the negative supply voltage, $(V_G = V_{SS})$. The (reverse) gate-source bias voltage $(-V_{GS})$ makes the source terminal $+V_{GS}$ above ground. So,

$$V_S = V_{SS} + V_{GS} \qquad\qquad \textbf{(10-14)}$$

When the voltage levels are understood, the type of circuit in Fig. 10-40(a) can be designed using a procedure similar to that for a voltage divider bias circuit. $I_{D(max)}$ is plotted on the maximum transfer characteristics, and V_{GS} is read from the horizontal axis. V_{GS} can also be determined by substituting $I_{D(max)}$, $I_{DSS(max)}$, and $V_{P(max)}$ into Eq. 10-8. Then, with the desired maximum voltage drops known, R_D and R_S are readily calculated. R_G is selected as 1 MΩ, or less, for the reasons already discussed.

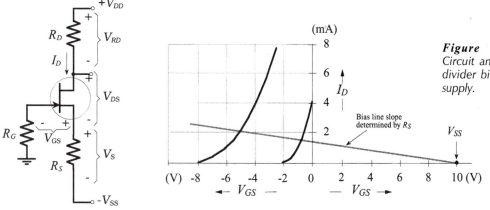

(a) Circuit using ± supply   (b) bias line for circuit using ± supply

Figure 10-40
Circuit and bias line for voltage divider bias using a plus/minus supply.

Analysis procedure for this circuit is also like that for a voltage divider bias circuit. One point on the bias line is plotted on the transfer characteristics at $V_G = V_{SS}$, as shown on Fig. 10-40(b), and the bias line is drawn at the slope determined by R_S. As always, the bias line intersections with the maximum and minimum characteristics gives the $I_{D(max)}$ and $I_{D(min)}$ levels.

Drain Feedback

The *drain feedback* bias circuit in Fig. 10-41 is essentially a voltage divider bias circuit. However, instead of V_G being derived from the supply voltage, the *FET* drain voltage (V_D) is divided by R_1 and R_2 to produce V_G. Modifying Eq. 10-7,

$$V_G = \frac{V_D \times R_2}{R_1 + R_2} \qquad \textbf{(10-15)}$$

The circuit is designed for a particular level of V_D and I_D, and the voltage drop across R_S helps to stabilise I_D, as in other voltage divider bias circuits. The voltage drop across R_D also provides feedback that helps to correct changes in I_D. When I_D is greater than the design value, V_{RD} is increased, V_D is reduced, and thus V_G is lowered to drive I_D back toward its original level. The two feedback effects (V_{RD} changes and V_{RS} changes) tend to keep $I_{D(min)}$ and $I_{D(max)}$ closer than in an ordinary voltage divider bias circuit.

Design procedure for this circuit is the same as for voltage divider bias with the exception that the voltage drop across R_1 is $(V_D - V_G)$ instead of $(V_{DD} - V_G)$. For circuit analysis a bias line should be drawn on the transfer characteristics, as always. This requires the equation relating V_{GS} and I_D.

$$V_{GS} = \frac{R_2\,(V_{DD} - I_D R_D)}{R_1 + R_2} - I_D R_S \qquad \textbf{(10-16)}$$

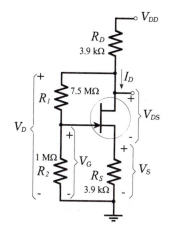

Figure 10-41
FET drain feedback bias circuit.

Example 10-10

Design the drain feedback circuit in Fig. 10-41 to use a *2N5458 FET* with a 20 V supply, and to have $I_{D(max)}$ = 1.5 mA and $V_{DS(min)}$ = 8 V.

Solution

Eq. 10-11, $R_D + R_S = \dfrac{V_{DD} - V_{DS(min)}}{I_{D(max)}} = \dfrac{20\,V - 8\,V}{1.5\,mA}$

$= 8\,k\Omega$

Select, $R_D = R_S = \dfrac{R_D + R_S}{2} = \dfrac{8\,k\Omega}{2}$

$= 4\,k\Omega$ (use 3.9 kΩ standard value)

From Appendix 1-11, $I_{DSS(max)}$ = 9 mA, and $V_{P(max)}$ = 7 V (for a 2N5458)

Eq. 10-8, $V_{GS(max)} = V_P\left[1 - \sqrt{(I_D/I_{DSS})}\right] = -7\,V\left[1 - \sqrt{(1.5\,mA/9\,mA)}\right]$

$= -4.14\,V$

$V_S = I_D R_S = 1.5\,mA \times 3.9\,k\Omega$

$= 5.85\,V$

$V_G = V_S + V_{GS(max)} = 5.85\,V - 4.14\,V$

$= 1.7\,V$

$V_D = V_{DD} - I_D R_D = 20\,V - (1.5\,mA \times 3.9\,k\Omega)$

$= 14.2\,V$

Select $R_2 = 1\,M\Omega$

From Eq. 10-15, $R_1 = \dfrac{(V_D - V_G) \times R_2}{V_G} = \dfrac{(14.2\,V - 1.7\,V) \times 1\,M\Omega}{1.7\,V}$

$= 7.4\,M\Omega$ (use 7.5 MΩ, ±5% standard value)

Constant Current Bias

As already discussed, the bias lines with the smallest slope give the closest levels of $I_{D(min)}$ and $I_{D(max)}$. In the two circuits in Fig. 10-42 the bipolar transistor (Q_2) keeps the *FET* source current constant. Consequently, as illustrated in Fig. 10-43, the bias line is drawn horizontally on the transfer characteristics, showing $I_{D(min)} = I_{D(max)}$, with different V_{GS} levels.

Both circuits provide bias voltage V_B to the base of Q_2, to give,

$$I_D = I_C \approx I_E = \frac{V_B - V_{BE}}{R_E} \qquad (10\text{-}17)$$

The circuit in Fig. 10-42(a) uses voltage divider bias to derive V_B from V_{DD}, and in Fig. 10-42(b) $V_B = V_{EE}$. In both cases, the *FET* gate is connected to the base of Q_2 via resistor R_G, and $V_{GS} = -V_{CB}$.

Design of each of the circuit in Fig. 10-42 simply involves

selection of appropriate voltage and current levels and calculation of the resistors. The emitter resistor voltage drop (V_E) should typically be a minimum of 5 V, but satisfactory results can be obtained with 3 V. It can, of course, be much larger than 5 V, as would normally be the case in Fig. 10-42(b). In both circuits,

$$V_B = V_E + V_{BE}$$

The level of V_B calculated in this way can be used to determine the ratio of R_1 and R_2 in Fig. 10-42(a). For the circuit in Fig. 10-42(b), V_{EE} dictates the V_B and V_E levels.

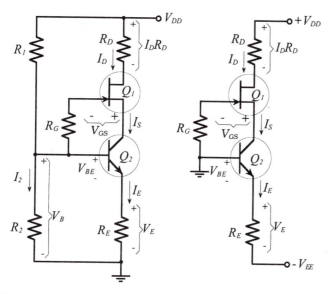

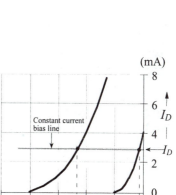

(a) Voltage divider bias with constant current source

(b) Circuit with a ± supply and a constant current source

Figure 10-42
FET circuits with BJT constant current bias circuits. In each case I_S is held constant at the I_C level.

The maximum V_{GS} level is determined by plotting I_D on the maximum transfer characteristic for the *FET*, and then reading $V_{GS(max)}$ at I_D. Alternatively, V_{GS} can be calculated from Eq. 10-8. The source terminal voltage is more positive than the gate. The equation for V_S,

$$V_S = V_B - V_{GS} \qquad \textbf{(10-18)}$$

Analysis of each of the circuits in Fig. 10-42 is very simple. Once V_B is determined, I_D is calculated using Eq. 10-17. $V_{GS(max)}$ and $V_{GS(min)}$ are found by drawing the bias line for I_D on the transfer characteristics, or by substituting I_D, V_P, and I_{DSS} into Eq. 10-8.

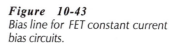

Figure 10-43
Bias line for FET constant current bias circuits.

Example 10-11

Design the constant current bias circuit in Fig. 10-42(b) to use a *2N5459 FET* with a ±20 V supply, and to have $I_{D(max)} = 3$ mA and $V_{DS(min)} = 9$ V.

Solution

$$V_E = V_{EE} - V_{BE} = 20 \text{ V} - 0.7 \text{ V}$$
$$= 19.3 \text{ V}$$

$$R_E = \frac{V_E}{I_D} = \frac{19.3 \text{ V}}{3 \text{ mA}}$$

$$= 6.4 \text{ k}\Omega \quad \text{(use 6.8 k}\Omega \text{ standard value)}$$

I_D becomes

$$\frac{V_E}{R_E} = \frac{19.3 \text{ V}}{6.8 \text{ k}\Omega}$$

$$= 2.83 \text{ mA}$$

From Appendix 1-11, $I_{DSS(max)} = 16$ mA, and $V_{P(max)} = 8$ V (for a 2N5459)

Eq. 10-8,
$$V_{GS(max)} = V_P[1 - \sqrt{(I_D/I_{DSS})}]$$
$$= -8 \text{ V}[1 - \sqrt{(2.83 \text{ mA}/16 \text{ mA})}]$$
$$= -4.6 \text{ V}$$

Eq. 10-18,
$$V_S = V_B - V_{GS} = 0 - (-4.6 \text{ V})$$
$$= +4.6 \text{ V}$$

$$V_{RD} = V_{DD} - V_{DS(min)} - V_S = 20 \text{ V} - 9 \text{ V} - 4.6 \text{ V}$$
$$= 6.4 \text{ V}$$

$$R_D = \frac{V_{RD}}{I_D} = \frac{6.4 \text{ V}}{2.83 \text{ mA}}$$

$$= 2.26 \text{ k}\Omega \quad \text{(use 2.2 k}\Omega \text{ standard value)}$$

Select $R_G = 1 \text{ M}\Omega$

Figure 10-44 shows the circuit voltage and current levels when the *FET* has $I_{DSS(max)}$ and $V_{P(max)}$, as used in the design. When the *FET* does not have the maximum parameters, only V_{GS} and V_{DS} should change.

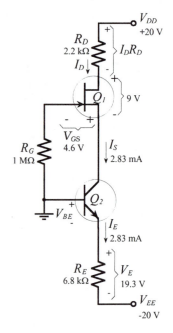

Figure 10-44
Constant current bias circuit designed in Ex. 10-11.

Practise Problems

10-8.1 Design the voltage divider bias circuit in Fig 10-40(a) to use $V_{DD} = \pm 18$ V. The circuit is to have $I_{D(max)} = 2.5$ mA and $V_{DS(min)} = 7$ V, and the *FET* has the transfer characteristics in Fig. 10-37.

10-8.2 The constant current bias circuit in Fig. 10-42(b) is to have a 2N5457 *FET*, $V_{DD} = 30$ V, and $V_{EE} = -12$ V. The bias conditions are to be $I_{D(max)} = 2$ mA and $V_{DS(min)} = 8$ V. Design the circuit.

10-8.3 Design a drain feedback bias circuit, as in Fig. 10-41, to have $I_{D(max)} = 2.5$ mA and $V_{DS(min)} = 9$ V. A 2N5457 *FET* is the be used, and the supply is $V_{DD} = 24$ V.

10-9   *Use of Universal Transfer Characteristic*

Universal Transfer Characteristic

A *universal transfer characteristic* for a *FET* is simply a transfer characteristic plotted with $I_{DSS} = 1$ and $V_P = 1$. Then, instead of the scales being calibrated in milliamps and volts, they are marked as the ratios I_D/I_{DSS} and V_{GS}/V_P. To construct the universal transfer characteristic, Eq. 9-1 is rewritten,

$$\frac{I_D}{I_{DSS}} = \left[1 - \frac{V_{GS}}{V_P}\right]^2 \qquad \textbf{(10-18)}$$

Now, by substituting convenient values for V_{GS}/V_P into Eq. 10-18 corresponding values of I_D/I_{DSS} are calculated, and these are used to plot the characteristic shown in Fig. 10-45.

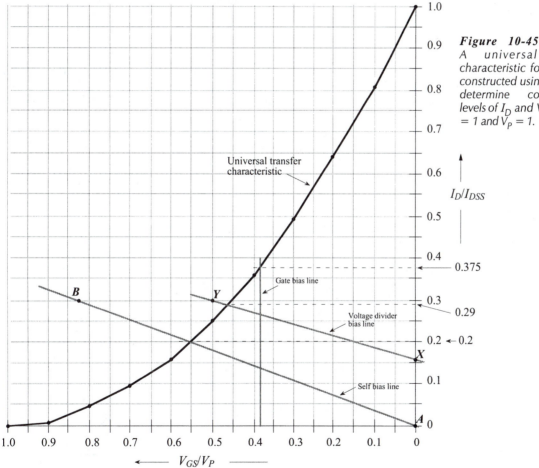

Figure 10-45
A universal transfer characteristic for a FET is constructed using Eq. 9-1 to determine corresponding levels of I_D and V_{GS} with $I_{DSS} = 1$ and $V_P = 1$.

The universal transfer characteristic can be applied to analyze or design a circuit using virtually any *JFET*. This saves the necessity of plotting the transfer characteristics for each device. The specified values of I_{DSS} and V_P must be known for the device. Then, the characteristic is used in the same way as an actual transfer characteristic, except that V_{GS} levels must be converted to V_{GS}/V_P ratios, and I_D levels must be converted to I_D/I_{DSS} before they can be plotted on the graph. Similarly, readings taken from the graph are converted back to V_{GS} and I_D levels by multiplying them by V_P and I_{DSS}, respectively.

Circuit Analysis

Consider the gate bias circuit analyzed in Ex. 10-3, and reproduced in Fig. 10-46. The gate voltage is given as -2.3 V, and the device used has $I_{DSS(max)}$ = 8 mA and $V_{P(max)}$ = -6 V. The bias line is drawn vertically on the universal transfer characteristic at,

$$\frac{V_{GS}}{V_P} = \frac{2.3\ V}{6\ V} = 0.38$$

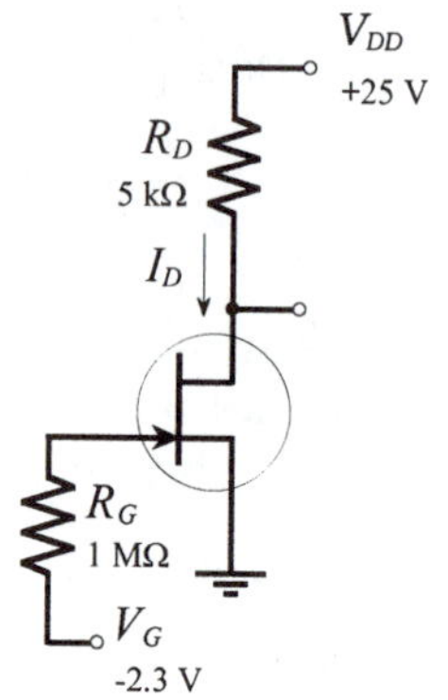

The gate bias line drawn (at V_{GS}/V_P = 0.38) on Fig. 10-45 intersects the universal characteristic at the point where I_D/I_{DSS} = 0.375,

giving, $I_{D(max)}$ = 0.375 I_{DSS} = 0.375 x 8 mA
            = 3 mA

This is the $I_{D(max)}$ level determined in Ex. 10-3.

The minimum I_D for a gate bias circuit can be determined from the universal transfer characteristic, by using the $I_{DSS(min)}$ and $V_{P(min)}$ values. Bias lines can also be drawn for analysis of self bias circuits and voltage divider bias circuits. Examples 10-12 and 10-13 demonstrate the analysis procedures.

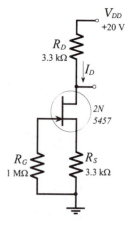

Figure 10-46
The gate bias circuit in Ex. 10-3 can be analyzed by use of the universal transfer characteristic.

Example 10-12

Use the *FET* universal transfer characteristic to determine $I_{D(max)}$ and $V_{DS(min)}$ in the self bias circuit shown in Fig. 10-47.

Solution

From Appendix 1-11, the *2N5457* has, $I_{DSS(max)}$ = 5 mA and $V_{P(max)}$ = 6 V

when I_D = 0, $V_{GS} = V_S = 0$

so, I_D/I_{DSS} = 0 and V_{GS}/V_P = 0

Plot point *A* on the universal transfer characteristic in Fig. 10-45 at

$$I_D/I_{DSS} = 0 \text{ and } V_{GS}/V_P = 0$$

when I_D = 1.5 mA, $V_{GS} = I_D \times R_S$ = 1.5 mA x 3.3 kΩ
                = 4.95 V

now, $\dfrac{I_D}{I_{DSS}} = \dfrac{1.5\ mA}{5\ mA} = 0.3$

and, $\dfrac{V_{GS}}{V_P} = \dfrac{4.95\ V}{6\ V} = 0.825$

Figure 10-47
Self bias circuit for Ex. 10-12.

Plot point B on the universal transfer characteristic at,

$$I_D/I_{DSS} = 0.3 \text{ and } V_{GS}/V_P = 0.825$$

Draw the self bias line through points A and B,and where the bias line intersects the universal transfer characteristic read,

$$I_D/I_{DSS} = 0.2$$

giving,

$$I_{D(max)} = 0.2\, I_{DSS} = 0.2 \times 5 \text{ mA}$$
$$= 1 \text{ mA}$$

$$V_{DS(min)} = V_{DD} - I_D(R_D + R_S)$$
$$= 20 \text{ V} - 1 \text{ mA}(3.3 \text{ k}\Omega + 3.3 \text{ k}\Omega)$$
$$= 13.4 \text{ V}$$

Example   10-13
Using the *FET* universal transfer characteristic, determine $I_{D(max)}$ and $V_{DS(min)}$ for the voltage divider bias circuit in Fig. 10-48.

Solution
From Appendix 1-11, the *2N5458* has, $I_{DSS(ma)} = 9$ mA and $V_{P(max)} = 7$ V

Eq. 10-7,
$$V_G = \frac{V_{DD} \times R_2}{R_1 + R_2} = \frac{22 \text{ V} \times 1 \text{ M}\Omega}{4.7 \text{ M}\Omega + 1 \text{ M}\Omega}$$
$$= 3.9 \text{ V}$$

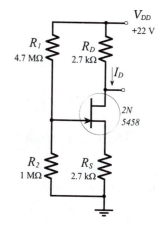

when $V_{GS} = 0$, $V_{GS}/V_P = 0$

and,
$$I_D R_S = V_G$$

so,
$$I_D = \frac{V_G}{R_S} = \frac{3.9 \text{ V}}{2.7 \text{ k}\Omega} = 1.4 \text{ mA}$$

and,
$$\frac{I_D}{I_{DSS}} = \frac{1.4 \text{ mA}}{9 \text{ mA}} = 0.16$$

Plot point X on the universal transfer characteristic in Fig. 10-45 at

$$I_D/I_{DSS} = 0.16 \text{ and } V_{GS}/V_P = 0$$

Figure   10-48
Voltage divider bias circuit for Ex. 10-13.

when $V_{GS}/V_P = 0.5$, $V_{GS} = 0.5\, V_P = 0.5 \times (-7 \text{ V})$
$$= -3.5 \text{ V}$$

from Eq. 10-6,
$$I_D R_S = V_G - V_{GS} = 3.9 \text{ V} - (-3.5 \text{ V})$$
$$= 7.4 \text{ V}$$

so,
$$I_D = \frac{I_D R_S}{R_S} = \frac{7.4 \text{ V}}{2.7 \text{ k}\Omega} = 2.7 \text{ mA}$$

$$\frac{I_D}{I_{DSS}} = \frac{2.7 \text{ mA}}{9 \text{ mA}} = 0.3$$

Plot point Y on the universal transfer characteristic at

$$I_D/I_{DSS} = 0.3 \text{ and } V_{GS}/V_P = 0.5$$

Draw the voltage divider bias line through points X and Y, and where the bias line intersects the universal transfer characteristic read,

$$I_D/I_{DSS} = 0.29$$

giving,

$$
\begin{aligned}
I_D &= 0.29 \, I_{DSS} = 0.29 \times 9 \text{ mA} \\
&= 2.61 \text{ mA}
\end{aligned}
$$

$$
\begin{aligned}
V_{DS} &= V_{DD} - I_D(R_D + R_S) \\
&= 22 \text{ V} - 2.61 \text{ mA}(2.7 \text{ k}\Omega + 2.7 \text{ k}\Omega) \\
&= 7.9 \text{ V}
\end{aligned}
$$

Practise   Problems

10-9.1   The self bias circuit in Fig. 10-47 has its *FET* changed to a *2N5458*. Use the *FET* universal transfer characteristic to determine $I_{D(max)}$ and $V_{DS(min)}$ for the circuit.

10-9.2   Using the *FET* universal transfer characteristic, analyze the circuit in Example 10-6 to determine $V_{DS(min)}$.

10-9.3   A gate bias circuit, as in Fig. 10-46, has a *2N5458*, $V_{DD} = 20$ V, $V_G = -4.5$ V, and $R_D = 4.7$ kΩ. Use the *FET* universal transfer characteristic to determine $I_{D(max)}$ and $V_{DS(min)}$.

10-10   MOSFET Biasing

DE-MOSFET Bias Circuits

DE-MOSFET bias circuits are similar to JFET bias circuits. Any of the *FET* bias circuits already discussed can be used to produce a negative V_{GS} level for an *n*-channel *MOSFET*, or a positive V_{GS} for a *p*-channel device. In this case, both devices would be operating in the depletion mode, just like JFETs. To operate an *n*-channel DE-MOSFET in the enhancement mode, the gate terminal must be made positive with respect to the source. A *p*-channel DE-MOSFET in the enhancement mode requires the gate to be negative with respect to the source.

Consider the four MOSFET bias circuits shown in Fig. 10-49, and assume that each device has the transfer characteristics in Fig. 10-50. In Fig. 10-49(a) the gate-source bias voltage is zero, so, the bias line is drawn on the transfer characteristics at $V_{GS} = 0$, as shown in Fig 10-50. The FET in Fig. 10-49(b) has a positive gate-source bias voltage, consequently, its bias line must be drawn vertically at $V_{GS} = +V_G$. The self bias circuit in Fig. 10-49(c) has its bias line drawn from $I_D = 0$ and $V_{GS} = 0$ at a slope determined by R_S, exactly as for a JFET self bias circuit. Similarly, the bias line for the voltage divider bias circuit in Fig. 10-49(d) is drawn from the point where $I_D = 0$ and $V_{GS} = +V_G$ with a slope set by R_S.

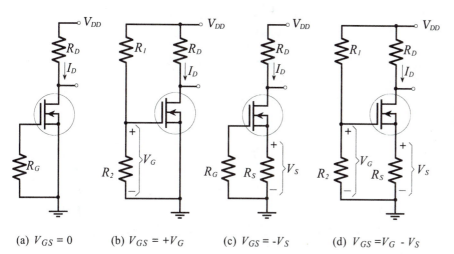

Figure 10-49
DE-MOSFET bias circuits.

(a) $V_{GS} = 0$ (b) $V_{GS} = +V_G$ (c) $V_{GS} = -V_S$ (d) $V_{GS} = V_G - V_S$

As always, $I_{D(min)}$ and $I_{D(max)}$ for any circuit are indicated by the bias line intersections with the maximum and minimum transfer characteristics. Fig. 10-50 shows that in some cases the *FET* is operating in depletion mode, and in other cases it is operating in enhancement mode. Analysis and design procedures for these circuits are essentially the same as for similar *JFET* bias circuits.

MOSFETs can also be used with a plus/minus supply voltage, as discussed in Section 10-8.

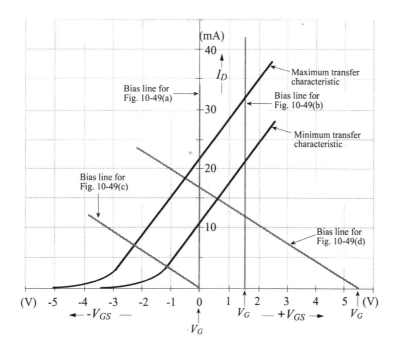

Figure 10-50
DE-MOSFET transfer characteristics and bias lines for the circuits in Fig. 10-49.

E-MOSFET Bias Circuits

Enhancement *MOSFETs* (such as the *VMOS* and *TMOS* devices discussed in Section 9-5) must have positive gate-source bias voltages in the case of *n*-channel devices, and negative V_{GS} levels for a *p*-channel *FET*. Thus, the gate bias circuit in Fig. 10-49(b) and the voltage divider bias circuit in Fig. 10-49(d) are suitable. In each case, the bias line is drawn exactly as already discussed.

Example 10-14

Determine I_D and V_{DS} for the bias circuit in Fig. 10-51, assuming that the device has the transfer characteristic in Fig. 10-52.

Solution

$$V_G = \frac{V_{DD} \times R_2}{R_1 + R_2} = \frac{40\ V \times 1\ M\Omega}{5.6\ M\Omega + 1\ M\Omega}$$

$$= 6.06\ V$$

Draw the bias line vertically on the transfer characteristic at $V_{GS} = 6.06\ V$, (see Fig. 10-52).
From the point where the bias line intersects the transfer characteristic,

$$I_D = 6.2\ A$$

$$V_{DS} = V_{DD} - I_D R_D = 40\ V - (6.2\ mA \times 4.7\ \Omega)$$

$$= 10.9\ V$$

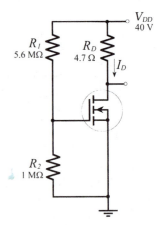

Figure   10-51
E-MOSFET gate bias circuit for Ex. 10-14.

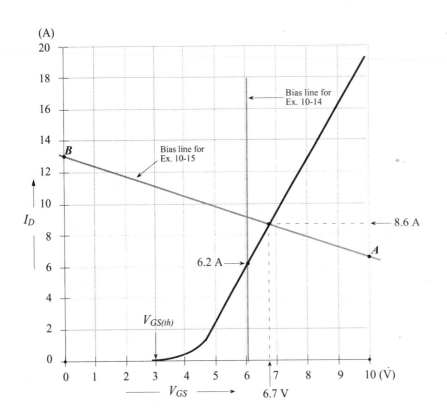

Figure   10-52
E-MOSFET transfer characteristic and bias lines.

The *drain-to-gate* bias circuit shown in Fig. 10-53 is uniquely suitable for an *E-MOSFET*. The *FET* gate is directly connected to the drain terminal via resistor R_G, so that $V_{GS} = V_{DS}$. Also,

$$V_{GS} = V_{DD} - I_D R_D \qquad \text{(10-19)}$$

If the drain current is higher than the design level, a higher voltage drop is produced across R_D, and this tends to reduce V_{GS}, and thus reduce I_D back toward the design level. Similarly, a lower than intended I_D level produces a lower $I_D R_D$ voltage drop, and results in a higher V_{GS} that drives I_D back up toward the desired current. Corresponding levels of I_D and V_{GS} can be calculated from Eq. 10-19. These are then plotted on the device transfer characteristics to draw the circuit bias line.

Example 10-15

Analyze the bias circuit in Fig. 10-53 to determine the typical levels of I_D and V_{DS}. Figure 10-52 shows the typical transfer characteristic for the device.

Solution

when $V_{GS} = 10$ V,

from Eq. 10-19, $I_D = \dfrac{V_{DD} - V_{GS}}{R_D} = \dfrac{20\,V - 10\,V}{1.5\,\Omega}$

$$= 6.7 \text{ A}$$

Plot *point A* on Fig. 10-52 at, $V_{GS} = 10$ V and $I_D = 6.7$ A

when $V_{GS} = 0$,

from Eq. 10-19, $I_D = \dfrac{V_{DD} - V_{GS}}{R_D} = \dfrac{20\,V - 0}{1.5\,\Omega}$

$$= 13 \text{ A}$$

Plot *point B* on Fig. 10-52 at, $V_{GS} = 0$ and $I_D = 13$ A

Draw the bias line for $R_D = 1.5\ \Omega$ through *points A* and *B*.

Where the bias line intersects the transfer characteristics,

$$I_D = 8.6 \text{ A and } V_{GS} = 6.7 \text{ V}$$

$$V_{DS} = V_{GS} = 6.7 \text{ V}$$

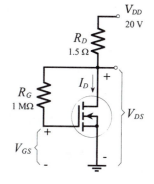

Figure 10-53
E-MOS drain-to-gate bias circuit.

Practise Problems

10-10.1 The *MOSFET* voltage divider bias circuit in Fig. 10-49(d) has $V_{DD} = 30$ V, $R_1 = 330$ kΩ, $R_2 = 150$ kΩ, and $R_D = R_S = 680$ Ω. The *FET* has the transfer characteristics in Fig. 10-50. Analyze the circuit to determine $V_{DS(min)}$ and $V_{DS(max)}$.

10-10.2 A drain-to-gate bias circuit, as in Fig. 10-53, has $V_{DD} = 35$ V, and a *MOSFET* with the transfer characteristics in Fig. 10-52. Calculate an approximate resistance for R_D, to give $V_{DS} = 5$ V.

10-11 Biasing FET Switching Circuits

JFET Switching

A field effect transistor in a switching circuit is normally in an *off* state with zero drain current, or in an *on* state with a very small drain-source voltage. When the *FET* is *off*, there is a drain-source leakage current so small that it can almost always be neglected. When the device is *on*, the drain-source voltage drop depends on the channel resistance ($r_{DS(on)}$) and the drain current (I_D).

$$V_{DS(on)} = I_D \, r_{DS(on)} \tag{10-20}$$

Field effect transistors designed specifically for switching applications have very low channel resistances. For example, the *2N4856* has $r_{DS(on)}$ = 25 Ω, (see Appendix 1-12). With low I_D levels, $V_{DS(on)}$ can be much smaller than the 0.2 V typical $V_{CE(sat)}$ for a *BJT*. This is an important advantage of a *FET* switch over a *BJT* switch.

Direct-Coupled JFET Switching Circuit

A direct-coupled *JFET* switching circuit is shown in Fig. 10-54(a), and the circuit waveforms are illustrated in Fig. 10-54(b). When V_i = 0, the *FET* gate and source voltages are equal, and there is no depletion region penetration into the channel. The output voltage is now $V_o = V_{DS(on)}$, as expressed by Eq. 10-20. When V_i exceeds the *FET* pinch-off voltage, the device is switched *off*, and the output voltage goes to V_{DD}, as illustrated.

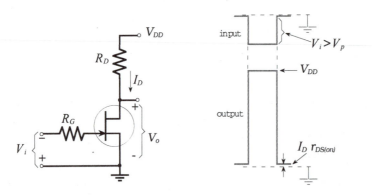

(a) Direct-coupled JFET switch   (b) Circuit waveforms

Figure 10-54
Direct-coupled JFET switching circuit, and the waveforms of the circuit input and output voltages.

Assuming that $V_{DS(on)}$ is very small, the drain current level is determined from,

$$V_{DD} \approx I_D R_D \tag{10-21}$$

Equation 10-21 can be used to determine R_D when V_{DD} and I_D are specified, or to calculate I_D when R_D is known. The I_D level can then be employed to determine $V_{DS(on)}$. The lowest drain current that can be used must be very much greater than the specified drain-source leakage current for the device.

To switch the *FET off*, V_i should exceed the maximum pinch-off voltage. However, V_i must not be so large that the drain-gate voltage ($V_{DG} = V_{DD} + V_i$) approaches the breakdown voltage. A rule-of-thumb is to select the input voltage 1 V larger than $V_{P(max)}$.

$$V_i = -(V_{P(max)} + 1 \text{ V}) \qquad \text{(10-22)}$$

The gate resistor (R_G) in the circuit in Fig. 10-54 is provided solely to limit any gate current in the event that the gate-source junctions become forward biased. The circuit might operate satisfactorily with R_G selected as 1 MΩ, however, high-value resistors can slow the switching speed of the circuit, so, quite small resistance values are often used for R_G.

Example 10-16
Design the *JFET* switching circuit in Fig. 10-54 to have $V_{DS(on)}$ not greater than 200 mV. A *2N4856 FET* is to be used with a 12 V supply.

Solution
From Appendix 1-12, $r_{DS(on)} = 25$ Ω and $V_{GS(off)} = 10$ V (max)

From Eq. 10-20, $\quad I_D = \dfrac{V_{DS(on)}}{r_{DS(on)}} = \dfrac{200 \text{ mV}}{25 \text{ Ω}}$

$\qquad\qquad\qquad = 8$ mA

From Eq. 10-21, $\quad R_D \approx \dfrac{V_{DD}}{I_D} = \dfrac{12 \text{ V}}{8 \text{ mA}}$

$\qquad\qquad\qquad = 1.5$ kΩ (standard value)

Eq. 10-22, $\qquad V_i = -(V_P + 1 \text{ V}) = -(10 \text{ V} + 1 \text{ V})$

$\qquad\qquad\qquad = -11$ V

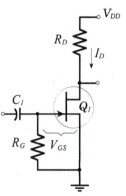

(a) Normally-*on* circuit

Capacitor-Coupled JFET Switching Circuits
Two capacitor-coupled *JFET* switching circuits are shown in Fig. 10-55. The *FET* in Fig. 10-55(a) is normally-*on* because it has $V_{GS} = 0$, and the device in Fig. 10-55(b) is normally-*off* with $-V_{GS}$ greater than the pinch-off voltage. In both circuits, the *FET* is switched *on* or *off* by a capacitor-coupled input pulse. The design procedure for these circuits uses the equations already discussed for the direct-coupled *JFET* switching circuit.

MOSFET Switching
Figure 10-56 shows two capacitor-coupled *MOSFET* switching circuits. In Fig. 10-56(a), the *FET* is biased *off* because $V_{GS} = 0$. A positive-going input signal is required to turn the device *on*. The *FET* in Fig. 10-56(b) is biased *on* by the positive V_{GS} provided by R_1 and R_2. In this case, a negative-going input voltage must be applied to turn the *FET off*. Equations 10-20 and 10-21 can be applied to these circuits to calculate I_D and $V_{DS(on)}$. To turn the

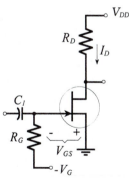

(b) Normally-*off* circuit

Figure 10-55
Normally-on and normally-off capacitor-coupled JFET switching circuits.

device *on* to a desired level of drain current, the transfer characteristics can be employed if they are available. Alternatively, Eq 9-5 can be used to estimate the required gate-source bias voltage, as explained in Section 9-5. To ensure that the *FET* is *off,* the gate-source voltage must be driven below the minimum threshold voltage for the device.

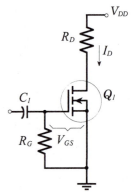

(a) Normally-*off* circuit

Example 10-17

The *MOSFET* in the switching circuit in Fig. 10-56(b) is to be biased *on* to have the smallest possible $V_{DS(on)}$. Determine suitable resistances for R_1 and R_2, and calculate $V_{DS(on)}$. The device has the transfer characteristics shown in Fig. 10-52, and its drain-source *on* resistance is 0.25 Ω.

Solution

From Eq. 10-21, $I_D \approx \dfrac{V_{DD}}{R_D} = \dfrac{50\ V}{10\ \Omega}$

$$= 5\ A$$

Referring to the transfer characteristics in Fig. 10-52,

at $I_D = 5\ A$, $V_{GS} \approx 5.7\ V$

Select $R_2 = 1\ M\Omega$

From Eq. 10-15, $R_1 = \dfrac{(V_D - V_{GS}) \times R_2}{V_G} = \dfrac{(50\ V - 5.7\ V) \times 1\ M\Omega}{5.7\ V}$

$= 7.7\ M\Omega$ (use 6.8 MΩ to make $V_{GS} > 5.7\ V$ to ensure that the *FET* is biased *on*)

Eq. 10-20, $V_{DS(on)} = I_D r_{DS(on)} = 5\ A \times 0.25\ \Omega$

$$= 1.25\ V$$

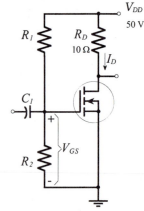

(b) Normally-*on* circuit

Figure 10-56
MOSFET   normally-on   and normally-off capacitor-coupled switching circuits.

Practise Problems

10-11.1   A normally-*off JFET* switching circuit, as in Fig. 10-55(b), is to use a 2N4861 *FET* with a 30 V supply, and is to have $I_{D(max)}$ = 10 mA. Calculate R_D, $V_{DS(on)}$, and the required V_G level.

10-11.2   The normally-*off MOSFET* switching circuit in Fig. 10-56(a) is to be employed to pass 15 A through a 3 Ω resistor. The device used has the transfer characteristics shown in Fig. 10-52, and $r_{DS(on)}$ = 0.2 Ω. Calculate approximate levels for the supply and input voltages.

Chapter-10 Review Questions

Section 10-1

10-1   Identify the components that constitute the *dc* load in a *FET* bias circuit. Explain the procedure for drawing the *dc* load line on the *FET* drain characteristics.

10-2   Explain the selection of a Q-point on a *FET dc* Load line, and discuss the limitations on the output voltage swing.

Section *10-2*
10-3   Sketch a gate bias circuit using an *n*-channel *JFET*. Identify the polarities of V_{DD}, V_{DS}, V_G, and V_{GS}. Also show the I_D direction. Briefly explain the circuit operation.

10-4   Repeat Question 10-3 for a gate bias circuit using a *p*-channel *JFET*.

10-5   Write equations for $V_{DS(max)}$ and $V_{DS(min)}$ for a *JFET* gate bias circuit.

10-6   State a typical maximum resistance for gate resistor R_G in a gate bias circuit. Briefly explain.

10-7   Explain why the device maximum and minimum transfer characteristics should be used in *FET* bias circuit analysis.

Section *10-3*
10-8   Sketch a self bias circuit using an *n*-channel *JFET*. Identify the polarities of V_{DD}, V_{DS}, V_S, V_G, and V_{GS}. Briefly explain the circuit operation.

10-9   Repeat Question 10-8 for a self bias circuit using a *p*-channel *JFET*.

10-10 Write equations for $V_{DS(max)}$ and $V_{DS(min)}$ for a *FET* self bias circuit. Also, write the equation used for drawing the circuit bias line on the transfer characteristics.

Section *10-4*
10-11 Sketch a voltage divider bias circuit using an *n*-channel *JFET*. Identify the polarities of V_{DD}, V_{DS}, V_S, V_G, and V_{GS}. Briefly explain the circuit operation.

10-12 Repeat Question 10-11 for a voltage divider bias circuit using a *p*-channel *JFET*.

10-13 Write equations for $V_{DS(max)}$ and $V_{DS(min)}$ for a *FET* voltage divider bias circuit. Also, write the equation used for drawing the circuit bias line on the transfer characteristics.

Section *10-5*
10-14 Compare the performance of the three basic *JFET* bias circuits in terms of the differences between $I_{D(max)}$ and $I_{D(min)}$ in each circuit and the predictability of circuit Q-points.

Section *10-6*
10-15 Briefly explain the procedure for testing a *FET* bias circuit. List common errors made when testing bias circuits.

10-16 List possible errors in a *JFET* bias circuit with $V_D \approx V_{DD}$ in the case of (a) gate bias, (b) self bias, (c) voltage divider bias.

10-17 List possible errors in a *JFET* bias circuit with $V_{DS} \approx 0$ in the case of (a) gate bias, (b) self bias, (c) voltage divider bias.

Section 10-7

10-18 Write the equation for calculating R_D in a gate bias circuit.

10-19 Explain why the *FET* maximum drain characteristics should be used when designing a *JFET* bias circuit.

10-20 Write equation for calculating R_D and R_S in a *JFET* self bias circuit.

10-21 Write equation for determining R_D, R_S, R_1 and R_2 in a *JFET* voltage divider bias circuit.

Section 10-8

10-22 Sketch an *n*-channel *JFET* voltage divider bias circuit using a plus-minus supply. Identify the polarities of V_{DD}, V_{DS}, V_S, V_G, and V_{GS}. Briefly explain the circuit operation.

10-23 Repeat Question 10-22 for a similar circuit using a *p*-channel *JFET*.

10-24 Sketch an *n*-channel *JFET* drain feedback bias circuit. Identify all voltage polarities, and explain the circuit operation.

10-25 Repeat Question 10-24 for a drain feedback bias circuit using a *p*-channel *JFET*.

10-26 Sketch an *n*-channel *JFET* constant current bias circuit using voltage divider bias at the gate. Identify all voltage polarities and current directions, and explain the circuit operation.

10-27 Repeat Question 10-26 for a constant current bias circuit using a *p*-channel *JFET*.

10-28 Sketch an *n*-channel *JFET* constant current bias circuit using a plus-minus supply. Identify all voltage polarities and current directions, and explain the circuit operation.

10-29 Repeat Question 10-28 for a constant current bias circuit using a *p*-channel *JFET*.

Section 10-9

10-30 Write the equation used for constructing a *JFET* universal transfer characteristic. Briefly explain the universal transfer characteristic and its application.

Section 10-10

10-31 Sketch *n*-channel *DE-MOSFET* gate bias circuits using (a) $V_{GS} = 0$, (b) $V_{GS} = +V_G$. Identify all voltage polarities and current directions, explain the operation of each circuit.

10-32 Draw the diagram of a self bias circuit using an *n*-channel *DE-MOSFET*. Identify all voltage polarities and current directions, and explain the circuit operation.

10-33 Draw a voltage divider bias circuit using an *n*-channel *DE-MOSFET*. Identify all voltage polarities and current directions, and explain the circuit operation.

10-34 Sketch approximate maximum and minimum transfer characteristics for a *DE-MOSFET*. On the transfer characteristics, draw typical bias lines for the circuits in Questions 10-31 through 10-33. Briefly explain.

10-35 Draw a circuit diagram for a drain-to-gate bias circuit using an *n*-channel *E-MOSFET*. Explain the circuit operation.

10-36 Draw a circuit diagram for a drain-to-gate bias circuit using a *p*-channel *E-MOSFET*. Explain.

Section 10-11

10-37 Sketch a circuit diagram for a direct-coupled switching circuit using an *n*-channel *JFET*. Sketch input and output waveforms, and explain the circuit operation.

10-38 Draw circuit diagrams for normally-*on* and normally-*off* capacitor-coupled *JFET* switching circuits. Explain the operation of each circuit.

10-39 Sketch circuit diagrams for normally-*on* and normally-*off* capacitor-coupled switching circuits using *E-MOSFETs*. Explain the operation of each circuit.

Chapter-10 Problems

Section 10-1

10-1 The circuit in Fig. 10-57 has a *JFET* with the characteristics in Fig. 10-58. Draw the *dc* load line for the circuit, and select a suitable gate voltage to give the maximum possible symmetrical output voltage swing.

10-2 Estimate the maximum possible symmetrical output voltage swing for the circuit in Problem 10-1 when V_G = -2 V.

10-3 A *JFET* circuit like the one in Fig. 10-57 is to have V_{DS} = 10 V and I_D = 2 mA. If V_{DD} = 18 V, draw the *dc* load line on the drain characteristics in Fig. 10-58, and determine the required resistance for R_D.

10-4 Using the drain characteristics in Fig. 10-58, draw the *dc* load line for the circuit in Fig. 10-59.

10-5 Determine the I_D and V_{DS} levels in the circuit in Problem 10-4 if the gate-source voltage is -1.5 V.

10-6 A *JFET* circuit as in Fig. 10-57 has R_D= 4.7 kΩ. The bias conditions are to be V_{DS} = 8 V and I_D = 1.5 mA. Draw the *dc* load line on the drain characteristics in Fig. 10-58, and determine a suitable supply voltage.

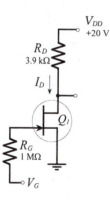

Figure 10-57

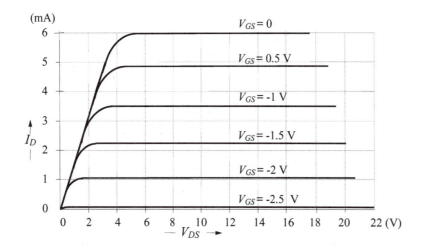

Figure   10-58

Section   10-2

10-7   The gate bias circuit in Fig. 10-60 has a *JFET* with the transfer characteristics in Fig. 10-61. Analyze the circuit to calculate $V_{DS(max)}$ and $V_{DS(min)}$.

10-8   Repeat Problem 10-7 using Eq. 9-1 instead of the transfer characteristics.

10-9   The *JFET* used in a gate bias circuit has the transfer characteristics in Fig. 10-61. If $R_D = 2.2$ kΩ, $V_{DD} = 20$ V, and $V_G = -2$ V, draw the bias line and calculate $V_{DS(max)}$ and $V_{DS(min)}$.

10-10   Repeat Problem 10-9 using Eq. 10-8 instead of the transfer characteristics.

10-11   Analyze the circuit in Fig. 10-62 to determine $V_{DS(max)}$ and $V_{DS(min)}$.

Section   10-3

10-12   Determine the maximum and minimum levels of V_{DS} for the *JFET* self bias circuit in Fig. 10-63. Assume that the device has the transfer characteristics in Fig. 10-61.

10-13   A *JFET* self bias circuit (as in Fig. 10-63) has $V_{DD} = 30$ V, $R_D = 4.7$ kΩ, $R_S = 820$ Ω, and $R_G = 1$ MΩ. If the *FET* is a *2N5457* calculate the minimum level of V_{DS}.

10-14   Recalculate $V_{DS(min)}$ for the circuit in Problem 10-13 when the device is changed to a *2N5458*.

Section   10-4

10-15   Assuming that the *JFET* in the voltage divider bias circuit in Fig. 10-64 has the transfer characteristics in Fig. 10-61, determine the maximum and minimum V_{DS} levels.

10-16   A *JFET* voltage divider bias circuit (as in Fig. 10-64) has $V_{DD} = 20$ V, $R_D = R_S = 2.7$ kΩ, $R_1 = 7.7$ MΩ, and $R_2 = 1$ MΩ. Using

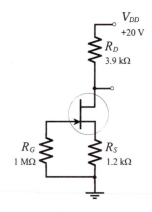

Figure   10-59

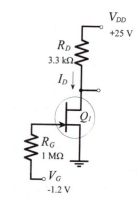

Figure   10-60

the *JFET* transfer characteristics in Fig. 10-61, determine the maximum and minimum V_{DS} levels.

10-17 A *JFET* with the transfer characteristics in Fig. 10-61 is connected in a voltage divider bias circuit with V_{DD} = 22 V. The circuit components are: $R_D = R_S$ = 3.9 kΩ, R_1 = 7.8 MΩ, and R_2 = 1 MΩ. Calculate $V_{DS(max)}$ and $V_{DS(min)}$.

10-18 A *2N5457 JFET* is used in a voltage divider bias circuit with V_{DD} = 20 V. The circuit components are: R_D = 4.7 kΩ, R_S = 3.9 kΩ, R_1 = 1.2 MΩ, and R_2 = 300 kΩ. Determine $V_{DS(min)}$.

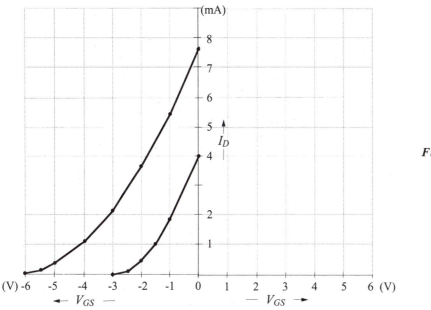

Figure   10-61

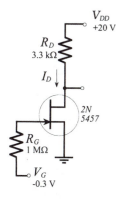

Figure   10-62

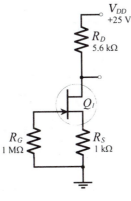

Figure   10-63

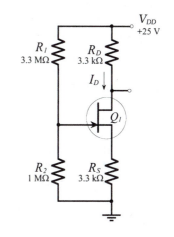

Figure   10-64

Section 10-7

10-19 A *JFET* gate bias circuit is to have $I_{D(max)}$ = 5.5 mA and $V_{DS(min)}$ = 7 V. The supply voltage is 25 V, and the *FET* transfer characteristics are shown in Fig. 10-61. Determine the required bias voltage and suitable resistor values.

10-20 Design a gate bias circuit using a *2N5457 JFET*. The circuit is to have $I_{D(max)}$ = 4.5 mA and $V_{DS(min)}$ = 7.5 V, and the supply voltage is to be 20 V.

10-21 A *JFET* self bias circuit is to have V_{DD} = 25 V, $I_{D(max)}$ = 2.5 mA, and $V_{DS(min)}$ = 7 V. Design the circuit using the *FET* transfer characteristics shown in Fig. 10-61.

10-22 Design a self bias circuit using a *2N5457 JFET* with V_{DD} = 20 V, $I_{D(max)}$ = 2 mA, and $V_{DS(min)}$ = 7.5 V.

10-23 A *JFET* voltage divider bias circuit is to have V_{DD} = 25 V, $I_{D(max)}$ = 2.5 mA, and $V_{DS(min)}$ = 7 V. Design the circuit using the device transfer characteristics shown in Fig. 10-61.

10-24 Design a *JFET* voltage divider bias circuit to have V_{DD} = 20 V, $I_{D(max)}$ = 2 mA, and $V_{DS(min)}$ = 7.5 V. Use a *2N5457 JFET*.

10-25 A *JFET* voltage divider bias circuit is to have V_{DD} = 22 V, $I_{D(max)}$ = 1.5 mA, and $V_{DS(min)}$ = 9 V. Design the circuit using the device transfer characteristics in Fig. 10-61.

10-26 A *JFET* with the device transfer characteristics in Fig. 10-61 is used in a voltage divider bias circuit with V_{DD} = 20 V and R_D = 4.7 kΩ. Determine suitable values for R_1, R_2, and R_S to maintain I_D within the limits of 1 mA to 1.3 mA.

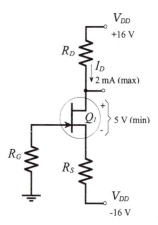

Figure   10-65

Section 10-8

10-27 Design the circuit shown in Fig. 10-65 using the transfer characteristics in Fig. 10-61.

10-28 Analyze the circuit designed for Problem 10-27 to determine the levels of $V_{DS(max)}$ and $V_{DS(min)}$.

10-29 A *JFET* circuit like the one in Fig. 10-65 is to have the following voltage and current levels: V_{DD} = 20 V, V_{SS} = -10 V, $I_{D(max)}$ = 1.5 mA, and $V_{DS(min)}$ = 8 V. Design the circuit to use a *2N5458 JFET*. Select suitable standard-value resistors.

10-30 Analyze the circuit designed for Problem 10-29 to determine $V_{DS(min)}$.

10-31 Design the circuit shown in Fig. 10-66 using the *JFET* transfer characteristics in Fig. 10-37. Select suitable standard-value resistors.

10-32 Analyze the circuit designed for Problem 10-31 to determine $V_{DS(max)}$ and $V_{DS(min)}$.

10-33 A *JFET* circuit like the one in Fig. 10-66 is to have the following voltage and current levels: V_{DD} = 22 V, $I_{D(max)}$ = 1.5 mA, and $V_{DS(min)}$ = 9 V. Design the circuit to use a *JFET* with

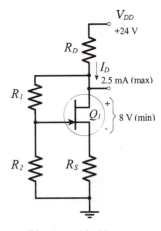

Figure   10-66

the transfer characteristics in Fig. 10-61. Select suitable standard-value resistors.

10-34 Analyze the circuit designed for Problem 10-33 to determine the maximum and minimum levels of V_{DS}.

10-35 Design the *JFET* circuit shown in Fig. 10-67 using the transfer characteristics in Fig. 10-61. Select suitable standard-value resistors.

10-36 Analyze the circuit designed for Problem 10-35 to determine all voltage and current levels.

Section 10-9

10-37 Use the *FET* universal transfer characteristic in Fig. 10-45 to determine $V_{DS(min)}$ for the gate bias circuit in Fig. 10-62.

10-38 Use the *FET* universal transfer characteristic in Fig. 10-45, to determine $V_{DS(min)}$ for the circuit in Fig. 10-63. The actual device transfer characteristics are shown in Fig. 10-61.

10-39 Assuming that a *2N5458 JFET* is connected in the voltage divider bias circuit in Fig. 10-64, determine $V_{DS(min)}$ using the *FET* universal transfer characteristic in Fig. 10-45.

Section 10-10

10-40 The *MOSFET* in the gate bias circuit in Fig. 10-68 has the transfer characteristics in Fig. 10-69. Draw the bias line and calculate $V_{DS(max)}$ and $V_{DS(min)}$.

10-41 Design the *MOSFET* gate bias circuit shown in Fig. 10-70 using the transfer characteristics in Fig. 10-69. Select suitable standard-value resistors.

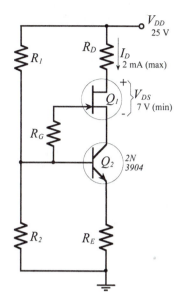

Figure 10-67

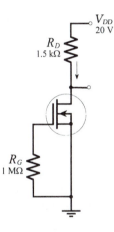

Figure 10-68

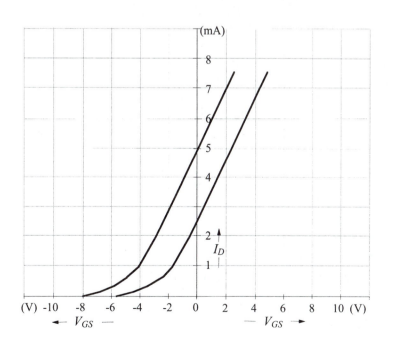

Figure 10-69

10-42   Analyze the circuit designed for Problem 10-41 to determine all voltage and current maximum and minimum levels.

10-43   Design the *MOSFET* voltage divider bias circuit shown in Fig. 10-71 using the transfer characteristics in Fig. 10-69. Select suitable standard-value resistors.

10-44   Analyze the circuit designed for Problem 10-43 to determine all voltage and current maximum and minimum levels.

10-45   Determine suitable resistors for the *MOSFET* drain-to-gate bias circuit in Fig. 10-72 using the transfer characteristics in Fig. 10-52. Calculate the drain-source voltage.

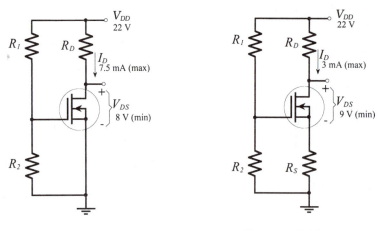

Figure 10-70   Figure 10-71

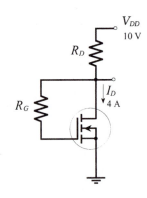

Figure 10-72

Section 10-11

10-46   Calculate V_{DS} levels for the direct-coupled *JFET* switching circuit in Fig. 10-73 when the *FET* is *off*, and when it is *on*. Appendix 1-12 gives the data sheet for the *2N4861*.

10-47   A direct-coupled *JFET* switching circuit (as in Fig. 10-73) is to be designed to use a *2N4861* and V_{DD} = 18 V. V_{DS} is not to exceed 100 mV when the device is *on*. Determine suitable resistor values and a suitable input voltage amplitude.

10-48   The *JFET* switching circuit in Fig. 10-74 is to have V_{DS} ≈ 50 mV when the device is *on*. Select a *FET* from the *2N4856* through *2N4861* range, calculate appropriate resistor values, and determine a suitable input voltage amplitude.

10-49   The switching circuit in Fig. 10-75 is required to pass 5 A through a 5 Ω resistance. Calculate a suitable $r_{DS(on)}$ value for the device. Also, determine a suitable input voltage amplitude using the transfer characteristic in Fig. 10-52.

10-50   The *MOSFET* in the switching circuit in Fig. 10-76 has the transfer characteristic in Fig. 10-52 and $r_{DS(on)}$ = 0.2 Ω. Calculate the required drain current and suitable values for resistors R_1 and R_2.

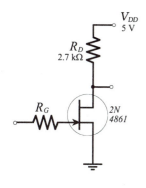

Figure 10-73

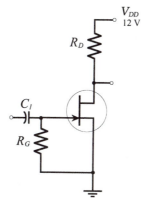

Figure 10-74

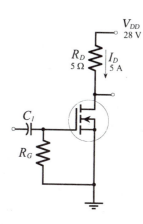

Figure 10-75

Practise Problem Answers

10-1.1	8 V
10-1.2	±5 V
10-2.1	8.4 V, 17.6
10-2.2	12.7 V
10-3.1	8.8 V, 16.1 V
10-4.1	14 V, 18.7 V
10-7.1	-1.7 V, 1 MΩ, 4.7 kΩ
10-7.2	1 MΩ, 3.9 kΩ, 1.5 kΩ
10-7.3	4.7 MΩ, 1 MΩ, 5.6 kΩ, 5.6 kΩ
10-8.1	1 MΩ, 3.3 kΩ, 8.3 kΩ
10-8.2	1 MΩ, 10 kΩ, 5.6 kΩ
10-8.3	(2.2 MΩ + 220 kΩ), 1 MΩ, 2.7 kΩ, 2.7 kΩ
10-9.1	12.4 V
10-9.2	10 V
10-9.3	14.1 V
10-10.1	9.6 V, 12.3 V
10-10.2	12 Ω
10-11.1	3.3 kΩ, 0.6 V, -5 V
10-11.3	48 V, 8.6 V

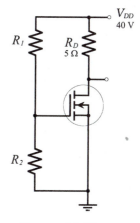

Figure 10-76

Chapter *11*

AC Analysis of FET Circuits

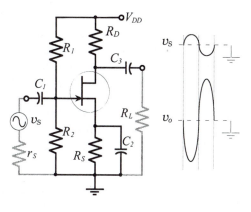

Chapter Contents

Objectives

You will be able to:

1   *Explain the need for coupling and bypass capacitors in FET circuits, and draw ac equivalent circuits for circuits containing capacitors.*

2   *Draw ac load lines for basic FET circuits.*

3   *Sketch ac equivalent circuits for various FET common-source, common-drain, and common-gate circuits.*

4   *Analyze various CS, CD, and CG FET circuits to determine input resistance, output resistance, and voltage gain.*

5   *Compare the performance of CS, CD, and CG circuits, and compare FET and BJT circuits.*

6   *Calculate the cutoff frequency for FET circuits.*

Introduction

Like *BJT* circuits, *FET* amplifiers have *ac* signals capacitor-coupled to the circuit input, and loads capacitor-coupled to the output terminals. Bypass capacitors are also employed both in *FET* and *BJT* circuits to ensure maximum *ac* voltage gains. Because of the *ac*-coupled load and the *ac* bypassed components, the *ac* load for a given circuit is different from the *dc* load, consequently, the *ac* load line is different from the *dc* load line.

There are three basic *FET* circuit configurations; common-source, common-drain, and common-gate. These are similar to the *BJT* common-emitter, common-collector, and common-base circuits, respectively. The common-source circuit is capable of voltage amplification, and it has a high input impedance. The common-drain circuit is a buffer amplifier, with a voltage gain of approximately one, high input impedance, and low output impedance. The common-gate circuit has voltage gain, low input impedance, and good high-frequency performance.

11-1   Coupling, Bypassing, and AC Load Lines

Coupling Capacitors

This subject is treated for *BJT* circuits in Section 6-1, and the discussion there applies equally to *FET* circuits. As explained, coupling capacitors are required at a circuit input to couple a signal source to the circuit without affecting the bias conditions. Similarly, loads are capacitor-coupled to the circuit output to avoid the change in bias conditions produced by direct coupling. Input and output coupling capacitors (C_1 and C_3) are shown in the *FET* circuit in Fig. 11-1.

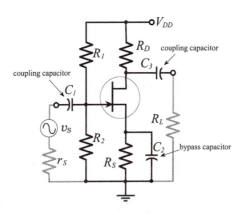

Figure 11-1
FET circuit with coupling and bypass capacitors.

Bypassing Capacitors

Bypass capacitors are also just as necessary in *FET* circuits as in *BJT* circuits. Bypass capacitor C_2 in Fig. 11-1 provides an *ac* short-circuit across resistor R_S. As will be shown, if C_2 is not present R_S substantially reduces the *ac* voltage gain of the circuit.

Figure 11-2 illustrates another situation where a bypassing capacitor is required. The *MOSFET* drain-to-gate bias circuit shown would have its voltage gain reduced by feedback from the drain to the gate via R_G (*ac* degeneration) if capacitor C_2 were not present. Splitting R_G into two equal resistors and *ac* shorting the junction to ground via C_2 eliminates the feedback.

AC Load Lines

Once again, this is a subject that applies equally to *BJT* and *FET* circuits, (see Section 6-2). The *dc* load for the *FET* circuit in Fig. 11-1 is $(R_D + R_S)$. With R_S *ac* bypassed and R_L absent, the *ac* load is R_D. With the capacitor-coupled load present in Fig. 11-1, the *ac* load is $R_D \| R_L$. The *dc* load line is drawn in the usual way, and the Q-point is marked, then the *ac* load line is drawn through the Q-point, as shown in Fig. 11-3.

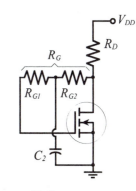

Figure 11-2
MOSFET circuit with bypass capacitor to eliminate ac degeneration.

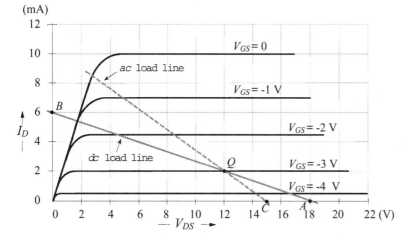

Figure 11-3
FET circuit dc and ac load lines.

Example 11-1

Draw the *dc* and *ac* load lines for the transistor circuit in Fig. 11-4 using the transistor common-emitter characteristics in Fig. 11-3, and identifying the Q-point current as $I_D = 2$ mA.

Solution

dc load line,

$$R_{L(dc)} = R_D + R_S = 1.5 \text{ k}\Omega + 1.5 \text{ k}\Omega$$
$$= 3 \text{ k}\Omega$$

The equation for V_{DS} is, $V_{DS} = V_{DD} - I_D (R_D + R_S)$

When $I_D = 0$, $V_{DS} = V_{DD} = 18$ V

Plot point A at, $I_D = 0$ and $V_{DS} = 18$ V

When $V_{DS} = 0$, $I_D = \dfrac{V_{DD}}{R_D + R_S} = \dfrac{18 \text{ V}}{3 \text{ k}\Omega}$

$$= 6 \text{ mA}$$

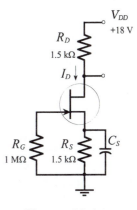

Figure 11-4
Circuit for Ex. 11-1.

Plot point B at, $I_D = 6$ mA and $V_{DS} = 0$
Draw the dc load line through points A and B.
Mark the Q-point on the dc load line at $I_D = 2$ mA.

ac load line,

$$R_{L(ac)} = R_D = 1.5 \text{ k}\Omega \text{ (when } R_S \text{ is bypassed)}$$

When I_D changes by $\quad \Delta I_D = 2$ mA,

$$\Delta V_{DS} = \Delta I_D \times R_D = 2 \text{ mA} \times 1.5 \text{ k}\Omega$$
$$= 3 \text{ V}$$

Plot point C at, $\Delta I_D = 2$ mA and $\Delta V_{DS} = 3$ V from the Q point.
Draw the ac load line through points C and Q.

Practise Problems

11-1.1 The circuit in Fig. 11-1 has $R_D = 2.2$ kΩ, $R_S = 1.8$ kΩ, $R_L = 15$ kΩ,
and $V_{DD} = 20$ V. Determine the dc and ac load resistances, and
draw both load lines on the characteristic in Fig. 11-3. Assume that
the Q-point drain current is 2.5 mA.

11-2   FET Models and Parameters

FET Equivalent Circuit

The complete equivalent circuit for a field effect transistor is shown
in Fig. 11-5(a). It is seen that the source terminal is common to
both input and out, so this is a *common-source* equivalent circuit.
Resistor R_{GS} between the gate and source terminals is the
resistance of the reverse-biased gate-source junction, and C_{gs} is
the junction capacitance. So, a signal applied to the input *sees*
R_{GS} in parallel with C_{gs}.

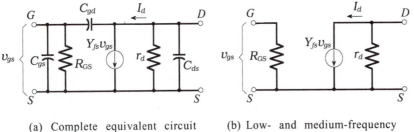

(a) Complete equivalent circuit   (b) Low- and medium-frequency
                                             equivalent circuit

Figure 11-5
*The complete equivalent circuit
for a FET includes capacitances
between terminals. For low- and
medium-frequency operations,
the capacitances are omitted.*

The output stage of the equivalent circuit is represented as a
current source ($Y_{fs} v_{gs}$) supplying current to the drain resistance
(r_d). Y_{fs} is the forward transfer admittance for the FET, and v_{gs} is
the ac signal voltage developed across the gate-source terminals,
so the ac drain current is ($Y_{fs} v_{gs}$). The drain-source capacitance
(C_{ds}) appears in parallel with r_d, and the gate-drain capacitance
(C_{gd}) is shown connected between the input and output stages.

For low- and medium-frequency operations the capacitances can be neglected, and the equivalent circuit is then as shown in Fig. 11-5(b). This is the *FET model* (or equivalent circuit) normally used in *ac* circuit analysis.

Equivalent Circuit Parameters

The parameters used in the equivalent circuit are discussed in Section 9-3. As explained, R_{GS} is a junction reverse resistance with a 10^9 Ω typical value for a *JFET*. Because its resistance is so high, R_{GS} is often regarded as an open circuit. Instead of R_{GS} being listed on a device data sheet, the gate-source reverse current is usually specified, and a value for R_{GS} can be calculated from this quantity.

The forward transfer admittance (Y_{fs} or g_{fs}) varies widely for different types of *FET*. For a small-signal or switching *JFET*, Y_{fs} typically ranges from 1000 μS to 5000 μS. For an *E-MOSFET*, Y_{fs} might be 2.9 S.

The drain resistance (r_d) shown in the equivalent circuit is the *ac* resistance offered between the drain and source terminals of the *FET* when operating in the pinch-off region of the drain characteristics. This quantity is usually defined in terms of an output admittance (Y_{os}) which equals $1/r_d$. Typical values of r_d range from 20 kΩ to 100 kΩ for a *JFET*.

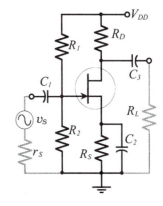

Figure 11-6
FET common-source amplifier circuit.

11-3 Common Source Circuit Analysis

Common Source Circuit

The circuit of a *FET* common-source amplifier is shown in Fig. 11-6. With the capacitors treated as *ac* short-circuits, the circuit input terminals are the gate and source, and the output terminals are the drain and the source. So, the source terminal is *common* to both input and output, and the circuit configuration is known as *common-source* (*CS*).

The current and voltage waveforms for the *CS* circuit in Fig. 11-6 are illustrated in Fig. 11-7. The 180° *phase shift* between the input and output waveforms can be understood by considering the effect of a positive-going input signal. An increase in v_s increases the *FET* gate-source voltage (V_{GS}), thus raising the level of I_D, and increasing the voltage drop across R_D. This produces a decrease in the level of the drain voltage (V_D), which is capacitor-coupled to the circuit output as a negative-going *ac* output voltage (v_o). Consequently, as v_s increases in a positive direction, v_o changes in a negative direction, as illustrated. Conversely, when v_s changes in a negative direction, the resultant decrease in V_{GS} reduces I_D and produces a positive-going output.

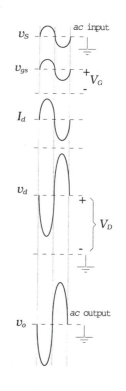

Figure 11-7
Voltage and current waveforms in a common-source circuit.

The circuit in Fig. 11-6 has an *input impedance* (Z_i), an *output impedance* (Z_o), and a *voltage gain* (A_v). Equations for these quantities can be determined by *ac* analysis of the circuit.

Common Source Equivalent Circuit

The first step in *ac* analysis of a *FET* (or *BJT*) circuit is to draw the *ac* equivalent circuit, by substituting *ac* short-circuits in place of the power supply and capacitors. When this is done for the circuit in Fig. 11-6, the *ac* equivalent circuit shown in Fig. 11-8(a) is created. For *ac* analysis, the *FET* equivalent circuit [from Fig. 11-5(b)] is now substituted in place of the device. This results in the CS *ac* equivalent circuit in Fig. 11-8(b).

The current directions and voltage polarities in Fig. 11-8(b) are those that occur when the instantaneous level of the input voltage is moving in a positive direction.

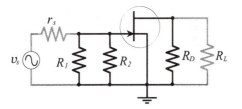

(a) *ac* equivalent circuit for *FET* CS circuit

Figure 11-8
A common-source ac equivalent circuit is drawn by first replacing the power supply and all capacitors with short-circuits, then the FET model is substituted in place of the device.

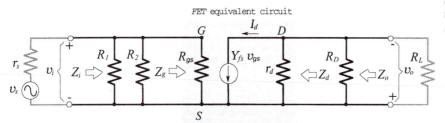

(b) Substitution of the *FET* equivalent circuit into the *CS* ac equivalent circuit

Input Impedance

Considering the input section of the equivalent circuit in Fig. 6-8(b) it is seen that the input impedance at the *FET* terminals is,

$$Z_g = R_{gs} \tag{11-1}$$

As already discussed, a typical value of R_{gs} for a low-current *JFET* is $10^9 \ \Omega$. At the circuit input terminals, resistors R_1 and R_2 are seen to be in parallel with Z_g. So, the circuit input impedance is,

$$Z_i = R_1 \| R_2 \| Z_g \tag{11-2}$$

Because Z_g is so large, the circuit input impedance is almost always determined by the bias resistors.

$$Z_i \approx R_1 \| R_2 \tag{11-3}$$

So, the input impedance of a *FET* circuit can usually be determined simply by calculating the equivalent resistance of the bias resistors. However, as already discussed, very high input resistance is one of the most important properties of a *FET*. In circumstances where it is necessary to achieve the highest possible input resistance, the signal source might be directly connected in series with the gate terminal, and in this case $Z_i \approx Z_g$.

In the case of the self biased circuit in Fig. 11-4, the circuit input impedance is equal to the gate bias resistor R_G.

Output Impedance

Looking into the output of the *CS* equivalent circuit in Fig. 11-8(b), the output impedance at the *FET* drain terminal is,

$$Z_d = r_d \qquad\qquad (11\text{-}4)$$

At the circuit output terminals, resistor R_D is in parallel with Z_d. So, the circuit output impedance is,

$$Z_o = R_D || r_d \qquad\qquad (11\text{-}5)$$

Because r_d is typically 100 kΩ, and R_D is usually much lower than 50 kΩ, the circuit output impedance is approximately equal to R_D. Using this information, the output impedance of a *CS* circuit can be discovered simply by reading the resistance of R_D.

$$Z_o \approx R_D \qquad\qquad (11\text{-}6)$$

Voltage Gain

Amplifier voltage gain is given by the equation,

$$A_v = \frac{v_o}{v_i}$$

From Fig. 11-8(b), $v_o = I_d\,(r_d||R_D||R_L)$

and $I_d = -Y_{fs}\,v_i$

so, $A_v = \dfrac{-Y_{fs}\,v_i\,(r_d||R_D||R_L)}{v_i}$

or, $A_v = -Y_{fs}(r_d||R_D||R_L) \qquad\qquad (11\text{-}7)$

The minus sign in Eq. 11-7 indicates that v_o is 180° out of phase with v_i. (When v_i increases, v_o decreases, and vice versa.) It is seen that the voltage gain of a common source amplifier is directly proportional to the Y_{fs} of the *FET*. If the appropriate Y_{fs} value and the resistance of R_D and R_L are known, the voltage gain of a *CS* circuit can be quickly estimated. For Y_{fs} = 5000 µS, R_D = 4.7 kΩ, and with $R_L \gg R_D$, and $r_d \gg R_D$, the voltage gain is -23.5.

The voltage gain of a *FET* common-source amplifier is typically about one tenth of the gain of a *BJT* common-emitter circuit. Low voltage gain, caused by the low Y_{fs} value, is the major disadvantage of *JFET* circuits compared to *BJT* circuits. The same is generally true of *MOSFET* circuits, except for *E-MOSFET* devices (*V-MOS* and *T-MOS*), which have much larger Y_{fs} values. However, *E-MOSFETs* use relatively large drain currents, and they are normally unsuitable for small-signal applications.

Summary of Typical CS Circuit Performance

Device input impedance	$Z_g = R_{gs}$
Circuit input impedance	$Z_i = R_1 \| R_2 \| Z_g \approx R_1 \| R_2$
Device output impedance	$Z_d = r_d$
Circuit output impedance	$Z_o = R_D \| r_d \approx R_D$
Circuit voltage gain	$A_v = -Y_{fs}(r_d \| R_D \| R_L)$

The common-source circuit has voltage gain, 180° phase shift, high input impedance, and relatively high output impedance.

Example 11-2

Calculate the input impedance, output impedance, and voltage gain for the common-source circuit in Fig. 11-9 using the *FET* typical parameters.

Solution

From the *2N5457* data sheet,

$$r_d = \frac{1}{Y_{os}} = \frac{1}{10\,\mu S}$$
$$= 100\ k\Omega$$

and, $Y_{fs} = 3000\,\mu S$

Eq. 11-3, $Z_i = R_1 \| R_2 = 1\ M\Omega \| 5.6\ M\Omega$
$= 848\ k\Omega$

Eq. 11-5, $Z_o = R_D \| r_d = 2.7\ k\Omega \| 100\ k\Omega$
$= 2.63\ k\Omega$

Eq. 11-7, $A_v = -Y_{fs}(r_d \| R_D \| R_L) = -3000\,\mu S\,(100\ k\Omega \| 2.7\ k\Omega)$
$= -7.9$

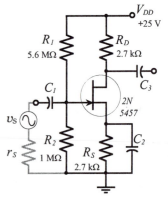

Figure 11-9
Circuit for Ex. 11-2.

Practise Problems

11-3.1   Calculate the typical input impedance, output impedance, and voltage gain for the circuit in Fig. 11-4, if the *FET* is a 2N5458.

11-3.2   A circuit similar to the one in Fig. 11-4 has R_G = 470 kΩ, R_D = 3.9 kΩ, and a *2N5457 FET*. Determine the typical input impedance, output impedance, and voltage gain for the circuit.

11-4 CS Circuit with Unbypassed Source Resistor

Equivalent Circuit

When an unbypassed source resistor (R_S) is present in a FET common-source circuit, as shown in Fig.11-10(a), it also appears in the ac equivalent circuit, [Fig. 11-10(b)]. In the complete equivalent circuit R_S must be shown connected between the FET source terminal and the circuit common input-output terminal, (Fig. 11-11). As with the previous equivalent circuit, the current directions and voltage polarities shown in Fig. 11-11 are those that occur when the instantaneous input voltage is positive-going. The presence of R_S without a bypass capacitor significantly affects the circuit voltage gain.

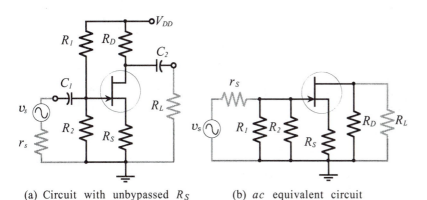

(a) Circuit with unbypassed R_S (b) ac equivalent circuit

Figure 11-10
FET common-source circuit with unbypassed source resistor, and ac equivalent circuit.

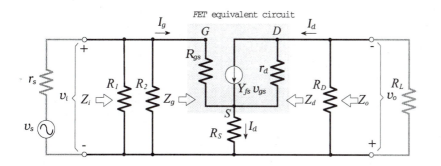

Figure 11-11
An unbypassed source resistor in a FET circuit must be included in the complete ac equivalent circuit.

Input Impedance

An equation for the input impedance at the FET gate can be determined from v_i and I_g. From Fig. 11-11,

$$v_i = v_{gs} + I_d R_S = v_{gs} + Y_{fs} v_{gs} R_S$$
$$= v_{gs}(1 + Y_{fs}R_S)$$

and, $$I_g = \frac{v_{gs}}{R_{gs}}$$

so,
$$Z_g = \frac{v_i}{I_g} = \frac{v_{gs}(1 + Y_{fs}R_S)}{v_{gs}/R_{gs}}$$

or,
$$\mathbf{Z_g = R_{gs}(1 + Y_{fs}R_S)} \tag{11-8}$$

Equation 11-8 gives the input impedance at the *FET* gate terminal. The circuit input impedance is again given by Eq. 11-2,

$$Z_i = R_1 \| R_2 \| Z_g$$

In this case, Z_g is much larger than $R_1 \| R_2$, so the circuit input impedance is determined by the gate bias resistors.

Eq. 11-3, $Z_i \approx R_1 \| R_2$

Output Impedance

To calculate the circuit output impedance, the *ac* signal voltage (v_s) is assumed to be zero, and an *ac* voltage (v_o) is applied at the output, (see Fig. 11-12(a)). The *ac* output current (I_d) is calculated in terms of v_o, then Z_o is determined as v_o divided by I_d. Figure 11-12(a) shows that when v_s is zero, the *ac* voltage across source resistor R_S is applied as a gate-source voltage.

$$v_{gs} = I_d R_S$$

Actually, $I_d R_S$ is divided across $r_s \| R_G$ and R_{gs}. However, $R_{gs} \gg r_s \| R_G$, so that all of $I_d R_S$ is effectively applied as a gate-source voltage. The v_{gs} produced in this way generates an *ac* drain current which opposes the drain current produced by v_o,

$$I = -Y_{fs}v_{gs} = -Y_{fs}(I_d R_S)$$

Converting the current generator [$(Y_{fs}v_{gs})$ in parallel with r_d] to a voltage generator [$(Y_{fs}v_{gs}r_d)$ in series with r_d] gives the equivalent circuit in Fig. 11-12(b). The equation for the voltage drops around this circuit is,

$$I_d(r_d + R_S) = v_o - (Y_{fs}I_d R_S r_d)$$

giving,
$$v_o = I_d(r_d + R_S + Y_{fs}R_S r_d)$$

and,
$$Z_d = \frac{v_o}{I_d} = r_d + R_S + Y_{fs}R_S r_d$$

or,
$$\mathbf{Z_d = r_d + R_S + Y_{fs}R_S r_d} \tag{11-9}$$

The circuit output impedance is,

Eq. 11-5, $Z_o = R_D \| Z_d$

Because the output impedance at the *FET* drain terminal is much larger than the drain resistor (R_D), the output impedance of the circuit with the unbypassed source resistor is still,

Eq. 11-6, $Z_o \approx R_D$

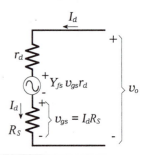

(a) Circuit output stage with input shorted

(b) Current source replaced with a voltage source

Figure 11-12
Analysis of the output impedance of a common-source circuit with an unbypassed source resistor.

Voltage Gain

From the derivation of the input impedance equation,

$$v_i = v_{gs}(1 + Y_{fs}R_S)$$

and neglecting r_d,

$$v_o = I_d(R_D \| R_L) = -Y_{fs}v_{gs}(R_D \| R_L)$$

so,

$$A_v = \frac{v_o}{v_i} = \frac{-Y_{fs}v_{gs}(R_D \| R_L)}{v_{gs}(1 + Y_{fs}R_S)}$$

giving,

$$A_v = \frac{-Y_{fs}(R_D \| R_L)}{1 + Y_{fs}R_S} \qquad \text{(11-10)}$$

Usually, $Y_{fs}R_S \gg 1$

so,

$$A_v \approx \frac{-(R_D \| R_L)}{R_E} \qquad \text{(11-11)}$$

The voltage gain of a *FET* common-source circuit with an unbypassed source resistor can be quickly estimated using Eq. 11-11. For the circuit in Fig. 11-10(a), with $R_D = 4.7$ kΩ, $R_S = 2.2$ kΩ, and $R_L \gg R_D$, $A_v \approx -2.1$.

Summary of Performance of CS Circuit with Unbypassed R_S

Device input impedance	$Z_g = R_{gs}(1 + Y_{fs}R_S)$
Circuit input impedance	$Z_i = R_G \| Z_g \approx R_1 \| R_2$
Circuit output impedance	$Z_o \approx R_D$
Circuit voltage gain	$A_v = \dfrac{-Y_{fs}(R_D \| R_L)}{1 + Y_{fs}R_S}$
Circuit voltage gain	$A_v \approx \dfrac{-(R_D \| R_L)}{R_S}$

The most significant feature of the performance of a *CS* circuit with an unbypassed source resistor is that its voltage gain is much lower than that for a *CS* circuit with R_S bypassed.

Example 11-3

For the common-source circuit in Fig. 11-13, calculate the gate input impedance, the drain output impedance, the circuit input and output impedances, and the voltage gain. Use the typical parameters for the *FET*.

Solution

From the *2N5458* data sheet,

$$r_d = \frac{1}{Y_{os}} = 100 \text{ k}\Omega$$

$$Y_{fs} = 4000 \,\mu S$$

and,

$$I_G = 1 \text{ nA with } V_{GS} = 15 \text{ V}$$

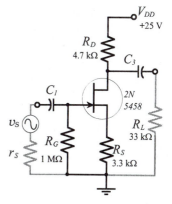

Figure 11-13
Circuit for Example 11-3.

so, $R_{gs} = \dfrac{V_{GS}}{I_G} = \dfrac{15\text{ V}}{1\text{ nA}}$

$= 15 \times 10^9\ \Omega$

Eq. 11-8, $Z_g = R_{gs}(1 + Y_{fs}R_S) = 15 \times 10^9\ \Omega[1 + (4\text{ mS} \times 3.3\text{ k}\Omega)]$

$= 2.13 \times 10^{11}\ \Omega$

from Eq. 11-3, $Z_i = R_G = 1\text{ M}\Omega$

Eq. 11-9, $Z_d = r_d + R_S + (Y_{fs}R_S r_d)$

$= 100\text{ k}\Omega + 3.3\text{ k}\Omega + (4\text{ mS} \times 3.3\text{ k}\Omega \times 100\text{ k}\Omega)$

$\approx 1.4\text{ M}\Omega$

Eq. 11-5, $Z_o = R_D\|Z_d = 4.7\text{ k}\Omega\|1.4\text{ M}\Omega$

$= 4.68\text{ k}\Omega$

Eq. 11-10, $A_v = \dfrac{-Y_{fs}\ (R_D\|R_L)}{1 + Y_{fs}R_S} = \dfrac{-4\text{ mS} \times (4.7\text{ k}\Omega\|33\text{ k}\Omega)}{1 + (4\text{ mS} \times 3.3\text{ k}\Omega)}$

$= -1.16$

or, Eq. 11-11, $A_v \approx \dfrac{-(R_D\|R_L)}{R_S} = \dfrac{4.7\text{ k}\Omega\|33\text{ k}\Omega}{3.3\text{ k}\Omega}$

$= -1.24$

Practise Problems

11-4.1   Calculate the gate input impedance, drain output impedance, the circuit input and output impedance, and the voltage gain for the circuit in Fig. 11-4, when the source bypass capacitor is removed. Assume that the *FET* is a *2N5458*.

11-4.2   Determine the items listed in Problem 11-4.1 for the circuit in Fig. 11-9 when the source bypass capacitor is removed.

11-5   Common Drain Circuit Analysis

Common Drain Circuit

The *FET* common-drain circuit shown in Fig. 11-14 has the output voltage developed across the source resistor (R_S). The external load (R_L) is capacitor-coupled to the source terminal of the *FET*, and the gate bias voltage (V_G) is derived from V_{DD} by means of voltage divider resistors R_1 and R_2. No resistor is connected in series with the drain terminal, and no source bypass capacitor is employed.

To understand the operation of the circuit in Fig. 11-14 note that the *dc* gate voltage (V_G) is a constant quantity, and the source voltage is,

$$V_S = V_G + V_{GS}$$

When an *ac* signal is applied to the gate via capacitor C_1, the gate

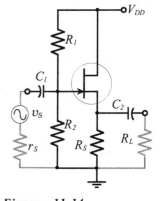

Figure 11-14
Common-drain circuit, also called a source follower. The ac input is applied to the FET gate, and the output is taken from the source terminal.

voltage is increased and decreased as the instantaneous level of the signal voltage rises and falls. Also, V_{GS} remains substantially constant, so the source voltage increases and decreases with the gate voltage. (See the waveforms in Fig. 11-15.) Thus, the *ac* output voltage is closely equal to the *ac* input voltage, and the circuit can be said to have unity gain. AC analysis of the circuit shows that v_o is slightly smaller than v_i. Because the output voltage at the source terminal follows the signal voltage at the gate, the common-drain circuit is also known as a *source follower*.

Common Drain Equivalent Circuit

As in the case of other circuits, the supply voltage and coupling capacitors in Fig. 11-14 must be replaced with short-circuits in order to study the circuit *ac* performance. This gives the common-drain *ac* equivalent circuit in Fig. 11-16(a). The input terminals of the *ac* equivalent circuit are seen to be the *FET* gate and drain, and the output terminals are the source and drain. Because the drain terminal is common to both input and output, the circuit configuration is named *common-drain (CD)*.

The completer *CD ac* equivalent circuit is drawn by substituting the *FET* model into the *ac* equivalent circuit, [Fig. 11-16(b)]. The indicated current directions and voltage polarities are, once again, those that are produced by a positive-going signal voltage.

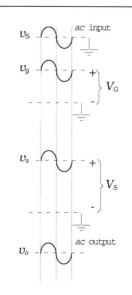

Figure 11-15
Voltage waveforms in a common-drain circuit.

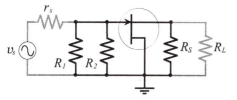

(a) *ac* equivalent circuit for *FET CD* circuit

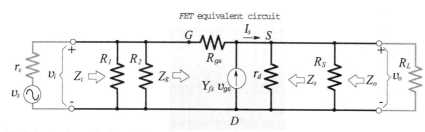

(b) Substitution of the *FET* equivalent circuit into the *CD* ac equivalent circuit

Figure 11-16
A common-drain ac equivalent circuit is drawn by first replacing the power supply and all capacitors with short-circuits, then the FET model is substituted in place of the device.

Input Impedance

Looking into the gate and drain terminals in Fig. 11-16(b), is similar to *looking into* the input of a common-source amplifier with an unbypassed source resistor. Therefore, the common-drain Z_g equation is derived in the same way as Eq. 11-8. One important difference is that the common-drain circuit has the external load resistor (R_l) in parallel with R_S.

$$Z_g = R_{gs}[1 + Y_{fs}(R_S \| R_L)] \qquad (11\text{-}12)$$

As discussed in Section 11-4, the equation for Z_g gives a very large value. This is important only when the signal source is connected directly in series with the gate terminal. With the signal capacitor-coupled to the input, the circuit input impedance is,

Eq. 11-2, $Z_i = R_1 \| R_2 \| Z_g$

which reduces to,

Eq. 11-3, $Z_i \approx R_1 \| R_2$

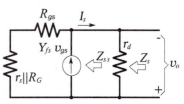

Output Impedance

In a common-drain circuit, any variation in the output voltage (v_o) has an effect on the *FET* gate-source voltage. To determine the impedance *looking into* the source terminal, the signal voltage [in Fig. 11-16(b)] is assumed to be zero, and the output current (I_s) is calculated in terms of v_o. The circuit is redrawn in Fig. 11-17 with resistor R_G representing the gate bias resistors, $[R_G = R_1 \| R_2$ in Fig. 11-16(a)]. With $v_s = 0$, v_o is divided across $R_G \| r_s$ and R_{gs}, (see Fig. 11-17). Because $R_{gs} \gg (R_G \| r_s)$, effectively all of v_o is applied as an *ac* gate-source voltage. So,

$$I_s = Y_{fs} v_o$$

and, $Z_{ss} = \dfrac{v_o}{I_s} = \dfrac{v_o}{Y_{fs} v_o}$ (see Fig. 11-17)

or, $Z_{ss} = 1/Y_{fs}$

also, $Z_s = Z_{ss} \| r_d$

so, $\boldsymbol{Z_s = (1/Y_{fs}) \| r_d}$ **(11-13)**

Figure 11-17
Analysis of the output impedance of a common-drain circuit.

Equation 11-13 gives the device output impedance. The circuit output impedance also involves R_S.

$$\boldsymbol{Z_o = (1/Y_{fs}) \| R_S \| r_d}$$ **(11-14)**

The drain resistance is usually much larger than R_S, and R_S is usually much larger than $1/Y_{fs}$, so that $Z_o \approx 1/Y_{fs}$. For $Y_{fs} = 5000$ µS, $Z_o = 200\ \Omega$.

Voltage Gain
From Fig. 11-16(b),

$$v_o = I_s(R_S \| R_L) = Y_{fs} v_{gs}(R_S \| R_L)$$

and, $v_i = v_{gs} + v_o = v_{gs} + Y_{fs} v_{gs}(R_S \| R_L)$

$$= v_{gs}[1 + Y_{fs}(R_S \| R_L)]$$

so, $A_v = \dfrac{v_o}{v_i} = \dfrac{Y_{fs} v_{gs}(R_S \| R_L)}{v_{gs}[1 + Y_{fs}(R_S \| R_L)]}$

or, $\boldsymbol{A_v = \dfrac{Y_{fs}(R_S \| R_L)}{1 + Y_{fs}(R_S \| R_L)}}$ **(11-15)**

As already discussed, the *ac* output voltage from a *CD* circuit is usually closely equal to the input voltage, so the voltage gain is normally taken as,

$$A_v \approx 1 \qquad\qquad (11\text{-}16)$$

Summary of CD Circuit Performance

Device input impedance	$Z_g = R_{gs}[1 + Y_{fs}(R_S \| R_L)]$
Circuit input impedance	$Z_i \approx R_1 \| R_2$
Device output impedance	$Z_s = (1/Y_{fs}) \| r_d$
Circuit output impedance	$Z_o = (1/Y_{fs}) \| R_S \| r_d \approx (1/Y_{fs}) \| R_S$
Circuit voltage gain	$A_v = \dfrac{Y_{fs}(R_S \| R_L)}{1 + Y_{fs}(R_S \| R_L)}$
Circuit voltage gain	$A_v \approx 1$

A common-drain circuit has a voltage gain approximately equal to 1, no phase shift between input and output, very high input impedance, and low output impedance. Because of its high Z_i, low Z_o, and unity gain, the *CD* circuit is usually used as a *buffer amplifier* between a high impedance signal source and a low impedance load.

Example 11-4

The common-drain circuit in Fig. 11-18 has a *FET* with $Y_{fs} = 3000\ \mu S$, $R_{gs} = 100\ M\Omega$, and $r_d = 50\ k\Omega$. Determine the input and output impedances at the device terminals, and the circuit input and output impedances. Also, calculate the circuit voltage gain.

Solution

Eq. 11-12,
$$\begin{aligned}Z_g &= R_{gs}[1 + Y_{fs}(R_S \| R_L)]\\ &= 100\ M\Omega[1 + 3000\ \mu S(5.6\ k\Omega \| 12\ k\Omega)]\\ &= 1.25 \times 10^9\ \Omega\end{aligned}$$

Eq. 11-2,
$$\begin{aligned}Z_i &\approx R_1 \| R_2 = 1.5\ M\Omega \| 1\ M\Omega\\ &= 600\ k\Omega\end{aligned}$$

Eq. 11-13,
$$\begin{aligned}Z_s &= (1/Y_{fs}) \| r_d = (1/3000\ \mu S) \| 50\ k\Omega\\ &= 326\ \Omega\end{aligned}$$

Eq. 11-14,
$$\begin{aligned}Z_o &= (1/Y_{fs}) \| R_S \| r_d = 326\ \Omega \| 5.6\ k\Omega \| 50\ k\Omega\\ &= 306\ \Omega\end{aligned}$$

Eq. 11-15,
$$A_v = \frac{Y_{fs}(R_S \| R_L)}{1 + Y_{fs}(R_S \| R_L)} = \frac{3000\ \mu S(5.6\ k\Omega \| 50\ k\Omega)}{1 + [3000\ \mu S(5.6\ k\Omega \| 50\ k\Omega)]}$$
$$= 0.94$$

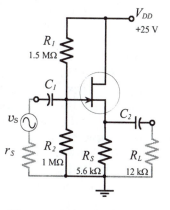

Figure 11-18
Circuit for Example 11-4.

11-6 Common-Gate Circuit Analysis

Common-Gate Circuit

The *FET* common-gate (*CG*) circuit shown in Fig. 11-19 uses voltage divider bias. The *ac* output is taken from the drain terminal, and an external load (R_L) is capacitor-coupled to the drain, exactly as in the case of a common-source circuit. Unlike a CS circuit, the *ac* input for the *CG* circuit is applied to the *FET* source terminal (via C_3), and a capacitor (C_1) is included to *ac* short-circuit the gate to ground. Because the gate is grounded, all of the *ac* input voltage appears across the gate-source terminals.

The current and voltage waveforms for the *CG* circuit shown in Fig. 11-20 illustrate the circuit operation. When the instantaneous level of the *ac* signal voltage increases, the source voltage increases while the gate voltage remains constant. Consequently, the gate-source voltage decreases, the drain current is reduced, and the output voltage at the drain terminal is increased. Similarly, when v_i decreases, v_{gs} increases, I_d increases, and v_d decreases. Thus, an increase in v_i causes an increase in v_o, and a decrease in v_i produces a decrease in v_o; so, there is no phase shift between the input and the output. (See the voltage and current waveforms in Fig. 11-20.)

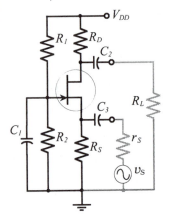

Figure 11-19
In a common-gate circuit the signal is capacitor-coupled to the FET source terminal, and the output is taken from the drain.

Common Gate Equivalent Circuit

As in the case of other circuits, the supply voltage and capacitors in Fig. 11-19 are replaced with short-circuits in order to study the circuit *ac* performance. This gives the common-gate *ac* equivalent circuit in Fig. 11-21(a). The input terminals of the *ac* equivalent circuit are seen to be the *FET* source and gate, and the output terminals are the drain and gate. The gate terminal is common to both input and output, so the circuit is named *common-gate*.

The completer *CG ac* equivalent circuit is drawn by substituting the *FET* model into the *ac* equivalent circuit, [Fig. 11-21(b)]. As always, the indicated current directions and voltage polarities are those that are produced by a positive-going signal voltage.

Input Impedance

'Looking into' the source and gate terminals of the *CG* circuit in Fig. 11-21(b), is similar to 'looking into' the output of a common-drain circuit. Therefore, the equation for the common-gate device input impedance (Z_s) is derived in the same way as the equation for Z_s in the common-drain circuit.

$$Z_s = 1/Y_{fs} \qquad\qquad (11\text{-}17)$$

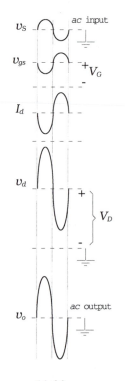

Figure 11-20
Voltage and current waveforms in a common-gate circuit.

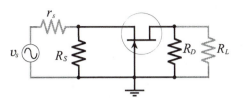

(a) *ac* equivalent circuit for *FET CG* circuit

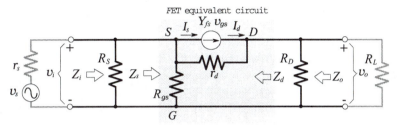

(b) Complete common-gate *ac* equivalent circuit

Figure   11-21
A common-gate ac equivalent circuit is drawn by first replacing the power supply and all capacitors with short-circuits, then the FET model is substituted in place of the device.

The circuit input impedance is R_S in parallel with the device input impedance.

so, $Z_i = (1/Y_{fs})\|R_S$ **(11-18)**

Using a typical Y_{fs} of 5000 µS, the input impedance of a common-gate circuit is around 200 Ω. Low input impedance is the major disadvantage of the common-gate circuit, because the signal voltage is divided across r_s and Z_s, (see Fig. 11-22).

$$v_i = \frac{v_s \times Z_i}{r_s + Z_i}$$ **(11-19)**

Output Impedance
The output of a common-gate circuit is taken from the drain terminal, as in the case of a common-source circuit. So, the common-gate output impedance is the same as the common-drain output impedance. At the drain terminal,

Eq. 11-4, $Z_d = r_d$

and the circuit output impedance is,

Eq. 11-5, $Z_o = R_D\|r_d$

Voltage Gain
From Fig. 11-21(b),

$$v_o = I_d(R_D\|R_L) = Y_{fs}v_i(R_D\|R_L)$$

and, $A_v = \dfrac{v_o}{v_i} = \dfrac{Y_{fs}v_i(R_D\|R_L)}{v_i}$

or, $A_v = Y_{fs}(R_D\|R_L)$ **(11-20)**

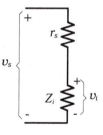
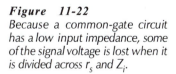

Figure   11-22
Because a common-gate circuit has a low input impedance, some of the signal voltage is lost when it is divided across r_s and Z_i.

Equation 11-20 is similar to Eq. 11-7 for the voltage gain of a common-source circuit, with the exception that there is no minus sign in Eq. 11-20. So, the voltage gain for a common-gate circuit is the same as the voltage gain for a common-source circuit with the same component values. As, already discussed, there is no phase shift between the input and output of a common-gate circuit, hence the absence of the minus sign in Eq. 11-20.

Effect of Unbypassed Gate Resistors

If capacitor C_1 is not present in the *CG* circuit in Fig. 11-19, the FET gate is *not ac* short-circuited to ground. So, a resistance ($R_G = R_1 \| R_2$) must be included in the *ac* equivalent circuit (see Fig. 11-23), and in the complete *CG ac* equivalent circuit. The presence of the unbypassed gate resistors affects the circuit input impedance and voltage gain. Analysis of the equivalent circuit shows that,

$$Z_i = (1/Y_{fs}) \frac{R_G + R_{gs}}{R_{gs}} \qquad \textbf{(11-21)}$$

and,

$$A_v = Y_{fs}(R_D \| R_L) \frac{R_{gs}}{R_G + R_{gs}} \qquad \textbf{(11-22)}$$

When $R_{gs} \gg R_G$ (which is usually the case), Eq. 11-21 gives $Z_i = 1/Y_{fs}$, and Eq. 11-22 gives $A_v = Y_{fs}(R_D \| R_L)$, which are the equations for a *CG* circuit with the gate bypassed to ground. The Z_i and A_v effects occur only in situations where very high value bias resistors are used in a *CG* circuit. Figure 11-24 shows a *CG* circuit that uses self bias. In this case, R_G is omitted because there is no signal applied at the gate input; the gate is directly grounded and no gate bypass capacitor is required.

Summary of CG Circuit Performance

With the gate bypassed to ground:

Device input impedance $\qquad Z_s = 1/Y_{fs}$

Circuit input impedance $\qquad Z_i = (1/Y_{fs}) \| R_S$

Device output impedance $\qquad Z_d = r_d$

Circuit output impedance $\qquad Z_o = R_D \| r_d$

Circuit voltage gain $\qquad A_v = Y_{fs}(R_D \| R_L)$

With the gate unbypassed:

Circuit input impedance $\qquad Z_i = (1/Y_{fs}) \dfrac{R_G + R_{gs}}{R_{gs}}$

Circuit voltage gain $\qquad A_v = Y_{fs}(R_D \| R_L) \dfrac{R_{gs}}{R_G + R_{gs}}$

A common-gate circuit has a voltage gain, no phase shift between input and output, low input impedance, and relatively high output impedance.

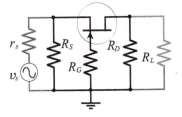

Figure 11-23
When the gate bypass capacitor is omitted in a CG circuit, the equivalent resistance of the gate bias resistors (R_G) must be included for circuit analysis.

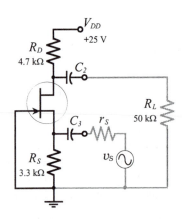

Figure 11-24
A common-gate circuit that uses self bias can have its gate directly grounded.

Example 11-5

The common-gate circuit in Fig. 11-24 has a *FET* with $Y_{fs} = 3000\ \mu S$, and $r_d = 50\ k\Omega$. Determine the device and circuit input and output impedances, and the circuit voltage gain. Also, calculate the overall voltage gain if the signal source impedance is $r_s = 600\ \Omega$.

Solution

Eq. 11-17, $\qquad Z_s = 1/Y_{fs} = 1/3000\ \mu S$
$\qquad\qquad\qquad = 333\ \Omega$

Eq. 11-18, $\qquad Z_i = (1/Y_{fs})\|R_S = 333\ \Omega\|3.3\ k\Omega$
$\qquad\qquad\qquad = 302\ \Omega$

Eq. 11-4, $\qquad Z_d = r_d = 50\ k\Omega$

Eq. 11-5, $\qquad Z_o = R_D\|r_d = 4.7\ k\Omega\|50\ k\Omega$
$\qquad\qquad\qquad = 4.3\ k\Omega$

Eq. 11-20, $\qquad A_v = Y_{fs}(R_D\|R_L) = 3000\ \mu S \times (4.7\ k\Omega\|50\ k\Omega)$
$\qquad\qquad\qquad = 12.9$

Combining Eq. 11-19 and 11-20, the overall voltage gain is

$$A_v = \frac{Y_{fs}(R_D\|R_L)\ Z_i}{r_s + Z_i} = \frac{12.9 \times 333\ \Omega}{600\ \Omega + 333\ \Omega}$$

$$= 4.6$$

Practise Problems

11-6.1 The common-gate circuit in Fig. 11-19 has the following components: $R_1 = 3.9\ M\Omega$, $R_2 = 2.2\ M\Omega$, $R_D = 3.3\ k\Omega$, $R_S = 1.5\ k\Omega$, and $R_L = 27\ k\Omega$. The *FET* has $Y_{fs} = 3.5\ mS$ and $r_d = 70\ k\Omega$, and the signal source has $r_s = 1\ k\Omega$. Calculate the device and circuit input and output impedances, the circuit voltage gain, and the overall voltage gain.

11-6.2 If the *FET* in Problem 11-6.1 has $R_{gs} = 10\ M\Omega$, determine the circuit input impedance and voltage gain when C_1 is removed.

11-7 Comparison of FET and BJT Circuits

CS, CD, and CG Circuit Comparison

Table 11-1 compares Z_i, Z_o, and A_v for CS, CD, and CG circuits. As already discussed, the CS circuit has voltage gain, high input impedance, high output impedance, and a 180° phase shift from input to output. The CD circuit has high input impedance, low output impedance, a voltage gain of approximately 1, and no phase shift. The CG circuit offers low input impedance, high output impedance, voltage gain, and no phase shift.

Table 11-1   Comparison of common-source, common-drain, and common-gate circuits.

Circuit configuration	Z_i	Z_o	A_v	Phase shift
CS	$\approx R_G$	$\approx R_D$	$-Y_{fs}(R_D\|R_L)$	180°
CD	$\approx R_G$	$\approx 1/Y_{fs}$	≈ 1	0
CG	$\approx 1/Y_{fs}$	$\approx R_D$	$Y_{fs}(R_D\|R_L)$	0

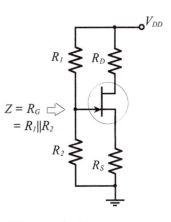

Figure 11-25
FET circuit impedance at the gate terminal.

Impedance at the FET Gate

Consideration of each type of circuit shows that the input or output impedance depends upon which device terminal is involved. In both the *CS* and *CD* circuits, the input signal is applied to the *FET* gate terminal, so Z_i is the impedance *looking into* the gate. Figure 11-25 shows that, for *CS* and *CD* circuits the gate input impedance is,

$$Z_g = R_{gs}$$

and the circuit input impedance is,

$$Z_i = R_G \| R_{gs}$$

where R_G is the equivalent resistance of the bias resistors. Because the reverse-biased gate-source resistance (R_{gs}) is so large, an unbypassed source resistor has no significant effect on Z_i at the gate. The circuit input impedance is largely determined by R_G in both the *CS* and *CD* circuits.

$$Z_i \approx R_G$$
$$= R_1 \| R_2 \text{ (in the case of voltage divider bias)}$$

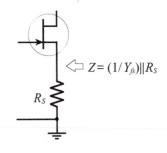

Figure 11-26
FET circuit impedance at the source terminal.

Impedance at the FET Source

The *FET* source is the output terminal for a *CD* circuit and the input terminal for a *CG* circuit. The device impedance in each case is the impedance *looking into* the source, (see Fig. 11-26).

$$Z_s = 1/Y_{fs}$$

The circuit impedance at the source terminal must include the source resistor.

$$Z_i \text{ or } Z_o = (1/Y_{fs})\|R_S$$

Impedance at the FET Drain

The output for *CS* and *CG* circuits is produced at the *FET* drain terminal. So, the impedance *looking into* the drain is the device output impedance (r_d). As illustrated in Fig. 11-27, the circuit output impedance for any circuit with the output taken from the *FET* drain terminal is,

$$Z_o = R_D \| r_d \approx R_D$$

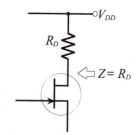

Figure 11-27
FET circuit impedance at the drain terminal.

Voltage gain

In a circuit with an unbypassed source resistor, the *ac* voltage at the source terminal follows the *ac* input at the gate, (see Fig. 11-28). A *CD* circuit (a source follower) has a voltage gain of approximately 1. A *CS* circuit with an unbypassed source resistor has v_i developed across R_S, and so,

$$v_o \approx v_i \times \frac{R_D}{R_S}$$

In a *CG* circuit, and in a *CS* circuit with R_S bypassed, the *ac* input (v_i) is developed across the gate-source terminals, and the *ac* output (v_o) is produced at the drain terminal. Thus, the voltage gain is the same for *CS* and *CG* circuits with similar component values and *FET* parameters. The *CS* voltage gain equation can be used for the *CG* circuit, with the omission of the minus sign that indicates *CS* phase inversion.

Although the voltages gains are equal for similar *CS* and *CG* circuits, the low input impedance of the *CG* circuit can substantially attenuate the signal voltage, and result in a low amplitude output. This is demonstrated in Example 11-6.

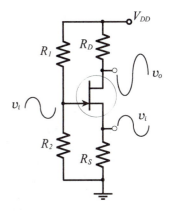

Figure 11-28
In a FET circuit that has an unbypassed R_s, v_i applied to the gate also appears at the source terminal.

Example 11-6

A 50 mV signal with a 600 Ω source resistance is applied to the circuit in Fig. 11-29. Calculate the *ac* output voltage when the conditions are: (a) *CS* circuit operation with the signal coupled to the gate, and R_S bypassed, (b) *CG* circuit operation with the signal applied to the source, and the gate bias resistors bypassed. The *FET* has a forward transfer admittance $Y_{fs} = 6000\,\mu S$.

Solution

(a) *CS circuit*

$$A_v = -Y_{fs}(R_D\|R_L) = -6000\,\mu S\,(2.7\,k\Omega\|33\,k\Omega)$$
$$\approx -15$$

$$Z_i = R_1\|R_2 = 100\,k\Omega\|47\,k\Omega$$
$$= 32\,k\Omega$$

$$v_i = \frac{v_s \times Z_i}{r_s + Z_i} = \frac{50\,mV \times 320\,k\Omega}{600\,\Omega + 320\,k\Omega}$$
$$= 49.1\,mV$$

$$v_o = A_v \times v_i = -15 \times 49.1\,mV$$
$$= 736\,mV$$

(b) *CG circuit*

$$A_v = Y_{fs}(R_D\|R_L) = 6000\,\mu S\,(2.7\,k\Omega\|33\,k\Omega)$$
$$= 15$$

$$Z_i = (1/Y_{fs})\|R_S = (1/6000\,\mu S)\|2.7\,k\Omega$$
$$= 157\,\Omega$$

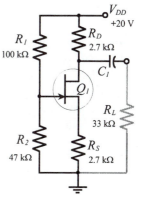

Figure 11-29
Circuit for Ex. 11-6.

$$v_i = \frac{v_s \times Z_i}{r_s + Z_i} = \frac{50 \text{ mV} \times 157 \text{ } \Omega}{600 \text{ } \Omega + 157 \text{ } \Omega}$$

$$= 10.4 \text{ mV}$$

$$v_o = A_v \times v_i = 15 \times 10.6 \text{ mV}$$

$$= 156 \text{ mV}$$

FET-BJT circuit Comparison

Table 11-2 compares Z_i, Z_o, and A_v for the basic FET and BJT circuits. The BJT CE and CB circuits have much higher voltage gains than the corresponding FET CS and CG circuits, while the FET CS and CD circuits have much higher input impedances than the BJT CE and CC circuits. Apart from these differences, BJTs and FETs can generally perform similar functions. High-frequency, fast-switching, and high-power devices of both types are available. In some switching applications, FETs have the advantage of a smaller voltage drop ($V_{DS(on)}$) than that across BJTs ($V_{CE(sat)}$). This is discussed in Section 10-11.

Table 11-2 Typical quantities for basic FET and BJT circuits

FET circuit	A_v	Z_i	Z_o	BJT circuit	A_v	Z_i	Z_o
CS	-25	500 kΩ	5 kΩ	CE	-250	5 kΩ	5 kΩ
CD	≈ 1	500 kΩ	200 Ω	CC	≈ 1	5 kΩ	20 Ω
CG	25	200 Ω	5 kΩ	CB	250	20 Ω	5 kΩ

Practise Problems

11-7.1 A FET circuit with voltage divider bias has the following component values: $R_1 = 560$ kΩ, $R_2 = 390$ kΩ, $R_D = 3.9$ kΩ, $R_S = 3.3$ kΩ, $R_L = 50$ kΩ. Assuming that the FET is a 2N5457 with typical parameters, calculate the impedances *looking into* the drain, source, and gate terminals.

11-8 Frequency Response of FET Circuits

Low-Frequency Response

The low frequency response of FET circuits is determined by exactly the same considerations as for BJT circuits. The lower cutoff frequency is normally set by a source bypass capacitor, and it can be affected by coupling capacitors. This is considered in detail in Chapter 12.

High-Frequency Response

Unlike BJTs, a device cutoff frequency is not normally specified for a *FET*. Instead, *FETs* intended for high-frequency operation have the parameters listed as measured at a specified high frequency. The low-frequency parameters are also normally listed. For example, a *2N5484 JFET* has the following parameters:

At f = 1 kHz	min	max
Output conductance (Y_{os})		50 µS
Forward transconductance (Y_{fs})	3000 µS	6000 µS

At f = 100 MHz		
Input admittance (Y_{is})		100 µS
Output conductance (Y_{os})		75 µS
Forward transconductance (Y_{fs})	2500 µS	

The device inter-terminal capacitances (C_{gs}, C_{gd}, and C_{ds}) are important quantities in determining the performance of a *FET* circuit, and these are used in the same way as the *BJT* capacitances. An input capacitance limited cutoff frequency ($f_{2(i)}$), and an output capacitance limited cutoff frequency ($f_{2(o)}$) can be calculated exactly as explained for *BJT* circuits in Section 8-4. The input capacitance is amplified by the Miller effect in the case of a *CS* circuit (an inverting amplifier). Equation 8-9 can be rewritten for the *FET* circuit,

$$C_{in} = C_{gs} + (1 + A_v)\, C_{gd} \qquad \textbf{(11-23)}$$

The *FET* capacitances are specified as the *input capacitance* (C_{iss}), the *reverse transfer capacitance* (C_{rss}), and the *output capacitance* (C_{oss}). The reverse transfer capacitance is another name for the gate-drain capacitance. The input capacitance is the sum of the gate-source capacitance and the gate-drain capacitance. The output capacitance is simply the drain-source capacitance.

$$C_{rss} = C_{gd} \qquad \textbf{(11-24)}$$

$$C_{iss} = C_{gs} + C_{gd} \qquad \textbf{(11-25)}$$

$$C_{oss} = C_{ds} \qquad \textbf{(11-26)}$$

Typical input capacitance values for a *2N5484* high-frequency *JFET* are: C_{rss} = 1 pF, C_{iss} = 5 pF, C_{oss} = 2 pF. Because, these are extremely small, the circuit performance can be easily affected by stray capacitance.

Example 11-7
Calculate the input capacitance limited cutoff frequency for the circuit in Fig. 11-30 when operated as a *CS* circuit with R_S bypassed. Assume that there is no additional stray capacitance at the input terminals.

Solution
Eq. 11-24, $C_{gd} = C_{rss} = 1$ pF

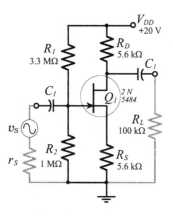

Figure 11-30
Circuit for Ex. 11-7.

From Eq. 11-25, $\qquad C_{gs} = C_{iss} - C_{rss} = 5 \text{ pF} - 1 \text{ pF}$
$$= 4 \text{ pF}$$

$$|A_v| = Y_{fs}[(1/Y_{os})\|R_D\|R_L]$$
$$= 2500 \ \mu S \ [(1/75 \ \mu S)\|5.6 \text{ k}\Omega\|100 \text{ k}\Omega]$$
$$= 9.5 \quad \text{(note that } A_v \text{ is a minimum value, because}$$
$$Y_{fs} \text{ is the minimum specified value.)}$$

Eq. 11-23, $\qquad C_{in} = C_{gs} + (1 + A_v)C_{gd} = 4 \text{ pF} + (1 + 9.5)1 \text{ pF}$
$$= 14.5 \text{ pF (note that } C_{in} \text{ is a minimum value,}$$
$$\text{because of the minimum } A_v.)$$

Eq. 11-2, $\qquad Z_i = R_1\|R_2\|(1/Y_{is}) = 3.3 \text{ M}\Omega\|1 \text{ M}\Omega\|(1/100 \ \mu S)$
$$= 9.87 \text{ k}\Omega$$

From Eq. 8-10, $\qquad f_{2(i)} = \dfrac{1}{2 \ \pi \ C_{in}(r_s\|Z_i)} = \dfrac{1}{2 \ \pi \times 14.5 \text{ pF} \times (600 \ \Omega\|9.87 \text{ k}\Omega)}$
$$= 19.4 \text{ MHz}$$

Practise Problems

11-8.1 The *FET* circuit in Fig. 11-30 is reconnected to function as a common-gate circuit with the gate terminal bypassed to ground and no source bypass capacitor. Calculate the input capacitance limited cutoff frequency.

11-8.2 The circuit in Ex. 11-7 has 5 pF of stray capacitance at the output terminal. Calculate the output capacitance limited cutoff frequency.

Chapter-11 Review Questions

Section 11-1

11-1 Sketch a practical common-source amplifier circuit using voltage divider bias. Show a capacitor-coupled signal source and capacitor-coupled load. Explain the need for coupling capacitors, and discuss the correct polarity for connecting the capacitors.

11-2 Explain what is meant by *ac* degeneration in a *FET* circuit. Show how *ac* degeneration is eliminated in voltage divider bias and self bias circuits.

11-3 Discuss the purpose of an *ac* load line for a *FET* circuit. State how *ac* load lines differ from *dc* load lines for circuits with bypassed source resistors, and for circuits with capacitor-coupled loads.

Section 11-2

11-4 Sketch the complete equivalent circuit for a field effect transistor connected in common-source configuration. Identify each component of the circuit and discuss its origin. Also, sketch a simplified form of the *FET* equivalent circuit for low- and medium-frequency applications.

11-5 Define each parameter in the *FET* low-frequency equivalent circuit, and state typical values.

Section 11-3

11-6 Sketch a practical common-source amplifier circuit using voltage divider bias. Draw *ac* voltage and current waveforms for the circuit, and explain its operation.

11-7 Show what happens to a *CS* circuit when the capacitors and *dc* supply are replaced with *ac* short circuits. Explain.

11-8 Draw the complete *ac* equivalent for a *CS* circuit with voltage divider bias, a bypassed source resistor, a capacitor-coupled signal source, and a capacitor-coupled load.

11-9 Draw the complete *ac* equivalent for a *CS* circuit with self bias, a bypassed source resistor, a capacitor-coupled signal source, and a capacitor-coupled load. Briefly explain.

11-10 Write equations for Z_i, Z_o, and A_v, for the *FET* circuit in Question 11-8.

11-11 Draw the complete *ac* equivalent for a *CS* circuit with gate bias, a bypassed source resistor, a capacitor-coupled signal source, and a capacitor-coupled load. Briefly explain.

11-12 List the performance characteristics for a *CS* circuit, and briefly discuss its usual applications.

Section 11-4

11-13 Draw the complete *ac* equivalent for a *CS* circuit with voltage divider bias, a capacitor-coupled signal source, a capacitor-coupled load, and *no* bypass capacitor across the source resistor. Explain the circuit.

11-14 Write equations for Z_i, Z_o, and A_v for the circuit described in Question 11-12.

11-15 Draw the complete *ac* equivalent for the circuit in Question 11-11 when the source bypass capacitor is removed.

Section 11-5

11-16 Sketch a practical common-drain amplifier circuit using voltage divider bias. Draw *ac* voltage and current waveforms for the circuit, and explain its operation.

11-17 Show what happens to a *CD* circuit when the capacitors and *dc* supply are replaced with *ac* short circuits. Explain.

11-18 Draw the complete *ac* equivalent for a *CD* circuit with voltage divider bias, a capacitor-coupled signal source, and a capacitor-coupled load. Explain the circuit.

11-19 Draw the complete *ac* equivalent for a *CD* circuit with self bias, a capacitor-coupled signal source, and a capacitor-coupled load. Explain.

11-20 Write equations for Z_i, Z_o, and A_v for the *FET* circuits in Questions 11-18 and 11-19.

11-21 List the performance characteristics for a *CD* circuit, and briefly discuss its usual applications.

Section 11-6

11-22 Sketch a practical common-gate amplifier circuit using voltage divider bias. Draw *ac* voltage and current waveforms for the circuit, and explain its operation.

11-23 Show what happens to a *CG* circuit when the capacitors and *dc* supply are replaced with *ac* short circuits. Explain.

11-24 Draw the complete *ac* equivalent for a *CG* circuit with voltage divider bias, a capacitor-coupled signal source, and a capacitor-coupled load. Explain the circuit.

11-25 Sketch a practical common-gate amplifier circuit using self bias. Draw the complete *ac* equivalent for the circuit.

11-26 Write equations for Z_i, Z_o, and A_v for the *FET* circuits in Questions 11-24 and 11-25.

11-27 List the performance characteristics for a *CG* circuit, and briefly discuss its usual applications.

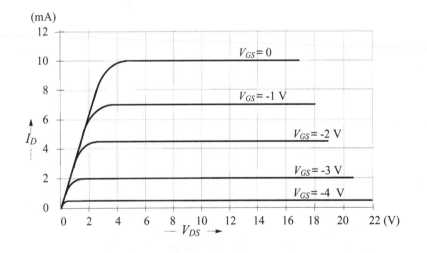

Figure   11-31

Section 11-7

11-28 Compare the performance of *CS*, *CD*, and *CG* circuits, and discuss typical applications of each type of circuit.

11-29 Write equations for the impedances 'seen looking into' the gate, source, and drain terminals of a *FET* in a circuit. Briefly explain each equation.

11-30   Construct a table to compare A_v, Z_i, and Z_o for basic *FET* and *BJT* circuits.

Chapter-11 Problems

Section 11-1

11-1   The drain characteristics for the circuit in Fig. 11-29 are shown in Fig. 11-31. Draw the *dc* and *ac* load lines for the circuit when it is operated in *CS* configuration with R_S bypassed. Assume that the *Q*-point drain current is 1.8 mA.

11-2   Draw the new *ac* load line for the circuit in Problem 11-1 when a 12 kΩ external load resistor (R_l) is capacitor-coupled to the *FET* drain terminal.

11-3   Using the device characteristics in Fig. 11-31, draw the *dc* and *ac* load lines for the *CS* circuit shown in Fig. 11-32. Take the *Q*-point drain current as 2 mA.

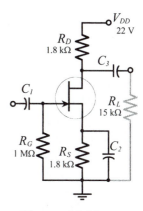

Figure   11-32

Section 11-3

11-4   The *FET* in the *CS* circuit in Fig. 11-33 has Y_{fs} = 6000 µS and r_d = 70 kΩ. Calculate the circuit input impedance, output impedance, and voltage gain.

11-5   Recalculate Z_i, Z_o, and A_v for the circuit in Problem 11-4 when the *FET* is replaced with one having Y_{fs} = 3500 µS, and r_d = 95 kΩ.

11-6   Determine the circuit input impedance, output impedance, and voltage gain for the *CS* circuit in Fig. 11-34.

11-7   Determine the circuit input impedance, output impedance, and voltage gain for the *CS* circuit in Fig. 11-35.

11-8   Determine the new values of input impedance, output impedance, and voltage gain for the *CS* circuit in Fig. 11-35 when a *2N5457 FET* is used and R_L is changed to 18 kΩ.

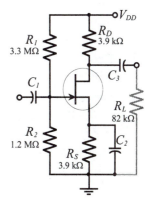

Figure   11-33

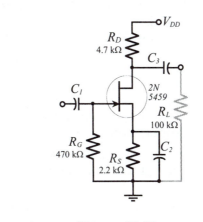

Figure   11-34   ***Figure   11-35***

11-9   A *CS* amplifier is to have a minimum voltage gain of 12. If R_L = 150 kΩ and a *2N5459 FET* is used, determine a suitable resistance for R_D. Also, calculate the typical gain.

Section 11-4

11-10   Calculate the new values of input impedance, output impedance, and voltage gain for the circuit in Problem 11-4 when the source bypass capacitor is removed.

11-11   Calculate the new values of input impedance, output impedance, and voltage gain for the circuit in Problem 11-7 when the source bypass capacitor is removed.

11-12   The source resistor in the circuit in Problem 11-4 is replaced with two series-connected resistors (R_{S1} and R_{S2}). R_{S2} is bypassed and R_{S1} is left unbypassed. Calculate a suitable resistance for R_{S1} to give a voltage gain of 5.

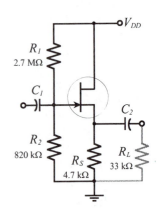

Figure   11-36

Section 11-5

11-13   The *FET* in the *CD* circuit in Fig. 11-36 has Y_{fs} = 6000 μS and r_d = 70 kΩ. Calculate the circuit input impedance, output impedance, and voltage gain.

11-14   Recalculate Z_i, Z_o, and A_v for the circuit in Problem 11-13 when the *FET* has Y_{fs} = 3500 μS and r_d = 95 kΩ.

11-15   Determine the circuit input impedance, output impedance, and voltage gain for the *CD* circuit in Fig. 11-37.

11-16   A *CD* circuit as in Fig. 11-36 has the following components: R_1 = 2.2 MΩ, R_2 = 1 MΩ, R_S = 6.8 kΩ, R_L = 150 kΩ. The *FET* parameters are Y_{fs} = 5000 μS, and r_d = 120 kΩ. Calculate the input impedance, output impedance, and voltage gain.

11-17   Determine the typical input impedance, output impedance, and voltage gain for the *CD* circuit in Problem 11-16 when a *2N5457 FET* is used and R_L is changed to 39 kΩ.

11-18   A *CD* amplifier is to have an output impedance of approximately 200 Ω. Select a suitable *FET* from the data sheet in Appendix 1-11. If R_S = 1 kΩ, calculate the typical, minimum, and maximum values of Z_o. Also, calculate the typical voltage gain for the circuit.

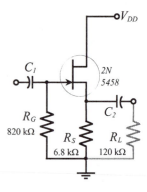

Figure   11-37

Section 11-6

11-19   The *FET* in the *CG* circuit in Fig. 11-38 has Y_{fs} = 6000 μS and r_d = 70 kΩ. Calculate the circuit input impedance, output impedance, and voltage gain.

11-20   A 120 mV *ac* signal with r_s = 300 Ω is applied as input to the *CG* circuit in Problem 11-19. Calculate the circuit *ac* output voltage.

11-21   Recalculate Z_i, Z_o, and A_v for the circuit in Problem 11-19 when the *FET* is replaced with one having Y_{fs} = 3500 μS, and r_d = 95 kΩ.

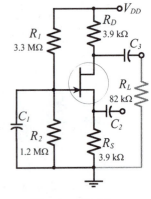

Figure   11-38

11-22 Determine the circuit input impedance, output impedance, and voltage gain for the *CG* circuit in Fig. 11-39.

11-23 Recalculate the values of input impedance, output impedance, and voltage gain for the circuit in Problem 11-22 when a *2N5457 FET* is used and R_L is changed to 18 kΩ.

11-24 Determine the input impedance, output impedance, and voltage gain for a *CG* circuit, as in Fig. 11-39, when a *2N5457 FET* is used and the resistors are; $R_S = 2.2$ kΩ, $R_D = 3.3$ kΩ, $R_L = 100$ kΩ.

11-25 An *ac* signal with $r_s = 600$ Ω is applied to the source input of the *CG* circuit in Problem 11-23. Calculate the required level of signal voltage if the *ac* output is to be 4 V.

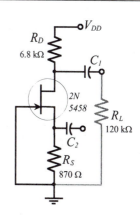

Figure   11-39

Section 11-7

11-26 The circuit in Fig. 11-40 uses a *FET* that has $Y_{fs} = 5000$ μS and $r_d = 100$ kΩ. Calculate the circuit impedances *looking into* each terminal.

11-27 A 30 mV ac signal with $r_s = 300$ Ω is applied to the circuit in Fig. 11-40. Determine the *ac* output voltage when the circuit is operated as; (a) a *CS* circuit with R_S bypassed, (b) a *CG* circuit with the gate bypassed to ground. The *FET* used has $Y_{fs} = 7000$ μS.

11-28 A *2N5458 FET* is used in a circuit with voltage divider bias and the following components: $R_1 = 560$ kΩ, $R_2 = 390$ kΩ, $R_D = 3.3$ kΩ, $R_S = 2.7$ kΩ, $R_L = 68$ kΩ. Calculate the impedances *looking into* the drain, source, and gate terminals. Use the *FET* typical parameters.

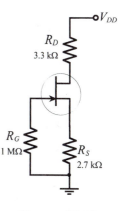

Figure   11-40

Section 11-8

11-29 The circuit in Fig. 11-40 has a *FET* with the following high-frequency parameters: $Y_{fs} = 5000$ μS, $Y_{is} = 67$ μS, $Y_{os} = 50$ μS, $C_{iss} = 7$ pF, $C_{rss} = 3$ pF, $C_{oss} = 3$ pF. Calculate the input capacitance limited cutoff frequency when the circuit is used as a *CS* amplifier with the source terminal bypassed to ground. The signal source has a 1 kΩ resistance.

11-30 Repeat Problem 11-29 for the circuit reconnected as a *CG* amplifier with the gate bypassed to ground.

11-31 The circuit in Problems 11-29 and 11-30 has 8 pF stray capacitance at the device drain terminal. Determine the output capacitance limited cutoff frequency.

11-32 The *CS* amplifier in Fig. 11-9 uses a *2N5484 JFET*. Calculate the input capacitance limited cutoff frequency for the circuit. Assume that $r_s = 4.7$ kΩ.

11-33 Recalculate the input capacitance limited cutoff frequency for the circuit in Problem 11-32 when it is connected to function as a *CG* circuit.

Practise Problem Answers

11-1.1   4 kΩ, 1.92 kΩ
11-3.1   1 MΩ, 1.48 kΩ, -5.5
11-3.2   470 kΩ, 3.75 kΩ, -11.25
11-4.1   1 MΩ, 1.5 kΩ, -1.3
11-4.2   848 kΩ, 2.7 kΩ, -0.89
11-5.1   194 kΩ, 232 Ω, 0.93
11-6.1   286 Ω, 240 Ω, 70 kΩ, 3.15 kΩ, 10.3, 2
11-6.2   274 Ω, 9.1
11-7.1   3.75 kΩ, 303 Ω, 230 kΩ
11-8.1   115 MHz, 6 MHz

Chapter *12*

Small Signal Amplifiers

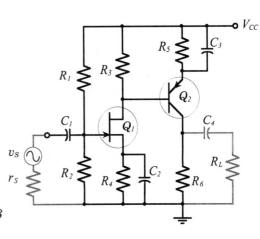

Chapter Contents

Objectives

You will be able to:
  Sketch the following small-signal amplifier circuits, explain the operation of each circuit, design each circuit for a given specification, and analyze each circuit to determine its performance:

1 *Single-stage BJT CE amplifier.*

2 *Single-stage FET CS amplifier.*

3 *Two-stage BJT CE capacitor-coupled amplifier.*

4 *Direct-coupled two-stage BJT amplifier.*

5 *Direct-coupled two-stage BJT amplifier circuits using complementary transistors.*

6 *Direct-coupled two-stage BJT amplifier with emitter-follower output stage.*

7 *DC feedback pair with CE output.*

8 *DC feedback pair with CC output.*

9 *Two-stage capacitor-coupled BIFET circuit.*

10 *Two-stage direct-coupled BIFET circuit.*

11 *Differential amplifier.*

12 *Common-base amplifier.*

13 *Cascode amplifier.*

Introduction

A *small-signal amplifier* accepts low-voltage *ac* inputs and produces amplified outputs. The *ac* outputs are also relatively low amplitude voltages, usually no larger than ±1 V. *BJT* and *FET* bias circuit design and *ac* analysis of single-stage circuits are treated in earlier chapters. So, this chapter covers the design of small-signal amplifier circuits to meet a given specification for voltage gain, load resistance, supply voltage, frequency response, etc.

A single-stage *BJT* circuit may be employed as a small-signal amplifier, but two cascaded stages give much greater amplification. For very high input impedance, a *FET* may be used as an input stage with a *BJT* as the second stage.

Calculation of circuit resistor values merely involves application of Ohm's law, after selecting suitable voltage and current levels throughout the circuit. Each capacitor value is determined in terms of the circuit lower cut-off frequency and the resistance in series with the capacitor.

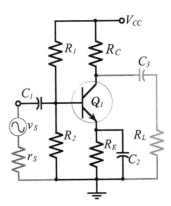

Figure 12-1
Common emitter circuit design commences with a specification for the circuit performance: voltage gain, load impedance, frequency response etc.

12-1   Single-Stage Common-Emitter Amplifier

Specification

Bias circuit design for the *CE* amplifier in Fig. 12-1 is explained in Chapter 5, and *ac* analysis of the circuit is covered in Chapter 6. Design of this circuit (or any other circuit) normally commences with a specification which might list: the supply voltage, minimum voltage gain, frequency response, signal source impedance, and the load impedance.

Selection of I_C, R_C, and R_E

Designing for a particular voltage gain requires the use of *ac negative feedback* to stabilize the gain, (see Chapter 13). The circuit shown in Fig. 12-1 has no provision for negative feedback, consequently, it is designed to achieve the largest possible voltage gain. From Eq. 6-15, the voltage gain of a *CE* circuit is,

$$A_v = \frac{-h_{fe}\,(R_C \| R_L)}{h_{ie}} \qquad \text{(12-1)}$$

Because A_v is directly proportional to $R_C \| R_L$, design for the greatest voltage gain would seem to require selection of the largest possible collector resistance. However, a very large value of R_C might make the collector current too small for satisfactory transistor operation. For most small-signal transistors, I_C should not be less than 500 µA. A good minimum I_C to aim for is 1 mA. Special low-noise transistors operate with much lower collector current levels.

The transistor h_{fe} value is related to I_C, so I_C might be selected high to give the largest h_{fe}, again to achieve the greatest A_v. But, a high I_C level results in a small R_C value for a given voltage drop (V_{RC}). So, a high I_C level might actually result in a low A_v value, although h_{fe} might be relatively large.

For a given level of I_C, the largest possible voltage drop across R_C gives the greatest R_C value, $(R_C = V_{RC}/I_C)$. To make V_{RC} as large as possible, V_{CE} and V_E should be held to a minimum, [see Fig. 12-2(a)]. The collector-emitter voltage should typically be around 3 V. This is large enough to ensure that the transistor operates linearly, and it also allows for a collector voltage swing of ±1 V which is usually adequate for a small-signal amplifier. Another consideration in selecting R_C is that there is nothing to be gained by making R_C larger than R_L. In fact, R_C should normally be very much smaller than R_L, so that R_L has little effect on the circuit voltage gain.

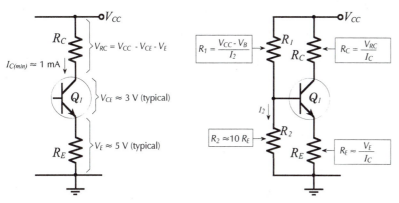
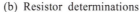
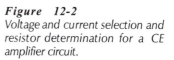

(a) Voltage and current selection   (b) Resistor determinations

Figure 12-2
Voltage and current selection and resistor determination for a CE amplifier circuit.

For good bias stability, the emitter resistor voltage drop (V_E) should be much larger than the base-emitter voltage (V_{BE}), (see Section 5-7). This is because $V_E = V_B - V_{BE}$, and when $V_E \gg V_{BE}$, V_E will be only slightly affected by any variation in V_{BE} (due to temperature change or other effects). Consequently, I_E and I_C remain fairly stable at $I_C \approx I_E = V_E/R_E$. A minimum V_E of 5 V gives good bias stability in most circumstances, [see Fig. 12-2 (a)]. With supply voltages less than 10 V, V_E might have to be reduced to 3 V to allow for reasonable levels of V_{CE} and V_{RC}. Normally, an emitter resistor voltage drop less than 3 V is likely to produce poor bias stability.

Once V_E, V_{CE}, and I_C are selected, V_{RC} is determined as,

$$V_{RC} = V_{CC} - V_{CE} - V_E$$

Then, R_C and R_E are calculated,

$$R_C = \frac{V_{RC}}{I_C} \text{ , and } R_E = \frac{V_E}{I_C} \quad \text{[see Fig. 12-2(b)]}$$

Bias Resistors

As discussed in Section 5-7, selection of the voltage divider current (I_2) as $I_C/10$ gives good bias stability and reasonably high input resistance. Where the input resistance is not important, I_2 may be made equal to I_C for excellent bias stability. The bias resistors are calculated as,

$$R_2 = \frac{V_B}{I_2} \text{ , and } R_1 \approx \frac{V_{CC} - V_B}{I_2}$$

Selecting $R_2 = 10 \, R_E$ gives $I_2 \approx I_C/10$, [Fig. 12-2(b)]. The precise level of I_2 can be calculated as $I_2 = V_B/R_2$, and this can be used in the equation for R_1.

Bypass Capacitor

All capacitors should be selected to have the smallest possible capacitance value, both to minimize the physical size of the circuit and for economy (large capacitors are the most expensive). Because each capacitor has its highest impedance at the lowest operating frequency, the capacitor values are calculated at the lowest signal frequency that the circuit is required to amplify. This frequency is the circuit *lower cutoff frequency*, or *low 3 dB frequency* (f_1), (see Fig. 12-3 reproduced from Fig. 8-5).

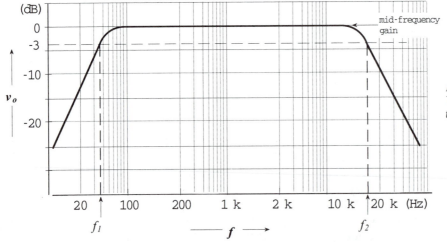

Figure 12-3
Typical frequency response for a transistor amplifier.

Bypass capacitor C_2 in Fig. 12-1 is normally the largest capacitor in the circuit, so C_2 is selected to set f_1 at the desired frequency. Equation 6-21 (for voltage gain) was developed for a *CE* circuit with an unbypassed emitter resistor (R_E). Rewriting the equation to include X_{C2} in parallel with R_E gives

$$A_v = \frac{-h_{fe}(R_C \| R_L)}{h_{ie} + (1 + h_{fe})(R_E \| X_{C2})}$$

Normally $R_E \gg X_{C2}$, so R_E can be omitted. Also, X_{C2} is capacitive,

so,

$$A_v = \frac{-h_{fe}(R_C \| R_L)}{\sqrt{\{h_{ie}^2 + [(1 + h_{fe})(X_{C2})]^2\}}}$$

When $h_{ie} = (1 + h_{fe}) X_{C2}$,

$$|A_v| = \frac{-h_{fe}(R_C \| R_L)}{h_{ie}\sqrt{(1^2 + 1^2)}} = \frac{\text{mid-frequency gain}}{\sqrt{2}}$$

$$= \text{(mid-frequency gain) - 3 dB}$$

Therefore, at f_1,

$$h_{ie} = (1 + h_{fe}) X_{C2}$$

or, $$\mathbf{X_{C2} = \frac{h_{ie}}{1 + h_{fe}}\ \textbf{at}\ f_1}\quad\textbf{(12-2)}$$

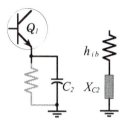

From Table 6-1 (in Section 6-3),

$$h_{ib} = \frac{h_{ie}}{1 + h_{fe}} = \text{(impedance 'looking into' the emitter)}$$

So, $$\mathbf{X_{C2} = h_{ib}\ \textbf{at}\ f_1}\qquad\qquad\textbf{(12-3)}$$

At f_1, X_{C2} = (impedance 'looking into' the emitter)
= (impedance in *series* with X_{C2}), (see Fig. 12-4)

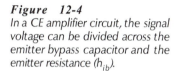

Figure 12-4
In a CE amplifier circuit, the signal voltage can be divided across the emitter bypass capacitor and the emitter resistance (h_{ib}).

Equations 12-2 and 12-3 give the smallest value for the bypass capacitor. When selecting a standard value for the capacitor, the next larger value should be chosen. This will give a cutoff frequency slightly lower than the f_1 value used in the calculations.

It is important to note that the emitter bypass capacitor is calculated in terms of the resistance 'seen looking into' the transistor emitter terminal (the resistance in series with C_2). The capacitor value is *not* determined in relation to the value of the emitter resistor (R_E). Example 12-1 demonstrates the results of correct and incorrect bypass capacitor selection.

Example 12-1

The circuit shown in Fig. 12-5 uses a transistor with $h_{fe} = 50$, $h_{ie} = 1\ k\Omega$, and $h_{ib} = 20\ \Omega$. The circuit lower cutoff frequency (f_1) is to be 100 Hz. Calculate the following quantities:
(a) The required capacitance for C_2.
(b) A_v with the emitter terminal completely bypassed to ground.
(c) A_v at $f = 100$ Hz, using the selected C_2 value.
(d) A_v at $f = 100$ Hz when C_2 is (incorrectly) selected as $X_{C2} = R_E/10$.

Solution

(a) C_2 determination:
Eq. 12-3, $$X_{C2} = h_{ib} = 20\ \Omega$$

$$C_2 = \frac{1}{2\pi f_1 X_{C2}} = \frac{1}{2 \times 100\ \text{Hz} \times 20\ \Omega}$$

$$= 79.6\ \mu F \quad \text{(use 80 } \mu F \text{ standard value)}$$

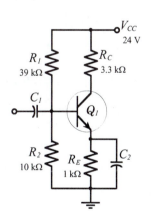

Figure 12-5
Common-emitter circuit for Example 12-1.

(b) Voltage gain:

Eq. 6-15,
$$A_v = \frac{-h_{fe}(R_C \| R_L)}{h_{ie}} = \frac{-50 \times 3.3 \text{ k}\Omega}{1 \text{ k}\Omega}$$

$$= -165$$

(c) A_v at 100 Hz:

at $f = 100$ Hz, $X_{C2} = 20 \ \Omega$

Eq. 12-1,
$$A_{v(1)} = \frac{-h_{fe}(R_C \| R_L)}{\sqrt{\{h_{ie}^2 + [(1 + h_{fe})X_{C2}]^2\}}}$$

$$= \frac{-50 \times 3.3 \text{ k}\Omega}{\sqrt{\{(1 \text{ k}\Omega)^2 + [(1 + 50) \times 20 \ \Omega]^2\}}}$$

$$= -116 = A_v / \sqrt{2}$$

(d) A_v at 100 Hz when $X_{C2} = R_E / 10$:

from Eq. 12-1,
$$A_{v(x)} = \frac{-h_{fe}(R_C \| R_L)}{\sqrt{\{h_{ie}^2 + [(1 + h_{fe})(R_E/10)]^2\}}}$$

$$= \frac{-50 \times 3.3 \text{ k}\Omega}{\sqrt{\{(1 \text{ k}\Omega)^2 + [(1 + 50) \times (1 \text{ k}\Omega/10)]^2\}}}$$

$$= -31.7$$

In Ex. 12-1 the circuit voltage gain is -165 when the emitter terminal is completely bypassed to ground. This is the mid-frequency gain (A_v) for the amplifier. With a correctly selected bypass capacitor, the voltage gain ($A_{v(1)}$) at a signal frequency of 100 Hz is -116. This is $1/\sqrt{2}$ times the mid-frequency gain, or 3 dB below the mid-frequency gain, which is to be expected because the bypass capacitor is selected to give a 3 dB gain reduction at 100 Hz. When the impedance of the bypass capacitor is *incorrectly* selected as $R_E/10$, the voltage gain ($A_{v(x)}$) at 100 Hz is much smaller than the required $A_v/\sqrt{2}$.

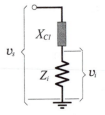

(a) The signal voltage is divided across X_{C1} and Z_i

Coupling Capacitors

The coupling capacitors (C_1 and C_3 in Fig. 12-1) should have a negligible effect on the frequency response of the circuit. Figure 12-6(a) illustrates the fact that X_{C1} and Z_i constitute a voltage divider. If X_{C1} is too large, the circuit *ac* input voltage (v_i) will be significantly smaller than the signal voltage (v_s). Similarly, X_{C3} and R_L attenuate the *ac* voltage at the transistor collector, so that the *ac* output voltage (v_o) can be smaller than the *ac* collector voltage (v_c), [Fig. 12-6(b)]. To minimize the effects of C_1 and C_3, the reactance of each coupling capacitor is selected to be approximately equal to one-tenth of the impedance in series with it at the lowest operating frequency for the circuit (f_1).

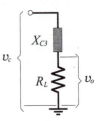

(b) The collector *ac* voltage is divided across X_{C3} and R_L

Figure 12-6
Input and output voltages can be divided across coupling capacitors and input or load resistances.

$$X_{C1} = \frac{Z_i + r_s}{10} \text{ at } f_1 \qquad\qquad \textbf{(12-4)}$$

$$X_{C3} = \frac{Z_o + R_L}{10} \text{ at } f_1 \qquad\qquad \textbf{(12-5)}$$

Usually, $R_L \gg Z_o$, and often $Z_i \gg r_s$, so that Z_o and r_s can be omitted in Equations 12-4 and 12-5. Once again, the equations give minimum capacitance values, so that the next larger standard values should always be selected for C_1 and C_3.

Equations 12-4 and 12-5 give an impression that approximately 10% of the signal and output voltages are lost across C_1 and C_3. This would be true if the quantities were resistive. However, X_{C1} and X_{C3} are capacitive while Z_i and R_L are usually resistive. So, when the actual loss is calculated, it is found to be only around 0.5% for each capacitor.

Another approach to the selection of coupling capacitors is to make $X_{C1} = Z_i$ at two octaves below f_1.

$$X_{C1} = Z_i \text{ at } f_1/4 \qquad\qquad \textbf{(12-6)}$$

The output coupling capacitor is then determined by making impedance of C_3 equal to R_L at two octaves below $f_1/4$.

$$X_{C3} = R_L \text{ at } f_1/16 \qquad\qquad \textbf{(12-7)}$$

When Equations 12-6 and 12-7 are used to determine the values of the coupling capacitors, it can be shown that the capacitor attenuation effects are less than 5% of A_v.

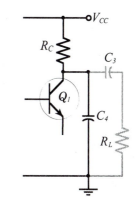

Shunting Capacitor

Sometimes an amplifier is required to have a particular *upper cutoff frequency*, (f_2 in Fig. 12-3). The transistor must be selected to have a much higher cutoff frequency than f_2, (see Section 8-3). The upper cutoff frequency for the circuit can then be set by connecting a small capacitor from the transistor collector terminal to ground, (C_4 in Fig. 12-7). The effect of shunt capacitance at the transistor collector is discussed in Section 8-4.

Figure 12-7
A shunting capacitor can be used at the output of a common-emitter circuit to select the upper 3 dB frequency.

Design Calculations

Example 12-2
Calculate suitable resistor values for the *CE* amplifier in Fig. 12-8.

Solution

$$R_C \ll R_L$$

select $$R_C = \frac{R_L}{10} = \frac{120 \text{ k}\Omega}{10}$$

$$= 12 \text{ k}\Omega \text{ (standard value)}$$

select $V_E = 5$ V, and $V_{CE} = 3$ V

$$V_{RC} = V_{CC} - V_{CE} - V_E = 24 \text{ V} - 3 \text{ V} - 5 \text{ V}$$
$$= 16 \text{ V}$$

$$I_C = \frac{V_{RC}}{R_C} = \frac{16 \text{ V}}{12 \text{ k}\Omega}$$
$$= 1.3 \text{ mA}$$

$$R_E = \frac{V_E}{I_C} = \frac{5 \text{ V}}{1.3 \text{ MA}}$$

       = 3.75 kΩ   (Use a 3.9 kΩ standard value to make I_C a little less than the design level, thus ensuring that V_{CE} is not less than 3 V.)

$$R_2 = 10 R_E = 10 \times 3.9 \text{ k}\Omega$$
$$= 39 \text{ k}\Omega \text{ (standard value)}$$

$$I_2 = \frac{V_E + V_{BE}}{R_2} = \frac{5 \text{ V} + 0.7 \text{ V}}{39 \text{ k}\Omega}$$
$$\approx 146 \, \mu A$$

$$R_1 = \frac{V_{CC} - V_B}{I_2} = \frac{24 \text{ V} - 5.7 \text{ V}}{146 \, \mu A}$$
$$\approx 125 \text{ k}\Omega \qquad \text{(use 120 k}\Omega \text{ standard value)}$$

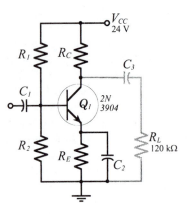

Figure 12-8
Common-emitter amplifier for Example 12-2.

Example 12-3
Determine suitable capacitor values for the *CE* amplifier in Ex. 12-2. The signal source resistance is 600 Ω, and the lower cutoff frequency is to be 100 Hz.

Solution
From Appendix 1-5, for the *2N3904:*
$$h_{fe} = 100, \text{ and } h_{ie} = 1 \text{ k}\Omega$$

Eq. 12-2, $$X_{C2} = \frac{h_{ie}}{1 + h_{fe}} = \frac{1 \text{ k}\Omega}{1 + 100}$$
$$\approx 10 \, \Omega$$

$$C_2 = \frac{1}{2 \pi f_1 X_{C2}} = \frac{1}{2 \pi \times 100 \text{ Hz} \times 10 \, \Omega}$$
$$= 159 \, \mu F \qquad \text{(use 180 } \mu F \text{ standard value)}$$

$$Z_i = R_1 \| R_2 \| h_{ie} = 120 \text{ k}\Omega \| 39 \text{ k}\Omega \| 1 \text{ k}\Omega$$
$$= 967 \, \Omega$$

from Eq. 12-4, $$C_1 = \frac{1}{2 \pi f_1 (Z_i + r_s)/10}$$

$$= \frac{1}{2\,\pi \times 100\text{ Hz} \times (967\ \Omega + 600\ \Omega)/10}$$

$$= 10.1\ \mu F \qquad \text{(use 10 } \mu F \text{ standard value)}$$

from Eq. 12-5, $\qquad C_3 = \dfrac{1}{2\,\pi\,f_1\,(R_C + R_L)/10}$

$$= \frac{1}{2\,\pi \times 100\text{ Hz} \times (12\text{ k}\Omega + 120\text{ k}\Omega)/10}$$

$$= 0.12\ \mu F \qquad \text{(standard value)}$$

The complete Common-source circuit designed in Examples 12-2 and 12-3 is shown in Fig. 12-9.

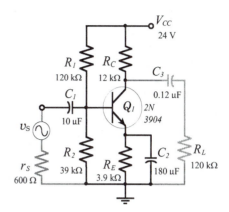

Figure 12-9
CE amplifier circuit designed in Examples 12-2 and 12-3.

Practise Problems

12-1.1 Determine suitable resistor values for a CE amplifier (see Fig. 12-9) that uses a 20 V supply and has a 68 kΩ capacitor-coupled load.

12-1.2 Determine suitable capacitor values for the circuit in Problem 12-1.1 to have a 40 Hz lower cutoff frequency and to use a 2N3903 transistor. Assume that $r_s = 500\ \Omega$.

12-2 Single-Stage Common-Source Amplifier

Specification

Bias circuit design for the common-source amplifier in Fig. 12-10 is treated in Chapter 10, and *ac* analysis of *FET* circuits is covered in Chapter 11. As with the common-emitter *BJT* circuit in Section 12-1, design commences with specification of the supply voltage, amplification, frequency response, load impedance, etc.

Selection of I_D, R_D, and R_S

The circuit shown in Fig. 12-10 has no provision for negative feedback, so, it should be designed to achieve the largest possible gain. From Eq. 11-7, the voltage gain of a *CS* circuit is,

$$A_v = -Y_{fs}\,(R_D \| R_L)$$

Because A_v is directly proportional to $R_D \| R_L$, design for the greatest voltage gain normally requires selection of the largest possible drain resistance. However, a very large value of R_D might make the drain current too small for satisfactory *FET* operation. Also, low I_D levels give small Y_{fs} values, which result in lower *ac* voltage gain. Furthermore, R_D should normally be much smaller than R_L, so that R_L has little effect on the circuit voltage gain.

For a given level of I_D, the largest possible voltage drop across R_D gives the greatest R_D value, $(R_D = V_{RD}/I_D)$. To make V_{RD} as large as possible, V_{DS} and V_S should be held to a minimum, [see Fig. 12-

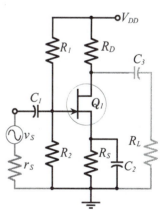
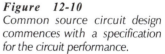

Figure 12-10
Common source circuit design commences with a specification for the circuit performance.

11(a)]. The drain-source voltage should typically be $V_{DS(min)} = (V_{p(max)} + 1\ V)$. This is large enough to ensure that the *FET* operates in the pinch-off region of its characteristics. It also allows for a drain voltage swing of ±1 V, which is usually adequate for a small-signal amplifier. If the gate-source bias voltage (V_{GS}) is other than zero, then, as illustrated in Fig. 12-11(a), the minimum V_{DS} should be calculated as,

$$V_{DS(min)} = V_{p(max)} + 1\ V - V_{GS} \qquad (12\text{-}8)$$

Recall from Chapter 10 that for good bias stability, the source resistor voltage drop (V_S) should be as large as possible. Where the supply voltage is small, V_S may be reduced to a minimum to allow for the minimum level of V_{DS}. A reasonable approach for most *FET* circuits is to calculate the the sum of V_S and V_{RD} from,

$$V_S + V_{RD} = V_{DD} - V_{DS(min)}$$

and then make,

$$V_S = V_{RD} = \frac{V_{DD} - V_{DS(min)}}{2} \qquad (12\text{-}9)$$

Once V_S, V_{RD}, and I_D are selected, R_D and R_S are calculated,

$$R_D = \frac{V_{RD}}{I_D}, \text{ and } R_S = \frac{V_S}{I_D} \text{ [see Fig. 12-11(b)]}$$

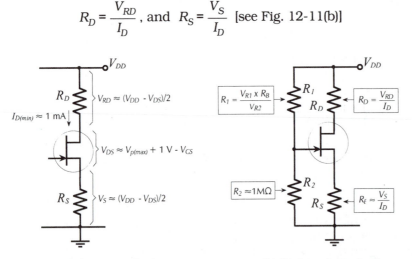

(a) Voltage and current selection   (b) Resistor determination

Figure 12-11
Voltage and current selection and resistor determination for a CS amplifier circuit.

As already discussed in Section 10-4, a bias line should be drawn upon the *FET* transfer characteristics to determine a suitable gate bias voltage (V_G). The selected maximum drain current ($I_{D(max)}$) is plotted on the maximum transfer characteristics for the *FET* used. The bias line is then drawn through this point with a slope of $1/R_S$. The gate bias voltage is read from the intersection of the bias line and the V_G scale. As an alternative to drawing the bias line, V_{GS} can be read from the transfer characteristic when $I_{D(max)}$ is plotted. Then,

$$V_G = V_S - V_{GS}$$

Instead of using the *FET* transfer characteristics, $V_{P(max)}$ and $I_{DSS(max)}$ can be substituted into Eq. 10-8 to calculate the V_{GS} level.

Bias Resistors

With a voltage divider bias circuit [as in Fig. 12-11(b)], R_2 is usually selected as 1 MΩ or less. Smaller resistance values may be used where a lower input impedance is acceptable. Larger resistance values may also be used, however, as discussed in Section 10-7, there are distinct disadvantages to using resistances higher than 1 MΩ. With R_2 determined, R_1 is calculated using R_2 and the ratio of V_{R1} to V_{R2}.

$$R_1 = \frac{V_{R1} \times R_2}{V_{R2}} \qquad \textbf{(12-10)}$$

Capacitors

As for a *BJT* capacitor-coupled circuit, coupling and bypass capacitors should be selected to have the smallest possible capacitance values. The largest capacitor in the circuit (source bypass capacitor C_2 in Fig 12-10) sets the circuit low 3 dB frequency (f_l). Equation 11-10 was developed for the voltage gain of a common-source circuit with an unbypassed source resistor (R_S). Rewriting the equation to include X_{C2} in parallel with R_S gives

$$A_v = \frac{-Y_{fs}(R_D\|R_L)}{1 + Y_{fs}(R_S\|X_{C2})}$$

Normally $R_S \gg X_{C2}$, so R_S can be omitted. Also, X_{C2} is capacitive,

so, $$A_v = \frac{-Y_{fs}(R_D\|R_L)}{\sqrt{[1^2 + (Y_{fs}X_{C2})^2]}} \qquad \textbf{(12-11)}$$

When $Y_{fs}X_{C2} = 1$,

so, $$|A_v| = \frac{-Y_{fs}(R_D\|R_L)}{\sqrt{(1^2 + 1^2)}} = \frac{\text{mid-frequency gain}}{\sqrt{2}}$$

$$= \text{(mid-frequency gain) - 3 dB}$$

Therefore, at f_l,

$$Y_{fs}X_{C2} = 1$$

or, $$X_{C2} = \frac{1}{Y_{fs}} \text{ at } f_1 \qquad \textbf{(12-12)}$$

From Section 11-7,

$$Z_S = \frac{1}{Y_{fs}} = \text{(impedance 'looking into' the } FET \text{ source)}$$

So, at f_l, $\quad X_{C2} = \text{(impedance 'looking into' the source terminal)}$
$\qquad\qquad = \text{(impedance in } series \text{ with } X_{C2}), \text{ (see Fig. 12-12)}$

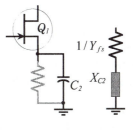

Figure 12-12
In a CS amplifier circuit, the signal voltage can be divided across the source bypass capacitor and the source resistance $(1/Y_{fs})$.

Equations 12-12 gives the smallest value for the source bypass capacitor. When selecting a standard value, the next larger capacitance value should be chosen. This will give a cutoff frequency slightly lower than the f_1 value used in the calculations.

As explained in Section 12-1, it is important to note that the bypass capacitor is calculated in terms of the resistance *seen looking into* the device terminal (the resistance in series with C_2). *The capacitance value is* **not** *determined in terms of the parallel resistor* (R_S).

The input and output coupling capacitors should have almost zero effect on the circuit frequency response. In Section 12-1 it is explained that X_{C1} in series with Z_i and X_{C3} in series with R_L constitute voltage dividers that can attenuate the *ac* input and output voltages. To minimize the attenuation, the reactance of each coupling capacitor is selected to be approximately one-tenth of the impedance in series with it at the lowest operating frequency for the circuit (f_1). Equations 12-4 and 12-5 apply once again for the calculation of the minimum C_1 and C_2 values.

Eq. 12-4, $$X_{C1} = \frac{Z_i + r_s}{10} \text{ at } f_1$$

Eq. 12-5, $$X_{C2} = \frac{Z_o + R_L}{10} \text{ at } f_1$$

As in *BJT* circuits, R_L is usually much larger than Z_o, and Z_i is often much larger than r_s, so Z_o and r_s can be omitted in Equations 12-4 and 12-5. As always, the equations give minimum capacitance values, so that the next larger standard values should always be selected for C_1 and C_3.

Alternatively, as discussed in Section 12-1, Equations 12-6 and 12-7 may be used for determining coupling capacitor values.

Design   Calculations

Example   12-4
Calculate suitable resistor values for the common-source circuit in Fig. 12-13.

Solution
Refer to the *2N5459* transfer characteristics in Fig. 9-19. A reasonable level of $I_{D(max)}$ might be 1.5 mA. Y_{fs} becomes smaller at lower I_D levels.

Eq. 10-8, $$V_{GS} = V_P[1 - \sqrt{(I_D/I_{DSS})}]$$

Using $V_{P(max)}$ and $I_{DSS(max)}$ from Appendix 1-11, (and from Fig. 9-19)

$$V_{GS} = 8 \text{ V } [1 - \sqrt{(1.5 \text{ mA}/16 \text{ mA})}]$$
$$= 5.6 \text{ V}$$

Eq. 12-8, $$V_{DS(min)} = V_{P(max)} + 1 \text{ V} - V_{GS} = 8 \text{ V} + 1 \text{ V} - 5.6 \text{ V}$$
$$= 3.4 \text{ V}$$

Eq. 12-9,
$$V_S = V_{RD} = \frac{V_{DD} - V_{DS(min)}}{2} = \frac{24 \text{ V} - 3.4 \text{ V}}{2}$$
$$= 10.3 \text{ V}$$

$$R_S = R_D = \frac{V_{RD}}{I_D} = \frac{10.3 \text{ V}}{1.5 \text{ mA}}$$

= 6.87 kΩ (use 6.8 kΩ standard value)

R_D should be $<< R_L$

6.8 kΩ $<<$ 120 kΩ

$$V_{R2} = V_G = V_S - V_{GS} = 10.3 \text{ V} - 5.5 \text{ V}$$
$$= 4.8 \text{ V}$$

$$V_{R1} = V_{DD} - V_G = 24 \text{ V} - 4.8 \text{ V}$$
$$= 19.2 \text{ V}$$

select
$$R_2 = 1 \text{ MΩ}$$

Eq. 12-10,
$$R_1 = \frac{V_{R1} \times R_2}{V_{R2}} = \frac{19.2 \text{ V} \times 1 \text{ MΩ}}{4.8 \text{ V}}$$

= 4 MΩ   (Use 4.7 MΩ standard value to ensure that V_G is slightly lower than the design level. This will give a lower level of I_D and a higher V_{DS} than the minimum design level.)

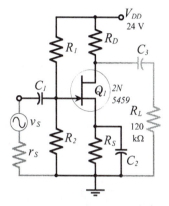

Figure 12-13
Common source amplifier circuit for Ex. 12-4.

Example 12-5
Determine suitable capacitor values for the common-source amplifier in Ex. 12-4. The signal source resistance is 600 Ω, and the lower cutoff frequency is to be 100 Hz.

Solution
From Appendix 1-11, for the *2N5459:*

$$Y_{fs(max)} = 6000 \text{ μS}$$

Eq. 12-12,
$$X_{C2} = \frac{1}{Y_{fs}} = \frac{1}{6000 \text{ μS}}$$

$$\approx 167 \text{ Ω}$$

$$C_2 = \frac{1}{2 \pi f_1 X_{C2}} = \frac{1}{2 \pi \times 100 \text{ Hz} \times 167 \text{ Ω}}$$

= 9.5 μF   (use 10 μF standard value)

$$Z_i = R_1 \| R_2 = 4.7 \text{ MΩ} \| 1 \text{ MΩ}$$
$$= 825 \text{ kΩ}$$

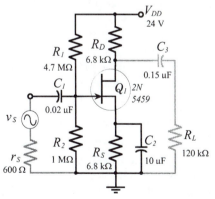

Figure 12-14
Common source circuit designed in Examples 12-4 and 12-5.

from Eq. 12-4, $C_1 = \dfrac{1}{2\pi f_1 (Z_i + r_s)/10}$

$= \dfrac{1}{2\pi \times 100\ \text{Hz} \times (825\ \text{k}\Omega + 600\ \Omega)/10}$

$= 0.019\ \mu\text{F}$ (use 0.02 μF standard value)

from Eq. 12-5, $C_3 = \dfrac{1}{2\pi f_1 (R_D + R_L)/10}$

$= \dfrac{1}{2\pi \times 100\ \text{Hz} \times (6.8\ \text{k}\Omega + 120\ \text{k}\Omega)/10}$

$= 0.13\ \mu\text{F}$ (use 0.15 μF standard value)

Practise Problems

12-2.1 A common-source amplifier (see Fig. 12-13) is to use a *2N5458 FET* and a 25 V supply. The external load is 100 kΩ. Determine suitable resistor values for the circuit.

12-2.2 Determine suitable capacitor values for the circuit in Problem 12-2.1 to give a 30 Hz lower cutoff frequency.

12-3   *Capacitor-Coupled Two-Stage CE Amplifier*

Circuit

A capacitor-coupled two-stage amplifier circuit is shown in Fig. 12-15. Each stage is similar to the single-stage circuit in Fig. 12-9. *Stage-1* is capacitor-coupled (via C_3) to the input of *Stage-2*. The signal is applied to the input of *Stage-1*, and the load is coupled to the output of *Stage-2*. The signal is amplified by *Stage-1*, and the output of *Stage-1* is amplified by *Stage-2*, so that the overall voltage gain is much greater than the gain of a single stage. As illustrated by the circuit waveforms in Fig. 12-16, the signal voltage is phase-shifted through 180° by *Stage-1*, and through a further 180° by *Stage-2*. Consequently, the overall phase shift from input to output is zero (or 360°).

Circuit Design

The simplest approach to design of this circuit is to make each stage identical. Then, when *Stage-2* has been designed, the components for *Stage-1* are selected as: $R_1 = R_5$, $R_2 = R_6$, $R_3 = R_7$, $R_4 = R_8$, $C_1 = C_3$, and $C_2 = C_4$. Note that C_3 is calculated in terms of the input resistance to *Stage-2*, which should be identical to that for *Stage-1* if the stages are otherwise identical. C_5 is calculated in the usual way for an output coupling capacitor.

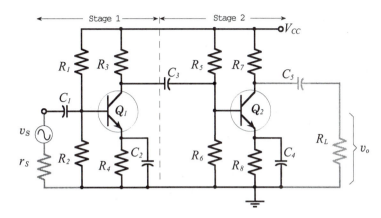

Figure 12-15
A two-stage capacitor-coupled CE amplifier consists of two similar CE circuits with capacitor-coupling between stages.

One way the design of a two-stage circuit differs from that of a single-stage circuit is in determination of emitter bypass capacitors C_2 and C_4. Consider Eq. 12-2, which was derived to calculate a bypass capacitor value to make the voltage gain of a single-stage circuit 3 dB down from the mid-frequency gain at the selected lower cutoff frequency (f_1).

$$X_C = \frac{h_{ie}}{1 + h_{fe}} \text{ at } f_1$$

If this equation is used to calculate the capacitance of C_2 and C_4 in the two-stage circuit, it is found that the gain of *Stage 1* is down by 3 dB at f_1, and the gain of *Stage 2* is also down by 3 dB at f_1, (see Fig. 12-17). This means that the overall gain of the circuit is down by a total of 6 dB at frequency f_1. For a 3 dB decrease in overall voltage gain at f_1, the bypass capacitors must be calculated to give a 1.5 dB reduction in the gain of each stage. A 1.5 dB reduction in stage gain at f_1, is achieved by making the reactance of the emitter bypass capacitor equal to 0.65 times the impedance in series with the capacitor. Thus, the equation for the bypass capacitor impedance now becomes,

$$\mathbf{X_{C2}} = \mathbf{X_{C4}} = \frac{\mathbf{0.65\ h_{ie}}}{\mathbf{1 + h_{fe}}} \textbf{ at } f_1 \qquad (12\text{-}13)$$

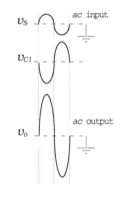

Figure 12-16
Voltage waveforms for a two-stage CE circuit.

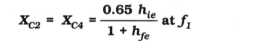
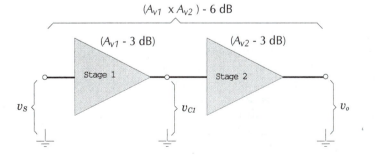

Figure 12-17
If the voltage gain of each stage of a two-stage circuit is down by 3 dB at f_1 the overall voltage gain is down by 6 dB.

Example 12-6

Design a two-stage capacitor-coupled amplifier, as shown in Fig. 12-15, to fulfil the specification used for Examples 12-2 and 12-3: 2N3904 transistors, $V_{CC} = 24$ V, $R_L = 120$ kΩ, $f_1 = 100$ Hz.

Solution

Except for the emitter bypass capacitors, each stage is designed exactly as in Examples 12-2 and 12-3. Therefore, $R_3 = R_7 = 12$ kΩ, $R_4 = R_8 = 3.9$ kΩ, $R_1 = R_5 = 120$ kΩ, $R_2 = R_6 = 39$ kΩ, $C_1 = C_3 = 10\ \mu$F, $C_5 = 0.12\ \mu$F.

From Appendix 1-5, for the 2N3904: $h_{fe} = 100$, and $h_{ie} = 1$ kΩ

Eq. 12-13, $$X_{C4} = \frac{0.65\ h_{ie}}{1 + h_{fe}} = \frac{0.65 \times 1\ \text{k}\Omega}{1 + 100}$$

$$\approx 6.4\ \Omega$$

$$C_2 = C_4 = \frac{1}{2\ \pi\ f_1\ X_{C4}} = \frac{1}{2\ \pi \times 100\ \text{Hz} \times 6.4\ \Omega}$$

$$= 249\ \mu\text{F} \qquad \text{(use 250}\ \mu\text{F standard value)}$$

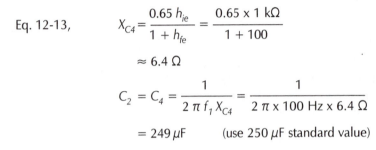

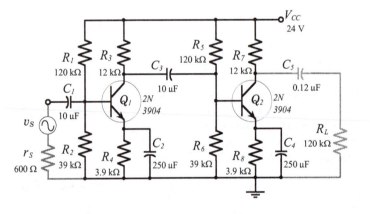

Figure 12-18
Two-stage capacitor-coupled CE circuit designed in Example 12-6.

Circuit Analysis

Ac analysis of a two-stage, capacitor-coupled circuit is similar to analysis of single-stage circuits, as covered in Chapter 6. The h-parameter equivalent circuit for the amplifier circuit in Fig. 12-15 is drawn in Fig. 12-19. Once again, this equivalent circuit is produced by replacing the supply voltage and all capacitors with short circuits, and substituting device h-parameter models for each transistor. The resultant equivalent circuit simply consists of two single-stage h-parameter equivalent circuits. As always, voltage polarities and current directions are indicated for an instantaneous positive-going signal voltage. The voltage polarities show a 180° phase shift between v_i and v_{c1}, and a further 180° phase shift between v_{c1} and v_o,

The performance equations for the two-stage circuit are readily derived from the equations for a single-stage circuit. Input and output impedance are exactly the same as for the single-stage circuit:

$$Z_i = R_1\|R_2\|h_{ie} \qquad \text{(12-14)}$$

$$Z_o \approx R_7 \qquad \text{(12-15)}$$

The voltage gain equation for the second stage is exactly the same as the single-stage gain equation. The capacitor-coupled load at the collector terminal of the first stage is the input impedance of the second stage. So the voltage gain equations are,

$$A_{v1} = \frac{-h_{fe1}\,(R_3\|Z_{i2})}{h_{ie1}} \qquad \text{(12-16)}$$

and,

$$A_{v2} = \frac{-h_{fe2}\,(R_7\|R_L)}{h_{ie2}} \qquad \text{(12-17)}$$

The overall voltage gain is found by multiplying the two individual stage gains:

$$A_v = A_{v1} \times A_{v2} \qquad \text{(12-18)}$$

Equations for current gain and power gain can also be derived from the single-stage equations, but these quantities are normally not important.

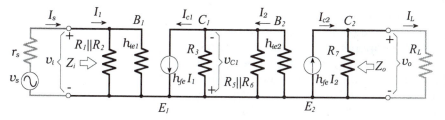

Figure 12-19
The h-parameter equivalent circuit for a two-stage CE amplifier is drawn by cascading the equivalent circuits for each stage.

Example 12-7

Analyze the two-stage amplifier circuit designed in Example 12-6 to determine its input impedance, output impedance, and minimum overall voltage gain.

Solution

Eq. 12-14, $Z_i = R_1\|R_2\|h_{ie} = 120\ \text{k}\Omega\|39\ \text{k}\Omega\|1\ \text{k}\Omega$

$ = 967\ \Omega$

Eq. 12-15, $Z_o \approx R_7 = 12\ \text{k}\Omega$

From Eq. 12-14, $Z_{i2} = R_5\|R_6\|h_{ie2} = 120\ \text{k}\Omega\|39\ \text{k}\Omega\|1\ \text{k}\Omega$

$\phantom{From Eq. 12-14,\quad Z_{i2}\ } = 967\ \Omega$

Eq. 12-16,
$$A_{v1} = \frac{-h_{fe1}\,(R_3\|Z_{i2})}{h_{ie1}} = \frac{-100 \times (12\ \text{k}\Omega\|967\ \Omega)}{1\ \text{k}\Omega}$$

$$\approx -89$$

Eq. 12-17,
$$A_{v2} = \frac{-h_{fe2}\,(R_7\|R_L)}{h_{ie2}} = \frac{-100 \times (12\ \text{k}\Omega\|120\ \text{k}\Omega)}{1\ \text{k}\Omega}$$

$$= -1090$$

Eq. 12-18,
$$A_v = A_{v1} \times A_{v2} = (-89) \times (-1090)$$

$$= 97\,000$$

Practise Problems

12-3.1   Determine suitable resistor values for a capacitor-coupled two-stage CE amplifier with a 15 V supply and a 82 kΩ external load.

12-3.2   Determine suitable capacitor values for the circuit in Problem 12-3.1. The circuit uses 2N3903 BJTs, the lower cutoff frequency is 70 Hz, and the signal source resistance is 1 kΩ.

12-3.3   Calculate the overall voltage gain for the circuit in Problems 12-3.1 and 12-3.2.

12-4   Direct-Coupled Two-Stage Circuits

Direct-Coupled Circuit

For economy, the number of components used in any circuit should be kept to a minimum. The use of direct coupling between stages is one way of eliminating components. Figure 12-20 shows a circuit that has the base of transistor Q_2 directly coupled to the collector of Q_1. Comparing this to the capacitor-coupled circuit in Fig. 12-15, it is seen that the two bias resistors for *Stage 2* and the interstage coupling capacitor have been eliminated. This is a saving of only three components, however, the total savings can be considerable when many similar circuits are to be manufactured.

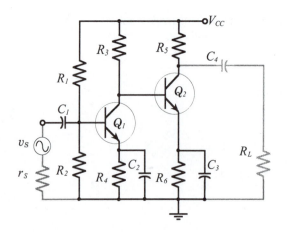

Figure 12-20
Two-stage direct-coupled CE amplifier. Because Q_2 base is directly coupled to Q_1 collector, the inter-stage coupling capacitor and the bias resistors for Q_2 are eliminated.

Circuit Design

The first step in the design of the circuit in Fig. 12-20 involves determining a suitable level of base bias voltage for Q_2. This is done by estimating satisfactory levels of V_{E1} and V_{CE1} for transistor Q_1. Then,

$$V_{B2} = V_{C1} = V_{E1} + V_{CE1}$$

$$V_{E2} = V_{B2} - V_{BE}$$

and,

$$V_{R5} = V_{CC} - V_{E2} - V_{CE2}$$

As discussed in Section 12-1, appropriate *dc* voltage levels for small-signal amplifiers are $V_{CE} = 3$ V and $V_E = 5$ V, except in the case of a very low supply voltage.

As for all amplifier circuits, the resistor at the collector of the output transistor (R_5 in Fig. 12-20) should be much smaller than the external load resistor. Once R_5 is selected, I_{C2} can be calculated using the voltage V_{R5} already determined. If I_{C2} looks too small for satisfactory operation of the transistor, a suitable current level should be selected and a new value of R_5 calculated. The collector current of Q_1 is determined by making I_{C1}, very much greater than the base current for Q_2. This is done to ensure that I_{B2} has a negligible effect on the bias conditions of Q_1. Normally, just making I_{C1} equal to I_{C2} is the simplest way to achieve the desired effect.

Example 12-8

Calculate suitable resistor values for a two-stage, direct-coupled amplifier, as in Fig. 12-20. The circuit is to use *2N3903* transistors, 14 V supply, 40 kΩ load resistance, and a signal source with a 600 Ω resistance.

Solution

Select, $V_{E1} = 5$ V, $V_{CE1} = V_{CE2} = 3$ V

$$V_{B2} = V_{C1} = V_{E1} + V_{CE1} = 5 \text{ V} + 3 \text{ V}$$
$$= 8 \text{ V}$$

$$V_{E2} = V_{B2} - V_{BE} = 8 \text{ V} - 0.7 \text{ V}$$
$$= 7.3 \text{ V}$$

$$V_{R5} = V_{CC} - V_{E2} - V_{CE2} = 14 \text{ V} - 7.3 \text{ V} - 3 \text{ V}$$
$$= 3.7 \text{ V}$$

Select, $$R_5 = \frac{R_L}{10} = \frac{40 \text{ k}\Omega}{10}$$
$$= 4 \text{ k}\Omega \text{ (use 3.9 k}\Omega \text{ standard value)}$$

$$I_{C2} = \frac{V_{R5}}{R_5} = \frac{3.7 \text{ V}}{3.9 \text{ k}\Omega}$$
$$= 949 \text{ }\mu\text{A}$$

$$R_6 = \frac{V_{E2}}{I_{C2}} = \frac{7.3 \text{ V}}{949 \text{ } \mu A}$$

$$= 7.7 \text{ k}\Omega \text{ (use 8.2 k}\Omega \text{ and recalculate } I_{C2})$$

$$I_{C2} = \frac{V_{E2}}{R_6} = \frac{7.3 \text{ V}}{8.2 \text{ k}\Omega}$$

$$= 890 \text{ } \mu A$$

$$I_{C1} >> I_{B2}, \text{ Select } I_{C1} = 1 \text{ mA}$$

$$V_{R3} = V_{CC} - V_{C1} = 14 \text{ V} - 8 \text{ V}$$

$$= 6 \text{ V}$$

$$R_3 = \frac{V_{R3}}{I_{C1}} = \frac{6 \text{ V}}{1 \text{ mA}}$$

$$= 6 \text{ k}\Omega \text{ (use 5.6 k}\Omega \text{ and recalculate } I_{C1}$$
$$\text{in order to keep } V_{B2} = 8 \text{ V })$$

$$I_{C1} = \frac{V_{R3}}{R_3} = \frac{6 \text{ V}}{5.6 \text{ k}\Omega}$$

$$= 1.07 \text{ mA}$$

$$R_4 = \frac{V_{E1}}{I_{C1}} = \frac{5 \text{ V}}{1.07 \text{ mA}}$$

$$\approx 4.7 \text{ k}\Omega \text{ (standard value)}$$

$$V_{R2} = V_{B1} = V_{E1} + V_{BE} = 5 \text{ V} + 0.7 \text{ V}$$

$$= 5.7 \text{ V}$$

$$V_{R1} = V_{CC} - V_{B1} = 14 \text{ V} - 5.7 \text{ V}$$

$$= 8.3 \text{ V}$$

$$R_2 = 10 \text{ } R_4 = 10 \times 4.7 \text{ k}\Omega$$

$$= 47 \text{ k}\Omega$$

$$I_2 = \frac{V_{B1}}{R_2} = \frac{5.7 \text{ V}}{47 \text{ k}\Omega}$$

$$= 121 \text{ } \mu A$$

Eq. 10-11, $$\qquad R_1 = \frac{V_{R1} \times R_2}{V_{R2}} = \frac{8.3 \text{ V} \times 47 \text{ k}\Omega}{5.7 \text{ V}}$$

$$= 68.4 \text{ k}\Omega \text{ (use 68 k}\Omega \text{ standard value)}$$

Example 12-9

Determine suitable capacitor values for the two-stage direct-coupled amplifier in Ex. 12-8. The lower cutoff frequency is to be 75 Hz.

Solution

From Appendix 1-5, for the 2N3903: $h_{fe(min)} = 50$, and $h_{ie(min)} = 1 \text{ k}\Omega$

$$Z_i = R_1 \| R_2 \| h_{ie} = 68\text{ k}\Omega \| 47\text{ k}\Omega \| 1\text{ k}\Omega$$
$$= 965\ \Omega$$

Eq. 12-4, $$X_{C1} = \frac{Z_i + r_s}{10} = \frac{965\ \Omega + 600\ \Omega}{10}$$
$$= 157\ \Omega$$

$$C_1 = \frac{1}{2\pi f_1 X_{C1}} = \frac{1}{2\pi \times 75\text{ Hz} \times 157\ \Omega}$$
$$= 13.5\ \mu F \text{ (use 15 } \mu F \text{ standard value)}$$

Eq. 12-13, $$X_{C2} = X_{C3} = \frac{0.65\ h_{ie}}{1 + h_{fe}} = \frac{0.65 \times 1\text{ k}\Omega}{1 + 50}$$
$$\approx 12.7\ \Omega$$

$$C_2 = C_3 = \frac{1}{2\pi f_1 X_{C2}} = \frac{1}{2\pi \times 75\text{ Hz} \times 12.7\ \Omega}$$
$$= 167\ \mu F \qquad \text{(use 180 } \mu F \text{ standard value)}$$

from Eq. 12-5, $$X_{C4} = \frac{R_5 + R_L}{10} = \frac{3.9\text{ k}\Omega + 40\text{ k}\Omega}{10}$$
$$= 4.39\text{ k}\Omega$$

$$C_4 = \frac{1}{2\pi f_1 X_{C4}} = \frac{1}{2\pi \times 75\text{ Hz} \times 4.39\text{ k}\Omega}$$
$$= 0.48\ \mu F \qquad \text{(use 0.47 } \mu F \text{ standard value)}$$

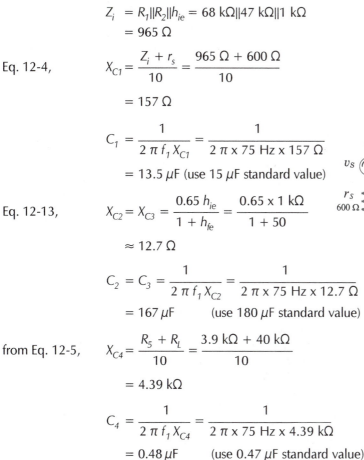
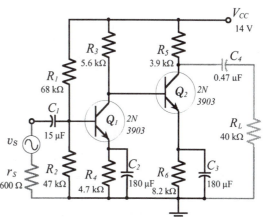
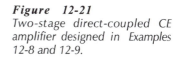

Figure 12-21
Two-stage direct-coupled CE amplifier designed in Examples 12-8 and 12-9.

Circuit Analysis

Analysis procedure for a two-stage direct-coupled circuit is similar to that for analysis of a two-stage capacitor-coupled circuit. The *h*-parameter equivalent for the circuit of Fig. 12-20 is exactly like Fig. 12-19, except that the bias resistors for the second stage are omitted. The voltage gain and impedance equations are as determined for the capacitor-coupled circuit. Because the component numbers differ slightly for the two circuits, care must be taken in substituting components into the equations. For example, from Eq. 12-17, the voltage gain of the second stage of the circuit in Fig. 12-20 is,

$$A_{v2} = \frac{-h_{fe2}\,(R_5 \| R_L)}{h_{ie2}}$$

Example 12-10
Calculate the minimum overall voltage gain for the circuit in Fig. 12-21.

Solution

Eq. 12-16,

$$A_{v1} = \frac{-h_{fe1}\,(R_3\|Z_{i2})}{h_{ie1}} = \frac{-h_{fe1}\,(R_3\|h_{ie2})}{h_{ie1}}$$

$$= \frac{-50 \times (5.6\ k\Omega\|1\ k\Omega)}{1\ k\Omega}$$

$$\approx -42.4$$

Eq. 12-17,

$$A_{v2} = \frac{-h_{fe2}\,(R_5\|R_L)}{h_{ie2}} = \frac{-50 \times (3.9\ k\Omega\|40\ k\Omega)}{1\ k\Omega}$$

$$= -177.7$$

Eq. 12-18,

$$A_v = A_{v1} \times A_{v2} = (-42.4) \times (-177.7)$$

$$= 7533$$

Use of Complementary Transistors

The direct-coupled, two-stage circuit illustrated in Fig. 12-22 is similar to that in Fig. 12-20, except that transistor Q_2 is a *pnp* device. The transistors are selected to have similar characteristics and parameters although one is *npn* and the other is *pnp*. This means that they are *complementary transistors*. Examination of Appendices 1-5 and 1-6 reveals that the *2N3903* and the *2N3905* are complementary, as are the *2N3904* and the *2N3906*.

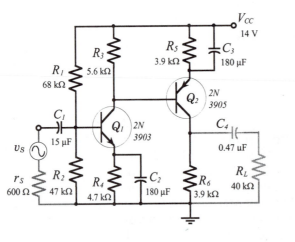

Figure 12-22
Direct-coupled two-stage CE amplifier using complementary transistors. Q_1 and Q_2 have similar parameters, although one is an npn transistor and the other is a pnp device.

Suppose the circuit in Fig. 12-22 is to be designed to use a 14 V supply, as in Example 12-8. Allowing $V_{E1} = 5$ V and $V_{CE1} = 3$ V, the base voltage for Q_2 (measured from $+V_{CC}$) is,

$$V_{B2} = V_{R3} = 6\ V$$

and

$$V_{R5} = V_{B2} - V_{BE} = 6\ V - 0.7\ V$$

$$= 5.3\ V$$

so,

$$V_{R6} = V_{CC} - V_{E2} - V_{CE2} = 14\ V - 5.3\ V - 3\ V$$

$$= 5.7\ V$$

The use of complementary transistors in Fig. 12-22 reduces the voltage drop across the emitter resistor of Q_2, compared to the situation in Fig. 12-20. This makes more voltage available for dropping across the Q_2 collector resistor.

The design procedure for the circuit using complementary transistors is very similar to the procedure for designing the non-complementary circuit in Fig. 12-20.

Practise Problems

12-4.1 A direct-coupled two-stage *CE* amplifier using complementary transistors (see Fig. 12-22) is to be designed to operate from a 25 V supply and to use a *2N3904* and a *2N3906* transistor. The capacitor-coupled load is 56 kΩ. Determine suitable resistor values.

12-4.2 Determine suitable capacitor values for the circuit in Problem 12-4.1 to give a 30 Hz lower cutoff frequency. Assume that $r_s = 300\ \Omega$.

12-4.3 Calculate the minimum overall voltage gain for the circuit in Practise Problems 12-4.1 and 12-4.2.

12-5 Two-Stage Circuit with Emitter Follower Output

Circuit

Figure 12-23 shows another direct-coupled circuit. This time the second stage is a common collector circuit, or emitter follower. *Stage 2* gives the circuit a very low output impedance, but has unity voltage gain. *Stage 1* still has substantial voltage gain.

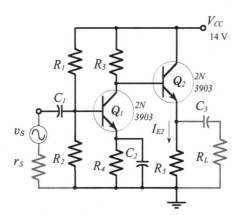

Figure 12-23
Two-stage direct-coupled CE amplifier consisting of a CE input stage and a CC (emitter follower) output stage for low output impedance.

The design procedure for this circuit is similar to the procedures already discussed. As with the circuit in Fig. 12-20, the Q_2 base bias voltage is the collector voltage of Q_1. A suitable level of emitter current I_{E2} is selected, and R_E is calculated to be $(V_{B2} - V_{BE})/I_{E2}$. Coupling capacitor C_1 is once again calculated from $X_{C1} = Z_i/10$.

The load resistor (R_L) may be relatively small, which means that coupling capacitor C_3 must be large. Therefore, both C_2 and C_3 are used to determine the low 3 dB frequency of the circuit by using the 0.65 factor employed in Eq. 12-14.

It is important to realize that the emitter follower in Fig. 12-23 is a small-signal circuit. In fact, this type of circuit functions best when the amplitude of the *ac* output voltage is much smaller than the *dc* base-emitter voltage of the output transistor (V_{BE}). With *ac* output voltages that approach or exceed V_{B2}, the base-emitter junction of Q_2 might become reverse biased when the output voltage moves rapidly in a negative direction. In this case, the circuit will not function correctly. Where large output voltages and low output impedance are required, a complementary emitter follower is used, (see Section 18-4).

With the second stage operating as a small-signal emitter follower, resistor R_5 in Fig. 12-23 does *not* have to be very much smaller than external load R_L. However, for satisfactory operation, Q_2 emitter current (I_{E2}) should be greater than the peak *ac* load current (i_p). The peak output current is calculated by dividing the desired peak output voltage v_p, by the load resistance; $i_p = v_p/R_L$.

Example 12-11

Determine suitable resistor values for the two-stage, amplifier circuit in Fig. 12-23. The circuit is to use *2N3903* transistors, the supply voltage is 20 V, the load resistance is 100 Ω, and the output voltage is to be ±100 mV.

Solution

$$i_p = \frac{v_p}{R_L} = \frac{100 \text{ mV}}{100 \text{ Ω}}$$

$$= 1 \text{ mA}$$

$$I_{E2} > i_p \quad \text{Select } I_{E2} = 2 \text{ mA}$$

Select, $\qquad V_{E1} = 5 \text{ V and } V_{CE1} = 3 \text{ V}$

$$V_{B2} = V_{C1} = V_{E1} + V_{CE1} = 5 \text{ V} + 3 \text{ V}$$

$$= 8 \text{ V}$$

$$V_{E2} = V_{B2} - V_{BE} = 8 \text{ V} - 0.7 \text{ V}$$

$$= 7.3 \text{ V}$$

$$R_5 = \frac{V_{E2}}{I_{E2}} = \frac{7.3 \text{ V}}{2 \text{ mA}}$$

$$= 3.65 \text{ kΩ (use 3.3 kΩ standard value)}$$

$$I_{C1} >> I_{B2}, \text{ select } I_{C1} = 1 \text{ mA}$$

$$V_{R3} = V_{CC} - V_{B2} = 20 \text{ V} - 8 \text{ V}$$

$$= 12 \text{ V}$$

$$R_3 = \frac{V_{R3}}{I_{C1}} = \frac{12\ V}{1\ mA}$$

$$= 12\ k\Omega\ \text{(standard value)}$$

$$R_4 = \frac{V_{E1}}{I_{C1}} = \frac{5\ V}{2\ mA}$$

$$= 5\ k\Omega\ \text{(use 4.7 k}\Omega\text{ standard value)}$$

$$V_{B1} = (I_{C1} \times R_4) + V_{BE} = (1\ mA \times 4.7\ k\Omega) + 0.7\ V$$

$$= 5.4\ V$$

Select, $R_2 = 10\ R_4 = 10 \times 4.7\ k\Omega$

$$= 47\ k\Omega$$

Eq 12-11, $R_1 = \dfrac{V_{R1} \times R_2}{V_{R2}} = \dfrac{(20\ V - 5.4\ V) \times 47\ k\Omega}{5.4\ V}$

$$= 127\ k\Omega\ \text{(use 120 k}\Omega\text{ standard value)}$$

Example 12-12

Calculate suitable capacitor values for the circuit designed in Example 12-11. The signal source resistance is 600 Ω, and the low 3-dB frequency is 150 Hz.

Solution

From Appendix 1-5, for the *2N3903:* $h_{fe(min)} = 50$, and $h_{ie(min)} = 1\ k\Omega$

$$Z_i = R_1 \| R_2 \| h_{ie} = 120\ k\Omega \| 47\ k\Omega \| 1\ k\Omega$$

$$= 971\ \Omega$$

$$X_{C1} = \frac{Z_i + r_s}{10} = \frac{971\ \Omega + 600\ \Omega}{10}$$

$$= 157\ \Omega$$

$$C_1 = \frac{1}{2\ \pi\ f_1\ X_{C1}} = \frac{1}{2\ \pi \times 150\ Hz \times 157\ \Omega}$$

$$= 6.8\ \mu F \qquad \text{(standard value)}$$

Eq. 12-14, $X_{C2} = \dfrac{0.65\ h_{ie}}{1 + h_{fe}} = \dfrac{0.65 \times 1\ k\Omega}{1 + 50}$

$$\approx 12.7\ \Omega$$

$$C_2 = \frac{1}{2\ \pi\ f_1\ X_{C2}} = \frac{1}{2\ \pi \times 150\ Hz \times 12.7\ \Omega}$$

$$= 83.6\ \mu F \quad \text{(use 82 } \mu F \text{ standard value)}$$

From Eqns. 6-24 and 6-25,

$$Z_o = \left[\frac{h_{ic} + R_3}{h_{fc}}\right] \| R_5 = \left[\frac{1\ k\Omega + 12\ k\Omega}{51}\right] \| 3.3\ k\Omega$$

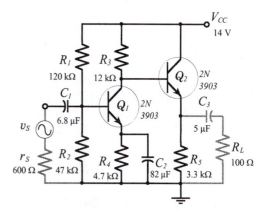

Figure 12-24
Two-stage direct-coupled CE amplifier designed in Examples 12-11 and 12-12.

$$= 237 \, \Omega$$

$$X_{C3} \approx 0.65 \, (R_L + Z_o) = 0.65 \, (100 \, \Omega + 237 \, \Omega)$$
$$= 219 \, \Omega$$

$$C_3 = \frac{1}{2 \pi f_1 X_{C3}} = \frac{1}{2 \pi \times 150 \, \text{Hz} \times 219 \, \Omega}$$

$$= 4.8 \, \mu F \qquad \text{(use 5 } \mu F \text{ standard value)}$$

Example 12-13
Analyze the two-stage transistor amplifier circuit in Fig. 12-24 to determine the minimum overall voltage gain.

Solution
From Eqn. 6-23 $Z_{i2} = h_{ic2} + h_{fc2}(R_5\|R_L) = 1 \, k\Omega + 51(3.3 \, k\Omega\|100 \, \Omega)$
$$= 5.95 \, k\Omega$$

Eq. 12-16, $A_{v1} = \dfrac{-h_{fe1} \, (R_3\|Z_{i2})}{h_{ie1}} = \dfrac{-50 \times (12 \, k\Omega\|5.95 \, k\Omega)}{1 \, k\Omega}$
$$= -199$$

From Eq. 6-27, $A_{v2} \approx 1$

From Eq. 12-18, $A_v = A_{v1} \times A_{v2} = -199 \times 1$
$$= -199$$

Practise   Problems

12-5.1   A two-stage direct-coupled *BJT* amplifier with an emitter-follower output is to be designed to operate from a 15 V supply. The capacitor-coupled load is 60 Ω. Using 2N3906 (*pnp*) transistors, determine suitable resistor values for the circuit.

12-5.2   Determine suitable capacitor values for the circuit in Problem 12-5.1 to give a 50 Hz lower cutoff frequency. Also, calculate the circuit minimum overall voltage gain. Assume that $r_s = 350 \, \Omega$.

12-6   DC Feedback Pair

DC Feedback Pair With Two Amplification Stages

The two-stage amplifier circuit shown in Fig. 12-25(a) is known as a *dc feedback pair*. The base of transistor Q_2 is directly connected to the collector of Q_1, and the base of Q_1 is biased from the emitter of Q_2 via resistor R_2. Comparing this circuit to the capacitor-coupled circuit of Fig. 12-15 and the direct-coupled circuit of Fig. 12-20, it is seen that a further saving in components has been effected: R_1, R_2, R_4, and C_2 in Fig. 12-20 are now replaced with the single resistor R_2 in Fig. 12-25.

To understand the bias arrangement for the *dc* feedback pair, note that there is only a *pn* junction voltage drop V_{BE2} between the Q_1 collector and the Q_2 emitter. This is illustrated in Fig. 12-25(b), where it is seen that the first stage of the circuit is essentially a collector-to-base bias circuit. *Stage 2* bias arrangement is similar to voltage divider bias, with resistor R_4 stabilizing the transistor emitter current. The base voltage for Q_2 is derived from the collector of Q_1; consequently, the bias stability of the second stage is no better than that of the first stage.

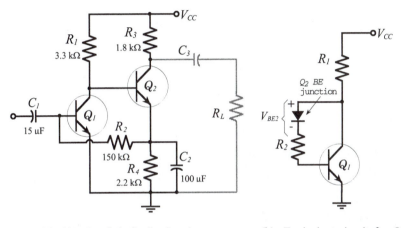

(a) Circuit of *dc* feedback pair   (b) Equivalent circuit for Q_1

Figure 12-25
In a dc feedback pair the base of Q_2 is directly connected to the collector of Q_1, and the base of Q_1 is biased via R_2 from the emitter of Q_2. This arrangement uses fewer components than a capacitor-coupled two-stage circuit.

Because collector-to-base bias is not as stable as voltage divider bias, the bias conditions in this circuit are not as stable as in the circuit of Fig. 12-20. However, the stability is adequate for most purposes, and the component savings can be a major advantage.

The *dc* feedback pair derives its name from the direct coupling between stages and from the feedback from the collector of Q_1 to its base. As explained for collector-to-base bias in Section 5-3, a change in the level of Q_1 collector voltage (V_{C1}), produces a change in base current (I_{B1}). The I_{B1} change in turn alters I_{C1}, and the voltage drop $I_{C1}R_1$ tends to push V_{C1} back towards its original level.

As well as stabilizing the circuit *dc* conditions, the feedback can produce *ac* degeneration, (see Section 6-1). Capacitor C_2 shorts the *ac* feedback to ground, to eliminate *ac* degeneration from *Stage 1*. As the emitter bypass capacitor, C_2 also eliminates *ac* degeneration from *Stage 2*.

The design approach to the *dc* feedback pair is similar to those already discussed. The output stage is designed exactly like the output stage for the direct-coupled circuit in Fig. 12-20. The input stage is designed as a collector-to-base bias circuit to have a collector voltage (V_{C1}) equal to the desired level of the *Stage 2* base voltage (V_{B2}).

The input impedance of the circuit is

$$Z_i = R_2 \| h_{ie} \qquad\qquad \textbf{(12-19)}$$

So, (From Eq. 12-4) the impedance of C_1 at f_1 is,

$$X_{C1} = \frac{(R_2 \| h_{ie}) + r_s}{10}$$

C_2 is by far the largest capacitor in the circuit, so it determines the circuit low 3 dB frequency:

Eq. 12-2, $X_{C2} = \dfrac{h_{ie}}{1 + h_{fe}}$ at f_1

C_3 is once again calculated from Eq. 12-5,

$$X_{C3} = \frac{Z_o + R_L}{10}$$

DC Feedback pair with Emitter Follower Output

Figure 12-26 shows a *dc* feedback pair that has an emitter follower output stage. In this case, a bypass capacitor cannot be used at the emitter of Q_2, otherwise the output would be *ac* shorted to ground. However, C_2 in Fig. 12-25 also serves the purpose of eliminating *ac* feedback from the collector of Q_1 to its base (via Q_2). So, now another method of removing this *ac* feedback must be employed. In Fig. 12-26, the base bias resistor for Q_1 is split into two resistors, and the junction of these two is *ac* grounded via capacitor C_2.

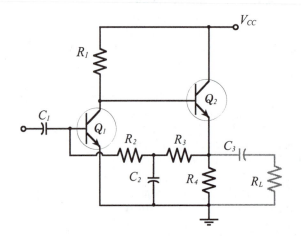

Figure 12-26
DC feedback pair with an emitter follower output stage. In this case, two series-connected feedback resistors (R_2 and R_3) are used, and the junction of the two is grounded via C_2 to eliminate ac degeneration.

The design of this circuit differs from other designs only in calculation of the capacitor values. In this case, X_{C2} should be very much smaller than R_3 at f_1. This is because the *ac* output voltage at the emitter of Q_2 is divided across R_3 and X_{C2} before being fed back to the Q_1 base. Capacitor values normally depend on the resistance in series with them, and in Fig. 12-26 R_3 is in series with C_2 for *ac* feedback. Because this capacitor is in the feedback path from output to input, it is calculated at f_1 from

$$X_{C2} = \frac{R_3}{100} \qquad\qquad \text{(12-20)}$$

The reasons for using a factor of 100, instead of the usual factor of 10, will be understood when negative feedback is studied, (see Chapter 13).

Normally, the largest capacitor in any circuit is employed to determine the circuit low 3 dB frequency (f_1). Where two capacitors are likely to be equally large, they are both used to set f_1. The largest capacitor is always the one in series with the lowest resistance. In the circuit in Fig. 12-26, the external load resistance is likely to be the smallest resistance in series with a capacitor. Consequently, C_3 is calculated from,

$$X_{C3} = (Z_o + R_L) \text{ at } f_1 \qquad\qquad \text{(12-21)}$$

Example 12-14

Design a *dc* feedback pair with an emitter follower output, as shown in Fig. 12-26. The circuit is to use *2N3904* transistors with a 12 V supply. The load resistance is 50 Ω, the lower cutoff frequency is 150 Hz, and the output voltage is to be ±50 mV. Assume that $r_s << Z_{in}$.

Solution

$$i_p = \frac{v_p}{R_L} = \frac{50 \text{ mV}}{50 \text{ Ω}}$$
$$= 1 \text{ mA}$$

$I_{E2} > i_p$ Select $I_{E2} = 2$ mA

Select, $V_{E2} = 5$ V

$$R_4 = \frac{V_{E2}}{I_{E2}} = \frac{5 \text{ V}}{2 \text{ mA}}$$
$$= 2.5 \text{ kΩ (use 2.2 kΩ standard value)}$$

$I_{C1} >> I_{B2}$, select $I_{C1} = 1$ mA

$$V_{R1} = V_{CC} - (V_{BE2} + V_{E2}) = 12 \text{ V} - (0.7 \text{ V} + 5 \text{ V})$$
$$= 6.3 \text{ V}$$

$$R_1 = \frac{V_{R1}}{I_{C1}} = \frac{6.3 \text{ V}}{1 \text{ mA}}$$
$$= 6.3 \text{ kΩ (use 5.6 kΩ and recalculate } I_{C1})$$

$$I_{C1} = \frac{6.3 \text{ V}}{5.6 \text{ kΩ}} = 1.13 \text{ mA}$$

From Appendix 1-5, for the *2N3904*: $h_{FE(min)} = 70$, $h_{fe} = 100$, and $h_{ie} = 1$ kΩ

so, $$I_{B1} = \frac{I_{C1}}{h_{FE}} = \frac{1.13 \text{ mA}}{70}$$

$$= 16.1 \ \mu A$$

$$R_2 + R_3 = \frac{V_{E2} - V_{BE}}{I_{B1}} = \frac{5 \text{ V} - 0.7 \text{ V}}{16.1 \ \mu A}$$

$$= 267 \text{ k}\Omega$$

select, $R_2 = 120$ kΩ, and $R_3 = 150$ kΩ

Eq. 12-19, $$Z_i = R_2 \| h_{ie} = 120 \text{ k}\Omega \| 1 \text{ k}\Omega$$

$$= 992 \ \Omega$$

$$X_{C1} = \frac{Z_i + r_s}{10} = \frac{992 \ \Omega}{10}$$

$$= 99.2 \ \Omega$$

$$C_1 = \frac{1}{2 \pi f_1 X_{C1}} = \frac{1}{2 \pi \times 150 \text{ Hz} \times 99.2 \ \Omega}$$

$$= 10.7 \ \mu F \qquad \text{(use 15 } \mu F \text{ standard value)}$$

Eq. 12-20, $$X_{C2} = \frac{R_3}{100} = \frac{150 \text{ k}\Omega}{100}$$

$$= 1.5 \text{ k}\Omega$$

$$C_2 = \frac{1}{2 \pi f_1 X_{C2}} = \frac{1}{2 \pi \times 150 \text{ Hz} \times 1.5 \text{ k}\Omega}$$

$$= 0.71 \ \mu F \quad \text{(use 0.82 } \mu F \text{ standard value)}$$

$$Z_o = \left[\frac{h_{ie2} + R_1}{h_{fe}}\right] \| R_4 = \left[\frac{1 \text{ k}\Omega + 5.6 \text{ k}\Omega}{100}\right] \| 2.2 \text{ k}\Omega$$

$$= 64 \ \Omega$$

Eq. 12-21, $$X_{C3} = R_L + Z_o = 50 \ \Omega + 64 \ \Omega$$

$$= 114 \ \Omega$$

$$C_3 = \frac{1}{2 \pi f_1 X_{C3}} = \frac{1}{2 \pi \times 150 \text{ Hz} \times 114 \ \Omega}$$

$$= 9.3 \ \mu F \qquad \text{(use 10 } \mu F \text{ standard value)}$$

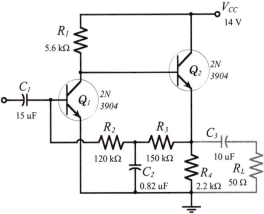

Figure   12-27
Two-stage dc feedback pair designed in Example 12-14.

Practise   Problems

12-6.1   A *dc* feedback pair as in Fig. 12-25 is to use $V_{CC} = 15$ V, $R_L = 47$ kΩ, and *2N3903* transistors. Determine suitable resistor values.

12-6.2   The circuit in Problem 12-6.1 to have $f_1 = 75$ Hz. Determine suitable capacitor values, and calculate the circuit minimum overall voltage gain. Assume that $r_s = 350 \ \Omega$.

12-7 BIFET Circuits

BJT-FET Considerations

Two-stage *BJT* circuits usually have relatively low input impedances. To increase Z_i, a field effect transistor may be used as the first stage. Circuits which are composed of *BJTs* and *FETs* are termed *BIFET* circuits. Since a *FET* circuit normally has much lower voltage gain than a *BJT* circuit, a *BIFET* circuit can be expected to have lower overall voltage gain than a two-stage *BJT* circuit. Two-stage *FET* circuits can be designed and constructed. However, the voltage gain of a two-stage *FET* circuit is substantially lower than that of a *BIFET* circuit, and there are no advantages to offset the lower voltage gain.

Capacitor-Coupled BIFET Amplifier

A two-stage, capacitor-coupled *BIFET* amplifier circuit is shown in Fig. 12-28. The input stage provides a very high input resistance and a voltage gain usually ranging from 5 to 15. *Stage 2* has the typical bipolar voltage gain of 200 to 500, depending upon transistor parameters and resistor values. Each stage is designed independently, as discussed in Sections 12-1 and 12-2. The two stages of the circuit shown in Fig. 12-28 are those designed in Examples 12-2 through 12-5. In this case, C_3 in Fig. 12-14 is omitted, and C_3 in Fig. 12-28 is the interstage coupling capacitor.

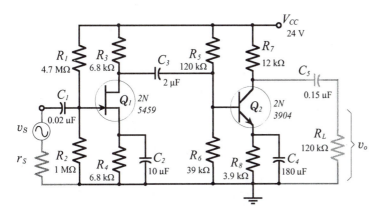

Figure 12-28
Capacitor-coupled two-stage BIFET amplifier. The FET input stage provides a high input impedance, and the BJT produces substantial voltage gain.

Direct-Coupled BIFET Amplifier

In the direct-coupled *BIFET* circuit in Fig. 12-29(a), transistor Q_1 is an *n*-channel *JFET* and Q_2 is a *pnp* bipolar transistor. An *npn* *BJT* could be used in the circuit, as illustrated in Fig. 12-29(b); however, this arrangement uses a large voltage drop across Q_2 emitter resistor (R_6), leaving a smaller voltage drop across the collector resistor (R_5). Q_2 can also be driven into saturation if the drain current of Q_1 is lower than the design level. For the circuit in Fig. 12-29(a), V_{CE2} is increased when I_{D1} is less than the $I_{D(max)}$ design level.

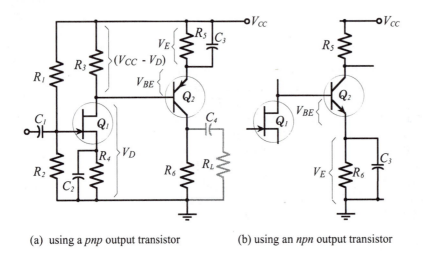

(a) using a *pnp* output transistor   (b) using an *npn* output transistor

Figure   12-29
Direct-coupled two-stage BIFET amplifier. When Q_1 is an n-channel FET, it is preferable that Q_2 be a pnp BJT.

Design   Calculations

Example   12-15

Determine suitable resistors for the BIFET amplifier in Fig. 12-29(a). The circuit is to use a *2N3906 BJT* and a *2N5486 FET*. The supply voltage is 20 V, the external load resistance is 80 kΩ, the circuit input impedance is to be approximately 500 kΩ, and the low 3 dB frequency is to be 150 Hz.

Solution

The *2N5486* has:

$$V_{p(max)} = -6 \text{ V},\ I_{DSS(max)} = 20 \text{ mA, and } Y_{fs} = 6000 \text{ } \mu\text{S (typically)}$$

Select, $I_{D(max)} \approx 2 \text{ mA}$

Eq. 10-8, $V_{GS} = V_P[\ 1 - \sqrt{(I_D/I_{DSS})}] = -6 \text{ V }[1 - \sqrt{(2 \text{ mA}/20 \text{ mA})}]$
                                    $\approx -4.1 \text{ V}$

Eq. 12-8, $V_{DS(min)} = V_{P(max)} + 1 \text{ V} - V_{GS} = 6 \text{ V} + 1 \text{ V} - 4.1 \text{ V}$
                        $= 2.9 \text{ V}$

$$V_{R3} + V_{R4} = V_{CC} - V_{DS} = 20 \text{ V} - 2.9 \text{ V}$$
$$= 17.1 \text{ V}$$

$$V_{R3} = V_{R4} = \frac{17.1 \text{ V}}{2}$$
$$\approx 8.5 \text{ V}$$

$$R_3 = R_4 = \frac{V_{R4}}{I_D} = \frac{8.5 \text{ V}}{2 \text{ mA}}$$

$\qquad\qquad = 4.25 \text{ k}\Omega$ (use 3.9 kΩ standard value, and recalculate V_{R3} and V_{R4})

$$V_{R3} = V_{R4} = I_D \times R_4 = 2 \text{ mA} \times 3.9 \text{ k}\Omega$$
$$= 7.8 \text{ V}$$

$$V_{R2} = V_G = V_{R4} - V_{GS} = 7.8 \text{ V} - 4.1 \text{ V}$$
$$= 3.7 \text{ V}$$

$$V_{R1} = V_{CC} - V_{R3} = 20 \text{ V} - 3.7 \text{ V}$$
$$= 16.3 \text{ V}$$

For $Z_i \approx 500 \text{ k}\Omega$, select $R_2 = 560 \text{ k}\Omega$

Eq. 12-10, $$R_1 = \frac{V_{R1} \times R_2}{V_{R2}} = \frac{16 \text{ V} \times 560 \text{ k}\Omega}{3.7 \text{ V}}$$

$$\approx 2.5 \text{ M}\Omega \text{ (use 2.7 M}\Omega \text{ standard value)}$$

Select, $$R_6 \approx \frac{R_L}{10} = \frac{80 \text{ k}\Omega}{10}$$

$$= 8 \text{ k}\Omega \text{ (use 8.2 k}\Omega \text{ standard value)}$$

$$V_{R5} = V_{R3} - V_{BE2} = 7.8 \text{ V} - 0.7 \text{ V}$$
$$= 7.1 \text{ V}$$

$$V_{R6} = V_{CC} - V_{R5} - V_{CE(min)} = 20 \text{ V} - 7.1 \text{ V} - 3 \text{ V}$$
$$= 9.9 \text{ V}$$

$$I_{C2} = \frac{V_{R6}}{R_6} = \frac{9.9 \text{ V}}{8.2 \text{ k}\Omega}$$

$$= 1.2 \text{ mA}$$

$$R_5 \approx \frac{V_{R5}}{I_{C2}} = \frac{7.1 \text{ V}}{1.2 \text{ mA}}$$

$$= 5.9 \text{ k}\Omega \text{ (use 6.8 k}\Omega \text{ standard value)}$$

Example 12-16

Determine suitable capacitor values for the BIFET direct-coupled amplifier in Example 12-15.

Solution

$$Z_i = R_1 \| R_2 = 2.7 \text{ M}\Omega \| 560 \text{ k}\Omega$$
$$= 464 \text{ k}\Omega$$

Eq. 12-4, $$X_{C1} = \frac{Z_i}{10} = \frac{464 \text{ k}\Omega}{10}$$

$$= 46.4 \text{ k}\Omega$$

$$C_1 = \frac{1}{2\pi f_1 X_{C1}} = \frac{1}{2\pi \times 150 \text{ Hz} \times 46.4 \text{ k}\Omega}$$

$$= 0.02 \, \mu F \qquad \text{(standard value)}$$

From Eq. 12-12,

$$X_{C2} = \frac{0.65}{Y_{fs}} = \frac{0.65}{6000 \, \mu S}$$

$$= 108 \, \Omega$$

$$C_2 = \frac{1}{2\pi f_1 X_{C2}} = \frac{1}{2\pi \times 150 \text{ Hz} \times 108 \, \Omega}$$

$$= 9.8 \, \mu F \quad \text{(use 10 } \mu F \text{ standard value)}$$

Eq. 12-13, $X_{C3} = \dfrac{0.65 \, h_{ie2}}{1 + h_{fe2}} = \dfrac{0.65 \times 2 \text{ k}\Omega}{1 + 100}$

$$\approx 13 \, \Omega$$

$$C_3 = \frac{1}{2\pi f_1 X_{C3}} = \frac{1}{2\pi \times 150 \text{ Hz} \times 13 \, \Omega}$$

$$= 81.6 \, \mu F \qquad \text{(use 82 } \mu F \text{ standard value)}$$

From Eq. 12-5, $X_{C4} = \dfrac{R_6 + R_L}{10} = \dfrac{8.2 \text{ k}\Omega + 80 \text{ k}\Omega}{10}$

$$= 8.8 \text{ k}\Omega$$

$$C_4 = \frac{1}{2\pi f_1 X_{C4}} = \frac{1}{2\pi \times 150 \text{ Hz} \times 8.8 \text{ k}\Omega}$$

$$= 0.12 \, \mu F \qquad \text{(standard value)}$$

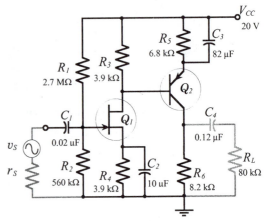

Figure 12-30
BIFET amplifier designed in Examples 12-15 and 12-16.

Example 12-17

Analyze the amplifier circuit in Fig. 12-30 to determine its minimum overall voltage gain.

Solution

For the *2N3906*, $Z_{i2} = h_{ie} = 2 \text{ k}\Omega$

Eq. 11-7, $A_{v1} = -Y_{fs} (R_3 \| Z_{i2}) = -6000 \, \mu S \times (3.9 \text{ k}\Omega \| 2 \text{ k}\Omega)$

$$= -7.9$$

From Eq. 12-17, $A_{v2} = \dfrac{-h_{fe2} (R_6 \| R_L)}{h_{ie2}} = \dfrac{-100 \times (8.2 \text{ k}\Omega \| 80 \text{ k}\Omega)}{2 \text{ k}\Omega}$

$$= -372$$

From Eq. 12-18, $A_v = A_{v1} \times A_{v2} = (-7.9) \times (-372)$

$$= 2939$$

12-8 Differential Amplifier

Differential Amplifier Circuit

The *differential amplifier* is widely applied in integrated circuitry, because it has both good bias stability and good voltage gain without the use of large bypass capacitors. Its use in integrated circuits is further studied in Chapter 14. Differential amplifiers can also be constructed as discrete component circuits.

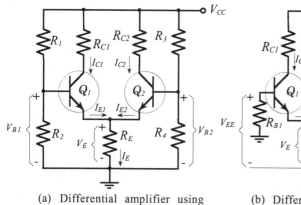

(a) Differential amplifier using single-polarity supply

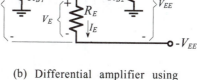

(b) Differential amplifier using plus/minus supply

Figure 12-31
A differential amplifier has two emitter-coupled transistors with a single emitter resistor. The emitter current tends to split evenly between Q_1 and Q_2.

Figure 12-31(a) shows that a basic differential amplifier circuit consists of two voltage divider bias circuits with a single emitter resistor. The circuit is also known as an *emitter-coupled amplifier*, because the transistors are coupled at the emitter terminals. If transistors Q_1 and Q_2 are assumed to be identical in all respects, and if $V_{B1} = V_{B2}$, then the emitter currents are equal, and the total emitter current is,

$$I_E = I_{E1} + I_{E2} \qquad (12\text{-}22)$$

also, $$V_E = V_B - V_{BE}$$

so, $$I_E = \frac{V_B - V_{BE}}{R_E}$$

and, $$I_{E1} = I_{E2} = \frac{I_E}{2}$$

Like the emitter current in a single-transistor voltage divider bias circuit, I_E in the differential amplifier remains virtually constant regardless of the transistor h_{FE} value. This results in, I_{E1}, I_{E2}, I_{C1}, and I_{C2} all remaining substantially constant, and the constant collector current levels keep V_{C1} and V_{C2} stable. So, the differential amplifier has the same excellent bias stability as a single-transistor voltage divider bias circuit.

The circuit of a differential amplifier using a plus-minus supply is shown in Fig. 12-31(b). In this case, the voltage across the emitter resistor is (V_{EE} - V_{BE}), as illustrated,

and, $$I_{E1} = I_{E2} = \frac{V_{EE} - V_{BE}}{2\,R_E}$$ **(12-23)**

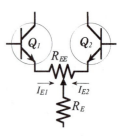

Figure 12-32
A potentiometer connected between the emitters of Q_1 and Q_2 allows I_{E1} and I_{E2} to be equalized.

The base resistors (R_{B1} and R_{B2}) are included to bias the transistor bases to ground while offering an acceptable input resistance to a signal applied to one of the bases. (See the emitter current bias circuit discussion in Section 5-8.) The transistor emitter currents (I_{E1} and I_{E2}) are exactly equal only if the devices are perfectly matched. To allow for some differences in transistor parameters, a small-value potentiometer (R_{EE}) is sometimes included between the emitters, (see Fig. 12-32). Adjustment of R_{EE} increases the resistance in series with the emitter of one transistor, and reduces the emitter resistance for the other transistor. This reduces the I_E for one transistor and increases it for the other, while the total emitter current remains constant.

AC Operation

Consider what happens when the *ac* input voltage (v_i) at the base of Q_1 is positive-going, as illustrated in Fig. 12-33. Q_1 emitter current (I_{E1}) increases. Also, I_{E2} decreases, because the total emitter current (I_{E1} + I_{E2}) remains constant. This means that I_{C1} increases and I_{C2} decreases, and consequently, V_{C1} falls and V_{C2} rises, as shown. So, the *ac* output voltage at Q_1 collector is in antiphase to v_i at Q_1 base, and the output at Q_2 collector is in phase with v_i.

Voltage Gain

The voltage gain of a single-stage amplifier with an unbypassed emitter resistor and no external load is given by Eq. 6-21,

$$A_v = \frac{-h_{fe}\,R_C}{h_{ie} + (1 + h_{fe})R_E}$$

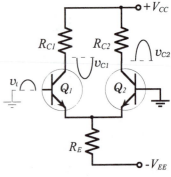

Figure 12-33
An positive-going ac input applied to the base of Q_1 produces a negative-going output at Q_{1C} and a positive-going output at Q_{2C}.

Refering to Fig. 12-34, it is seen that the resistance *looking into* the emitter of Q_2 is h_{ib}, so $h_{ib}\|R_E$ behaves like an unbypassed resistor in series with the emitter of Q_1. Neglecting R_E because it is very much larger than h_{ib}, the voltage gain from the base of Q_1 to its collector is,

$$A_v = \frac{-h_{fe}\,R_C}{h_{ie} + (1 + h_{fe})h_{ib}}$$

this reduces to, $$A_v = \frac{-h_{fe}\, R_C}{2\, h_{ie}}$$ **(12-24)**

Equation 12-24 gives the voltage gain from one input terminal to one output of a differential amplifier. It is seen to be half the voltage gain of a similar single-transistor *CE* amplifier with R_E bypassed; but note that *the differential amplifier requires no bypass capacitor.* This is an important advantage, because bypass capacitors are usually large and expensive.

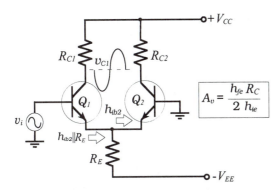

$$A_v = \frac{h_{fe}\, R_C}{2\, h_{ie}}$$

Figure 12-34
The voltage gain of a differential amplifier is half the gain of a common-emitter circuit using similar components. Note the absence of an emitter bypass capacitor.

Another way to contemplate the operation of the differential amplifier circuit is to think of the input voltage being equally divided between Q_1 base-emitter and Q_2 base-emitter. This is illustrated in Fig. 12-35 where it is seen that (for a positive-going input) $v_i/2$ is applied *positive on the base* of Q_1, while the other half of v_i appears *positive on the emitter* of Q_2. Thus, for v_i at Q_{1B}, transistor Q_1 behaves as a common-emitter circuit, and because Q_2 receives the input at its emitter, Q_2 behaves as a common-base circuit. Consequently,

$$v_{C1} = \frac{v_i}{2} \times \frac{-h_{fe}\, R_C}{h_{ie}}$$

and, $$v_{C2} = \frac{v_i}{2} \times \frac{h_{fe}\, R_C}{h_{ie}}$$

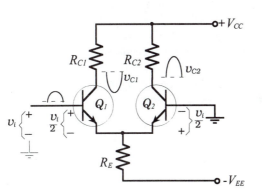

Figure 12-35
A signal at the base of Q_1 is divided equally between the BE junctions of Q_1 and Q_2. Half of v_i appears at Q_1 base-emitter as a common-emitter circuit, and the other half is applied to Q_2 base-emitter as a common-base circuit.

Input and Output Impedances

From Eq. 6-20, the input impedance at the base of a *CE* circuit with an unbypassed emitter resistor is,

$$Z_b = h_{ie} + (1 + h_{fe})R_E$$

Referring to Fig. 12-34, the differential amplifier has $h_{ib}\|R_E$ as an unbypassed resistor in series with the emitter of Q_1. Neglecting R_E (because $R_E \gg h_{ib}$), the input resistance at Q_{1B} is,

$$Z_b = h_{ie} + (1 + h_{fe})h_{ib}$$

This reduces to,

$$\mathbf{Z_b = 2\ h_{ie}} \qquad\qquad \textbf{(12-25)}$$

or, $Z_b = 2\ (Z_b$ for a single-stage *CE* circuit)

Note that there are usually bias resistors in parallel with Z_b, so that (from Eq. 6-12) the circuit input impedance is,

$$Z_i = R_B\|Z_b$$

As in the case of *CE* and *CB* circuits, the output impedance at the transistor collector terminals is given by Eq. 6-14;

$$Z_o = R_C\|(1/h_{oe})$$

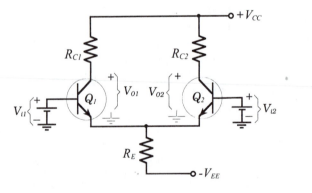

Figure 12-36
When dc voltages are applied to the two inputs of a differential amplifier, the outputs at the transistor collectors are proportional to the differences between the input voltages.

DC Amplification

When one transistor base is grounded in a differential amplifier, and an input is applied to the other one, as already discussed, v_i is amplified to produce the outputs at the collector terminals. In this case v_i is the *voltage difference* between the two base terminals. Figure 12-36 shows a differential amplifier with *dc* input voltages V_{i1} and V_{i2} applied to the transistor bases. If the voltage gain from the base to the collector is A_v, the *dc* voltage changes at the collectors are;

$$\Delta V_{o1} = A_v(V_{i2} - V_{i1})$$

and, $$\Delta V_{o2} = A_v(V_{i1} - V_{i2})$$

It is seen that the differential amplifier can be employed as a *direct-coupled amplifier*, or *dc amplifier*. The term *difference amplifier* is also used for this circuit.

Design Calculations

Design procedures for a differential amplifier are similar to those for voltage divider bias circuits. Because there is no bypass capacitor in a differential amplifier, one of the coupling capacitors determines the circuit lower cutoff frequency (f_1). The capacitor with the smallest resistance in series with it is normally the largest capacitor, and in the case of a differential amplifier this is usually the input coupling capacitor. So, the input coupling capacitor determines the circuit lower cut-off frequency.

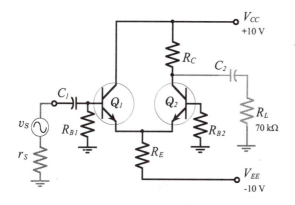

Figure 12-37
Differential amplifier circuit with capacitor-coupled signal and load.

Consider the capacitor-coupled differential amplifier in Fig. 12-37. The circuit uses a plus-minus supply, and a single collector resistor (R_C). No output is taken from Q_1 collector, so there is no need for a collector resistor. R_C is selected in the usual way for a small-signal amplifier; $R_C \ll R_L$. The collector-emitter voltage should be a minimum of 3 V, as always. Then, I_C is calculated from R_C and the selected voltage drop across R_C.

The total emitter current is determined as,

$$I_E \approx I_{C1} + I_{C2} \approx 2\, I_C$$

and, $$R_E = \frac{V_{EE} - V_{BE}}{I_E} \qquad \textbf{(12-26)}$$

The base bias resistors are determined from Eq. 5-17,

$$R_B = \frac{V_{BE}}{10\, I_{B(max)}}$$

As discussed, capacitor C_1 sets the lower cutoff frequency. So,

$$X_{C1} = Z_i \text{ at } f_1 \qquad \textbf{(12-27)}$$

and C_2 is determined from Eq. 12-5,

$$X_{C2} = R_L/10 \text{ at } f_1$$

Example   12-18

Determine suitable resistor values for the differential amplifier circuit in Fig. 12-37. The transistor parameters are $h_{FE} = 60$, $h_{fe} = 60$, and $h_{ie} = 1.4$ kΩ.

Solution

$$R_{C2} \ll R_L$$

select

$$R_{C2} \approx \frac{R_L}{10} \approx \frac{70 \text{ k}\Omega}{10}$$

$$= 7 \text{ k}\Omega \text{ (use 6.8 k}\Omega \text{ standard value)}$$

select

$$V_{CE(min)} = 3 \text{ V}$$

$$V_{RC2} = V_{CC} + V_{BE} - V_{CE(min)} = 10 \text{ V} + 0.7 \text{ V} - 3 \text{ V}$$

$$= 7.7 \text{ V}$$

$$I_C = \frac{V_{RC2}}{R_C} = \frac{7.7 \text{ V}}{6.8 \text{ k}\Omega}$$

$$= 1.13 \text{ mA}$$

Eq. 12-26,

$$R_E = \frac{V_{EE} - V_{BE}}{I_E} \approx \frac{10 \text{ V} - 0.7 \text{ V}}{2 \times 1.13 \text{ mA}}$$

$$= 4.1 \text{ k}\Omega \text{ (use 4.7 k}\Omega \text{ standard value)}$$

From Eq. 5-17,

$$R_B = \frac{V_{BE}}{10 \, I_{B(max)}} = \frac{V_{BE}}{10 \, (I_C / h_{FE})} = \frac{0.7 \text{ V}}{10 \times (1.1 \text{ mA}/60)}$$

$$= 3.8 \text{ k}\Omega \text{ (use 3.9 k}\Omega \text{ standard value)}$$

$$R_{B1} = R_{B2} = 3.9 \text{ k}\Omega$$

Example   12-19

Determine suitable capacitor values for the amplifier in Ex. 12-18 to give $f_1 = 60$ Hz. Also, calculate the circuit voltage gain. Assume that $r_s \ll Z_i$.

Solution

Eq. 12-25,

$$Z_b = 2 \, h_{ie} = 2 \times 1.4 \text{ k}\Omega$$

$$= 2.8 \text{ k}\Omega$$

$$Z_i = R_B \| Z_b = 3.9 \text{ k}\Omega \| 2.8 \text{ k}\Omega$$

$$= 1.6 \text{ k}\Omega$$

from Eq. 12-27,

$$C_1 = \frac{1}{2 \pi f_1 \, Z_i} = \frac{1}{2 \pi \times 60 \text{ Hz} \times 1.6 \text{ k}\Omega}$$

$$= 1.66 \, \mu F \text{ (use 1.8 } \mu F \text{ standard value)}$$

from Eq. 12-5,

$$C_2 = \frac{1}{2 \pi f_1 \, (R_L / 10)} = \frac{1}{2 \pi \times 60 \text{ Hz} \times (70 \text{ k}\Omega / 10)}$$

$$= 0.38 \, \mu F \text{ (use 0.39 } \mu F \text{ standard value)}$$

from Eq. 12-24, $A_v = \dfrac{h_{fe}\,(R_{C2}\|R_L)}{2\,h_{ie}} = \dfrac{60 \times (6.8\ k\Omega\|70\ k\Omega)}{2 \times 1.4\ k\Omega}$

$\qquad\qquad = 133$

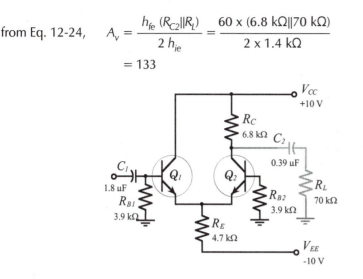

Figure 12-38
Capacitor-coupled   differential amplifier circuit designed in Examples 12-18 and 12-19.

Practise Problems

12-8.1 The differential amplifier in Fig. 12-31(a) is to use a *2N3904* transistor, a 20 V supply, and to have a 100 kΩ load capacitor-coupled to Q_2 collector. Determine suitable resistor values.

12-8.2 Determine suitable coupling capacitor values for the circuit in Problem 12-9.1 to give a 75 Hz lower cutoff frequency. Also, calculate the overall voltage gain. Assume that $r_s = 200\ \Omega$.

12-9   Small-Signal High-Frequency Amplifiers

Common Base Amplifier

A practical *CB* amplifier (using a plus/minus supply) is shown in Fig. 12-39. [Note the polarity of the input coupling capacitor (C_1).] The *CB* circuit is analysed in Section 6-7, where it is found to have a very low input impedance, as well as a voltage gain and output impedance similar to a *CE* circuit. Because the *CB* circuit has no phase inversion between the input and output terminals, there is no Miller-effect amplification of the collector-base capacitance, (see Sections 8-3 and 8-4). Consequently, a *CB* circuit is capable of operating at much higher frequencies than a *CE* circuit.

Design procedure for a *CB* circuit is exactly the same as that for a *CE* circuit, (see Section 12-1). Resistor calculation is performed just as if a *CE* circuit were being designed. The output coupling capacitor (C_2 in Fig. 12-39) is determined in the usual way. Because of the very low input impedance of a *CB* circuit, the input coupling capacitor (C_1) is the largest capacitor in the circuit. So, C_1 determines the circuit low cutoff frequency, and its capacitance is calculated like an emitter bypass capacitor in a *CE* circuit.

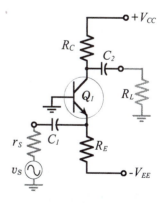

Figure 12-39
A capacitor-coupled CB circuit can be designed in exactly the same way as a CE circuit.

Cascode Amplifier

The very low input impedance of a *CB* circuit (typically 25 Ω) is a major disadvantage. To increase the circuit input impedance while retaining its high frequency performance, the *cascode* amplifier (Fig. 12-40) uses a *CE* input stage driving a *CB* output stage. Transistor Q_1 and its associated components in Fig. 12-40 operates as a *CE* input stage, while the circuit of Q_2 functions as a *CB* output stage.

The circuit *dc* voltage and current levels are illustrated in Fig. 12-41. It is seen that I_{E1} is set by V_{B1} and R_E. Collector current I_{C1} approximately equals I_{E1}, and I_{E2} is the same current as I_{C1}. So, I_{C2} approximately equals I_{E1}. The current levels remain constant so long as V_{CE1} and V_{CE2} are large enough for transistor operation.

A positive-going *ac* signal at the base of Q_1 produces an increase in I_{E1}, and this results in an I_{C2} increase. Thus, the voltage drop across R_C is increased, producing a decrease in the output voltage. So, there is an *ac* phase inversion between the input and output.

The input impedance at the emitter of Q_2 constitutes the load for the collector of Q_1. Therefore, from Eq. 6-15, the voltage gain from the circuit input to Q_1 collector is,

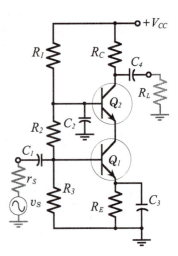

Figure   12-40
A cascode amplifier is like a CB circuit with an additional input transistor (Q_1) that behaves like a CE stage with a gain of one.

$$A_{v1} = \frac{-h_{fe} \times (Z_i \text{ to } Q_2)}{h_{ie}} = \frac{-h_{fe} \times [h_{ie}/(1 + h_{fe})]}{h_{ie}}$$

$$\approx -1$$

With an input stage gain of -1, there is no significant Miller effect at the circuit input.

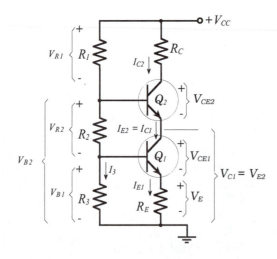

Figure   12-41
DC Voltage levels in a cascode circuit.

From Eq. 6-32, the voltage gain produced by the Q_2 *CB* circuit is,

$$A_{v2} = \frac{h_{fb} \times (R_C \| R_L)}{h_{ib}}$$

Converting to *CE* parameters, the overall voltage gain for the cascode amplifier is the same as that for a *CE* circuit;

$$A_v \approx \frac{-h_{fe} \times (R_C\|R_L)}{h_{ie}}$$

Design Calculations

The design procedure for a cascode amplifier is easily derived from the *CE* design process. The voltage across R_E should typically be 5 V, and the V_{CE} for each transistor should be a minimum of 3 V. The emitter voltage for Q_2 is,

$$V_{E2} = (V_{E1} + V_{CE1})$$

The transistor base voltages are,

$$V_{B1} = (V_{E1} + V_{BE})$$

and, $$V_{B2} = (V_{E2} + V_{BE})$$

When the voltage and current levels are selected, the resistor values are easily calculated.

   The base bypass capacitor (C_2) in Fig. 12-40 should have an impedance equal to one tenth of the h_{ie} of Q_2. This ensures that the impedance looking into Q_2 emitter terminal is not significantly affected by R_1 and R_2. Also, that no portion of the input signal (at Q_2 base) is developed at Q_1 base.

Example 12-20

Determine suitable resistor values the cascode amplifier circuit in Figs. 12-40 and 12-41 if $V_{CC} = 20$ V and $R_L = 90$ kΩ. The transistor parameters are $h_{fe} = 50$, $h_{ie} = 1.2$ kΩ, and $h_{ib} = 24$ Ω.

Solution

select $$R_C \approx \frac{R_L}{10} \approx \frac{90 \text{ k}\Omega}{10}$$

$$= 9 \text{ k}\Omega \text{ (use 8.2 k}\Omega \text{ standard value)}$$

select $$V_{CE1(min)} = V_{CE2(min)} = 3 \text{ V}$$

and, $$V_E = 5 \text{ V}$$

$$V_{RC} = V_{CC} - V_{CE1} - V_{CE2} - V_E = 20 \text{ V} - 3 \text{ V} - 3 \text{ V} - 5 \text{ V}$$
$$= 9 \text{ V}$$

$$I_C = \frac{V_{RC}}{R_C} = \frac{9 \text{ V}}{8.2 \text{ k}\Omega}$$

$$= 1.1 \text{ mA}$$

Eq. 12-27, $$R_E \approx \frac{V_E}{I_C} = \frac{5 \text{ V}}{1.1 \text{ mA}}$$

$$= 4.5 \text{ k}\Omega \text{ (use 4.7 k}\Omega \text{ standard value)}$$

select
$$R_3 = 10\,R_E = 10 \times 4.7\ \text{k}\Omega$$
$$= 47\ \text{k}\Omega$$

$$V_{B1} = V_E + V_{BE} = 5\ \text{V} + 0.7\ \text{V}$$
$$= 5.7\ \text{V}$$

$$I_3 = \frac{V_{B1}}{R_3} = \frac{5.7\ \text{V}}{47\ \text{k}\Omega}$$
$$= 121\ \mu\text{A}$$

$$V_{B2} = V_E + V_{CE1} + V_{BE2} = 5\ \text{V} + 3\ \text{V} + 0.7\ \text{V}$$
$$= 8.7\ \text{V}$$

$$V_{R2} = V_{B2} - V_{B1} = 8.7\ \text{V} - 5.7\ \text{V}$$
$$= 3\ \text{V}$$

$$R_2 = \frac{V_{R2}}{I_3} = \frac{3\ \text{V}}{121\ \mu\text{A}}$$
$$= 24.8\ \text{k}\Omega \quad (\text{use } 22\ \text{k}\Omega + 2.7\ \text{k}\Omega)$$

$$R_1 = \frac{V_{CC} - V_{B2}}{I_3} = \frac{20\ \text{V} - 8.7\ \text{V}}{121\ \mu\text{A}}$$
$$= 93.4\ \text{k}\Omega \quad (\text{use } 47\ \text{k}\Omega + 47\ \text{k}\Omega)$$

Example 12-21

Determine suitable capacitor values for the cascode circuit in Example 12-20 to give a 25 Hz low cutoff frequency. Assume that $r_s \ll Z_i$.

Solution

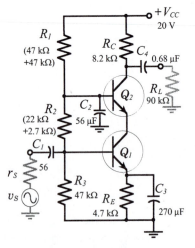

$$Z_i = h_{ie1}\|R_3\|R_2 = 1.2\ \text{k}\Omega\|47\ \text{k}\Omega\|24.7\ \text{k}\Omega$$
$$= 1.1\ \text{k}\Omega$$

from Eq. 12-4,
$$C_1 = \frac{1}{2\pi f_1\,(Z_i/10)} = \frac{1}{2\pi \times 25\ \text{Hz} \times (1.1\ \text{k}\Omega/10)}$$
$$= 57.9\ \mu\text{F} \quad (\text{use } 56\ \mu\text{F standard value})$$

$$C_2 = \frac{1}{2\pi f_1\,(h_{ie2}/10)} = \frac{1}{2\pi \times 25\ \text{Hz} \times (1.2\ \text{k}\Omega/10)}$$
$$= 53\ \mu\text{F} \quad (\text{use } 56\ \mu\text{F standard value})$$

from Eq. 12-2,
$$C_3 = \frac{1}{2\pi f_1\,h_{ib}} = \frac{1}{2\pi \times 25\ \text{Hz} \times 24\ \Omega}$$
$$= 265\ \mu\text{F} \quad (\text{use } 270\ \mu\text{F standard value})$$

from Eq. 12-5,
$$C_4 = \frac{1}{2\pi f_1\,(R_C + R_L)/10} = \frac{1}{2\pi \times 25\ \text{Hz} \times (99\ \text{k}\Omega/10)}$$
$$= 0.64\ \mu\text{F} \quad (\text{use } 0.68\ \mu\text{F standard value})$$

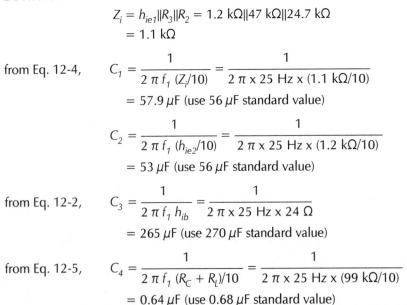

Figure 12-42
Cascode amplifier designed in Examples 12-20 and 12-21.

Differential Amplifier as a High-Frequency Amplifier

The differential amplifier discussed in Section 12-8 can have the same high-frequency performance as a *CB* circuit. The differential amplifier circuit shown in Fig. 12-43, has zero voltage gain from Q_1 base to Q_1 collector, because there is no collector resistor. Consequently, there is no Miller effect on the input capacitance. Also, the amplifier input is applied (via Q_1) to the emitter of Q_2, as in the case of a *CB* circuit. So, there is no Miller effect associated with Q_2. The input impedance (at the base of Q_1) is reasonably high, like a *CE* circuit and unlike the case of a *CB* amplifier.

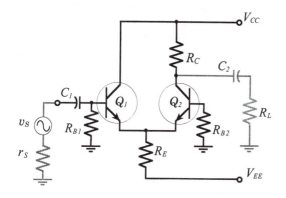

Figure 12-43
A differential amplifier can have a good high frequency response when the input stage has a voltage gain of one, because of the absence of a collector resistor.

Practise Problems

12-9.1 A cascode amplifier circuit as in Fig. 12-40 is to use *2N3904 BJTs* and a 16 V supply. The external load resistor is 33 kΩ. Determine suitable resistor values for the circuit.

12-9.2 Assuming that $r_s << Z_i$, determine suitable capacitor values for the circuit in Problem 12-9.1 to give a 100 Hz lower cutoff frequency. Also, calculate the circuit minimum voltage gain.

12-10 Amplifier Testing

Preparation

Transistor amplifiers and other circuits should be tested in a methodical fashion; otherwise the results obtained may be useless. The circuit should first be carefully constructed in bread-board form, placing the components to resemble the circuit diagram as closely as possible. This makes testing and troubleshooting much easier than when the layout is untidy. Figure 12-24(a) shows a *CE* circuit, and Fig. 12-24(b) illustrates a suitable circuit layout on a plug-in type breadboard.

Instability

After the circuit has been constructed, the power supply should be set to the appropriate voltage and then connected and switched *on*. An oscilloscope should be connected to monitor the output of the amplifier to check that the circuit is not oscillating. If the circuit is oscillating, the oscillations must be stopped before proceeding further. For example, *dc* voltage measurements made on an oscillating circuit have absolutely no value.

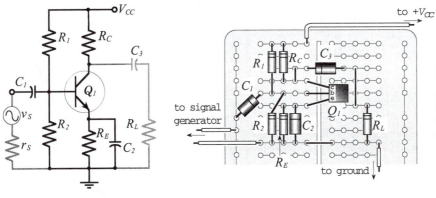

(a) Amplifier circuit diagram

(b) Breadboarded circuit

Figure 12-44
When constructing a circuit for laboratory testing, the layout of the breadboard should resemble the circuit diagram as closely as possible.

Amplifier instability can be the result of incorrect design or poor circuit layout. It can also be caused by feedback along the conductors from the power supply to the circuit. To stabilize an unstable amplifier, commence by connecting a 0.01 μF (or 0.1 μF, if necessary) decoupling capacitor from the positive supply line to ground, (Fig. 12-45). Where a plus-minus supply is used, connect capacitors from each supply line to ground. *The capacitors must be installed right on the breadboard.* It is useless to place them at power supply terminals.

If the circuit is still unstable, small shunt capacitors should be connected from the transistor collector terminals to ground, or between collector and base, (Fig. 12-45). Start with 50 pF capacitors for this purpose, and increase the capacitance in steps to about 300 pF, as necessary. Normally, these capacitors should not be part of the final circuit. They are usually installed temporarily (if they are required) only until the source of the circuit instability is located.

DC Voltage Measurements

Once it is established that the circuit is stable, the next step is to measure the *dc* voltage levels at all transistor terminals. An electronic voltmeter should be used, and every *dc* voltage level should be measured with respect to ground. Some electronic voltmeters have their common terminal capacitor-coupled to ground. So, it is best to keep the voltmeter common terminal grounded at all times. If the *dc* voltages are not satisfactory, they must be corrected before proceeding further. Even when the voltage levels are not exactly as designed, it might be possible to proceed.

The measured *dc* voltage levels should at least show that every *BJT* base-emitter junction is forward biased by the appropriate amount (approximately 0.7 V for silicon transistors and 0.3 V for

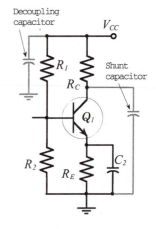

Figure 12-45
Oscillations may be eliminated in an unstable circuit by using power supply decoupling capacitors and/or by employing small shunt capacitors from the transistor collector to ground.

germanium devices). The measurements should also show a minimum of 2 V for each collector-emitter voltage drop. *BJT* and *FET* bias circuit trouble-shooting procedures are discussed in Sections 5-6 and 10-6, respectively.

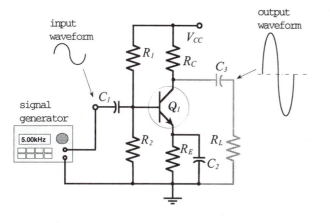

Figure 12-46
For ac testing, a signal generator should be connected at the circuit input, and an oscilloscope should be used to monitor the input and output waveforms.

AC Testing

When satisfactory *dc* levels are established throughout the circuit, *ac* tests may proceed. Connect a signal generator to the input, and connect a two-channel oscilloscope to simultaneously monitor the input and output voltage waveforms, (Fig. 12-46). Adjust the signal frequency to somewhere around the middle of the amplifier bandwidth, and adjust the signal amplitude to give maximum undistorted output voltage. If the signal generator amplitude cannot be set to a low enough level to give an undistorted output, a resistive voltage divider should be employed to further reduce the signal amplitude, (see Fig. 12-47).

The amplifier input and output voltage amplitudes (v_i and v_o) can be measured and the voltage gain can be calculated as $A_v = v_o/v_i$. When a two-stage circuit is being tested, one oscilloscope probe may be moved to monitor the output voltage (v_{o1}) of the first stage. The gain of the first stage is $A_{v1} = v_{o1}/v_i$, and that of the second stage is $A_{v2} = v_o/v_{o1}$.

With the oscilloscope monitoring the input and output waveforms, the frequency response of the circuit may be investigated. The signal voltage level should be maintained constant, and the signal frequency adjusted in convenient steps. At each frequency, v_o should be measured and A_v calculated. The upper and lower cutoff frequencies are found where v_o drops to 0.707 of its normal (mid-frequency) level with v_i still at it normal level.

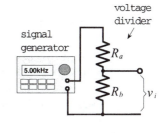

Figure 12-47
A resistive voltage divider might be required to reduce the signal generator amplitude to a low enough level for the circuit input voltage.

Input and Output Impedances

The amplifier input and output impedances should normally be measured with the signal frequency at the middle of the bandwidth. The simplest way to measure the input impedance is to connect a resistor (not a decade box) in series with the circuit input terminal and the signal source, [see Fig. 12-48(a)]. When the *ac* output voltage drops to half its normal level, the series-connected resistor equals the input impedance. If the output voltage change is other than half, the input resistance can be calculated using the voltage divider equation.

The output impedance of a circuit may be determined by connecting a resistor as a capacitor-coupled load at the circuit output, [Fig. 12-48(b). The *ac* output voltage is once again halved when the load resistance equals the circuit output impedance.

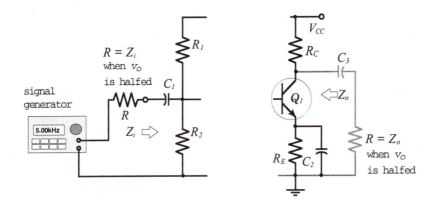

(a) Input resistance measurement   (b) Measurement of output resistance

Figure 12-48
Circuit input impedance is measured by connecting a resistor in series with the circuit input. Output resistance is measured by the use of a capacitor-coupled load resistor.

Section 12-10 Review

12-10.1   Prepare a point form list of the correct test procedure for a two-stage transistor amplifier circuit.

Practise Problem

12-10.1   Refer to the circuit designed in Examples 12-2 and 12-3. Calculate the percentage change in output voltage when a 1 kΩ resistor is connected in series with the signal source and the circuit input. Also, determine the effect on the output voltage when a 27 kΩ resistor is connected in place of the capacitor-coupled load.

Chapter-12 Review Questions

Section 12-1

12-1   Discuss the factors involved in the selection of I_C, R_C, and R_E for a single-stage common-emitter *BJT* amplifier circuit using voltage divider bias. Also, explain the process for determining suitable bias resistors.

12-2   Derive the equation for calculating the value of the emitter bypass capacitor for a single-stage common-emitter *BJT* amplifier using voltage divider bias.

12-3   Write an equation for calculating the coupling capacitors in a capacitor-coupled common-emitter amplifier. Show that, when this equation is used, the coupling capacitors have very little effect on the performance of the circuit.

12-4   Show how a capacitor may be used to obtain a desired upper cutoff frequency in a transistor common-emitter amplifier. Derive the equation for determining the capacitor value.

Section 12-2

12-5   Discuss the factors involved in the selection of I_D, R_D, and R_S for a single-stage common-source *FET* amplifier circuit using voltage divider bias. Also, explain the process for determining suitable bias resistors.

12-6   Derive the equation for calculating the value of the source bypass capacitor for a single-stage common-source amplifier using voltage divider bias.

Section 12-3

12-7   Sketch the circuit of a two-stage capacitor-coupled common-emitter *BJT* amplifier. Briefly explain its operation

12-8   Discuss the lower cutoff frequency for each stage of a two-stage common-emitter amplifier and the lower cutoff frequency for the complete circuit. Also, write equations for calculating the bypass capacitor values for each stage.

12-9   Write equations for calculating the coupling capacitors in a two-stage capacitor-coupled common-emitter amplifier. Explain.

12-10   Sketch the *h*-parameter equivalent circuit for a two-stage common-emitter amplifier. Write equations for Z_i, Z_o, A_{v1}, A_{v2}, and A_v.

Section 12-4

12-11   Sketch the circuit of a two-stage direct-coupled common-emitter amplifier using *npn BJTs*. Discuss the advantages of direct coupling between stages.

12-12   Sketch the *h*-parameter equivalent circuit for the amplifier in Review Question 12-11. Write equations for Z_i, Z_o, and A_v for the circuit.

12-13   Sketch a two-stage, direct-coupled, *BJT* amplifier using an *npn* input stage and a *pnp* output stage. Explain the advantages of using complementary transistors.

Section 12-5

12-14   Sketch the circuit of a two-stage, direct-coupled, *BJT* amplifier using a common-emitter input stage and an emitter follower output stage. Discuss the circuit operation.

12-15 Write equations for Z_i, Z_o, and A_v for the circuit in Review Question 12-14.

Section 12-6

12-16 Sketch the circuit of a *dc* feedback pair with two stages of amplification. Explain the circuit operation, and discuss its advantages and disadvantages.

12-17 Sketch the circuit of a *dc* feedback pair with an amplifier input stage and an emitter follower stage. Explain the capacitor bypassing for this circuit.

Section 12-7

12-18 Sketch the circuit of a two-stage, capacitor-coupled amplifier using an *FET* common-source input stage and a *BJT* common-emitter output stage. Discuss the advantages of this type of *BIFET* circuit.

12-19 Write equations for Z_i, Z_o, and A_v for the circuit in Review Question 12-18.

12-20 Sketch the circuit of a two-stage, direct-coupled amplifier using an *n*-channel *FET* common-source input stage and a *pnp BJT* common-emitter output stage. Explain the advantages of this arrangement.

Section 12-8

12-21 Draw the circuit diagram for a *BJT* differential amplifier: (a) using a single-polarity supply, (b) using a plus/minus supply. Explain the *dc* and *ac* operation of each circuit, and discuss the advantages and disadvantages of differential amplifiers.

12-22 Write equations voltage gain and impedance equations for a differential amplifier circuit.

Section 12-9

12-23 Sketch the circuit of a single-stage, capacitor-coupled common base amplifier, and then show the modifications required to convert it to a cascode amplifier. Explain the operation of the cascode amplifier.

12-24 Derive an equation for a cascode amplifier voltage gain. Also, write equations for input and output impedance.

12-25 Sketch a differential amplifier circuit for high-frequency use. Explain its advantages in this application.

Section 12-10

12-26 Briefly explain the measures that should be taken to stop a laboratory-constructed circuit from oscillating.

12-27 List the correct procedure for testing a laboratory-constructed amplifier circuit for voltage gain, frequency response, input impedance, and output impedance.

Chapter-12 Problems

Section 12-1

12-1   A single-stage, *BJT*, common-emitter amplifier with voltage divider bias is to be designed to use a *2N3903* transistor. A 15 V supply is to be used and 70 kΩ load resistor is to be capacitor-coupled to the output. Sketch the circuit and determine suitable resistor values.

12-2   Determine suitable capacitor values for the circuit in Problem 12-1 to give a 60 Hz lower cutoff frequency and a 60 kHz upper cutoff frequency. The signal source resistance is 300 Ω.

12-3   Analyze the circuit designed for Problems 12-1 and 12-2 to determine A_v, Z_i, and Z_o.

12-4   A single-stage common-emitter amplifier using a *2N3906* transistor with voltage divider bias is to be designed to have V_{CC} = 18 V, and R_L = 56 kΩ. Sketch the circuit and determine suitable resistor values.

12-5   The circuit in Problem 12-5 to have f_1 = 80 Hz and f_2 = 75 kHz. The signal source resistance is 500 Ω. Determine suitable capacitor values.

12-6   Calculate A_v, Z_i, and Z_o for the circuit designed for Problems 12-4 and 12-5.

12-7   A single-stage common-emitter amplifier with voltage divider bias is to use a *2N3904* transistor with V_{CC} = 12 V. The circuit is to be designed to have $Z_o ≈ 3.9$ kΩ. Sketch the circuit and determine suitable resistor values.

12-8   The circuit in Problem 12-7 is to have f_1 = 50 Hz, R_L = 47 kΩ, and a signal source resistance r_s = 600 Ω. Determine suitable capacitor values.

12-9   Determine A_v, Z_i, and Z_o for the circuit designed for Problems 12-7 and 12-8.

Section 12-2

12-10 A single-stage common-source amplifier with voltage divider bias is to use a *2N5457 FET* with V_{DD} = 22 V and R_L = 70 kΩ. Sketch the circuit and determine suitable resistor values.

12-11 Determine suitable capacitor values for the circuit in Problem 12-10 to give a 60 Hz lower cutoff frequency and a 60 kHz upper cutoff frequency. The signal source resistance is 300 Ω.

12-12 Analyze the circuit designed for Problems 12-10 and 12-11 to determine A_v, Z_i, and Z_o.

12-13 A single-stage common-source amplifier using a *2N5458 FET* with voltage divider bias is to have to have V_{DD} = 25 V, and R_L = 56 kΩ. Sketch the circuit and determine suitable resistor values.

12-14 The circuit in Problem 12-13 to have f_1 = 75 Hz and f_2 = 90 kHz. The signal source resistance is 500 Ω. Determine suitable capacitor values.

12-15 Calculate A_v, Z_i, and Z_o for the circuit designed for Problems 12-13 and 12-14.

12-16 A single-stage common-source amplifier with voltage divider bias is to use a *2N5459 FET* with V_{DD} = 20 V. The circuit is to be designed to have Z_i ≈ 500 kΩ and Z_o ≈ 5 kΩ. Draw the circuit diagram and determine suitable resistor values.

12-17 The circuit in Problem 12-16 is to have f_1 = 80 Hz, R_L = 47 kΩ, and a signal source resistance r_S = 600 Ω. Determine suitable capacitor values.

12-18 Determine A_v, Z_i, and Z_o for the circuit designed for Problems 12-16 and 12-17.

Section 12-3

12-19 A two-stage, capacitor-coupled, common-emitter amplifier is to be designed to have Z_o ≈ 3.9 kΩ. The circuit is to use *2N3903* transistors with V_{CC} = 12 V. Draw the complete circuit diagram and calculate suitable resistor values.

12-20 The circuit in Problem 12-19 to have f_1 = 100 Hz, R_L = 47 kΩ, and r_S = 600 Ω. Determine suitable capacitor values.

12-21 Calculate A_v, Z_i, and Z_o for the circuit designed for Problems 12-19 and 12-20.

12-22 A two-stage, capacitor-coupled, common-emitter amplifier is to be designed to use *2N3906 pnp* transistors with V_{CC} = 18 V and R_L = 60 kΩ. Draw the circuit diagram and calculate suitable resistor values.

12-23 The circuit in Problem 12-22 to have f_1 = 120 Hz and f_2 = 82 kHz. The signal source resistance is r_S = 600 Ω. Determine suitable capacitor values.

12-24 Calculate A_v, Z_i, and Z_o for the circuit designed for Problems 12-22 and 12-23.

Section 12-4

12-25 A two-stage, direct-coupled, common-emitter amplifier is to be designed to use *2N3903* transistors with V_{CC} = 16 V. The capacitor-coupled load is R_L = 75 kΩ, and the signal source resistance is r_s = 200 Ω. Draw the circuit diagram and calculate suitable resistor values.

12-26 Determine suitable capacitor values for the circuit in Problem 12-25. The lower cutoff frequency is to be 90 Hz.

12-27 Calculate A_v, Z_i, and Z_o for the circuit designed for Problems 12-25 and 12-26.

12-28 A two-stage, direct-coupled, common-emitter amplifier is to be designed to use a *2N3904 npn* transistor as *stage 1*, and a *2N3906 pnp* device as *stage 2*. The circuit is to have V_{CC} = 17 V and R_L = 85 kΩ. Draw the circuit diagram and calculate suitable resistor values.

12-29 The circuit in Problem 12-28 to have f_1 = 55 Hz. Determine suitable capacitor values if r_s = 600 Ω.

12-30 Calculate A_v, Z_i, and Z_o for the circuit designed for Problems 12-28 and 12-29.

12-31 A two-stage, direct-coupled, common-emitter amplifier is to use a *2N3905 pnp* transistor as *stage 1*, and a *2N3903 npn* device as *stage 2*. The circuit is to have V_{CC} = 11 V, R_L = 75 kΩ, f_1 = 80 Hz, and r_s = 300 Ω. Sketch the circuit diagram and determine suitable resistor and capacitor values.

Section 12-5

12-32 A two-stage, direct-coupled, common-emitter amplifier with an emitter follower output is to be designed to use *2N3904* transistors. The circuit is to have V_{CC} = 16 V, R_L = 65 Ω, f_1 = 50 Hz, and v_o ≈ ±75 mV. Sketch the circuit and determine suitable resistor and capacitor values. Take r_s = 300 Ω.

12-33 Analyze the circuit designed for Problem 12-32 to calculate the overall voltage gain.

12-34 Design a two-stage, direct-coupled, common-emitter amplifier with an emitter follower output. The circuit is to use *2N3906 BJTs* with V_{CC} = 18 V. The load is R_L = 85 Ω, the lower cutoff frequency is f_1 = 65 Hz, the output voltage is v_o ≈ ±55 mV, and r_s = 100 Ω. Draw the circuit diagram and determine suitable resistor and capacitor values.

12-35 Determine the overall voltage gain for the amplifier circuit in Problem 12-34.

Section 12-6

12-36 A *dc* feedback pair with two stages of amplification is to be designed to use *2N3904* transistors. The circuit is to have V_{CC} = 18 V, R_L = 50 kΩ, f_1 = 75 Hz, and r_s = 100 Ω. Sketch the circuit and determine suitable component values.

12-37 Analyze the circuit designed for Problem 12-36 to calculate the overall voltage gain.

12-38 A *dc* feedback pair with two stages of amplification is to be designed to use *2N3906 BJTs* and to have V_{CC} = 15 V, R_L = 60 kΩ, f_1 = 150 Hz, and r_s = 220 Ω. Draw the circuit diagram and determine suitable component values.

12-39 A *dc* feedback pair with an emitter follower output stage is to use *2N3904 BJTs*, to have V_{CC} = 14 V, R_L = 40 Ω, r_s = 200 Ω, f_1 = 60 Hz, and v_o ≈ ±65 mV. Determine suitable component values.

12-40 Analyze the circuit designed for Problem 12-39 to calculate A_v, Z_i, and Z_o.

12-41 A *dc* feedback pair with an emitter follower output stage is to use *2N3906 BJTs*. The circuit is to have V_{CC} = 9 V, R_L = 100 Ω, and f_1 = 75 Hz, and v_o ≈ ±75 mV. Sketch the circuit and determine suitable resistor and capacitor values.

12-42 Calculate A_v, Z_i, and Z_o for the circuit in Problem 12-41.

Section 12-7

12-43 A two-stage, capacitor-coupled, *BIFET* amplifier using a common-source input stage and a common-emitter output stage is to be designed to use a *2N5459 FET* and a *2N3904 BJT*. The external load resistance is 120 kΩ, the supply is 22 V, and the circuit input impedance is 800 kΩ. Draw the circuit diagram and calculate suitable resistor values.

12-44 The circuit in Problem 12-43 is to have f_1 = 140 Hz. Assuming that r_s = 220 Ω, determine suitable capacitor values.

12-45 Calculate A_v, Z_i, and Z_o for the circuit designed for Problems 12-43 and 12-44.

12-46 A two-stage, direct-coupled, *BIFET* amplifier using a common-source input stage with a *2N5457 FET* and a common-emitter output stage with a *2N3906 pnp BJT* is to be designed. The external load resistance is 100 kΩ, the supply is 25 V, and the circuit input impedance is to be approximately 300 kΩ. Draw the circuit diagram and calculate suitable resistor values.

12-47 The circuit in Problem 12-46 to have f_1 = 80 Hz. Determine suitable capacitor values. Assume that r_s = 800 Ω.

12-48 Calculate A_v, Z_i, and Z_o for the circuit designed for Problems 12-46 and 12-47.

12-49 Design a two-stage, capacitor-coupled, *BIFET* amplifier circuit using a *2N5486 FET* with self bias for Q_1 and a voltage divider biased *2N3904 BJT* for Q_2. The circuit specification is: V_{CC} = 23 V, R_L = 200 kΩ, f_1 = 80 Hz. For the *2N5486*, Y_{fs} = 6000 μS typical, 8000 μS maximum, $I_{DSS(max)}$ = 20 mA, and $V_{P(max)}$ = -6 V. Assume $r_s \ll Z_i$.

12-50 Calculate A_v, Z_i, and Z_o for the circuit in Problem 12-49.

12-51 Redesign the circuit in Problem 12-49 to use a *2N3905 pnp BJT* for the second stage with direct coupling between stages.

Section 12-8

12-52 A differential amplifier is to be designed to use *2N3904 BJTs*. The external capacitor-coupled load resistance is 120 kΩ, and the supply is ±12 V. Draw the circuit diagram and calculate suitable resistor values.

12-53 Determine suitable capacitor values for the circuit in Problem 12-52 to give $f_1 = 100$ Hz. Assume that $r_s = 600$ Ω.

12-54 Determine A_v, Z_i, and Z_o for the circuit designed for Problems 12-52 and 12-53.

12-55 A differential amplifier is to use *2N3903 BJTs* with a 15 V supply. The external load resistance is 66 kΩ, the signal source resistance is 600 Ω, and the lower cutoff frequency is to be 75 Hz. Draw the circuit diagram and determine suitable resistor and capacitor values. Assume $r_s \ll Z_i$.

12-56 Determine A_v, Z_i, and Z_o for the circuit in Problem 12-55.

Section 12-9

12-57 Design a common-base *BJT* amplifier to use a *2N3904* transistor. The specification is: $V_{CC} = 9$ V, $R_L = 27$ kΩ, and $f_1 = 75$ Hz. Sketch the circuit and determine suitable resistor and capacitor values.

12-58 Calculate A_v, Z_i, and Z_o for the circuit in Problem 12-57.

12-59 Design a cascode amplifier to use *2N3904* transistors. The specification is: $V_{CC} = 19$ V, $R_L = 47$ kΩ, and $f_1 = 85$ Hz. Draw the circuit diagram and calculate suitable resistor and capacitor values.

12-60 Analyze the circuit in Problem 12-59 to determine A_v, Z_i, and Z_o.

Section 12-10

12-61 The output voltage of a transistor amplifier under test falls from ±2 V to ±1.5 V when a 4.7 kΩ resistor is connected in series with the 600 Ω source resistance and the circuit input. Calculate the circuit input resistance.

12-62 Determine the output resistance of a transistor amplifier if v_o changes from ±1 V to ±1.2 V when the capacitor-coupled 60 kΩ load resistance is disconnected.

12-63 The voltage gain of the first stage of a two-stage *BJT* amplifier drops from 300 to 66 when the second stage is connected. The first stage uses a 6.8 kΩ collector resistor. Calculate the input impedance of the second stage.

Practise Problem Answers

12-1.1 150 Ω, 33 kΩ, 6.8 kΩ, 3.3 kΩ
12-1.2 27 μF, 200 μF, 0.56 μF
12-2.1 3.3 MΩ, 1 MΩ, 8.2 kΩ, 8.2 kΩ
12-2.2 0.068 μF, 30 μF, 0.5 μF
12-3.1 100 kΩ, 56 kΩ, 8.2 kΩ, 5.6 kΩ, 100 kΩ, 56 kΩ, 8.2 kΩ, 5.6 kΩ
12-3.2 15 μF, 180 μF, 25 μF, 180 μF, 0.25 μF
12-3.3 14 616
12-4.1 (150 kΩ + 15 kΩ), 56 kΩ, 5.6 kΩ, 5.6 kΩ, 1.8 kΩ, 5.6 kΩ

12-4.2   47 μF, 1000 μF, 500 μF, 1 μF

12-4.3   37 338

12-5.1   47 kΩ, (82 kΩ + 1.5 kΩ), 4.7 kΩ, 6.8 kΩ, 2.7 kΩ

12-5.2   15 μF, 250 μF, 39 μF, -182

12-6.1   10 kΩ, 220 kΩ, 4.7 kΩ, 3.3 kΩ

12-6.2   33 μF, 180 μF, 0.68 μF, 9737

12-7.1 $C_2 = 15\,\mu$F, $C_3 = 2\,\mu$F, $C_4 = 250\,\mu$F

12-7.2   5541

12-8.1   56 kΩ, 22 kΩ, 10 kΩ, 2.2 kΩ, 10 kΩ, 56 kΩ, 22 kΩ

12-8.2   1.2 μF, 0.2 μF, 454

12-9.1   (39 kΩ + 3.3 kΩ), (15 kΩ + 2.2 kΩ), 33 kΩ, 3.3 kΩ, 3.3 kΩ

12-9.2   18 μF, 18 μF, 180 μF, 0.47 μF, -300

12-10.1   -39%, -24%

Chapter *13*

Amplifiers with Negative Feedback

Chapter Contents

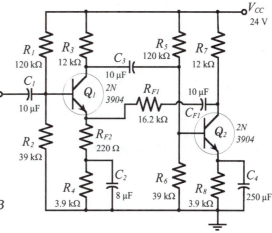

Objectives

You will be able to:

1   Using diagrams, explain series voltage negative feedback. List its effects on amplifier performance, and write voltage gain and impedance equations.

2   Sketch the following amplifier circuits using series voltage negative feedback, and explain the operation of each:
  • Two-stage, BJT capacitor-coupled
  • Complementary BJT direct-coupled
  • DC feedback pair
  • Two-stage BIFET
  • Differential-input

3   Analyze each of the circuit types listed in item 3 to determine its performance characteristics.

4   Design each of the circuit types listed in item 3 to fulfil a given specification.

5   Explain emitter current feedback. List its effects, and write voltage gain and impedance equations.

6   Design and analyze emitter current feedback circuits.

7   Explain p arallel current negative feedback. List its effects, and write current gain and impedance equations.

8   Design and analyze parallel current negative feedback circuits.

9   Explain the effects of negative feedback on circuit bandwidth, distortion, and phase shift, and make related calculations.

Introduction

Negative feedback is produced by feeding a portion of an amplifier output back to the input where it behaves as an additional signal. The feedback quantity is applied in opposition to the signal, so that the effective input is reduced. This results in stabilized amplifier gain, extended bandwidth, reduced distortion, and modified input and output impedances. *Series voltage feedback* increases input impedance; *parallel current feedback* reduces input impedance. The most important advantage of negative feedback is that it produces amplifiers with stable, predictable, voltage gains.

13-1   Series Voltage Negative Feedback

Negative Feedback Concept

In a negative feedback amplifier, a small portion of the output voltage is fed back to the input. When the feedback voltage is applied in series with the signal voltage, the arrangement is *series voltage feedback*. The instantaneous polarity of the feedback voltage is normally opposite to the signal voltage polarity, (they are in *series-opposition*). So, the feedback voltage is *negative* with respect to the signal voltage; hence the term *negative feedback*.

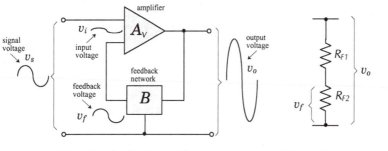

(a) Negative feedback amplifier   (b) Feedback network

Figure   13-1
In a negative feedback amplifier a portion of the output is fed back to the input. The instantaneous polarity of the feedback voltage (v_f) is negative with respect to the signal voltage (v_s).

Consider the illustration in Fig. 13-1(a). An amplifier with two input terminals and one output is shown (in triangular representation). The amplifier has a voltage gain (A_v), and its output voltage (v_o) is applied to a feedback network that reduces v_o by a factor (B) to produce a feedback voltage (v_f). The feedback network may be as simple as the resistive voltage divider shown in Fig. 13-1(b). At the amplifier input, the instantaneous level of v_f is applied negative with respect to v_s, so that the amplifier input terminal voltage is,

$$v_i = v_s - v_f$$

Because the amplifier input voltage is lower than the signal voltage, the output voltage is lower than that produced when negative feedback is not used. This means, of course, that the overall voltage gain (v_o/v_i) is reduced by negative feedback. However, as will be demonstrated, the stability of the voltage gain is greatly improved with negative feedback.

Voltage gain

Consider the feedback amplifier illustrated in Fig. 13-2, and recall that v_i is amplified to produce v_o, and that v_o is divided by the feedback network to produce v_f. Also, that v_f is applied (along with v_s) to the amplifier input. It is seen that there is a *closed loop* from the amplifier input to the output, and then back to the input. Because of this closed loop, the overall voltage gain with negative feedback is termed the *closed-loop gain* (A_{CL}).

When the feedback network is disconnected, the loop is opened, and the gain without feedback is referred to as the *open-loop gain* (A_v). Alternative symbols sometimes used for the open-loop gain are A_{OL} and $A_{v(OL)}$.

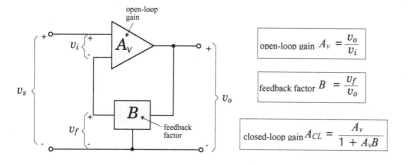

$$\text{open-loop gain } A_v = \frac{v_o}{v_i}$$

$$\text{feedback factor } B = \frac{v_f}{v_o}$$

$$\text{closed-loop gain } A_{CL} = \frac{A_v}{1 + A_v B}$$

Figure 13-2
Open-loop voltage gain, feedback factor, and closed-loop voltage gain for a negative feedback amplifier.

The overall voltage gain (closed-loop gain) of the amplifier in Fig. 13-2 is,

$$A_{CL} = \frac{v_o}{v_s}$$

and the *feedback factor* is,

$$B = \frac{v_f}{v_o} \qquad\qquad \textbf{(13-1)}$$

The input voltage is,

$$v_i = v_s - v_f$$

so,

$$v_i = v_s - B v_o$$

also,

$$v_o = A_v v_i = A_v(v_s - B v_o)$$
$$= A_v v_s - A_v B v_o$$

or,

$$v_o(1 + A_v B) = A_v v_s$$

giving,

$$\frac{v_o}{v_s} = \frac{A_v}{1 + A_v B}$$

So, the equation for overall voltage gain with negative feedback is,

$$A_{CL} = \frac{A_v}{1 + A_v B} \qquad\qquad \textbf{(13-2)}$$

Analysis of any transistor amplifier not using negative feedback shows that the voltage gain has a wide range, depending on the actual h_{fe} values of the transistors used. Consequently, the open-loop gain of a negative feedback amplifier can vary significantly. However, as demonstrated in Ex. 13-1, the closed loop gain can be a very stable quantity.

Example 13-1

Calculate the closed-loop gain for the negative feedback amplifier shown in Fig. 13-3. Also, calculate the closed loop gain when the open-loop gain is changed by $\pm 50\%$. Note that $A_v = 100\,000$ and $B = 1/100$.

Solution

when $A_v = 100\,000$,

Eq. 13-2, $A_{CL} = \dfrac{A_v}{1 + A_v B} = \dfrac{100\,000}{1 + (100\,000/100)}$

$= 99.9$

when $M = 150\,000$,

Eq. 13-2, $A_{CL} = \dfrac{A_v}{1 + A_v B} = \dfrac{150\,000}{1 + (150\,000/100)}$

$= 99.93$

when $M = 50\,000$,

Eq. 13-2, $A_{CL} = \dfrac{A_v}{1 + A_v B} = \dfrac{50\,000}{1 + (50\,000/100)}$

$= 99.8$

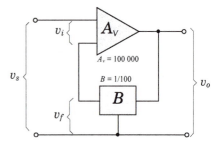

Figure 13-3
Negative feedback amplifier for Ex. 13-1.

Example 13-1 shows that when the open-loop gain changes by $\pm 50\%$ the closed-loop gain of the negative feedback amplifier remains stable within $\pm 0.1\%$. Thus,

series-voltage negative feedback stabilizes amplifier voltage gain.

Amplifier gain stabilization is the most important advantage of negative feedback amplifiers. Amplifiers that are required to have stable voltage gain are always designed as negative feedback amplifiers. Examination of Eq. 13-2 reveals that if $A_v B \gg 1$, then

$$A_{CL} \approx \frac{1}{B} \qquad\qquad \textbf{(13-3)}$$

From Eq. 13-3, it is seen that to design a negative feedback amplifier with a particular closed-loop gain, it is only necessary to design the feedback network to give $A_{CL} \approx 1/B$. For example, for $A_{CL} = 100$, $B \approx 1/100$. It must be remembered that for Eq. 13-3 to be correct, $A_v B \gg 1$, or

$$A_v \gg \frac{1}{B}$$

which means that,

$$A_v \gg A_{CL}$$

Thus, for satisfactory operation of a negative feedback amplifier,

**the open-loop voltage gain must be much greater than the
required closed-loop gain.**

In general, for A_{CL} to be stable within ±1%, A_v should be equal to
or greater than $100\,A_{CL}$.

Input Impedance

The input impedance of a *BJT* amplifier without negative feedback
is typically the impedance *looking into* the base of a transistor. If no
negative feedback is present in the amplifier in Fig. 13-4, the input
impedance is,

$$Z_b = \frac{v_i}{i_i}$$

and this gives,

$$i_i = \frac{v_i}{Z_b}$$

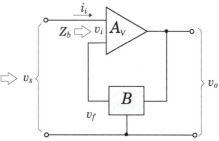

With negative feedback, the input impedance is,

$$Z_i = \frac{v_s}{i_i} = \frac{v_s \times Z_b}{v_i} \qquad (1)$$

and,

$$v_i = v_s - v_f$$
$$= v_s - Bv_o$$
$$= v_s - A_v Bv_i$$

Figure 13-4
*The input impedance of an
amplifier is increased by a factor of
$(1 + A_v B)$ when series voltage
negative feedback is employed.*

or,

$$v_i(1 + A_v B) = v_s \qquad (2)$$

Substituting for v_s from Eq. 2 into Eq. 1 gives,

$$\mathbf{Z_i = (1 + A_v B)Z_b} \qquad \textbf{(13-4)}$$

where Z_i is the input impedance with negative feedback, and Z_b is
the input impedance without negative feedback. It is seen that

**series voltage feedback increases the input impedance of
an amplifier by a factor of $(1 + A_v B)$.**

This is the same factor involved in gain reduction.

Input Impedance with Bias Resistors

The bias resistors at the input of a transistor circuit (*BJT* or *FET*)
are not normally affected by series voltage negative feedback, (see
Fig. 13-5). This is because the bias resistors are outside the
feedback loop. The circuit input resistance in Fig 13-5 is,

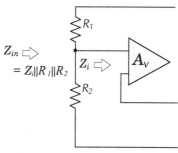

$$\mathbf{Z_{in} = Z_i \| R_1 \| R_2} \qquad \textbf{(13-5)}$$

Figure 13-5
*The bias resistors at the input of
an amplifier are not affected by
negative feedback.*

Example 13-2

The input impedance of the amplifier in Fig. 13-6 is $Z_b = 1\ k\Omega$ when negative feedback is not used, and the open-loop voltage gain is 100 000. Calculate the circuit input impedance with negative feedback.

Solution

From Eq. 13-1,
$$B = \frac{v_f}{v_o} = \frac{R_{F2}}{R_{F1} + R_{F2}} = \frac{560\ \Omega}{560\ \Omega + 56\ k\Omega}$$
$$= \frac{1}{101}$$

Eq. 13-4,
$$Z_i = (1 + A_v B)Z_b = \left[1 + \left(\frac{100\ 000}{101}\right)\right] \times 1\ k\Omega$$
$$= 991\ k\Omega$$

Eq. 13-5,
$$Z_{in} = Z_i\|R_1\|R_2 = 991\ k\Omega\|68\ k\Omega\|33\ k\Omega$$
$$= 21.7\ k\Omega$$

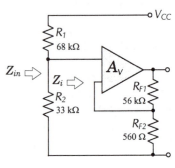

Figure 13-6
Negative feedback amplifier for Ex. 13-2.

Output Impedance

To derive an equation for the output impedance of a negative feedback amplifier, refer to Fig. 13-7. This is similar to the circuit of Fig. 13-1, but showing the internal impedance at the amplifier output terminal. The input terminals are short-circuited, so that v_f is the only voltage at the amplifier input. The output impedance without feedback is identified as Z_c, (the output impedance at a BJT collector terminal, normally equal to $1/h_{oe}$).

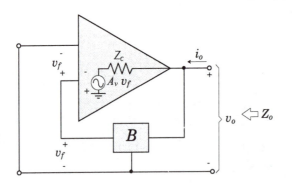

Figure 13-7
The output impedance of an amplifier is reduced by a factor of $(1 + A_v B)$ by series voltage negative feedback.

The output voltage (v_o) produces the feedback voltage (v_f), and this generates a voltage $A_v v_f$ in series with Z_c, as illustrated. Assuming an instantaneous positive polarity for v_o, the output current (i_o) occurs in the direction shown.

Writing an equation for the voltage drops around the output circuit of the negative feedback amplifier,

$$v_o = i_o Z_c - A_v v_f$$

or,

$$i_o Z_c = v_o + A_v v_f$$
$$= v_o + A_v B v_o$$
$$= v_o(1 + A_v B)$$

The output impedance of the negative feedback amplifier is,

$$Z_o = \frac{v_o}{i_o}$$

so,

$$Z_o = \frac{Z_c}{1 + A_v B} \qquad\qquad \text{(13-6)}$$

In Eq. 13-6, Z_o is the output impedance with negative feedback, and Z_c is the output impedance without negative feedback. So,

> **series voltage feedback reduces the output impedance of an amplifier by a factor of $(1 + A_v B)$.**

This is the same factor involved in the reduction of amplifier gain, and in the increase of the input impedance.

Output Impedance with a Collector Resistor

The collector resistor in the second stage of a two-stage circuit is not normally affected by series voltage negative feedback, (see Fig. 13-8). This is because (like the case of input bias resistors) the collector resistor is outside the feedback loop.

Recall from Section 6-4 that, for a *CE* circuit, $Z_c = 1/h_{oe}$, and $Z_o = R_C \| (1/h_{oe})$. Because $1/h_{oe}$ is usually much larger than R_C, Z_o is usually taken as equal to R_C when there is no feedback involved. However, with negative feedback, the collector output impedance is much lower than $1/h_{oe}$.

From Eq. 13-6, $\qquad Z_o = \dfrac{1/h_{oe}}{1 + A_v B}$

So, the circuit output impedance should be calculated as,

$$Z_{out} = R_C \| Z_o \qquad\qquad \text{(13-7)}$$

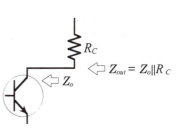

Figure 13-8
The output impedance at the collector of Q_2 is $R_C \| Z_o$. Resistor R_C is not affected by negative feedback.

Example 13-3

The amplifier circuit in Fig. 13-9 (reproduced from Fig. 12-21) is designed and analyzed in Examples 12-8 through 12-10. The input impedance at the base of Q_1 is $Z_b = 1$ kΩ, and Q_2 has $1/h_{oe} = 50$ kΩ. Calculate the circuit input and output impedances when negative feedback is used with a feedback factor of 1/100.

Solution

Eq. 13-4, $\qquad Z_i = (1 + A_v B)Z_b = \left[1 + \left(\dfrac{7533}{100}\right)\right] \times 1 \text{ kΩ}$

$$= 76 \text{ kΩ}$$

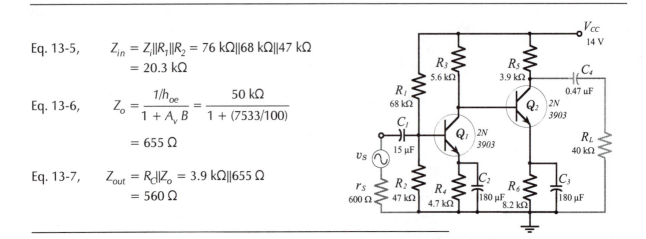

Eq. 13-5, $Z_{in} = Z_i \| R_1 \| R_2 = 76\ \text{k}\Omega \| 68\ \text{k}\Omega \| 47\ \text{k}\Omega$

$\qquad = 20.3\ \text{k}\Omega$

Eq. 13-6, $Z_o = \dfrac{1/h_{oe}}{1 + A_v\,B} = \dfrac{50\ \text{k}\Omega}{1 + (7533/100)}$

$\qquad = 655\ \Omega$

Eq. 13-7, $Z_{out} = R_C \| Z_o = 3.9\ \text{k}\Omega \| 655\ \Omega$

$\qquad = 560\ \Omega$

Figure 13-9
Amplifier circuit for Ex. 13-3.

Practise   Problems

13-1.1   A *BJT* amplifier with an open-loop gain of 350 000 has transistor parameters of $h_{ie} = 1.5\ \text{k}\Omega$, $h_{fe} = 100$, and $1/h_{oe} = 73\ \text{k}\Omega$. The input bias resistors are $R_1 = 82\ \text{k}\Omega$ and $R_2 = 39\ \text{k}\Omega$, and the output collector resistor is $R_C = 5.6\ \text{k}\Omega$. Negative feedback is applied with $B = 1/135$. Calculate Z_{in} and Z_{out} without negative feedback and with negative feedback. Also, determine the closed-loop gain.

13-2   Two-Stage CE Amplifier with Series Voltage Negative Feedback

Negative Feedback Amplifier Circuit

A two-stage, capacitor-coupled *BJT* amplifier is shown in Fig. 13-10. This is the same two-stage *CE* circuit discussed in Section 12-3 with the addition of feedback components R_{F2}, R_{F1}, and C_{F1}. These components constitute the feedback network referred to in Section 13-1. The output voltage is divided across R_{F2} and R_{F1} to produce a feedback voltage in series with the signal at the base of Q_1.

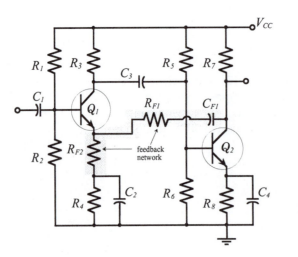

Figure 13-10
Two-stage capacitor-coupled BJT amplifier using series voltage negative feedback. R_{F1}, R_{F2}, and C_{F1} constitute the feedback network.

C_{F1} (in Fig. 13-10) is a *dc* blocking capacitor to prevent the *dc* voltage at Q_2 collector from affecting the Q_1 bias conditions. C_{F1} behaves as an open-circuit to *dc* and a short-circuit to *ac*. Consequently, C_{F1} is not included in the mid-frequency *ac* equivalent circuit of the feedback network in Fig. 13-11. Capacitor C_2 behaves as a short-circuit at middle and higher frequencies, so C_2 and R_4 (shorted by C_2) are also absent from the mid-frequency *ac* equivalent circuit.

The signal voltage (v_s) is applied between Q_1 base and ground, as illustrated in Fig. 13-11, and the output voltage (v_o) is developed between Q_2 collector and ground. The feedback voltage (v_f) is developed across resistor R_{F2}, between the Q_1 emitter terminal and ground. The input voltage (v_i) appears across the Q_1 base-emitter terminals as the difference between v_s and v_f.

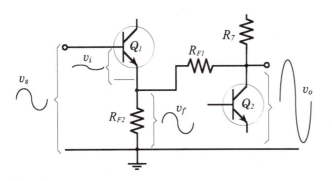

Figure 13-11
Mid-frequency ac equivalent circuit for the feedback network in the circuit of Fig. 13-10.

From previous studies of a two-stage amplifier circuit (Section 12-3), it is known that the circuit output voltage is in phase with the signal voltage. When the instantaneous level of v_s is positive-going, v_o is also positive-going, and consequently v_f is in *phase-opposition* with v_s. When v_s goes up (positively), v_f also goes up, and consequently, the voltage (v_i) at the base-emitter of Q_1 is reduced from v_s to,

$$v_i = v_s - v_f$$

Note that the Q_1 bias resistors (R_1 and R_2) in Fig. 13-10 are outside the feedback loop, and are unaffected by feedback, (as discussed in Section 13-1). The impedance *looking-into* the transistor base is within the feedback loop, so it is altered by feedback. Also, recall that the Q_2 collector resistor (R_7) is unaffected by negative feedback, and that the impedance *looking-into* the transistor collector terminal is changed by negative feedback.

The feedback factor for the circuit in Figs. 13-10 and 13-11 is,

$$B = \frac{R_{F2}}{R_{F1} + R_{F2}} \qquad \text{(13-8)}$$

From Eq. 13-3,
$$A_{CL} \approx \frac{R_{F1} + R_{F2}}{R_{F2}} \qquad \text{(13-9)}$$

Usually $R_{F1} \gg R_{F2}$, so $\qquad A_{CL} \approx \dfrac{R_{F1}}{R_{F2}}$ $\qquad\qquad$ **(13-10)**

Example 13-4

Determine the voltage gain, input impedance, and output impedance for the amplifier in Fig. 13-12. Without the feedback network, the circuit has an open-loop voltage gain of 58 000, an input impedance at Q_1 base of $Z_b = 1$ kΩ, and a Q_2 collector resistance of $1/h_{oe} = 85$ kΩ.

Solution

Eq. 13-8, $\qquad B = \dfrac{R_{F2}}{R_{F1} + R_{F2}} = \dfrac{220\ \Omega}{16.2\ \text{k}\Omega + 220\ \Omega}$

$\qquad\qquad\qquad = \dfrac{1}{74.6}$

Eq. 13-2, $\qquad A_{CL} = \dfrac{A_v}{(1 + A_vB)} = \dfrac{58\,000}{1 + (58\,000/74.6)}$

$\qquad\qquad\qquad = 74.5$

or, using Eq. 13-9,

$\qquad A_{CL} \approx \dfrac{R_{F1} + R_{F2}}{R_{F2}} = \dfrac{16.2\ \text{k}\Omega + 220\ \Omega}{220\ \Omega}$

$\qquad\qquad\qquad = 74.6$

Eq. 13-4, $\qquad Z_i = Z_b(1 + A_vB) = 1\ \text{k}\Omega[1 + (58\,000/74.6)]$

$\qquad\qquad\qquad = 778\ \text{k}\Omega$

Eq. 13-5, $\qquad Z_{in} = Z_i\|R_1\|R_2 = 778\ \text{k}\Omega\|120\ \text{k}\Omega\|39\ \text{k}\Omega$

$\qquad\qquad\qquad = 28.4\ \text{k}\Omega$

Eq. 13-6, $\qquad Z_o = \dfrac{1/h_{fe}}{(1 + A_vB)} = \dfrac{85\ \text{k}\Omega}{1 + (58\,000/74.6)}$

$\qquad\qquad\qquad = 109\ \Omega$

Eq. 13-7, $\qquad Z_{out} = R_7\|Z_o = 12\ \text{k}\Omega\|109\ \Omega$

$\qquad\qquad\qquad = 108\ \Omega$

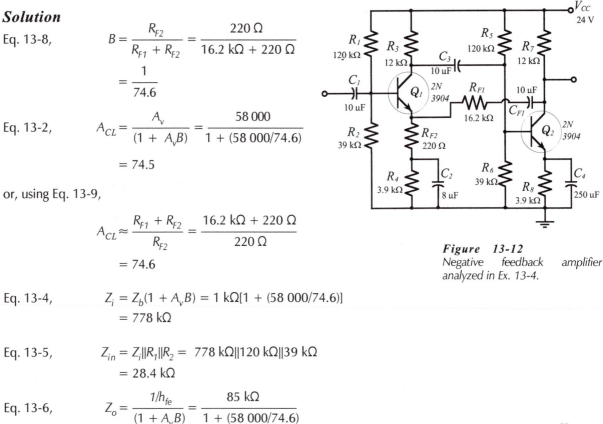

Figure 13-12
Negative feedback amplifier analyzed in Ex. 13-4.

Negative Feedback Amplifier Design

A negative feedback amplifier, like the one shown in Fig. 13-12, is best designed first as an amplifier without feedback. The feedback component values are then determined, and the circuit is modified as necessary.

Like R_L, R_{F1} is an additional load at the collector of the second stage in Fig. 13-12. As illustrated in Fig. 13-13, the voltage gain of

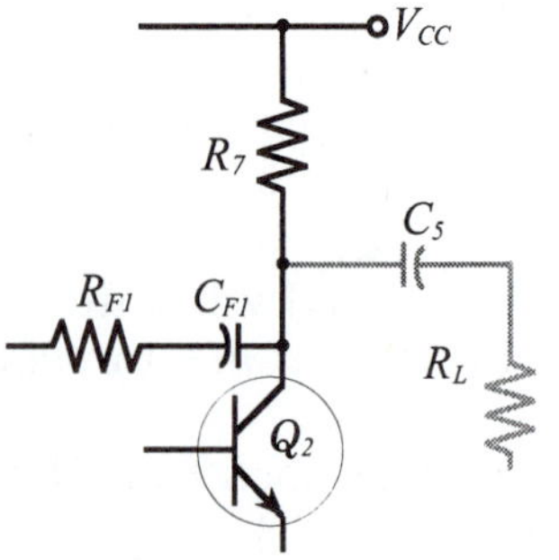

Figure 13-13
The circuit open-loop gain can be reduced by the presence of feedback resistor R_{F1}.

the second stage is proportional to the total load $(R_L \| R_7 \| R_{F1})$, consequently, R_{F1} has the effect of reducing the second stage gain. A large open-loop gain is required for good gain stability, so R_{F1} should be selected as large as possible.

Resistor R_{F2} is an unbypassed resistor in series with the emitter of Q_1. As such, it would seem to reduce the first stage voltage gain, and thus reduce the circuit open-loop gain. However, because v_f is applied to Q_1 emitter, the effective open-loop gain is the voltage gain from the BE terminals of Q_1 to the collector of Q_2. So, R_{F2} does *not alter the overall open-loop gain*. (R_{F2} is omitted when calculating the circuit open-loop gain.)

Another consideration for R_{F2} is that it is in parallel with the impedance (Z_{e1}) at the emitter of Q_1, (see Fig. 13-14). Without negative feedback, Z_{e1} would usually equal the transistor h_{ib} value, which is typically 26 Ω when I_C = 1 mA. With negative feedback, Z_{e1} can be shown to range from about 2.5 kΩ to 30 kΩ, depending upon A_v and B. Resistor R_{F2} should be selected much smaller than Z_{e1}. However, R_{F2} should be as large as possible to give the largest possible value for R_{F1}; (from Eq. 13-10, $R_{F1} \approx A_{CL} \times R_{F2}$).

A reasonable starting point for feedback network design is to select R_{F2} much lower than $Z_{e1}/10$; in the range of 100 Ω to 470 Ω. R_{F2} could also alter the *dc* bias conditions for Q_1. Consequently, it might be necessary to reduce R_4, so that R_4 and R_{F2} add up to the original resistance selected for R_4 (or R_E), (see Fig. 13-14).

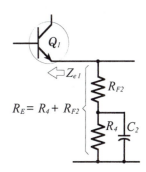

Figure 13-14
Resistor R_{F2} is in parallel with the impedance Z_{e1} 'looking into' the emitter of Q_1. The total (dc) emitter resistance (R_E) is also increased by the presence of R_{F2}.

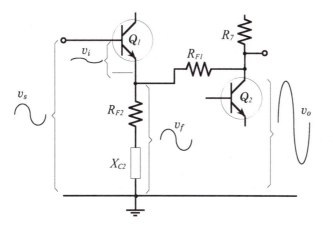

Figure 13-15
Low-frequency ac equivalent circuit for the feedback network in the circuit of Fig. 13-12. Note that X_{C2} is in series with R_{F2}.

Figure 13-15 shows a low-frequency *ac* equivalent circuit for the feedback network. Note that capacitor C_{F1} is still taken as an *ac* short circuit, so X_{CF1} is assumed to be very much smaller than R_{F1}. The impedance of capacitor C_2 is show, and because X_{C2} is assumed to be very much smaller than resistor R_4, R_4 is omitted.

From Fig. 13-15, v_o is divided across R_{F1} and $(R_{F2} + jX_{C2})$ to give v_f, so the feedback factor is,

$$B = \frac{R_{F2} - jX_{C2}}{(R_{F2} - jX_{C2}) + R_{F1}}$$

Assuming that $(R_{F2} - jX_{C2}) << R_{F1}$,

$$B \approx \frac{R_{F2} - jX_{C2}}{R_{F1}}$$

From Eq. 13-3,

$$A_{CL} \approx \frac{R_{F1}}{R_{F2} - jX_{C2}}$$

or,

$$|A_{CL}| \approx \frac{R_{F1}}{\sqrt{(R_{F2}{}^2 + X_{C2}{}^2)}}$$

When $X_{C2} = R_{F2}$,

$$|A_{CL}| \approx \frac{R_{F1}}{\sqrt{2} \, R_{F2}} \approx 0.707 \times \text{(mid-frequency gain)}$$

To set a desired frequency lower cutoff frequency (f_1),

$$\boldsymbol{X_{C2} = R_{F2} \text{ at } f_1} \qquad\qquad \textbf{(13-11)}$$

As already mentioned, X_{CF1} must be very much smaller than R_{F1} at the low cutoff frequency for the amplifier. It can be shown that, if $X_{CF1} = R_{F1}$ at f_1, the closed-loop gain would actually increase at frequencies below f_1. C_{F1} is normally calculated from,

$$\boldsymbol{X_{CF1} = \frac{R_{F1}}{100} \text{ at } f_1} \qquad\qquad \textbf{(13-12)}$$

Capacitor C_{F1} can often be eliminated. If R_{F1} is large enough, direct connection of R_{F1} from Q_{2C} to Q_{1E} in Fig. 13-12 will have a negligible effect on the circuit bias conditions. In fact, with a direct-coupled circuit, direct connection of the feedback resistors might enhance the circuit bias stability.

The circuit input impedance is increased by negative feedback, so input coupling capacitor C_1 can be recalculated as a smaller capacitor, $(X_{C1} = Z_{in}/10 \text{ at } f_1)$. Recall that the bias resistors are not affected by feedback. Regardless of how high the circuit input impedance might become, C_1 should normally be selected as a minimum of 0.1 µF, to ensure that it is much larger than any stray capacitance at the circuit input.

The emitter bypass capacitor C_4 (in Fig. 13-12) was calculated (in Ex. 12-6) to have $X_{C4} = 0.65 \, h_{ib2}$ at f_1. So, X_{C4} will reduce the circuit open-loop gain (A_v) at f_1. However, even with this gain reduction, A_v should still be much larger than A_{CL}, so C_4 can normally be left as calculated for an amplifier without feedback.

Example 13-5
Modify the two-stage capacitor-coupled *BJT* amplifier designed in Example 12-6 to use series voltage feedback, as in Fig. 13-16. The closed-loop gain is to be 75, and the lower cutoff frequency is to be 100 Hz.

Solution

Select $\qquad R_{F2} = 220\ \Omega \qquad (R_{F2} < Z_{E1})$

R_4 becomes, $\qquad R_4 = $ (original R_4) $- R_{F2} = 3.9\ k\Omega - 220\ \Omega$

$\qquad\qquad\qquad = 3.68\ k\Omega$ (use $3.9\ k\Omega$ standard value)

From Eq. 13-9, $\qquad R_{F1} = (A_{CL} - 1)\ R_{F2} = (75 - 1) \times 220\ \Omega$

$\qquad\qquad\qquad = 16.3\ k\Omega$ (use $15\ k\Omega + 1.2\ k\Omega$)

Eq. 13-11, $\qquad X_{C2} = R_{F2}$ at f_1

$$C_2 = \frac{1}{2\ \pi\ f_1\ R_{F2}}$$

$$= \frac{1}{2\ \pi \times 100\ Hz \times 220\ \Omega}$$

$$= 7.2\ \mu F \text{ (use } 8\ \mu F)$$

Eq. 13-12, $\qquad X_{CF1} = \dfrac{R_{F1}}{100}$ at f_1

$$C_{F1} = \frac{1}{2\ \pi\ f_1\ R_{F1}/100}$$

$$= \frac{1}{2\ \pi \times 100\ Hz \times 16.2\ k\Omega/100}$$

$$= 9.8\ \mu F \text{ (use } 10\ \mu F)$$

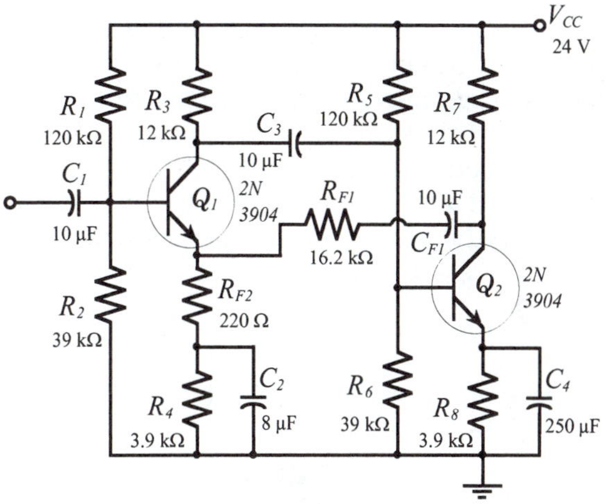

Figure 13-16
Negative feedback amplifier for Example 13-5. Note that this is the circuit analyzed in Ex. 13-4.

Recall from Chapter 12 that, when designing an amplifier without feedback, the collector resistor (R_{C2}) for the output transistor is usually selected as equal to, or less than, $R_L/10$. This is because without negative feedback $Z_o \approx R_{C2}$. Negative feedback normally reduces Z_o to a value much smaller than R_{C2} (see Ex. 13-4). So, R_{C2} (R_7 in Fig. 13-16) does not have to be selected as $R_L/10$ when designing a negative feedback amplifier. Instead, any convenient level of V_{RC} and I_C can be used for calculating a suitable R_{C2} value.

Practise Problems

13-2.1 Modify the circuit in Fig. 12-18 to use series voltage negative feedback (as in Fig. 13-16) to give $A_{CL} = 100$, and $f_1 = 50$ Hz.

13-2.2 Analyze the negative feedback circuit in Problem 13-2.1 to determine the input and output impedances. The transistor parameters are $h_{ie} = 1\ k\Omega$, $h_{oe} = 40\ \mu S$, and $h_{fe} = 100$.

13-3   *More Amplifiers with Series Voltage Negative Feedback*

Complementary Direct-Coupled Circuit with Feedback

The two-stage, direct-coupled *BJT* amplifier in Fig. 13-17 is reproduced from Fig. 12-22, and modified to include feedback components R_{F1} and R_{F2}. The feedback network design procedure is exactly as discussed in Section 13-2. Other direct-coupled circuits can be converted into negative feedback amplifiers by following the same procedure. Similarly, *BIFET* circuits, both direct-coupled and capacitor-coupled, can be designed for overall negative feedback. As always, the best approach is to first design the circuit as a non-feedback amplifier, then determine the feedback component values.

Note that there is no coupling-capacitor for the feedback network in the circuit shown in 13-17. This is because (as discussed in Section 13-2) the omission of the capacitor can have a negligible effect on the circuit *dc* conditions if feedback resistor R_{F1} is large enough.

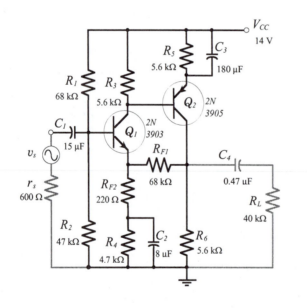

Figure 13-17
Complementary, direct-coupled amplifier using series-voltage negative feedback. The amplifier closed-loop voltage gain is,
$$A_{CL} = (R_{F1} + R_{F2})/R_{F2}.$$

Example 13-6

Modify the direct-coupled amplifier in Fig. 12-22 to use series-voltage negative feedback as in Fig 13-17. The amplifier closed-loop voltage gain is to be approximately 300, and the lower cutoff frequency is to be 75 Hz.

Solution

$$R_{F2} < Z_{e1}$$

Select $R_{F2} = 220\ \Omega$

R_4 becomes, R_4 = (original R_4) - R_{F2} = 4.7 kΩ - 220 Ω

$\qquad\qquad\qquad$ = 4.48 kΩ   (use 4.7 kΩ standard value)

From Eq. 13-9, $R_{F1} = (A_{CL} - 1) R_{F2} = (300 - 1) \times 220 \; \Omega$

$\qquad\qquad\qquad \approx 66 \; k\Omega$ (use 68 kΩ standard value)

Eq. 13-11, $X_{C2} = R_{F2}$ at f_1

$$C_2 = \frac{1}{2 \pi f_1 R_{F2}} = \frac{1}{2 \pi \times 100 \; Hz \times 220 \; \Omega}$$

$$= 7.2 \; \mu F \quad (\text{use } 8 \; \mu F \text{ standard value})$$

Analysis of the circuit in Ex. 13-6 reveals that the transistor *dc* collector currents are approximately 1 mA, and that the direct current through R_{F1} is around 4.4 μA.

DC Feedback Pair Using Negative Feedback

Figure 13-18(a) shows a *dc* feedback pair with *ac* feedback components C_{F1}, R_{F1}, and R_{F2}. This is the most economical of all two-stage *BJT* amplifier circuits, because it has the smallest quantity of components. It can have just as high a voltage gain (open-loop and closed-loop) as any other two-stage circuit. As a negative feedback amplifier, its input resistance is normally higher than that for a *BJT* circuit using voltage divider bias, (R_2 is usually larger than the parallel resistance of voltage divider resistors). As always, the feedback components are determined after the circuit is designed for the largest possible open-loop voltage gain.

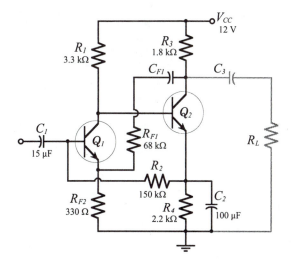

(a) DC feedback pair using series-voltage negative feedback

(b) Circuit modification for predictable low 3 dB frequency

Figure 13-18
DC feedback pair as a negative feedback amplifier. The closed-loop voltage gain is,
$A_{CL} = (R_{F1} + R_{F2})/R_{F2}$.

The lower cutoff frequency for the circuit in Fig. 13-18(a) is unpredictable. The modification shown in Fig. 13-18(b) gives a definite low 3 dB frequency, (when $X_{C4} = R_{F2}$). In this case, emitter resistor R_5 affects the circuit *dc* bias conditions, so it must be designed into the circuit from the start. As in other circuits, it might be possible to omit the feedback network coupling capacitor, (C_{F1} in Fig. 13-18). The direct current through R_{F1} should be calculated to ensure that it does not significantly affect the bias conditions of the circuit.

BIFET Two-Stage Circuit With Negative Feedback

A two-stage, direct-coupled, *BIFET* circuit with series-voltage negative feedback is shown in Fig. 13-19. This is a modified version of the circuit in Fig. 12-30, designed in Examples 12-15 and 12-16.

The open-loop voltage gain of a *BIFET* circuit is typically 50 to 100 times smaller than that for a two-stage *BJT* amplifier, (see Ex. 12-17). Therefore, *BIFET* negative feedback amplifiers must be designed for relative small closed-loop voltage gains, usually a maximum of around 30. The single major advantage of *BIFET* circuits, very high input impedance, is largely unchanged by negative feedback. The design procedure for the feedback network in the *BIFET* circuit is exactly as already discussed.

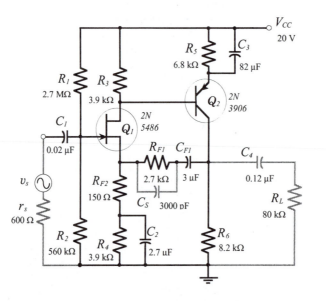

Figure 13-19
BIFET circuit using series-voltage negative feedback. The closed-loop voltage gain is,
$$A_{CL} = (R_{F1} + R_{F2})/R_{F2}.$$

Setting the Circuit Upper Cutoff Frequency

The upper cutoff frequency of any of the series voltage negative feedback amplifiers already discussed can be set by the simple process of connecting a capacitor (C_S) across feedback resistor R_{F1}, as illustrated in Fig. 13-19. It can be shown that f_2 occurs when $X_{CS} = R_{F1}$, so long as the upper cutoff frequency without C_S is much higher than the desired f_2 value. Therefore, to determine C_S,

$$X_{CS} = R_{F1} \text{ at } f_2 \qquad\qquad (13\text{-}13)$$

13-4   Two-Stage Differential Amplifier with Negative Feedback

Two-Stage Differential Amplifier

The circuit shown in Fig. 13-20 has a differential-amplifier input stage with *npn* BJTs, and a direct-coupled *pnp* transistor differential amplifier second stage. (The differential amplifier is discussed in detail in Section 12-8.) The circuit uses a plus/minus supply voltage, and the base terminals of Q_1 and Q_2 are biased to ground via resistors R_1 and R_6, respectively. Transistors Q_3 and Q_4 have the bases directly connected to the collector terminals of Q_1 and Q_2. So, bias voltage for Q_3 and Q_4 bases is provided by the voltage drops across the Q_1 and Q_2 collector resistors, (R_2 and R_4).

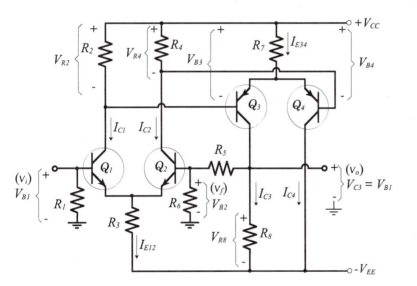

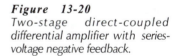

Figure   13-20
Two-stage   direct-coupled differential amplifier with series-voltage negative feedback.

The amplifier input terminal is the base of Q_1, and the output is taken from Q_3 collector. No collector resistor is provided for Q_4, because no output or feedback is taken from Q_4. Note that the feedback network (R_5 and R_6) is connected from the output at the collector of Q_3 back to the base of Q_2. Because the emitter of Q_2 is directly coupled to Q_1 emitter, applying the feedback voltage (v_f) to Q_2 base is similar to applying it to Q_1 emitter.

DC Bias Conditions

The circuit in Fig. 13-20 is designed to have the *dc* bias voltage at Q_2 base equal to Q_1 base bias voltage; in this case, ground level. This means that the *dc* voltage at the collector of Q_3 must also equal the Q_1 base voltage, because Q_3 collector is directly connected to Q_2 base via R_5. Consider what would happen if V_{C3} (at the output terminal) is not exactly equal to V_{B1} (at the input).

Suppose V_{C3} goes lower than its normal level:

- V_{B2} is reduced below the level of V_{B1}, and this reduces I_{C2} and increases I_{C1}.
- The increased level of I_{C1} causes an increased voltage drop across R_2, and the reduced level of I_{C2} decreases the voltage drop across R_4. Thus V_{B3} is increased and V_{B4} is decreased.
- This raises the level of I_{C3} and lowers I_{C4}. So, V_{R8} is increased (by the I_{C3} increase) to drive V_{C3} back up to its normal voltage.

Similarly, if V_{C3} somehow drifts to a higher than normal level:

- V_{B2} is raised above the level of V_{B1}, thus increasing I_{C2} and reducing I_{C1}.
- The decreased level of I_{C1} and increased level of I_{C2} produces voltage drops across R_2 and R_4 that reduce V_{B3} and increase V_{B4}.
- This change reduces I_{C3}, thus decreasing V_{R8} to drive V_{C3} back down to its normal voltage.

It is seen that there is *dc negative feedback* that stabilizes the circuit bias conditions.

AC Operation

Consider the circuit waveforms shown in Fig. 13-21. A positive-going input signal (v_i) at the base of Q_1 produces a positive-going output (v_o) at Q_3 collector. As in the case of the *dc* bias conditions, the instantaneous *ac* voltage at Q_2 base follows the instantaneous level of the *ac* signal voltage at Q_1 base.

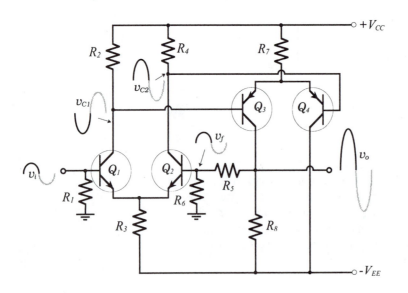

Figure 13-21
AC waveforms in a two-stage differential amplifier with series voltage negative feedback.

With $v_{b2} = v_{b1}$, or $v_i = v_f$,

$$v_f = B v_o = v_o \times \frac{R_6}{R_5 + R_6}$$

so,

$$v_o = v_f \times \frac{R_5 + R_6}{R_6} = v_i \times \frac{R_5 + R_6}{R_6}$$

giving,

$$A_{CL} = \frac{R_5 + R_6}{R_6}$$

Resistors R_5 and R_6 in Figs. 13-20 and 13-21 are comparable to R_{F1} and R_{F2}, respectively, in the other negative feedback circuits already discussed. So, the closed-loop gain equation is the same equation that applies in the case of all series-voltage negative feedback amplifiers. As always, the equation is true only when the open-loop gain is very much larger than the closed-loop gain.

The input and output impedances for a circuit with a differential input stage are calculated exactly as discussed for other negative feedback circuits.

Note that there are no bypass capacitors in the circuit in Figs. 13-20 and 13-21. If the signal and load are capacitor coupled to the circuit, the coupling capacitors determine the lower cutoff frequency. If the signal and load are direct coupled, then the circuit is a *dc amplifier*; one that amplifies direct voltage signals.

Circuit Design

The design procedure for the input stage is similar to that discussed in Section 12-8. The second stage components are determined in essentially the same way as the input stage components, bearing in mind that the second stage is *upside down* compared to the input stage. Feedback network resistor R_6 at the base of Q_2 is selected equal to bias resistor R_1 at Q_1 base. This is to equalize the resistances at the bases of Q_1 and Q_2, and thus equalize any voltage drops due to I_{B1} and I_{B2}. R_5 is calculated from the specified closed loop gain and the R_6 resistance.

Example 13-7

The two-stage differential amplifier in Fig. 13-22 is to use a ± 10 V supply, and have a closed-loop gain of 33. Select suitable direct current and voltage levels, and calculate resistor values for the circuit. Assume that the transistors have $h_{FE} = 100$.

Solution

Select,

$$I_{C1} = I_{C2} = I_{C3} = I_{C4} \approx 1 \text{ mA}$$

From Eq. 5-17,

$$R_1 \approx \frac{V_{BE}}{10\, I_{C1}/h_{FE}} \approx \frac{0.7\text{ V}}{10 \times 1\text{ mA}/100}$$

$$= 7 \text{ k}\Omega \quad \text{(use 6.8 k}\Omega \text{ standard value)}$$

$$R_3 \approx \frac{V_{EE} - V_{BE}}{I_{C1} + I_{C2}} = \frac{10\ V - 0.7\ V}{2\ mA}$$

$$= 4.65\ k\Omega\ \text{(use 4.7 k}\Omega\text{ standard value)}$$

Select, $V_{CE1} \approx 3\ V$

$V_{R2} = V_{CC} + V_{BE1} - V_{CE1} = 10\ V + 0.7\ V - 3\ V$

$= 7.7\ V$

$$R_2 = R_4 = \frac{V_{R2}}{I_{C1}} = \frac{7.7\ V}{1\ mA}$$

$$= 7.7\ k\Omega\quad \text{(use 6.8 k}\Omega\text{ standard value)}$$

$$R_7 = \frac{V_{R2} - V_{BE}}{I_{C3} + I_{C4}} = \frac{7.7\ V - 0.7\ V}{2\ mA}$$

$$= 3.5\ k\Omega\quad \text{(use 3.3 k}\Omega\text{ standard value)}$$

$$R_8 = \frac{V_{EE}}{I_{C3}} = \frac{10\ V}{1\ mA}$$

$$= 10\ k\Omega\quad \text{(standard value)}$$

Select, $R_6 = R_1 = 6.8\ k\Omega$

From Eq. 13-9, $R_5 = (A_{CL} - 1)\ R_6 = (33 - 1) \times 6.8\ k\Omega$

$= 217.6\ k\Omega\ \text{(use 220 k}\Omega\text{ standard value)}$

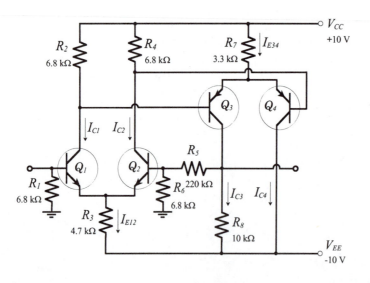

Figure 13-22
Amplifier circuit for Ex. 13-7.

Example 13-8

Assuming an open-loop gain of 25 000, $h_{ie} = 2\ k\Omega$, $h_{ib} = 25\ \Omega$, and $1/h_{oe} = 100\ k\Omega$, for the circuit in Ex. 13-7, calculate A_{CL}, Z_{in}, and Z_{out}.

Solution

From Eq. 13-8,
$$B = \frac{R_6}{R_5 + R_6} = \frac{6.8 \text{ k}\Omega}{220 \text{ k}\Omega + 6.8 \text{ k}\Omega}$$
$$= \frac{1}{33.4}$$

Eq. 13-2,
$$A_{CL} = \frac{A_v}{1 + A_v B} = \frac{25\,000}{1 + (25\,000/33.4)}$$
$$= 33.4$$

Eq. 13-4,
$$Z_i = 2\,h_{ie}(1 + A_v B) = 2 \text{ k}\Omega[1 + (25\,000/33.4)]$$
$$= 3 \text{ M}\Omega$$

Eq. 13-5,
$$Z_{in} = Z_i \| R_1 = 3 \text{ M}\Omega \| 6.8 \text{ k}\Omega$$
$$\approx 6.8 \text{ k}\Omega$$

Eq. 13-6,
$$Z_o = \frac{1/h_{oe}}{1 + A_v B} = \frac{100 \text{ k}\Omega}{1 + (25\,000/33.4)}$$
$$= 133 \ \Omega$$

Eq. 13-7,
$$Z_{out} = R_8 \| Z_o = 10 \text{ k}\Omega \| 133 \ \Omega$$
$$= 131 \ \Omega$$

Modification for Reduced Z_{out}

Although the application of series-voltage negative feedback substantially reduces the output impedance of a circuit, further reduction of Z_{out} is sometimes required. Figure 13-23 shows the circuit of Fig. 13-22 with the addition of an emitter follower output stage (Q_5). Note that the feedback network is now connected at the emitter follower output terminal.

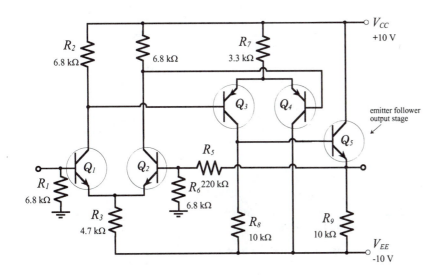

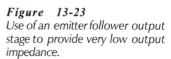

Figure 13-23
Use of an emitter follower output stage to provide very low output impedance.

The output impedance at Q_5 emitter terminal in Fig. 13-23 is found from the derivation of Eq. 6-24 (for a common-collector circuit).

$$Z_e = \frac{h_{ic} + (Z \text{ in series with } Q_5 \text{ base})}{h_{fc}}$$

This gives,

$$Z_e = \frac{h_{ic} + R_8}{h_{fc}} \qquad\qquad \textbf{(13-14)}$$

This quantity is modified by negative feedback according to (a rewritten) Eq. 13-6,

$$Z_o = \frac{Z_e}{1 + A_v B}$$

Then, as always, the circuit output impedance is given by Eq. 13-7. which must also be rewritten for the circuit in Fig. 13-23.

$$Z_{out} = Z_o \| R_9$$

An emitter follower output stage can be added to any of the amplifiers previously discussed. However, the combination of an emitter follower output stage with a differential amplifier has a particular significance that will be explained in Chapter 14.

Example 13-9

Calculate the output impedance for the circuit modification in Fig. 13-23, assuming that $h_{ic} = h_{ie} = 2$ kΩ, and $h_{fc} = h_{fe} = 100$. Also assume that $A_v = 25\,000$ and $B = 1/33.4$, as for Ex. 13-7 and 13-8.

Solution

Eq. 13-14,
$$Z_e = \frac{h_{ic} + R_4}{h_{fc}} = \frac{2 \text{ k}\Omega + 10 \text{ k}\Omega}{100}$$
$$= 120 \ \Omega$$

Eq. 13-6,
$$Z_o = \frac{Z_e}{1 + A_v B} = \frac{120 \ \Omega}{1 + (25\,000/33.4)}$$
$$= 0.16 \ \Omega$$

Eq. 13-7,
$$Z_{out} = R_5 \| Z_o = 10 \text{ k}\Omega \| 0.16 \ \Omega$$
$$= 0.16 \ \Omega$$

Practise Problems

13-4.1 A two-stage differential amplifier with negative feedback, as in Fig. 13-22, is to use 2N3904 and 2N3906 BJTs with a ±15 V supply. The closed-loop gain is to be 50. Determine suitable resistor values.

13-4.2 Calculate the minimum open loop gain for the circuit in Problem 13-4.1. Also, determine Z_{in} and Z_{out}.

13-5   *Emitter Current Feedback*

Emitter Current Feedback Circuit

Emitter current feedback is produced by connecting an unbypassed resistor in series with the emitter terminal of a transistor, as shown in Fig. 13-24(a). The effects of an unbypassed emitter resistor in a *CE* circuit have been analyzed in Section 6-5, where it is shown that the circuit impedances and voltage gains are;

Eq. 6-20, $Z_b = h_{ie} + R_{E1}(1 + h_{fe})$

Eq. 6-12, $Z_{in} = R_1 \| R_2 \| Z_b$

Eq. 6-14, $Z_{out} = (1/h_{oe}) \| R_C \approx R_C$

Eq. 6-21, $A_v = \dfrac{-h_{fe}(R_C \| R_L)}{h_{ie} + R_{E1}(1 + h_{fe})}$

Eq. 6-22, $A_v \approx \dfrac{R_C \| R_L}{R_{E1}}$

Equation 6-20 shows that the input impedance at the base of a transistor with an unbypassed emitter resistor is considerably increased above its normal value without feedback (h_{ie}). This raises the circuit input impedance, as defined by Eq. 6-12. Therefore,

emitter current feedback increases the circuit input impedance.

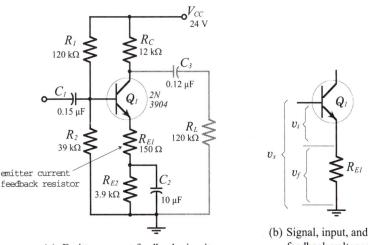

(a) Emitter current feedback circuit

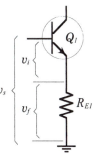

(b) Signal, input, and feedback voltages

Figure   13-24
Single-stage CE   amplifier using emitter current feedback. The voltage gain is $A_v \approx (R_C \| R_L)/R_{E1}$.

Figure 13-24(b) shows that the circuit feedback loop is from the transistor base to the emitter, then back to the base again. The collector resistor is outside the feedback loop. The circuit output impedance remains approximately equal to R_C, (see Eq. 6-14). So, *emitter current feedback has no effect on the circuit output impedance,* (see Section 6-5).

The precise voltage gain for the circuit in Fig. 13-24(a) is given by Eq. 6-21, and Eq. 6-22 can be used for calculation of the approximate voltage gain. Equation 6-22 is derived from Eq. 6-21 by assuming that $h_{ie} \ll R_{E1}(1 + h_{fe})$. Using typical values of $R_{E1} = 100\ \Omega$, $h_{fe} = 100$,

$$R_{E1}(1 + h_{fe}) \approx 10\ \text{k}\Omega$$

This is ten times the typical value for h_{ie}. So, Eq. 6-22 gives a reasonably close approximation to the circuit voltage gain.

It is apparent that,

emitter current feedback stabilizes single-stage amplifier voltage gain.

Example 13-10

The emitter current feedback circuit in Fig. 13-24 uses a transistor with $h_{fe(min)} = 100$, $h_{fe(max)} = 400$, $h_{ie(min)} = 1\ \text{k}\Omega$, and $h_{ie(max)} = 5\ \text{k}\Omega$. Calculate the precise value of the circuit voltage gain at the h_{fe} extremes, then calculate the approximate voltage gain using Eq. 6-22.

Solution

Eq. 6-21, $\quad A_v = \dfrac{h_{fe}(R_C \| R_L)}{h_{ie} + R_{E1}(1 + h_{fe})}$

$$A_{v(max)} = \frac{h_{fe(max)}(R_C \| R_L)}{h_{ie(max)} + R_{E1}(1 + h_{fe(max)})} = \frac{400(12\ \text{k}\Omega \| 120\ \text{k}\Omega)}{5\ \text{k}\Omega + 150\ \Omega(1 + 400)}$$

$$= 67$$

$$A_{v(min)} = \frac{h_{fe(min)}(R_C \| R_L)}{h_{ie(min)} + R_{E1}(1 + h_{fe(min)})} = \frac{100(12\ \text{k}\Omega \| 120\ \text{k}\Omega)}{1\ \text{k}\Omega + 150\ \Omega(1 + 100)}$$

$$= 67.5$$

Eq. 6-22, $\quad A_v \approx \dfrac{R_C \| R_L}{R_{E1}} = \dfrac{12\ \text{k}\Omega \| 120\ \text{k}\Omega}{150\ \Omega}$

$$= 72.7$$

Circuit Design

As in the case of other negative feedback amplifiers, an emitter current feedback circuit should first have its resistor values calculated as for a circuit without feedback. Resistor R_{E1} is then determined from Eq. 6-22, and R_{E2} is reduced to its original value minus R_{E1}, if R_{E1} is large enough to affect the circuit bias conditions. The input and output coupling capacitors are calculated in the usual way, remembering that Z_{in} is increased by the presence of R_{E1}. The equation for the emitter bypass capacitor (Eq. 12-3) is rewritten to take R_{E1} into account. Equation 12-3 was derived from the fact that X_{C2} must equal the impedance in series with C_2 at the lower cutoff frequency for the circuit. For the circuit in Fig. 13-24(a),

$$X_{C2} = (h_{ib} + R_{E1}) \text{ at } f_1$$

usually, $R_{E1} \gg h_{ib}$, so the capacitor can be calculated from,

$$\mathbf{X_{C2} = R_{E1} \text{ at } f_1} \qquad\qquad \textbf{(13-15)}$$

Example 13-11

Modify the *CE* amplifier circuit designed in Examples 12-2 and 12-3 to use emitter current feedback to give $A_v = 70$. The lower cutoff frequency is $f_1 = 100$ Hz, and the signal source resistance is $r_s = 600\ \Omega$.

Solution

From Eq. 6-22, $R_{E1} \approx \dfrac{R_C \| R_L}{A_v} = \dfrac{12\ \text{k}\Omega \| 120\ \text{k}\Omega}{70}$

$\approx 156\ \Omega$ (use $150\ \Omega$ standard value)

$R_{E2} = R_E - R_{E1} = 3.9\ \text{k}\Omega - 150\ \Omega$
$= 3.75\ \text{k}\Omega$ (use $3.9\ \text{k}\Omega$)

Eq. 6-20, $Z_{b(min)} = h_{ie} + R_{E1}(1 + h_{fe}) = 1\ \text{k}\Omega + 150\ \Omega(1 + 100)$
$= 16.2\ \text{k}\Omega$

Eq. 6-12, $Z_{in} = R_1 \| R_2 \| Z_b = 120\ \text{k}\Omega \| 39\ \text{k}\Omega \| 16.2\ \text{k}\Omega$
$= 10.4\ \text{k}\Omega$

Eq. 12-4, $C_1 = \dfrac{1}{2\,\pi\,f_1\,(Z_{in} + r_s)/10}$

$= \dfrac{1}{2\,\pi \times 100\ \text{Hz} \times (10.4\ \text{k}\Omega + 600\ \Omega)/10}$

$= 1.5\ \mu\text{F}$ (standard value)

Eq. 13-15, $X_{C2} = R_{E1} \text{ at } f_1$

$C_2 = \dfrac{1}{2\,\pi\,f_1\,R_{E1}} = \dfrac{1}{2\,\pi \times 100\ \text{Hz} \times 150\ \Omega}$

$= 10.6\ \mu\text{F}$ (use $10\ \mu\text{F}$ standard value)

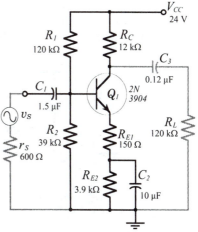

Figure 13-25
Emitter current feedback circuit for Ex. 13-11.

Two-Stage Circuit with Emitter Current Feedback

All of the two-stage amplifiers discussed in Ch. 12 can be designed to use emitter current feedback for voltage gain stabilization and increased input impedance. The design procedures are essentially those already discussed.

Consider the two-stage, capacitor-coupled circuit in Fig. 13-26. The voltage gain for *stage-1* is,

From Eq. 6-22, $$A_{v1} \approx \frac{R_3 \| Z_{in2}}{R_4}$$ (13-16)

and for stage-2, $$A_{v2} \approx \frac{R_8 \| R_L}{R_9}$$ (13-17)

The overall voltage gain is,

$$A_v = A_{v1} \times A_{v2}$$

When designing a two-stage circuit with emitter current feedback, the gain of each stage is determined as,

$$A_{v1} = A_{v2} = \sqrt{A_v}$$ (13-18)

Although the two stages can be designed for identical gains, $R_3 \| Z_{i2}$ is unlikely to equal $R_8 \| R_L$. So, R_4 and R_9 are likely to be unequal.

For the reasons discussed in Section 12-3, The emitter bypass capacitors are determined from,

$$X_{C2} = 0.65\ R_4\ \text{at}\ f_1$$ (13-19)

and, $$X_{C4} = 0.65\ R_9\ \text{at}\ f_1$$ (13-20)

The circuit input impedance is, once again, given by Eqs. 6-20 and 6-12, and the output impedance is simply the collector resistor in the second stage, (see Eq. 6-14).

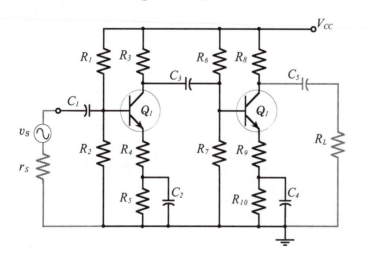

Figure 13-26
Two-stage, capacitor-coupled, BJT amplifier using emitter current feedback in each stage.

Example 13-12

The circuit in Fig. 13-25 is to be modified for use as each stage of the circuit in Fig. 13-26. The overall voltage gain is to be 1000, and the lower cutoff frequency is to be 100 Hz. Determine suitable emitter resistor values. Assume transistor parameters of $h_{ie} = 1\ k\Omega$ and $h_{fe} = 100$.

Solution

Eq. 13-18, $A_{v1} = A_{v2} = \sqrt{A_v} = \sqrt{1000}$

$= 31.6$

From Eq. 13-17, $R_9 \approx \dfrac{R_8 \| R_L}{A_{v2}} = \dfrac{12\ k\Omega \| 120\ k\Omega}{31.6}$

$\approx 345\ \Omega$ (use $330\ \Omega$ standard value)

$R_{10} = R_E - R_9 = 3.9\ k\Omega - 330\ \Omega$

$= 3.57\ k\Omega$ (use $3.9\ k\Omega$)

A_{v2} becomes, $A_{v2} \approx \dfrac{R_8 \| R_L}{R_9} = \dfrac{12\ k\Omega \| 120\ k\Omega}{330\ \Omega}$

≈ 33

so, $A_{v1} \approx \dfrac{A_v}{A_{v2}} = \dfrac{1000}{33}$

$= 30.3$

From Eq. 6-20, $Z_{i2} = h_{ie} + R_9(1 + h_{fe})$

$= 1\ k\Omega + 330\ \Omega(1 + 100)$

$= 34.3\ k\Omega$

From Eq. 6-12, $Z_{in2} = R_6 \| R_7 \| Z_{i2} = 120\ k\Omega \| 39\ k\Omega \| 34.3\ k\Omega$

$= 15.8\ k\Omega$

From Eq. 13-16, $R_4 \approx \dfrac{R_3 \| Z_{in2}}{A_{v1}} = \dfrac{12\ k\Omega \| 15.8\ k\Omega}{30.3}$

$\approx 225\ \Omega$ (use $220\ \Omega$ standard value)

$R_5 = R_E - R_4 = 3.9\ k\Omega - 220\ \Omega$

$= 3.68\ k\Omega$ (use $3.9\ k\Omega$)

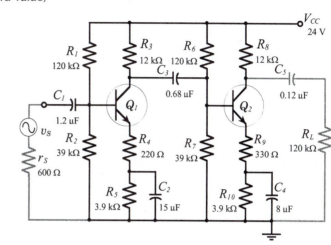

Figure   13-27
Emitter current feedback amplifier circuit for Ex. 13-12 and 13-13.

Example   13-13

Calculate suitable capacitor values for the two-stage circuit in Ex. 13-12.

Solution

Eq. 6-20, $Z_{i1} = h_{ie} + R_4(1 + h_{fe}) = 1\ k\Omega + 220\ \Omega(1 + 100)$

$= 23.2\ k\Omega$

Eq. 6-12, $Z_{in1} = R_1 \| R_2 \| Z_{i1} = 120\ k\Omega \| 39\ k\Omega \| 23.2\ k\Omega$

$= 13\ k\Omega$

Eq. 12-4, $C_1 = \dfrac{1}{2\pi f_1\ (Z_{in1} + r_s)/10} = \dfrac{1}{2\pi \times 100\ Hz \times (13\ k\Omega + 600\ \Omega)/10}$

$= 1.2\ \mu F$ (standard value)

Eq. 13-19, $X_{C2} = 0.65\,R_4$ at f_1

$$C_2 = \frac{1}{2\,\pi\,f_1\,0.65\,R_4} = \frac{1}{2\,\pi \times 100\text{ Hz} \times 0.65 \times 220\ \Omega}$$

$$= 11.1\ \mu\text{F}\ \text{ (use 15 }\mu\text{F standard value)}$$

Eq. 12-4, $C_3 = \dfrac{1}{2\,\pi\,f_1\,(Z_{in2} + R_3)/10} = \dfrac{1}{2\,\pi \times 100\text{ Hz} \times (15.8\text{ k}\Omega + 12\text{ k}\Omega)/10}$

$$= 0.57\ \mu\text{F}\ \text{ (use 0.68 }\mu\text{F standard value)}$$

Eq. 13-20, $X_{C4} = 0.65\,R_9$ at f_1

$$C_4 = \frac{1}{2\,\pi\,f_1\,0.65\,R_9} = \frac{1}{2\,\pi \times 100\text{ Hz} \times 0.65 \times 330\ \Omega}$$

$$= 7.4\ \mu\text{F}\ \text{ (use 8 }\mu\text{F standard value)}$$

Eq. 12-5, $C_5 = \dfrac{1}{2\,\pi\,f_1\,(R_8 + R_L)/10} = \dfrac{1}{2\,\pi \times 100\text{ Hz} \times (12\text{ k}\Omega + 120\text{ k}\Omega)/10}$

$$= 0.12\ \mu\text{F}\ \text{ (standard value)}$$

Other two-stage circuits can be designed to use emitter current bias, (for example, see Fig. 13-28). However, series voltage feedback produces more stable (closed-loop) voltage gains than single-stage emitter current feedback. Series voltage feedback also produces reduced output impedance, as well as increased input impedance. Emitter current feedback increases the circuit input impedance, but leaves the output impedance unaffected. Consequently, overall series voltage negative feedback is normally preferable to single-stage emitter current feedback.

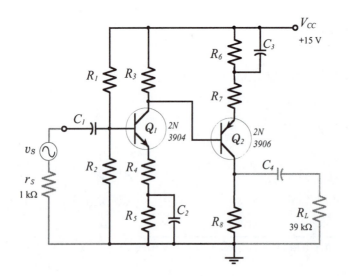

Figure 13-28
Two-stage direct-coupled BJT amplifier circuit using emitter current feedback.

13-6 Parallel Current Negative Feedback

Current Feedback Circuit

In series-voltage feedback, a portion of the output voltage is fed back in series with the signal source. In *parallel current feedback*, a portion of the output current is fed back in parallel with the signal source. Just as series voltage feedback stabilizes the voltage gain of a circuit, so parallel current feedback stabilizes the current gain.

Consider the *dc* feedback pair circuit in Fig. 13-29. An unbypassed resistor (R_{F2}) is connected in the emitter circuit of transistor Q_2, and Q_1 bias resistor (R_{F1}) is connected at the emitter terminal of Q_2. With this arrangement, any *ac* voltage developed across R_{F2} is applied to R_{F1}.

Look at the instantaneous polarities of the input voltage (v_i) and feedback voltage (v_f), as illustrated in the *ac* equivalent circuit in Fig. 13-30. When v_i is moving in a positive direction, v_{c1} is negative-going. Because the emitter of Q_2 follows the voltage (v_{c1}) at its base, v_f (across R_{F2}) is also negative-going at this time.

Now look at the instantaneous *ac* current directions indicated in Fig. 13-30. When v_i goes positive, the signal current (i_s) flows into the circuit, and an input current (i_i) flows into the base of Q_1, as shown. Because v_f goes negative when v_i is positive-going, a feedback current (i_f) flows in R_{F1} in the direction indicated, (from Q_1 base toward Q_2 emitter). Thus, some of the signal current is diverted away from the base of Q_1. This means that the output current (i_{c2}) is less than it would be if there was no current feedback, consequently, the circuit current gain is reduced.

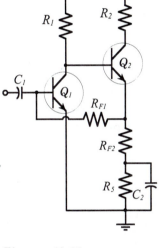

Figure **13-29**
DC feedback pair with parallel current negative feedback. The circuit current gain is
$$A_{i(CL)} \approx (R_{F1} + R_{F2})/R_{F2}$$

Current Gain

To determine the effect of negative feedback on the circuit current gain, first investigate the current gain without feedback, (the open-loop current gain). Suppose that R_{F2} in Fig. 13-30 has a bypass capacitor connected across it, so that there is no negative feedback. Assuming that the resistance of R_{F1} is very much larger than the Q_1 input impedance (h_{ie1}), virtually all of i_s enters the base of Q_1, (see Fig. 13-30).

So, $i_s \approx i_1$

and $i_{c1} \approx h_{fe1} i_1$

i_{c1} divides between R_1 and Z_{i2}. Using the current divider equation,

$$i_{b2} = \frac{i_{c1} \times R_1}{R_1 + Z_{i2}}$$

$$i_o = i_{c2} = h_{fe2} \, i_{b2}$$

or,
$$i_o = \frac{h_{fe2} \, i_{c1} \, R_1}{R_1 + Z_{i2}} = \frac{h_{fe1} \, h_{fe2} \, i_i \, R_1}{R_1 + Z_{i2}}$$

The open-loop current gain is,

$$A_i = \frac{i_o}{i_i}$$

$$A_i = \frac{h_{fe1} \, h_{fe2} \, R_1}{R_1 + Z_{i2}} \qquad \textbf{(13-21)}$$

Now derive an equation for the current gain with negative feedback. Referring to Fig. 13-30, it is seen that i_o (in the emitter circuit of Q_2) divides between R_{F1} and R_{F2}, giving

$$i_f = \frac{i_o \times R_{F2}}{R_{F1} + R_{F2}} = i_o B$$

where
$$B = \frac{R_{F2}}{R_{F1} + R_{F2}}$$

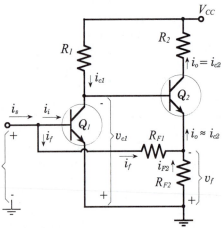

Also,
$$i_s = i_i + i_f = i_i + i_o B$$

and,
$$i_o = A_i \, i_i$$

so,
$$i_s = i_i + (A_i \, i_i \, B) = i_i(1 + A_i B)$$

The closed-loop current gain is,

$$A_{i(CL)} = \frac{i_o}{i_s} = \frac{A_i \, i_i}{i_i(1 + A_i B)}$$

$$A_{i(CL)} = \frac{A_i}{1 + A_i B} \qquad \textbf{(13-22)}$$

If $A_i B \gg 1$, then

$$A_{i(CL)} \approx \frac{1}{B} \approx \frac{R_{F1} + R_{F2}}{R_{F2}} \qquad \textbf{(13-23)}$$

Figure 13-30
AC equivalent circuit for the current feedback circuit Fig. 13-29, showing ac voltages and currents.

It is seen that,

> ***parallel current negative feedback stabilizes circuit current gain.***

It should be noted that, in the current gain equations, the *ac* output current is taken as the current in the collector circuit of transistor Q_2. How this current divides between a capacitor-coupled external load (R_L) and the Q_2 collector resistor (R_3) depends on the values of R_L and R_3.

Output and Input Impedances

In the circuit shown in Figs. 13-29 and 13-30, the Q_2 collector resistor (R_2) is outside the feedback loop. Therefore, like the case of emitter current feedback, the output impedance of this circuit is largely unaffected by feedback. (It can be shown that the impedance *looking into* the collector of Q_2 is increased by current feedback.) So, the circuit output impedance is, (from Eq. 6-14),

$$Z_o \approx R_2$$

Assuming that $R_{F1} \gg h_{ie1}$, the circuit input impedance without feedback is,

$$Z_i \approx h_{ie1} = \frac{v_i}{i_s}$$

As already shown, with negative feedback,

$$i_s = i_i + i_f = i_i(1 + A_i B)$$

so,

$$Z_i = \frac{v_i}{i_s} = \frac{v_i}{i_i(1 + A_i B)}$$

or,

$$Z_i = \frac{h_{ie1}}{1 + A_i B} \qquad \textbf{(13-24)}$$

Therefore,

> *parallel current negative feedback reduces circuit input impedance by a factor of* $(1 + A_i B)$.

Although R_{F1} is a bias resistor, it is also part of the feedback network, so it is *not* in parallel with the circuit input impedance.

Another way of looking at the circuit input impedance is in terms of the the feedback resistor R_{F1} and the voltage gain of the first stage. It can be shown that,

$$Z_i = h_{ie1} \| \frac{R_{F1}}{A_{v1}} \qquad \textbf{(13-25)}$$

Circuit Design

As with other negative feedback circuits, design of a parallel current negative feedback amplifier is approached by first ignoring the *ac* negative feedback components. Resistors R_{F2} and R_5 in Fig. 13-29 are calculated as a single emitter resistor $(R_E = R_{F2} + R_5)$ to fulfil the desired *dc* conditions. Also, R_{F1} is calculated (as for other *dc* feedback bias circuits) to provide the required base current to Q_1. R_{F2} is then calculated from Eq. 13-23. Usually, the resistance of R_{F1} is very large, and this results in a calculated resistance for R_{F2} larger than $(R_{F2} + R_5)$! In this case, the base resistor for Q_1

should be made up of two resistors one of which is bypassed; see $(R_2 = R_{F1} + R_7)$ in Fig. 13-31. R_{F1} is the unbypassed portion of R_2, and this is calculated in relation to R_{F2}.

Bypass capacitor C_2 (in Fig. 13-31) is calculated to give $X_{C2} = R_{F2}$ at the desired lower cutoff frequency. Because the signal is derived from a current source, the impedance of the input coupling capacitor (C_1) at f_1 should be much smaller than the (normally-high) impedance of the signal source. Capacitor C_3 is determined by making X_{C3} very much smaller than R_{F1} at f_1.

Example 13-14

Analyze the circuit in Fig. 13-31 to determine the current gain and the input impedance. Assume transistor parameters of $h_{fe} = 100$, and $h_{ie} = 2$ kΩ.

Solution

From Eq. 6-20, $\quad Z_{i2} = h_{ie2} + (1 + h_{fe2})R_4 = 2$ k$\Omega + (1 + 100)100$ Ω

$$= 12.1 \text{ k}\Omega$$

Open-loop current gain,

Eq. 13-21, $\qquad A_i = \dfrac{h_{fe1} h_{fe2} R_1}{R_1 + Z_{i2}} = \dfrac{100 \times 100 \times 5.6 \text{ k}\Omega}{5.6 \text{ k}\Omega + 12.1 \text{ k}\Omega}$

$$= 3164$$

From Eq. 13-23, $\quad B = \dfrac{R_4}{R_6 + R_4} = \dfrac{100 \ \Omega}{2.2 \text{ k}\Omega + 100 \ \Omega}$

$$= \dfrac{1}{23}$$

Closed-loop current gain,

Eq. 13-22, $\qquad A_{i(CL)} = \dfrac{A_i}{1 + A_i B} = \dfrac{3164}{1 + (3164/23)}$

$$= 22.8$$

Eq. 13-24, $\qquad Z_i = \dfrac{h_{ie1}}{1 + A_i B} = \dfrac{2 \text{ k}\Omega}{1 + (3164/23)}$

$$= 14.6 \ \Omega$$

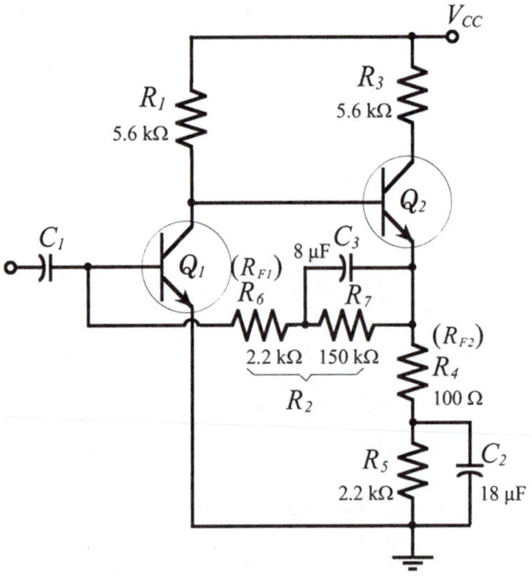

Figure 13-31
Current feedback circuit with a portion of R_2 bypassed to minimise R_4. The current gain is $A_{i(CL)} \approx (R_6 + R_4)/R_4$

Practise Problems

13-6.1 Modify the dc feedback pair circuit shown in Fig. 12-25(a) to give $A_{i(CL)} = 55$ and $f_1 = 80$ Hz. Assume $r_s = 12$ kΩ.

13-6.2 Assuming transistor parameters of $h_{ie} = 1$ kΩ and $h_{fe} = 50$, calculate Z_i and $A_{i(CL)}$ for the circuit in Problem 13-6.1.

13-7   Additional Effects of Negative Feedback

Decibels of Feedback

Negative feedback can be measured in decibels. A statement that 40 dB of feedback has been applied to an amplifier means that the amplifier gain has been reduced by 40 dB, (that is, by a factor of 100). Thus,

$$A_{CL} = A_v - 40 \text{ dB} = \frac{A_v}{100}$$

Bandwidth

Consider the typical gain-frequency response of an amplifier, as illustrated in Fig. 13-32. Without negative feedback, the amplifier open-loop (A_v) gain falls off to its lower 3 dB frequency ($f_{1(OL)}$), as illustrated. This is usually due to the impedance of bypass capacitors increasing as the frequency decreases. Similarly, the open-loop upper cutoff frequency ($f_{2(OL)}$) is produced by transistor cutoff, by shunting capacitance, or by a combination of both. As discussed in Section 8-2, the circuit open-loop bandwidth is,

$$BW_{OL} = f_{2(OL)} - f_{1(OL)}$$

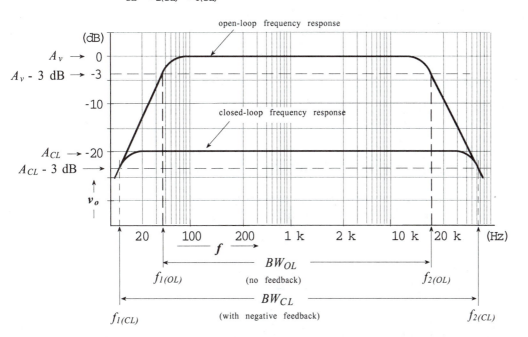

Now look at the typical frequency response for the same amplifier when negative feedback is used. The closed-loop gain (A_{CL}) is much smaller than the open-loop gain, and A_{CL} does not begin to fall off (at high or low frequencies) until A_v (open-loop) falls substantially. Consequently, $f_{1(CL)}$ is much lower than $f_{1(OL)}$, and $f_{2(CL)}$ is much higher than $f_{2(OL)}$. So, the circuit bandwidth with negative feedback (the closed-loop bandwidth) is much greater than the bandwidth without negative feedback.

Figure   13-32
Amplifier frequency response with and without negative feedback. Negative feedback extends the amplifier bandwidth.

$$BW_{CL} = f_{2(CL)} - f_{1(CL)}$$

From Eq. 13-2,

$$A_{CL} = \frac{A_v}{1 + A_v B}$$

It can be shown that there is a 90° phase shift associated with the open-loop gain at frequencies below $f_{1(OL)}$ and above $f_{2(OL)}$. Thus, Eq. 13-2 must be rewritten as,

$$A_{CL} = \frac{-jA_v}{1 - jA_v B}$$

or, $$|A_{CL}| = \frac{A_v}{\sqrt{[1 + (A_v B)^2]}}$$

When $A_v = 1/B$,

$$|A_{CL}| = \frac{1/B}{\sqrt{[1 + 1]}} = \frac{A_{CL}}{\sqrt{2}}$$

$$= A_{CL} - 3 \text{ dB}$$

Thus, for a negative feedback amplifier designed to have the widest possible bandwidth, the cutoff frequencies would occur when the open-loop gain falls to the equivalent of $1/B$. Thus, $f_{2(CL)}$ occurs when,

$$\mathbf{A_v = 1/B \approx A_{CL}} \tag{13-26}$$

So, for example, the cutoff frequencies for a negative feedback amplifier designed for a closed-loop gain of 100 would occur when the open-loop gain falls to 100. It is seen that,

negative feedback increases amplifier bandwidth.

The upper cutoff frequency for an amplifier is usually greater than 20 kHz, and the lower cutoff frequency is around 100 Hz, or lower. So, $f_2 \gg f_1$, and consequently,

$$BW = f_2 - f_1 \approx f_2$$

This means that the amplifier bandwidth is essentially equal to the upper cutoff frequency.

Now refer to Fig. 13-32 once again. The amplifier gain multiplied by the upper cutoff frequency is a constant quantity. This is known as the *gain-bandwidth product*. Therefore,

$$A_{CL} \times f_{2(CL)} = A_v \times f_{1(OL)}$$

or, $$\mathbf{f_{2(CL)} = \frac{A_v \, f_{1(OL)}}{A_{CL}}} \tag{13-27}$$

So, the closed-loop upper cut-off frequency for a negative feedback amplifier can be calculated from the open-loop upper cutoff frequency, the open-loop gain, and the closed-loop gain.

Harmonic Distortion

Harmonic Distortion occurs when a transistor or other device is driven beyond the linear range of its characteristics. Figure 13-33 shows the I_C/V_{BE} characteristics of a transistor biased to a base-emitter voltage V_B. As illustrated, when the base-emitter voltage changes ($\pm v_{be}$) are very small, the collector current changes by equal positive and negative amounts ($\pm i_c$). When $\pm v_{be}$ is large, the change in $+i_c$ is greater than the change in $-i_c$. This is because of the nonlinearity in the I_C/V_{BE} characteristics. The result is *harmonic distortion* (or *nonlinear distortion*) in the waveform of i_c and in the amplifier output.

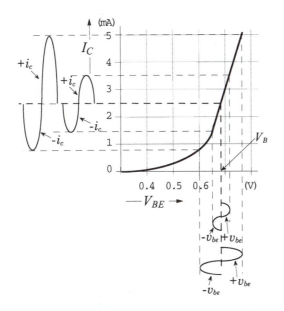

Figure 13-33
Harmonic distortion in the output of an amplifier is caused by nonlinearity in transistor characteristics. Negative feedback reduces this type of distortion by a factor of $(1 + A_v B)$.

The distorted waveform can be shown to consist of a *fundamental* frequency waveform and a number of smaller amplitude *harmonic* components, (see Fig. 13-34). The fundamental waveform is the amplified signal, or amplifier output voltage (v_o), and the harmonics are unwanted voltage components (v_H). The *harmonic distortion* is the rms value of v_H expressed as a percentage of the rms value of v_o. The harmonics (generated within the feedback loop) are reduced by negative feedback by a factor of $(1 + A_v B)$. So that,

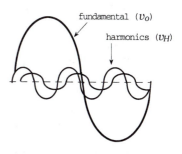

fundamental (v_O)

harmonics (v_H)

$$\% \textbf{ distortion with } \textit{NFB} = \frac{\% \textbf{ distortion without } \textit{NFB}}{(1 + A_v B)} \quad \textbf{(13-28)}$$

So,

Figure 13-34
Harmonic distortion in an amplifier is reduced by negative feedback.

> *negative feedback reduces harmonic distortion.*

Attenuation Distortion

Attenuation distortion occurs when different frequencies are amplified by different amounts. This type of distortion is the result of the amplifier open-loop gain being frequency-dependent, (see Fig. 13-35). When negative feedback is used, the closed-loop gain is a constant quantity largely independent of signal frequency within the circuit bandwidth. Like harmonic distortion, attenuation distortion is reduced by a factor of $(1 + A_v B)$ by negative feedback. Equation 13-28 applies. Thus,

negative feedback reduces attenuation distortion.

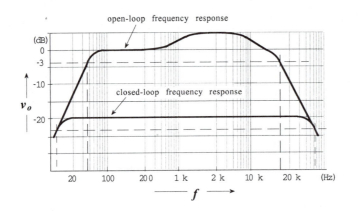

Figure 13-35
Attenuation distortion, resulting from different signal frequencies being amplified by different amounts, is reduced by negative feedback.

Phase Shift

Input signals to an amplifier go through phase-shifts from the input to the output. The waveforms in Fig. 13-36 show that, in a two-stage circuit, there is a 180° phase shift from v_s to v_{o1}, and a further 180° shift from v_{o1} to v_{o2}. In some cases, an amplifier can produce different phase shifts for different signal frequencies. For example, higher frequency signals might be shifted by more than 180° at each stage, resulting in an undesirable phase difference (ø) between input and output, as illustrated.

Audio (and other) signals at an amplifier input are usually made up of a combination of component waveforms with different amplitudes and different frequencies. So, any variation in amplifier-introduced phase shift will create distortion in the output waveform that is not present in the input.

To investigate the effect of negative feedback on amplifier phase shift, assume that the open-loop gain has a phase shift angle of ø, and that the closed-loop phase shift is $ø_{CL}$. Then,

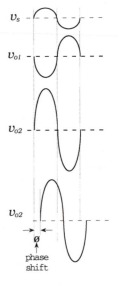

Figure 13-36
Undesirable phase shifts in amplifier ac waveforms are reduced by negative feedback.

$$\text{open-loop gain} = A_v \underline{/ø}$$

and,

$$A_{CL}\underline{/ø_{CL}} = \frac{A_v\underline{/ø}}{1 + A_v B\underline{/ø}} = \frac{A_v B \sin ø}{1 + A_v B \cos ø + jA_v B \sin ø}$$

giving,

$$ø_{CL} = ø - \tan^{-1}\frac{A_v B \sin ø}{1 + A_v B \cos ø} \qquad \textbf{(13-29)}$$

So,

negative feedback reduces amplifier phase shift

by an angle of,

$$\tan^{-1} \frac{A_v B \sin \o}{1 + A_v B \cos \o}$$

Example 13-15

A negative feedback amplifier has an open-loop gain of 60 000 and a closed-loop gain of 300. If the open-loop upper cutoff frequency is 15 kHz, estimate the closed-loop upper cutoff frequency. Also, calculate the total harmonic distortion with feedback if there is 10% harmonic distortion without feedback.

Solution

Eq. 13-27,

$$f_{2(CL)} = \frac{A_v \; f_{1(OL)}}{A_{CL}} = \frac{60\,000 \times 15 \text{ kHz}}{300}$$

$$= 3 \text{ MHz}$$

Eq. 13-28,

$$\% \text{ distortion with } NFB = \frac{\% \text{ distortion without } NFB}{(1 + A_v B)} = \frac{10\%}{1 + (60\,000/300)}$$

$$= 0.48\%$$

Noise

Circuit *noise* generated within the feedback loop of an amplifier is reduced by a factor of $(1 + A_v B)$ in the same way that unwanted harmonics are reduced. It should be noted that only the noise generated within the feedback loop is reduced by feedback. Noise produced by bias resistors outside the feedback loop will not be affected by feedback.

Negative feedback reduces circuit noise.

Circuit Stability

In Chapter 16, it is explained that, for a circuit to oscillate, the *loop phase shift* must be 360° when the *loop gain* is 1. The loop phase shift is the phase shift around the loop from the amplifier input to the output, and back via the feedback network to the input. Similarly, the loop gain is the product of the amplifier open-loop gain and the feedback network attenuation, $(A_v B)$.

The loop phase shift of an amplifier can approach 360° as the gain falls off at the high end of the bandwidth. In this case, the amplifier might oscillate, and it is said to be *unstable.* The usual method of combating this kind of instability is to include small shunting capacitors from the transistor collector terminals to ground. These tend to cause the loop gain to fall below 1 before the loop phase shift approaches 360°.

Amplifier instability and compensation techniques are further covered in Ch. 15.

Practise   Problems

13-7.1   The amplifier in Ex. 13-15 has a 15° open-loop phase shift at high signal frequencies. Calculate the phase shift with negative feedback.

13-7.2   A negative feedback amplifier with a closed loop gain of 250 has a 4 MHz upper cutoff frequency. Calculate the open-loop cutoff frequency if the open-loop gain is 200 000.

Chapter-13   Review   Questions

Section   13-1

13-1   Draw a sketch to illustrate the principle of series voltage negative feedback, and briefly explain. Also, list the major effects of negative feedback on an amplifier.

13-2   Derive an equation for the voltage gain of an amplifier that uses series voltage negative feedback.

13-3   Derive an equation for the input impedance of an amplifier that uses series voltage negative feedback.

13-4   Derive an equation for the output impedance of an amplifier that uses series voltage negative feedback.

13-5   Discuss the effect of input bias resistors on the input impedance of a negative feedback amplifier, and the effect of a collector resistor at the output of a *BJT* negative feedback amplifier.

Section   13-2

13-6   Sketch the circuit of a two-stage, capacitor-coupled, *BJT* amplifier that uses series voltage negative feedback. Briefly explain how the feedback operates, and write an equation for the voltage gain in terms of the feedback components.

13-7   Briefly discuss the procedure for determining the negative feedback components for the circuit in Question 13-6

13-8   For the circuit in Question 13-6, write equations for the input impedance and output impedance in terms of the circuit components and transistor parameters.

Section   13-3

13-9   Sketch the circuit of a direct-coupled, two-stage *BJT* amplifier that uses series voltage negative feedback. Q_1 should be an *npn* device, and Q_2 should be a *pnp* transistor. Briefly explain the feedback operation, and write the equation for voltage gain in terms of the circuit components.

13-10   For the circuits in Question 13-9, write equations for Z_{in} and Z_{out} in terms of the transistor parameters and the circuit components.

13-11 Briefly discuss the design procedures for the circuit in Questions 13-9.

13-12 Sketch the circuit of a *dc* feedback pair that uses series voltage negative feedback. Explain the operation of the feedback network, and write an equation for the voltage gain in terms of the circuit components.

13-13 For the circuits in Questions 13-12, write equations for Z_{in} and Z_{out} in terms of the transistor parameters and the circuit components.

13-14 Briefly discuss the design procedures for the circuit in Questions 13-12.

13-15 Sketch the circuit of a direct-coupled, two-stage, *BIFET* amplifier that uses series voltage negative feedback. Explain the operation of the circuit, and write an equation for the voltage gain in terms of the circuit components.

13-16 For the circuits in Questions 13-15, write equations for Z_{in} and Z_{out} in terms of the transistor parameters and the circuit components.

13-17 Briefly discuss the design procedures for the circuit in Questions 13-15.

Section 13-4

13-18 Sketch the circuit of a two-stage, direct-coupled negative feedback amplifier that uses a two-transistor, emitter-coupled input stage. Explain the operation of the circuit.

13-19 Write equations for $A_{v(CL)}$, Z_{in}, and Z_{out} for the circuit in Questions 13-18.

13-20 Briefly discuss the design procedures for the circuit in Questions 13-18.

Section 13-5

13-21 Sketch the circuit of a single-stage common emitter amplifier that uses emitter current feedback. Briefly explain the operation of the circuit.

13-22 For the single-stage circuit sketched for Question 13-21, write equations for $A_{v(CL)}$, Z_{in}, and Z_{out}.

13-23 Briefly explain the design procedure for the circuit in Question 13-21.

13-24 Sketch a two-stage, capacitor-coupled, *BJT* amplifier that uses emitter current feedback in each stage. Briefly explain the operation of the circuit.

13-25 Discuss the design procedure for the two-stage circuit in Question 13-24.

13-26 Sketch the circuit of a direct-coupled, two-stage, *BJT* amplifier that uses emitter current feedback. Q_1 should be an *npn* device, and Q_2 should be a *pnp* transistor.

13-27 Write equations for $A_{v(CL)}$, Z_{in}, and Z_{out} for the circuit in Question 13-26.

Section 13-6

13-28 Sketch a direct-coupled *dc* feedback pair amplifier that uses parallel current negative feedback. Briefly explain the operation of the circuit, and discuss its performance characteristics.

13-29 Write equations for $A_{i(CL)}$, Z_{in}, and Z_{out} for the circuit in Question 13-28.

13-30 Discuss the design procedure for the two-stage circuit in Question 13-28.

Section 13-7

13-31 Using illustrations, explain the effects of negative feedback on the bandwidth of an amplifier. Discuss the effect of open-loop gain reduction on the closed-loop gain.

13-32 Draw sketches to explain how harmonic distortion occurs in an amplifier. Discuss the effects of negative feedback on harmonic distortion.

13-33 Explain attenuation distortion, and discuss the effects of negative feedback on attenuation distortion.

13-34 Explain how phase shift distortion occurs in an amplifier, and discuss how negative feedback affects phase shift.

13-35 Discuss the effect of negative feedback on circuit noise.

13-36 Briefly discuss amplifier stability with negative feedback.

Chapter-13 Problems

Section 13-1

13-1 A two-stage *BJT* amplifier has an open-loop voltage gain of 224 000 when the transistors have a maximum h_{fe} value, and 14 000 when the transistor h_{fe} values are a minimum. Calculate the closed-loop gain for both cases when negative feedback is used with a feedback factor of 1/125.

13-2 The amplifier in Problem 13-1 has an input impedance of 1.2 kΩ at the base of the input transistor when negative feedback is not used. Voltage divider bias with R_1 = 150 kΩ and R_2 = 56 kΩ is used at the input transistor. Calculate the circuit input impedance (with negative feedback) at both h_{fe} extremes.

13-3 The output transistor in the amplifier in Problem 13-1 has $1/h_{oe}$ = 80 kΩ, and a 5.6 kΩ collector resistor. Calculate the circuit output impedance at both h_{fe} extremes.

13-4   The closed loop gain of a negative feedback amplifier is measured as 99.7 when the feedback factor is 1/100, and as 297 when a feedback factor of 1/300 is used. Determine the open-loop gain.

Section 13-2

13-5   The circuit in Fig. 13-16 is to be modified to have an overall voltage gain of 50 and a lower cutoff frequency of 150 Hz. Calculate the new values for R_{F1}, R_{F2}, C_2 , and C_{F1}.

13-6   Analyze the negative feedback amplifier in Problem 13-5 to determine the input and output impedances. Assume that the transistor parameters are h_{ie} = 800 Ω, h_{fe} = 60, and $1/h_{oe}$ = 90 kΩ. The open-loop gain is 97 000.

13-7   A two-stage, capacitor-coupled, *BJT* amplifier as in Fig. 12-15 has the following components: R_1 = 150 kΩ, R_2 = 56 kΩ, R_3 = 4.7 kΩ, R_4 = 5.6 kΩ, R_5 = 150 kΩ, R_6 = 56 kΩ, R_7 = 4.7 kΩ, R_8 = 5.6 kΩ, R_L = 47 kΩ, C_1 = 10 μF, C_2 = 180 μF, C_3 = 10 μF, C_4 = 180 μF. Modify the circuit to use series voltage negative feedback. The voltage gain is to be 180, and the required lower cutoff frequency is 90 Hz.

13-8   Calculate the input and output impedances for the circuit in Problem 13-7. Take the transistor parameters as: h_{ie} = 1.2 kΩ, h_{fe} = 90, and $1/h_{oe}$ = 75 kΩ.

Section 13-3

13-9   A two-stage, direct-coupled, *BJT* amplifier as in Fig. 12-20 has the following components: R_1 = 100 kΩ, R_2 = 39 kΩ, R_3 = 3.9 kΩ, R_4 = 3.9 kΩ, R_5 = 2.7 kΩ, R_6 = 3.9 kΩ, R_L = 33 kΩ, C_1 = 8 μF, C_2 = 150 μF, C_3 = 150 μF. Modify the circuit to use series voltage negative feedback to give a gain of 75, and a lower cutoff frequency of 60 Hz.

13-10   Calculate the input and output impedances for the circuit in Problem 13-9 if the transistor parameters are: h_{ie} = 1 kΩ, h_{fe} = 120, and $1/h_{oe}$ = 100 kΩ.

13-11   The two-stage, direct-coupled, *BJT* amplifier designed in Exs. 12-8 and 12-9 (and analyzed in Ex. 12-10) is to be converted into a negative feedback amplifier with a voltage gain of 120, and a lower cutoff frequency of 20 Hz. Make the necessary modifications and determine the new component values.

13-12   A two-stage, direct-coupled amplifier with complementary transistors is to be designed to use negative feedback, as in Fig. 13-17. The overall voltage gain is to be 150, and the lower 3 dB frequency is to be 200 Hz. Calculate suitable values for R_{F1}, R_{F2}, and C_2.

13-13   A *dc* feedback pair using negative feedback (as in Fig. 13-18) is to have $A_{v(CL)}$ = 80, f_1 = 120 Hz, and f_2 = 50 kHz. Calculate suitable component values for the feedback network.

13-14   Calculate Z_{in} and Z_{out} for the circuit in Fig. 13-18. Use the component values shown, and transistor parameters of: h_{ie} = 1.2 kΩ, h_{fe} = 90, and $1/h_{oe}$ = 95 kΩ.

13-15   A two-stage, negative feedback, *BIFET* amplifier, as in Fig. 13-19, is to have $A_{v(CL)}$ = 19, and f_1 = 400 Hz. Calculate suitable values for R_{F1}, R_{F2}, and C_{F1}.

13-16   Calculate the direct current through the feedback networks in the circuits in Figures 13-17 and 13-19 when the feedback network is directly connected instead of capacitor coupled.

Section 13-4

13-17   The circuit in Fig. 13-21 is to be designed to have $A_{v(CL)}$ = 175. Using V_{CC} = ±12.5 V, select suitable current and voltage levels, and calculate resistor values. Use *2N3904* and *2N3906* transistors.

13-18   Calculate the minimum values of $A_{v(CL)}$, Z_{in}, and Z_{out} for the circuit in Problem 13-17.

13-19   A two-stage differential amplifier circuit (as in Fig. 13-21) is to be designed to amplify a 10 mV input to 750 mV. The amplifier supply is ±9 V, and the transistors to be used have h_{FE} = h_{fe} = 150, h_{ie} = 1.5 kΩ, and $1/h_{oe}$ = 150 kΩ. determine suitable resistor values.

13-20   Calculate Z_i and Z_o for the circuit in Problem 13-19.

Section 13-5

13-21   Design a single-stage *BJT* common emitter amplifier with emitter current feedback. The circuit specification is: Q_1 = 2N3906, V_{CC} = -18 V, R_L = 150 kΩ, A_v = 45, f_1 = 50 Hz.

13-22   Analyze the circuit designed for Problem 13-21 to determine A_v, Z_{in}, Z_{out}, and f_1.

13-23   A common emitter amplifier as in Fig. 12-1 has the following components: R_1 = 120 kΩ, R_2= 33 kΩ, R_C= 6.8 kΩ, R_E = 3.3 kΩ, and R_L = 68 kΩ. Modify the circuit to use emitter current feedback. The voltage gain is to be 23, and the required lower cutoff frequency is 60 Hz. The transistor parameters are: h_{ie} = 1.2 kΩ, h_{fe} = 90, and $1/h_{oe}$ = 95 kΩ.

13-24   Calculate A_v, Z_{in}, and Z_{out} for the circuit in Problem 13-23.

13-25   Two stages of the circuit in Problem 13-23 are to be connected in cascade to give an overall voltage gain of approximately 1600 and a lower cutoff frequency of 50 Hz. Determine the necessary modifications.

13-26   An amplifier is to be designed using two stages of the circuit designed for Problem 13-21. The overall voltage gain is to be 900 and the lower cutoff frequency is to be 100 Hz. Determine the necessary modifications.

13-27 Analyze the circuit designed for Problem 13-25 to determine A_v, Z_{in}, and Z_{out}.

13-28 Analyze the circuit designed for Problem 13-26 to determine A_v, Z_{in}, and Z_{out}.

13-29 A *dc* feedback pair [as in Fig 12-25(a)] is to be modified to use emitter current feedback. The circuit specification is: V_{CC} = 15 V, R_L = 80 kΩ, A_v = 400, f_1 = 70 Hz. Design the circuit to use *2N3904 BJTs*.

13-30 Analyze the circuit designed for Problem 13-29 to determine A_v, Z_{in}, and Z_{out}.

Section 13-6

13-31 A *dc* feedback pair is to be designed to use parallel current negative feedback, as in Fig. 13-29. Using V_{CC} = 22 V, and transistors with h_{FE} = h_{fe} = 75 and h_{ie} = 1.5 kΩ, design the circuit to have I_C = 0.9 mA, A_i = 22, f_1 = 200 Hz.

13-32 Calculate A_i and Z_{in} for the circuit in Problem 13-31.

13-33 A *dc* feedback pair as in Fig. 12-25(a) has the following components: R_1 = 12 kΩ, R_2 = 330 kΩ, R_3 = 4.7 kΩ, R_4 = 2.7 kΩ, and C_2 = 240 μF. Determine the necessary modification to give A_i = 30 and f_1 = 75 Hz.

13-34 Analyze the circuit designed for Problem 13-33 to determine A_i and Z_{in}. Assume h_{fe1} = h_{fe2} = 100, and h_{ie1} = h_{ie2} = 1 kΩ.

Section 13-7

13-35 The amplifier in Example 13-4 has 5% total harmonic distortion in its output waveform when negative feedback is not used. Calculate the new distortion content with negative feedback.

13-36 If the amplifier in Example 13-4 has a 17° open-loop phase shift at high frequencies, determine the phase shift with negative feedback.

13-37 The negative feedback amplifier in Examples 13-7 and 13-8 has a 30 kHz open-loop upper cutoff frequency. Calculate the upper cutoff frequency with negative feedback.

13-38 A negative feedback with B = 1/105 produces a 2 V output when the input is 20 mV. Distortion content in the output is 1%. Find the new signal level to give 2 V output when the negative feedback is disconnected. Also, determine the new distortion content in the output.

Practise Problem Answers

13-1.1   1.42 kΩ, 5.2 kΩ, 26.3 kΩ, 28 Ω, 134.9
13-2.1   100 Ω, 3.9 kΩ, 33 μF, 10 kΩ, 33 μF
13-2.2   27.7 kΩ, 52 Ω
13-3.1   330 Ω, 68 kΩ, 171 μA

13-3.2   150 Ω, 2.7 kΩ, 5.6 μF, 30 μF, 3000 pF
13-4.1   4.7 kΩ, 10 kΩ, 6.8 kΩ, 10 kΩ, (220 kΩ + 10 kΩ), 4.7 kΩ, 4.7 kΩ, 15 kΩ
13-4.2   4.7 kΩ, 12.4 Ω
13-5.1   (68 kΩ + 10 kΩ), 47 kΩ, 4.7 kΩ, 120 Ω, 4.7 kΩ, 2.7 kΩ, 150 Ω, 3.9 kΩ
13-5.2   2.7 μF, 27 μF, 22 μF, 0.47 μF
13-6.1   100 Ω, 5.6 kΩ, 1.8 μF, 20 μF, 39 μF
13-6.2   61 Ω, 878
13-7.1   0.08°
13-7.2   5 kHz

Chapter *14*

IC Operational Amplifiers and Basic Op-Amp Circuits

Chapter Contents

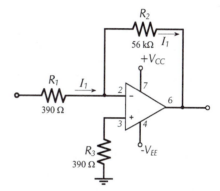

Objectives

You will be able to:

1 *Sketch and briefly explain an operational amplifier circuit symbol and the basic input and output stages. Identify all terminals.*

2 *List and discuss the most important op-amp parameters.*

3 *Design op-amp bias circuits, and show how to null the output offset.*

4 *Sketch the following linear op-amp circuits in direct-coupled and capacitor-coupled configuration, and explain the operation of each:*
   • *Voltage follower*
   • *Noninverting amplifier*
   • *Inverting amplifier*

5 *Analyze and design circuits of the type listed in item 4 above.*

6 *Sketch the following linear op-amp circuits and explain the operation of each:*
   • *Summing amplifier*
   • *Difference amplifier*
   • *Instrumentation amplifier*

7 *Analyze and design circuits of the type listed in item 6 above.*

8 *Sketch the following nonlinear op-amp circuits and explain the operation of each:*
   • *Zero crossing detectors*
   • *Inverting Schmitt trigger circuits*
   • *Noninverting Schmitt triggers*

9 *Analyze and design circuits of the type listed in item 8 above.*

Introduction

An operational amplifier is a high-gain amplifier circuit with two high-impedance input terminals and one low-impedance output. The inputs are identified as *inverting input* and *noninverting input*. The basic circuit consists of a differential amplifier input stage and a low impedance emitter-follower output stage. The voltage gains of integrated circuit (*IC*) operational amplifiers are extremely high, typically 200 000. Because of this high voltage gain, externally connected resistors must be employed to provide negative feedback. Design of *IC* op-amp circuits involves determination of suitable values for the external components. The design calculations are usually much simpler than discrete-component circuit design.

14-1   Integrated Circuit Operational Amplifiers

Circuit Symbol and Packages

Figure 14-1(a) shows the triangular circuit symbol for an operational amplifier (op-amp). As illustrated, there are two input terminals, one output terminal, and two supply terminals. The inputs are identified as *the inverting input* (- sign) and *the noninverting input* (+ sign). A positive-going voltage at the inverting input produces a negative-going (inverted) voltage at the output terminal. Conversely, a positive-going voltage at the noninverting input generates an output which is also positive-going (noninverted). Plus-minus supplies are normally used with op-amps, so the supply terminals are identified as $+V_{CC}$ and $-V_{EE}$.

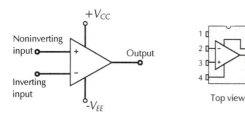

(a) Op-amp circuit symbol   (b) Terminal connections for *DIL* package   (c) Terminal connections for metal-can package

Figure 14-1
Operational amplifier circuit symbol and terminal connections. The + and - signs identify the noninverting and inverting input terminals, respectively.

Two typical op-amp packages are illustrated in Figs. 14-1 (b) and (c). For both the *dual-in-line* (*DIL*) plastic package and the metal can package, terminals 2 and 3 are the inverting and noninverting inputs respectively, terminal 6 is the output, and terminals 7 and 4 are the + and - supply terminals.

Basic Op-Amp Circuit

The basic circuit of an *IC* operational Amplifier is essentially as shown in Fig. 13-23, and further illustrated in Figs. 14-2(a) and (b). The differential-amplifier input stage has two (inverting and noninverting) high-impedance input terminals, and the emitter follower output stage has a low impedance. An intermediate stage (between the input and output stages) is largely responsible for the op-amp high voltage gain. Figure 14-2(a) shows a *BJT* input stage, while Fig. 14-2(b) has a *FET* input stage. Appropriately, op-amps that exclusively use *BJTs* are termed bipolar op-amps, and those that use a *FET* input stage with *BJT* additional stages are referred to as *BIFET* op-amps.

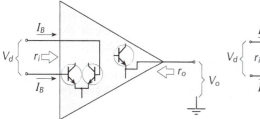

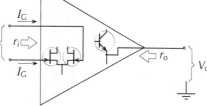

Figure 14-2
Input and output stages for bipolar and BIFET operational amplifiers. The FET gate current is very much smaller than the BJT base current.

(a) Basic input and output stages   (b) Basic input and output stages
    for a bipolar op-amp   for a *BIFET* op-amp

Important Parameters

Some of the most important parameters for the *741* operational amplifier (one of the most commonly used) are listed in the data sheet portion shown in Fig. 14-3.

The *large signal voltage gain* (A_v), also identified as the *open-loop gain* ($A_{v(OL)}$), for the *741* is listed as as 50 000 minimum and 200 000 typical. This means, for example, that a signal applied as a voltage difference between the two input terminals is amplified at least by a factor 50 000. The typical amplification is 200 000, and the maximum amplification can be considerably greater. This is similar to the situation with the h_{fe} of a *BJT*, where the manufacturer specifies minimum, typical, and maximum values.

The *input bias current* (I_{IB}) is the base current for the op-amp input stage transistors. In Fig. 14-3 I_{IB} is listed as 80 nA typical, and 500 nA maximum.

Ideally, the input stage transistors should be perfectly matched, so that zero voltage difference between the two input (base) terminals should produce a zero output voltage. In practise, there is always some difference between the base-emitter voltages of the transistors, and this results in an *input offset voltage* (V_{IO}), that might be as large as 5 mV for a 741 op-amp. Similarly, because of mismatch of input transistors, there is also an *input offset current* (I_{IO}). For the 741 I_{IO} is a maximum of 200 nA.

The *input resistance* (r_i) and the *output resistance* (r_o) are the resistances at the op-amp terminals when no feedback is involved. Virtually all op-amp applications use negative feedback, so these (resistance) quantities are modified in practical circuits.

The *dc* supply voltage for electronic circuits can vary when the supply current changes, and in some cases there is an *ac ripple voltage* superimposed on the *dc.* The supply voltage changes can be passed to the output of an amplifier. Ideally, the supply voltage variations should have no effect on the circuit output. How good the amplifier is at attenuating supply voltage variations is defined by the *power supply voltage rejection ratio (PSRR)*. This is usually specified in decibel form.

Common mode inputs are voltage changes that are applied simultaneously to both input terminal of an operational amplifier. Common mode input voltages can be passed to the output. The *common mode rejection ratio (CMRR)* (expressed in *dB*) defines how good the amplifier is at attenuating common mode input voltages.

Electrical characteristics (V_S = ±15 V, T_A = 25°C unless otherwise specified)

Parameters	Conditions	Min	Typ	Max	Unit
Large signal voltage gain (A_v)	$R_L \geq 2$ kΩ, V_{out} = ±10 V	50 000	200 000		
Input bias current (I_{IB})			80	500	nA
Input offset voltage (V_{IO})	$R_S \leq 10$ kΩ		1.0	5.0	mV
Input offset current (I_{IO})			20	200	nA
Input resistance (r_i)		0.3	2		MΩ
Output resistance (r_o)			75		Ω

Figure 14-3
Portion of data sheet for a 741 op-amp.

Practise Problems

14-1.1   Determine the following parameter values from the *LM108* op-amp data sheet in Appendix 1-14: large signal voltage gain, input bias current, input offset voltage, input resistance.

14-1.2   Repeat Problem 14-1.1 for the *LF353* op-amp in Appendix 1-15.

14-2   Biasing Operational Amplifiers

Biasing Bipolar Op-amps

Like other electronic devices, operational amplifiers must be correctly biased if they are to function properly. As already discussed, the inputs of an operational amplifier are the base terminals of the transistors in a differential amplifier. Base currents must flow into these terminals for the transistors to operate. Consequently, the input terminals must be directly

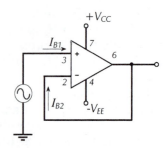

Figure 14-4
Op-amp circuit with one input terminal grounded via a signal generator, and the other input directly connected to the output.

connected to suitable *dc* bias voltage sources.

The most appropriate *dc* bias voltage level for op-amp inputs is approximately halfway between the + and - supply voltages. One of the two input terminals is usually connected in some way to the op-amp output to facilitate negative feedback. Where a +/- supply is used, the other input might be biased directly to ground via a signal source [see Fig. 14-4]. Base current I_{B1} flows into the op-amp noninverting input via the signal source, while I_{B2} flows from the output into the inverting input, as illustrated.

Figure 14-5 shows a situation in which one input is connected via resistor R_1 to ground, and the other is connected via R_2 to the op-amp output. Once again base current for the input stage transistors flows into both input terminals. R_1 and R_2 should have equal resistance values, so that voltage drops $I_{B1}R_1$ and $I_{B2}R_2$ are approximately equal. Any difference in these two voltage drops appears as an op-amp *dc* input voltage which may be amplified to produce a *dc* offset at the output.

If very small resistance values are selected for R_1 and R_2 in the circuit in Fig. 14-5, then a very small input resistance (R_1) would be offered to a capacitor-coupled signal source, [see Fig. 14-5]. On the other hand, if R_1 and R_2 are very large, the voltage drops $I_{B1}R_1$ and $I_{B2}R_2$ might be ridiculously large. An acceptable maximum voltage drop across these resistors must be very much smaller than the typical V_{BE} level for a forward-biased base-emitter junction.

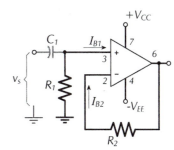

Figure 14-5
Op-amp circuit with one input grounded via resistor R_1, and the other input connected to the output via R_2.

Select, $V_{R1} \approx V_{BE}/10$

and, $R_{1(max)} = V_{R1}/I_{B(max)}$

or, $$R_{1(max)} = \frac{V_{BE}}{10\ I_{B(max)}}$$ **(14-1)**

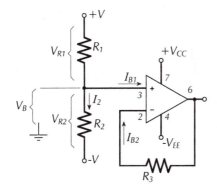

Figure 14-6
Op-amp circuit using a voltage divider bias circuit.

Figure 14-6 shows a voltage divider (R_1 and R_2) providing an input terminal bias level from the the supply voltages. The voltage divider current I_2 should be selected to be very much larger than the input bias current. This is to ensure that I_B has a negligible

is selected as,

$$I_{2(min)} = 100\ I_{B(max)} \qquad \textbf{(14-2)}$$

Then R_1 and R_2 are simply calculated as, V_{R1}/I_2 and V_{R2}/I_2 respectively.

Resistor R_3 in Fig. 14-6 should be selected approximately equal to $R_1\|R_2$ to minimize any difference between $I_{B1}(R_1\|R_2)$ and $I_{B1}R_3$ that could behave as a *dc* input voltage.

Example 14-1

Calculate the maximum resistance for R_1 and R_2 in Fig. 14-7(a) if a *741* op-amp is used. Also, determine suitable resistances for R_1, R_2, and R_3 in Fig. 14-7(b) to give $V_B = 0$ when the supply voltages are ±15 V.

Solution

For Fig. 14-7(a),

$$V_{BE} \approx 0.7\ V \quad \text{(for op-amps with BJT input stages)}$$

From the 741 data sheet,

$$I_{B(max)} = 500\ nA$$

Eq. 14-1,
$$R_{1(max)} = \frac{V_{BE}}{10\ I_{B(max)}} = \frac{0.7\ V}{10 \times 500\ nA}$$

$$= 140\ k\Omega \text{ (use 120 k}\Omega \text{ standard value)}$$

For Fig. 14-7(b),

Eq. 14-2,
$$I_2 = 100\ I_{B(max)} = 100 \times 500\ nA$$
$$= 50\ \mu A$$

$$V_{R1} = V_{R2} = 15\ V$$

$$R_1 = R_2 = \frac{V_{R1}}{I_2} = \frac{15\ V}{50\ \mu A}$$

$$= 300\ k\Omega \text{ (use 270 k}\Omega \text{ standard value)}$$

$$R_3 = R_1\|R_2 = 270\ k\Omega\|270\ k\Omega$$
$$= 135\ k\Omega \text{ (use 120 k}\Omega \text{ standard value)}$$

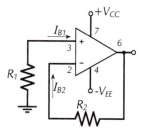

(a) Grounded input circuit

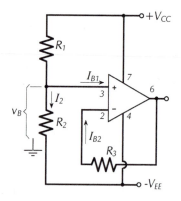

(b) Voltage divider bias

Figure 14-7
Op-amp circuits for Ex. 14-1.

A single-polarity supply voltage can be used with an operational amplifier. For example, a 741 could use a +30 V (as illustrated in Fig. 14-8) instead of a ±15 V supply. Resistors R_1 and R_2 are normally selected to set $V_B = V_{CC}/2$.

The input offset voltage referred to in Section 14-1, and the resultant output offset, can be reduced or eliminated in the circuit shown in Fig. 14-6 by making one of the resistors adjustable. Similarly, if R_3 in Fig. 14-8 is a variable resistor, it could be adjusted to reduce the op-amp output offset voltage. The process is termed *voltage offset nulling*.

Figure 14-9 shows another method of voltage offset nulling with a 741 op-amp. A 10 kΩ potentiometer is connected to the differential amplifier input stage via terminals 1 and 5, as illustrated. With the moving contact connected to $-V_{EE}$, the potentiometer can be adjusted to null the output offset voltage.

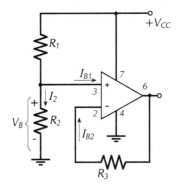

Figure 14-8
Op-amp circuit using a single-polarity supply.

Biasing BIFET Op-amps

The input bias current for a *BIFET* op-amp is typically 50 pA, which is very much smaller that that for a *Bioplar* op-amp. So, the bias resistor selection method already discussed would produce very high resistor values. This is undesirable because electric charges can accumulate at the *FET* gates, and thus make the bias levels unstable. Also, stray capacitance becomes more effective with high value bias resistors, possibly resulting in unwanted circuit oscillations. To combat these effects, the largest resistor in a *BIFET* op-amp bias circuit should normally not exceed 1 MΩ.

Example 14-2
Determine suitable resistor values for the type of circuit shown in Fig. 14-7(b) when a *BIFET* op-amp is used with a ±9 V supply.

Solution

Select, $R_1 = R_2 = 1 \text{ M}\Omega$

$R_3 = R_1 \| R_2 = 1 \text{ M}\Omega \| 1 \text{ M}\Omega$

$= 500 \text{ k}\Omega$ (use 470 kΩ standard value)

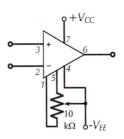

Figure 14-9
Output offset nulling can be accomplished in a 741 op-amp by the use of a 10 kΩ potentiometer connected to terminals 1 and 5.

Practise Problems

14-2.1 A *741* op-amp is used with a +24 V supply in the type of circuit shown in Fig. 14-8. Determine suitable bias resistor values.

14-2.2 Repeat Problem 14-2.1 to bias the input to +10 V.

14-3 Voltage Follower Circuits

Direct-Coupled Voltage Follower
The *IC* operational amplifier can be employed for an infinite variety of applications. The very simplest application is the direct-coupled *voltage follower* shown in Fig. 14-10(a). The output terminal is connected directly to the inverting input terminal, the signal is applied to the noninverting input, and the load is directly-coupled to the output.

Figure 14-10(b) shows the basic op-amp circuit connected as a voltage follower. With the differential amplifier input stage, V_2 (which equals V_o) must equal V_i. When V_2 is higher or lower than

V_i, the voltage difference is amplified to move the output back to equality with the input. Suppose V_o was slightly higher than V_i. The voltage at Q_2 base (terminal 2) would be higher than that at Q_1 base (terminal 3), therefore I_{C2} would be increased above its normal level. This would cause an increase in the voltage drop across R_C, thus reducing V_{B3} and driving V_o back to equality with V_i. Similarly, when V_o is lower than V_i, Q_2 base voltage is lower than that at Q_1 base, I_{C2} is reduced, thus reducing V_{RC}, increasing V_{B3}, and driving V_o back up toward V_i.

When the input voltage (at terminal 3) is increased or decreased, the feedback effect causes the output to almost perfectly follow the input. The actual difference between the input and output voltage is easily calculated from the output voltage level and the op-amp open-loop voltage gain.

Like an emitter follower, the voltage follower has a high input impedance, a low output impedance, and a voltage gain of 1. The voltage follower performance is superior to that of the emitter follower. As demonstrated in Example 14-3, the typical difference between input and output voltage with a voltage follower is only 5 μV when the input amplitude is 1 V. With an emitter follower, there is a 0.7 V dc voltage difference between input and output. The voltage follower also has a much higher input impedance and a much lower output impedance than the emitter follower. The actual values of Z_{in} and Z_o can be calculated from the negative feedback equations derived in Chapter 13.

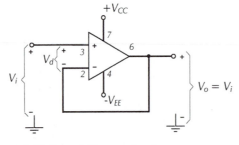

(a) Voltage follower circuit

Figure 14-10
In a voltage follower circuit, the output terminal is directly connected to the inverting input terminal. As the input voltage changes, the output follows the input to keep the inverting input terminal voltage equal to the noninverting terminal voltage.

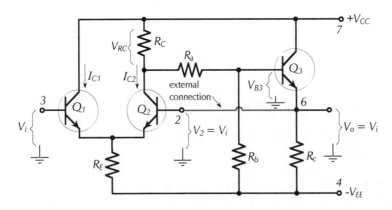

(b) Basic op-amp circuit connected as a voltage follower

Example 14-3

Calculate the typical difference between the input and output voltage for a voltage follower using a 741 op-amp with a 1 V input, as in Fig. 14-11. Also, determine typical Z_{in} and Z_{out} values.

Solution

From Appendix 1-13, typical parameters are:

$$A_v = 200\ 000,\ r_i = 2\ M\Omega,\ r_o = 75\Omega$$

$$V_d = V_o - V_i$$

$$V_d = \frac{V_o}{A_v} = \frac{1\ V}{200\ 000}$$

$$= 5\ \mu V$$

from Eq. 13-4, $Z_i = (1 + A_v B)r_i = [1 + (200\ 000 \times 1)] \times 2\ M\Omega$

$$= 400\ 000\ M\Omega$$

from Eq. 13-6, $Z_o = \dfrac{r_o}{1 + A_v B} = \dfrac{75\ \Omega}{1 + (200\ 000 \times 1)}$

$$= 0.375 \times 10^{-3}\ \Omega$$

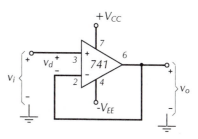

Figure   14-11
Voltage   follower   circuit   for Example 14-3.

Capacitor-Coupled Voltage Follower

When a voltage follower is to have capacitor-coupled input and output terminals, the noninverting input must be grounded via a resistor, [see (R_1) Fig. 14-12]. As discussed, the resistor is required for passing bias current to the noninverting input terminal. It also offers an input resistance to the signal source, rather than a short-circuit. Because the load is capacitor-coupled, the *dc* output offset voltage might seem to be unimportant. However, the *dc* offset voltage at the op-amp output terminal can limit the amplitude of the output voltage. So, a resistor (R_2) equal to R_1 should be included in series with the inverting input terminal to equalize the $I_B R_B$ voltage drops and thus minimize output offset voltage, as already discussed.

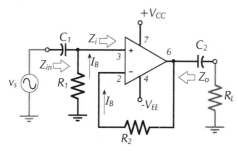

Figure   14-12
A   capacitor-coupled   voltage follower must have the op-amp noninverting terminal grounded via a resistor to provide a path for input bias current.

Design of the capacitor-coupled circuit in Fig. 14-12 involves calculation of R_1, C_1, and C_2. The circuit input impedance is $R_1 \| Z_i$, however, Z_i is always much larger than R_1. So, the circuit input impedance is simply taken as,

$$Z_{in} = R_1 \qquad\qquad \text{(14-3)}$$

Normally, The load resistance R_L is much smaller R_1, and consequently, the smallest capacitor values are calculated when C_2 is selected to set f_1. Therefore,

$$X_{C2} = (Z_o + R_L) \text{ at } f_1 \qquad\qquad \text{(14-4)}$$

Capacitor C_1 should then be calculated from, $X_{C1} = (Z_{in} + r_s)/10$ at f_1, so that it has no significant effect on the circuit lower cutoff frequency.

$$X_{C1} = (R_1 + r_s)/10 \text{ at } f_1 \qquad\qquad \text{(14-5)}$$

Coupling capacitor equations are explained in Section 12-1.

Example 14-4

Design a capacitor-coupled voltage follower using a 741 op-amp. The circuit lower cutoff frequency is to be 70 Hz, the load resistance is 4 kΩ, and the signal source resistance is much smaller than R_1.

Solution

From the 741 data sheet,

$$I_{B(max)} = 500 \text{ nA}$$

Eq. 14-1, $\qquad R_{1(max)} = \dfrac{V_{BE}}{10\, I_{B(max)}} = \dfrac{0.7 \text{ V}}{10 \times 500 \text{ nA}}$

$$= 140 \text{ k}\Omega \text{ (use 120 k}\Omega \text{ standard value)}$$

From Eq. 14-4, $\quad C_2 = \dfrac{1}{2\pi f_1 R_L} = \dfrac{1}{2\pi \times 70 \text{ Hz} \times 4 \text{ k}\Omega}$

$$\approx 0.56\ \mu\text{F (standard value)}$$

From Eq. 14-5, $\quad C_1 = \dfrac{1}{2\pi f_1 R_1/10} = \dfrac{1}{2\pi \times 70 \text{ Hz} \times 120 \text{ k}\Omega/10}$

$$\approx 0.19\ \mu\text{F (use 0.2}\ \mu\text{F standard value)}$$

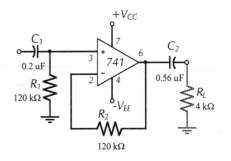

Figure 14-13
Capacitor-coupled voltage follower for Example 14-4.

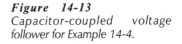

Practise Problems

14-3.1 A capacitor-coupled voltage follower with Z_{in} = 22 kΩ and R_L = 10 kΩ is to have f_1 = 50 Hz. Determine suitable capacitor values.

14-3.2 Determine coupling capacitor values for the circuit in part (b) of Ex. 14-1, if r_s = 600 Ω, R_L = 8.2 kΩ, and f_1 = 20 Hz.

14-3.3 Calculate the maximum dc offset voltage that might be produced at the output of the circuit designed in part (a) of Ex. 14-1 when R_2 is replaced with a short circuit.

14-4   Noninverting Amplifiers

Direct-Coupled Noninverting Amplifier

The noninverting amplifier circuit in Fig. 14-14 behaves similarly to a voltage follower circuit with one major difference. Instead of all of the output voltage being fed directly back to the inverting input terminal (as in a voltage follower), only a portion of V_o is fed back. The output voltage is divided by resistors R_2 and R_3, and the voltage across R_3 is applied to the inverting input terminal. As in the case of the voltage follower, the output voltage changes as necessary to keep the inverting input terminal voltage equal to that at the noninverting input. Thus, the voltage V_{R3} always equals V_i, and the output voltage is then determined by the resistances of R_2 and R_3.

Because the signal voltage is applied to the op-amp noninverting input terminal, the output always has the same polarity as the input. A positive-going input produces a positive-going output, and vice versa. Thus, the input is not inverted (at the output), and the circuit is identified as a *noninverting amplifier.*

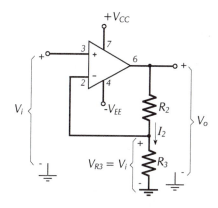

Figure   14-14
Op-amp noninverting amplifier circuit. The signal is applied to the noninverting input terminal. The circuit voltage gain is,
$$A_{CL} = (R_2 + R_3)/R_3$$

The voltage divider current (I_2) is always selected to be very much larger than the operational amplifier input bias current, and

$$\boldsymbol{V_{R3} = V_i = I_2\,R_3} \qquad\qquad \textbf{(14-6)}$$

Also, in Fig. 14-14 it is seen that,

$$\boldsymbol{I_2 = \frac{V_o}{R_2 + R_3}} \qquad\qquad \textbf{(14-7)}$$

The circuit voltage gain is,

$$A_{CL} = \frac{V_o}{V_i} = \frac{I_2\,(R_2 + R_3)}{I_2\,R_3}$$

or, $$A_{CL} = \frac{R_2 + R_3}{R_3} \qquad \textbf{(14-8)}$$

Figure 14-15 shows a noninverting amplifier circuit with a resistor (R_1) connected in series with the noninverting input terminal. As in the case of other op-amp circuits, this is done to equalize the resistor voltage drops produced by input bias current. In this case, R_1 is selected to be approximately equal to the resistance 'seen looking out of' the inverting input terminal. Thus,

$$R_1 = R_2 \| R_3 \qquad \textbf{(14-9)}$$

The input and output impedances for a noninverting amplifier are easily determined from the negative feedback equations in Chapter 13.

Design of a noninverting amplifier mostly involves determining suitable voltage divider resistors (R_2 and R_3). So, as discussed in Section 14-2, design commences with selection of the voltage divider current to be much larger than the op-amp input bias current.

Example 14-5

Design a direct-coupled noninverting amplifier (as in Fig. 14-15) to use a 741 op-amp. The output voltage is to be 2 V when the input is 50 mV.

Solution

From the 741 data sheet,

$$I_{B(max)} = 500 \text{ nA}$$

From Eq. 14-2, $\quad I_{2(min)} = 100\, I_{B(max)} = 100 \times 500 \text{ nA}$
$$= 50\,\mu A$$

From Eq. 14-6, $\quad R_3 = \dfrac{V_i}{I_2} = \dfrac{50 \text{ mV}}{50\,\mu A}$
$$= 1 \text{ k}\Omega \quad \text{(standard resistor value)}$$

From Eq. 14-7, $\quad R_2 + R_3 = \dfrac{V_o}{I_2} = \dfrac{2 \text{ V}}{50\,\mu A}$
$$= 40 \text{ k}\Omega$$

$$R_2 = (R_2 + R_3) - R_3 = 40 \text{ k}\Omega - 1 \text{ k}\Omega$$
$$= 39 \text{ k}\Omega \quad \text{(standard value)}$$

From Eq. 14-9, $\quad R_1 = R_2 \| R_3 = 39 \text{ k}\Omega \| 1 \text{ k}\Omega$
$$\approx 1 \text{ k}\Omega \quad \text{(use 1 k}\Omega)$$

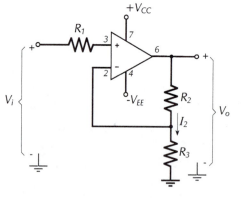

Figure 14-15
Noninverting amplifier circuit with a resistor R_1 in series with the input. R_1 is selected to be approximately equal to $R_2 \| R_3$.

Example 14-6
Calculate typical input and output impedances for the noninverting amplifier designed in Example 14-5.

Solution

From Appendix 1-13, typical parameters are:

$$A_v = 200\,000, \ r_i = 2 \ \text{M}\Omega, \ r_o = 75\Omega$$

Also,
$$B = \frac{R_3}{R_2 + R_3} = \frac{1 \ \text{k}\Omega}{39 \ \text{k}\Omega + 1 \ \text{k}\Omega}$$

$$= \frac{1}{40}$$

from Eq. 13-4,
$$Z_i = (1 + A_v B) \, r_i = [1 + \frac{200\,000}{40}] \times 2 \ \text{M}\Omega$$

$$\approx 10\,000 \ \text{M}\Omega$$

from Eq. 13-6,
$$Z_o = \frac{r_o}{1 + A_v B} = \frac{75 \ \Omega}{1 + (200\,000/40)}$$

$$\approx 0.015 \ \Omega$$

Capacitor-Coupled Noninverting Amplifier
When a noninverting amplifier is to have a signal capacitor-coupled to its input, the op-amp noninverting input terminal must be grounded via a resistor to provide a path for the input bias current. This is illustrated in Fig. 14-16, where R_1 allows for the passage of I_{B1}. As in the case of the capacitor-coupled voltage follower, the input resistance is essentially equal to R_1 for the capacitor-coupled noninverting amplifier.

Resistors R_1, R_2, and R_3 in the capacitor-coupled circuit are determined exactly as for a direct-coupled noninverting amplifier. The capacitor values are calculated in the same way as for a capacitor-coupled voltage follower, using Eqs. 14-4 and 14-5.

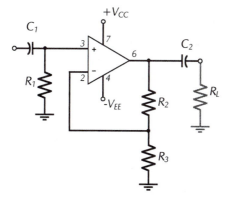

Figure 14-16
A capacitor-coupled noninverting amplifier circuit must have the noninverting input terminal grounded via a resistor to provide a path for the input bias current.

Example 14-7

Calculate the voltage gain and lower cutoff frequency for the capacitor-coupled noninverting amplifier circuit in Fig. 14-17.

Solution

Eq. 14-8,
$$A_{CL} = \frac{R_2 + R_3}{R_3} = \frac{56\ k\Omega + 2.2\ k\Omega}{2.2\ k\Omega}$$

$$= 26.5$$

From Eq. 14-4,
$$f_1 = \frac{1}{2\pi C_2 R_L} = \frac{1}{2\pi \times 8.2\ \mu F \times 600\ \Omega}$$

$$= 32.3\ Hz$$

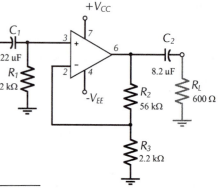

Figure 14-17
Capacitor-coupled noninverting amplifier circuit for Ex. 14-7.

Practise Problems

14-4.1 A direct-coupled noninverting amplifier using a 741 op-amp is to have a 100 mV input voltage amplitude and a voltage gain of 40. Select suitable resistor values.

14-4.2 Design a capacitor-coupled noninverting amplifier using a 741 op-amp to have a ± 5 V output amplitude when the input is ± 250 mV. The load resistance is $R_L = 820\ \Omega$, and the lower cutoff frequency is to be $f_1 = 20$ Hz.

14-4.3 Redesign the circuit in Problem 14-4.1 to use a *BIFET* op-amp.

14-5   Inverting Amplifiers

Direct-Coupled Inverting Amplifier

The circuit in Fig. 14-18 is termed an *inverting amplifier* because, with V_i applied via R_1 to the inverting input terminal, the output goes negative when the input goes positive, and vice versa. Note that the noninverting input terminal is grounded via resistor R_3. With the noninverting terminal grounded, the voltage at the op-amp inverting input terminal remains close to ground. Any increase or decrease in the (very small) difference voltage between the two input terminals is amplified by the op-amp open-loop gain and fed back via R_2 and R_1 to correct the change. Because the inverting input terminal is not grounded but remains close to ground, the inverting input terminal in this application is termed a *virtual ground* or *virtual earth.*

The circuit input current in Fig. 14-18 can be calculated as,

$$I_1 = \frac{V_{R1}}{R_1}$$

The input voltage is applied to one end of R_1, and the other end of R_1 is at ground level. Consequently,

$$V_{R1} = V_i$$

and, $$I_1 = \frac{V_i}{R_1}$$ **(14-10)**

I_1 is always selected to be very much larger than the op-amp input bias current (I_B). Consequently, virtually all of I_1 flows through resistor R_2, (see Fig.14-18), and the voltage drop across R_2 is,

$$V_{R2} = I_1 R_2$$

The left side of R_2 is connected to the op-amp inverting input terminal, which, as discussed, is always at ground level. This means that the right side of R_2 (the output terminal) is V_{R2} below ground. So,

$$V_o = -I_1 R_2$$ **(14-11)**

From Eq. 14-10, $V_i = I_1 R_1$

So, the circuit voltage gain is,

$$A_{CL} = \frac{V_o}{V_i} = \frac{-I_1 R_2}{I_1 R_1}$$

or, $$A_{CL} = \frac{-R_2}{R_1}$$ **(14-12)**

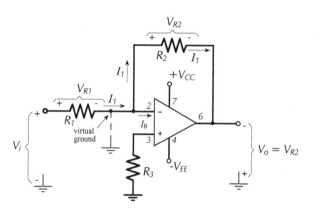

Figure 14-18
Op-amp inverting amplifier. The signal voltage is applied via resistor R_1 to the inverting input terminal. The circuit voltage gain is $A_{CL} = -R_2/R_1$.

If the input of the inverting amplifier is grounded, the circuit is seen to be exactly the same as noninverting amplifier with R_3 as the input resistor and zero input voltage. Thus, negative feedback occurs (as in a noninverting amplifier) to maintain the op-amp inverting input terminal at the same voltage level as the noninverting input terminal.

The output impedance of an inverting amplifier is calculated in exactly the same way as for a noninverting amplifier, (using Eq. 13-6). Like the case of the noninverting amplifier, the output impedance is very low for an inverting amplifier.

The input impedance of an inverting amplifier is easily determined by recalling that the right side of R_1 (in Fig. 14-18) is always at ground level, and that the signal is applied to the left side of R_1. Therefore,

$$Z_i = R_1 \qquad\qquad (14\text{-}13)$$

Design of an inverting amplifier is very simple. Voltage divider current I_1 is selected very much larger than the op-amp input bias current. Resistors R_1 and R_2 are calculated from Eqs. 14-10 and 14-11, or 14-12, and R_3 is selected approximately equal to $R_1\|R_2$.

Example 14-8

Design a direct-coupled inverting amplifier, as in Fig. 14-19, to use a 741 op-amp. The input voltage amplitude is 20 mV and the voltage gain is to be 144.

Solution

For the 741, $I_{B(max)} = 500\,\text{nA}$

select, $I_{1(min)} = 100\,I_{B(max)} = 100 \times 500\,\text{nA}$
$$= 50\,\mu\text{A}$$

From Eq. 14-10, $R_1 = \dfrac{V_i}{I_1} = \dfrac{20\,\text{mV}}{50\,\mu\text{A}}$

$$= 400\,\Omega \quad \text{(use 390 }\Omega \text{ standard value)}$$

From Eq. 14-12, $R_2 = A_{CL}\,R_1 = 144 \times 390\,\Omega$

$$= 56.2\,\text{k}\Omega \quad \text{(use 56 k}\Omega \text{ standard value)}$$

$$R_3 = R_1\|R_2 = 390\,\Omega\|56\,\text{k}\Omega$$

$$\approx 390\,\Omega \quad \text{(use 390 }\Omega \text{)}$$

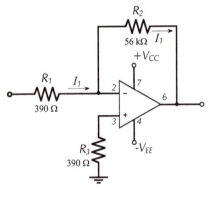

Figure 14-19
Inverting amplifier for Ex. 14-8.

Capacitor-Coupled Inverting Amplifier

A capacitor-coupled inverting amplifier circuit is shown in Fig. 14-20. In this case, the bias current to the op-amp inverting input terminal flows (from the output) via resistor R_2, so that the input coupling capacitor does not interrupt the input bias current. A voltage drop $I_B R_2$ is produced at the inverting input by the bias current flow, and this must be equalized by the $I_B R_3$ voltage drop at the noninverting input. Therefore, for the capacitor-coupled noninverting circuit,

$$R_3 = R_2 \qquad\qquad (14\text{-}14)$$

Apart from the selection of R_3, the circuit is designed exactly as the direct-coupled circuit, and the capacitor values are calculated in the same way as for a capacitor-coupled noninverting amplifier.

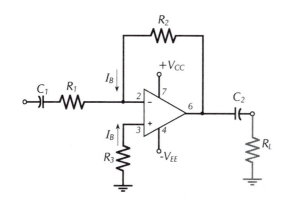

Figure 14-20
Capacitor-coupled inverting amplifier circuit. Note that the flow of bias current is not interrupted by capacitor C_1.

Practise Problems

14-5.1   A direct-coupled inverting amplifier using a 741 op-amp is to have a 300 mV input voltage amplitude and a voltage gain of 15. Select suitable resistor values.

14-5.2   The circuit in Problem 14-5.1 is to have a 1 kΩ capacitor-coupled load and a 600 Ω capacitor coupled signal source. Determine suitable capacitor values for a 20 Hz lower cutoff frequency.

14-5.3   Redesign the circuit in Problem 14-5.1 to use a *BIFET* op-amp.

14-6   Summing Amplifier

The *summing amplifier* circuit in Fig. 14-21 is simply a direct-coupled inverting amplifier with two inputs applied to two resistors (R_1 and R_1). With two input voltages (V_{i1} and V_{i2}) there are two input currents (I_1 and I_2). Also, because the op-amp inverting input terminal behaves as a virtual ground (as for an inverting amplifier), the input currents are,

$$I_1 = \frac{V_{i1}}{R_1} \qquad I_2 = \frac{V_{i2}}{R_2}$$

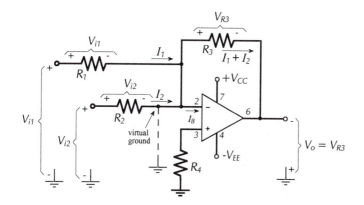

Figure 14-21
An op-amp summing amplifier amplifies the sum of two, or more, input voltages.

All of I_1 and I_2 flows through resistor R_3, giving

$$V_o = - (I_1 + I_2) R_3$$

or,

$$V_o = [\frac{V_{i1}}{R_1} + \frac{V_{i2}}{R_2}] R_3$$

With $R_1 = R_2$,

$$\boldsymbol{V_o = \frac{- R_3}{R_1}(V_{i1} + V_{i2})} \qquad \textbf{(14-15)}$$

When $R_3 = R_1 = R_2$, the output voltage is the direct (inverted) sum of the two inputs. When R_3 is greater than R_1 and R_2, the output is an amplified version of the sum of the inputs.

A summing amplifier is not limited to two inputs. There can be almost any number of inputs, and the output remains the sum of the inputs. Summing amplifiers are designed in the same way as ordinary inverting amplifiers.

Example 14-9

Design a three-input summing amplifier, as in Fig. 14-22, to use a *BIFET* op-amp and to have a voltage gain of 3. Calculate the resistor currents and the output voltage when all three inputs are 1 V.

Solution

Select, $R_4 = 1 M\Omega$

From Eq. 14-12, $R_1 = R_2 = R_3 = \dfrac{R_4}{A_{CL}} = \dfrac{1 M\Omega}{3}$

$\qquad\qquad\qquad = 333 k\Omega$ (use 330 kΩ standard value)

$\qquad R_5 = R_1\|R_2\|R_3\|R_4 = 330\ k\Omega\|330\ k\Omega\|330\ k\Omega\|1\ M\Omega$

$\qquad\qquad = 99\ k\Omega$ (use 100 kΩ standard value)

$\qquad I_1 = I_2 = I_3 = \dfrac{V_i}{R_1} = \dfrac{1\ V}{330\ k\Omega}$

$\qquad\qquad = 3.03\ \mu A$

$\qquad I_4 = I_1 + I_2 + I_3 = 3.03\ \mu A + 3.03\ \mu A + 3.03\ \mu A$

$\qquad\qquad = 9.09\ \mu A$

$\qquad V_o = -I_4 R_4 = -9.09\ \mu A \times 1\ M\Omega$

$\qquad\qquad = -9.09\ V$

or, from Eq 14-15,

$$V_o = \frac{-R_4}{R_1}(V_{i2} + V_{i2} + V_{i3}) = \frac{-1\ M\Omega}{330\ k\Omega}(1\ V + 1\ V + 1V)$$

$\qquad\qquad = -9.09\ V$

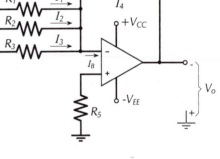

Figure 14-22
Three-input summing amplifier for Example 14-9.

14-7 Difference Amplifier

Circuit Operation

A *difference amplifier* amplifies the difference between two inputs. The circuit shown in Fig. 14-23 is a combination of inverting and noninverting amplifiers. Resistors R_1, R_2, and the op-amp constitutes an inverting amplifier for a voltage (V_{i1}) applied to R_1. The same components (R_1, R_2, and the op-amp) also function as a noninverting amplifier for a voltage (V_{R4}) at the noninverting input terminal. It is seen that V_{R4} is derived from input voltage V_{i2} by the voltage divider R_3 and R_4. To understand the circuit operation, consider the output produced by each input voltage when the other input is zero:

with $V_{i2} = 0$, $V_{o1} = \dfrac{-R_2}{R_1} \times V_{i1}$

with $V_{i1} = 0$, $V_{o2} = \dfrac{R_1 + R_2}{R_1} \times V_{R4}$

and, $V_{R4} = \dfrac{R_4}{R_3 + R_4} \times V_{i2}$

so, $V_{o2} = \dfrac{R_1 + R_2}{R_1} \times \dfrac{R_4}{R_3 + R_4} \times V_{i2}$

with $R_3 = R_1$, and $R_4 = R_2$,

$$V_{o2} = \dfrac{R_2}{R_1} \times V_{i2}$$

When both inputs are present,

$$V_o = V_{o2} + V_{o1}$$

$$= \dfrac{R_2}{R_1} V_{i2} + \dfrac{-R_2}{R_1} V_{i1}$$

giving, $$\mathbf{V_o = \dfrac{R_2}{R_1} (V_{i2} - V_{i1})} \qquad\qquad \textbf{(14-16)}$$

When $R_2 = R_1$, the output voltage (as calculated by Eq. 14-16) is

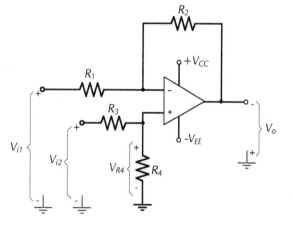

Figure 14-23
An op-amp difference amplifier amplifies the difference between two input voltages.

the direct difference between the two inputs. With R_2 greater than R_1, the output becomes an amplifier version of $(V_{i2} - V_{i1})$.

Input resistances

Consider the input portion of the difference amplifier circuit reproduced in Fig. 14-24. The resistance at input *terminal 1* is the same as the input impedance for an inverting amplifier: $Z_{i1} = R_1$. The input resistance at the op-amp noninverting input terminal, is very high (as in the case of a noninverting amplifier), and this is in parallel with resistor R_4. So, the input impedance at *terminal 2* in Fig. 14-24 is $Z_{i2} = R_3 + R_4$.

Equation 14-16 was derived by assuming that $R_3 = R_1$, and $R_4 = R_2$. It can be shown that the same result is obtained when the *ratio* R_4/R_3 equals R_2/R_1, so that the actual resistor values do not have to be equal. For equal resistances at the two input terminals

select, $$R_3 + R_4 = R_1 \qquad (14\text{-}17)$$

Then, calculate the resistances of R_3 and R_4 from,

$$\frac{R_4}{R_3} = \frac{R_2}{R_1} \qquad (14\text{-}18)$$

A simple rule-of-thumb can be used for determining suitable resistance values for R_3 and R_4 when the two input resistances do not have to be exactly equal. Select, $R_4 = R_2/A_{CL}$; which always makes $R_4 = R_1$. Then, calculate R_3 as, $R_3 = R_1/A_{CL}$.

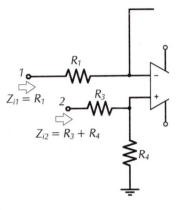

Figure 14-24
Difference amplifier circuit input impedances.

Example 14-10

A difference amplifier is to be designed to amplify the difference between two voltages by a factor of 10. The inputs each approximately equal 1 V. Determine suitable resistor values for a circuit using a 741 op-amp.

Solution

Select,

$$I_1 \approx 100\,I_B = 100 \times 500 \text{ nA}$$
$$= 50\,\mu A$$

$$R_1 = \frac{V_{i1}}{I_1} = \frac{1 \text{ V}}{50\,\mu A}$$
$$= 20 \text{ k}\Omega \text{ (use 18 k}\Omega \text{ standard value)}$$

$$R_2 = A_{CL}\,R_1 = 10 \times 18 \text{ k}\Omega$$
$$= 180 \text{ k}\Omega \text{ (standard values)}$$

$$R_4 = R_1 = 18 \text{ k}\Omega$$

$$R_3 = \frac{R_1}{A_{CL}} = \frac{18 \text{ k}\Omega}{10}$$
$$= 1.8 \text{ k}\Omega \text{ (standard value)}$$

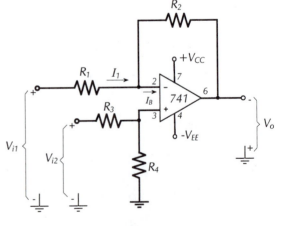

Figure 14-25
Difference amplifier circuit for Ex. 14-10.

Common-Mode Voltages

A common-mode input voltage is a signal voltage (*dc* or *ac*) applied to both input terminals at the same time. This is illustrated in Fig. 14-26, where V_{i1}, V_{i2}, and the common-mode voltage (V_n) are all represented as inputs from *dc* sources. As shown, the input voltages at *terminals 1* and *2* are changed from V_{i1} and V_{i2} to ($V_n + V_{i1}$) and ($V_n + V_{i2}$).

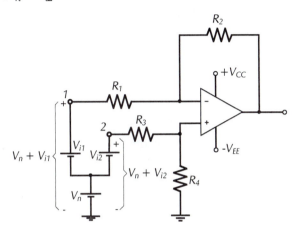

Figure 14-26
Common-mode input voltage applied to a difference amplifier.

Equation 14-16 shows that the output voltage is the amplified difference between the two input voltages. So,

$$V_o = \frac{R_2}{R_1}[(V_{i2} + V_n) - (V_{i1} + V_n)]$$

$$= \frac{R_2}{R_1}(V_{i2} - V_{i1})$$

This shows that the common-mode input is completely cancelled. However, recall that the gain equation depends upon the resistor ratios (R_4/R_3 and R_2/R_1) being equal. If the ratios are not exactly equal, one input will experience a larger amplification than the other. Also, the common-mode voltage at one input terminal will be amplified by a larger amount than that applied to the other input terminal. In this case, common-mode inputs will not be completely cancelled.

Because it is difficult to perfectly match resistor ratios (especially for standard-value components), some common-mode output voltage is almost certain to be produced where a common-mode input exists. Figure 14-27 shows a circuit modification for minimizing common-mode outputs from a difference amplifier. Resistor R_4 is made up of a fixed-value resistor and a small-value adjustable resistor connected in series. This provides adjustment of the ratio R_4/R_3 to match R_2/R_1, so that common-mode outputs can be nulled to zero.

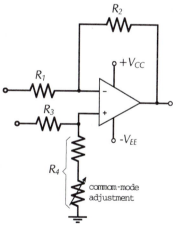
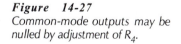

Figure 14-27
Common-mode outputs may be nulled by adjustment of R_4.

14-8   Instrumentation Amplifier

Circuit Operation

At first glance, the *instrumentation amplifier* circuit in Fig. 14-28 looks complex, but when considered section by section it is found to be quite simple. First, note that the second stage (consisting of op-amp A_3 and resistors R_4 through R_7) is a difference amplifier that operates exactly as described in Section 14-7. Next, look at A_1 and resistors R_1 and R_2; this is a noninverting amplifier. Similarly, A_2 combined with resistors R_2 and R_3 constitutes another noninverting amplifier. Because the first stage circuits share a single resistor, their operation is slightly different from the usual noninverting amplifier operation.

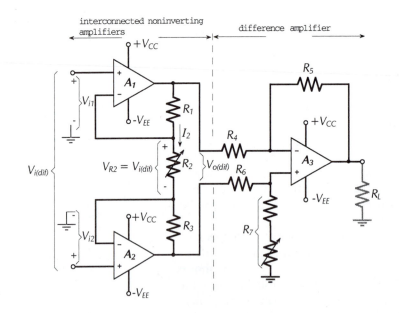

Figure 14-28
An instrumentation amplifier has two interconnected noninverting amplifiers as the first stage, and a difference amplifier as the second stage.

The first stage accepts a *differential input voltage* ($V_{i(dif)}$), and produces a *differential output voltage* ($V_{o(dif)}$). The differential input could be the difference between two grounded inputs (V_{i1} and V_{i2}), as illustrated. But, is often a differential ungrounded input voltage derived, for example, from two voltage monitoring electrodes connected to a human body for medical purposes. In this case, there is often a large common mode input voltage, which can be shown to pass to the output of the first stage without amplification.

The difference amplifier second stage accepts $V_{o(dif)}$ from the first stage as an input, and produces an output to a grounded load, as shown on the circuit diagram. As explained in Section 14-7, the difference amplifier tends to reject common-mode voltages, and the circuit can also have an adjustment for reducing common-mode outputs to zero.

The instrumentation amplifier is now seen to be a circuit with two high-impedance input terminals, and one low-impedance output. The differential input voltage is amplified and converted to a single-ended output, and common-mode inputs are attenuated.

Voltage Gain

Recall that, with a noninverting amplifier, the feedback voltage to the op-amp inverting input terminal always equals the input voltage to the noninverting input terminal. Therefore, the voltage at the $R_1 R_2$ junction equals V_{i1}, and that at the $R_2 R_3$ junction equals V_{i2}. Consequently, the voltage drop across R_3 equals the difference between the two input voltages, which also means that V_{R2} equals the differential input voltage $(V_{i(dif)})$. The current through R_2 can now be calculated as,

$$I_2 = \frac{V_{i(dif)}}{R_2}$$

The voltage drop across R_1, R_2, and R_3 is the differential output voltage of the first stage $(V_{o(dif)})$.

$$V_{o(dif)} = I_2 (R_1 + R_2 + R_3)$$

$$= \frac{V_{i(dif)}}{R_2}(R_1 + R_2 + R_3)$$

The (closed-loop) voltage gain of the *differential input-differential output* first stage is,

$$A_{CL1} = \frac{V_{o(dif)}}{V_{i(dif)}} = \frac{R_1 + R_2 + R_3}{R_2}$$

Normally, R_1 and R_3 are always equal. So, the first stage gain can be written as,

$$A_{CL1} = \frac{2 R_1 + R_2}{R_2} \qquad (14\text{-}19)$$

From Eq. 14-16, the second-stage gain is,

$$A_{CL2} = \frac{R_5}{R_4}$$

The overall voltage gain is,

$$A_{CL} = A_{CL1} \times A_{CL2}$$

The second stage is often designed for a gain of one, so that the overall voltage gain can be calculated from Eq. 14-19. Note that, as shown in Fig. 14-28, R_2 can be a variable resistor for adjustment of the circuit overall voltage gain.

Example 14-11

Calculate the overall voltage gain for the instrumentation amplifier in Fig. 14-29. Also, determine the current and voltage levels throughout the circuit when a +1 V common-mode input (V_{cm}) is present along with the ±10 mV signals.

Solution

From Eqs. 14-16 and 14-19,

$$A_{CL} = \frac{2R_1 + R_2}{R_2} \times \frac{R_5}{R_4}$$

$$= \frac{(2 \times 33\ k\Omega) + 300\ \Omega}{300\ \Omega} \times \frac{15\ k\Omega}{15\ k\Omega}$$

$$= 221$$

at the junction of R_1 and R_2,

$$V_b = V_{i1} + V_{cm} = 10\ mV + 1\ V$$

$$= +1.01\ V$$

at the junction of R_2 and R_3,

$$V_c = V_{i2} + V_{cm} = -10\ mV + 1\ V$$

$$= +0.99\ V$$

current through R_2,

$$I_2 = \frac{V_b - V_c}{R_2} = \frac{1.01\ V - 0.99\ V}{300\ \Omega}$$

$$= 66.67\ \mu A$$

at the output of A_1,

$$V_a = V_b + (I_2 \times R_1) = 1.01\ V + (66.67\ \mu A \times 33\ k\Omega)$$

$$= +3.21\ V$$

at the output of A_2,

$$V_d = V_c - (I_2 \times R_3) = 0.99\ V - (66.67\ \mu A \times 33\ k\Omega)$$

$$= -1.21\ V$$

at the junction of R_6 and R_7,

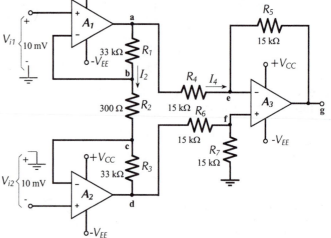

Figure 14-29
Instrumentation amplifier circuit for Ex.14-11.

$$V_f = V_d \times \frac{R_7}{R_6 + R_7} = -1.21 \text{ V} \times \frac{15 \text{ k}\Omega}{15 \text{ k}\Omega + 15 \text{ k}\Omega}$$

$$= -0.605 \text{ V}$$

at the junction of R_4 and R_5,

$$V_e = V_f = -0.605 \text{ V}$$

current through R_4,

$$I_4 = \frac{V_a - V_e}{R_2} = \frac{3.21 \text{ V} - (-0.605 \text{ V})}{15 \text{ k}\Omega}$$

$$= 254.3 \text{ }\mu\text{A}$$

at the output of A_3,

$$V_g = V_e - (I_4 \times R_5) = -0.605 \text{ V} - (254.3 \text{ }\mu\text{A} \times 15 \text{ k}\Omega)$$
$$= -4.42 \text{ V}$$

Practise   Problems

14-8.1 An instrumentation amplifier is to be designed to produce a 4.75 V output when the differential input is 50 mV. Using *741* op-amps, determine suitable resistor values.

14-9   Voltage Level Detectors

Op-amps in Switching Applications

Operational amplifiers are often used in circuits in which the output is switched between the positive and negative saturation voltages, $+V_{o(sat)}$ and $-V_{o(sat)}$. The actual voltage change that occurs is known as the *output voltage swing*. For many op-amps (such as a *741*, for example), the output saturation voltages are typically the supply voltage levels minus 1 V. Thus, as illustrated in Fig. 14-30, the typical output voltage swing is,

$$\Delta V \approx (V_{CC} - 1 \text{ V}) - (V_{EE} + 1 \text{ V}) \qquad \qquad \textbf{(14-20)}$$

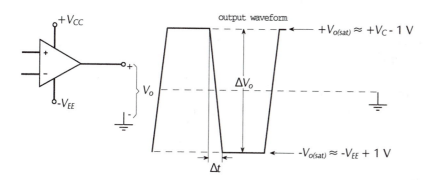

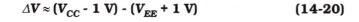

Figure 14-30
An operational amplifier used in a switching application produces an output voltage that switches between positive and negative saturation levels.

For many op-amps the output can be switched from one supply level to the other (there is no 1 V drop), so that

$$\Delta V \approx V_{CC} - V_{EE} \qquad \textbf{(14-21)}$$

This is referred to as *rail-to-rail* operation.

The switching speed, or rate-of change, of the op-amp output voltage is termed the *slew rate (SR)*. Refering to Fig. 14-30,

$$\textbf{SR} = \frac{\Delta V}{\Delta t} \qquad \textbf{(14-22)}$$

Suppose the output changes from -10 to +10 V; the voltage change is $\Delta V = 20$ V. If the transition time (or *rise time*, see Section 8-5), is $\Delta t = 1$ μs, the slew rate is 20 V/μs.

Another integrated circuit known as a *voltage comparator*[1] is often substituted in place of an operational amplifier in switching applications. Voltage comparators are similar to op-amps in that they have two (inverting and noninverting) input terminals and one output. However, comparators are designed exclusively for switching, and they have slew rates much faster that those available with op-amps.

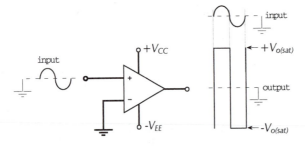

(a) Noninverting zero crossing detector

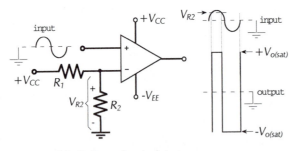

(b) Voltage level detector

Figure 14-31
Operation amplifiers can be connected to function as zero crossing detectors, or as voltage level detectors.

Zero Crossing Detector

The *zero crossing detector* circuit in Fig. 14-31(a) is seen to be simply an operational amplifier with the inverting input grounded and the signal applied to the noninverting input. When the input is above ground level the output is saturated at its positive

---

[1] David A. Bell, *Solid State Pulse Circuits,* 4th ed. (1997), p. 193.

maximum, and when the input is below ground the output is at its negative maximum level. This is illustrated by the input and output waveforms which show that the output voltage changes from one extreme to the other each time the input voltage crosses zero. The input waveform could have any shape (sinusoidal, pulse, ramp, etc.), and the output will always be a rectangular-type wave.

The actual input voltage that causes the output to switch is not precisely zero, but some very small voltage above or below zero, depending upon the op-amp open-loop gain.

If the op-amp non-inverting input is grounded and the signal is applied to the inverting input in Fig. 14-31(a), the output is negative when the input is above ground, and vice versa. Because of the waveform inversion, this circuit is often termed an *inverter*.

The circuit in Fig. 14-31(b) has the op-amp inverting input biased via a voltage divider (R_1 and R_2). The bias voltage level (V_{R2}) could be positive or negative. In the circuit illustrated, the bias voltage is a positive quantity. The waveforms show that the output voltage changes when the input voltage crosses the bias level. This circuit is appropriately named a *voltage level detector*.

Example 14-12

For the zero crossing detector in Fig. 14-32 determine the typical output voltage swing, and the typical input voltage level above and below ground at which the output switches. Also, calculate the rise time of the output voltage.

Solution

Eq. 14-20, $\Delta V \approx (V_{CC} - 1\text{ V}) - (V_{EE} + 1\text{ V}) \approx (15\text{ V} - 1\text{ V}) - (-15\text{ V} + 1\text{ V})$

$\approx 28\text{ V}$

From Appendix 1-13,

$A_v = 200\,000$ (typical), and $SR = 0.5$ V/μs

$$V_i = \frac{V_o}{A_v} = \frac{\pm 14\text{ V}}{200\,000}$$

$$= \pm 70\,\mu V$$

From Eq. 14-22,

$$\Delta t = \frac{\Delta V}{SR} = \frac{28\text{ V}}{0.5\text{ V/}\mu s}$$

$$= 56\,\mu s$$

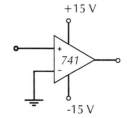

+15 V

741

-15 V

Figure 14-32
Zero crossing detector circuit for Ex.14-12.

Practise Problems

14-9.1   A *353* op-amp with a $\pm$18 V supply is used as a zero crossing detector. Determine the typical output voltage swing, and the typical input voltage level required to switch the output. Also, calculate the rise time of the output voltage. Appendix 1-15 shows a partial specification for the *353*.

14-10   *Schmitt Trigger Circuits*

Inverting Schmitt Trigger

A *Schmitt trigger* circuit is a fast-operating voltage level detector. When the input voltage arrives at a level determined by the circuit components, the output voltage switches rapidly between its maximum positive level and its maximum negative level.

An op-amp inverting Schmitt trigger circuit is shown in Fig. 14-33 together with input and output waveforms. At first glance the circuit looks like a noninverting amplifier. But note that (unlike a noninverting amplifier) the input voltage (V_i) is applied to the inverting input terminal, and the feedback voltage goes to the noninverting input. The waveforms show that the output switches rapidly from the positive saturation ($+V_{o(sat)}$) voltage to the negative saturation level ($-V_{o(sat)}$) when the input exceeds a certain positive level; the *upper trigger point* (*UTP*). Similarly, the output voltage switches from low to high when the input goes below a negative triggering point; the *lower trigger point* (*LTP*).

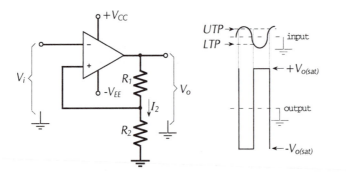

Figure 14-33
A Schmitt trigger circuit is a fast-operating voltage level detector that switches between negative and positive output voltage levels. In an inverting Schmitt trigger circuit the output goes negative when the input goes positive, and vice versa.

Note that after V_i has increased to the *UTP* and V_o has switched to $-V_{o(sat)}$, the output remains at $-V_{o(sat)}$ even when V_i falls below the *UTP*. Switch over from $-V_{o(sat)}$ to $+V_{o(sat)}$ does not occur until V_i = *LTP*. Similarly, after V_i has been reduced to the *LTP* and V_o has switched to $+V_{o(sat)}$, the output remains at $+V_{o(sat)}$ when V_i is increased above the *LTP*. Switch-over from $+V_{o(sat)}$ to $-V_{o(sat)}$ does not occur again until V_i = *UTP*.

Triggering Points

If the output voltage to the circuit in Fig. 14-33 is high, the voltage at the noninverting terminal is,

$$V_{R2} = \frac{+V_o \times R_2}{R_1 + R_2}$$

If the input voltage (at the inverting input terminal) is below V_{R2} (at the noninverting input), the output voltage is kept at its high positive level. For the output to switch to its low level, the input voltage must exceed V_{R2} by a very small amount (approximately 70 μV for a *741* op-amp). So, the *UTP* essentially equals V_{R2}.

$$UTP = \frac{+V_o \times R_2}{R_1 + R_2} \qquad \text{(14-23)}$$

When the output is negative, the *LTP* can be calculated as,

$$LTP = \frac{-V_o \times R_2}{R_1 + R_2} \qquad \text{(14-24)}$$

Input/Output Characteristic

A graph of output voltage (V_o) versus input voltage (V_i) can be plotted for an inverting Schmitt trigger circuit, as shown in Fig. 14-34. This is the circuit input/output characteristic. To understand the characteristic, consider the voltages at each of the numbered points on the graph:

- At *point 1*, $V_o = +V_{o(sat)}$, and $V_i < LTP$, thus keeping $V_o = +V_{o(sat)}$.
- From *point 1* through *points 2* and *3*, V_o remains at $+V_{o(sat)}$ as V_i is increased through the *LTP* and through zero, until $V_i = UTP$.
- From *point 3* to *point 4*, V_o switches rapidly from $+V_{o(sat)}$ to $-V_{o(sat)}$ when $V_i = UTP$.
- From *point 4* to *point 5*, V_o remains at $-V_{o(sat)}$ as V_i is increased above the *UTP*.
- From *point 5* through *points 4* and *6*, V_o remains at $-V_{o(sat)}$ as V_i is reduced through the *UTP* and through zero, until $V_i = LTP$.
- From *point 6* to *point 2*, V_o switches rapidly from $-V_{o(sat)}$ to $+V_{o(sat)}$ when $V_i = LTP$.
- From *point 2* to *point 1*, V_o remains at $+V_{o(sat)}$ as V_i is reduced below the *LTP*.

The difference between the *UTP* and the *LTP* is termed *hysteresis*. Some applications require a small amount of hysteresis, and for other applications a large amount of hysteresis is essential.

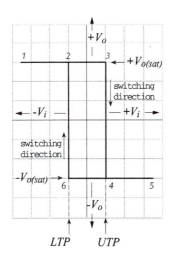

Figure 14-34
output/input characteristics for an inverting Schmitt trigger circuit.

Circuit Design

Design procedure for a Schmitt trigger circuit is similar to op-amp amplifier design. A voltage divider current (I_2 in Fig. 14-33) is selected much larger than the op-amp input bias current. The resistor values are then calculated as,

$$R_2 = \frac{UTP}{I_2} \qquad \text{(14-25)}$$

and,

$$R_1 = \frac{V_{o(sat)} - UTP}{I_2} \qquad \text{(14-26)}$$

Example 14-13

Using a *741* op-amp, calculate resistor values for the Schmitt trigger circuit in Fig. 14-35 to give triggering points of ± 5 V.

Solution

Select, $I_1 \approx 100\, I_B = 100 \times 500$ nA

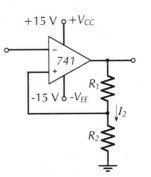

$$= 50 \, \mu A$$

Eq. 14-25, $R_2 = \dfrac{UTP}{I_2} = \dfrac{5 \text{ V}}{50 \, \mu A}$

$$= 100 \text{ k}\Omega \text{ (standard value)}$$

Eq. 14-26, $R_1 = \dfrac{V_{o(sat)} - UTP}{I_2} \approx \dfrac{(15 \text{ V} - 1 \text{ V}) - 5 \text{ V}}{50 \, \mu A}$

$$= 180 \text{ k}\Omega \text{ (standard value)}$$

Figure 14-35
Inverting Schmitt trigger circuit for Ex. 14-13.

Adjusting the Trigger Points

Many Schmitt trigger circuit applications require *UTP* and *LTP* levels that are not equal in magnitude. This is usually achieved by the use of diodes, as illustrated in Fig. 14-36.

The circuit shown in Fig. 14-36(a) simply has a diode (D_1) connected in series with resistor R_1. The diode is forward biased only when the op-amp output is a positive quantity. At this time, the *UTP* is V_{R2}, as before. When V_o is negative, D_1 is reverse biased, making I_2 equal zero. Consequently, there is no voltage drop across R_2, and so the noninverting terminal is grounded via R_2. This gives a zero level for the *LTP*. Thus, this circuit has a positive *UTP* and a zero voltage *LTP*.

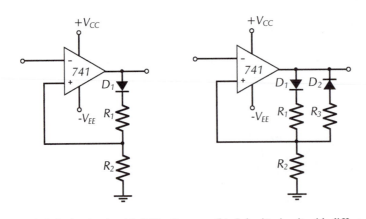

(a) Schmitt circuit with *LTP* = 0   (b) Schmitt circuit with different
                                                    UTP and *LTP* levels

Figure 14-36
Diodes can be used with inverting Schmitt trigger circuits to produce different levels of positive and negative triggering voltages.

Figure 14-36(b) shows a circuit with two different-level trigger points. When V_o is positive, D_1 is forward biased and D_2 is reversed, and the *UTP* is set by resistors R_1 and R_2. With V_o negative, D_2 is forward biased and D_1 is reversed. The *LTP* is now determined by the resistances of R_3 and R_2.

The diode forward voltage drop (V_F) must be accounted for when calculating the trigger points for both of the circuits in Fig. 14-36. This is done simply by replacing V_o with $(V_{o(sat)} - V_F)$ in Eqs. 14-25 and 14-26. Another important design consideration is that the voltage divider current (I_2) should normally be a minimum of 100 µA for satisfactory diode operation.

Noninverting Schmitt Trigger

A noninverting Schmitt trigger circuit is shown in Fig. 14-37. This circuit looks like an inverting amplifier, but note that (unlike an inverting amplifier) the inverting input is grounded and the noninverting input is connected to the junction of R_1 and R_2. The waveforms in Fig. 14-37 show that V_o switches rapidly from $-V_{o(sat)}$ to $+V_{o(sat)}$ when V_i arrives at the UTP, and that V_o switches back to $-V_{o(sat)}$ when V_i falls to the LTP.

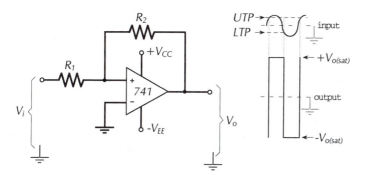

Figure 14-37
Noninverting Schmitt trigger circuit. The output goes positive when the input arrives at the UTP, and negative when V_i = LTP.

Consider the situation when V_i = UTP and V_o has just switched to $+V_{o(sat)}$. The voltage at the junction of R_1 and R_2 is pulled up far above the ground level voltage at the op-amp inverting input terminal. So, the positive voltage at the noninverting input keeps the output at its positive saturation level. To switch the output to $-V_{o(sat)}$, the voltage at the junction of R_1 and R_2 must be pulled down to the (ground level) voltage at the inverting input terminal. This occurs when V_i is at the LTP, [see Fig. 14-38(a)]. So, switching occurs when one end of R_1 (the right end) is at ground, and the other (left) end is at V_i = LTP; that is, when V_{R1} = LTP. Similarly, switching in the other direction occurs when V_{R1} = UTP, [Fig. 14-38(b)]. So, at the trigger points,

$$V_i = UTP = LTP = I_2 R_1$$

Figure 14-38(a) and (b) show that the output voltage is at one of its saturation levels at the instant of triggering. This means that one end of R_2 is at ground (left end), and the other (right) end is at $V_{o(sat)}$. So, $V_{R2} = V_{o(sat)}$ when triggering occurs.

giving,
$$I_2 = \frac{V_{o(sat)}}{R_2}$$

also,
$$UTP = I_2 R_1$$

so,
$$\boldsymbol{UTP = V_{o(sat)} \times \frac{R_1}{R_2}}$$ **(14-27)**

and, $$LTP = V_{o(sat)} \times \frac{R_1}{R_2}$$ (14-28)

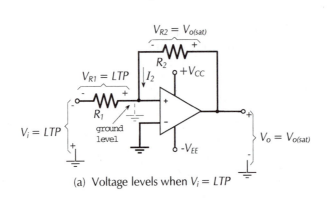

(a) Voltage levels when $V_i = LTP$

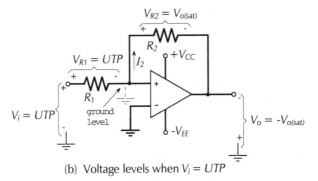

(b) Voltage levels when $V_i = UTP$

Figure 14-38
At the instant of switching in a noninverting Schmitt trigger circuit, the op-amp noninverting input terminal voltage must be at ground level.

Design procedure for a noninverting Schmitt trigger circuit is just as simple as for the inverting circuit. Voltage divider current I_2 is again selected much larger than the op-amp input bias current. Then the resistor values are,

$$R_1 = \frac{UTP}{I_2}$$ (14-29)

and, $$R_2 = \frac{V_{o(sat)}}{I_2}$$ (14-30)

Noninverting Schmitt trigger circuits can be designed for different upper and lower trigger point voltages by the use of diodes, as in the case of the inverting circuit. Figure 14-39(a) and (b) shows two possible circuits. The diode forward voltage drop (V_F) must be included in the *UTP* and *LTP* calculations.

Example 14-14
Determine the upper and lower trigger points for the noninverting Schmitt trigger circuit shown in Fig. 14-39(b). Assume that the op-amp used can be rail-to-rail operated.

Solution

With rail-to-rail operation,

$$V_{o(sat)} = \pm V_{CC} = \pm 15 \text{ V}$$

UTP calculation (including V_F):

From Eq. 14-30, $I_2 = \dfrac{V_{o(sat)} - V_F}{R_2} = \dfrac{15 \text{ V} - 0.7 \text{ V}}{150 \text{ k}\Omega}$

$$= 95.3 \text{ } \mu A$$

From Eq. 14-29, $UTP = I_2 R_1 = 95.3 \text{ } \mu A \times 27 \text{ k}\Omega$

$$= 2.57 \text{ V}$$

LTP calculation (including V_F):

From Eq. 14-30, $I_3 = \dfrac{V_{o(sat)} - V_F}{R_3} = \dfrac{15 \text{ V} - 0.7 \text{ V}}{120 \text{ k}\Omega}$

$$= 119 \text{ } \mu A$$

From Eq. 14-29, $LTP = I_3 R_1 = 119 \text{ } \mu A \times 27 \text{ k}\Omega$

$$= 3.2 \text{ V}$$

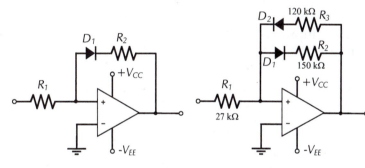

(a) Noninverting Schmitt circuit
with *LTP* = 0

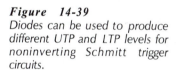

(b) Noninverting Schmitt circuit with
different *UTP* and LTP levels.

Figure 14-39
Diodes can be used to produce different UTP and LTP levels for noninverting Schmitt trigger circuits.

Practise Problems

14-10.1 Using a *741* op-amp with a ±12 V supply, design an inverting Schmitt trigger circuit [as in Fig. 14-36(b)] to have *UTP* = 3 V and *LTP* = -2 V.

14-10.2 Design a noninverting Schmitt trigger circuit to have approximately ±1 V trigger points. Use an *353* op-amp with a ±15 V supply.

Chapter-14 Review Questions

Section 14-1

14-1 Sketch the circuit symbol for an operational amplifier, identify all terminals, and briefly explain.

14-2   Sketch the input and output stages of a bipolar operational amplifier, explain the circuit operation, and discuss the input and output impedances.

14-3   Repeat Question 14-2 for a *BIFET* operational amplifier.

14-4   List the most important parameters for an operational amplifier, and state typical values for each parameter.

Section 14-2

14-5   Draw a complete circuit diagram to show an operational amplifier with a signal source capacitor-coupled to the noninverting input, and 100% feedback from the output to the inverting input terminal. Explain the requirement for all components.

14-6   Draw a complete circuit diagram for an operational amplifier with a plus/minus supply, its noninverting input terminal biased to a positive voltage level, and 100% feedback from the output to the inverting input terminal. Explain the requirement for all components, and discuss the component current levels.

14-7   An operational amplifier is to use a single-polarity supply, and have its noninverting input terminal biased to a positive voltage level. Also, 100% feedback is to be provided from the output to the inverting input terminal. Draw the circuit diagram, and explain the function of each component.

14-8   Explain output voltage offset nulling, and show how it can be accomplished in a *741* op-amp.

Section 14-3

14-9   Draw a circuit diagram for a direct-coupled voltage follower, identifying all terminals. Also, sketch a basic op-amp (internal) circuit with the external connection for voltage follower operation. Explain the circuit operation.

14-10 Discuss the performance of a voltage follower, and compare it to an emitter follower.

14-11 Sketch a capacitor-coupled voltage follower circuit, and explain the effects of capacitor-coupling on the circuit input impedance.

Section 14-4

14-12 Draw a circuit diagram for a direct-coupled noninverting amplifier. Identify all terminals and explain the circuit operation. Also, derive the closed-loop voltage gain equation, and write equations for the input and output impedances.

14-13 Sketch a basic operational amplifier (internal) circuit with the external connection for noninverting amplifier operation. Explain how the circuit amplifies a signal.

14-14 Sketch a capacitor-coupled noninverting amplifier circuit. Explain the circuit operation, and discuss the effects of capacitor-coupling on the circuit input impedance.

Section 14-5

14-15 Sketch a circuit diagram for a direct-coupled inverting amplifier. Identify all terminals and explain the circuit operation. Also, derive the closed-loop voltage gain equation, and write equations for the input and output impedances.

14-16 Sketch a circuit for capacitor-coupled inverting amplifier. Explain the amplifier operation, and discuss the effects of capacitor-coupling on the circuit input impedance.

Section 14-6

14-17 Sketch a three-input summing amplifier circuit and explain its operation.

14-18 For a three-input summing amplifier, derive an equation for the output voltage in terms of the input voltage and the circuit components.

Section 14-7

14-19 Sketch the circuit of an op-amp difference amplifier. Explain its operation, and derive an equation for the output voltage in terms of the input and the circuit components.

14-20 Identify the input resistances for the difference amplifier in Question 14-19. Also, discuss the effects of a common-mode input voltage, and show how common-mode voltages can be nulled.

Section 14-8

14-21 Draw the circuit diagram for an instrumentation amplifier, identify each section of the circuit, and explain its operation.

14-22 Derive an equation for the overall voltage gain of the instrumentation amplifier circuit drawn for Question 14-21. Discuss the effects of common mode input voltages.

Section 14-9

14-23 Sketch circuit diagrams for an op-amp used as (a) a zero-crossing detector, (b) as a voltage level detector. Show typical input and output waveforms and briefly explain. Define slew rate, and show its effect on an op-amp output waveform.

Section 14-10

14-24 Draw the circuit diagram of an op-amp inverting Schmitt trigger circuit. Sketch typical input and output waveforms, and explain the circuit operation. Write equations for the circuit upper and lower trigger points.

14-25 Sketch typical output/input characteristics for an inverting Schmitt trigger circuit. Identify the upper and lower trigger points, and explain the characteristics.

14-26 Draw circuit diagrams for inverting Schmitt trigger circuits with different levels of positive and negative trigger points. Explain.

14-27 Draw the circuit diagram of an op-amp noninverting Schmitt trigger circuit. Sketch typical input and output waveforms, and explain the circuit operation. Write equations for the circuit upper and lower trigger points.

14-28 Draw circuit diagrams for noninverting Schmitt trigger circuits with different levels of positive and negative trigger points. Explain.

Chapter-14 Problems

Section 14-1

14-1 Referring to the *741* operational amplifier specification in Appendix 1-13, determine typical values of the following parameters: input resistance, input bias current, output resistance, output short-circuit current, supply current, power dissipation.

14-2 Referring to Appendix 1-14, determine typical values of the following parameters for a *LM308* operational amplifier: input resistance, input bias current, supply current, power supply rejection ratio.

14-3 From the *LF353A* op-amp specification in Appendix 1-15, determine typical values of the following parameters: supply current, common-mode rejection ratio, output voltage swing, input offset voltage.

Section 14-2

14-4 A *741* op-amp with a ±9 V supply uses voltage divider bias to provide +3 V at the noninverting input terminal. There is also 100% feedback from the output to the inverting input. Determine suitable resistor values.

14-5 Determine resistor values for a *741* op-amp with a +18 V supply, +9 V voltage divider bias at the noninverting input terminal, and 100% feedback from the output to the inverting input.

14-6 Redesign the circuit in Problem 14-4 to use a *LM108* op-amp. Note that, because the *LM108* has a very low bias current, it should be treated like a *BIFET* op-amp.

14-7   Redesign the circuit in Problem 14-5 to use a *LF353* op-amp. Then, investigate the effect of substituting a *741* in place of the *LF353* without changing the resistor values.

Section 14-3

14-8   An operational amplifier with an open-loop of 200 000 is used as a voltage follower. Calculate the precise level of the output voltage if the input voltage is exactly 8 V.

14-9   Calculate the input and output resistance of a voltage follower circuit that uses an op-amp with the following parameters: $A_{v(OL)}$ = 100 000, r_i = 1 MΩ, r_o = 50 Ω.

14-10   Design a capacitor-coupled voltage follower to use a *741* op-amp. The load resistance is 2 kΩ, the signal source resistance is 600 Ω, and the lower cutoff frequency is to be 100 Hz.

14-11   A capacitor-coupled voltage follower uses an op-amp with a 1 μA input bias current. The load resistance is 1 kΩ, and the signal source resistance is 400 Ω. Select suitable component values for the circuit to give a 120 Hz lower cutoff frequency.

Section 14-4

14-12   Design a direct-coupled noninverting amplifier to produce a ±5 V output when the input is ±75 mV. Use a *741* op-amp.

14-13   A direct-coupled noninverting amplifier that uses an op-amp with $I_{B(max)}$ = 750 nA is to have a maximum output of ±10 V and a voltage gain of 120. Design the circuit.

14-14   A direct-coupled noninverting amplifier (as in Fig. 14-15) uses a *BIFET* op-amp and has the following resistor values: R_1 = 22 kΩ, R_2 = 1 MΩ, R_3 = 22 kΩ. Calculate the circuit voltage gain.

14-15   Redesign the circuit in Problem 14-12 to use an *LF353 BIFET* op-amp.

14-16   The noninverting amplifier designed for Problem 14-12 is to be modified to operate as a capacitor-coupled circuit with f_1 = 80 Hz, R_L = 100 kΩ, and r_s = 300 Ω. Calculate suitable capacitor values.

14-17   Calculate the input impedance of the direct-coupled circuit in Problem 14-12, and of the capacitor-coupled circuit in Problem 14-16.

14-18   A capacitor-coupled noninverting amplifier is to be designed to produce a 7 V output from a 70 mV input. The supply voltage is ±18 V, the lower cutoff frequency is 60 Hz, the load is 82 kΩ, and the signal source resistance is 600 Ω. Design the circuit to have an input impedance of 12 kΩ.

14-19   Design a direct-coupled noninverting amplifier using a *BIFET* op-amp with a 100 000 open-loop voltage gain. The closed-loop gain is to be 45, and the maximum output is to be ±2 V.

14-20   Modify the circuit for Problem 14-19 to make it capacitor-coupled with $R_L = 2$ kΩ and $f_1 = 150$ Hz.

Section 14-5

14-21   A direct-coupled inverting amplifier using a *741* op-amp is to have a 45 mV input signal and a voltage gain of 200. Calculate suitable resistor values and determine the circuit input impedance.

14-22   A direct-coupled inverting amplifier is to have a 1 kΩ input impedance, and is to produce a ±3.3 V output from a ±100 mV input. Determine suitable resistor values.

14-23   Redesign the circuit in Problem 14-21 to use a *BIFET* op-amp.

14-24   The inverting amplifier designed for Problem 14-21 is to be modified to operate as a capacitor-coupled circuit with $f_1 =$ 50 Hz, $R_L = 150$ kΩ, and $r_s = 300$ Ω. Determine suitable capacitor values.

14-25   A capacitor-coupled inverting amplifier is to have a 680 Ω input impedance, and a voltage gain of 49. The load resistance is 12 kΩ, the lower cutoff frequency is to be 40 Hz. Determine suitable component values.

14-26   The inverting amplifier designed for Problem 14-22 is to function as a capacitor-coupled circuit with $f_1 = 30$ Hz, $R_L =$ 33 kΩ, and $r_s = 600$ Ω. Determine suitable capacitor values.

14-27   A capacitor-coupled inverting amplifier is to use a *741* op-amp with a +30 V single-polarity supply, and have $R_L = 5$ kΩ, $A_{CL} = 50$, $V_o = ±3$ V, and $f_1 = 75$ Hz. Sketch the circuit and determine suitable component values.

Section 14-6

14-28   Design a two-input summing amplifier for a voltage gain of 10 and input levels of 100 mV. Use an operational amplifier with $I_{B(max)} = 1$ μA and $A_{v(OL)} = 300\ 000$.

14-29   The circuit designed for Problem 14-28 has inputs of 60 mV and 80 mV. Analyze the circuit to determine all voltage and current levels.

14-30   A three-input summing amplifier using a *741* op-amp is to be designed for a voltage gain of 1. The input voltages are approximately 1 V. Calculate suitable component values.

Section 14-7

14-31   The difference amplifier circuit in Fig. 14-27 has the following components: $R_1 = 15$ kΩ, $R_2 = 33$ kΩ, $R_3 = 6.8$ kΩ,

R_4 = (12 kΩ +2.7 kΩ + 500 Ω variable). The input voltages are: V_{i1} (to R_1) = 1.47 V and V_{i2} (to R_3) = 1.23 V. Calculate V_o when R_4 is adjusted to 14.96 kΩ.

14-32 A difference amplifier is to be designed for a voltage gain of 100 and input levels of 100 mV. Using an operational amplifier with $I_{B(max)}$ = 1 µA and $A_{v(OL)}$ = 300 000, determine suitable resistor values.

14-33 If the difference amplifier in Problem 14-31 has a 500 mV common-mode input, calculate the common-mode output voltage; (a) when R_4 is set to its maximum resistance, (b) when R_4 is at its minimum value, (c) when R_4 = 14.96 kΩ.

Section 14-8

14-34 Design an instrumentation amplifier to produce a 3 V output when the inputs are +450 mV and +150 mV. Use *741* op-amps with a ±12 V supply.

14-35 Analyze the circuit designed for Problem 14-34 to determine all current and voltage levels.

Section 14-9

14-36 A zero-crossing detector uses an op-amp with V_{CC} = ±9 V, SR = 3 V/µs, and $A_{v(OL)}$ = 100 000. Calculate the typical output voltage swing, the output rise time, and the input voltage level required to switch the output.

Section 14-10

14-37 Using a *741* op-amp with V_{CC} = ±18 V, design an inverting Schmitt trigger circuit to have ±2 V triggering points.

14-38 Plot the output/input characteristics for the circuit in Problem 14-37.

14-39 Design a noninverting Schmitt trigger circuit to fulfil the specification in Problem 14-37.

14-40 An inverting Schmitt trigger circuit is to be designed to have *UTP* = 0 V and *LTP* = -1.5 V. Sketch the circuit and determine suitable component values if a *741* op-amp is used with a ±15 V supply.

14-41 An noninverting Schmitt trigger circuit is to be designed to have *UTP* = 1 V and *LTP* = -0.5 V. Sketch the circuit and determine suitable component values if a *LF353* op-amp is used with a ±12 V supply.

14-42 An noninverting Schmitt trigger circuit using a *741* op-amp with a ±10 V supply is to have trigger points adjustable from ±0.5 V to ±0.75 V. Sketch the circuit and determine suitable component values.

Practise Problem Answers

14-1.1	300 000, 0.8 nA, 0.7 mV, 70 MΩ
14-1.2	100 000, 50 pA, 5 mV, 10^{12} Ω
14-2.1	220 kΩ, 220 kΩ, 100 kΩ
14-2.2	180 kΩ, (220 kΩ + 33 kΩ), 100 kΩ
14-3.1	1.5 μF, 0.33 μF
14-3.2	0.68 μF, 1 μF
14-3.3	60 mV
14-4.1	1.8 kΩ, (68 kΩ + 2.2 kΩ), 1.8 kΩ
14-4.2	4.7 kΩ, (82 kΩ + 6.8 kΩ), 4.7 kΩ, 18 μF, 10 μF
14-4.3	180 kΩ, 1 MΩ, (220 kΩ + 33 kΩ)
14-5.1	5.6 kΩ, (82 kΩ + 2.2 kΩ), 5.6 kΩ
14-5.2	8 μF, 15 μF
14-5.3	(33 kΩ + 33 kΩ), 1 MΩ, 68 kΩ
14-6.1	4.7 kΩ, 4.7 kΩ, 47 kΩ, 2.2 kΩ
14-6.2	-10.6 V, 2.27 μA, 4.55 μA, 3.79 μA, 10.6 μA
14-7.1	+7.5 V, -33.9 μA, -33.9 μA, 75.76 μA, 1.36 V
14-7.2	25 x 10^{-3}
14-7.3	(100 kΩ + 68 kΩ), 1 MΩ, 27 kΩ, (100 kΩ + 68 kΩ)
14-8.1	47 kΩ, 1 kΩ, 47 kΩ, 47 kΩ, 47 kΩ, 47 kΩ, 47 kΩ
14-9.1	33 V, ±330 μV, 2.5 μS
14-10.1	(68 kΩ + 27 kΩ), 39 kΩ, (150 kΩ + 12 kΩ)
14-10.2	(56 kΩ + 18 kΩ), 1 MΩ

Chapter *15*

Operational Amplifier Frequency Response and Compensation

Chapter Contents

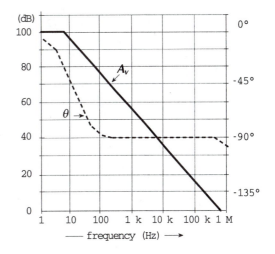

Objectives

You will be able to:

1 *Show how feedback can produce instability in op-amp circuits.*

2 *Sketch and explain typical gain/ frequency and phase/frequency response graphs for uncompensated and compensated operational amplifiers.*

3 *Define: loop gain, loop phase shift, phase margin.*

4 *Discuss compensation methods for stabilizing op-amp circuits, and calculate component values for compensating circuits.*

5 *Explain how the bandwidth of an op-amp circuits is affected by closed-loop gain.*

6 *Define: Gain-bandwidth product, slew rate, full-power bandwidth.*

7 *Determine the bandwidths of various op-amp circuits using; frequency response graphs, gain-bandwidth product, and slew rate.*

8 *Explain how stray and load capacitance can affect op-amp circuit stability, sketch appropriate compensating circuits, and calculate suitable component values.*

9 *List precautions that should be observed to ensure op-amp circuit stability.*

Introduction

Signals applied to operational amplifiers experience phase shifts as they pass from input to output. These phase shifts are greatest at high frequencies, and at some particular frequency the total loop phase shift (from the inverting input terminal to the output and back to the input via the feedback network) can add up to 360°. When this occurs, the amplifier circuit can go into a state of unwanted oscillation. The conditions that produce oscillation are that the loop voltage gain be greater than or equal to unity when the loop phase shift approaches 360°. Measures taken to combat instability include the use of capacitors and resistors to reduce the total phase shift. Most operational amplifiers have compensating components included in the circuitry to ensure stability.

15-1 Operational Amplifier Circuit Stability

Loop Gain and Loop Phase Shift

Consider the inverting amplifier circuit and waveforms in Fig. 15-1(a). The signal voltage voltage (v_s) is amplified by a factor R_2/R_1, and phase shifted through -180°. The circuit is redrawn in Fig. 15-1(b) to illustrate the fact that the output voltage (v_o) is divided by the feedback network to produce the feedback voltage (v).

For an *ac* voltage (v) at the op-amp inverting input terminal [in Fig. 15-1(b)], the amplified output is $v_o = A_v v$, as shown. The output is divided by the feedback factor [$B = R_1/(R_1 + R_2)$], and fed back to the input. An additional -180° of phase shift can occur within the op-amp at high frequencies, and this causes v to be in-phase with v_o, as illustrated. Thus, the feed back voltage can be exactly equal to and in phase with the voltage (v) at the inverting input. In this case, the circuit is supplying its own *ac* input voltage, and a state of continuous oscillation exists.

Because of the feedback network, high-frequency oscillations can occur in many operational amplifier circuits, and when this happens the circuit is termed *unstable*. Measures taken to combat circuit instability are referred to as *frequency compensation*.

Two conditions normally have to be fulfilled for a circuit to oscillate; the *loop gain* must be equal to or greater than 1, and the *loop phase shift* should equal 360°. The loop gain is the voltage gain around the loop from the inverting input terminal to the amplifier output, and back to the input via the feedback network. The loop phase shift is the total phase shift around the loop from the inverting input terminal to the amplifier output, and back to the input via the feedback network.

The gain from the inverting input terminal to the output is the op-amp open-loop gain (A_v). For the feedback network, the gain from the amplifier output back to the input is actually an attenuation. So,

loop gain = (amplifier gain) x (feedback network attenuation)
            = $A_v B$

Assuming that the feedback network is purely resistive, it adds nothing to the loop phase shift. The loop phase shift is essentially the amplifier phase shift. The phase shift from the inverting input terminal to the output is normally -180°. (The output goes negative when the input goes positive, and vice versa.) At high frequencies there is additional phase shift caused by circuit capacitances, and the total can approach -360°. When this occurs, the circuit is virtually certain to oscillate. Most currently-available operational amplifiers have internal *compensating components* to prevent oscillations. In some cases, compensating components must be connected externally to stabilize a circuit.

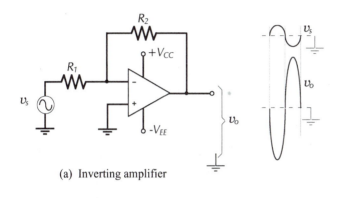

(a) Inverting amplifier

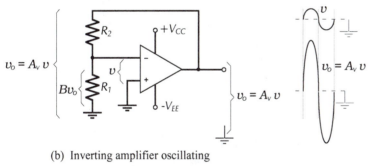

(b) Inverting amplifier oscillating

Figure 15-1
Because an inverting amplifier consists of an operational amplifier and a feedback network, the circuit can supply its own ac input (v derived from v_o), and a state of continuous oscillations can occur.

Uncompensated Gain and Phase Response

A straight-line approximation of the gain/frequency response graph for a typical operational amplifier *without any compensating components* is shown in Fig. 15-2. Note that the overall open-loop voltage gain (A_v) initially falls off (from its 100 dB level) at 6 dB/octave (-20 dB/decade) from f_{p1} (*pole frequency #1*). From f_{p2}, the rate of decline is 12 dB/octave (-40 dB/decade), and from f_{p3}, the fall-off rate of A_v is 18 dB/octave (-60 dB/decade).

The phase/frequency response graph in Fig. 15-2 shows that the phase shift (θ) is approximately -45° at f_{p1}, -135° at f_{p2}, and -225° at f_{p3}. This open-loop phase shift is in addition to the -180° phase shift that normally occurs from the op-amp inverting input terminal to the output. Thus, the total loop phase shift (ϕ_L) at f_{p1} is (-45° -180°) = -225°; at f_{p2}, ϕ_L = (-135° -180°) = -315°; and at f_{p3}, ϕ_L = (-225° -180°) = -405°.

As already discussed, oscillations occur when the loop gain is equals or exceeds 1 and the loop phase shift is 360°. In fact, the phase shift does not have to be exactly 360° for oscillation to occur. A phase shift of 330° at $A_vB = 1$ makes the circuit unstable. To avoid oscillations, the total loop phase shift must not be greater than 315° when $A_vB = 1$. The difference between 360° and the actual loop phase shift at $A_vB = 1$ is referred to as the *phase margin* (ϕ_m). Thus, for circuit stability, the phase margin should be a minimum of,

$$\phi_m = 360° - 315° = 45°$$

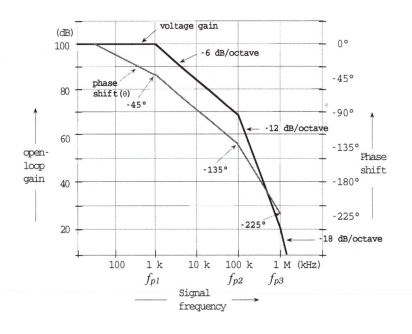

Figure 15-2
The three stages of an op-amp (internal) circuit each has its own gain/frequency response with a 6 dB/octave fall-off, and its own phases shift/ frequency response with a maximum phase shift of 90°. These responses combine to give the overall op-amp A_v/f and θ/f responses.

Compensated Op-amp Gain and Phase Response

The open-loop gain/frequency and phase/frequency responses for two internally compensated operational amplifiers are shown in Figs. 15-3 and 15-4. The *741* frequency response graphs in Fig. 15-3 shows that the gain starts at 100 dB and falls by 20 dB/decade over most of its frequency range. The phase shift remains -90° or less for most of the frequency range. The open-loop gain falls off to 1 (0 dB) at a frequency of approximately 800 kHz. The *741* is known as a general purpose operational amplifier for use in relatively low frequency applications.

The *AD843* frequency response in Fig. 15-4 shows an open-loop gain of 90 dB at low frequencies, falling off at 20 dB per decade to 34 MHz at $A_v = 1$. Instead of the open-loop phase shift, the phase margin is plotted versus frequency. The phase margin is close to 90° over much of the frequency range, starts to become smaller around 3.4 MHz, and falls to approximately 40° at $f = 34$ MHz.

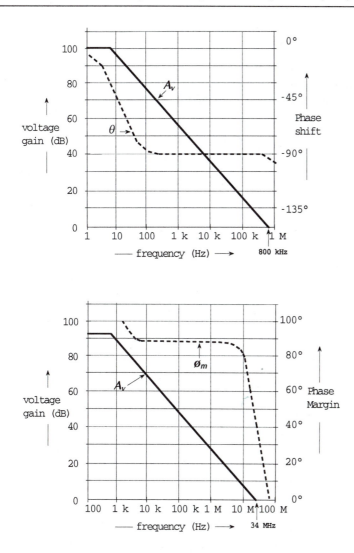

Figure 15-3
Approximate gain/frequency and phase/frequency responses for a 741 op-amp.

Figure 15-4
Approximate gain/frequency and phase-margin/frequency response for an AD843 op-amp.

Amplifier Stability and Gain

From Eq. 13-3, the overall voltage gain of an amplifier with negative feedback is,

$$A_{CL} \approx \frac{1}{B}$$

and the loop gain is,

$$A_v B \approx \frac{A_v}{A_{CL}}$$

So, the loop gain $(A_v B)$ equals 1 when

$$A_{CL} \approx A_v$$

This is one of the conditions required for circuit oscillation. To determine if oscillation will occur in a given circuit, it is necessary to first find the frequency at which $A_{CL} \approx A_v$, then determine the op-amp phase margin at that frequency.

Example 15-1

The inverting amplifier in Fig. 15-5 is to be investigated for stability. Determine the frequency at which the loop gain equals 1 and estimate the phase margin if the operational amplifier is: (a) one with the gain/frequency characteristics in Fig. 15-2, (b) a *741*, (c) an *AD843*.

Solution

(a) Refer to the A_v/f and θ/f graphs reproduced in Fig. 15-6 (from Fig. 15-2).

$$A_{CL} = \frac{R_2}{R_1} = \frac{560 \text{ k}\Omega}{1.8 \text{ k}\Omega}$$

$$= 311$$

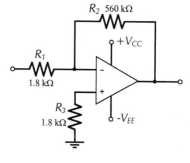

or, $A_{CL} = 20 \log 311$

$$\approx 50 \text{ dB}$$

Figure 15-5
Inverting amplifier circuit for Example 15-1.

Draw a horizontal line on the frequency response graph at $A_v = A_{CL} = 50$ dB, (Fig. 15-6). Draw a vertical line where the horizontal line intersects the A_v/f graph. The frequency at this point is identified as f_2.

$$f_2 \approx 150 \text{ kHz (logarithmic scale)}$$

From the θ/f graph, the op-amp phase shift at f_2 is,

$$\theta \approx -165°$$

The loop phase shift is,

$$\emptyset_L = \theta - 180° = -165° - 180°$$

$$= -345°$$

and, $\emptyset_m = 360° - \emptyset_L = 360° - 345°$

$$= 15°$$

Because the phase margin is less than 45°, the circuit is likely to be unstable.

(b) For a *741*

A horizontal line at 50 dB on the frequency response in Fig. 15-3 gives,

$$f_2 \approx 1.5 \text{ kHz and } \theta \approx -90°$$

$$\emptyset_L = -90° - 180°$$

$$= -270°$$

and, $\emptyset_m = 360° - 270°$

$$= 90° \text{ (stable circuit)}$$

(c) For an *AD843*

A horizontal line at 50 dB on the frequency response in Fig. 15-4 gives,

$$f_2 \approx 90 \text{ kHz}$$

and, $\emptyset_m \approx 90°$ (stable circuit)

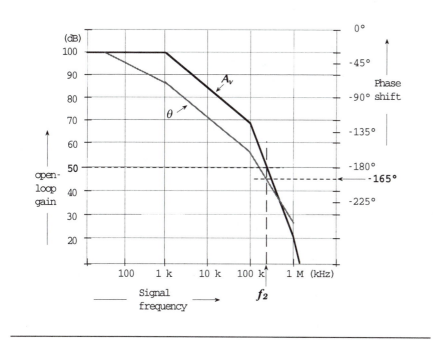

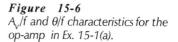

Figure 15-6
A$_v$/f and θ/f characteristics for the op-amp in Ex. 15-1(a).

A circuit with the frequency response in Fig. 15-6 and with A_{CL} = 50 dB was shown to be unstable. If the amplifier had A_{CL} = 70 dB, reconsideration shows that it is stable. That is, an amplifier with a high closed-loop gain is more likely to be stable than one with the lower gain. Low gain amplifiers are more difficult to stabilize than high gain circuits. The voltage follower (with a closed-loop gain of 1) can be one of the most difficult circuits to stabilize.

Some internally compensated op-amps are specified as being stable to closed-loop gains as low as 5. In this case, external compensating components must be used with lower gain circuits.

Practise Problems

15 -1.1 Investigate the stability of an inverting amplifier with a closed-loop gain of 60 dB if the operational amplifier is; (a) one with the gain/frequency characteristics in Fig. 15-2, (b) a *741*, (c) a *AD843*.

15-2 Frequency Compensation Methods

Phase-Lag and Phase-Lead Compensation

Lag compensation and *lead compensation* are two methods often employed to stabilize op-amp circuits. The phase-lag network in Fig. 15-7(a) introduces additional phase lag at some low frequency where the op-amp phase shift is still so small that additional phase lag has no effect. It can be shown that at frequencies where $X_{C1} \gg R_2$, the voltage v_2 lags v_1 by as much as 90°. At higher frequencies where $X_{C1} \ll R_2$ no significant phase lag occurs, and the lag network merely introduces some attenuation. The effect of this attenuation is that the A_v/f graph is moved to the left, as illustrated in Fig. 15-7(b). Thus, the frequency (f_{x1}) at which $A_vB = 1$ [for a given closed-loop gain (A_{CL})] is moved to a lower frequency (f_{x2}), as shown. Because f_{x2} is less than f_{x1}, the phase shift at f_{x2} is less than that at f_{x1}, and the circuit is likely to be stable.

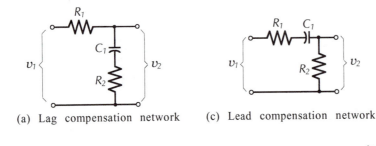

(a) Lag compensation network   (c) Lead compensation network

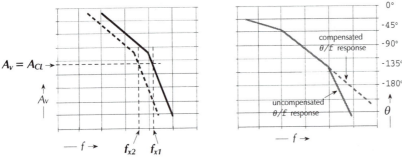

(b) Effect of lag compensation   (d) Effect of lead compensation

Figure 15-7
A phase-lag network reduces an amplifier open-loop gain, so that the phase shift where $A_vB = 1$ is too small for instability. A phase-lead network cancels phase lag.

The network in Fig 15-7(c) introduces a phase lead. In this network, when $X_{C1} \gg R_1$, the voltage v_2 leads v_1. This phase lead cancels some of the unwanted phase lag in the operational amplifier θ/f graph, [see Fig. 15-7(d)], thus rendering the circuit more stable. Phase-lag and phase-lead networks are both used internally to compensate op-amp circuits. Both types of circuit can also be used externally.

Manufacturer's Recommended Compensation
Most currently-available operational amplifiers contain internal compensating components, and do not require additional external components. Some have internal compensating resistors, and need only a capacitor connected externally to complete a compensating network. For those that require compensation, *IC* manufacturers list recommended component values and connection methods on the op-amp data sheet. An example of this is illustrated in Fig. 15-8 for the *LM108*.

When selecting standard value compensating capacitors the next larger values should be used. This is termed *over-compensation* and it results in better amplifier stability, but it also produces a smaller circuit bandwidth.

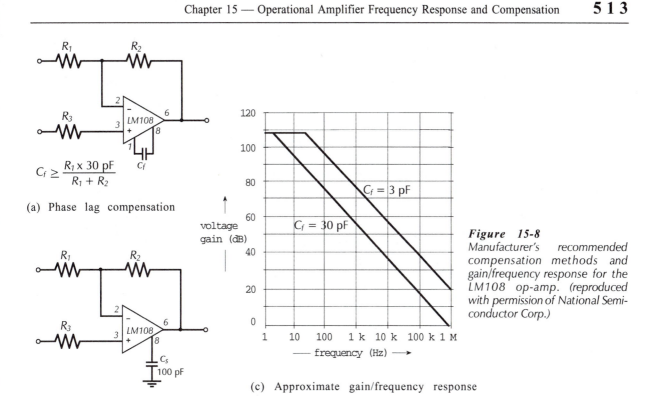

$$C_f \geq \frac{R_1 \times 30 \text{ pF}}{R_1 + R_2}$$

(a) Phase lag compensation

(b) Alternate phase lag compensation

(c) Approximate gain/frequency response

Figure 15-8
Manufacturer's recommended compensation methods and gain/frequency response for the LM108 op-amp. (reproduced with permission of National Semiconductor Corp.)

Example 15-2

The inverting amplifier in Fig. 15-9 is required to amplify a 200 mV input by a factor of 4.5. Determine suitable component values.

Solution

Because the *LM108* has a very low input bias current (see Appendix 1-14), it should be treated as a *BIFET* op-amp.

Select $R_2 = 1 \text{ M}\Omega$

$$R_1 = \frac{R_2}{A_{CL}} = \frac{1 \text{ M}\Omega}{4.5}$$

$= 222 \text{ k}\Omega$ (use 220 kΩ standard value)

$$R_3 = R_1 \| R_2 = 220 \text{ k}\Omega \| 1 \text{ M}\Omega$$

$= 180 \text{ k}\Omega$ (standard value)

From Fig. 15-8,

$$C_f = \frac{R_1 \times 30 \text{ pF}}{R_1 + R_2} = \frac{220 \text{ k}\Omega \times 30 \text{ pF}}{220 \text{ k}\Omega + 1 \text{ M}\Omega}$$

$= 5.4 \text{ pF}$ (use 10 pF standard value for over-compensation)

Connect C_f between terminals 1 and 8, as shown in Fig. 15-9.

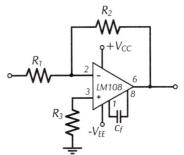

Figure 15-9
Op-amp circuit for Ex. 15-2.

Miller-Effect Compensation

Miller effect (discussed in Section 8-3) involves connecting a capacitor between the output and input terminals of an inverting amplifier. Miller-effect compensation of an op-amp circuit is very simple, and it is often the only external method available for stabilizing a circuit where the op-amp is internally compensated. A capacitor (C_f) is connected across the feedback resistor, as shown in Fig. 15-10(a) and (b). The capacitor value is calculated to have an impedance equal the feedback resistor value at the desired signal cutoff frequency (f_2).

$$X_{Cf} = R_f \text{ at } f_2 \qquad \qquad (15\text{-}1)$$

This reduces the closed-loop by 3 dB at the selected frequency. So long as the op-amp is stable at this frequency, the circuit will not oscillate. The op-amp used should have an upper cutoff frequency much higher than f_2.

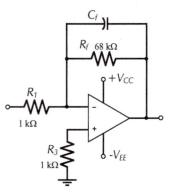

(a) Inverting amplifier with Miller-effect compensation

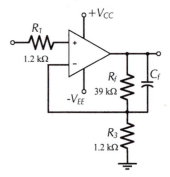

(b) Noninverting amplifier with Miller-effect compensation

Figure 15-10
Miller-effect frequency compensation for amplifier circuits.

Example 15-3

Calculate a suitable Miller-effect capacitor to stabilize the circuit in Fig. 15-10(a) at $f_2 = 35$ kHz.

Solution

From Eq. 15-1, $C_f = \dfrac{1}{2\pi f_2 R_f} = \dfrac{1}{2\pi \times 35 \text{ kHz} \times 68 \text{ k}\Omega}$

$$\approx 67 \text{ pF (use 68 pF standard value)}$$

Practise Problems

15-2.1 The components of the lag and lead compensation network in Fig. 15-7 are: $R_1 = 6.8$ kΩ, $R_2 = 390\,\Omega$, and $C_1 = 500$ pF. Calculate the approximately phase lag and phase lead at a frequency of 50 kHz.

15-2.2 Calculate a suitable Miller-effect capacitor to stabilize the circuit in Fig. 15-10(b) at $f_2 = 50$ kHz.

15-3 Op-amp Circuit Bandwidth and Slew Rate

Low Cutoff Frequency

Operational amplifiers are direct-coupled internally, so where they are employed in direct-coupled applications, the circuit lower cutoff frequency (f_1) is zero. In capacitor-coupled circuits, the lower cutoff frequency is determined by the selection of coupling capacitors. The circuit high cutoff frequency (f_2) is, of course, dependent on the frequency response of the operational amplifier.

High Cutoff Frequency

In Section 13-7 it is shown that for a negative feedback amplifier, the high cutoff frequency occurs when the amplifier open-loop gain approximately equals the circuit closed-loop gain:

Eq. 13-26, $A_v \approx A_{CL}$

So, the circuit high cutoff frequency (f_2) can be found simply by drawing a horizontal line at $A_v \approx A_{CL}$ on the op-amp open-loop gain/ frequency response graph. Because the op-amp low cutoff frequency is zero (as explained above), The circuit bandwidth is,

$$BW = f_2 - f_1$$
$$= f_2$$

Consequently, the op-amp high cutoff frequency is often referred to as the circuit bandwidth.

The frequency response graphs published on manufacturer's data sheets are typical for each particular type of operational amplifier. Like all typical device characteristics, the precise frequency response differs from one op-amp to another. All frequencies derived from the response graphs should be taken as typical quantities. The process of determining circuit cutoff frequency from the op-amp frequency response graph is demonstrated in Example 15-4.

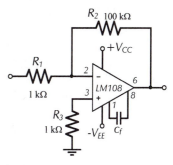

Figure 15-11
Amplifier circuit for Example 15-4.

Example 15-4

Determine the typical upper cutoff frequency for the inverting amplifier in Fig. 15-11 when the compensating capacitor (C_f) value is; (a) 30 pF, (b) 3 pF. The A_v/f graph for the *LM108* is shown in Fig. 15-12.

Solution

$$A_{CL} = \frac{R_2}{R_1} = \frac{100 \text{ k}\Omega}{1 \text{ k}\Omega}$$

$$= 100 = 40 \text{ dB}$$

f_2 occurs at $A_v = A_{CL} = 40$ dB

(a) For $C_f = 30$ pF:
Draw a horizontal line on the A_v/f graph at $A_v = 40$ dB. Where the line cuts the A_v/f characteristic for $C_f = 30$ pF read,

$$f_2 \approx 8 \text{ kHz}$$

(b) For $C_f = 3$ pF:
Where the $A_v = 40$ dB line cuts the A_v/f characteristic for $C_f = 3$ pF read,

$$f_2 \approx 80 \text{ kHz}$$

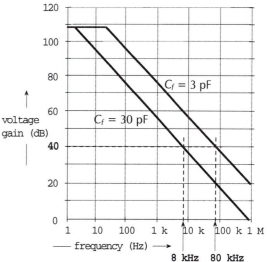

Figure 15-12
Straight line approximation of LM108 gain/frequency response.

Gain-Bandwidth Product

The *gain-bandwidth produce (GBW)*, or *unity-gain bandwidth*, of an operational amplifier is the open-loop gain at a given frequency multiplied by the frequency. Referring to the A_v/f response for the *741* reproduced in Fig. 15-13, it is seen that at $A_{v(a)} = 10^4$, the frequency is $f_{(a)} \approx 80$ Hz. Thus,

$$GBW = A_{v(a)} \times f_{(a)} = 10^4 \times 80 \text{ Hz}$$
$$= 8 \times 10^5$$

Similarly, at $A_{v(b)} = 10$, $f_{(b)} \approx 80$ kHz, again giving $GBW = 8 \times 10^5$. Also, at $A_{v(c)} = 1$, $f_{(c)} \approx 800$ kHz, once more giving $GBW = 8 \times 10^5$. This last determination explains the term *unity-gain bandwidth*, because the GBW is simply equal to the frequency at which $A_v = 1$.

Because the high cutoff frequency for an op-amp circuit occurs when the closed-loop gain equals the open-loop gain, the circuit upper cutoff frequency can be calculated by dividing the gain-bandwidth product by the closed-loop gain;

$$f_2 = \frac{GBW}{A_{CL}} \qquad\qquad \textbf{(15-2)}$$

It is important to note that Eq. 15-2 applies only to operational amplifiers that have a gain/frequency response that falls off to the unity-gain frequency at 20 dB/decade. Where the A_v/f response falls off at some other rate, Eq. 15-2 cannot be used.

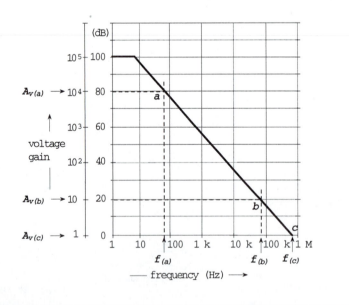

Figure 15-13
The gain-bandwidth product (GBW) for an operational amplifier can be used to determine the cutoff frequency for any given closed-loop gain.

Example 15-5

Using the gain-bandwidth product, determine the cutoff frequencies for the circuit in Ex. 15-4 (reproduced in Fig. 15-14), when the compensating capacitor is (a) $C_f = 30$ pF, (b) $C_f = 3$ pF.

Solution

(a) For $C_f = 30$ pF:

Referring to the *LM108* A_v/f graph for $C_f = 30$ pF in Fig. 15-12,

$$GBW = f \text{ at } A_v = 1$$
$$\approx 800 \text{ kHz}$$

Eq. 15-2, $$f_2 = \frac{GBW}{A_{CL}} = \frac{800 \text{ kHz}}{100}$$
$$= 8 \text{ kHz}$$

(b) For $C_f = 3$ pF:

Referring to the *LM108* A_v/f graph for $C_f = 3$ pF in Fig. 15-12,

at $A_v = 20$ dB $= 10$, $f \approx 800$ kHz

$$GBW = f \times A_v = 800 \text{ kHz} \times 10$$
$$= 8 \text{ MHz}$$

Eq. 15-2, $$f_2 = \frac{GBW}{A_{CL}} = \frac{8 \text{ MHz}}{100}$$
$$= 80 \text{ kHz}$$

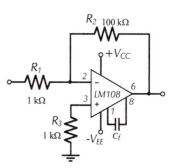

Figure   15-14
Amplifier circuit for Example 15-5.

Full-Power BW and Slew Rate

The A_v/f response graphs, upper cutoff frequencies, and *GBW* specified on op-amp data sheets normally refer to the operational amplifier performance as a small-signal amplifier. In this case, the measurements are usually made *for output amplitudes not exceeding 100 mV peak-to-peak*. Where an amplifier circuit has to produce a large output voltage, the op-amp *full-power bandwidth (f_p)* must be used. The *AD843* operational amplifier, for example, is specified as having a typical unity gain bandwidth of 34 MHz for an output amplitude of 90 mV p-to-p, and a typical full power bandwidth of 3.9 MHz when the output amplitude is 20 V p-to-p.

The op-amp slew rate (*SR*) (see Section 14-9) can be used to calculate the full-power bandwidth for a given output amplitude. For a sinusoidal voltage waveform, the fastest rate-of-change of voltage occurs at the point where the waveform crosses from its negative half-cycle to its positive half-cycle, and vice versa. This is illustrated in Fig. 15-15(a). It can be shown that the voltage rate-of-change at this point is,

$$\Delta V/\Delta t = 2\pi f V_p \text{ (volts/second)}$$

The maximum rate-of-change of the waveform is limited by the maximum slew rate of the op-amp used. Where the waveform amplitude or frequency is higher than the limits imposed by the slew rate, distortion will occur as illustrated in Fig. 15-15(b).

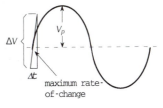

(a) Sine wave maximum
rate-of-change

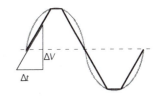

(b) Sine wave distortion
caused by the slew rate

Figure   15-15
The op-amp slew-rate limits the upper cutoff frequency of an op-amp circuit, and limits the output amplitude at a given frequency.

The SR can be equated to the sine wave rate-of-change,

$$SR = 2\pi f_p V_p \qquad \qquad \textbf{(15-3)}$$

where f_p is the slew-rate limited frequency, or full-power bandwidth, and V_p is the peak level of the circuit output voltage. Equation 15-3 can be used to determine the full-power bandwidth of an op-amp circuit for a given output voltage amplitude. Sometimes Eq. 15-3 gives an f_p value greater than that determined from the A_v/f graph or the GBW product. In these cases, the circuit bandwidth is still dictated by the A_v/f graph or the GBW product.

Example 15-6

(a) Calculate the full-power bandwidth for an *AD843* op-amp circuit (Fig. 15-16), given a 1 V peak input and op-amp slew rate of 250 V/μs.

(b) Determine the maximum peak output voltage obtainable from a *741* op-amp circuit with a 100 kHz signal frequency. ($SR = 0.5$ V/μs for a 741.)

Solution

(a) For the *AD843*:

$$V_{o(p)} = \frac{R_2 + R_3}{R_3} \times V_{i(p)} = \frac{39\ k\Omega + 4.7\ k\Omega}{4.7\ k\Omega} \times 1\ V$$

$$= 9.3\ V$$

From Eq. 15-3, $\qquad f_p = \dfrac{SR}{2\pi V_p} = \dfrac{250\ V/\mu s}{2\pi \times 9.3\ V}$

$$\approx 4.2\ MHz$$

(b) For a *741*:

From Eq. 15-3, $\qquad V_p = \dfrac{SR}{2\pi f_p} = \dfrac{0.5\ V/\mu s}{2\pi \times 100\ kHz}$

$$= 0.79\ V$$

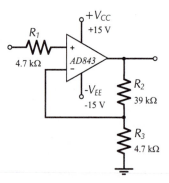

Figure 15-16
Amplifier circuit for Example 15-6.

Practise Problems

15-3.1 Determine the typical upper cutoff frequency for an inverting amplifier with a closed-loop gain of 15 using a *741* op-amp. The A_v/f graph for the *741* is shown in Fig. 15-13.

15-3.2 Using the gain-bandwidth product, calculate the cutoff frequencies for an inverting amplifier with a closed-loop gain of 30 when the op-amp used is (a) *741*, (b) an *AD843*.

15-3.3 Calculate the full-power bandwidth for an *LF353* op-amp circuit, with a 14 V peak-to-peak output voltage.

15-4   Stray Capacitance Effects

Stray capacitance (C_s) at the input terminals of an operational amplifier effectively introduces an additional phase-lag network in the feedback loop, (see Fig. 15-17), thus making the op-amp circuit unstable. Stray capacitance problems can be avoided by good circuit construction techniques that keep the stray to a minimum. The effects of stray capacitance also depend upon the resistor values used in the feedback network. High resistance values make it easier for small stray capacitances to produce phase lag. With low-resistances, small stray capacitances normally have little effect on the circuit stability.

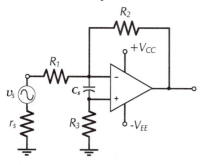

(a) Stray capacitance (C_s) at amplifier input

(b) C_s and its series resistance

Figure 15-17
Stray capacitance can cause instability in an op-amp circuit by introducing additional phase-lag in the feedback network.

Analysis of an RC phase lag circuit shows that the capacitor voltage lags the input voltage by 45° when the capacitor impedance (X_c) equal the series resistance (R). Also, when $X_c = 10$ R, the phase lag is approximately 10°, and it is this 10° of additional phase lag that might make the circuit oscillate if its phase margin is already close to the minimum for stability. If the phase margin is known to be large at the frequency where $A_v B = A_{CL}$ (the frequency at which the circuit is likely to oscillate), the stray capacitance might be unimportant. Where the phase margin is small, for circuit stability the op-amp input stray capacitance should normally be much less than,

$$C_s = \frac{1}{2\pi f(10\ R)} \tag{15-4}$$

where R is the equivalent resistance in series with the stray capacitance. In Fig. 15-17, $R = R_3 + (R_1 + r_s)\|R_2$.

From Eq. 15-4 it is seen that (as already mentioned) the larger resistor values the smaller the stray capacitance that can produce circuit instability. If the signal source is disconnected from the circuit, R becomes equal to $(R_2 + R_3)$, which is much larger than $[R_3 + (r_s + R_1)\|R_2]$. In this situation, extremely small stray capacitance values can make the circuit unstable.

Miller-effect compensation can be used to compensate for stray capacitance at an op-amp input, as shown in Fig. 15-18. To eliminate the phase shift introduced by the stray capacitance the division of the output voltage produced by C_S and C_2 in series should be equal to the division produced by R_1 and R_2. Therefore,

$$\frac{X_{CS}}{X_{C2}} = \frac{R_1}{R_2}$$

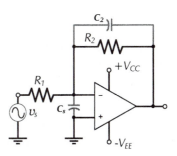

Figure 15-18
Use of Miller-effect compensation for stray capacitance at the input terminals of an op-amp.

This gives, $$C_2 R_2 = C_S R_1 \qquad \qquad \textbf{(15-5)}$$

Note that Eq. 15-5 does not allow for r_s or R_3 in Fig. 15-17. Where r_s is not very much smaller than R_1, it must be added to R_1. Also, resistor R_3 could be bypassed with another capacitor to reduce the total series resistance.

Example 15-7

Calculate the op-amp input terminal stray capacitance that might cause instability in the circuit of Fig. 15-19 if the amplifier cutoff frequency is 800 kHz. Also, determine a suitable Miller-effect compensating capacitor value.

Solution

Stray capacitance:

Eq. 15-4,

$$C_s = \frac{1}{2 \pi f \times 10[(r_s + R_1) \| R_2]}$$

$$= \frac{1}{2 \pi \times 800 \text{ kHz} \times 10 \, [(600 \, \Omega + 1 \text{ k}\Omega) \| 10 \text{ k}\Omega]}$$

$$= 14.4 \text{ pF}$$

Compensation:

Eq. 15-5,

$$C_2 = \frac{C_s (r_s + R_1)}{R_2} = \frac{14.4 \text{ pF} \times (600 \, \Omega + 1 \text{ k}\Omega)}{10 \text{ k}\Omega}$$

$$= 2.3 \text{ pF}$$

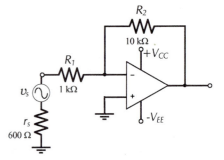

Figure 15-19
Op-amp amplifier circuit for Example 15-7.

Practise Problems

15-4.1 Determine the op-amp input stray capacitance that might cause instability in an inverting amplifier with $R_1 = 1.8 \text{ k}\Omega$, $R_2 = 560 \text{ k}\Omega$, and $f_2 = 600 \text{ kHz}$; (a) when the signal source is open-circuited, (b) when $r_s = 600 \, \Omega$, and R_1 and R_2 are reduced by a factor of 10.

15-4.2 Determine a suitable Miller-effect compensating capacitor value for the circuit in part (b) of Problem 15-4.1.

15-5   Load Capacitance Effects

Capacitance connected at the output of an operational amplifier is termed *load capacitance* (C_L). Figure 15-20 shows that C_L is in series with the op-amp output resistance (r_o), so C_L and r_o constitute a phase-lag circuit in the feedback network. As in the case of stray capacitance, another 10° of phase lag introduced by C_L and r_o could cause circuit instability where the phase margin is already small. The equation for calculating the load capacitance that might cause instability is similar to that for stray capacitance;

$$C_L = \frac{1}{2 \pi f (10 r_o)} \qquad \qquad \textbf{(15-6)}$$

In Eq. 15-6 f is the frequency at which $A_v B = A_{CL}$. If r_o is reduced, Eq 15-6 gives a larger C_L value. Thus, *an op-amp with a low output resistance can tolerate more load capacitance than one with a higher output resistance.*

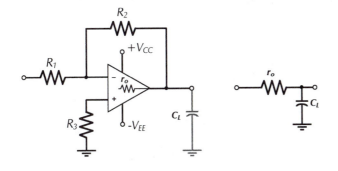

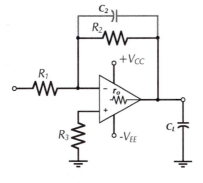

Figure 15-20
Load capacitance at an op-amp output can cause instability by introducing additional phase lag in the feedback network.

One method often used to counter instability caused by load capacitance is shown in Fig. 15-21(a). A resistor (R_x), usually ranging from 12 Ω to 400 Ω, is connected in series with the load capacitance. The presence of R_x (with R_2 connected at the op-amp output) can severely reduce the phase lag produced by r_o and C_L. However, R_x also has the undesirable effect of increasing the circuit output impedance to approximately the resistance of R_x.

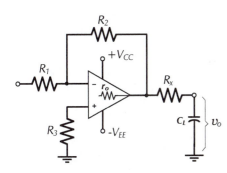

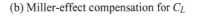

(a) Use of R_x to reduce phase lag

(b) Miller-effect compensation for C_L

Figure 15-21
Compensation methods for load capacitance.

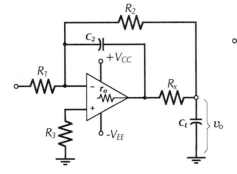

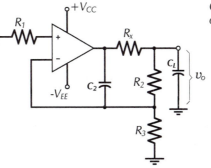

(c) Inverting amplifier with combination of R_x and C_2 for C_L compensation

(d) Noninverting amplifier with R_x C_2 combination for C_L compensation

A Miller-effect capacitor (C_2) connected across feedback resistor R_2 may be used to compensate for the load capacitance, [see Fig. 15-21(b)]. In this case, C_2 introduces some phase-lead in the feedback network to counter the phase-lag. The equation for calculating a suitable capacitance for C_2 is, once again, similar to that for stray capacitance;

$$C_2 R_2 = C_L r_o \qquad (15\text{-}7)$$

A modified form of Miller-effect compensation for load capacitance is shown in Fig. 15-21(c). An additional resistor (R_x) is included in series with C_L to reduce the phase lag, as discussed. But now, R_2 is connected at the junction of R_x and C_L, so that (because of feedback) R_x has no significant effect on the circuit output impedance. Also, C_2 is connected from the op-amp output terminal to the inverting input. With this arrangement, Eq. 15-7 is modified to,

$$C_2 R_2 = C_L(r_o + R_x) \qquad (15\text{-}8)$$

It should be noted from Equations 15-7 and 15-8 that, as for stray capacitance, smaller resistance values for R_2 give larger, more convenient, compensating capacitor values.

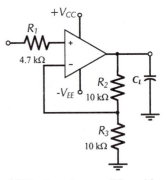

(a) Noninverting amplifier with load capacitance (C_L)

Example 15-8

Calculate the load capacitance that might cause instability in the circuit of Fig. 15-22(a) if the amplifier cutoff frequency is 2 MHz and its output resistance is 25 Ω. Also, determine a suitable compensating capacitor value for the circuit as modified in Fig. 15-22(b) with a 0.1 μF load capacitance.

Solution

Load capacitance:

Eq. 15-6, $\quad C_L = \dfrac{1}{2\pi f\,(10\,r_o)} = \dfrac{1}{2\pi \times 2\text{ MHz} \times 10 \times 25\ \Omega}$

$\quad\quad\quad\quad = 318\text{ pF}$

Compensation:

Eq. 15-8, $\quad C_2 = \dfrac{C_L\,(r_o + R_x)}{R_2} = \dfrac{0.1\ \mu F \times (25\ \Omega + 25\ \Omega)}{10\text{ k}\Omega}$

$\quad\quad\quad\quad = 500\text{ pF (standard value)}$

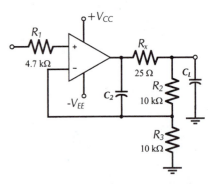

(b) Amplifier compensated with R_x and C_2

Figure 15-22
Circuits for Example 15-8.

Practise Problems

15-5.1 Calculate the load capacitance that might cause instability in the circuit in Ex. 15-7 if the op-amp output resistance is 20 Ω. Determine a suitable Miller-effect compensating capacitor value.

15-5.2 The circuit is Problem 15-5.1 is modified as in Fig. 15-21(c) with $R_x = 5\,R_o$ and $C_L = 0.5\ \mu F$. Calculate the required C_2 value.

15-6 Circuit Stability Precautions

Power Supply Decoupling

Feedback along supply lines is another source of op-amp circuit instability. This can be minimized by connecting 0.01 µF high-frequency capacitors from each supply terminal to ground (see Fig. 15-23). The capacitors must be connected as close as possible to the *IC* terminals. Sometimes larger-value capacitors are required.

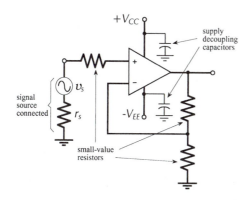

Figure 15-23
For op-amp circuit stability, keep resistor values to a minimum, use the recommended compensating components, bypass the supply terminals to ground, and keep the signal source connected.

Stability Precautions

The following precautions should be observed for circuit stability:

1. Where low-frequency performance is required, use an internally compensated op-amp. Alternatively, use Miller-effect compensation to give the lowest acceptable cutoff frequency.

2. Use small-value resistors in the feedback network, if possible, instead of using the largest possible resistor values.

3. With an op-amp that must be compensated, use the methods and components recommended by the *IC* manufacturer.

4. Keep all component leads as short as possible, and take care with component placement. A resistors connected to an op-amp input terminal should have the resistor body placed close to the input terminal.

5. Use 0.01 µF capacitors (or 0.1 µF capacitors if necessary) to bypass the supply terminals of op-amp (or groups of op-amps) to ground. Connect these capacitors close to the *ICs*.

6. Always have a signal source connected to a circuit being tested. Alternatively, ground the circuit input. With an open-circuited input, very small stray capacitances can cause instability.

7. Do not connect oscilloscopes or other instrument at the op-amp input terminals. Instrument input capacitance can cause instability.

8. If a circuit is unstable after all of the above precautions have been observed, reduce the value of all circuit resistors (except compensating resistors). Also, reduce the signal source resistance if possible.

Chapter-15 Review Questions

Section 15-1

15-1   Show how feedback in an op-amp inverting amplifier can produce instability. Explain the conditions necessary for oscillations to occur in an op-amp circuit.

15-2   Show how feedback in an op-amp noninverting amplifier circuit can produce instability.

15-3   Define: loop gain, loop phase shift, phase margin.

15-4   Sketch typical gain/frequency response and phase/frequency response graphs for an uncompensated operational amplifier. Identify the pole frequencies and rates of fall of voltage gain, and show the typical phase shift at each pole frequency.

15-5   Sketch typical gain/frequency response and phase/frequency response graphs for a compensated operational amplifier. Briefly explain.

15-6   Derive an equation for the open-loop gain of an operational amplifier when the loop gain equals 1.

Section 15-2

15-7   Sketch a lag compensation circuit. Explain its operation and show how it affects the frequency and phase response graphs of an operational amplifier.

15-8   Sketch a lead compensation circuit. Explain its operation and show how it affects the frequency and phase response graphs of an operational amplifier.

15-9   Show how Miller-effect compensation can be applied to an op-amp circuit. Briefly explain.

Section 15-3

15-10   Define bandwidth, gain-bandwidth product, and full-power bandwidth for an operational amplifier. Explain the circuit conditions that apply in each case.

15-11   Define slew-rate, and explain its effect on the output waveform from an operational amplifier.

Section 15-4

15-12   Discuss the effects of stray capacitance at the input terminals of an operational amplifier. Explain the precautions that should be observer to deal with input stray capacitance problems.

15-13   Show how Miller-effect compensation can be used to counter the effects of op-amp input stray capacitance.

Section 15-5

15-14   Discuss the effects of op-amp load capacitance.

15-15   Show how op-amp instability due to load capacitance can be countered by means of: an additional resistor, Miller-effect compensation, a combination of both.

Section 15-6

15-16   List precautions that should be observed for operational amplifier circuit stability. Briefly explain in each case.

Chapter-15 Problems

Section 15-1

15-1   Investigate the stability of the circuit in Fig. 15-11, if the *LM108* is replaced with an op-amp that has the gain/frequency and phase/frequency responses in Fig. 15-2.

15-2   Investigate the stability of the inverting amplifier circuit in Fig. 15-10(a) if the op-amp used is (a) a *741*, (b) an *AD843*. Use the response graphs in Fig. 15-3 and 15-4.

15-3   Investigate the stability of an amplifier circuit with A_{CL} = 70 dB if the op-amp has the gain/frequency and phase/frequency responses in Fig. 15-2.

Section 15-2

15-4   The phase-lag network in Fig. 15-7(a) has: R_1 = 8.2 kΩ, R_2 = 470 Ω, and C_1 = 3300 pF. Calculate the approximate phase lag at 7.5 kHz and at 750 kHz.

15-5   The phase-lead network in Fig. 15-7(c) has: R_1 = 560 Ω, R_2 = 27 kΩ, and C_1 = 1000 pF. Calculate the approximate phase lead at 6 kHz and 300 kHz.

15-6   If the circuit in Fig. 15-22(a) uses an *LM108* op-amp, determine suitable compensation capacitor values. Refer to the *LM108* information in Fig. 15-8.

15-7   Calculate a suitable Miller-effect compensating capacitor to stabilize the circuit in Fig. 15-19 at an 80 kHz cutoff frequency.

15-8   The circuit in Fig. 15-10(b) is to be stabilized at f_2 = 50 kHz. Determine the value of a suitable Miller-effect compensating capacitor.

Section 15-3

15-9   Determine the bandwidth of the circuit in Fig. 15-10(a) if the op-amp has the gain/frequency response graph in Fig. 15-2. Also, determine the bandwidth of the circuit in Fig 15-10(b) if the op-amp is a *741*.

15-10   Find the upper cutoff frequency for the circuit in Fig. 15-19 if the op-amp is (a) a *741*, (b) an *AD843*. Use the response graphs in Fig. 15-3 and 15-4.

15-11   Using the gain-bandwidth product, determine the upper cutoff frequencies for the circuits in Problem 15-10.

15-12   Use the gain-bandwidth product to determine the upper cutoff frequencies for the circuits in Figs. 15-10(a) and (b) if they both have *LM108* op-amps with C_f = 30 pF.

15-13   The circuits in Examples 14-5 and 14-8 use *741* op-amps. Use the gain-bandwidth product to determine the upper cutoff frequency for each circuit.

15-14   If the circuit in Fig. 15-11 has the *LM108* replaced with an *LF353*, use the gain-bandwidth product to determine the upper cutoff frequency.

15-15   Calculate the full-power bandwidth for an amplifier using *741* op-amp if the output voltage is to be (a) 5 V peak-to-peak, (b) 1 V peak-to-peak.

15-16   Recalculate the full power bandwidth in each case in Problem 15-15 if the *741* is replaced with an *LF353*.

15-17   Calculate the full-power bandwidth for the circuit in Example 14-5 if the peak output is to be 2 V. Also, determine the maximum peak output voltage that can be produced by the circuit at the cutoff frequency calculated in Problem 15-13.

15-18   Calculate the slew-rate limited cutoff frequency for the circuit in Example 14-8 if the peak input is 20 mV. Also, determine the maximum peak output voltage at the circuit cutoff frequency calculated in Problem 15-13.

Section 15-4

15-19   A circuit as in Fig. 15-10(a) with C_f removed has a cutoff frequency of 600 kHz. Determine the op-amp input stray capacitance that might cause instability; (a) when the signal source is open-circuited, (b) when a 300 Ω signal source is connected.

15-20   Determine the input stray capacitance that might make the circuit in Fig. 15-10(b) become unstable when a 300 Ω signal source is connected. Assume that the circuit cutoff frequency is 30 kHz and that C_f is removed.

15-21   Calculate the Miller-effect capacitor value required to compensate for 250 pF of input stray capacitance in the circuitry of Problem 15-19(b).

15-22   Calculate the Miller-effect capacitor value required to compensate for 90 pF of stray capacitance in the circuit of Problem 15-20.

15-23   An inverting amplifier (as in Fig. 15-19) uses a *LF353* op-amp, and has; r_s = 600 Ω, R_1 = 220 kΩ, R_2 = 2.2 MΩ, f_2 = 18 kHz. Calculate the input stray capacitance that might make the circuit unstable; (a) when the signal source is connected, (b) when the signal source is open-circuited.

15-24   Repeat Problem 15-23 when R_1 and R_2 are each reduced by a factor of 10.

Section 15-5

15-25   Determine the load capacitance that might cause instability in the circuit in Fig. 15-10(a) with C_f removed, if the circuit cutoff frequency is 600 kHz, and $r_o = 100\ \Omega$.

15-26   An inverting amplifier (as in Fig. 15-19) uses an op-amp with a 300 Ω output resistance, and has; $R_1 = 220\ k\Omega$, $R_2 = 2.2\ M\Omega$. Calculate the load capacitance that might make the circuit unstable.

15-27   Calculate the Miller-effect capacitor value required to compensate for a 0.1 μF load capacitance in the circuit of Fig. 15-10(a) if $f_2 = 600$ kHz.

15-28   Calculate the Miller-effect capacitor value required to compensate for the load capacitance in Problem 15-26.

15-29   Determine the load capacitance that might cause instability in the circuit in Fig. 15-22(a), if the cutoff frequency is 400 kHz, and the op-amp has $r_o = 150\ \Omega$.

15-30   The circuit in Problem 15-29 is rearranged as in Fig. 15-22(b) with $R_x = 10\ r_o$. Calculate the required C_2 value to compensate for a 5000 pF load capacitance.

Practise Problem Answers

15-1.1	30°, 90°, 90°
15-2.1	-48.5°, 41.5°
15-2.2	82 pF
15-3.1	50 kHz
15-3.2	27 kHz, 1.13 MHz
15-3.3	296 kHz
15-4.1	0.05 pF, 34.5 pF
15-4.2	0.5 pF
15-5.1	995 pF, 2 pF
15-5.2	6000 pF

Chapter *16*
Signal Generators

Chapter   Contents

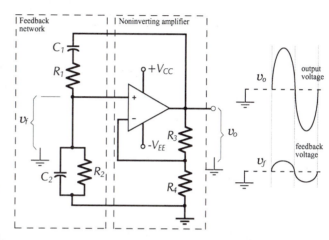

Objectives
You will be able to:

1 Draw the following types of sine wave oscillator circuits, and explain the operation of each: Phase shift, Colpitts, Hartley, Wein bridge

2 Analyze each of the above oscillator circuits to determine the oscillation frequency.

3 Design each of the above oscillator circuits to produce a specified output frequency.

4 Sketch oscillator amplitude stabilization circuits and explain their operation.

5 Design circuits to limit oscillator outputs to a specified amplitude.

6 Draw square wave and triangular wave generator circuits. Sketch the circuit waveforms, and explain the operation of each circuit.

7 Analyze square and triangular waveform generators to determine the amplitude and frequency of the output waveform.

8 Design square and triangular waveform generators to produce a specified output amplitude and frequency.

9 Explain piezoelectric crystals, sketch the crystal equivalent circuit, and the crystal impedance/frequency graph.

10 Show how crystals may be used for oscillator frequency stabilization, and design crystal-controlled oscillators.

Introduction

A sinusoidal oscillator usually consists of an amplifier and a phase-shifting network. The amplifier receives the output from the network, amplifies it, phase shifts it by 180°, and applies it to the network input. The network phase shifts the amplifier output by a further 180°, and attenuates it before feeding it back to the amplifier input. When the amplifier gain equals the inverse of the network attenuation, and the amplifier phase shift equals the network phase shift, the circuit is amplifying an input to produce an output which is attenuated to become the input. The circuit is generating its own input signal, and a state of oscillation exists.

Some signal generators produce square or triangular waveforms. These normally use nonlinear circuits and resistor/capacitor charging circuits.

16-1   Phase Shift Oscillators

Op-Amp Phase Shift Oscillator

Figure 16-1 shows the circuit of a *phase shift oscillator*, which consists of an inverting amplifier and an *RC* phase-shifting network. The amplifier phase-shifts its input by -180°, and the *RC phase-lead* network phase-shifts the amplifier output by a +180°, giving a total loop phase shift of zero. The attenuated feedback signal (at the amplifier input) is amplified to reproduce the output. In this condition the circuit is generating its own input signal, consequently, it is oscillating. The output and feedback voltage waveforms in Fig. 16-1 illustrate the circuit operation.

For a state of oscillation to be sustained in any sinusoidal oscillator circuit, certain conditions, known as the *Barkhausen criteria*, must be fulfilled:

> *The loop gain must be equal to (or greater than) one.*
> *The loop phase shift must be zero.*

The *RC* phase-lead network in Fig. 16-1 consists of three equal-value resistors and three equal-value capacitors. Resistor R_1 functions as the last resistor in the *RC* network and as the amplifier input resistor. A *phase-lag* network would give a total loop phase shift of -360°, and so it would work just as well as the phase lead network.

The frequency of the oscillator output depends upon the component values in the *RC* network. The circuit can be analyzed to show that the phase shift is 180° when

$$X_c = \sqrt{6}\, R$$

This gives an oscillation frequency,

$$f = \frac{1}{2\,\pi\,R\,C\,\sqrt{6}} \qquad\qquad \textbf{(16-1)}$$

As well as phase shifting the amplifier output, the *RC* network attenuates the output. It can be shown that, when the required 180° phase shift is produced, the feedback factor (B) is always 1/29. This means that the amplifier must have a closed-loop voltage gain (A_{CL}) of at least 29 to give a loop gain ($B\,A_{CL}$) of one; otherwise the circuit will not oscillate. For example, if the amplifier output voltage is 10 V, the feedback voltage is,

$$v_f = B\,v_o = 10\ \text{V}/29$$

To reproduce the 10 V output, v_f must be amplified by 29,

$$v_o = A_{CL}\,v_f = 29 \times (10\ \text{V}/29)$$
$$= 10\ \text{V}$$

If the amplifier voltage gain is much greater than 29, the output waveform will be distorted. When the gain is slightly greater than 29, a reasonably pure sine wave output can be expected. The gain is usually designed to be just over 29 to ensure that the circuit oscillates. The output voltage amplitude normally peaks at $\pm(V_{CC} - 1\ \text{V})$, unless a rail-to-rail op-amp is used (see Section 14-9).

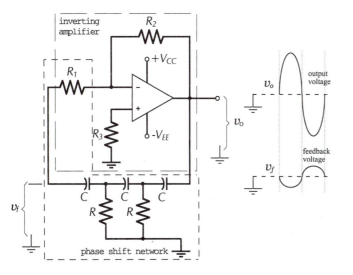

Figure 16-1
A phase shift oscillator consists of an inverting amplifier and an RC phase shifting feedback network. The RC network attenuates the output and phase shifts it by 180°. The amplifier amplifies the network output and phase shifts it through a further 180°.

Circuit Design

Design of a phase shift oscillator begins with design of the amplifier to have a closed-loop gain just greater than 29. The resistor values for the RC network are then selected equal to the amplifier input resistor (R_1), and the capacitor values are calculated from Eq. 16-1. In some cases, this procedure might produce capacitor values not much larger than stray capacitance. So, alternatively, the design might start with selection of convenient capacitor values. Equation 16-1 is then used to calculate the resistance of R (and R_1). Finally, R_2 is selected to give the required amplifier gain.

Example 16-1

Using a *741* op-amp with a ± 10 V supply, design the phase shift oscillator in Fig. 16-2 to produce a 1 kHz output frequency.

Solution

Select

$$I_1 \approx 100 \times I_{B(max)} = 100 \times 500 \text{ nA}$$
$$= 50 \,\mu\text{A}$$

$$v_o \approx \pm(V_{CC} - 1 \text{ V}) \approx \pm(10 \text{ V} - 1 \text{ V})$$
$$\approx \pm 9 \text{ V}$$

$$v_i = \frac{v_o}{A_{CL}} = \frac{\pm 9 \text{ V}}{29}$$
$$= \pm 0.31 \text{ V}$$

$$R_1 = \frac{v_i}{I_1} = \frac{0.31 \text{ V}}{50 \,\mu\text{A}}$$
$$= 6.2 \text{ k}\Omega \text{ (use 5.6 k}\Omega \text{ standard value)}$$

$$R_2 = A_{CL} R_1 = 29 \times 5.6 \text{ k}\Omega$$
$$\approx 162 \text{ k}\Omega \text{ (use 180 k}\Omega \text{ to give } A_{CL} > 29)$$

$$R_3 = R_2 = 180 \text{ k}\Omega \text{ (the } dc \text{ path through } R_1$$
$$\text{is interrupted by } C \,)$$

$$R = R_1 = 5.6 \text{ k}\Omega$$

From Eq. 16-1,
$$C = \frac{1}{2\pi R f \sqrt{6}} = \frac{1}{2\pi \times 5.6 \text{ k}\Omega \times 1 \text{ kHz} \times \sqrt{6}}$$
$$\approx 0.01 \,\mu\text{F (standard value)}$$

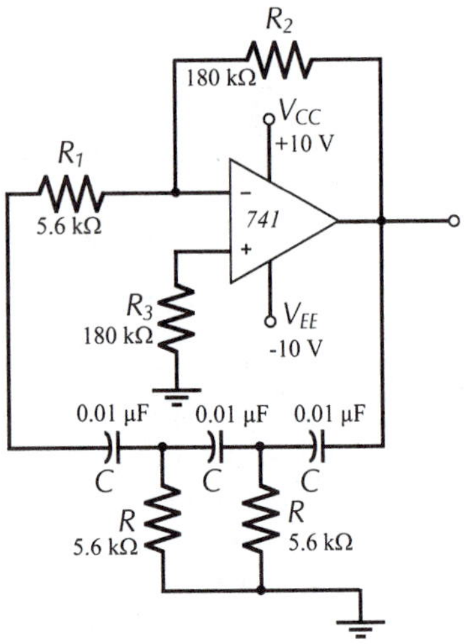

Figure 16-2
Phase shift oscillator circuit for Example 16-1.

Although the *741* op-amp used in Ex. 16-1 is likely to be quite suitable for the particular circuit, some care should always be taken when selecting an operational amplifier. It should be recalled (from Ch. 15) that when a large output voltage swing is required, the op-amp full-power bandwidth is involved. This must be considered when selecting an operational amplifier for an oscillator circuit.

BJT Phase Shift Oscillator

A phase shift oscillator using a single *BJT* amplifier is shown in Fig. 16-3. Once again, the amplifier and phase shift network each produce 180° of phase shift, the *BJT* amplifies the network output, and the network attenuates the amplifier output.

First thoughts about this circuit (in comparison to the op-amp phase shift oscillator) would suggest that a *BJT* amplifier with a voltage gain of 29 is required. An attempt to design such a circuit reveals that in many cases the amplifier output is overloaded by the phase shift network, or else the network output is overloaded by the amplifier input. The problem can be solved by including an emitter follower in the circuit. However, the circuit can function satisfactorily without any additional components if the transistor is treated as a current amplifier, rather than as a voltage amplifier. In this case, circuit analysis gives,

$$f = \frac{1}{2 \pi R C \sqrt{(6 + 4R_C/R)}} \tag{16-2}$$

and,
$$h_{fe(min)} = 23 + \frac{29 R}{R_C} + \frac{4 R_C}{R} \tag{16-3}$$

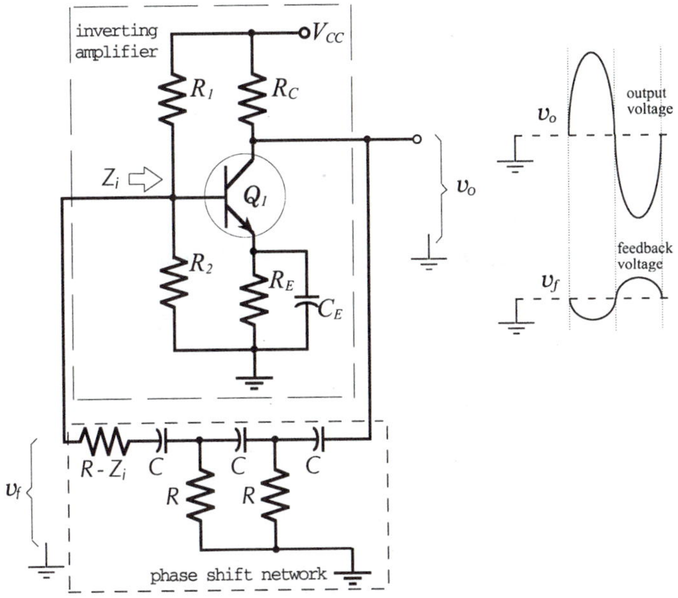

Figure 16-3
Phase shift oscillator using an inverting BJT amplifier and an RC feedback network.

The circuit oscillates only if the *BJT* current gain is equal to (or larger than) the minimum value determined from Eq. 16-3. With $R = R_C$, a minimum h_{fe} of 56 is required to sustain circuit oscillation. With $R = 10 R_C$, $h_{fe(min)}$ must be greater than 300. The output waveform is likely to be distorted if h_{fe} is substantially greater than the calculated $h_{fe(min)}$. Because h_{fe} varies widely from one transistor to another, R_C should be partially adjustable to minimize distortion. Note that in Fig. 16-3, the amplifier input resistance (Z_i) constitutes part of the last resistor in the phase shift network.

Circuit Design

BJT phase shift oscillator design should be approached by first selecting R equal to or greater than the estimated amplifier Z_i. Then, R_C is selected equal to R, C is calculated from Eq. 16-2, and the rest of the component values are determined for the circuit *dc* conditions. The impedance of C_E should be much lower than $h_{ie}/(1 + h_{fe})$ at the oscillating frequency.

Practise Problems

16-1.1   Using a *BIFET* op-amp with rail-to-rail operation, design a phase shift oscillator to produce a 6.5 kHz, ±12 V output.

16-1.2   Design a *BJT* phase shift oscillator, as in Fig. 16-3, to produce a 900 Hz waveform with a 10 V peak-to-peak amplitude. Assume that the BJT has $h_{fe(min)} \approx 60$ and $h_{ie} \approx 1.5$ kΩ.

16-2   Colpitts Oscillators

Op-Amp Colpitts Oscillator

The *Colpitts oscillator* circuit show in Fig. 16-4 is similar to the op-amp phase shift oscillator, except that an *LC* network is used to produce the necessary phase shift in the feedback voltage. In this case, the *LC* network acts as a filter that passes the oscillating frequency and blocks all other frequencies. The filter circuit resonates at the required oscillating frequency. For resonance,

$$X_L = X_{CT}$$

where X_{CT} is the total capacitance in parallel with the inductor. This gives the resonance frequency (and oscillating frequency) as,

$$f = \frac{1}{2\pi\sqrt{(L_1 C_T)}} \qquad\qquad \textbf{(16-4)}$$

Capacitors C_1 and C_2 are connected in series across L_1; so,

$$C_T = \frac{C_1 C_2}{C_1 + C_2} \qquad\qquad \textbf{(16-5)}$$

Consideration of the *LC* network shows that its attenuation (from the amplifier output to input) is due to the voltage divider effect of L and C_1. This gives,

$$B = \frac{X_{C1}}{X_{L1} - X_{C1}}$$

It can be shown that the required 180° phase shift occurs when

$$X_{C2} = X_{L1} - X_{C1}$$

and this gives,

$$B = \frac{X_{C1}}{X_{C2}} = \frac{C_2}{C_1}$$

As in the case of all oscillator circuits, the loop gain must be a minimum of one to ensure oscillation. Therefore,

$$A_{CL(min)} B = 1$$

or, $$A_{CL(min)} = \frac{C_1}{C_2}$$ **(16-6)**

When deriving the above equations, it was assumed that the inductor coil resistance is very much smaller than the inductor impedance; that is, that the coil Q factor $(\omega L/R)$ is large. This must be taken into considered when selecting an inductor. It was also assumed that the amplifier input resistance is much greater than the impedance of C_1 at the oscillating frequency. Because of the inductor resistance and the amplifier input resistance, and because of stray capacitance effects when the oscillator operates at a high-frequency, the amplifier voltage gain usually has to be substantially larger than C_1/C_2.

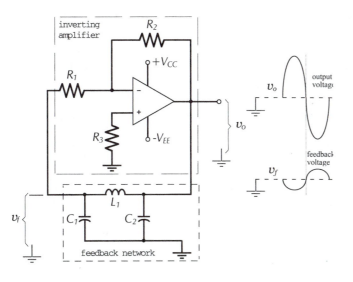

Figure 16-4
A Colpitts oscillator consists of an inverting amplifier and an LC phase shifting feedback network.

Circuit Design
Colpitts oscillator design can commence with selection of the smallest capacitor (C_2) much larger than stray capacitance, or with selection of a convenient value of L. To keep the amplifier input voltage to a fairly low level, the feedback network is often designed to attenuate the output voltage by a factor of 10. This requires that $C_1/C_2 \approx 10$. (It should be recalled that large A_{CL} values require larger op-amp bandwidths.) Also, X_{C2} should be much larger than the amplifier output impedance. Using the desired oscillating frequency, L can be calculated from Eq. 16-4. Amplifier input resistor R_1 must be large enough to avoid overloading the feedback network, $(R_1 \gg X_{C1})$. Resistor R_2 is determined from A_{CL} and R_1.

Example 16-2
Design the Colpitts oscillator in Fig. 16-5 to produce a 40 kHz output frequency. Use a 100 mH inductor and an op-amp with a ±10 V supply.

Solution

From Eq. 16-4,

$$C_T = \frac{1}{4\pi^2 f^2 L} = \frac{1}{4\pi^2 \times (40\ \text{kHz})^2 \times 100\ \text{mH}}$$

$$= 153.8\ \text{pF}$$

For $C_1 \approx 10\ C_2$,

$$C_1 \approx 10\ C_T = 10 \times 153.8\ \text{pF}$$

$$\approx 1538\ \text{pF (use 1500 pF standard value)}$$

From Eq. 16-5,

$$C_2 = \frac{1}{(1/C_T) - (1/C_1)} = \frac{1}{(1/158.3\ \text{pF}) - (1/1500\ \text{pF})}$$

$$= 177\ \text{pF (use 180 pF standard value)}$$

$$X_{C2} = \frac{1}{2\pi f C_2} = \frac{1}{2\pi \times 40\ \text{kHz} \times 180\ \text{pF}}$$

$$= 22\ \text{k}\Omega$$

$$X_{C2} \gg Z_o \text{ of the amplifier}$$

$$X_{C1} = \frac{1}{2\pi f C_1} = \frac{1}{2\pi \times 40\ \text{kHz} \times 1500\ \text{pF}}$$

$$= 2.65\ \text{k}\Omega$$

$$R_1 \gg X_{C1}$$

Select

$$R_1 = 10\ X_{C1} = 10 \times 2.65\ \text{k}\Omega$$

$$= 26.5\ \text{k}\Omega \text{ (use 27 k}\Omega\text{ standard value)}$$

From Eq. 16-6,

$$A_{CL(min)} = \frac{C_1}{C_2} = \frac{1500\ \text{pF}}{180\ \text{pF}}$$

$$= 8.33$$

$$R_2 = A_{CL}\ R_1 = 8.33 \times 27\ \text{k}\Omega$$

$$= 225\ \text{k}\Omega \text{ (use 270 k}\Omega\text{ standard value)}$$

$$R_3 = R_1 \| R_2 = 27\ \text{k}\Omega \| 270\ \text{k}\Omega$$

$$= 24.5\ \text{k}\Omega \text{ (use 27 k}\Omega\text{ standard value)}$$

The op-amp full-power bandwidth (f_p) must be a minimum of 40 kHz when $v_o \approx \pm 9$ V and $A_{CL} = 8.33$.

from Eq. 15-2,

$$f_2 = A_{CL} \times f_p = 8.33 \times 40\ \text{kHz}$$

$$= 333\ \text{kHz}$$

from Eq. 15-3,

$$SR = 2\pi f_p\ v_p = 2\pi \times 40\ \text{kHz} \times 8\ \text{V}$$

$$= 2\ \text{V}/\mu\text{s}$$

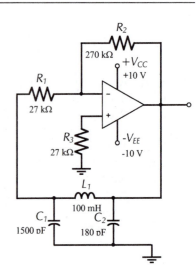

Figure 16-5
Colpitts oscillator circuit for Example 16-2.

BJT Colpitts Oscillator

A Colpitts oscillator using a single *BJT* amplifier is shown in Fig. 16-6(a). This is the basic circuit, and its similarity to the op-amp Colpitts oscillator is fairly obvious. A more complex version of the circuit is shown in Fig. 16-6(b). Components Q_1, R_1, R_2, R_E, and C_E in (b) are unchanged from (a), but collector resistor R_C is replaced with inductor L_1. A radio frequency choke (*RFC*) is included in series with V_{CC} and L_1. This allows *dc* collector current (I_C) to pass, but offers a very high impedance at the oscillating frequency, so that the top of L_1 is *ac* isolated from V_{CC} and ground. The output of the *LC* network (L_1, C_1, C_2) is coupled to via C_c to the amplifier input. The circuit output voltage (v_o) is derived from a secondary winding (L_2) coupled to L_1. As in the case of the *BJT* phase shift oscillator, the transistor current gain is important. Circuit analysis gives Eq. 16-4 for frequency, and for current gain,

$$h_{fe(min)} = \frac{C_1}{C_2} \qquad\qquad \textbf{(16-7)}$$

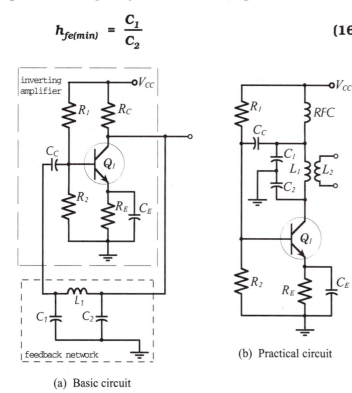

(a) Basic circuit

(b) Practical circuit

Figure 16-6
Colpitts oscillator using an inverting BJT amplifier and an LC feedback network.

Practise Problems

16-2.1 Design a Colpitts oscillator circuit to produce a 12 kHz, ±10 V output. Use a *741* op-amp.

16-2.2 Design the oscillator in Fig. 16-6(a) to produce a 20 kHz, 6 V p-to-p output. Use a 10 mH inductor and assume that the *BJT* has $h_{1b} \approx$ 26 Ω and $h_{ie} \approx$ 1.5 kΩ.

16-3   Hartley Oscillators

Op-Amp Hartley Oscillator

The *Hartley oscillator* circuit is similar to the Colpitts oscillator, except that the feedback network consists of two inductors and a capacitor instead of two capacitors and an inductor. Figure 16-7(a) shows the Hartley oscillator circuit, and Fig. 16-7(b) illustrates the fact that L_1 and L_2 may be wound on a single core so that there is mutual inductance (M) between the two windings. In this case, the total inductance is,

$$L_T = L_1 + L_2 + 2\,M \qquad\qquad (16\text{-}8)$$

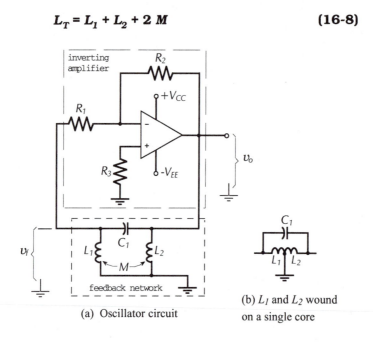

Figure 16-7
Hartley oscillator circuit using an op-amp inverting amplifier and an LC feedback network.

(a) Oscillator circuit

(b) L_1 and L_2 wound on a single core

Oscillation occurs at the feedback network resonance frequency ;

$$f = \frac{1}{2\,\pi\,\sqrt{(C_1\,L_T)}} \qquad\qquad (16\text{-}9)$$

The attenuation of the feedback network is,

$$B = \frac{X_{L1}}{X_{L1} - X_{C1}}$$

It can be shown that the required 180° phase shift occurs when

$$X_{L2} = X_{L1} - X_{C1}$$

The loop gain must be a minimum of one, giving;

$$A_{CL(min)} = \frac{L_2}{L_1} \qquad\qquad (16\text{-}10)$$

Design procedure for a Hartley oscillator circuit is similar to that for a Colpitts oscillator.

Example 16-3

Design the Hartley oscillator in Fig. 16-8 to produce a 100 kHz output frequency with an amplitude of approximately ±8 V. For simplicity, assume that there is no mutual inductance between L_1 and L_2.

Solution

$$V_{CC} \approx v_o + 1\ \text{V} = \pm(8\ \text{V} + 1\ \text{V})$$
$$\approx \pm 9\ \text{V}$$

$$X_{L2} >> Z_o \text{ of the amplifier}$$

select $$X_{L2} \approx 1\ \text{k}\Omega$$

$$L_2 = \frac{X_{L2}}{2\,\pi\,f} = \frac{1\ \text{k}\Omega}{2\,\pi \times 100\ \text{kHz}}$$

$$= 1.59\ \text{mH} \quad \text{(use 1.5 mH standard value)}$$

select $$L_1 \approx \frac{L_2}{10} = \frac{1.5\ \text{mH}}{10}$$

$$= 150\ \mu\text{H} \quad \text{(standard value)}$$

$$L_T = L_1 + L_2 = 1.5\ \text{mH} + 150\ \mu\text{H} \ \text{(assuming } M = 0\text{)}$$
$$= 1.65\ \text{mH}$$

From Eq. 16-9, $$C_1 = \frac{1}{4\,\pi^2\,f^2\,L_T} = \frac{1}{4\,\pi^2 \times (100\ \text{kHz})^2 \times 1.65\ \text{mH}}$$

$$= 1535\ \text{pF} \quad \text{(use 1500 pF with additional parallel capacitance, if necessary)}$$

$$C_1 >> \text{stray capacitance}$$

$$X_{L1} = 2\,\pi\,f\,L_1 = 2\,\pi \times 100\ \text{kHz} \times 150\ \mu\text{H}$$
$$= 94.2\ \Omega$$

$$R_1 >> X_{L1}$$

Select $$R_1 = 1\ \text{k}\Omega \ \text{(standard value)}$$

From Eq. 16-10,

$$A_{CL(min)} = \frac{L_2}{L_1} = \frac{1.5\ \text{mH}}{150\,\mu\text{H}}$$

$$= 10$$

$$R_2 = A_{CL}\,R_1 = 10 \times 1\ \text{k}\Omega$$
$$= 10\ \text{k}\Omega \ \text{(standard value)}$$

$$R_3 = R_1\|R_2 = 1\ \text{k}\Omega\|10\ \text{k}\Omega$$
$$= 909\ \Omega \ \text{(use 1 k}\Omega \text{ standard value)}$$

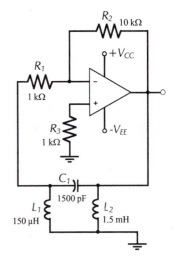

Figure 16-8
Hartley oscillator circuit for Example 16-3.

The op-amp full-power bandwidth (f_p) must be a minimum of 100 kHz when $v_o \approx \pm 8$ V and $A_{CL} = 10$.

from Eq. 15-2, $f_2 = A_{CL} \times f = 10 \times 100$ kHz
$$= 1 \text{ MHz}$$

from Eq. 15-3, $SR = 2 \pi f_p v_p = 2 \pi \times 100$ kHz $\times 8$ V
$$= 5 \text{ V/}\mu s$$

BJT Hartley Oscillator

Figure 16-9 shows the circuit of a Hartley oscillator using a *BJT* amplifier. The basic circuit in Fig. 16-9(a) is similar to the op-amp Hartley oscillator, and its operation is explained in the same way as for the op-amp circuit. Note that coupling capacitors C_2 and C_4 are required to avoid *dc* grounding the transistor base and collector terminals through L_1 and L_2.

In the practical *BJT* Hartley oscillator circuit shown in Fig. 16-9(b) L_1, L_2, and C_1 constitute the phase shift network. In this case, the inductors are directly connected in place of the transistor collector resistor (R_C). The circuit output is derived from the additional inductor winding (L_3). The radio frequency choke (*RFC*) passes the direct collector current, but *ac* isolates the upper terminal of L_1 from the power supply. Capacitor C_2 couples the output of the feedback network back to the amplifier input. Capacitor C_4 at the *BJT* collector in Fig. 16-9(a) is not required in Fig. 16-9(b), because L_2 is directly connected to the collector terminal. The junction of L_1 and L_2 must now be capacitor coupled to ground (via C_3) instead of being direct coupled.

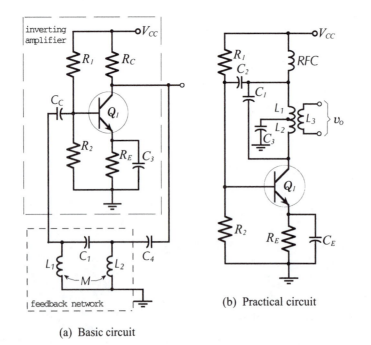

Figure 16-9
Hartley oscillator consisting of a BJT inverting amplifier and an LC feedback network.

(a) Basic circuit

(b) Practical circuit

Practise Problems

16-3.1 A Hartley oscillator circuit using a *741* op-amp is to produce a 7 kHz, ±10 V output. Determine suitable component values.

16-3.2 Analyze the *BJT* Hartley oscillator in Fig. 16-9(b) to determine the oscillating frequency. Some of the component values are: $L_1 = L_2 = 4.7$ mH, $C_1 = 600$ pF, $C_2 = C_3 = 0.03~\mu F$. The mutual inductance between L_1 and L_2 is 100 μH.

16-4   Wein bridge Oscillator

The *Wein bridge* is an *ac* bridge that balances only at a particular supply frequency. In the *Wein bridge oscillator* (Fig. 16-10), a Wein bridge circuit is used as a feedback network between the amplifier output and input. The bridge is made up of all of the resistors and capacitors. The operational amplifier together with resistors R_3 and R_4 constitute a noninverting amplifier. The feedback network from the amplifier output to its noninverting input terminal is made up of components C_1, R_1, C_2 and R_2.

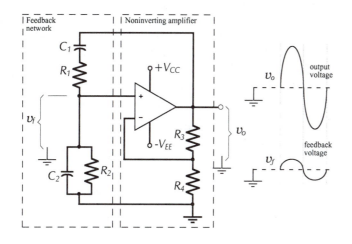
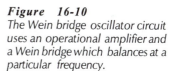

Figure 16-10
The Wein bridge oscillator circuit uses an operational amplifier and a Wein bridge which balances at a particular frequency.

At the balance frequency of the Wein bridge, the feedback voltage is in phase with the amplifier output. This (in-phase) voltage is amplified to reproduce the output. At all other frequencies, the bridge is off balance; that is, the feedback and output voltages do not have the correct phase relationship to sustain oscillations. The Barkhausen requirement for zero loop phase shift is fulfilled in this circuit by the amplifier and feedback network both having zero phase shift at the oscillation frequency.

Analysis of the bridge circuit shows that balance is obtained when two equations are fulfilled:

$$\frac{R_3}{R_4} = \frac{R_1}{R_2} + \frac{C_2}{C_1} \qquad\qquad \textbf{(16-11)}$$

and,
$$2\pi f = \frac{1}{\sqrt{(R_1\, C_1\, R_2\, C_2)}}$$
(16-12)

If $R_1 C_1 = R_2 C_2$, Eq.16-12 yields,

$$f = \frac{1}{2\pi\, R_1\, C_1}$$
(16-13)

For simplicity, the components are often selected as, $R_1 = R_2$ and $C_1 = C_2$, causing Eq. 16-11 to give,

$$R_3 = 2\, R_4$$
(16-14)

In this case, the amplifier closed-loop gain is, $A_{CL} = 3$.

Sometimes it is preferable to have an amplifier voltage gain substantially greater than 3, then the relationship between the component values is determined by Equations 16-11 and 16-12.

Design of a Wein bridge oscillator can be commenced by selecting a current level for each arm of the bridge. This should be much larger than the op-amp input bias current. Resistors R_3 and R_4 can then be calculated using the estimated output voltage and the closed-loop gain. After that, the other component values can be determined from the above equations.

An alternative design approach is to start by selecting a convenient value for the smallest capacitor in the circuit. The other component values are then calculated from the equations.

Example 16-4
Design the Wein bridge oscillator in Fig. 16-11 to produce a 100 kHz, ±9 V output. Design the amplifier to have a closed-loop gain of 3.

Solution

$$V_{CC} \approx \pm(V_o + 1\text{ V}) = \pm(9\text{ V} + 1\text{ V})$$
$$= \pm10\text{ V}$$

for $A_{Cl} = 3$, $R_1 = R_2$ and $C_1 = C_2$

also, $R_3 = 2\, R_4$

select, $C_1 = 1000$ pF (standard value)

$C_2 = C_1 = 1000$ pF

From Eq. 16-13, $R_1 = \dfrac{1}{2\,\pi\, f\, C_1} = \dfrac{1}{2\,\pi \times 100\text{ kHz} \times 1000\text{ pF}}$

$= 1.59$ kΩ (use 1.5 kΩ standard value)

$R_2 = R_1 = 1.5$ kΩ

select, $R_4 \approx R_2 = 1.5$ kΩ (standard value)

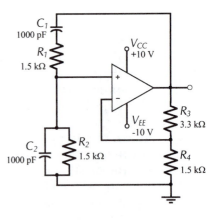

Figure 16-11
Wein bridge oscillator circuit for Example 16-4.

$$R_3 = 2\,R_4 = 2 \times 1.5\ \text{k}\Omega$$
$$= 3\ \text{k}\Omega\ (\text{use } 3.3\ \text{k}\Omega\ \text{standard value})$$

The op-amp must have a minimum full-power bandwidth (f_p) of 100 kHz when $v_o \approx \pm 9$ V and $A_{CL} = 3$.

from Eq. 15-2, $\quad f_2 = A_{CL} \times f = 3 \times 100$ kHz
$$= 300\ \text{kHz}$$

from Eq. 15-3, $\quad SR = 2\,\pi\,f_p\,v_p = 2\,\pi \times 100\ \text{kHz} \times 9\ \text{V}$
$$\approx 5.7\ \text{V/}\mu\text{s}$$

Practise Problems

16-4.1 Resistors R_1 and R_2 in Fig. 16-11 are switched to (a) 15 kΩ and (b) 5.6 kΩ. Calculate the new oscillating frequency in each case.

16-4.2 A Wein bridge oscillator using an op-amp is to produce a 15 kHz, ± 14 V output. Design the circuit with the amplifier having $A_{CL} = 11$.

16-5 Oscillator Amplitude Stabilization

Output Amplitude

For all of the oscillator circuits discussed, the output voltage amplitude is determined by the amplifier maximum output swing. The output waveform may also be distorted by the amplifier output saturation limitations. To minimize distortion and reduce the output voltage to an acceptable level, *amplitude stabilization* circuitry must be employed. Amplitude stabilization operates by ensuring that oscillation is not sustained if the output exceeds a predetermined level.

Diode Stabilization Circuit for a Phase Shift Oscillator

The phase shift oscillator discussed in Section 16-1 must have a minimum amplifier gain of 29 for the circuit to oscillate. Consider the oscillator circuit in Fig. 16-12(a) that has part of resistor R_2 bypassed by series-parallel connected diodes. When the output amplitude is low, the diodes do not become forward biased, and so they have no effect on the circuit. At this time, the amplifier voltage gain is, $A_{CL} = R_2/R_1$. As always for a phase shift oscillator, A_{CL} is designed to exceed the critical value of 29. When the output amplitude becomes large enough to forward bias either D_1 and D_2, or D_3 and D_4, resistor R_4 is short-circuited, and the amplifier gain becomes, $A_{CL} = R_5/R_1$. This is designed to be too small to sustain oscillations. So, this circuit cannot oscillate with a high-amplitude output, however it can (and does) oscillate with a low-amplitude output.

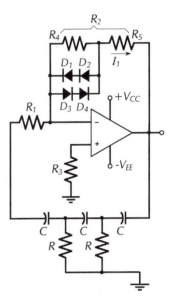

(a) Phase shift oscillator with amplitude stabilization

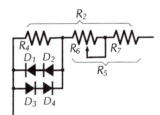

(b) Use of adjustable resistor for distortion control

Figure 16-12
The output amplitude of a phase shift oscillator can be limited by using diodes to modify the amplifier gain.

In designing the amplitude stabilization circuit, the inverting amplifier is designed in the usual manner with one important difference. The current (I_1) used in calculating the resistor values must be selected large enough to forward bias the diodes into the near-linear region of their characteristics. This usually requires a minimum current around 1 mA. Resistor R_1 is calculated using,

$$R_1 = \frac{v_o/29}{I_1} \tag{16-15}$$

and R_2 is determined as,

$$R_2 = 29\ R_1 \tag{16-16}$$

The diodes should become forward biased just when the output voltage is at the desired maximum level. At this time, I_1 produces a voltage drop of 2 V_F across R_4,

so,

$$R_4 \approx \frac{2\ V_F}{I_1} \tag{16-17}$$

and,

$$R_5 = R_2 - R_4$$

The resultant component values should give $(R_4 + R_5)/R_1$ slightly greater than 29, and R_5/R_1 less than 29.

Some distortion of the waveform can occur if $(R_4 + R_5)/R_1$ is much larger than 29, however, attempts to make the gain close to 29 can cause the circuit to stop oscillating. Making a portion of R_5 adjustable, as illustrated in Fig. 16-12(b), provides for gain adjustment to give the best possible output waveform. Typically, R_6 should be approximately 40% of the calculated value of R_5, and R_7 should be 80% of R_5. This gives a ±20% adjustment of R_5.

The diodes selected should be low-current switching devices. The diode reverse breakdown voltage should exceed the circuit supply voltage, and the maximum reverse recovery time $(t_{rr(max)})$ should be around one-tenth of the time period of the oscillation frequency,

$$t_{rr(max)} = \frac{T}{10} \tag{16-18}$$

Example 16-5

Design the phase shift oscillator in Fig. 16-13 to produce a 5 kHz, ±5 V output waveform.

Solution

Select $I_1 \approx 1$ mA when $v_{o(peak)} = 5$ V

Eq. 16-15, $R_1 = \dfrac{v_o/29}{I_1} = \dfrac{5\ V/29}{1\ mA}$

                   $= 170\ \Omega$ (use 150 Ω)

Eq. 16-16, $\quad R_2 = 29\,R_1 = 29 \times 150\ \Omega$

$\qquad\qquad \approx 4.4\ k\Omega$

Eq. 16-17, $\quad R_4 = \dfrac{2\,V_F}{I_1} = \dfrac{2 \times 0.7\ V}{1\ mA}$

$\qquad\qquad = 1.4\ k\Omega$ (use 1.5 kΩ standard value)

$\qquad R_5 = R_2 - R_4 = 4.4\ k\Omega - 1.5\ k\Omega$

$\qquad\qquad = 2.9\ k\Omega$

$\qquad R_6 = 0.4\,R_5 = 0.4 \times 2.9\ k\Omega$

$\qquad\qquad = 1.16\ \Omega$ (use 1 kΩ adjustable)

$\qquad R_7 = 0.8\,R_5 = 0.8 \times 2.9\ k\Omega$

$\qquad\qquad = 2.32\ k\Omega$ (use 2.7 kΩ standard value)

$\qquad R_3 \approx R_2 = 4.4\ k\Omega$ (use 4.7 kΩ standard value)

$\qquad R = R_1 = 150\ \Omega$

From Eq. 16-1, $\quad C = \dfrac{1}{2\,\pi\,R\,f\,\sqrt{6}} = \dfrac{1}{2\,\pi \times 150\ \Omega \times 5\ kHz \times \sqrt{6}}$

$\qquad\qquad = 0.087\ \mu F$ (use 0.082 μF standard value)

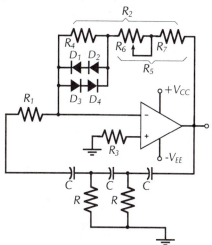

Figure 16-13
Amplitude-controlled phase shift
oscillator circuit for Ex. 16-5.

Diode Stabilization Circuit for a Wein Bridge Oscillator

Figures 16-14, and 16-15 show two output amplitude stabilization methods that can be used with a Wein bridge oscillator. These can also be applied to other oscillator circuits, because they all operate by limiting the amplifier voltage gain.

The circuit in Fig. 16-14 uses diodes and operates in the same way as the amplitude control for the phase-shift oscillator. Resistor R_6 becomes shorted by the diodes when the output amplitude exceeds the design level, thus rendering the amplifier gain too low to sustain oscillations.

FET Stabilization Circuit for a Wein Bridge Oscillator

The circuit in Fig. 16-15 is slightly more complex than the diode circuit, however, like other circuits, it stabilizes the oscillator output amplitude by controlling the amplifier gain. The channel resistance (r_{DS}) of the p-channel FET (Q_1) is in parallel with resistor R_4. Capacitor C_3 ensures that Q_1 has no effect on the amplifier dc conditions. The amplifier voltage gain is,

$$A_{CL} = \frac{R_3 + R_4 \| r_{DS}}{R_4 \| r_{DS}} \qquad (16\text{-}19)$$

The FET gate-source bias voltage is derived from the amplifier ac output. The output voltage is divided across resistors R_5 and R_6,

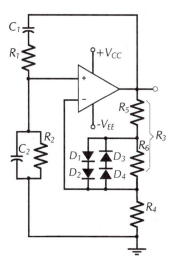

Figure 16-14
Wein bridge oscillator with its output amplitude stabilized by a diode circuit that modifies the amplifier gain.

and rectified by diode D_1. Capacitor C_4 smoothes the rectified waveform to give the *FET dc* bias voltage (V_{GS}). The polarity shown on the circuit diagram reverse biases the gate-source of the *p*-channel device. When the output amplitude is low, V_{GS} is low and this keeps the *FET* drain-source resistance (r_{DS}) low. When the output gets larger, V_{GS} is increased causing r_{DS} to increase. The increase in r_{DS} reduces A_{CL}, thus preventing the circuit from oscillating with a high output voltage level. It is seen that the *FET* is behaving as a *voltage variable resistance* (VVR).

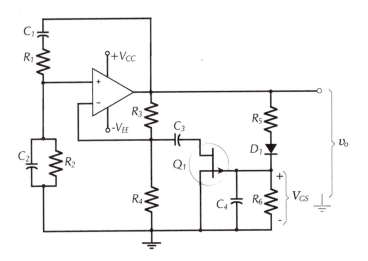

Figure 16-15
Wein bridge oscillator with the output amplitude stabilized by a FET voltage controlled resistance circuit.

Design of a FET Stabilization Circuit

To design the *FET* amplitude stabilization circuit, knowledge of a possibly suitable *FET* is required; in particular, the channel resistance at various gate-source voltages must be known.

Suppose the circuit is to oscillate when $r_{DS} = 6$ kΩ at $V_{GS} = 1$ V, and that the peak output is to be $V_{o(pk)} = 6$ V. Resistors R_5 and R_6 should be selected to give $V_{GS} = 1$ V when $V_{o(pk)} = 6$ V, allowing for V_F across the diode. Capacitor C_4 smoothes the half-wave rectified waveform, and discharges via R_6 during the time interval between peaks of the output waveform. The capacitance of C_4 is calculated to allow perhaps a 10% discharge during the time period of the oscillating frequency. The voltage divider current (I_5) should be a minimum of around 100 µA for satisfactory diode operation.

C_3 is a coupling capacitor; its impedance at the oscillating frequency should be much smaller than the r_{DS} of the *FET*. Resistors R_3 and R_4 are calculated using Eq. 16-19 to give the required amplifier voltage gain when $r_{DS} = 6$ kΩ.

Example 16-6

Design the *FET* output amplitude stabilization circuit in Fig. 16-16 to limit the output amplitude of the Wein bridge oscillator in Example 16-4 to ±6 V. Assume that the $r_{ds} = 500$ Ω at $V_{GS} = 1$ V, and 800 Ω at $V_{GS} = 3$ V.

Solution

Select
$$R_4 \approx r_{DS} \text{ at } V_{GS} = 1 \text{ V}$$
$$\approx 600 \ \Omega \text{ (use 560 } \Omega)$$
$$R_4 \| r_{DS} = 560 \ \Omega \| 600 \ \Omega$$
$$\approx 290 \ \Omega$$

for $A_{CL} = 3$,
$$R_3 = 2 \ (R_4 \| r_{DS}) = 2 \times 290 \ \Omega$$
$$= 640 \ \Omega \ \text{ (use 680 } \Omega)$$

Select,
$$I_5 \approx 200 \ \mu A \text{ when } V_{o(peak)} = 6 \text{ V}$$

$$R_6 = \frac{V_{GS}}{I_5} = \frac{1 \text{ V}}{200 \ \mu A}$$
$$= 5 \text{ k}\Omega \text{ (use 4.7 k}\Omega)$$

$$R_5 = \frac{V_{o(peak)} - (V_{GS} + V_{D1})}{I_5} = \frac{6 \text{ V} - (1 \text{ V} + 0.7 \text{ V})}{200 \ \mu A}$$
$$= 21.5 \text{ k}\Omega \text{ (use 22 k}\Omega)$$

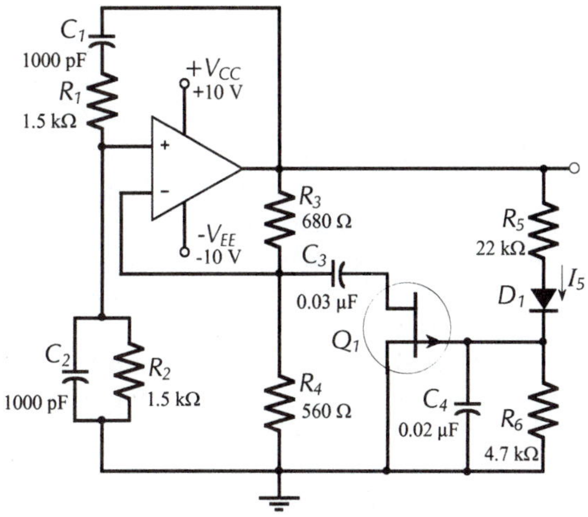

Figure 16-16
Wein bridge oscillator circuit for Example 16-6.

C_4 discharge voltage,
$$\Delta V_C = 0.1 \ V_{GS} = 0.1 \times 1 \text{ V} = 0.1 \text{ V}$$

C_4 discharge time,
$$T = \frac{1}{f} = \frac{1}{100 \text{ kHz}}$$
$$= 10 \ \mu s$$

$$I_C \approx \frac{V_{R6}}{R_6} \approx I_5 \approx 200 \ \mu A$$

$$C_4 = \frac{I_C \ T}{\Delta V_C} = \frac{200 \ \mu A \times 10 \ \mu s}{0.1 \text{ V}}$$
$$= 0.02 \ \mu F \text{ (standard value)}$$

$$X_{C3} = \frac{r_{DS}}{10} \text{ at the oscillating frequency}$$

$$C_3 = \frac{1}{2 \ \pi \ f \ r_{DS}/10} = \frac{1}{2 \ \pi \times 100 \text{ kHz} \times 500 \ \Omega/10}$$
$$= 0.032 \ \mu F \text{ (use 0.03 } \mu F)$$

D_1 should be a low current switching diode with a $t_{rr} \ll T$.

Practise Problems

16-5.1 An op-amp phase shift oscillator is to produce a 3.3 kHz, ±7 V output. Design the circuit to use diode amplitude stabilization.

16-5.2 Modify the Wein bridge oscillator circuit in Ex. 16-4 to stabilize the output amplitude to ±5 V. Use the diode circuit in Fig. 16-14.

16-6   Square Wave Generator

A *square wave generator* can be constructed by adding a resistor and capacitor to an inverting Schmitt trigger circuit (see Section 14-10). Figure 16-17 shows the circuit, which is also known as an *astable multivibrator*. The operational amplifier together with resistors R_2 and R_3 constitute the inverting Schmitt trigger circuit. Capacitor C_1 controls the voltage at the Schmitt input, and resistor R_1 charges and discharges C_1 from the Schmitt output.

The circuit waveforms in Fig. 16-17 illustrate the square wave generator operation. When the output is *high* (at the op-amp output saturation level), current flows through R_1 charging C_1 positively until v_{C1} equals the Schmitt *UTP*. The Schmitt output then switches to the op-amp negative saturation level. Current now commences to flow out of the capacitor via R_1, causing v_{C1} to decrease until it arrives at the *LTP* of the Schmitt circuit. At this point, the Schmitt output switches to its positive level once again, and the cycle recommences.

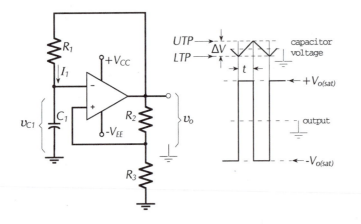

Figure 16-17
Square wave generator consisting of an inverting Schmitt trigger circuit and a series RC circuit.

Design of this very simple square wave generator involves design of the Schmitt trigger circuit, and determination of suitable R_1 and C_1 values for the trigger voltage levels and the required charge and discharge times. Selecting the *UTP* and *LTP* very much smaller than the op-amp output levels keeps the voltage drop across R_1 approximately constant. This means that the capacitor charging current is also maintained fairly constant, and so the simple constant-current equation can be used for the capacitor;

$$C_1 = \frac{I_1 \times t}{\Delta V} \tag{16-20}$$

In Eq. 16-20, I_1 is the average charging current to the capacitor (through R_1), t is the charging time (see Fig. 16-17), and ΔV is the capacitor voltage change between the *UTP* and *LTP*, as illustrated. Charging current I_1 should be selected much larger than the op-amp input bias current; then R_1 and C_1 are calculated. Alternatively, a convenient value of C_1 can first be selected. The level of I_1 is then determined from Eq. 16-20.

If a Schmitt trigger circuit with a large difference between the *UTP* and *LTP* is used, the capacitor changing equation[1] is,

$$e_c = E - (E - E_o)\varepsilon^{-t/RC} \qquad\qquad (16\text{-}21)$$

Example 16-7
Design the square wave generator in Fig. 16-18 to produce a 1 kHz square wave with an amplitude of approximately ±14 V. Use a *741* op-amp.

Solution

$$V_{CC} \approx \pm(V_o + 1\text{ V}) \approx \pm(14\text{ V} + 1\text{ V})$$
$$\approx \pm15\text{ V}$$

$$V_{R3} = UTP = -LTP << V_{CC}$$

Select
$$V_{R3} \approx 0.5\text{ V}$$

Select
$$I_2 \approx 100 \times I_{B(max)} = 100 \times 500\text{ nA}$$
$$= 50\ \mu A$$

$$R_3 = \frac{V_{R3}}{I_2} = \frac{0.5\text{ V}}{50\ \mu A}$$
$$= 10\text{ k}\Omega \quad (\text{standard value})$$

$$R_2 = \frac{V_o - V_{R3}}{I_2} = \frac{14\text{ V} - 0.5\text{ V}}{50\ \mu A}$$
$$= 270\text{ k}\Omega \quad (\text{standard value})$$

$$t = \frac{T}{2} = \frac{1}{2f} = \frac{1}{2 \times 1\text{ kHz}}$$
$$= 0.5\text{ ms}$$

$$\Delta V = UTP - LTP = 0.5\text{ V} - (-0.5\text{ V})$$
$$= 1\text{ V}$$

Select
$$C_1 = 0.1\ \mu F \text{ (convenient value)}$$

From Eq. 16-20,
$$I_1 = \frac{C_1\ \Delta V}{t} = \frac{0.1\ \mu F \times 1\text{ V}}{0.5\text{ ms}}$$
$$= 200\ \mu A$$

$$V_{R1(ave)} \approx 14\text{ V}$$

$$R_1 = \frac{V_{R1}}{I_1} = \frac{14\text{ V}}{200\ \mu A}$$
$$= 70\text{ k}\Omega \ (\text{ use } 68\text{ k}\Omega \text{ standard value})$$

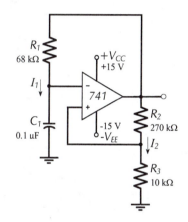

Figure   16-18
Square wave generator circuit for Example 16-7.

---

[1] David A. Bell, *Solid State Pulse Circuits 4th ed.,* (1997) p. 33.

Practise Problems

16-6.1 The square wave generator designed in Example 16-7 is to be modified to make the output frequency adjustable. Determine the maximum and minimum values for R_1 to produce a frequency range of 500 Hz to 5 kHz.

16-6.2 A Schmitt trigger circuit with ±0.8 V trigger points and a ±9 V supply is to be used in a 9 kHz square wave generator (as in Fig. 16-17). Determine suitable R_1 and C_1 values.

16-7   Triangular Wave Generator

Integrator

The circuit in Fig. 16-19 is an *integrator*; its output amplitude can be shown to be directly proportional to the area of the input pulse (amplitude x time). The circuit is simiar to an op-amp inverting amplifier, except that capacitor C_1 replaces the resistor usually connected between the output and inverting input terminals. As in the case of the inverting amplifier, the op-amp inverting input terminal remains at ground level (a virtual ground) because the noninverting input is grounded. The output voltage depends upon the capacitor charge;

$$V_o = V_{C1}$$

If the capcitor charge is $V_{C1} = 0$, then the output is $V_o = 0$. If $V_{C1} = -1$ V (negative on the right, as illustrated), $V_o = -1$ V. If $V_{C1} = +1$ V (positive on the right), $V_o = +1$ V.

The circuit input current is,

$$I_1 = \frac{V_i}{R_1}$$

With V_i constant, I_1 is a constant current flowing into C_1. The capacitor constant current charging equation can be used to calculate the capacitor voltage;

$$V_{C1} = \frac{I_1 t}{C_1}$$

or, $$V_o = \frac{I_1 t}{C_1} \qquad\qquad (16\text{-}22)$$

In Eq. 16-22, t is the time duration of I_1, or the input pulse width, as illustrated in Fig. 16-19.

When the input voltage is positive $(+V_i)$, I_1 is a positive quantity flowing through R_1 and into C_1 in the direction shown on the circuit diagram. This causes C_1 to charge, positive on the left, negative on the right, as illustrated. With the input voltage constant, I_1 is a constant current, and V_o increases constantly in a negative direction, as shown. When the polarity of the input voltage is reversed (to $-V_i$), the direction of I_1 is reversed, and the charging direction of C_1 is also reversed. Thus, V_o commences to grow in a positive direction. So, a square wave input to the integrating circuit produces a triangular wave output, as illustrated.

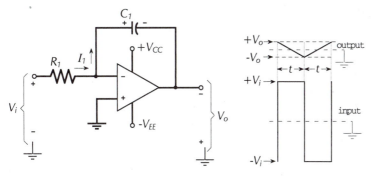

Figure 16-19
An integrator circuit produces a triangular waveform output from a square wave input.

Integrator Combined with Schmitt Trigger

Figure 16-20 shows an integrator combined with a noninverting Schmitt trigger circuit, (see Section 14-10). As will be explained, this combination constitutes a *triangular waveform generator*. The Schmitt trigger output is applied to the integrator input, and the integrator output functions as the Schmitt circuit input. The waveforms illustrate the operation of the circuit.

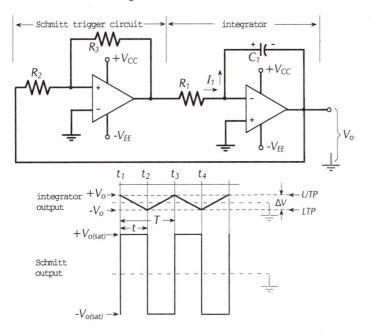

Figure 16-20
Triangular waveform generator consisting of a noninverting Schmitt trigger circuit and an integrator. The Schmitt output is the integrator input, and the integrator output is applied as the Schmitt input.

During the time from instant t_1 to instant t_2, the Schmitt output is positive (at $+V_{o(sat)}$), and the integrater output is changing at a constant rate in a negative-going direction. The output change is ΔV; from $+V_o$ to $-V_o$. The Schmitt circuit is designed to have upper and lower trigger points (*UTP* and *LTP*) equal to the desired levels of $+V_o$ and $-V_o$. Thus, when the integrator output arrives at the *LTP*, the Schmitt output (the integrator *input*) switches from $+V_{o(sat)}$ to $-V_{o(sat)}$. The integrator input is now a constant negative voltage, so that the integrator output direction is reversed. From t_2 to t_3, the integrator output increases linearly from $-V_o$ to $+V_o$, that is from the Schmitt *LTP* to its *UTP*. At the *UTP* the Schmitt output reverses again, causing the integrator output to reverse direction once more. The cycle repeats again and again producing a triangular waveform at the integrator output terminal.

The frequency of the triangular output wave can be varied by altering the charging rate of capacitor C_1. This is done by making R_1 adjustable, so that I_1 can be increased or decreased. Reducing the resistance of R_1 increases the level of I_1 and causes C_1 to be charged faster. This reduces the time (t) between $+V_o$ and $-V_o$, and thus increases the output frequency. An increase in the resistance of R_1 reduces I_1, increases t, and results in a lower output frequency. The output amplitude of the triangular wave can be varied by altering the Schmitt *UTP* and *LTP*. This can be done by making one of the Schmitt resistors (R_2 or R_3) partially adjustable.

Circuit Design

A waveform generator is normally designed to produce a specified output amplitude and frequency. This means that the Schmitt Circuit trigger points must be equal to the required positive and negative output peaks of the triangular wave, (see Section 14-10). Also, the integrator has to accept the Schmitt output as its input, and its capacitor charging time should equal half the time period of the specified output frequency. Resistor R_1 might be made partially variable to precisely set the output to the required frequency.

Example 16-8

Design a triangular waveform generator to produce a ± 3 V, 500 Hz output. Use *741* op-amps with a ± 9 V supply.

Solution

Integrator design:

$$V_i \approx \pm(V_{CC} - 1\text{ V}) = \pm(9\text{ V} - 1\text{ V})$$
$$= \pm 8\text{ V}$$

$$\Delta V = +V_o - (-V_o) = 3\text{ V} - (-3\text{V})$$
$$= 6\text{ V}$$

$$I_1 >> I_{B(max)}\text{ for the op-amp}$$

select, $$I_1 = 1\text{ mA}$$

$$R_1 = \frac{V_i}{I_1} = \frac{8\text{ V}}{1\text{ mA}}$$
$$= 8\text{ k}\Omega\text{ (use 8.2 k}\Omega$$

$$t = \frac{1}{2f} = \frac{1}{2 \times 500 \text{ Hz}}$$

$$= 1 \text{ ms}$$

from Eq. 16-22, $C_1 = \dfrac{I_1 t}{\Delta V} = \dfrac{1 \text{ mA} \times 1 \text{ ms}}{6 \text{ V}}$

$$= 0.16 \ \mu\text{F} \text{ (use 0.15 standard value)}$$

Schmit design:

select, $I_2 = 1 \text{ mA}$

$$R_2 = \frac{UTP}{I_2} = \frac{3 \text{ V}}{1 \text{ mA}}$$

$$= 3 \text{ k}\Omega \text{ (use 3.3 k}\Omega)$$

$$R_3 = \frac{V_i}{I_2} = \frac{8 \text{ V}}{1 \text{ mA}}$$

$$= 8 \text{ k}\Omega \text{ (use 8.2 k}\Omega)$$

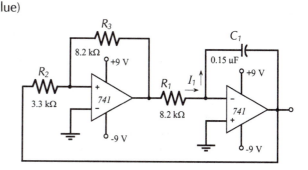

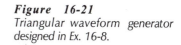

Figure 16-21
Triangular waveform generator designed in Ex. 16-8.

Practise Problems

16-7.1 A triangular waveform generator circuit (as in Fig. 16-21) has the following component values: $R_1 = 4.7 \text{ k}\Omega$, $R_2 = 3.9 \text{ k}\Omega$, $R_3 = 22$ kΩ, $C_1 = 0.05 \ \mu\text{F}$. If the supply is $V_{CC} = \pm12 \text{ V}$, determine the amplitude and frequency of the output.

16-7.2 Design a triangular wave generator to have $V_o = \pm2.5 \text{ V}$, and an output frequency adjustable from 200 Hz to 400 Hz. Use *741* op-amps with $V_{CC} = \pm15 \text{ V}$.

16-8   Oscillator Frequency Stabilization

Frequency Stability

The output frequency of oscillator circuits is normally not as stable as required for a great many applications. The component values all have tolerances, so that the actual oscillating frequency may easily be 10% higher or lower than the desired frequency. However, by making a capacitor or resistor partially adjustable, the frequency of most oscillator circuits can be set fairly precisely. Such adjustments may not be convenient in production circuits.

A further problem is that component values vary with the changes in temperature, and this causes changes in oscillating frequency. The component temperature changes might be due to variations in the ambient temperature, or the result of component power dissipation. A well-designed circuit (of any type) normally uses as little power as possible to avoid component heating and to reduce the power supply load. Oscillator frequency stability can be dramatically improved by the use of *piezoelectric crystals*.

Piezo Electric Crystals

If a mechanical stress is applied to a wafer of quartz crystal, a voltage proportional to the pressure appears at the surfaces of the crystal, [see Fig. 16-22(a)]. Also, the crystal vibrates, or *resonates*, when an alternating voltage with the natural resonance frequency of the crystal is applied to its surfaces, [Fig. 16-22(b)]. All materials with this property are termed *piezoelectric*. Because the crystal resonance frequency is extremely stable, piezoelectric crystals are used to stabilize the frequency of oscillators.

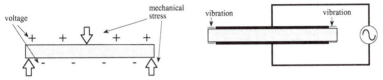

(a) Crystal with mechanical stress   (b) Crystal with alternating voltage

Figure 16-22
A piezoelectric crystal under stress produces a surface voltage. It also vibrates when an ac voltage is applied to its surfaces.

Quartz crystals for electronics applications are cut from the natural material in several different shapes. The crystals are ground to precise dimensions, and silver or gold electrodes are plated on opposite sides for electrical connections. The crystal is usually mounted inside a vacuum-sealed glass envelope or in a hermetically-sealed metal can. Figure 16-23 shows a typical crystal contained in a metal can enclosure. Crystals are also available in surface-mount and other types of enclosures.

Crystals Equivalent Circuit

The electrical equivalent circuit for a crystal is shown in Fig. 16-24(a). The crystal behaves as a series *RLC* circuit (R_s, L_s, C_s) in parallel with the capacitance of the connecting terminals (C_p). The series *RLC* components are referred to as the *motional resistance* (R_s), the *motional inductance* (L_s), and the *motional capacitance* (C_s), because they represent the piezoelectric performance of the crystal.

C_p is sometimes referred to as a *parasitic capacitance*. Because of the presence of C_p, the crystal has a *parallel resonance frequency* (f_p) when C_p resonates with the series circuit reactance, as well as a *series resonance frequency* (f_s), when L_s and C_s resonate. At series resonance the device impedance is reduced to R_s, and at parallel resonance the impedance is very high.

Figure 16-24(b) shows that, like all *RLC* circuits, the impedance of the crystal equivalent circuit is capacitive below the series resonance frequency and inductive above f_s. At frequencies greater than f_s, the series *RLC* circuit becomes inductive until it resonates with the parallel capacitance at f_p. The impedance of the complete circuit then becomes capacitive (with increasing frequency) as the reactance of C_p is reduced. The two resonance frequencies (f_s and f_p) are very close together.

A measure of the quality of a resonance circuit is the ratio of reactance to resistance, termed the *Q factor*. Because the resistive component of the crystal equivalent circuit is relatively small,

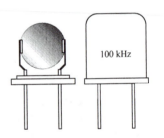

100 kHz

Figure 16-23
Electronic crystal contained in a metal can.

crystals have very large Q factors. Crystal Q factors range approximately from 2000 to 100 000, compared to a maximum of about 400 for an actual LC circuit. Resonance frequencies of available crystals are typically 10 kHz to 200 MHz.

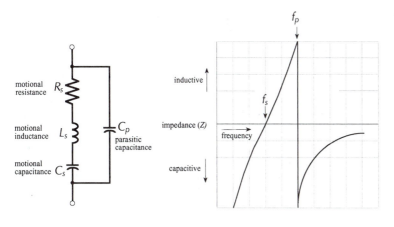

(a) Crystal equivalent circuit   (b) Crystal impedance/frequency graph

Figure 16-24
The equivalent circuit for a piezoelectric crystal is a series RLC with a parallel capacitance. The circuit has two resonance frequencies: (f_s) and (f_p).

When a series LRC circuit is operating at its resonance frequency the inductive and capacitive reactances cancel each other, and the power supplied is dissipated in the resistance. If the power dissipation increases the temperature of a crystal, the resonance frequency can drift by a small amount. Most crystals maintain their frequency to within a few cycles of the resonance frequency at 25°C. For greater frequency stability, crystals are sometimes enclosed in an insulated, thermostatically controlled, *crystal oven*.

Crystals Control of Oscillators

The frequency of an oscillator may be stabilized by using a crystal operating at either its series or parallel resonance frequency. The circuit is then usually referred to as a *crystal oscillator*.

In many circuits the crystal is connected in series with the feedback network. The crystal offers a low impedance at its series resonance and a high impedance at all other frequencies, so that oscillation occur only at the crystal series resonance frequency. The *Pierce oscillator* in Fig. 16-25 would appear to be such a circuit, however, it actually operates as a Colpitts oscillator. Capacitors C_1 and C_2 are present at the input and output of the inverting amplifier [compare to Fig. 16-6(a)]. A Colpitts oscillator also has an inductor connected between the amplifier input and output. In Fig. 16-25 the crystal behaves as an inductance by operating at a frequency slightly above f_s, (see Fig. 16-24).

The circuit in Fig. 16-25 is often used as a square wave generator or *clock oscillator* for digital circuit applications. In this situation, the amplifier gain is made as large as possible, so that Q_1 is driven between saturation and cutoff to produce a square wave output.

Figure 16-26 shows one method of modifying a Wein Bridge oscillator for frequency stabilization. The crystal is connected in

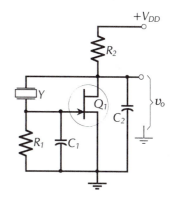

Figure 16-25
Pierce oscillator circuit using a crystal that operates at its series resonance frequency.

series with C_1 and R_1, and in this situation it offers a low resistance at its series resonance frequency (f_s), and a relatively high impedance at all other frequencies. Feedback voltage from the output via C_1 and R_1 is severely attenuated at all frequencies except f_s, so the circuit can oscillate only at f_s.

A crystal-controlled Wein bridge oscillator circuit is first designed without the crystal. A crystal is selected with f_s equal to the desired frequency. Then, the circuit is modified by subtracting the crystal series resonance resistance (R_s) from the calculated value of the series resistor, $(R_1$ in Fig. 16-26). Alternatively, circuit design could start by selection of a crystal. R_1 is then made larger than R_s, and $R_2 = (R_1 + R_s)$. C_1 and C_2 are calculated from f_s and R_2.

The crystal power dissipation must be kept below the specified maximum. Too much power can overheat the crystal and cause drift in the resonance frequency. Also, like other devices, too much power dissipation can destroy a crystal. Typical maximum crystal drive powers are 1 mW to 10 mW, but manufacturers usually recommend operating at 1/10 of the specified maximum.

The *dc* insulation resistance of crystals ranges from 100 MΩ to 500 MΩ. This allows them to be directly connected into a circuit without the need for coupling capacitors. Care must be taken in any crystal oscillator circuit to ensure that the crystal does not interrupt the flow of a necessary direct current.

The typical frequency stability of oscillators that do not use crystals is around 1 in 10^4. This means, for example, that the frequency of a 1 MHz oscillator might be 100 Hz higher or lower than 1 MHz. Using a crystal, the frequency stability can be improved to better than 1 in 10^6, which gives a ±1 Hz variation in the output of a 1 MHz oscillator.

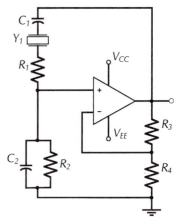

Figure 16-26
Use of a series-connected crystal to stabilize the frequency of a Wein bridge oscillator circuit.

Example 16-9

Redesign the Wein bridge oscillator in Example 16-4 to use a 100 kHz crystal with $R_S = 1.5$ kΩ.

Solution

select
$$R_1 = 2 R_s = 2 \times 1.5 \text{ k}\Omega$$
$$= 3 \text{ k}\Omega \text{ (use 2.7 k}\Omega \text{ standard value)}$$

$$R_2 = R_1 + R_s = 2.7 \text{ k}\Omega + 1.5 \text{ k}\Omega$$
$$= 4.2 \text{ k}\Omega \text{ (use 3.9 k}\Omega \text{ standard value)}$$

From Eq. 16-13,
$$C_1 = \frac{1}{2 \pi f R_2} = \frac{1}{2 \pi \times 100 \text{ kHz} \times 3.9 \text{ k}\Omega}$$
$$= 408 \text{ pF (use 390 pF standard value)}$$

$$R_4 \approx R_2 = 3.9 \text{ k}\Omega \text{ (standard value)}$$

$$R_3 = 2 R_4 = 2 \times 3.9 \text{ k}\Omega$$
$$= 7.8 \text{ k}\Omega \text{ (use 8.2 k}\Omega \text{ standard value)}$$

Example 16-10

The Pierce oscillator in Fig. 16-27 has a crystal with $f_s = 1$ MHz and $R_s = 700$ Ω. Calculate the inductance offered by the crystal at the circuit oscillating frequency. Also, estimate the power dissipated in the crystal.

Solution

eq. 16-5,
$$C_T = \frac{C_1 \times C_2}{C_1 + C_2} = \frac{1000 \text{ pF} \times 100 \text{ pF}}{1000 \text{ pF} + 100 \text{ pF}}$$

$$= 90.9 \text{ pF}$$

at resonance,
$$X_L = X_{CT}$$

$$2 \pi f L = \frac{1}{2 \pi f C_T}$$

or,
$$L = \frac{1}{(2 \pi f)^2 C_T} = \frac{1}{(2 \pi \times 1 \text{ MHz})^2 \times 90.9 \text{ pF}}$$

$$= 279 \, \mu H$$

$$i_p = \frac{V_{DD}}{R_1 + R_2 + R_s} = \frac{5 \text{ V}}{1 \text{ M}\Omega + 10 \text{ k}\Omega + 700 \, \Omega}$$

$$\approx 5 \, \mu A$$

$$P_D = (0.707 \, i_p)^2 \, R_s = (0.707 \times 5 \, \mu A)^2 \times 700 \, \Omega$$

$$= 9 \text{ nW}$$

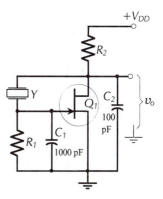

Figure 16-27
Crystal oscillator circuit for Example 16-10.

Practise Problems

16-8.1 A 40 kHz crystal with $R_s = 3$ kΩ is to be used with the Colpitts oscillator in Fig. 16-5. Determine the necessary circuit modifications, and estimate the peak power dissipation in the crystal.

16-8.2 A Pierce oscillator (as in Fig. 16-27) has a crystal with $L = 500 \, \mu H$ at $f_s = 1$ MHz. Select suitable capacitor values and calculate the minimum amplifier voltage gain.

16-8.3 Design a 50 kHz, crystal controlled, Wein bridge oscillator to use a crystal with $R_s = 2$ kΩ. The available supply voltage is $V_{CC} = \pm 12$ V. Specify the op-amp bandwidth.

Chapter-16 Review Questions

Section 16-1

16-1 State the Barkhausen criteria for a sinewave oscillator, and explain why they must be fulfilled to sustain oscillations.

16-2 Draw the circuit diagram of an op-amp phase shift oscillator. Sketch the circuit waveforms, and briefly explain the oscillator operation.

16-3   Write the oscillating frequency equation for a phase shift oscillator. Discuss the phase shift network attenuation and the amplifier gain requirements.

16-4   Sketch the circuit diagram for a phase shift oscillator that uses a single-stage *BJT* amplifier. Briefly explain the circuit operation.

Section 16-2

16-5   Draw the circuit diagram of an op-amp Colpitts oscillator. Sketch the oscillator waveforms, and briefly explain the circuit operation.

16-6   Write the frequency equation for a Colpitts oscillator. Discuss the phase shift network attenuation and the amplifier gain requirements.

16-7   Sketch circuit diagrams for a Colpitts oscillator using a single-stage *BJT* amplifier. Briefly explain the circuit operation.

Section 16-3

16-8   Draw the circuit diagram of an op-amp Hartley oscillator. Sketch the oscillator waveforms, and explain the circuit operation.

16-9   Write the frequency equation for a Hartley oscillator. Discuss the phase shift network attenuation and the amplifier gain requirements.

16-10 Sketch circuit diagrams for a Hartley oscillator using a single-stage *BJT* amplifier. Briefly explain the circuit operation.

Section 16-4

16-11 Draw the circuit diagram of an op-amp Wein bridge oscillator. Sketch the oscillator waveforms, and explain the circuit operation.

16-12 Write the frequency equation for a Wein bridge oscillator. Discuss the phase shift network attenuation and the amplifier gain requirements.

Section 16-5

16-13 Show how the output amplitude of a phase shift oscillator can be stabilized by means of a diode circuit. Explain the circuit operation.

16-14 Sketch a diode amplitude stabilization circuit for a Wein bridge oscillator, and explain its operation.

16-15 Draw a *FET* circuit diagram for stabilizing the output amplitude of a Wein Bridge oscillator. Explain the circuit operation.

Section 16-6

16-16 Draw the circuit diagram of a square wave generator that uses an inverting Schmitt trigger circuit. Sketch the circuit waveforms, and explain its operation.

Section 16-7

16-17 Sketch an op-amp integrating circuit together with the circuit waveforms. Explain the circuit operation.

16-18 Draw the circuit diagram of a triangular waveform generator that uses an integrator circuit and a noninverting Schmitt trigger. Sketch the waveforms and explain the circuit operation.

16-19 Discuss how the output amplitude and frequency can be made adjustable in a triangular wave generator.

Section 16-8

16-20 Describe a piezoelectric crystal as used with electronic circuits. Sketch the crystal equivalent circuit and impedance/frequency graph. Explain the behaviour of electronic crystals.

16-21 Show how piezoelectric crystals are employed for oscillator stabilization. Explain.

Chapter-16 Problems

Section 16-1

16-1 Design a phase shift oscillator to have a 3 kHz output frequency. Use a *741* op-amp with $V_{CC} = \pm 12$ V.

16-2 A phase shift oscillator is to use three 0.05 μF capacitors and an op-amp with $V_{CC} = \pm 9$ V. Design the circuit to have f = 7 kHz. Select a suitable operation amplifier.

16-3 Redesign the circuit in Problem 16-1 to use a single-stage *BJT* amplifier. Use a *2N3904* transistor with $V_{CC} = 15$ V.

16-4 The phase shift oscillator circuit in Fig. 16-1 has the following component values: R_1 = 3.9 kΩ, R_2 = 120 kΩ, R_3 = 120 kΩ, R = 3.9 kΩ, C = 0.025 μF. Calculate the circuit oscillating frequency.

16-5 Determine new capacitor values for the phase shift oscillator in Problem 16-1 to switch its frequency to (a) 700 Hz, (b) 5 kHz.

Section 16-2

16-6 Using a *741* op-amp with $V_{CC} = \pm 10$ V, design a Colpitts oscillator to have a 30 kHz oscillating frequency.

16-7 Design a Colpitts oscillator to have f = 55 kHz, and to use a 20 mH inductor and a *741* op-amp with $V_{CC} = \pm 18$ V.

16-8 The Colpitts oscillator circuit in Fig. 16-4 has: R_1 = 18 kΩ, R_2 = 180 kΩ, R_3 = 18 kΩ, L = 75 mH, C_1 = 3000 pF, and C_2 = 300 pF. Calculate the circuit oscillating frequency.

16-9 Design a 20 kHz Colpitts oscillator using a 10 mH inductor and a single-stage *BJT* amplifier. Use a *2N3904* transistor with V_{CC} = 20 V.

Section 16-3

16-10 Design a 6 kHz Hartley oscillator circuit to use a *741* op-amp, and an inductor with L_1 = 10 mH and L_2 = 100 mH.

16-11 Calculate the oscillating frequency for the Hartley oscillator in Fig. 16-7 if the components are: R_1 = 3.3 kΩ, R_2 = 56 kΩ, R_3 = 3.3 kΩ, L_1 = 3 mH, L_2 = 50 mH, and C_1 = 1500 pF.

16-12 An op-amp Hartley oscillator has two inductors with L_1 = 5 mH, L_2 = 40 mH, and a total inductance of L_T = 50 mH. Determine the capacitor value to give f = 2.25 kHz. Also, calculate the required amplifier voltage gain.

Section 16-4

16-13 Design a 15 kHz Wein bridge oscillator to use an op-amp with V_{CC} = ±14 V. Specify the operational amplifier.

16-14 A Wein bridge oscillator uses two 5000 pF capacitors and a *741* op-amp with V_{CC} = ±12 V. Complete the circuit design to produce a 9 kHz output frequency.

16-15 The Wein bridge oscillator in Fig. 16-10 has the following components: R_1 = R_2 = R_4 = 5.6 kΩ, R_3 = 12 kΩ, C_1 = C_2 = 2700 pF. Calculate the oscillating frequency.

16-16 The oscillator in Problem 16-15 has its capacitors changed to 0.05 µF, and resistors R_1 and R_2 are adjustable from 4.7 kΩ to 6.7 kΩ. Calculate the maximum and minimum output frequencies.

Section 16-5

16-17 The phase shift oscillator in Problem 16-1 is to have its output amplitude stabilized to ±7 V. Design a suitable diode amplitude stabilization circuit.

16-18 Design a diode amplitude stabilization circuit to limit the output of the Wein bridge oscillator in Example 16-4 to a maximum of ±4 V.

16-19 Design a 3 kHz Wein bridge oscillator using a diode circuit to stabilize the output to ±10 V. A ±20 V supply is to be used, and a suitable op-amp is to be selected.

16-20 Modify the circuit for Problem 16-19 to use a *FET* stabilization circuit, as in Fig. 16-15. Use a *FET* with r_{DS} = 200 Ω at V_{GS} = 3 V, and r_{DS} = 600 Ω at V_{GS} = 5 V.

Section 16-6

16-21 The square wave generator circuit in Fig. 16-17 has the following quantities: R_1 = 33 kΩ, R_2 = 56 kΩ, R_3 = 2.2 kΩ, C_1 = 0.15 µF, V_{CC} = ±12 V. A rail-to-rail op-amp is used. Determine the output amplitude and frequency.

16-22 Design a square wave generator circuit (as in Fig. 16-17) to produce a 3 kHz, ±9 V output. Select a suitable op-amp.

16-23 A square wave generator circuit is to use a *BIFET* op-amp with V_{CC} = ±18 V. Design the circuit to produce an output frequency adjustable from 500 Hz to 5 kHz.

16-24 Modify the circuit designed for problem 16-22 to make the output adjustable from 2.5 kHz to 3.5 kHz.

Section 16-7

16-25 Design a triangular waveform generator to produce a ±1 V, 1 kHz output. Use *741* op-amps with V_{CC} = ±12 V.

16-26 Modify the circuit designed for Problem 16-25 to make the output adjustable from 500 Hz to 1.5 kHz.

16-27 Determine the amplitude and frequency of the output from a triangular waveform generator (as in Fig. 16-20) that has the following components and supply voltage: R_1 = 10 kΩ, R_2 = 3.9 kΩ, R_3 = 22 kΩ, C_1 = 0.5 µF, V_{CC} = ±12 V.

16-28 Determine the output frequency range from the circuit in Problem 16-27 if R_1 is replaced with a 4.7 kΩ resistor in series with a 10 kΩ potentiometer.

Section 16-8

16-29 The oscillating frequency of the Hartley oscillator circuit in Fig. 16-8 is to be stabilized by means of a 100 kHz crystal with R_S = 700 Ω. Determine the necessary modifications, and estimate the peak power dissipation in the crystal.

16-30 Select suitable capacitor values for a Pierce oscillator (as in Fig. 16-27) that uses a 3 MHz crystal. The crystal has L = 390 µH at the oscillating frequency.

16-31 A 200 kHz crystal with R_S = 700 Ω is used in the Wein bridge oscillator circuit in Fig. 16-26. Design the circuit using A_{CL} ≈ 3 for the amplifier. Assume that V_{CC} = ±15 V and that the output is limited to ±5 V.

Practise Problem Answers

16-1.1 10 kΩ, 10 kΩ, 330 kΩ, 1000 pF

16-1.2 V_{CC} = 18 V, R_1 = (56 kΩ + 2.2 kΩ), R_2 = 27 kΩ, R_C = 3.3 kΩ, $(R - Z_{in})$ = 1.8 kΩ, R = 3.3 kΩ, C = (0.15 µF‖2000 pF), C_E = 75 µF

16-2.1 12 kΩ, 120 kΩ, 12 kΩ, 0.01 µF, 220 mH, 1000 pF, ±12 V

16-2.2 V_{CC} = 15 V, R_1 = (68 kΩ + 8.2 kΩ), R_2 = 47 kΩ, R_C = 4.7 kΩ, R_E = 4.7 kΩ, C_1 = 0.06 µF, C_2 = 6000 pF, C_C = 0.05 µF, C_E = 3 µF

16-3.1 $\pm$12 V, 1 kΩ, 10 kΩ, 1 kΩ, 2.2 mH, 22 mH, 0.02 μF
16-3.2 66.3 kHz
16-4.1 10.6 kHz, 28.4 kHz
16-4.2 10 kΩ, 2.2 kΩ, 27 kΩ, 2.2 kΩ, 1000 pF, 5000 pF
16-5.1 220 Ω, 5.6 kΩ, 1.5 kΩ, 2 kΩ, 3.9 kΩ, 220 Ω, 0.082 μF
16-5.2 1.5 kΩ, 2.2 kΩ, 1.5 kΩ
16-6.1 12 kΩ, 150 kΩ
16-6.2 0.1 μF, 2.7 kΩ
16-7.1 $\pm$1.42 V, 6.13 kHz
16-7.2 (6.8 kΩ + 10 kΩ pot.), 0.5 μF, 2.2 kΩ, 12 kΩ
16-8.1 Crystal in series with R_1 = 22 kΩ, 2.8 μW
16-8.2 270 pF, 62 pF, 4.4
16-8.3 3.9 kΩ, 5.6 kΩ, 560 pF, 560 pF, 12 kΩ, 5.6 kΩ, 150 kHz

Chapter *17*
Linear and Switching Voltage Regulators

Chapter Contents

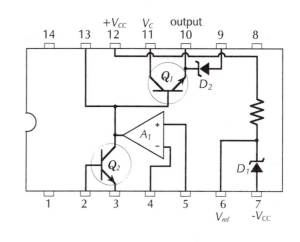

Objectives

You will be able to:

1 Sketch transistor series regulator circuits, and explain their operation.

2 Show how regulator circuits may be improved by the use of error amplifiers, additional series-pass transistors, preregulation etc., and how the output voltage may be adjusted.

3 Design transistor series regulators circuits to fulfil a given specification.

4 Analyze transistor series regulators to determine source effect, load effect, line regulation, load regulation, and ripple reduction.

5 Sketch operational amplifier series regulator circuits, and explain their operation.

6 Design op-amp series regulators circuits to fulfil a given specification.

7 Sketch and explain the basic circuit of a 723 IC voltage regulator, and design circuits using 723 ICs.

8 Sketch IC regulator block diagrams, and determine external component values for various applications.

9 Sketch the block diagram and waveforms for a switching regulator, and explain its operation.

10 Sketch circuits and waveforms for step-down, step-up, and inverting converters, and explain their operation.

11 Design LC filter circuit for various switching converters.

12 Show how an IC controller circuit is used with switching converters, and calculate values for the externally-connected components.

Introduction

Almost all electronic circuits require a direct voltage supply. This is usually derived from the standard industrial or domestic *ac* supply by transformation, rectification, and filtering. The resultant *raw dc* is not stable enough for most purposes, and it usually contains an unacceptably large *ac* ripple waveform. Voltage regulator circuit are employed to render the voltage more constant and to attenuate the ripple.

Unregulated power supplies and Zener diode regulators are discussed in Chapter 3. Zener diode regulators are normally used only where the load current does not exceed 25 mA. A transistor operating as an emitter follower circuit may be connected to a Zener diode regulator to supply larger load currents. The regulator circuit performance is tremendously improved when an *error amplifier* is included, to detect and amplify the difference between the output and the voltage reference source, and to provide feedback to correct the difference.

A variety of voltage regulator circuits are available in integrated circuit form.

17-1   Transistor Series Regulator

Basic   Circuit

When a low-power Zener diode is used in the simple regulator circuit described in Section 3-6, the load current is limited by the maximum diode current. A high-power Zener used in such a circuit can supply higher levels of load current, but much power is wasted when the load is light. The emitter follower regulator shown in Fig. 17-1 is an improvement on the simple regulator circuit because it draws a large current from the supply only when required by the load. In Fig. 17-1(a), the circuit is drawn in the form of the common collector amplifier (emitter follower) discussed in Section 6-6. In Fig. 17-1(b), the circuit is shown in the form usually referred to as a *series regulator*. Transistor Q_1 is termed a *series-pass transistor*.

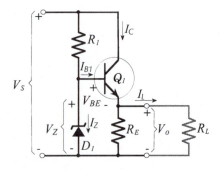

(a) Emitter follower voltage regulator

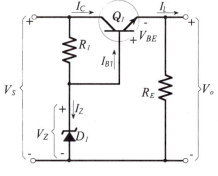

(b) Series voltage regulator circuit

Figure   17-1
To supply a large output current from a Zener diode voltage regulator, a transistor (Q_1) is connected as an emitter follower. This converts the circuit into a series voltage regulator.

The output voltage (V_o) from the series regulator is $(V_Z - V_{BE})$, and the maximum load current $(I_{L(max)})$ can be the maximum emitter current that Q_1 is capable of passing. For a *2N3055* transistor (specification in Appendix 1-8), I_L could approach 15 A. When I_L is zero, the current drawn from the supply is approximately $(I_Z + I_{C(min)})$, where $I_{C(min)}$ is the minimum collector current to keep Q_1 operational. The Zener diode circuit $(R_1$ and $D_1)$ has to supply only the base current of the transistor. The series voltage regulator is, therefore, much more efficient than a simple Zener diode regulator.

Regulator with Error Amplifier

A series regulator using an additional transistor as an *error amplifier* is shown in Fig. 17-2. The error amplifier improves the line and load regulation of the circuit, (see Section 3-5). The amplifier also makes it possible to have an output voltage greater than the Zener diode voltage. Resistor R_2 and diode D_1 are the Zener diode reference source. Transistor Q_2 and its associated components constitute the error amplifier, that controls the series-pass transistor (Q_1). The output voltage is divided by resistors R_3 and R_4, and compared to the Zener voltage level (V_Z). C_1 is a large-value capacitor, usually 50 µF to 100 µF, connected at the output to suppress any tendency of the regulator to oscillate.

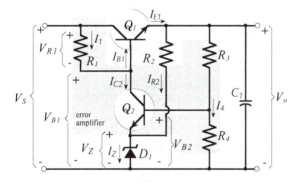

Figure 17-2
Series voltage regulator circuit with an error amplifier. The error amplifier improves the regulator line and load regulation, and gives an output voltage greater than the Zener diode voltage.

When the circuit output voltage changes, the change is amplified by transistor Q_2 and fed back to the base of Q_1 to correct the output voltage level. Suppose that the circuit is designed for $V_o = 12$ V, and that the supply voltage is $V_S = 18$ V. A suitable Zener diode voltage in this case might be $V_Z = 6$ V. For this V_Z level, the base voltage of Q_2 must be, $V_{B2} = V_Z + V_{BE2} = 6.7$ V. So, resistors R_3 and R_4 are selected to give $V_{B2} = 6.7$ V and $V_o = 12$ V. The voltage at the base of Q_1 is, $V_{B1} = V_o + V_{BE1} = 12.7$ V. Also, $V_{R1} = V_S - V_{B1} = 5.3$ V. The current through R_1 is largely the collector current of Q_2.

Now suppose the output voltage drops slightly for some reason. When V_o decreases, V_{B2} decreases. Because the emitter voltage of Q_2 is held at V_Z, any decrease in V_{B2} appears across the base-emitter of Q_2. A reduction in V_{BE2} causes I_{C2} to be reduced. When I_{C2} falls, V_{R1} is reduced, and the voltage at the base of Q_1 rises $(V_{B1} = V_S - V_{R1})$ causing the output voltage to increase. Thus, a decrease in V_o produces a feedback effect which causes V_o to increase back toward its normal level. Taking the same approach, a rise in V_o above its normal level produces a feedback effect which pushes V_o down again toward its normal level.

When the input voltage changes, the voltage across resistor R_1 changes in order to keep the output constant. This change in V_{R1} is produced by a change in I_{C2}, which itself is produced by a small change in V_o. Therefore, a supply voltage change (ΔV_S) produces a small output voltage change (ΔV_o). The relationship between ΔV_S and ΔV_o depends upon the amplification of the error amplifier. Similarly, when the load current (I_l) changes, I_{B1} alters as necessary to increase or decrease I_{E1}. The I_{B1} variation is produced by a change in I_{C2} which, once again, is the result of an output voltage variation ΔV_o.

Series Regulator Performance

The performance of a series regulator without an error amplifier (Fig. 17-1) is similar to that of a Zener diode regulator, (see Section 3-6), except in the case of the load effect. The series-pass transistor tends to improve the regulator load effect by a factor equal to the transistor h_{FE}.

The error amplifier in the regulator in Fig 17-3 (reproduced from Fig. 17-2) improves all aspects of the circuit performance by an an amount directly related to the amplifier voltage gain (A_v). When V_S changes by ΔV_S, the output change is,

$$\Delta V_o = \frac{\Delta V_S}{A_v} \qquad \textbf{(17-1)}$$

If ΔV_S is produced by a variation in the *ac* supply voltage, the power supply source effect is reduced by a factor of A_v. ΔV_S might also be the result of an increase or decrease in load current that causes a change in the average level of the *dc* supply voltage. Thus, the load effect of the power supply is reduced by a factor of A_v.

Now consider the effect of supply voltage ripple on the circuit in Fig. 17-3. The ripple waveform appears at the collector of transistor Q_1. If there was no negative feedback, it would also be present at Q_1 base and at the regulator output. However, like supply voltage changes, the input ripple is reduced by a factor of A_v when it appears at the output. The ripple rejection ratio is calculated as the decibel ratio of the input and output ripple voltages.

Example 17-1

The supply voltage for the regulator in Fig. 17-3 has $V_S = 21$ V on no load, and $V_S = 20$ V when $I_{L(max)} = 40$ mA. The output voltage is $V_o = 12$ V, and the regulator has an error amplifier with a gain of 100. Calculate the source effect, load effect, line regulation, and load regulation for the complete power supply. Also determine the ripple rejection ratio in decibels.

Solution

Eq. 3-20,   Source Effect $= \Delta V_o$ for ($\Delta V_S = 10\%$)

Eq. 17-1,
$$\Delta V_o = \frac{\Delta V_S}{A_v} = \frac{10\% \text{ of } 21 \text{ V}}{100}$$
$$= 21 \text{ mV}$$

Eq. 3-22,   Load Effect $= \Delta V_o$ for $\Delta I_{L(max)}$

Eq. 17-1,
$$\Delta V_o = \frac{\Delta V_S}{A_v} = \frac{21 \text{ V} - 20 \text{ V}}{100}$$
$$= 10 \text{ mV}$$

Eq. 3-21,   Line regulation $= \dfrac{(\text{Source Effect}) \times 100\%}{E_o}$
$$= \frac{21 \text{ mV} \times 100\%}{12 \text{ V}}$$
$$= 0.175\%$$

Eq. 3-23,   Load regulation $= \dfrac{(\text{Load Effect}) \times 100\%}{E_o}$
$$= \frac{10 \text{ mV} \times 100\%}{12 \text{ V}}$$
$$= 0.08\%$$

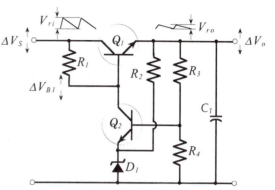

Figure 17-3
Voltage variations at the input of a regulator with an error amplifier are reduced by a factor equal to the amplifier voltage gain.

ripple rejection $= 20 \log (1/A_v) = 20 \log (1/100)$
$$= -40 \text{ dB}$$

Regulator Design

To design a series regulator circuit (as in Fig. 17-3), the Zener diode is selected to have V_Z less than the output voltage. A Zener diode voltage approximately equal to 0.75 V_o is usually suitable. Appropriate current levels are chosen for each resistor, and the resistor values are calculated using Ohm's law. Transistor Q_1 is selected to pass the required load current and to survive the necessary power dissipation. A heat sink (see Section 8-8) is normally required for the series-pass transistor in a regulator that supplies large load currents. As discussed, a large capacitor is usually connected across the output to ensure amplifier *ac* stability, (C_1 in Fig. 17-3).

The difference between the regulator input and output voltages is the collector-emitter voltage of the series-pass transistor (Q_1), and this voltage must be large enough to keep the transistor operational. The minimum level of V_{CE1} (known as the *dropout voltage*) occurs at the lowest point in the ripple waveform of (rectified and filtered) *raw dc* input. If V_{CE1} is too small for correct operation at this point, a large-amplitude ripple waveform appears at the regulator output.

Example 17-2

Design the voltage regulator circuit in Fig. 17-4 to produce $V_o = 12$ V and $I_{L(max)} = 40$ mA. The supply voltage is $V_s = 20$ V.

Solution

Select,

$$V_Z \approx 0.75\ V_o = 0.75 \times 12\ V$$
$$= 9\ V$$

For D_1, use a *1N757* Zener diode with $V_Z = 9.1$ V (see Appendix 1-4)

For minimum D_1 current,

Select

$$I_{R2} = 10\ mA$$

$$R_2 = \frac{V_o - V_Z}{I_{R2}} = \frac{12\ V - 9.1\ V}{10\ mA}$$
$$= 290\ \Omega \quad \text{(use 270 } \Omega \text{ standard value)}$$

$$I_{E1(max)} \approx I_{L(max)} + I_{R2} = 40\ mA + 10\ mA$$
$$= 50\ mA$$

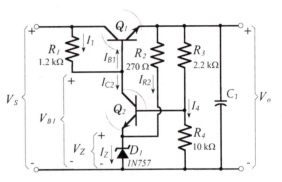

Specification for Q_1,

$$V_{CE1(max)} \approx V_S = 20\ V$$

$$I_{C1(max)} \approx I_{E(max)} = 50\ mA$$

$$P_{D(max)} = (V_S - V_o) \times I_{E1(max)} = (20\ V - 12\ V) \times 50\ mA$$
$$= 400\ mW$$

Figure 17-4
Voltage regulator circuit for Example 17-2.

Assuming $h_{FE1(min)} = 50$,

$$I_{B1(max)} = \frac{I_{E1(max)}}{h_{FE1(min)}} = \frac{50\ mA}{50}$$
$$= 1\ mA$$

$$I_{C2} > I_{B1(max)}$$
select, $$I_{C2} = 5\ mA$$

$$R_1 = \frac{V_S - V_{B1}}{I_{C2} + I_{B1}} = \frac{20\ V - (12\ V + 0.7\ V)}{5\ mA + 1\ mA}$$
$$= 1.21\ k\Omega \quad \text{(use 1.2 } k\Omega \text{ standard value)}$$

$$I_Z = I_{E2} + I_{R2} = 5\ mA + 10\ mA$$
$$= 15\ mA$$

$$I_4 \gg I_{B1(max)}$$
select, $$I_4 = 1\ mA$$

$$R_4 = \frac{V_Z + V_{BE2}}{I_4} = \frac{9.1\ V + 0.7\ V}{1\ mA}$$

$$= 9.8\ k\Omega \quad (\text{use } 10\ k\Omega \text{ standard value})$$

$$R_3 = \frac{V_o - V_{R4}}{I_4} = \frac{12\ V - 9.8\ V}{1\ mA}$$

$$= 2.2\ k\Omega \quad (\text{standard value})$$

Practise   Problems

17-1.1   A 12 V *dc* power supply has an 18 V input (V_S) from a rectifier and filter circuit. There is a 1 V drop in V_S from no load to full load. If the regulator has an error amplifier with $A_v = 70$, calculate the source effect, load effect, line regulation, and load regulation.

17-1.2   A voltage regulator circuit as in Fig. 17-4 has $V_S = 25$ V, and is to produce $V_o = 15$ V with $I_{L(max)} = 60$ mA. Design the circuit and specify the series transistor. Assume $h_{FE1(min)} = 100$.

17-2 Improving Regulator Performance

Error Amplifier Gain

The performance of a regulator is dependent on the voltage gain of the error amplifier, (see Eq. 17-1). A higher gain amplifier gives better line and load regulation. So, anything that improves the amplifier voltage gain will improve the regulator performance. Two possibilities to increase A_v are: using as transistor with a high h_{FE} value for Q_2, and using the highest possible resistance value for R_1.

Output Voltage Adjustment

Regulator circuits such as the one designed in Example 17-2 are unlikely to have V_o exactly as specified. This is because of component tolerances, as well as, perhaps, not finding standard values close to the calculated values. Hence, some form of adjustment is required to enable the output voltage to be set to the desired level. In Fig. 17-5 the potentiometer (R_5) connected between resistors R_3 and R_4 provides output adjustment. The maximum output voltage level is produced when the potentiometer moving contact is at the bottom of R_5:

$$V_{o(max)} = \frac{R_3 + R_4 + R_5}{R_4} \times V_{B2} \qquad \textbf{(17-2)}$$

Minimum V_o occurs when the moving contact is at the top of R_5:

$$V_{o(min)} = \frac{R_3 + R_4 + R_5}{R_4 + R_5} \times V_{B2} \qquad \textbf{(17-3)}$$

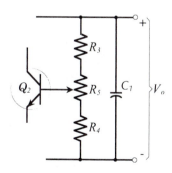

Figure   17-5
Use of a potentiometer to provide regulator output voltage adjustment.

Example 17-3
Modify the voltage regulator circuit in Ex. 17-2 (as in Fig. 17-6) to make V_o adjustable from 11 V to 13 V.

Solution

select $\qquad I_{4(min)} = 1$ mA

For $V_o = 11$ V, (moving contact at top of R_5)

$$R_3 = \frac{V_o - V_{B2}}{I_{4(min)}} = \frac{11 \text{ V} - 9.8 \text{ V}}{1 \text{ mA}}$$

$$= 1.2 \text{ k}\Omega \text{ (standard value)}$$

$$R_4 + R_5 = \frac{V_{B2}}{I_{4(min)}} = \frac{9.8 \text{ V}}{1 \text{ mA}}$$

$$= 9.8 \text{ k}\Omega$$

For $V_o = 13$ V, (moving contact at bottom of R_5)

I_4 becomes, $\qquad I_4 = \dfrac{V_o}{R_3 + R_4 + R_5} = \dfrac{13 \text{ V}}{1.2 \text{ k}\Omega + 9.8 \text{ k}\Omega}$

$$= 1.18 \text{ mA}$$

$$R_4 = \frac{V_{B2}}{I_4} = \frac{9.8 \text{ V}}{1.18 \text{ mA}}$$

$$= 8.3 \text{ k}\Omega \text{ (use 8.2 k}\Omega \text{ standard value)}$$

$$R_5 = (R_4 + R_5) - R_4 = 9.8 \text{ k}\Omega - 8.2 \text{ k}\Omega$$

$$= 2.6 \text{ k}\Omega \text{ (use 2.5 k}\Omega \text{ standard value potentiometer)}$$

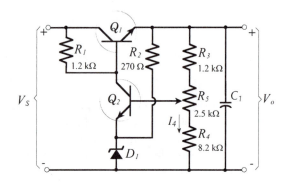

Figure 17-6
Adjustable output regulator circuit for Example 17-3.

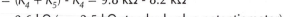

High Output Current Circuit
For the circuit in Fig. 17-6 to function correctly, the collector current of Q_2 must be larger than the maximum base current flowing into Q_1. In regulators that supply a large output current, I_{B1} can be too large for I_{C2} to control. In this case, an additional transistor (Q_3) should be connected at the base of Q_1 to form a *Darlington Circuit* (or *Darlington Pair*), as in Fig. 17-7. Transistor Q_1 is usually a high power *BJT* requiring a heat sink (see Section 8-8), and Q_3 is a low power device. The Q_3 base current is,

$$I_{B3} = \frac{I_L}{h_{FE1} \times h_{FE3}} \qquad (17\text{-}4)$$

The current gain for the Darlington is ($h_{FE1} \times h_{FE3}$), and the level of I_{B3} is low enough that I_{C2} may easily be made much greater than I_{B3}. Note resistor R_6 in Fig. 17-7, which is included to provide a suitable minimum Q_3 operating current when I_L is very low.

Darlington transistors consisting of a pair of (low-power and high-power) *BJTs* fabricated together and packaged as a single device are available. These are usually referred to as *power Darlingtons*. The *2N6039* is an *npn* power Darlington with an h_{FE} specified as 750 minimum, 18 000 maximum. Resistor R_6 in Fig. 17-7 is not required when a power Darlington is used.

Example 17-4

Modify the voltage regulator circuit in Ex. 17-2 to change the load current to 200 mA. Use a Darlington circuit (as in Fig. 17-7), and assume that $h_{FE1} = 20$, and that $h_{FE3} = 50$. Specify transistor Q_1.

Solution

$$I_{E1(max)} \approx I_{L(max)} + I_{R2} = 200 \text{ mA} + 10 \text{ mA}$$
$$= 210 \text{ mA}$$

$$I_{B1(max)} = \frac{I_{E1(max)}}{h_{FE1}} = \frac{210 \text{ mA}}{20}$$
$$= 10.5 \text{ mA}$$

$$I_{B3(max)} = \frac{I_{B1(max)}}{h_{FE3}} = \frac{10.5 \text{ mA}}{50}$$
$$= 0.21 \text{ mA}$$

$$I_{C2} > I_{B3(max)}$$

select, $\quad I_{C2} = 1 \text{ mA}$

$$R_1 = \frac{V_S - V_{B3}}{I_{C2} + I_{B3}} = \frac{20 \text{ V} - (12 \text{ V} + 0.7 \text{ V} + 0.7 \text{ V})}{1 \text{ mA} + 0.21 \text{ mA}}$$
$$= 5.45 \text{ k}\Omega \quad \text{(use 5.6 k}\Omega \text{ standard value)}$$

select, $\quad I_6 = 0.5 \text{ mA}$

$$R_6 = \frac{V_o + V_{BE1}}{I_6} = \frac{12 \text{ V} + 0.7 \text{ V}}{0.5 \text{ mA}}$$
$$= 25.4 \text{ k}\Omega \quad \text{(use 22 k}\Omega \text{ standard value)}$$

Specification for Q_1,

$$V_{CE1(max)} \approx V_S = 20 \text{ V}$$

$$I_{C1(max)} \approx I_{E1(max)} = 210 \text{ mA}$$

$$P_{D(max)} = (V_S - V_o) \times I_{E1(max)} = (20 \text{ V} - 12 \text{ V}) \times 210 \text{ mA}$$
$$= 1.68 \text{ W}$$

Preregulation

Refer to Fig. 17-8, which is a partially reproduction of Fig. 17-2. Note that resistor R_1 is supplied from the regulator input, and consider what happens when the supply voltage drops by 1 V. If I_{C2}

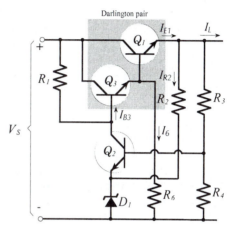

Figure 17-7
An additional transistor (Q_3) connected at the base of Q_1 to constitute a Darlington circuit allows a voltage regulator to supply a higher output current.

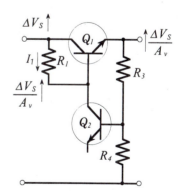

Figure 17-8
A change in a regulator supply voltage (ΔV_S) produces an output change ($\Delta V_o = \Delta V_S / A_v$).

does not change, the voltage across R_1 remains constant, and the 1 V drop also occurs at the base of Q_1 and at the output of the regulator. This does not happen, of course. Instead, I_{C2} changes to reduce the voltage across R_1 and thus keep the output voltage close to its normal level. The change in I_{C2} is produced by a small change in the output voltage. If R_1 is connected to a *constant-voltage source* instead of the input, the change in I_{C2} would not be required when V_S changes, and consequently, the output voltage change would not occur.

Now look at Fig. 17-9(a) where R_1 is shown connected to another Zener diode voltage source (R_7 and D_2). This arrangement is called a *preregulator*. When V_S changes, the change in V_{Z2} that occurs is negligible compared to ΔV_S. Thus, virtually no change is required in I_{C2} and V_o. A preregulator substantially improves the line and the load regulation of a regulator circuit.

The minimum voltage drop across R_1 should typically be 3 V. (Small values of R_1 give low amplifier voltage gain.) Also, a minimum of perhaps 6 V is required across R_7 to keep a reasonably constant current level through D_2. The voltage (V_{Z2}) for D_2 is usually a relatively high voltage for a Zener diode. It may be necessary to use two diodes in series to give the desired voltage.

Figure 17-9(b) shows another preregulator circuit that has R_7 and D_2 connected across transistor Q_1. This gives an R_1 supply voltage of $V_1 = (V_o + V_{Z2})$.

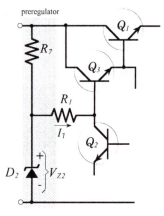

(a) Preregulator at the input

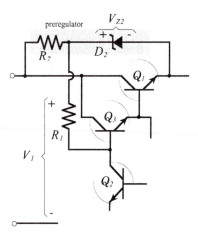

(b) Preregulator connected between input and output

Figure 17-9
A preregulator circuit improves the performance of a voltage regulator.

Example 17-5

Determine suitable component values for a preregulator circuit as in Fig. 17-9(a) for the voltage regulator modified in Ex. 17-4.

Solution

select $\quad V_{R1(min)} = 3\text{ V}$

$$R_1 = \frac{V_{R1}}{I_{C2} + I_{B3}} = \frac{3\text{ V}}{1\text{ mA} + 0.21\text{ mA}}$$

$$= 2.5\text{ k}\Omega \quad (\text{use } 2.7\text{ k}\Omega \text{ standard value})$$

$$V_{Z2} = V_o + V_{BE1} + V_{BE3} + V_{R1}$$

$$= 12\text{ V} + 0.7\text{ V} + 0.7\text{ V} + 3\text{ V}$$

$$= 16.4\text{ V} \quad (\text{Use a } IN966A \text{ with } V_Z = 16\text{ V. Alternatively,}$$
$$\text{use a } 1N753 \text{ and a } 1N758 \text{ connected in}$$
$$\text{series. See Appendix 1-4})$$

$$I_{R7} \gg I_1 \text{ and } I_{Z2} > (I_{ZK} \text{ for Zener diode } D_2)$$

select $\quad I_{R7} = 5\text{ mA}$

$$R_2 = \frac{V_S - V_{Z2}}{I_{R7}} = \frac{20\text{ V} - 16\text{ V}}{5\text{ mA}}$$

$$= 800\text{ }\Omega \quad (\text{use } 820\text{ }\Omega \text{ standard value})$$

Constant Current Source

The constant-current source shown in Fig. 17-10 may be used in place of resistor R_1 as an alternative to a preregulator circuit. This arrangement passes all the required current to Q_3 base and Q_2 collector, but behaves as a very high resistance ($1/h_{oe4}$) at the collector of transistor Q_2. Consequently, it increases the error amplifier voltage gain and results in a substantial improvement in the regulator performance. To design the constant current source, V_{CE4} should typically be a minimum of 3 V. The various voltage drops and current levels are then easily determined for component selection.

Differential Amplifier

Figure 17-11 shows a regulator that uses a differential amplifier, or *difference amplifier*, (see Section 12-8). In this circuit, Zener diode D_2 is the voltage reference source, and D_1 is used only to provide an appropriate voltage level at the emitter of transistor Q_2. V_o is divided to provide V_{B6}, and V_{B6} is compared to V_{Z2}. Any difference in these two voltages is amplified by Q_5, Q_6, Q_2, and the associated components, and then applied to the base of Q_3 to change the output in the required direction.

The performance of the regulator is improved by the increased voltage gain of the error amplifier. However, the performance is also improved in another way. In the regulator circuit in Figs. 17-2 and 17-7, when I_{C1} changes, the current through D_1 is also changed, ($I_{Z1} = I_{R2} + I_{E2}$). This causes the Zener voltage to change by a small amount, and this change in V_{Z1} produces a change in output voltage. In the circuit in Fig. 17-11, the current through the reference diode (D_2) remains substantially constant because it is supplied from the regulator output. Therefore, there is no change in output voltage due to a change in the reference voltage.

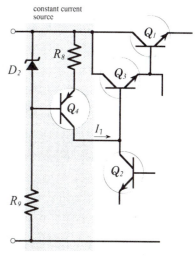

constant current source

Figure 17-10
A constant-current source can be used as an alternative to a preregulator circuit to improve regulator performance.

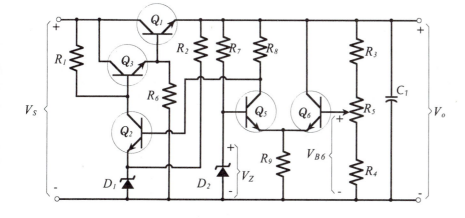

Figure 17-11
Use of a differential amplifier (Q_5 and Q_6 etc.,) improves the regulator performance by improving the gain of the error amplifier, and by using D_2 as a stable voltage reference source.

Example 17-6

Design the differential amplifier stage for the regulator in Fig. 17-11. Use the Q_1, Q_2, Q_3 stage already designed in Examples 17-2 and 17-4. The output voltage is to be adjustable from 10 V to 12 V.

Solution

$$V_{C5} = V_{B2} = 9.8 \text{ V}$$

select $$V_{CE5} = 3 \text{ V}$$

$$V_{R9} = V_{C5} - V_{CE5} = 9.8 \text{ V} - 3 \text{ V}$$
$$= 6.8 \text{ V}$$

$$V_{Z2} = V_{R9} + V_{BE5} = 6.8 \text{ V} + 0.7 \text{ V}$$
$$= 7.5 \text{ V} \text{ (use a } 1N755 \text{ Zener diode)}$$

$$I_{C5} >> I_{B2}$$
select $$I_{C5} = 1 \text{ mA}$$

$$R_8 = \frac{V_o - V_{C2}}{I_{C5}} = \frac{12 \text{ V} - 9.8 \text{ V}}{1 \text{ mA}}$$
$$= 2.2 \text{ k}\Omega \text{ (standard value)}$$

$$I_{R9} \approx 2 \times I_{C5} = 2 \text{ mA}$$

$$R_9 = \frac{V_{R9}}{I_{R9}} = \frac{6.8 \text{ V}}{2 \text{ mA}}$$
$$= 3.4 \text{ k}\Omega \text{ (use 3.3 k}\Omega \text{ standard value)}$$

$$I_{Z2} >> I_{B5} \text{ and } I_{Z2} > (I_{ZK} \text{ for the Zener diode)}$$
select $$I_{Z2} = 10 \text{ mA}$$

$$R_7 = \frac{V_o - V_{Z2}}{I_{Z2}} = \frac{12 \text{ V} - 7.5 \text{ V}}{10 \text{ mA}}$$
$$= 450 \text{ }\Omega \text{ (use 470 }\Omega \text{ standard value)}$$

$$I_4 >> I_{B6}$$
select $$I_4 = 1 \text{ mA}$$

$$V_{B6} = V_{Z2} = 7.5 \text{ V}$$

when $V_o = 11$ V, (moving contact at the top of R_5)

$$R_3 = \frac{V_o - V_{B6}}{I_4} = \frac{11 \text{ V} - 7.5 \text{ V}}{1 \text{ mA}}$$
$$= 3.5 \text{ k}\Omega \text{ (use 3.3 k }\Omega \text{ standard value)}$$

I_4 becomes, $$I_4 = \frac{V_o - V_{B6}}{R_3} = \frac{11 \text{ V} - 7.5 \text{ V}}{3.3 \text{ mA}}$$
$$= 1.06 \text{ mA}$$

$$R_4 + R_5 = \frac{V_{B6}}{I_4} = \frac{7.5 \text{ V}}{1.06 \text{ mA}}$$
$$= 7.07 \text{ k}\Omega$$

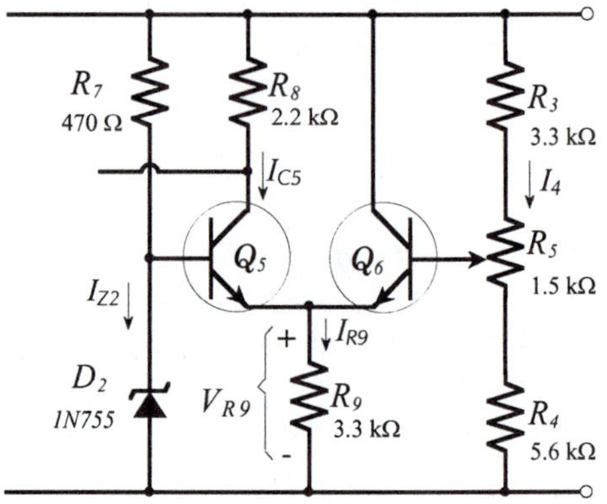

Figure 17-12
Differential amplifier circuit for Example 17-6.

When $V_o = 13$ V, (moving contact at the bottom of R_5)

I_4 becomes, $$I_4 = \frac{V_o}{R_3 + R_4 + R_5} = \frac{13 \text{ V}}{3.3 \text{ k}\Omega + 7.07 \text{ k}\Omega}$$

$$= 1.25 \text{ mA}$$

$$R_4 = \frac{V_{B6}}{I_4} = \frac{7.5 \text{ V}}{1.25 \text{ mA}}$$

$$= 6.25 \text{ k}\Omega \text{ (use 5.6 k}\Omega \text{ standard value)}$$

$$R_5 = (R_4 + R_5) - R_4 = 7.07 \text{ k}\Omega - 5.6 \text{ k}\Omega$$

$$= 1.47 \text{ k}\Omega \text{ (use 1.5 k}\Omega \text{ standard value potentiometer)}$$

Practise Problems

17-2.1 A voltage regulator circuit as in Fig. 17-6 uses a 6.2 V Zener diode. Determine suitable resistor values for R_3, R_4, and R_5 to produce an output adjustable from 9 V to 12 V.

17-2.2 A voltage regulator has an 18 V supply, a 10 V output, and a 150 mA load current. The circuit uses Darlington connected transistors, as in Fig. 17-7. Calculate suitable resistor values, and specify transistor Q_1. Assume that $h_{FE1} = 20$ and $h_{FE3} = 50$.

17-2.3 Determine suitable components for the constant current circuit in Fig. 17-10. Assume that the supply voltage is 20 V, the output is 12 V, and that I_{B3} is 100 μA.

17-3 Current Limiting

Short-Circuit Protection

Power supplies used in laboratories are subject to overloads and short circuits. Short-circuit protection by means of current limiting circuits is necessary in such equipment to prevent the destruction of components when an overload occurs. Transistor Q_7 and resistor R_{10} in Fig. 17-13(a) constitute a *current limiting circuit*. When the load current (I_L) flowing through resistor R_{10} is below the normal maximum level, the voltage drop V_{R10} is not large enough to forward bias the base-emitter junction of Q_7. In this case, Q_7 has no effect on the regulator performance.

When the load current reaches the selected maximum ($I_{L(max)}$), V_{R10} biases Q_7 on. Current I_{C7} then produce a voltage drop across resistor R_1 that drives the output voltage down to near zero.

The voltage/current characteristic of the regulator is shown in Fig. 17-13(b). It is seen that output voltage remains constant as the load current increases up to $I_{L(max)}$. Beyond $I_{L(max)}$, V_o drops to zero, and a short-circuit current (I_{SC}) slightly greater than $I_{L(max)}$ flows at the output. Under this circumstance, series pass transistor Q_1 is carrying all of the short-circuit current and has virtually all of the supply voltage developed across its terminals.

So, the power dissipation in Q_1 is,

$$P_1 = V_S \times I_{SC} \qquad (17\text{-}5)$$

Obviously, Q_1 must be selected to survive this power dissipation.

Design of the current limiting circuit in Fig. 17-13 is very simple. Assuming that Q_7 is a silicon transistor, it should begin to conduct when $V_{R10} \approx 0.5$ V. Therefore, R_{10} is calculated as,

$$R_{10} \approx \frac{0.5 \text{ V}}{I_{L(max)}} \qquad (17\text{-}6)$$

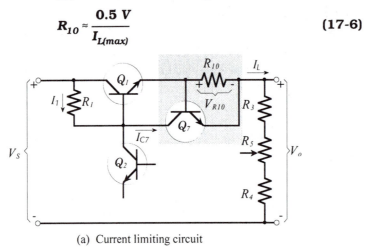

(a) Current limiting circuit

Figure 17-13
Regulator overload protection circuit. When I_L increases to a selected maximum level, V_{R10} causes Q_7 to conduct, and this pulls Q_1 base down causing V_o to be reduced to near zero.

(b) V_o/I_o characteristic of current limiting circuit

Fold-Back Current Limiting

A problem with the simple short-circuit protection method just discussed is that there is a large amount of power dissipation in the series pass transistor while the regulator remains short-circuited. The foldback current limiting circuit in Fig. 17-14(a) minimizes this transistor power dissipation. The graph of V_o/I_L in Fig. 17-14(b) shows that the regulator output voltage remains constant until $I_{L(max)}$ is approached. Then the current reduces (or folds back) to a lower short-circuit current level (I_{SC}). The lower level of I_{SC} produces a lower power dissipation in Q_1.

To understand how the circuit in Fig. 17-14 operates, first note that when the output is shorted, V_o equals zero. Consequently, the voltage drop across resistor R_{11} is almost zero. The voltage across R_{10} is (I_{SC} x R_{10}), and (as in the simple short-circuit protection circuit) this is designed to just keep transistor Q_3 biased *on*. When the regulator is operating normally with I_L less than $I_{L(max)}$, the voltage drop across R_{11} is,

$$V_{R11} \approx \frac{V_o \times R_{11}}{R_{11} + R_{12}}$$

To turn Q_7 *on*, the voltage drop across R_{10} must become larger than V_{R11} by enough to forward bias the base-emitter junction of Q_7. Using $V_{BE7} \approx 0.5$ V, and making $V_{R11} \approx 0.5$ V gives

$$I_{L(max)} R_{10} = 0.5 \text{ V} + 0.5 \text{ V} = 1 \text{ V}$$

With $I_{SC} R_{10} = 0.5$ V,

$$I_{L(max)} \approx 2 I_{SC} \quad \text{[see Fig. 17-14(b)]}$$

If V_{R11} is selected as 1 V,

$$I_{L(max)} R_{10} = 0.5 \text{ V} + 1 \text{ V} = 1.5 \text{ V}$$

and,

$$I_{L(max)} \approx 3 I_{SC}$$

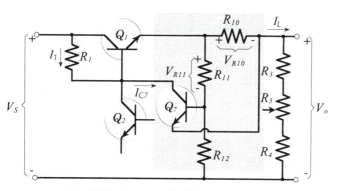

(a) Foldback current limiting circuit

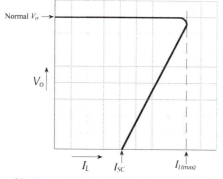

(b) Characteristic of foldback current limiting circuit

Figure 17-14
Foldback current limiting. The short-circuit current (I_{SC}) is less than $I_{L(max)}$, thus minimizing the power dissipation in Q_1.

The fold-back current limiting circuit is designed by first calculating R_{10} to give the desired level of I_{SC}. Then, the voltage divider resistors (R_{11} and R_{12}) are calculated to provide the necessary voltage drop across R_{11} for the required relationship between I_{SC} and $I_{L(max)}$.

Example 17-7

Design a foldback current limiting circuit for a voltage regulator with a 12 V output. The maximum output current is to be 200 mA, and the short-circuit current is to be 100 mA.

Solution

$$I_{SC} = 100 \text{ mA}$$

select

$$V_{R10} \approx 0.5 \text{ V at short-circuit}$$

$$R_{10} = \frac{V_{R10}}{I_{SC}} = \frac{0.5 \text{ V}}{100 \text{ mA}}$$

$$= 5 \ \Omega \ \text{(use 4.7 } \Omega \text{ standard value)}$$

at $I_{L(max)}$,

$$V_{R10} = I_{L(max)} \times R_{10} = 200 \text{ mA} \times 4.7 \ \Omega$$
$$= 0.94 \text{ V}$$

$$V_{R11} = (I_{L(max)} \ R_{10}) - 0.5 \text{ V}$$
$$= 0.44 \text{ V}$$

$$I_{11} >> I_{B7}$$

select

$$I_{11} = 1 \text{ mA}$$

$$R_{11} = \frac{V_{R11}}{I_{11}} = \frac{0.44 \text{ V}}{1 \text{ mA}}$$

$$= 440 \ \Omega \ \text{(use 470 } \Omega \text{ standard value)}$$

$$R_{12} = \frac{V_o + V_{R10} - V_{R11}}{I_{11}} = \frac{12 \text{ V} + 0.94 \text{ V} - 0.44 \text{ V}}{1 \text{ mA}}$$

$$= 12.5 \text{ k}\Omega \ \text{(use 12 k}\Omega \text{ standard value)}$$

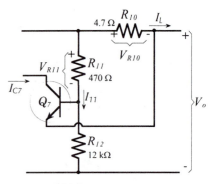

Figure 17-15
Foldback current limiting circuit for Example 17-7.

Practise Problems

17-3.1 A short-circuit protection circuit, as in Fig 17-13, is to be designed to limit the output of a 15 V regulator to 400 mA. Select a suitable value for R_{10}, and specify Q_1. Assume that $V_S = 25$ V.

17-3.2 Modify the circuit designed for Problem 17-3.1 to convert it to foldback current limiting with a 150 mA short-circuit current.

17-4   Op-amp Voltage Regulators

Voltage Follower Regulator

Refer once again to the voltage regulator circuit in Fig. 17-11. The complete error amplifier has two input terminals at the bases of Q_5 and Q_6 and one output at the collector of Q_2. Transistor Q_6 base is an inverting input and Q_5 base is a noninverting input. The error amplifier circuit is essentially an operational amplifier. Thus, *IC* operational amplifiers with their extremely high open-loop voltage gain are ideal for use as error amplifiers in *dc* voltage regulator circuits. Normally, an internally compensated op-amp (such as the *741*) is quite suitable for most voltage regulator applications.

A simple voltage follower regulator circuit is illustrated in Fig. 17-16. In this circuit, the op-amp output voltage always follows the voltage at the noninverting terminal, consequently, V_o remains constant at V_Z. The only design calculations are those required for design of the Zener diode voltage reference circuit (R_1 and D_1), and for the specification of Q_1.

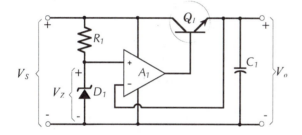

Figure 17-16
Operational amplifier connected as a voltage follower regulator. The output voltage is held constant at V_Z.

Adjustable Output Regulator

The circuit in Fig. 17-17 is that of a variable-output, highly stable *dc* voltage regulator. As in the transistor circuit in Fig. 17-11, the reference diode in Fig. 17-17 is connected at the amplifier noninverting input, and the output voltage is divided and applied to the inverting input. The operational amplifier positive supply terminal has to be connected to regulator supply voltage. If it were connected to the regulator output, the op-amp output voltage (at Q_2 base) would have to be approximately 0.7 V higher than its positive supply terminal, and this is impossible.

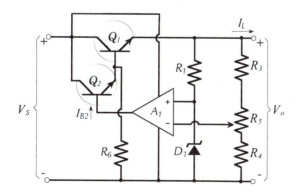

Figure 17-17
Highly stable, adjustable output, voltage regulator using an operational amplifier.

Design of the regulator circuit in Fig. 17-17 involves selection of R_1 and D_1, design of the voltage divider network (R_3, R_4, and R_5), and specification of transistors Q_1 and Q_2. Clearly, an op-amp voltage regulator is more easily designed than a purely transistor regulator circuit.

Example 17-8

Design the voltage regulator circuit in Fig. 17-18 to give an output voltage adjustable from 12 V to 15 V. The maximum output current is to be 250 mA, and the supply voltage is 20 V. Assuming $h_{FE1} = 20$ and $h_{FE2} = 50$, estimate the op-amp maximum output current.

Solution

$$V_Z \approx 0.75\, V_{o(min)} = 0.75 \times 12\ \text{V}$$
$$= 9\ \text{V}$$

Use a *1N757* diode with $V_Z = 9.1\ \text{V}$

$$I_Z >> (\text{op-amp}\ I_{B(max)}),\ \text{and}\ I_Z > (I_{ZK}\ \text{for the diode})$$

Select

$$I_Z = 10\ \text{mA}$$

$$R_1 = \frac{V_{o(min)} - V_Z}{I_{Z1}} = \frac{12\ \text{V} - 9.1\ \text{V}}{10\ \text{mA}}$$

$$= 290\ \Omega\ (\text{use}\ 270\ \Omega\ \text{standard value})$$

$$I_{3(min)} >> (\text{op-amp}\ I_{B(max)})$$

select

$$I_{3(min)} = 1\ \text{mA}$$

when $V_o = 12\ \text{V}$, (moving contact at top of R_5)

$$R_3 = \frac{V_o - V_Z}{I_{3(min)}} = \frac{12\ \text{V} - 9.1\ \text{V}}{1\ \text{mA}}$$

$$= 2.9\ \text{k}\Omega\ (\text{use}\ 2.7\ \text{k}\Omega\ \text{standard value})$$

$$R_4 + R_5 = \frac{V_Z}{I_{3(min)}} = \frac{9.1\ \text{V}}{1\ \text{mA}}$$

$$= 9.1\ \text{k}\Omega$$

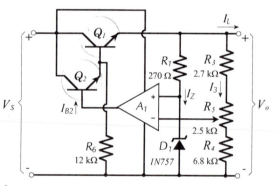

Figure 17-18
Op-amp voltage regulator for Example 17-8.

When $V_o = 15\ \text{V}$, (moving contact at bottom of R_5)

I_3 becomes,

$$I_3 = \frac{V_o}{R_3 + R_4 + R_5} = \frac{15\ \text{V}}{2.7\ \text{k}\Omega + 9.1\ \text{k}\Omega}$$

$$= 1.27\ \text{mA}$$

$$R_4 = \frac{V_Z}{I_3} = \frac{9.1\ \text{V}}{1.27\ \text{mA}}$$

$$= 7.16\ \text{k}\Omega\ (\text{use}\ 6.8\ \text{k}\Omega\ \text{standard value})$$

$$R_5 = (R_4 + R_5) - R_4 = 9.1\ \text{k}\Omega - 6.8\ \text{k}\Omega$$

$$= 2.3 \text{ k}\Omega \text{ (use 2.5 k}\Omega \text{ potentiometer)}$$

select

$$I_{R6} = 1 \text{ mA}$$

$$R_6 \approx \frac{V_o}{I_{R6}} = \frac{12 \text{ V}}{0.5 \text{ mA}}$$

$$= 24 \text{ k}\Omega \text{ (use 22 k}\Omega \text{ standard value)}$$

op-amp output current,

$$I_{B2} = \frac{I_{L(max)}}{h_{FE1} \times h_{FE2}} = \frac{250 \text{ mA}}{20 \times 50}$$

$$= 0.25 \text{ mA}$$

Current Limiting with an Op-amp Regulator

When a large output current is to be supplied by an operational amplifier voltage regulator, one of the current limiting circuits described in Section 17-3 may be used with one important modification. Figure 17-19 shows the modification. A resistor (R_{13}) must be connected between the op-amp output terminal and the junction of Q_{2B} and Q_{7C}. When an overload causes the regulator output voltage to go to zero, the op-amp output goes high (close to V_S) as it attempts to return V_o to its normal level. Consequently, because the op-amp normally has a very low output resistance, R_{13} is necessary to allow I_{C7} to drop the voltage at Q_2 base to near ground level. The additional resistor at the op-amp output is calculated as $R_{13} \approx V_S/I_{C7}$. Resistor R_{13} must not be so large that an excessive voltage drop occurs across it when the regulator is supposed to be operating normally.

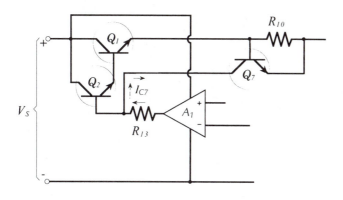

Figure 17-19
When current limiting is used with an op-amp regulator, a resistor (R_{13}) must be included in series with the op-amp output.

Practise Problems

17-4.1 Design an op-amp voltage regulator circuit as in Fig. 17-17) to produce an output adjustable from 15 V to 18 V, with a 300 mA maximum load current. The supply voltage is 25 V.

17-4.2 Design a fold-back current limiting circuit for the regulator in Problem 17-4.1, to give $I_{L(max)} = 300$ mA and $I_{SC} = 200$ mA when $V_o = 15$ V.

17-5 IC Linear Voltage Regulators

723 IC Regulator

The basic circuit of a *723* IC voltage regulator in a dual-in-line package is shown in Fig. 17-20. This IC has a voltage reference source (D_1), an error amplifier (A_1), a series pass transistor (Q_1), and a current limiting transistor (Q_2), all contained in one small package. An additional Zener diode (D_2) is included for voltage dropping in some applications. The IC can be connected to function as a positive or negative voltage regulator with an output voltage ranging from 2 V to 37 V, and output current levels up to 150 mA. The maximum supply voltage is 40 V, and the line and load regulations are each specified as 0.01%. A partial specification for the *723* regulator is given in Appendix 1-16.

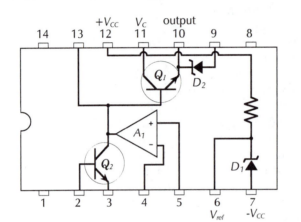

Figure 17-20
The 723 IC voltage regulator contains a reference diode (D_1), and error amplifier (A_1), a series-pass transistor (Q_1), a current limiting transistor (Q_2), and a voltage-dropping diode (D_2).

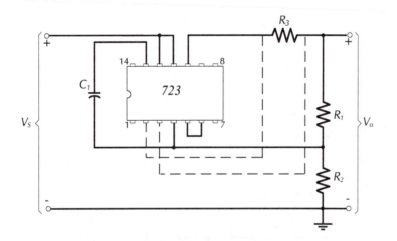

Figure 17-21
Voltage regulator circuit using a 723 IC regulator. The output voltage range is 7 V to 37 V.

Figure 17-21 shows a *723* connected to function as a positive voltage regulator. The complete arrangement (including the internal circuitry shown in Fig. 17-20) is similar to the op-amp regulator circuit in Fig. 17-17. One difference between the two circuits is the 100 pF capacitor (C_1) connected to the error amplifier output and its inverting input terminal. This capacitor is used instead of a large capacitor at the output terminals to prevent the regulator from oscillating. (Sometimes both capacitors are required.) By appropriate selection of resistors R_1 and R_2 in Fig. 17-21, the regulator output can be set to any level between 7.15 V (the reference voltage) and 37 V. A potentiometer can be included between R_1 and R_2 to make the output voltage adjustable.

The dashed lines (in Fig. 17-21) show connections for simple (non-foldback) current limiting. Fold-back current limiting can also be used with the *723*.

It is important to note that, as for all linear regulator circuits, the supply voltage at the lowest point on the ripple waveform should be at least 3 V greater than the regulator output; otherwise a high-amplitude output ripple might occur. The total power dissipation in the regulator should be calculated to ensure that it does not exceed the specified maximum. The specification lists 1.25 W as the maximum power dissipation at a free air temperature of 25°C for a *DIL* package. This must be derated at 10 mW/°C for higher temperatures. For a metal can package, $P_{D(max)} =$ 1 W at 25°C free air temperature, and the derating factor is 6.6 mW/°C. An external series-pass transistor may be Darlington-connected to (internal transistor) Q_1, to enable a *723* regulator to handle larger load current. This is illustrated in Fig. 17-22.

A regulator output voltage less than the 7.15 V reference level can be obtained by using a voltage divider across the reference source (terminals 6 and 7 in Fig. 17-20). Terminal 5 is connected to the reduced reference voltage, instead of to terminal 6.

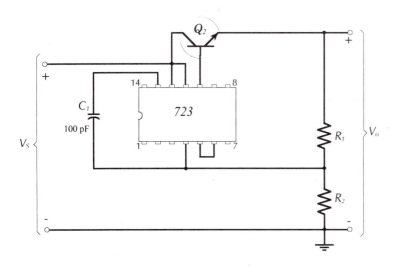

Figure 17-22
An external series-pass transistor may be added to a 723 IC voltage regulator to supply higher load currents than the 723 can normally handle.

Example 17-9
The regulator circuit in Fig. 17-21 is to have an output of 10 V. Calculate resistor values for R_1 and R_2, select a suitable input voltage, and determine the maximum load current that may be supplied if $P_{D(max)} = 1000$ mW.

Solution

$$I_2 >> \text{(error amplifier input bias current)}$$

Select $\quad\quad\quad I_2 = 1 \text{ mA}$

$$V_{R2} = V_{ref} = 7.15 \text{ V}$$

$$R_2 = \frac{V_{ref}}{I_2} = \frac{7.15 \text{ V}}{1 \text{ mA}}$$

$$= 7.15 \text{ k}\Omega \quad \text{(use 6.8 k}\Omega \text{ standard value,}$$
$$\text{and recalculate } I_2\text{)}$$

I_2 becomes, $\quad\quad I_2 = \dfrac{V_{ref}}{R_2} = \dfrac{7.15 \text{ V}}{6.8 \text{ k}\Omega}$

$$= 1.05 \text{ mA}$$

$$R_1 = \frac{V_o - V_{ref}}{I_1} = \frac{10 \text{ V} - 7.15 \text{ V}}{1.05 \text{ mA}}$$

$$\approx 2.85 \text{ K}\Omega \text{ (use 2.7 k}\Omega \text{ standard value)}$$

For satisfactory operation of the series pass transistor,

select $\quad\quad V_S - V_o = 5 \text{ V}$

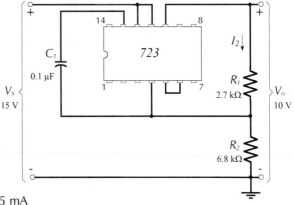

$$V_S = V_o + 5 \text{ V} = 10 \text{ V} + 5 \text{ V}$$
$$= 15 \text{ V}$$

The internal circuit current is,

$$I_{(standby)} + I_{ref} \approx 25 \text{ mA}$$

The *723* internal power dissipation on no-load is,

$$P_i = V_S \times (I_{(standby)} + I_{ref}) \; = 15 \text{ V} \times 25 \text{ mA}$$
$$= 375 \text{ mW}$$

Figure 17-23
Regulator circuit for Ex. 17-9.

Maximum power dissipated in the (internal) series-pass transistor,

$$P_D = \text{(specified } P_{D(max)}) - P_i = 1000 \text{ mW} - 375 \text{ mW}$$
$$= 625 \text{ mW}$$

Maximum load current,

$$I_{L(max)} = \frac{P_D}{V_S - V_o} = \frac{625 \text{ mW}}{5 \text{ V}}$$

$$= 125 \text{ mA}$$

LM317 and LM337 IC Regulators

The *LM317* and *LM337 IC* regulators are three-terminal devices which are extremely easy to use. The *317* is a positive voltage regulator [Fig. 17-24(a)], and the *337* is a negative voltage regulator

[Fig. 17-24(b)]. In each case, *input* and *output* terminals are provided for supply and regulated output voltage, and an adjustment terminal (*ADJ*) is included for output voltage selection. The output voltage range is 1.2 V to 37 V, and the maximum load current ranges from 300 mA to 2 A, depending on the device package type. Typical line and load regulations are specified as 0.01%/(volt of V_o), and 0.3%/(volt of V_o), respectively.

The internal reference voltage for the *317* and *337* regulators is typically 1.25 V, and V_{ref} appears across the *ADJ* and *output* terminals. Consequently, the regulator output voltage is,

$$V_o = \frac{R_1 + R_2}{R_1} \times V_{REF} \qquad (17\text{-}7)$$

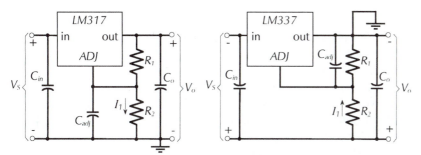

(a) *LM317* positive voltage regulator   (b) *LM337* negative voltage regulator

Figure 17-24
Application of the *LM317* and *LM337* integrated circuit voltage regulators. The internal reference voltage appears across resistor R_1.

To determine suitable values for R_1 and R_2 for a desired output voltage, first select the voltage divider current (I_1) to be much larger than the current that flows in the *ADJ* terminal of the device. This is specified as 100 μA maximum on the device data sheet. The resistors are calculated using the relationship in Eq. 17-7.

Note the capacitors included in the regulator circuits. Capacitor C_{in} is necessary only when the regulator is not located close to the power supply filter circuit. C_{in} eliminates the oscillatory tendencies that can occur with long connecting leads between the filter and regulator. Capacitor C_o improves the transient response of the regulator and ensures *ac* stability, and C_{adj} improves the ripple rejection ratio.

Figure 17-25 shows the terminal connections for *LM317* and *LM337* regulators in *221A*-type packages.

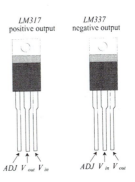

Figure 17-25
Terminal connections for *LM317* and *LM337* IC regulators contained in 221A packages.

Example 17-10
An *LM317* regulator is to provide a 6 V output from a 15 V supply. The load current is 200 mA. Determine suitable resistance values for R_1 and R_2, (in Fig. 17-24) and calculate the regulator power dissipation.

Solution

$$I_1 \gg I_{ADJ}$$

Select $\qquad I_1 = 1\ mA$

$$R_1 = \frac{V_{ref}}{I_1} = \frac{1.25\ V}{1\ mA}$$

$$= 1.25\ k\Omega \quad (use\ 1.2\ k\Omega\ standard\ value)$$

$$R_2 = \frac{V_o - V_{ref}}{I_1} = \frac{6\ V - 1.25\ V}{1\ mA}$$

$$= 4.75\ k\Omega\ (use\ 4.7\ k\Omega\ standard\ value)$$

$$P_D = (V_S - V_o) \times I_{L(max)} = (15\ V - 6\ V) \times 200\ mA$$
$$= 1.8\ W$$

LM340 Regulators

LM340 devices are three-terminal positive voltage regulators with fixed output voltages ranging from a low of 5 V to a high of 23 V. The regulator is selected for the desired output voltage and then simply provided with a voltage (V_S) from a power supply filter circuit, as illustrated in Fig. 17-26. Here again, capacitor C_1 is required only when the regulator is not located close to the filter.

The *LM340* data sheet specifies the regulator performance for an output current of 1 A. The tolerance on the output voltage is ±2%, the line regulation is *0.01% per output volt*, and the load regulation is *0.3% per amp of load current*. The IC includes current limiting, and a thermal shutdown circuit that protects against excessive internal power dissipation. As with all series regulators, a heat sink must be used when high power dissipation is involved.

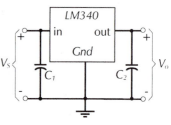

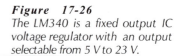

Figure 17-26
The LM340 is a fixed output IC voltage regulator with an output selectable from 5 V to 23 V.

Practise Problems

17-5.1 Design the regulator circuit in Fig. 17-22 to have an 18 V, 200 mA output. Include short circuit protection, and specify Q_2.

17-5.2 Using an *LM317*, design a 9 V, 150 mA voltage regulator. Select a supply voltage, and calculate the regulator power dissipation.

17-6 Switching Regulator Basics

Switching Regulator Operation

A switching regulator can be thought of as similar to a linear regulator, but with the series-pass transistor operating as a switch that is either *off*, or switched *on* (in a saturated state). The output voltage from the switch is a pulse waveform which is smoothed into a *dc* voltage by the action of an *LC* filter.

Switching regulators can be classified as:

- *step-down converter* (output voltage lower than input)
- *step-up converter* (output higher than input)
- *inverting converter* (output polarity opposite to input)

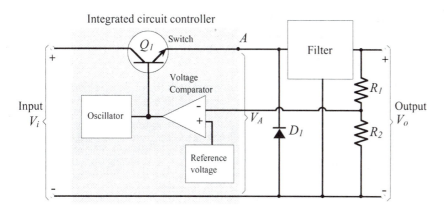

Figure 17-27
A switching regulator has a switch (contained in an IC controller), a filter, and a diode. The switch converts the dc input into a pulse waveform, and the filter smoothes the pulse wave into a direct voltage with a ripple waveform.

The basic block diagram of a step-down switching regulator in Fig. 17-27 consists of a transistor *switch* (Q_1) (also termed a *power switch*), an *oscillator*, a *voltage comparator*, a *voltage reference source*, a *diode* (D_1), and a *filter*. The switch, oscillator, comparator, and reference source are all usually contained within an integrated circuit *controller*, as illustrated. The filter usually consists of an inductor and capacitor. The operation of the regulator is as follows:

* The *dc* input voltage (V_i) is converted into a pulse waveform (V_A) by the action of the switch (Q_1) turning *on* and *off*. This is illustrated by the waveforms in Fig. 17-28.
* The oscillator switches Q_1 *on*, causing current to flow to the filter, and the output voltage to rise.
* The voltage comparator compares the output voltage (divided by R_1 and R_2) to the reference voltage, and it holds Q_1 *on* until V_o equals V_{ref}. Then, Q_1 is turned *off* again.
* The pulse waveform (V_A) at the filter input is produced by Q_1 turning *on* and *off*.
* The filter smoothes the pulse waveform to produce a *dc* output voltage (V_o) with a *ripple* waveform (V_r).
* The ripple waveform is the result of the filter capacitor charging via the filter inductor during t_{on}, and then discharging to the load during t_{off} via D_1. (The is further explained in Section 17-7.)

In the operation described above, the controller can be thought of as a *pulse width modulator*; the *on* time of Q_1 (*pulse width* of its output) is increased or decreased as necessary to supply the required output current. Other systems involve control of the switch *off* time.

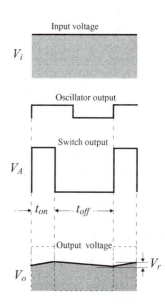

Input voltage

Oscillator output

Switch output

Output voltage

Figure 17-28
Switching regulator waveforms. The switch output is a pulse waveform, and the filter output is a dc voltage with a ripple.

Comparison of Linear and Switching Regulators

The power dissipated in the series-pass transistor in a linear regulator is wasted power. This is not very important when the load current is lower than 500 mA. With high current levels, the regulator efficiency becomes important, and there can also be serious heat dissipation problems.

In a switching regulator, the power dissipation in the switching transistor (whether it is *on* or *off*) is very much smaller than in the series-pass transistor of a linear regulator with a similar output voltage and load current. So, a switching regulator is more efficient than a linear regulator.

The approximate efficiency of a linear regulator can be calculated by assuming that the only wasted power (P_D) is that dissipated in the series-pass transistor, [see Fig. 17-29(a)].

$$P_i \approx P_o + (V_{CE} \times I_o) \qquad \text{(17-8)}$$

To estimate the efficiency of a switching regulator, it should be noted that D_1 (in Fig. 17-27) is biased *off* when Q_1 is *on*, and that D_1 is *on* when Q_1 is *off*. So, the power dissipated in the switching transistor and diode can be taken as the total wasted power, [Fig. 17-29(b)]. The Q_1 and D_1 power dissipations can be calculated in terms of the actual current level in each device and the *on* and *off* times. This analysis shows that P_D for each device is simply [(average output current) x (device voltage drop)]. So,

$$P_i \approx P_o + I_o (V_{CE(sat)} + V_F) \qquad \text{(17-9)}$$

As discussed, there is much less power dissipation in the transistor and diode in a switching regulator than in the series-pass transistor of a linear regulator with the same output conditions. So, a lower power transistor can be used in the switching regulator. Also, with a switching regulator there is usually no need for the heat sink normally required with a series regulator. Example 17-11 compares switching and linear regulators with similar load requirements.

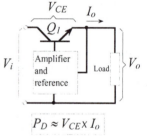

$$\boxed{P_D \approx V_{CE} \times I_o}$$

(a) Linear regulator

Example 17-11
Calculate the efficiencies of a linear regulator and a switching regulator that each supply 10 V, 1 A loads. The series-pass transistor of the linear regulator has $V_{CE} = 7$ V. The switching regulator has $V_{CE(sat)} = V_F = 1$ V.

Solution

$$P_o = V_o \times I_o = 10\,V \times 1\,A$$
$$= 10\,W$$

Linear Regulator

Eq. 17-8,
$$P_i \approx P_o + (V_{CE} \times I_o) = 10\,W + (7\,V \times 1\,A)$$
$$= 17\,W$$

$$\text{Efficiency} = \frac{P_o \times 100\%}{P_i} = \frac{10\,W \times 100\%}{17\,W}$$
$$= 59\%$$

Switching Regulator

Eq. 17-9,
$$P_i \approx P_o + I_o(V_{CE(sat)} + V_F) = 10\,W + 1\,A\,(1\,V + 1\,V)$$
$$= 12\,W$$

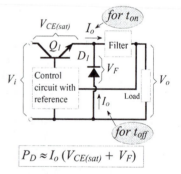

$$\boxed{P_D \approx I_o\,(V_{CE(sat)} + V_F)}$$

(b) Switching regulator

Figure 17-29
Comparison of power dissipated in linear and switching regulators.

$$Efficiency = \frac{P_o \times 100\%}{P_i} = \frac{10\,W \times 100\%}{12\,W}$$

$$= 83\%$$

The efficiency calculations in this example are approximate, because power dissipations in other parts of the circuits are neglected.

As well as efficiency, there are other considerations in the choice between linear and switching regulators. The output ripple voltage with a switching regulator is substantially larger than with a linear regulator. The line effect can be similar with both types of regulator, but the load effect is usually largest with a switching regulator. For low levels of output power, a switching regulator is usually more expensive than a linear regulator. A linear regulator is usually the best choice for output power levels up to 10 W. A switching regulator should be considered when the output is above 10 W. Table 17-1 compares the two voltage regulator types.

Table 17-1   *Comparison of linear and switching regulator performance.*

	Linear Regulator	Switching Regulator
Typical Efficiency	30% to 70%	85%
Line Effect	< 10 mV	< 10 mV
Load Effect	< 10 mV	> 50 mV
Ripple Voltage	< 10 mV (120 Hz)	100 mV (10 kHz to 200 kHz)

Practise Problems

17-6.1 Calculate the approximate efficiencies of linear and switching voltage regulators which each supply a 15 V, 750 mA load. The input voltage to both regulators is $V_i = 21$ V. The switching regulator uses a *FET* with a drain-source resistance of $r_{DS(on)} = 0.6\,\Omega$ as the power switch, and a diode with $V_F = 0.7$ V

17-7 Step-Down, Step-Up, and Inverting Converters

Step-Down Converter

A *step-down* switching regulator, or *step-down converter* (also termed a *buck* converter), produces a *dc* output voltage lower than its input voltage. The basic circuit arrangement for a step-down converter is shown in Fig. 17-30. Note the presence of the *catch diode* (D_1). This is normally reverse biased when Q_1 is *on*, but becomes forward biased when Q_1 switches *off*.

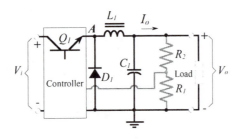

Figure 17-30
Step-down converter circuit.

Figure 17-31(a) illustrates the situation when Q_1 is *on* (in saturation). Inductor current (I_ℓ) flows from V_i, producing an inductor voltage drop (V_ℓ) which is **+** on the left, **-** on the right. At this time D_1 is reverse biased. When Q_1 switches *off* [Fig. 17-31(b)], the inductor has stored energy, and it opposes any change in current level. In order to maintain inductor current flow (with Q_1 *off*), the inductor voltage reverses, becoming **-** on the left, **+** on the right, as illustrated. The reversed polarity of V_ℓ forward biases D_1 to provide a path for I_ℓ. If D_1 was not present in the circuit, the inductor voltage would become large enough to break down the junctions of Q_1. So, D_1 *catches* V_ℓ and stops it from increasing; hence the name *catch diode*.

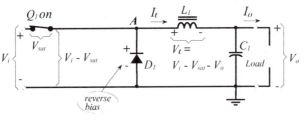

(a) Circuit conditions with Q_1 on

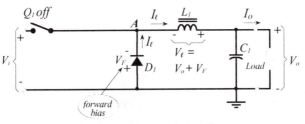

(b) Circuit conditions with Q_1 off

Figure 17-31
In a step-down converter, inductor current I_ℓ flows while Q_1 is on, producing a voltage drop (V_ℓ). When Q_1 switches off, V_ℓ reverses and D_1 becomes forward biased to keep I_ℓ flowing.

Step-Down Converter Equations

Figure 17-32 shows the circuit waveforms during the Q_1 *on* and *off* times (t_{on} and t_{off}). These are; switch voltage (V_A) at *point A* (Q_1 emitter), inductor voltage (V_ℓ), inductor current (I_ℓ), and output voltage (V_o). Equations for calculating component values can be derived by considering the circuit conditions and the waveforms.

The input voltage (V_i) appears at the collector of transistor Q_1. When Q_1 is *on*, the voltage across the inductor and capacitor (in series) is ($V_i - V_{sat}$), and the inductor voltage is ($V_i - V_{sat} - V_o$), [see Fig. 17-31(a)]. When Q_1 switches *off*, the inductor voltage becomes

$(V_o + V_F)$, [Fig. 17-31(b)]. Typically, $V_F = 0.7$ V, and V_{sat} ranges from 0.2 V to 1.5 V, depending on the transistor and the load current. The energy supplied *to the inductor* during t_{on} is proportional to [t_{on} x $(V_i - V_{sat} - V_o)$], and the energy supplied *by the inductor* to the output during t_{off} is proportional to [t_{off} x $(V_o + V_F)$]. Because the inductor input energy must equal its output energy,

$$t_{on}(V_i - V_{sat} - V_o) = t_{off}(V_o + V_F)$$

giving,

$$\frac{t_{on}}{t_{off}} = \frac{V_o + V_F}{V_i - V_{sat} - V_o} \qquad \textbf{(17-10)}$$

From Fig. 17-32,

$$t_{on} + t_{off} = T = 1/f \qquad \textbf{(17-11)}$$

The inductor is charged from the supply via Q_1 during t_{on}, and it discharges to the load and the output capacitor via D_1 during t_{off}. When the output current is a maximum ($I_{o(max)}$), the inductor current (I_L) can be allowed to change from zero to a peak level.

$$I_p = 2\,I_{o(max)} \qquad \textbf{(17-12)}$$

This is the absolute maximum inductor current change ($\Delta I = I_p$) that can occur when the circuit is operating correctly.

An equation for the inductance may be derived from a knowledge of the inductor voltage and the inductor current change during t_{on}.

$$L = \frac{e_L \Delta t}{\Delta I}$$

Giving,

$$L_{1(min)} = \frac{(V_i - V_{sat} - V_o)\,t_{on}}{I_p} \qquad \textbf{(17-13)}$$

The output voltage waveform (Fig. 17-32) is also the capacitor voltage waveform. As illustrated, the capacitor voltage increases while I_L is greater than I_o, and decreases during the time that I_L is less than I_o. While I_L is less than I_o, the average inductor current is $\Delta I/4$ below I_o, and the capacitor must supply this current to the load. The time involved is $T/2$, and the capacitor voltage change is the peak-to-peak output ripple voltage (V_r). So, an equation for the minimum capacitance of C_1 can be derived as,

$$C_1 = \frac{I\,t}{\Delta V} = \frac{(\Delta I/4) \times T/2}{V_r}$$

or,

$$C_{1(min)} = \frac{I_p}{8\,f\,V_r} \qquad \textbf{(17-14)}$$

It should be noted that Equations 17-13 and 17-14 give minimum inductor and capacitor values. The use of component values larger than the minimum will result in a lower ripple voltage at the regulator output.

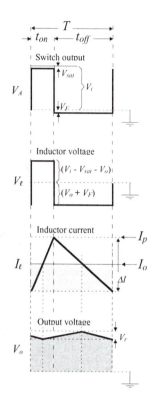

Figure 17-32
Step-down converter waveforms.

When selecting components, the capacitor *equivalent series resistance* (*ESR*) and the inductor *winding resistance* (R_W) must be taken into account. *ESR* can affect the output ripple voltage. The maximum *ESR* should usually be limited to ($0.1\, V_r/\Delta I$). If R_W is too large, it will produce an excessive voltage drop across the inductor. The maximum winding resistive voltage drop should typically not exceed 0.2 V. Also, the peak level of the inductor current must be passed without saturating the core.

Resistance values for R_1 and R_2 (in Fig. 17-30) depend on the output and reference voltages. This is treated in Section 17-8.

Example 17-12

A switching regulator with a 30 V input is to have a 12 V, 500 mA output with a 100 mV maximum ripple. The switching frequency is to be 50 kHz. Calculate the minimum filter components values. Assume that $V_{sat} = 1$ V.

Solution

$$T = \frac{1}{f} = \frac{1}{50\ \text{kHz}}$$

$$= 20\ \mu s$$

Eq. 17-10,
$$\frac{t_{on}}{t_{off}} = \frac{V_o + V_F}{V_i - V_{sat} - V_o} = \frac{12\ \text{V} + 0.7\ \text{V}}{30\ \text{V} - 1\ \text{V} - 12\ \text{V}}$$

$$= 0.75$$

or,
$$t_{on} = 0.75\ t_{off}$$

Eq. 17-11,
$$T = t_{on} + t_{off} = 0.75\ t_{off} + t_{off}$$

or,
$$t_{off} = \frac{T}{1.75} = \frac{20\ \mu s}{1.75}$$

$$\approx 11.4\ \mu s$$

$$t_{on} = T - t_{off} = 20\ \mu s - 11.4\ \mu s$$

$$= 8.6\ \mu s$$

Eq. 17-12,
$$I_p = 2\,I_{o(max)} = 2 \times 500\ \text{mA}$$
$$= 1\ \text{A}$$

Eq. 17-13,
$$L_{1(min)} = \frac{(V_i - V_{sat} - V_o)\,t_{on}}{I_p} = \frac{(30\ \text{V} - 1\ \text{V} - 12\ \text{V}) \times 8.6\ \mu s}{1\ \text{A}}$$

$$= 146\ \mu H$$

Eq. 17-14,
$$C_{1(min)} = \frac{I_p}{8\,f\,V_r} = \frac{1\ \text{A}}{8 \times 50\ \text{kHz} \times 100\ \text{mV}}$$

$$= 25\ \mu F$$

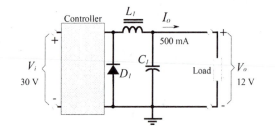

Figure 17-33
Step-down converter circuit for Example 17-12.

Step-Up Converter

A *step-up converter*, or *boost converter*, produces a *dc* output voltage higher than its supply voltage. In the circuit shown in Fig. 17-34(a), L_1 is directly connected to the supply, and D_1 is in series with L_1 and C_1. The collector of Q_1 is connected to the junction of L_1 and D_1 (*point A*), and its emitter is grounded.

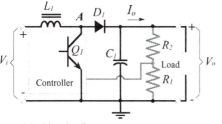

(a) Circuit of step-up converter

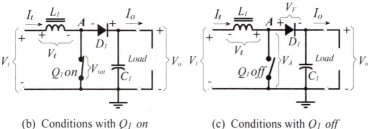

(b) Conditions with Q_1 on   (c) Conditions with Q_1 off

Figure 17-34

In a step-up converter circuit, the Inductor current I_ℓ flows while Q_1 is on, producing V_ℓ. When Q_1 switches off, V_ℓ reverses and D_1 becomes forward biased. The output voltage is $(V_i + V_\ell - V_F)$.

When Q_1 is *on* [Figure 17-34(b)], D_1 is reverse biased, and

$$V_\ell = V_i - V_{sat}$$

When Q_1 is *off* [Figure 17-34(c)], V_ℓ reverses polarity to keep I_ℓ flowing. The output voltage is now,

$$V_o = V_i + V_\ell - V_F$$

So, the output voltage is larger than the input, and its actual level can be set by selection of R_1 and R_2 [in Fig. 17-34(a)].

The voltage at *point A* has the pulse waveform (V_A) illustrated in Fig. 17-35. During t_{on} (Q_1 on time), $V_A = V_{sat}$, and during t_{off}, $V_A = (V_i + V_\ell)$. The inductor current increases to I_p during t_{on}, and decreases during t_{off} as L_1 discharges to C_1 and the load. The output voltage (V_o across C_1) decreases during t_{on}, because D_1 is reverse biased and C_1 is supplying all the load current. During t_{off}, D_1 is forward biased and C_1 is recharged from L_1. V_o increases while I_ℓ is greater than I_o, and decreases when I_ℓ is less than I_o.

From the circuit conditions and the waveforms, the following equations can be derived:

$$\frac{t_{on}}{t_{off}} = \frac{V_o + V_F - V_i}{V_i - V_{sat}}$$

(17-15)

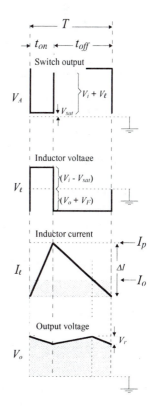

Figure 17-35
Step-up converter circuit waveforms.

$$I_p = \frac{2\,T\,I_{o(max)}}{t_{off}} \qquad\qquad \textbf{(17-16)}$$

$$L_{1(min)} = \frac{(V_i - V_{sat})\,t_{on}}{I_p} \qquad\qquad \textbf{(17-17)}$$

$$C_{1(min)} = \frac{t_{on}\,I_{o(max)}}{V_r} \qquad\qquad \textbf{(17-18)}$$

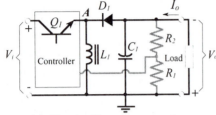

(a) Circuit of inverting converter

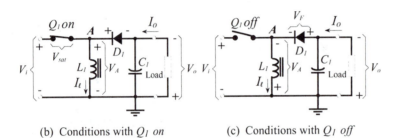

(b) Conditions with Q_1 on   (c) Conditions with Q_1 off

Inverting Converter

Figure 17-36(a) shows the circuit of an *inverting converter*, (also termed a *flyback converter*). This circuit produces a negative *dc* output voltage from a positive supply voltage. In this case, L_1 is connected between ground and the emitter terminal of Q_1, (*point A*). Diode D_1 is in series with Q_1 emitter and C_1. Note the polarity of D_1 and output capacitor C_1.

Figure 17-36(b) shows that when Q_1 is *on*, D_1 is reverse biased, and

$$V_A = V_\ell = V_i - V_{sat}$$

From Fig. 17-36(c), when Q_1 is *off* V_ℓ reverses to keep I_ℓ flowing. Diode D_1 is now forward biased, giving

$$V_o = -(V_\ell - V_F)$$

The output voltage is negative, and once again the actual output voltage level is set by selection of R_1 and R_2.

The voltage at *point A* in Fig. 17-36 is also the inductor voltage, and it has the pulse waveform shown in Fig. 17-37. During t_{on}, $V_A = (V_i - V_{sat})$; and during t_{off}, $V_A = -(V_o + V_F)$. As in other switching converters, the inductor current increases to I_p during t_{on} and

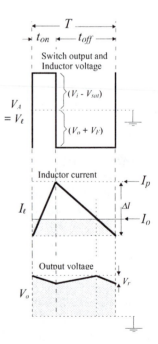

Figure 17-36
In an inverting converter circuit, the Inductor current I_ℓ flows while Q_1 is on, producing V_ℓ. When Q_1 switches off, V_ℓ reverses and D_1 becomes forward biased. The output voltage is $-(V_\ell - V_F)$.

Figure 17-37
Inverting converter circuit waveforms.

decreases during t_{off}, discharging to C_1 and the load. V_o decreases during t_{on} (D_1 reverse biased) as C_1 supplies all of I_o. Then, C_1 is recharged from L_1 during t_{off} (D_1 forward biased), and V_o increases while I_ℓ exceeds I_o, and decreases while I_ℓ is less than I_o.

For the inverting converter,

$$\frac{t_{on}}{t_{off}} = \frac{V_o + V_F}{V_i - V_{sat}} \qquad \textbf{(17-19)}$$

Equations 17-16, 17-17, and 17-18 apply for the inverting converter, as well as for the step-up converter.

Practise Problems

17-7.1 A switching regulator with $V_i = 20$ V and $f = 30$ kHz is to produce a 5 V, 1 A output with a 200 mV ripple. Assuming that $V_{sat} = 0.5$ V and $V_F = 0.7$ V, determine minimum filter component values.

17-7.2 A switching regulator is to have $V_i = 5$ V, $V_o = 9$ V, $I_{o(max)} = 600$ mA, $f = 30$ kHz, and $V_r = 200$ mV. Determine minimum filter component values. Assume that $V_{sat} = 0.5$ V and $V_F = 0.7$ V.

17-8 IC Controller for Switching Regulators

Functional Block Diagram

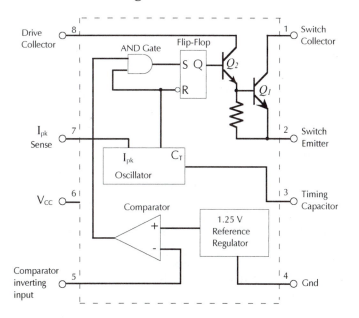

Figure 17-38
Motorola MC34063 switching regulator controller functional block diagram.
(Courtesy of Motorola Inc.)

The functional block diagram of a *MC34063* integrated circuit controller is shown in Fig. 17-38. This *IC* is designed to be used as a variable *off* time switching regulator. The components parts of the diagram in Fig. 17-38 are:

- *Switching transistor* (Q_1) controlled by transistor Q_2.
- *Set-reset flip-flop* with input terminals S and R, and output Q.

- *AND gate* that controls Q_1 and Q_2 by controlling the flip-flop.
- *Oscillator* that generates a square wave output at the desired switching frequency. The oscillator frequency is set by an externally-connected timing capacitor (C_T) at terminal *3*.
- *Current limiter* terminal (I_{pk} *sense*). This input to the oscillator permits Q_1 to be turned *off* via the reset terminal of the flip-flop.
- *Comparator* for comparing the output and reference voltages.
- *Reference voltage source* (1.25 V).

Step-Down Converter Using an MC34063

Figure 17-39 shows an *MC34063* connected to function as a step-down converter. The positive terminal of the supply is connected to terminal *6*, and the negative (grounded) terminal is connected to terminal *4*. Note that a 100 µF capacitor (C_2) is connected from terminal *6* to ground (right at the *IC* terminals) to smoothe supply-line pulses that result from fast load current changes.

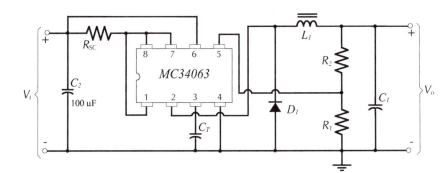

Figure 17-39
Step-down switching converter circuit using an MC34063 controller

The supply voltage (V_i) is connected (via R_{SC}) to the collectors of Q_1 and Q_2 at terminals *8* and *1*, (see Fig. 17-38). Resistor R_{SC} (connected to terminal *7*) senses the peak supply current (I_S) and provides a voltage drop to turn Q_1 *off* when the current exceeds the maximum design level. The equation for the current-limiting resistor is given on the device data sheet as,

$$R_{SC} = \frac{0.33 \text{ V}}{I_P} \qquad (17\text{-}20)$$

The output voltage (V_o) is divided across resistors R_1 and R_2 and applied to the inverting input of the voltage comparator to compare it to the reference voltage (V_{ref}). The voltage across R_1 must be equal to the V_{ref} when the circuit is operating correctly, and the voltage across ($R_1 + R_2$) equals V_o. As in all voltage divider designs, the resistor current should be much larger than the device input bias current; the comparator input current in this case.

The timing capacitor (C_T), connected from terminal 3 to ground, is selected for the desired *on* time of the switching transistor. An equation for the capacitance of C_T is given on the *IC* data sheet;

$$C_T = 4.8 \times 10^{-5} \times t_{on} \qquad (17\text{-}21)$$

Variable Off Time Modulator

Switching converter circuits are designed to supply a particular maximum load current. For a converter circuit using an *MC34063* controller, the circuit waveforms are typically as shown in Fig. 17-40(a) when supplying *full load current*. Transistor Q_1 (in Fig. 17-38) is switched *on* at the start of the each cycle of the oscillator square wave output. Also, t_{on} and t_{off} add up to the time period (T) of the oscillator, as illustrated. When the *load current is low*, capacitor C_1 discharges more slowly than when supplying full load current, [see Fig. 17-40(b)]. So, V_{R1} (in Fig. 17-39) has not dropped below V_{ref} by the beginning of the next cycle of the oscillator square wave. Consequently, Q_1 is *not* switched *on* again at this point, and so the pulse waveform frequency at Q_1 emitter is now lower than the oscillator frequency. A switching converter controller designed to operate in this way is known as a *variable off-time modulator.*

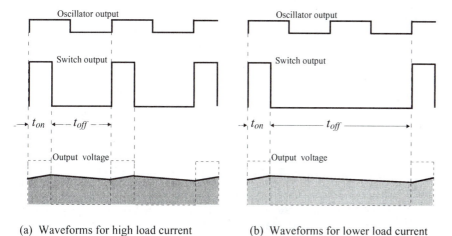

(a) Waveforms for high load current   (b) Waveforms for lower load current

Figure   17-40
Waveforms for high-current and low-current conditions in a variable off-time controller for a switching regulator.

Example   17-13

A *MC34063* controller is to be used with the step-down switching regulator designed in Ex. 17-12, as in Fig. 17-39. Determine suitable component values for R_1, R_2, R_{SC}, and C_T.

Solution

$$I_1 >> [\text{comparator input bias current } (I_B)]$$

$$I_B = -400 \text{ nA (from the } IC \text{ data sheet)}$$

Select, $I_1 = 1 \text{ mA}$

$$R_1 = \frac{V_{ref}}{I_1} = \frac{1.25 \text{ V}}{1 \text{ mA}}$$

$$= 1.25 \text{ k}\Omega \text{ (use 1.2 k}\Omega \text{ standard value)}$$

I_1 becomes, $$I_1 = \frac{V_{ref}}{R_1} = \frac{1.25 \text{ V}}{1.2 \text{ k}\Omega}$$

$$= 1.04 \text{ mA}$$

$$R_2 = \frac{V_o - V_{ref}}{I_1} = \frac{12 - 1.25 \text{ V}}{1.04 \text{ mA}}$$

$$= 10.34 \text{ k}\Omega \ [\text{use } (10 \text{ k}\Omega + 330 \ \Omega) \text{ standard values}]$$

Eq. 17-20, $$R_{SC} = \frac{0.33 \text{ V}}{I_p} = \frac{0.33 \text{ V}}{1 \text{ A}}$$

$$= 0.33 \ \Omega \ (\text{special low-value resistor})$$

Eq. 17-21, $$C_T = 4.8 \times 10^{-5} \times t_{on} = 4.8 \times 10^{-5} \times 8.6 \ \mu s$$

$$= 413 \text{ pF (use 430 pF standard value)}$$

Catch Diode Selection

The diode power dissipation can be calculated approximately from the output current and the diode voltage drop;

$$P_{D1} \approx I_o \times V_F \tag{17-22}$$

The reverse recovery time (t_{rr}) of the diode should be less than one tenth of the minimum diode *on* or *off* time, which ever is smaller;

$$t_{rr} < 0.1 \ (t_{on(min)} \text{ or } t_{off(min)}) \tag{17-23}$$

Diode Snubber

A *diode snubber* is a series *RC* circuit connected across the catch diode to suppress high frequency *ringing* that can be produced by resonance of the inductance and capacitance of components and connecting leads. (Termed *parasitic* inductance and capacitance.) A diode snubber circuit is shown in Fig. 17-41. The capacitance of C_s is usually selected in the range of four to ten times the diode junction capacitance (C_D), and then the resistance of R_s is calculated from C_s and the ringing frequency (f_r).

$$C_s = (4 \text{ to } 10) \times C_D \tag{17-24}$$

$$R_s \, C_s = \frac{1}{f_r} \tag{17-25}$$

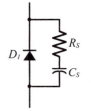

Figure 17-41
A series RC circuit connected in parallel with a diode (a diode snubber circuit) is often used to suppress high frequency ringing.

High Power Converters

When the load current of a switching regulator is too high for the controller internal transistor, an externally-connected device can be employed. Figure 17-42(a) illustrates how a high-power *BJT* (Q_3) should be connected for use with a step-down converter, and Fig. 17-42(b) shows how an external *FET* should be connected. Resistor R_B at the base of Q_3 in Fig. 17-42(a) ensures that Q_3 is biased *off* when Q_1 switches *off*. For the *p*-channel *MOSFET* (Q_3) In Fig. 17-43(b), resistors R_C and R_E are selected to give a gate-channel voltage (V_{RC}) that will switch Q_3 *on* when Q_1 is *on*.

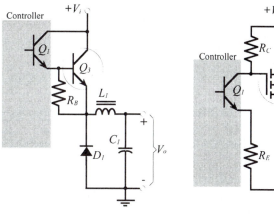

(a) Use of external *BJT*

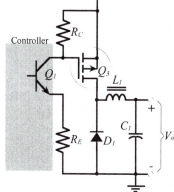

(b) External *MOSFET* connection

Figure 17-42
When the peak current is too high for an IC switching converter controller, an external power device can be connected to pass higher current levels.

Practise Problems

17-8.1 The step-down switching regulator designed for Problem 17-7.1 is to use an *MC34063* controller. Determine suitable values for the additional components.

17-8.2 The step-up switching regulator designed for Problem 17-7.2 is to use an *MC34063* controller. Determine suitable values for the additional components.

Chapter-17 Review Questions

Section 17-1

17-1 Sketch the circuit of an emitter follower voltage regulator. Explain the operation of the circuit, and discuss the effect of the transistor on the performance of the regulator.

17-2 Sketch the circuit of a series regulator with a one-transistor error amplifier. Explain the operation of the circuit, and discuss the effect of the error amplifier on the performance of the regulator.

Section 17-2

17-3 Show how the regulator circuit in Question 17-2 should be modified to produce an adjustable output voltage. Explain.

17-3 Show how the regulator in Question 17-2 should be modified to handle a large load current. Explain.

17-4 Show how voltage regulator performance may be improved by the use of a preregulator. Explain how the circuit performance is improved.

17-5 Show how a constant-current source may be used instead of a preregulator to improve the performance of a voltage regulator. Explain the operation of the circuit, and discuss how the performance of the regulator is improved.

17-6   Sketch the complete circuit of a transistor voltage regulator that uses a differential amplifier. Explain the operation of the circuit and discuss its performance.

Section 17-3

17-7   Sketch a simple current limiting circuit for short-circuit protection on a voltage regulator. Explain the operation of the circuit, and sketch the V_o/I_L characteristic for the regulator. Discuss the effects of current limiting on the series-pass transistor.

17-8   Sketch a foldback current limiting circuit for a voltage regulator. Sketch the V_o/I_L characteristic, and explain the operation of the circuit.

17-9   Write equations for power dissipation in the series pass transistor in voltage regulators using (a) simple current limiting, (b) foldback current limiting. Compare the two current limiting methods.

Section 17-4

17-10   Sketch the circuit of a voltage follower regulator using an *IC* operational amplifier. Explain the circuit operation.

17-11   Compare the performance of an *IC* op-amp voltage regulator to the performance of an emitter follower voltage regulator.

17-12   Sketch the circuit of a series regulator that uses an *IC* operational amplifier as an error amplifier. Explain the circuit operation.

17-13   For an *IC* op-amp series regulator, write equations for V_o in terms of V_Z. Briefly discuss the supply voltage requirements.

Section 17-5

17-14   Sketch the basic circuit of a *723 IC* voltage regulator. Briefly explain.

17-15   Sketch a *723 IC* voltage regulator connected to function as a positive voltage regulator with V_o greater than V_{ref}. Write an equation for V_o in terms of V_{ref}. Briefly discuss the required supply voltage.

17-16   Sketch a regulator circuit that uses an *LM317 IC* positive voltage regulator. Briefly explain.

17-17   Sketch a regulator circuit that uses an *LM337 IC* negative voltage regulator. Briefly explain.

17-18   Sketch a regulator circuit that uses an *LM340 IC* fixed voltage regulator. Briefly explain.

Section 17-6

17-19   Draw a basic block diagram and waveforms for a switching regulator. Explain its operation.

17-20 Compare switching regulators and linear regulators.

Section 17-7

17-21 Sketch the basic circuit for a step-down switching regulator. Draw current and voltage waveforms, and explain the circuit operation.

17-22 Repeat Question 17-21 for a step-up switching regulator.

17-23 Repeat Question 17-21 for an inverting switching regulator.

Section 17-8

17-24 Sketch the functional block diagram for an *MC34063* controller for a switching regulator. Briefly discuss each item in the block diagram.

17-25 Sketch the circuit of a step-up switching regulator using an *MC34063* in a similar form to Fig. 17-39.

17-26 Sketch the circuit of an inverting switching regulator using an *MC34063* in a similar form to Fig. 17-39.

17-27 Sketch a diode snubber circuit, and briefly explain.

17-28 Draw circuits to show how an external *BJT* and an external *FET* should be connected for use with step-down converters. Briefly explain.

Chapter-17 Problems

Section 17-1

17-1 Using a 15 V supply and a transistor with $h_{FE(min)}$ = 50, design a voltage follower regulator circuit (as in Fig. 17-1) to give a 9 V output with a 100 mA maximum load current.

17-2 Calculate the line regulation, load regulation, and ripple rejection ratio for the circuit designed for Problem 17-1.

17-3 A regulator circuit as in Fig. 17-1 has: V_Z = 7 V, R_1 = 560 Ω, R_E = 8.2 kΩ, R_L = 180 Ω. V_S = 19 V when I_L = 0 , and V_S = 18 V when I_L = 35 mA. The transistor has h_{FE} = 60. Calculate the Q_1 emitter current and the Zener diode current when the load is connected and when the load is disconnected.

17-4 Determine the line and load regulations for the regulator in Problem 17-3. Assume that Z_Z = 5 Ω.

17-5 Measurements on a voltage regulator with V_S = 25 V and V_o = 18 V give the following results: (when I_L changes from zero to 200 mA, ΔV_S = 2 V, and ΔV_o = 10 mV), (when ΔV_S = ±2.5 V, ΔV_o = ±12.5 mV). Calculate the source effect, load effect, line regulation, and load regulation.

17-6 Determine the voltage gain of the error amplifier in Problem 17-5, and estimate the output ripple voltage amplitude if there is a 2 V peak-to-peak input ripple.

17-7   A series regulator circuit as in Fig. 17-2 is to have a 15 V output and a 50 mA maximum load current. Select a suitable minimum supply voltage and design the circuit.

17-8   Design a voltage regulator circuit as in Fig. 17-2 to produce a 9 V output with a 30 mA maximum load current.

17-9   For the regulator designed for Problem 17-7, calculate the approximate line regulation, load regulation, and ripple reduction, if V_S drops by 1 V from zero to full load current.

17-10   Determine the approximate line regulation, load regulation, and ripple reduction for the regulator designed for Problem 17-8. Assume that $\Delta V_S = 1$ V when $\Delta I_L = $ full load current.

Section 17-2

17-11   Modify the circuit designed for Problem 17-7 to make the output adjustable from 12 V to 15 V.

17-12   Modify the regulator designed for Problem 17-8 to make the output adjustable from 8 V to 10 V.

17-13   Modify the regulator designed for Problem 17-7 to supply a maximum load current of 210 mA. Assume that all transistors used have $h_{FE} = 30$.

17-14   Modify the regulator designed for Problem 17-8 to supply a maximum load current of 300 mA. Assume that $h_{FE1} = 20$ and $h_{FE3} = 100$.

17-15   Design a series voltage regulator with a one-transistor error amplifier to provide an output adjustable from 15 V to 18 V. The load current is to be 250 mA. Assume that all transistors have $h_{FE} = 50$.

17-16   Calculate the line and load regulations for the modified regulator circuit in Problem 17-14. Assume that $\Delta V_S = 1$ V when $\Delta I_L = $ full load current.

17-17   Design a preregulator for the circuit designed for Problem 17-7 and modified for Problem 17-13.

17-18   Determine the approximate line regulation, load regulation, and ripple reduction for the modified regulator in Problem 17-17. Assume that $\Delta V_S = 1$ V when $\Delta I_L = $ full load current.

17-19   The preregulator in the circuit referred to in Problem 17-17 is to be replaced with a constant-current source. Design the constant-current source.

17-20   Design a differential amplifier as in Fig. 17-12 to use with the regulator designed for Problem 17-15.

Section 17-3

17-21   Design a current limiting circuit as in Fig. 17-13(a), to limit the maximum load current to approximately 220 mA for the regulator in Problem 17-13. Calculate the power dissipation in the series pass transistor at I_{SC}.

17-22 Design a foldback current limiting circuit. as in Fig. 17-14(a), to set $I_{L(max)}$ to approximately 220 mA and I_{SC} to approximately 150 mA for the regulator in Problem 17-13. Calculate the power dissipation in the series-pass transistor at $I_{L(max)}$ and at I_{SC}.

17-23 The current limiting circuit in Fig. 17-14(a) has R_{10} = 1.6 Ω, R_{11} = 2.7 kΩ, and R_{12} = 47 kΩ. If the normal output level is 20 V, calculate $I_{L(max)}$ and I_{SC}.

Section 17-4

17-24 Design an op-amp voltage regulator to have V_o = 15 V and $I_{L(max)}$ = 120 mA. Use a *741 IC* operational amplifier.

17-25 Design a voltage regulator using a *741 IC* operational amplifier to have V_o adjustable from 9 V to 12 V and to deliver a maximum load current of 60 mA.

17-26 Design an op-amp series voltage regulator to provide an output voltage adjustable from 15 V to 18 V. The load current is to be 300 mA. Use a *741* op-amp, and assume that the transistors all have h_{FE} = 60.

17-27 For the regulator designed for Problem 17-25, calculate the approximate line regulation, load regulation, and ripple reduction, if V_S drops by 1 V from zero to full load current.

Section 17-5

17-28 Calculate R_1, R_2, and R_3 for the *723 IC* positive voltage regulator circuit in Fig. 17-21. V_o is to be 25 V, and $I_{L(max)}$ is to be approximately 55 mA. Select a suitable supply voltage.

17-29 A regulator circuit using a *723 IC* is to be designed to provide V_o adjustable from 15 to 20 V. Design the circuit, select a suitable input voltage, and calculate the maximum load current that can be supplied.

17-30 The *LM317* positive voltage regulator in Fig. 17-24(a) is to produce an 8 V output of with $I_{L(max)}$ = 100 mA. Calculate suitable resistances for R_1 and R_2, select an appropriate supply voltage, and determine the device power dissipation.

17-31 The *LM317* negative voltage regulator in Fig. 17-24(b) is to produce a 12 V output with $I_{L(max)}$ = 80 mA. Calculate suitable resistances for R_1 and R_2, select an appropriate supply voltage, and determine the *IC* power dissipation.

Section 17-6

17-32 Calculate the approximate efficiency of the linear regulator in Problem 17-24. Also calculate the approximate efficiency of a switching regulator with similar supply and load conditions. Assume that the switching transistor has $V_{(sat)}$ = 0.5 V, and that the switching diode has V_F = 0.7 V.

17-33 Calculate the approximate efficiency of the *IC* series regulator in Problem 17-30. A switching regulator with similar supply and load conditions uses a *FET* with $r_{DS(on)}$ = 10 Ω, and a switching diode with V_F = 0.7 V. Calculate the approximate efficiency of the switching regulator.

Section 17-7

17-34 Determine minimum filter component values for a step-down switching regulator to have; f = 28 kHz, V_i = 30 V, V_o = 15 V, I_o = 300 mA, and V_r = 250 mV. Assume that $V_{(sat)}$ = V_F = 1 V.

17-35 A step-up switching regulator that is to have; f = 28 kHz, V_i = 12 V, V_o = 30 V, I_o = 150 mA, and V_r = 250 mV. Determine suitable minimum filter component values. Assume that $V_{(sat)}$ = V_F = 1 V.

17-36 Determine minimum filter component values for an inverting switching regulator to have; f = 28 kHz, V_i = 30 V, V_o = -12 V, I_o = 300 mA, and V_r = 250 mV. Assume that $V_{(sat)}$ = V_F = 1 V.

Section 17-8

17-37 The step-down switching regulator in Problem 17-34 is to use an *MC34063 IC* controller. Determine suitable values for the additional components.

17-38 Calculate values for the additional components for the step-up switching regulator in Problem 17-35 when an *MC34063 IC* controller is used.

17-39 Calculate the additional component values for the inverting switching regulator in Problem 17-36 to use an *MC34063 IC* controller.

Practise Problem Answers

17-1.1   25.7 mV, 14.3 mV, 0.21%, 0.12%
17-1.2   1.5 kΩ, 470 Ω, (3.9 kΩ + 390 Ω), 10 kΩ, *1N758*,
        (25 V, 70 mA, 700 mW)
17-2.1   2.2 kΩ, 5.6 kΩ, 2 kΩ
17-2.2   5.6 kΩ, 270 Ω, 1.8 kΩ, 8.2 kΩ, 22 kΩ, (18 V, 160 mA, 1.28 W)
17-2.3   *1N749*, 3.3 kΩ, 1.5 kΩ
17-3.1   1.25 Ω, (25 V, 400 mA, 10 W)
17-3.2   3.3 Ω, 820 Ω, 15 kΩ
17-4.1   470 Ω, 15 kΩ, 4.7 kΩ, 8.2 kΩ, 2 kΩ, *1N758*
17-4.2   2.5 Ω, 220 Ω, (12 kΩ + 1.8 kΩ), 2.2 kΩ
17-5.1   10 kΩ, 6.8 kΩ, 2.5 Ω, 23 V, (200 mA, 23 V, 4.6 W)
17-5.2   270 Ω, 1.5 kΩ, 12 V, 0.45 W
17-6.1   71%, 93%
17-7.1   60 μH, 47 μF
17-7.2   30 μH, 60 μF
17-8.1   1.2 kΩ, (3.3 kΩ + 470 Ω), 0.165 Ω, 360 pF
17-8.2   1.2 kΩ, (6.8 kΩ + 1 kΩ), 0.18 Ω, 560 pF

Chapter *18*
Audio Power Amplifiers

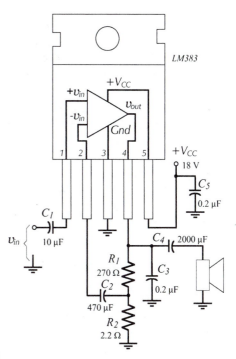

Chapter Contents

Objectives
You will be able to:

1 Sketch and explain Class-A, Class-B, and Class-AB transformer-coupled power amplifier circuits.

2 Design and analyze transformer-coupled power amplifiers, and draw the circuit dc and ac load lines.

3 Sketch and explain the basic circuits of Class-AB capacitor-coupled and direct-coupled power amplifiers.

4 Show how capacitor-coupled and direct-coupled power amplifiers should be modified for: high load currents, output current limiting, the use of overall negative feedback.

5 Explain: complementary and quasi-complementary emitter followers, V_{BE} multiplier, and supply decoupling.

6 Draw and explain complete circuits for direct- and capacitor-coupled power amplifiers using: BJT driver stages, op-amp drivers, BJT output stages, and MOSFET output stages.

7 Design and analyze the types of circuit listed in item 6 above.

8 Draw and explain complementary common-source power amplifiers, and design and analyze such circuits.

9 Explain the applications of various integrated circuit power amplifiers.

Introduction

A *power amplifier*, or *large-signal amplifier*, develops relatively large output voltages across low impedance loads. Audio amplifiers are large-signal amplifiers that supply *ac* output power to speakers. Power amplifiers may be categorized as *Class-A* circuits in which the output transistor is biased to the center of its load line, or as *Class-B* or *Class-AB* amplifiers in which two output transistors are biased at or close to cutoff. The amplifier load may be transformer-coupled, capacitor-coupled, or direct-coupled. Direct coupling usually gives the best performance, but plus-and-minus supply voltages are required. The output stage of the amplifier may use power *BJTs* or power *MOSFETs*. *IC* operational amplifiers may also be used in power amplifiers, and complete power amplifiers are available as integrated circuits.

18-1   Transformer-Coupled Class-A Amplifier

Class A Circuit

Instead of capacitor coupling, a transformer may be used to *ac* couple amplifier stages while providing *dc* isolation between stages. The resistance of the transformer windings is normally very small, so that there is no effect on the transistor bias conditions.

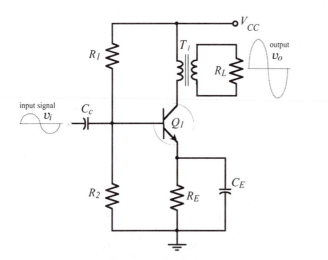

Figure 18-1
Transformer-coupled Class-A circuit using emitter current bias. Q_1 is biased to produce maximum equal positive and negative V_{CE} changes. The transistor ac load depends on R_L and the square of the transformer turn ratio.

Figure 18-1 shows a load resistance (R_L) transformer-coupled to a transistor collector. The low resistance of the transformer primary winding allows any desired level of (*dc*) collector current to flow, while the transformer core couples all variations in I_C to R_L via the secondary winding. This circuit is an emitter current bias circuit with voltage divider resistors R_1 and R_2 determining the transistor base voltage (V_B), and resistor R_E setting the emitter current level.

The circuit in Fig. 18-1 is referred to as a *class-A amplifier*, which is defined as one that has the *Q*-point (bias point) approximately at the center of the *ac* load line. This enables the circuit to produce maximum equal positive and negative changes in V_{CE}.

DC and AC Loads

The total *dc* load for transistor Q_1 in the circuit in Fig. 18-1 is the sum of the emitter resistor (R_E) and the transformer primary winding resistance (R_{PY}).

$$R_{L(dc)} = R_E + R_{PY} \qquad \text{(18-1)}$$

Consider the transformer illustrated in Fig. 18-2. N_1 is the number of turns on the primary winding, and N_2 is the number of secondary turns. The primary *ac* voltage and current are v_1 and i_1, and the secondary quantities are v_2 and i_2. The (secondary) load resistance can be calculated as,

$$R_L = \frac{v_2}{i_2}$$

The *ac* load resistance measured at the transformer primary terminals (r_L) is calculated as,

$$r_L = \frac{v_1}{i_1}$$

From basic transformer theory,

$$\frac{v_1}{v_2} = \frac{N_1}{N_2} \quad \text{and} \quad \frac{i_1}{i_2} = \frac{N_2}{N_1}$$

These equations give,

$$v_1 = \frac{N_1}{N_2} v_2 \quad \text{and} \quad i_1 = \frac{N_2}{N_1} i_2$$

Substituting for v_1 and i_1 in the equation for r_L,

$$r_L = \left[\frac{N_1}{N_2}\right]^2 R_L \qquad \text{(18-2)}$$

The load resistance calculated in this way is termed the *reflected load*, or the *referred load*; meaning that R_L is reflected or referred from the transformer secondary to the primary as r_L. The total *ac* load at the transistor collector is the sum of the referred load and the transformer primary winding resistance,

$$r_{L(ac)} = r_L + R_{PY} \qquad \text{(18-3)}$$

Example 18-1

Draw the *dc* and *ac* load lines for the circuit in Fig. 18-3 on the transistor common emitter characteristics in Fig. 18-4. The transformer has $R_{PY} = 40\,\Omega$, $N_1 = 74$, and $N_2 = 14$.

Solution

Q point:

$$V_B = V_{CC} \times \frac{R_2}{R_1 + R_2} = 13\ V \times \frac{3.7\ k\Omega}{4.7\ k\Omega + 3.7\ k\Omega}$$

$$\approx 5.7\ V$$

$$I_C \approx I_E = \frac{V_B - V_{BE}}{R_E} = \frac{5.7\ V - 0.7\ V}{1\ k\Omega}$$

$$\approx 5\ mA$$

$$V_{CE} = V_{CC} - I_C(R_{PY} + R_E) = 13\ V - 5\ mA\ (40\ \Omega + 1\ k\Omega)$$

$$\approx 8\ V$$

Plot the Q-point on the characteristics at $I_C = 5$ mA and $V_{CE} = 8$ V.

dc load line:

$$V_{CE} = V_{CC} - I_C(R_{PY} + R_E)$$

When $I_C = 0$, $V_{CE} = V_{CC} = 13$ V

Plot point *A* on the characteristics at $I_C = 0$ and $V_{CE} = 13$ V.

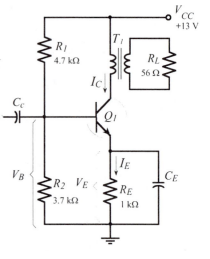

Figure 18-3
Transformer-coupled   Class-A amplifier for Example 18-1.

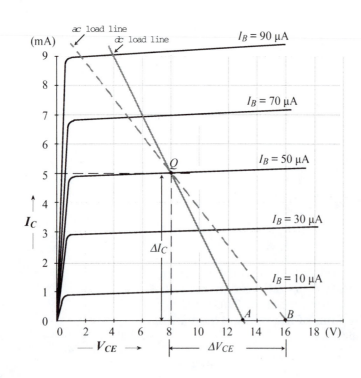

Figure 18-4
The dc load line for a transformer-coupled amplifier is drawn through the Q-point at a slope of $1/R_{L(dc)}$. The ac load line is drawn through the Q-point at a slope of $1/r_{L(ac)}$.

Draw the *dc* load through point *A* and point *Q*.

ac load line:

Eq. 18-2,
$$r_L = \left[\frac{N_1}{N_2}\right]^2 R_L = \left[\frac{74}{14}\right]^2 \times 56\ \Omega$$
$$= 1565\ \Omega$$

Eq. 18-3,
$$r_{L(ac)} = r_L + R_{PY} = 1565\ \Omega + 40\ \Omega$$
$$\approx 1.6\ k\Omega$$

When I_C changes by $\Delta I_C = 5$ mA,

$$\Delta V_{CE} = \Delta I_C \times r_{L(ac)} = 5\ \text{mA} \times 1.6\ k\Omega$$
$$= 8\ V$$

Measure ΔI_C and ΔV_{CE} from the *Q*-point on the characteristics to give point *B* at $V_{CE} = 16$ V and $I_C = 0$. Draw the *ac* load line through points *Q* and *B*.

Collector Voltage Swing

The *ac* load line drawn in Ex. 18-1 is reproduced in Fig. 18-5 to show the effect of an input signal. When the input causes I_B to increase from 50 μA (at I_{BQ}) to 90 μA, the current and voltage become $I_C \approx 9$ mA and $V_{CE} \approx 1.6$ V, (point *C* on the *ac* load line). The changes are: $\Delta I_C = +4$ mA, and $\Delta V_{CE} = -6.4$ V. When the input causes I_B to decrease (from I_{BQ}) to 10 μA, I_C changes from 5 mA to 1 mA, and the V_{CE} changes from 8 V to 14.4 V, (point *D*). The current and voltages changes are now: $\Delta I_C = -4$ mA and $\Delta V_{CE} = +6.4$ V.

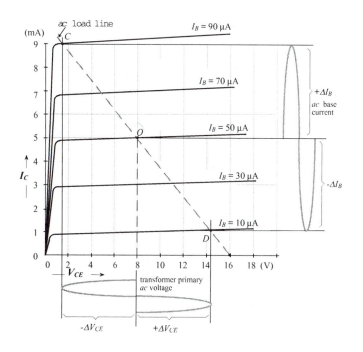

Figure 18-5
A transformer-coupled Class-A amplifier has its Q-point at the center of the ac load line. The ac voltage applied to the transformer primary is $\pm\Delta V_{CE}$, which is produced by $\pm\Delta I_B$.

It is seen that an I_B change of ±40 µA produces a ±4 mA I_C change and a ±6.4 V change in V_{CE}. The V_{CE} variation appears at the primary winding of transformer T_1, (Fig. 18-3), and the I_C variation flows in the primary winding.

Note that, although V_{CC} = 13 V, the transistor V_{CE} can actually go to 16 V. This is due to the inductive effect of the transformer primary winding. The transistor used in this type of circuit should have a minimum breakdown voltage approximately equal to 2 V_{CC}.

Efficiency of a Class A Amplifier

Power is delivered to an amplifier from the *dc* power supply. The amplifier converts the *dc* power into *ac* power delivered in the load, (see Fig. 18-6). Some of the input power is dissipated in the transistor or in other components. This is wasted power. The efficiency (η) of a power amplifier is a measure of how good the amplifier is at converting the *dc* input (supply) power (P_i) into *ac* output power (P_o) dissipated in the load.

$$\eta = \frac{P_o}{P_i} \times 100\% \qquad\qquad (18\text{-}4)$$

The *dc* supply power is,

$$P_i = V_{CC} \times I_{ave}$$

In the case of a class A amplifier, $I_{ave} = I_{CQ}$.

So, $P_i = V_{CC} \times I_{CQ}$

Refer again to the class A circuit in Fig. 18-6, and assume that $V_E \ll V_{CC}$. In this case, V_{CEQ} is approximately equal to V_{CC}, and the peak voltage developed across the transformer primary approaches ±V_{CC} if the transistor is driven to cutoff and saturation. Also, the peak current developed in the transformer windings approaches ±I_{CQ}. Thus, the maximum *ac* power delivered to the transformer primary can be calculated as,

$$P_o' = V_{rms} \times I_{rms} = (V_p/\sqrt{2}) \times (I_p/\sqrt{2})$$

giving, $\boldsymbol{P_o' = 0.5\, V_p I_p}$ (18-5)

Using the highest possible current and voltage levels, and assuming that the transformer is 100% efficient,

$$P_o = 0.5\, V_{CC} I_{CQ}$$

The maximum theoretical efficiency for a Class-A transformer-coupled power amplifier can now be determined as,

Eq. 18-4, $\eta = \dfrac{P_o}{P_i} \times 100\% = \dfrac{0.5\, V_{CC} I_{CQ}}{V_{CC} I_{CQ}} \times 100\%$

                  = 50%

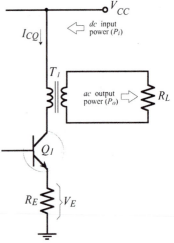

Figure 18-6
A power amplifier converts dc input (supply) power into ac output power.

In a practical Class-A transformer-coupled power amplifier circuit, 50% efficiency is never approached. Any practical calculation of power amplifier efficiency must take the output transformer efficiency (η_t) into account.

$$\eta_t = \frac{P_o}{P_o{'}} \times 100\% \qquad\qquad \text{(18-6)}$$

A typical transformer efficiency might be 80%. There is also power dissipation in the transistor emitter resistor and in the bias circuit. The practical maximum efficiency for a Class-A power amplifier is usually around 25%. This means, for example, that 4 W of *dc* supply power must be provided to deliver 1 W of *ac* output power to the load.

Example 18-2

Calculate the maximum efficiency of the Class-A amplifier circuit in Ex. 18-1, (reproduced in Fig. 18-7). Assuming that the transformer has an 80% efficiency.

Solution

$$P_i = V_{CC} \times I_{CQ} = 13\ \text{V} \times 5\ \text{mA}$$
$$= 65\ \text{mW}$$

$$V_p \approx V_{CEQ} = 8\ \text{V}$$

$$I_p \approx I_{CQ} = 5\ \text{mA}$$

Eq. 18-5, $\quad P_o{'} = 0.5\ V_p\,I_p = 0.5 \times 8\ \text{V} \times 5\ \text{mA}$
$$= 20\ \text{mW}$$

Eq. 18-6, $\quad P_o = \eta_t\,P_o{'} = 0.8 \times 20\ \text{mW}$
$$= 16\ \text{mW}$$

Eq. 18-4, $\quad \eta = \dfrac{P_o}{P_i} \times 100\% = \dfrac{16\ \text{mW}}{65\ \text{mW}} \times 100\%$

$$= 24.6\%$$

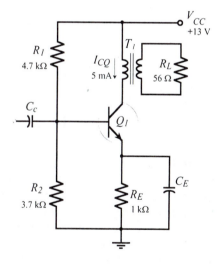

Figure 18-7
Class-A power amplifier for Example 18-2.

Practise Problems

18-1.1 The circuit in Ex. 18-1 has the quantities changed to: V_{CC} = 15 V, R_1 = 3.9 kΩ, R_2 = 2.2 kΩ, R_E = 1.5 kΩ, R_{PY} = 33 Ω, R_L = 100 Ω , N_1 = 118, N_2 = 20. Draw the new *dc* and *ac* load lines on Fig. 18-4.

18-1.2 Determine the maximum efficiency for the amplifier in Problem 18-1.1. Assume that the transformer efficiency is 75%.

18-2 Transformer-Coupled Class-B and Class-AB Amplifiers

Class B Amplifier

The inefficiency of Class-*A* amplifiers is largely due to the transistor bias conditions. In a Class-*B* amplifier, the transistors are biased to cutoff, so that there is no transistor power dissipation when there is no input signal. This gives the Class-*B* amplifier a much greater efficiency than the Class-*A* circuit.

The output stage of a Class-*B* transformer-coupled amplifier is shown in Fig. 18-8. Transformer T_2 couples load resistor R_L to the collector circuits of transistors Q_2 and Q_3. Note that the supply is connected to the center-tap of the transformer primary, and that Q_2 and Q_3 have grounded emitters. The transistor bases are grounded via resistors R_1 and R_2, so that both are biased *off*.

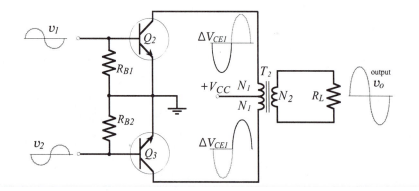

Figure 18-8
In a Class-B power amplifier, two transistors operate in push-pull. When the input signal turns Q_2 on, Q_3 is turned off, and vice versa.

The input signals applied to the transistor bases consist of two separate sine waves which are identical, except that they are in antiphase. When v_1 is going positive, v_2 is going negative, so that Q_3 is being biased further *off* as Q_2 is being biased *on*. As the collector current in Q_2 increases from zero, it produces a half sine wave of voltage across the upper half of the transformer primary, as illustrated. When the positive half-cycle of input signal to Q_2 base begins to go negative, the signal at Q_3 base is commencing to go positive. Thus, as Q_2 becomes biased *off* again, Q_3 is biased *on*, and a half-cycle of voltage waveform is generated across the lower half of the transformer primary.

The two half-cycles in the separate sections of the transformer primary produce a magnetic flux in the transformer core that flows first in one direction and then in the opposite direction. This flux links with the secondary winding and generates a complete sine wave output which is passed to the load.

In the Class-*B* circuit, the two output transistors are said to be operating in *push-pull*. The push-pull action is best illustrated by drawing the *ac* load line on the *composite characteristics* for Q_2 and Q_3. The composite characteristics are created by drawing the Q_2

characteristics in the normal way, and presenting the Q_3 characteristics upside down. This is illustrated in Fig. 18-9.

Example 18-3

Draw the ac load lines for the circuit in Fig. 18-8 on the transistor composite characteristics in Fig. 18-9. The transformer has $R_{PY} = 40\ \Omega$, $N_1 = 74$, and $N_2 = 14$, and the supply voltage is $V_{CC} = 16$ V.

Solution

Q-point : $V_B = 0$ V and $I_C = 0$ mA

$$V_{CE} = V_{CC} - I_C(R_{PY} + R_F) = 16\ V - 0$$
$$= 16\ V$$

Plot the Q-point on the characteristics at $I_C = 0$ mA and $V_{CE} = 16$ V.

ac load line:

·Eq. 18-2, $r_L = \left[\dfrac{N_1}{N_2}\right]^2 R_L = \left[\dfrac{74}{14}\right]^2 \times 56\ \Omega$

$$= 1565\ \Omega$$

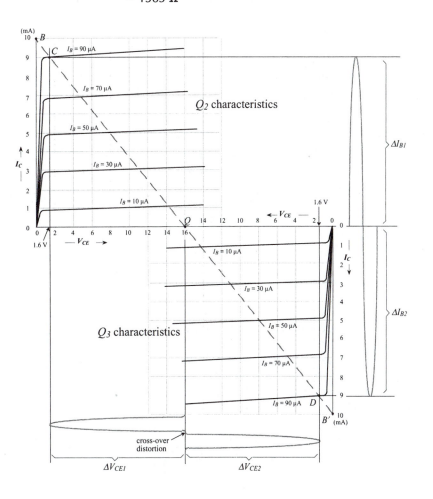

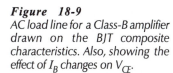

Figure 18-9
AC load line for a Class-B amplifier drawn on the BJT composite characteristics. Also, showing the effect of I_B changes on V_{CE}.

Eq. 18-3, $r_{L(ac)} = r_L + R_{py} = 1565\ \Omega + 40\ \Omega$
$$\approx 1.6\ k\Omega$$

When I_C changes by $\Delta I_C = 10$ mA,

$$V_{CE} = \Delta I_C \times r_{L(ac)} = 10\ \text{mA} \times 1.6\ k\Omega$$
$$= 16\ V$$

Measure ΔI_C and V_{CE} from the Q-point to give points B and B' at $V_{CE} = 16$ V and $I_C = 0$. Draw the ac load line through points Q, B, and B'.

Now consider the effect of a signal applied to the bases of Q_2 and Q_3 in the circuit in Fig. 18-8. When I_{B1} is increased from zero to 90 µA, Q_3 remains *off* and V_{CE1} falls to 1.6 V, (point C on the composite characteristics in Fig. 18-9). At this point the voltage across the upper half of the transformer primary in is,

$$V_{py} = V_{CC} - V_{CE} = 16\ V - 1.6\ V$$
$$= 14.4\ V$$

When I_{B2} is increased from 0 to 90 µA, Q_2 is *off*, and the Q_3 current and voltage conditions move to point D on the ac load line. This produces 14.4 V across the lower half of the transformer primary. Thus, a full sine wave is developed at the transformer output. When no signal is present, both transistors remain *off* and dissipate zero power. Power is dissipated only while each device is conducting. The wasted power is considerably less with the Class-B amplifier than with a Class-A circuit.

Cross-Over Distortion

The waveform delivered to the transformer primary and the resultant output are not perfectly sinusoidal in the Class-B circuit. *Cross-over distortion* is produced in the output waveform, as illustrated in Fig. 18-9 and 18-10, due to the fact that the transistors do not begin to turn on until the input base-emitter voltage is about 0.5 V for a silicon device, or 0.15 V for a germanium transistor. To eliminate this effect, the transistors are partially biased *on* instead of being biased *off*. With this modification, the Class-B amplifier becomes a *Class-AB amplifier*.

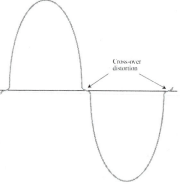

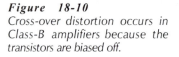

Figure   18-10
Cross-over distortion occurs in Class-B amplifiers because the transistors are biased off.

Class AB Amplifier

Figure 18-11 shows a Class-AB transformer-coupled output stage with a Class-A transformer-coupled *driver stage*. The output transformer (T_2) has a center-tapped primary winding, with each half of the winding constituting a load for one of the output transistors (Q_2 and Q_3). Resistors R_4 and R_5 bias Q_2 and Q_3 partially *on*, and resistors R_6 and R_7 limit the emitter (and collector) currents to the desired bias levels. Transformer T_1 together with transistor Q_2 and the associated components comprise a Class-A stage. The secondary of T_1 is center-tapped to provide the necessary antiphase signals to Q_2 and Q_3.

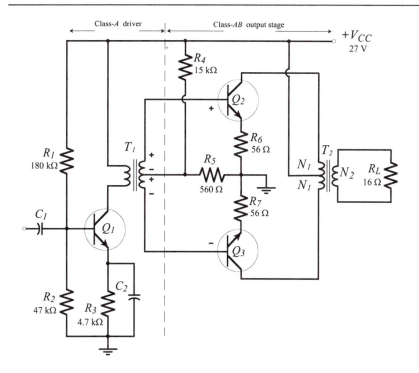

Figure 18-11
Class-AB transformer-coupled output stage with a Class-A driver stage. Transistors Q$_2$ and Q$_3$ are biased to a low I_C level.

When the instantaneous polarity of T_1 output is positive at the top, Q_2 base voltage is positive and Q_3 base voltage is negative, as illustrated. At this time Q_2 is *on* and Q_3 is *off*. When the polarity reverses at T_1 output, the base of Q_3 becomes positive and that of Q_2 becomes negative. The output stage functions exactly as for a Class-*B* circuit, except that each device commences to conduct just before the signal to its base becomes positive. This eliminates the transistor turn-on delay that creates crossover distortion in a Class-*B* amplifier.

The Class-*A* portion of the circuit in Fig. 18-11 is referred to as a *driver stage,* simply because it provides the input signals to drive the Class-*AB* output stage. The input power handled by the driver stage is very much smaller than the circuit output power, so that the inefficiency of the Class-*A* stage is unimportant.

Efficiency of Class-B and Class-AB Amplifiers

For a Class-*B* amplifier, the *dc* supply power (see Fig. 18-12) is calculated as,

$$P_{i(dc)} = V_{CC} \times I_{ave}$$

or, $$P_{i(dc)} = V_{CC} \times 0.636\ I_p \qquad (18\text{-}7)$$

The *ac* power input to the transformer primary is given by Eq. 18-5,

$$P_o' = 0.5\ V_p I_p \approx 0.5\ V_{CC} I_p$$

Assuming a 100% efficiency for the output transformer,

$$P_o = P_o' \approx 0.5\ V_{CC} I_p$$

So, the maximum theoretical efficiency for a Class-B transformer-coupled power amplifier is,

Eq. 18-4, $\eta = \dfrac{P_o}{P_i} \times 100\% = \dfrac{0.5\, V_{CC} I_p}{0.636\, V_{CC} I_p} \times 100\%$

$= 78.6\%$

Once again, the efficiency of a practical amplifier is lower than the theoretical efficiency. Some power is wasted in the transistors and emitter resistors, and the transformer is never 100% efficient.

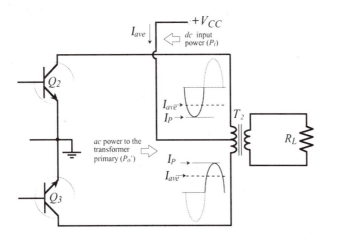

Figure 18-12
The dc input power to a Class-B amplifier is $P_i = (V_{CC} \times I_{ave})$, and the ac power to the transformer primary is $P_o' = (0.5\, V_{CC} I_p)$.

The efficiency of a Class-AB power amplifier is typically a little less than that of a Class-B circuit, because of the additional small amount of power wasted in keeping the output transistors biased in a low-current *on* state. Class-B and Class-AB power amplifiers are employed more often than Class-A circuits, because of their greater efficiency,

---

Example 18-4
Calculate the power delivered to the load in the Class-AB amplifier Fig. 18-13, (reproduced from Fig. 18-11). Assume that T_2 has a 79% efficiency, and that there is a 0.5 V drop across Q_2 and across Q_3 at peak output voltage. Also, assume that $N_1 = 60$, $N_2 = 10$, and that the primary winding resistance is small enough to neglect.

Solution
Referred load:

Eq. 18-2, $r_L = [\dfrac{N_1}{N_2}]^2 \times R_L = [\dfrac{60}{10}]^2 \times 16\,\Omega$

$= 576\,\Omega$

Total *ac* load in series with each of Q_2 and Q_3:

$R_L' = r_L + R_6 + R_{PY} = 576\,\Omega + 56\,\Omega + 0$

$= 632\,\Omega$

Peak primary current:

$$I_p = \frac{V_{CC} - V_{CE}}{R_L'} = \frac{27\text{ V} - 0.5\text{ V}}{632\ \Omega}$$

$$= 41.9\text{ mA}$$

Peak primary voltage:

$$V_p = V_{CC} - V_{CE} - (I_p R_6)$$

$$= 27\text{ V} - 0.5\text{ V} - (41.9\text{ mA} \times 56\ \Omega)$$

$$= 24.15\text{ V}$$

Power delivered to primary:

Eq. 18-5, $P_o' = 0.5\, V_p\, I_p$

$$= 0.5 \times 24.15\text{ V} \times 41.9\text{ mA}$$

$$= 506\text{ mW}$$

Power delivered to the load:

$$P_o = P_o' \times \text{(transformer efficiency)}$$

$$= 506\text{ mW} \times 0.79$$

$$= 400\text{ mW}$$

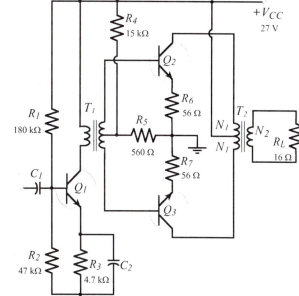

Figure 18-13
Class-AB transformer-coupled amplifier for Example 18-4.

Practise Problems

18-2.1   Calculate the bias currents in the circuit in Fig. 18-13, and using the results from Ex. 18-4 estimate the overall efficiency of the amplifier.

18-2.2   Draw the *dc* and *ac* load lines for the output stage in the circuit in Fig. 18-13 on blank composite characteristics.

18-3   Transformer-Coupled Amplifier Design

Design of a transformer-coupled amplifier commences with the load resistance and output power specification. A signal voltage amplitude may also be stated, as well as the upper and lower cutoff frequencies for the amplifier. If a supply voltage is given, then the design must determine a specification for the transformer. Where an available transformer is to be employed, the supply voltage is calculated to suit the transformer. The maximum levels of V_{CE}, I_C, and P_D must be calculated for each transistor,

The power delivered to the transformer primary is determined from the Eq. 18-6,

$$P_o' = \frac{P_o}{\eta_t}$$

The power delivered to the primary can also be expressed as,

$$P_o' = \frac{(V_{rms})^2}{r_L}$$

where V_{rms} is the *rms* primary voltage, and r_L is the *ac* resistance offered by the transformer primary (the referred resistance). Using peak voltages, the equation becomes,

$$P_o' = \frac{(V_p/\sqrt{2})^2}{r_L} = \frac{V_p^2}{2\,r_L}$$

This gives the equation for r_L:

$$r_L = \frac{V_p^2}{2\,P_o'} \qquad\qquad \textbf{(18-8)}$$

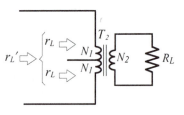

Figure 18-14
The input resistance at each half of the center-tapped primary winding of an output transformer is r_L, and the input resistance to the whole primary is $r_L' = 4\,r_L$.

Using Eq. 18-2, r_L can be calculated in terms of the load resistance and the transformer turn ratio:

$$r_L = \left[\frac{N_1}{N_2}\right]^2 R_L$$

The resistance seen when "looking into" the whole winding of a transformer with a center-tapped primary (see Fig. 18-14) is,

$$r_L' = \left[\frac{2\,N_1}{N_2}\right]^2 R_L = 4\left[\frac{N_1}{N_2}\right]^2 R_L$$

or, $\qquad\qquad$ $r_L' = 4\,r_L$ $\qquad\qquad\qquad$ **(18-9)**

The transformer can now be specified in terms of the output power (P_o), the load resistance (R_L), and the referred resistance (r_L').

Referring to Fig. 18-15, note that when Q_2 is *off* and Q_3 is *on* a voltage with a peak value of approximately $+V_{CC}$ is induced in the half of the transformer primary connected to Q_2. The induced voltage occurs because the other half of the primary has $+V_{CC}$ applied to it via Q_3, and because both windings are on the same magnetic core. The induced voltage is superimposed upon the supply, so that the voltage that appears at the collector of Q_2 is,

$$V_{CE(max)} = 2\,V_{CC} \qquad\qquad \textbf{(18-10)}$$

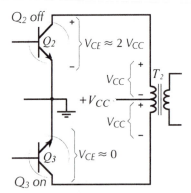

Figure 18-15
With Q_1 off and Q_2 on, the voltage across Q_1 collector-emitter is approximately $2\,V_{CC}$.

To determine the peak transistor current, the equation for power delivered to the transformer primary is rewritten. From Eq. 18-5,

$$I_p = \frac{2\,P_o'}{V_p} \qquad\qquad \textbf{(18-11)}$$

The power dissipated in the two output transistors is the difference between the *dc* supply power to the amplifier and the *ac* power delivered to the transformer primary:

$$2\,P_T = P_{i(dc)} - P_o'$$

Each transistor is *on* for half of each cycle of the input signal, so the power dissipated in each transistor is half of $2 P_T$.

$$P_T = 0.5 \, (P_{i(dc)} - P_o') \qquad\qquad (18\text{-}12)$$

The transistors are specified in terms of the device power dissipation (P_T), the peak current (I_p), and the maximum collector-emitter voltage ($V_{CE(max)}$). The transistors must also be operated below the maximum power dissipation curve, (see Section 8-7).

Example 18-5

The Class-B circuit in Fig. 18-16 is to dissipate 4 W in the 16 Ω load. Specify the output transformer and transistors. Assume an 80% transformer efficiency.

Solution

Eq. 18-6,
$$P_o' = \frac{P_o}{\eta_t} = \frac{4\text{ W}}{0.8}$$
$$= 5\text{ W}$$

$$V_p \approx V_{CC} = 30\text{ V}$$

Eq. 18-8,
$$r_L = \frac{V_p^2}{2\,P_o'} = \frac{(30\text{ V})^2}{2 \times 5\text{ W}}$$
$$= 90\ \Omega$$

$$r_L' = 4\,r_L = 4 \times 90\ \Omega$$
$$= 360\ \Omega$$

Transformer specification:
$$P_o = 4\text{ W},\ R_L = 16\ \Omega,\ r_L' = 360\ \Omega\ \text{center-tapped}$$

Eq. 18-10,
$$V_{CE(max)} = 2\,V_{CC} = 2 \times 30\text{ V}$$
$$= 60\text{ V}$$

Eq. 18-11,
$$I_p = \frac{2\,P_o'}{V_p} = \frac{2 \times 5\text{ W}}{30\text{ V}}$$
$$= 333\text{ mA}$$

Eq. 18-7,
$$P_{i(dc)} = V_{CC} \times 0.636\,I_p = 30\text{ V} \times 0.636 \times 333\text{ mA}$$
$$= 6.35\text{ W}$$

Eq. 18-12,
$$P_T = 0.5\,(P_{i(dc)} - P_o') = 0.5\,(6.35\text{ W} - 5\text{ W})$$
$$= 0.68\text{ W}$$

Transistor specification:
$$P_T = 0.68\text{ W},\ V_{CE(max)} = 60\text{ V},\ I_p = 333\text{ mA}$$

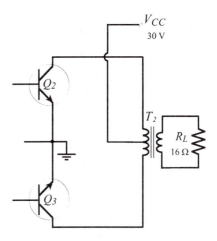

Figure 18-16
Class-B power amplifier output stage for Ex. 18-5.

Poor frequency response is one disadvantage of transformer-coupled amplifiers, both at the low and high ends of the audio frequency range. This can be improved by the use of overall negative feedback. However, substantial improvement in frequency response can be achieved by eliminating transformers from the circuit. The alternatives are capacitor-coupling and direct-coupling of the amplifier load.

Practise Problems

18-3.1   A Class-A transformer-coupled amplifier is to produce a peak current of 15 mA in a 200 Ω load. The available transformer has R_L = 200 Ω, r_L = 320 Ω, and efficiency = 90%. Determine a suitable supply voltage and specify the transistor.

18-3.2   A Class-B transformer-coupled amplifier with V_{CC} = 25 V is to supply 6 W to a 12 Ω load. Assuming a 75% transformer efficiency, specify the transistors and transformer.

18-4   Capacitor-Coupled and Direct-Coupled Output Stages

Complementary Emitter Follower

Two *BJTs* connected to function as emitter followers are shown in Fig. 18-17. Although one is *npn* and the other is *pnp*, the devices are selected to have similar parameters, so they are *complementary transistors*. The circuit is termed a *complementary emitter follower*.

A single-transistor emitter follower is essentially a small-signal circuit, because large signals can reverse bias the transistor base-emitter junction when the input polarity is opposite to the transistor V_{BE} polarity. An *npn* emitter follower might not correctly reproduce the negative-going portion of a large signal, while a *pnp* emitter follower might not reproduce the positive-going portion. Complementary emitter followers have similar signals applied simultaneously to both device bases, as illustrated. Transistor Q_1 conducts during the positive half-cycle of the signal, and it *pulls* the output voltage up to follow the input. During this time, Q_2 base-emitter junction is reverse biased. For the duration of the negative half-cycle of the input, Q_1 base-emitter junction is reversed and Q_2 conducts, pulling the output down to follow the input. Thus, the complementary emitter follower is a large-signal circuit with the low output impedance typical of emitter followers.

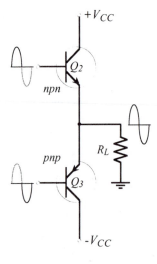

Figure 18-17
A complementary emitter follower uses an npn transistor and a pnp transistor that have similar characteristics.

Capacitor-Coupled Class-AB Output Stage

The basic circuit of a Class-AB amplifier using a complementary emitter follower output stage and a capacitor-coupled load is shown in Fig. 18-18. The circuit is termed a *complementary symmetry amplifier*. Transistor Q_1 and resistors R_1, R_2, R_C, and R_{E1} comprise a common emitter amplifier stage that produces all of the circuit voltage gain. The output of Q_1 is developed across R_C and applied to the bases of Q_2 and Q_3. Capacitor C_o *ac* couples R_L, and *dc* isolates R_L to keep it from affecting the circuit bias conditions.

The total voltage drop (V_B) across diodes D_1 and D_2 and resistor R_B forward biases the base-emitter junctions of Q_2 and Q_3 to avoid cross-over distortion. Emitter resistors R_{E2} and R_{E3} help limit the quiescent current through Q_2 and Q_3. Adjustment of the bias voltage (V_B) is provided by variable resistor R_B. The diodes have voltage drops (V_D) that approximately match the output transistor V_{BE} levels. Also, V_D does not change significantly when the diode current changes, so, the diodes behave like bypassed resistors. The diodes and output transistors can be *thermally-coupled* by mounting D_1 and Q_2 on a single heat sink, and D_2 and Q_3 on a single heat sink. In this case, V_D follows the V_{BE} level changes with temperature, thus stabilizing the transistor bias conditions over a wide temperature range. The junction of R_{E2} and R_{E3} must be biased to $V_{CC}/2$, so that the output voltage to C_o can swing by equal amounts in positive- and negative-going directions.

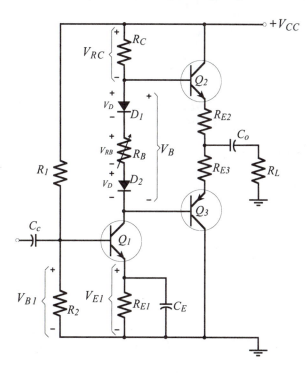

Figure 18-18
A Class-AB capacitor-coupled power amplifier that uses a complementary emitter follower as the output stage is known as a complementary symmetry amplifier.

Class-AB Capacitor-Coupled Amplifier Design

Design of the type of circuit shown in Fig. 18-18 is largely a matter of selecting appropriate resistor voltage drops and current levels, and then applying Ohm's law to calculate the resistor values. The peak output voltage (V_P) and peak output current (I_P) can be determined from Eqs. 18-8 and 18-11, respectively. Those equations were developed for the power delivered to a transformer primary, but they apply equally to power delivered to any load resistor.

The voltage drops across R_{E2} and R_{E3} when the peak output current is flowing are typically selected as 5% to 10% of the peak output voltage. This is illustrated in Fig. 18-19(a) and (b) where the output capacitor is represented as an *ac* short-circuit. So,

$$R_{E2} = R_{E3} \approx 0.05 \ R_L \text{ to } 0.1 \ R_L \qquad \textbf{(18-13)}$$

It should be remembered that R_{E2} and R_{E3} are included to help stabilize the transistor quiescent currents at a level that eliminates cross-over distortion in the output waveform. For the type of amplifier circuit in Fig. 18-18, without overall negative feedback, it is best to select the emitter resistors as large as possible. Smaller emitter resistors can be used in circuits with *dc* and *ac* negative feedback, (see Sections 18-5 and 18-6).

When the output is at its negative-going peak, V_{CE1} should be 1 V minimum, to ensure that Q_1 does not go into saturation. Also, V_{E1} should typically be 3 V, (see Section 5-7). So, the minimum level of V_{C1} is typically 4 V, [Fig. 18-19(b)]. Similarly, when the output is at its positive-going peak, there must be an appropriate minimum voltage drop across resistor R_C, [Fig. 18-19(a)]. It is not acceptable to set a 1 V minimum for V_{RC}, because the current through R_C would be too small for the required peak base current to Q_2. So, it is best to select,

$$V_{RC(min)} = V_{C1(min)} = 4 \text{ V} \qquad \textbf{(18-14)}$$

The minimum current through R_C ($I_{RC(min)}$) should typically be selected 1 mA larger than the peak base current for the output transistors. R_C is calculated from $V_{RC(min)}$ and $I_{RC(min)}$.

Referring to Fig. 18-19(a), the supply required to produce the positive output peak voltage is,

$$V_+ = V_P + V_{RE2} + V_{BE2} + V_{RC(min)}$$

Also, from Fig. 18-19(b), the negative output peak requires,

$$V_- = V_P + V_{RE3} + V_{BE3} + V_{C1(min)}$$

and, $\qquad V_+ = V_-$

So, the total supply voltage is,

$$V_{CC} = 2(V_P + V_{RE2} + V_{BE2} + V_{RC(min)}) \qquad \textbf{(18-15)}$$

The voltage drop across the diodes and R_B should just bias Q_2 and Q_3 on for Class-*AB* operation, (see Fig. 18-20). The current through R_B is the Q_1 quiescent current (I_{CQ1}), and this is calculated from R_C and the *dc* voltage drop across R_C. $V_{RC(dc)}$ equals $V_{C1(dc)}$, and the sum of them equals ($V_{CC} - V_B$), (see Fig. 18-20). So,

$$V_{RC(dc)} = V_{C1(dc)} = 0.5(V_{CC} - V_B) \qquad \textbf{(18-16)}$$

The resistance of R_B is now calculated from $V_{RC(dc)}$ and I_{CQ1}. Resistors R_1, R_2, and R_{E1} are determined in the usual manner for an emitter current bias circuit.

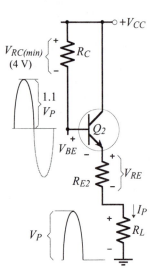

(a) Positive output peak

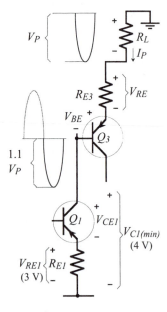

(a) Negative output peak

Figure 18-19
Supply voltage determination for a Class-AB capacitor-coupled output stage.

The output transistors should be specified in terms of their maximum voltage, current, and power dissipation. The maximum V_{CE} for Q_2 and Q_3 (in Fig. 18-18) is the total supply voltage (V_{CC}). Maximum current for Q_2 and Q_3 is the peak load current plus the selected quiescent current (I_{CQ3}). This normally 1.1 I_P. Transistor power dissipation is calculated by determining the *dc* power delivered to the output stage from the power supply, and then subtracting the *ac* load power. The remainder is halved to find the power dissipated in each transistor. Equation 18-12 applies; $P_T = 0.5 (P_i - P_o)$. Recall that the transistors must be operated within the safe operating area of the characteristics, (see Section 8-7).

With a capacitor-coupled load, current is drawn from the power supply during the positive half-cycle of the output, but not during the negative half cycle. The capacitor acts as an energy reservoir to supply load current when the output is negative-going. Consequently, the supply current has a half-wave rectified waveform (see Fig. 18-21), and so,

$$I_{ave} = 0.5 \times 0.636 \ 1.1 \ I_P$$
$$= 0.35 \ I_P$$

The *dc* supply power to the output stage is,

$$P_i = V_{CC} \times I_{ave}$$

or, $\qquad$ **$P_i = 0.35 \ V_{CC} \ I_p$** $\qquad\qquad$ **(18-17)**

As always, with the exception of the capacitor selected to set the circuit low 3 dB frequency (f_1), each capacitor impedance is selected as one tenth of the impedance in series with the capacitor. Where there is no overall negative feedback, the capacitor with the lowest-value series-connected impedance is normally selected to set f_1. For the circuit shown in Fig. 18-18, the impedance looking into the emitter of Q_1 is h_{ib1}, and this is in series with C_E. If h_{ib1} is smaller than R_L, then at f_1:

$$X_{CE} \approx h_{ib1}, \ X_{CC} \approx 0.1 \ Z_i, \ X_{CO} \approx 0.1 \ R_L$$

Most power amplifiers typically have $R_L = 8 \ \Omega$ or $16 \ \Omega$. So, the load-coupling capacitor normally sets the circuit low 3 dB frequency.

$$X_{CO} = R_L \text{ at } f_1 \qquad\qquad \textbf{(18-18)}$$

Example 18-6
The Class-*AB* amplifier in Fig. 18-22 (reproduced from Fig. 18-18) is to deliver 1 W to a 50 Ω load. Determine the required supply voltage, and calculate resistor values for R_C, R_B, R_{E2}, and R_{E3}. Assume $h_{FE(min)} = 50$ for Q_2 and Q_3.

Solution

from Eq. 18-8, $\quad V_p = \sqrt{(2 \ R_L \ P_o)} = \sqrt{(2 \times 50 \ \Omega \times 1 \ W)}$
$$= 10 \text{ V}$$

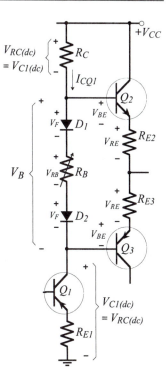

Figure 18-20
The voltage drops across D_1, D_2, and R_B bias Q_2 and Q_3 on for Class-AB operation.

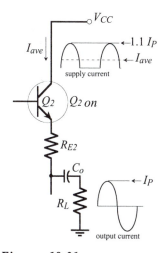

Figure 18-21
The supply current for the output stage of a capacitor-coupled Class-AB power amplifier has a half-wave rectified waveform.

$$I_P = \frac{V_P}{R_L} = \frac{10 \text{ V}}{50 \text{ }\Omega}$$

$$= 200 \text{ mA}$$

Eq. 18-13, $R_{E2} = R_{E3} = 0.1 \, R_L = 0.1 \times 50 \text{ }\Omega$

$$= 5 \text{ }\Omega \text{ (use 4.7 }\Omega\text{ standard value)}$$

select, $I_{CQ2} = 0.1 \, I_P = 0.1 \times 200 \text{ mA}$

$$= 20 \text{ mA}$$

$$V_B = V_{BE2} + I_{CQ2} \, (R_{E2} + R_{E3}) + V_{BE3}$$

$$= 0.7 \text{ V} + 20 \text{ mA} \, (4.7 \text{ }\Omega + 4.7 \text{ }\Omega) + 0.7 \text{ V}$$

$$= 1.6 \text{ V}$$

Eq. 18-14, $V_{C1(min)} = V_{RC(min)} = 4 \text{ V}$

$$I_{B2(max)} = \frac{I_P}{h_{FE2(min)}} = \frac{200 \text{ mA}}{50}$$

$$= 4 \text{ mA}$$

select, $I_{RC(min)} = I_{B2(max)} + 1 \text{ mA} = 4 \text{ mA} + 1 \text{ mA}$

$$= 5 \text{ mA}$$

$$R_C = \frac{V_{RC(min)}}{I_{RC(min)}} = \frac{4 \text{ V}}{5 \text{ mA}}$$

$$= 800 \text{ }\Omega \text{ (use 680 }\Omega\text{ standard value)}$$

Eq. 18-15, $V_{CC} = 2 \, (V_P + V_{RE2} + V_{BE2} + V_{RC(min)})$

$$= 2 \, [10 \text{ V} + 1 \text{ V} + 0.7 \text{ V} + 4 \text{ V}]$$

$$= 31.4 \text{ V} \text{ (use 32 V)}$$

Eq. 18-16, $V_{RC(dc)} = 0.5 \, (V_{CC} - V_B) = 0.5 \, (32 \text{ V} - 1.6 \text{ V})$

$$= 15.2 \text{ V}$$

$$I_{C1(dc)} = \frac{V_{RC(dc)}}{R_{C1}} = \frac{15.2 \text{ V}}{680 \text{ }\Omega}$$

$$= 22.4 \text{ mA}$$

$$R_B = \frac{V_B - V_{D1} - V_{D2}}{I_{C1(dc)}} = \frac{1.6 \text{ V} - 0.7 \text{ V} - 0.7 \text{ V}}{22.4 \text{ mA}}$$

$$= 8.9 \text{ }\Omega \text{ (use 20 }\Omega\text{ standard value variable resistor to allow for } \pm \text{ adjustment)}$$

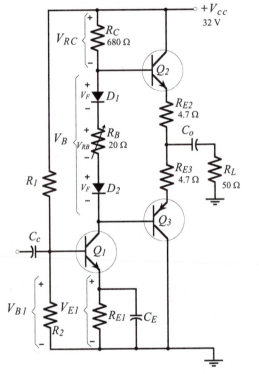

Figure 18-22
Capacitor-coupled   Class-AB power amplifier for Example 18-6.

Example 18-7

Specify the output transistors for the circuit in Example 18-6.

Solution

$$V_{CE(max)} = V_{CC} = 32 \text{ V}$$

$$
\begin{aligned}
I_{C(max)} &= 1.1\, I_p = 1.1 \times 200 \text{ mA} \\
&= 220 \text{ mA}
\end{aligned}
$$

Eq. 18-17, $\begin{aligned}P_{i(dc)} &= 0.35\, V_{CC} I_p = 0.35 \times 32 \text{ V} \times 200 \text{ mA} \\ &= 2.24 \text{ W}\end{aligned}$

from Eq. 18-12, $\begin{aligned}P_T &= 0.5\,(P_{i(dc)} - P_o) = 0.5\,(2.24 \text{ W} - 1 \text{ W}) \\ &= 0.62 \text{ W}\end{aligned}$

Example 18-8

Calculate capacitor values for C_E and C_o for the circuit in Example 18-6 if the lower cutoff frequency is to be 50 Hz. Assume $h_{ib1} = 20 \ \Omega$.

Solution

Because $h_{ib1} < R_L$, C_E sets f_1:

$$
\begin{aligned}
C_E &\approx \frac{1}{2\pi f_1 h_{ib1}} = \frac{1}{2 \times \pi \times 50 \text{ Hz} \times 20 \ \Omega} \\
&= 159 \ \mu F \text{ (use 180 } \mu F \text{ standard value)}
\end{aligned}
$$

$$
\begin{aligned}
C_o &\approx \frac{1}{2\pi f_1\, 0.1\, R_L} = \frac{1}{2 \times \pi \times 50 \text{ Hz} \times 0.1 \times 50 \ \Omega} \\
&= 637 \ \mu F \text{ (use 680 } \mu F \text{ standard value)}
\end{aligned}
$$

Direct-Coupled Class-AB Output Stage

The output capacitor in a capacitor-coupled power amplifier is a large expensive component that should be eliminated if possible. Figure 18-23 shows a Class-*AB* amplifier circuit with a direct-coupled load. In this case, the supply voltages must be positive and negative quantities, $+V_{CC}$ and $-V_{EE}$ as shown, so that the *dc* voltage at the output is zero. This is necessary to avoid a power-wasting direct current through the load. Apart from the positive/negative supply, the direct-coupled circuit operates in the same way as the capacitor-coupled amplifier.

With a few exceptions, the design procedure for a direct-coupled circuit is essentially the same as for a capacitor coupled circuit. Equation 18-15 gives the total supply voltage, and this must be halved to give $+V_{CC}$ and $-V_{EE}$ for the direct-coupled amplifier. The total supply voltage must be used to calculate P_i in Eq. 18-17. Also, the transistor maximum V_{CE} is the total supply voltage.

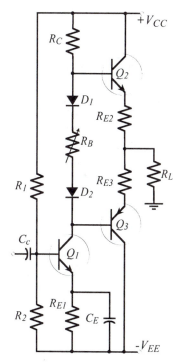

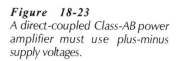

Figure 18-23
A direct-coupled Class-AB power amplifier must use plus-minus supply voltages.

Practise Problems

18-4.1 A Class-*AB* power amplifier, as in Fig. 18-22, is to dissipate 200 mW in a 60 Ω load. Calculate V_{CC}, and specify the output transistors.

18-4.2 Determine R_C and R_B for the amplifier in Problem 18-4.1.

18-5 Modifications to Improve Power Amplifier Performance

Darlington-Connected Output Transistors

High-power transistors usually have low current gains, so relatively large base currents must flow into Q_2 and Q_3 in the circuits in Figs. 18-22 and 18-23 to supply a high load current. This means that the quiescent current through Q_1 must be large, and consequently resistor R_C is small. The small value of R_C keeps the amplifier voltage gain low. To improve on this situation, *Darlington-connected* output transistors may be used, as illustrated in Fig. 18-24(a). Transistors Q_4 and Q_5 in Fig. 18-24(a) are low-power devices that supply base current to output transistors Q_2 and Q_3, respectively. Note the four biasing diodes in Fig. 18-24(a) to bias the four transistor base-emitter junctions.

When peak load current (I_p) flows, the peak base current to Q_4 and Q_5 is,

$$I_B = \frac{I_P}{h_{fe2} \times h_{fe4}}$$

This reduced base current allows $I_{RC(min)}$ to be smaller, giving a larger resistance for R_C, and resulting in a larger voltage gain.

Resistors R_8 and R_9 in Fig. 18-24(a) are included to bias Q_2 and Q_3 *off* when Q_4 and Q_5 are in cutoff. The largest possible resistance values should normally be selected for R_8 and R_9. When Q_2 and Q_3 are *off*, the collector-base leakage current I_{CBO} flows in R_8 and R_9, [see Fig. 18-24(b)]. The voltage drop across the resistors ($I_{CBO}R_8$) should be much smaller than the normal transistor base-emitter voltage. Selecting $I_{CBO}R_8$ equal to 0.01 V normally gives satisfactory resistor values. R_8 and R_9 are not required when power Darlingtons are used, as illustrated in Fig. 18-25.

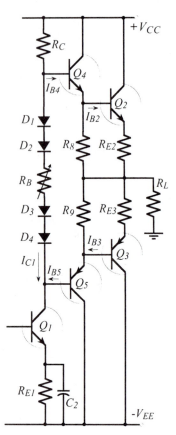

(a) Darlington-connected output stage

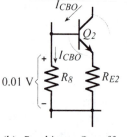

(b) R_8 biases Q_2 off

Figure 18-24
Darlington-connected output transistors reduce the current to be supplied by R_C and Q_1, and allow R_C to be increased for a larger voltage gain.

Example 18-9

The Class-*AB* amplifier in Example 18-6 is to be redesigned to use power Darlingtons, as in Fig. 18-25. Assuming that the power Darlingtons have $h_{FE(min)} = 2000$, and $V_{BE} = 1.4$ V. Determine new values for: V_{CC}, R_C, and R_B.

Solution

from Ex. 18-6, $V_p = 10$ V, $I_p = 200$ mA, $I_{CQ2} = 20$ mA, $R_{E1} = R_{E2} = 4.7$ Ω,
$V_{E1} = 3$ V, $V_{C1(dc)} = V_{RC(dc)} = 15.2$ V, $V_{RC(min)} = 4$ V

$$V_B = V_{BE2} + I_{CQ2} (R_{E2} + R_{E3}) + V_{BE3}$$
$$= 1.4 \text{ V} + 20 \text{ mA} (4.7 \text{ Ω} + 4.7 \text{ Ω}) + 1.4 \text{ V}$$
$$= 3 \text{ V}$$

$$V_{CC} = V_{RC(dc)} + V_{C1(dc)} + V_B = 15.2 \text{ V} + 15.2 \text{ V} + 3 \text{ V}$$
$$= 33.4 \text{ V}$$

use, $V_{CC}/V_{EE} = \pm 17$ V

$$I_{B2(max)} = \frac{I_P}{h_{FE2(min)}} = \frac{200 \text{ mA}}{2000}$$

$$= 100 \, \mu A$$

select, $\quad I_{RC(min)} = 1 \text{ mA}$

$$R_C = \frac{V_{RC(min)}}{I_{RC(min)}} = \frac{4 \text{ V}}{1 \text{ mA}}$$

$$= 4 \text{ k}\Omega \text{ (use 3.9 k}\Omega \text{ standard value)}$$

$$I_{C1(dc)} = \frac{V_{RC(dc)}}{R_{C1}} = \frac{15.2 \text{ V}}{3.9 \text{ k}\Omega}$$

$$= 3.9 \text{ mA}$$

$$R_B = \frac{V_B - (4 \times V_D)}{I_{C1(dc)}} = \frac{3 \text{ V} - (4 \times 0.7 \text{ V})}{3.9 \text{ mA}}$$

$$\approx 51 \, \Omega \text{ (use a 100 } \Omega \text{ variable}$$
$$\text{resistor for adjustment)}$$

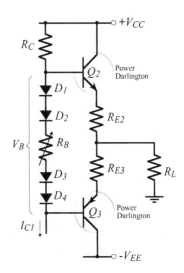

Figure 18-25
Class-AB output stage using power Darlingtons.

Quasi-Complementary Output Stage

The *quasi-complementary* circuit was originally developed because complementary high-power transistors were not readily available. Despite the fact that such transistors are now available, the quasi-complementary circuit is still widely used.

Consider the arrangement in Fig. 18-26. Q_3 is a high-power *npn* transistor, and Q_5 is a low-power *pnp* device. When Q_5 base current (I_{B5}) flows, the collector current of Q_5 behaves (largely) as base current (I_{B3}) for transistor Q_3. This produces a Q_3 collector current flow (I_{C3}), which combines with I_{E5} to constitute a current flow in the load. Because $I_{C3} \gg I_{E5}$, the output current can be taken to be approximately I_{C3};

$$I_{C3} = h_{FE3} \times I_{B3}$$

or, $\qquad I_{C3} \approx h_{FE3} \times h_{FE5} \times I_{B5}$

This is the same as the current gain with Darlington-connected transistors.

Now note that when a negative-going input voltage is applied to the base of Q_5 a negative-going output occurs, because the emitter of Q_5 (and the collector of Q_3) follow the input voltage. Thus, the combination of transistors Q_3 and Q_5 in Fig. 18-26 behave as a high-power *pnp* emitter follower, just like Darlington-connected transistors Q_3 and Q_5 in Fig. 18-24(a). Because transistors Q_2 and Q_3 in Fig. 18-26 are both *npn* devices, they can be the same type of transistor. This eliminates any problem with finding suitable complementary high-power transistors. Resistor R_9 in Fig. 18-26 ensures that Q_3 is biased off when Q_5 goes into cutoff.

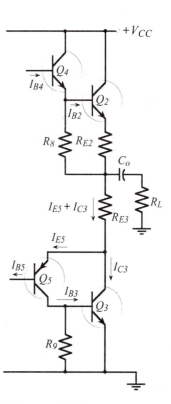

Figure 18-26
A quasi-complementary output stage allows output transistors to be the same type; Q_2 and Q_3 are both npn transistors.

Output Current Limiting

Because the output transistors can be destroyed by excessive current flow, output current limiting circuits are often included in a power amplifier. Figure 18-27 shows the typical arrangement for a current limiting circuit. Emitter resistors R_{E2} and R_{E3} are each made up of two components (R_A and R_B), as illustrated. The current limiting transistors (Q_4 and Q_5) are connected as shown, so that the voltage drop across R_{B2} and R_{B3} (produced by I_L) can turn Q_4 and Q_5 on. This occurs only when I_L is at the selected $I_{L(max)}$ level. When Q_4 turns on, it pulls the base of Q_2 down, so that Q_2 cannot supply current in excess of $I_{L(max)}$. Similarly, turning Q_5 on causes the base of Q_3 to be pulled *up* toward its emitter, thus again limiting the output current to $I_{L(max)}$.

As already discussed for voltage regulator current limiting circuits, Q_4 and Q_5 typically turn on when $V_{BE} \approx 0.5\,\text{V}$, (see Section 17-3). The component values shown in Fig. 18-27 limit the peak output current to approximately 1 A. *Catch diodes* D_3 and D_4 in Fig. 18-27 are usually included with current limiting, to protect the output transistors from the excessive voltage levels generated in an inductive load when the current growth is limited. The diodes prevent the load voltage from exceeding the supply voltage level.

V_{BE} Multiplier

Figure 18-28 shows an alternative to diode biasing for the output stage transistors in a Class-*AB* amplifier. This circuit is known as a V_{BE} *multiplier* because it produces a bias voltage ($V_B = V_{CE6}$) which is a multiple of the V_{BE} of transistor Q_6. Referring to the circuit, note that I_{10} is the current through the the voltage divider (R_{10}, R_{11}, and R_{12}) that biases Q_6. Because I_{10} is much smaller than I_{C6}, the Q_1 collector current (I_{C1}) approximately equals I_{C6}.

$$I_{10} = \frac{V_{BE6}}{R_{11} + R_{12}}$$

$$V_B = I_{10}(R_{10} + R_{11} + R_{12})$$

or, $$V_B = \frac{V_{BE6}(R_{10} + R_{11} + R_{12})}{R_{11} + R_{12}} \qquad \textbf{(18-19)}$$

It is seen that the base bias voltage for the output stage transistors can be set by suitable selection of the V_{BE} multiplier resistors. The bias voltage remains constant when I_{C1} changes. Also, it can also be shown that the changes produced in V_B by temperature variations closely match the total V_{BE} temperature changes in the four output stage transistors.

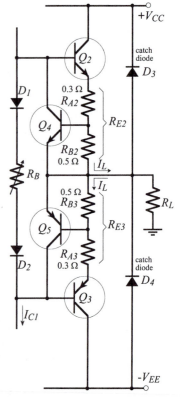

Figure 18-27
Current limiting transistors (Q_4 and Q_5) turn on when the load current is at the selected maximum. This stops further current increase in Q_2 and Q_3.

Example 18-10

Design the V_{BE} multiplier in Fig. 18-28 to have $V_B = 3.2\,\text{V}$ adjustable by ± 0.5 V when $I_{C1} = 5$ mA.

Solution

$$V_{B(min)} = 3.2 \text{ V} - 0.5 \text{ V} = 2.7 \text{ V}$$
$$V_{B(max)} = 3.2 \text{ V} + 0.5 \text{ V} = 3.7 \text{ V}$$

select

$$I_{10} \approx 0.1 \, I_{C1} = 0.1 \times 5 \text{ mA}$$
$$\approx 500 \, \mu A$$

for $V_{CE} = 3.2 \text{ V},$

$$R_{10} = \frac{V_{CE} - V_{BE}}{I_{10}} = \frac{3.2 \text{ V} - 0.7 \text{ V}}{500 \, \mu A}$$
$$= 5 \text{ k}\Omega \text{ (use 4.7 k}\Omega \text{ standard value)}$$

for $V_{CE} = 3.7 \text{ V},$

$$I_{10(max)} = \frac{V_{CE(max)} - V_{BE}}{R_{10}} = \frac{3.7 \text{ V} - 0.7 \text{ V}}{4.7 \text{ k}\Omega}$$
$$= 638 \, \mu A$$

for $V_{CE} = 2.7 \text{ V},$

$$I_{10(min)} = \frac{V_{CE(min)} - V_{BE}}{R_{10}} = \frac{2.7 \text{ V} - 0.7 \text{ V}}{4.7 \text{ k}\Omega}$$
$$= 426 \, \mu A$$

$$R_{11} + R_{12} = \frac{V_{BE}}{I_{10(min)}} = \frac{0.7 \text{ V}}{426 \, \mu A}$$
$$= 1.6 \text{ k}\Omega$$

$$R_{11} = \frac{V_{BE}}{I_{10(max)}} = \frac{0.7 \text{ V}}{638 \, \mu A}$$
$$= 1.1 \text{ k}\Omega \text{ (use 1 k}\Omega \text{ standard value)}$$

$$R_{12} = (R_{11} + R_{12}) - R_{11} = 1.6 \text{ k}\Omega - 1 \text{ k}\Omega$$
$$= 600 \, \Omega \text{ (use a 750 }\Omega \text{ standard value potentiometer)}$$

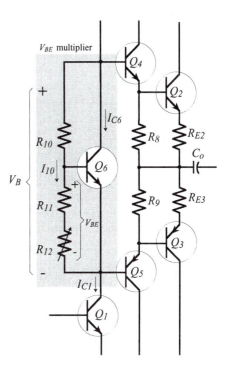

Figure 18-28
A V_{BE} multiplier is an alternative to diode biasing for the output transistors.

Power Supply Decoupling

High-power amplifiers require high supply current levels, so unregulated power supplies are often employed to avoid the power wasted in a series regulator. The high ripple voltage that occurs with unregulated supplies can be amplified to appear at speaker outputs as very unpleasant *power supply hum*. Supply *decoupling* components C_D and R_{15} are employed as shown in Fig. 18-29(a) to combat hum. Capacitive impedance X_{CD} forms an *ac* voltage divider with resistor R_{15}, so that the ripple amplitude is attenuated, as illustrated. The resistance of R_{15} is usually selected approximately equal to emitter resistor R_{E1}, and X_{CD} is made very much smaller than R_{15} at the ripple frequency (f_r). If $X_{CD} = R_{15}/100$ at f_r, the ripple voltage will be attenuated approximately by a factor of 100. The *dc* voltage drop across R_{15} must be taken into account when calculating V_{CC}.

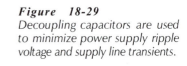

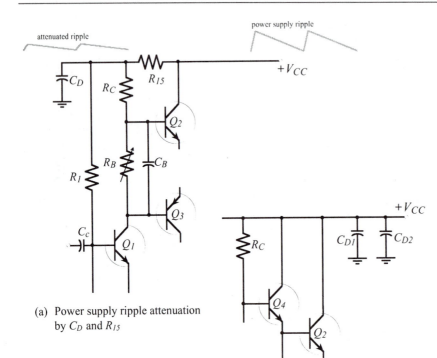

(a) Power supply ripple attenuation by C_D and R_{15}

Figure   18-29
Decoupling capacitors are used to minimize power supply ripple voltage and supply line transients.

(b)   Power supply decoupling by C_{D1} and C_{D2}

Figure 18-29(b) shows power supply decoupling capacitors (C_{D1} and C_{D2}) without any series connected resistors. These are often used even when the supply voltages are regulated and hum is not a problem. When a sudden high output current is switched *on* or *off*, the current change can produce short-lived (spike-type) voltage drops (*transient*) on the supply lines. These transients may be amplified to produce output distortion if they are allowed to appear at the supply lines for the first or second stages of a circuit. Capacitor C_{D1} is a relatively high-capacitance component that might normally be expected to perform the necessary decoupling. However, such capacitors usually offer a relatively high impedance to high-frequency variations or fast transients. Consequently, the high-frequency low-capacitance component C_{D2} is required to ensure satisfactory decoupling. *It is very important that decoupling capacitors be located right on circuit boards, as close as possible to the terminals of the circuit to be decoupled.*

Increased Voltage Gain and Negative Feedback
The Q_1 collector resistor (R_C) in the circuits discussed so far has its resistance limited by the need to supply base current to the *npn* output transistor. The resistance of R_C also dictates the Q_1 collector current level. This is shown by the $I_{C1(dc)}$ calculation in Ex. 18-6. A larger resistance for R_C would give a greater voltage gain, which is desirable for negative feedback. In Section 13-7 it is shown that negative feedback reduces distortion and increases circuit bandwidth, so its use is necessary in all power amplifiers.

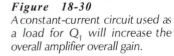

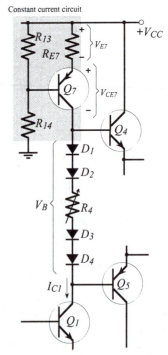

Figure   18-30
A constant-current circuit used as a load for Q_1 will increase the overall amplifier overall gain.

In Fig. 18-30 R_C is replaced by a constant current circuit (see Section 17-2) constituted by transistor Q_7 together with emitter resistor R_{E7}, and base bias resistors R_{13} and R_{14}. The minimum level of V_{CE7} and the R_{E7} voltage drop are selected to equal the Q_1 levels. This allows the voltage to the output stage to swing positively by the same amount as it can go negatively. The constant current circuit offers a high ac load resistance $(1/h_{oe7})$ for the Q_1 stage, to give the highest possible voltage gain.

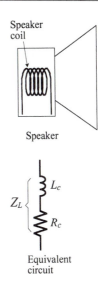

Speaker coil

Speaker

Load Compensation

All design and analysis calculations for power amplifiers assume a resistive load with a given resistance value (R_L). For audio amplifiers, the load is usually the coil of a speaker which, as illustrated in Fig. 18-31, combines coil inductance L_c and winding resistance R_C. The load impedance is, $Z_L = R_c + j\,(2\pi f\,L_c)$, and clearly Z_L increases (from a low of R_c at dc) as the signal frequency increases. An 8 Ω speaker might offer an impedance of 8 Ω only at a frequency around 400 Hz. The fact that the load is inductive means that the load current lags the load voltage, and typically the phase angle could be as high as 60°. Similarly, when capacitive loads are involved the load current can lead the load voltage. The phase difference between load current and voltage can put stress on the output transistors, so output compensating components are often included to minimize the phase difference.

Figure 18-32 shows the typical arrangement of the compensating components. Inductor L_x and its parallel-connected resistor R_x are usually recommended by device and IC manufacturers for isolating a capacitive load. Capacitor C_o and series-connected resistor R_o help to correct the lagging phase angle of an inductive load.

Figure 18-31
A speaker coil has resistance and inductance, so its impedance varies with signal frequency.

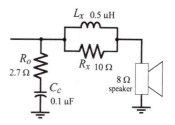

Figure 18-32
Typical arrangement of components to compensate for inductive and capacitive loads.

Practise Problems

18-5.1 A direct-coupled Class-*AB* power amplifier output stage, as in Fig. 18-24(a) is to deliver 5 W to a 8 Ω load. The output transistors are power Darlingtons with $h_{FE} = 3000$ and $V_{BE} = 1.5$ V. Calculate the required V_{CC} and the resistance values for R_C and R_B.

18-5.2 Design a V_{BE} multiplier to provide a (4 ± 0.7) V bias for a Class-*AB* output stage. The bias current (I_{C1}) is 2 mA.

18-5.3 The constant current circuit in Fig. 18-30 has $V_{CC} = 25$ V and $I_{C1} = 3$ mA. Determine suitable resistance values for R_{13}, R_{14}, and R_{E7}. Also, calculate the resistance for R_4 to give $V_B = 3.5$ V ± 0.4 V.

18-6 BJT Power Amplifier with Differential Input Stage

Amplifier Circuit

The direct-coupled amplifier in Fig. 18-33 has a differential-amplifier input stage constituted by transistors Q_1 and Q_2, (see Section 12-8). It also has an intermediate stage (Q_3) with a constant current load (Q_4). Both pairs of output stage transistors (Q_5 and Q_7) and (Q_6 and Q_8) are in quasi-complementary configuration, instead of the usual complementary form. This arrangement helps to minimize the required supply voltage by removing the V_{BE} of the power transistors from the V_{CC} equation.

$$V_{CC} = V_P + V_{R14} + V_{BE5} + V_{CE3(min)} + V_{R9} \qquad \textbf{(18-20)}$$

$$V_{EE} = -(V_P + V_{R15} + V_{BE6} + V_{CE4(min)} + V_{R11})$$
$$= -V_{CC}$$

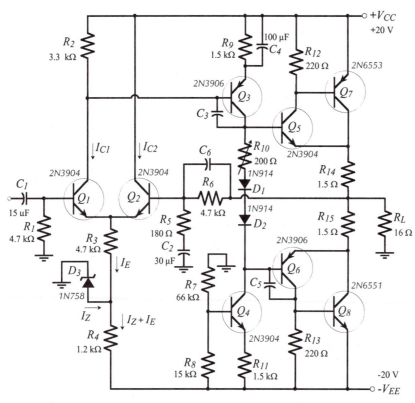

Figure 18-33
Complementary symmetry Class-AB power amplifier with a quasi-complementary output stage, a differential input stage, and overall negative feedback. The closed-loop gain is
$$A_{CL} = (R_5 + R_6)/R_5.$$

The differential input stage facilitates negative feedback (*NFB*), and the whole circuit functions like an operation amplifier. Q_1 base is the noninverting input, Q_2 base is the inverting input, and the junction of R_{14} and R_{15} is the output terminal. There is 100% dc *NFB* provided from the output via R_6 to Q_2 base. This keeps the output *dc* voltage at the same level as Q_1 base, (at ground). With C_2 behaving as an *ac* short-circuit, the *ac NFB* is divided by R_5 and R_6 to give a closed loop gain; $A_{CL} = (R_5 + R_6)/R_5$.

Zener diode D_3 and resistor R_4 decouple the power supply ripple on the negative supply line. The dynamic impedance of D_1 combined with R_4 functions as an *ac* voltage divider to attenuate the ripple at the emitters of Q_1 and Q_2. Ripple at this point is amplified just like an input signal. Capacitors C_3 and C_5 are frequency-compensating components, (see Section 15-2).

Amplifier Design

The design procedure for the output and intermediate stages of the circuit in Fig. 18-33 is similar to procedures already discussed. Design of the differential stages is very simple. The *dc* collector currents for Q_1 and Q_2 should be larger than the peak base current for Q_3, and the voltage drop across R_2 is ($V_{E3} + V_{BE3}$). Zener voltage V_{Z3} is any convenient level, usually around 0.5 V_{EE}. Resistor R_3 is calculated to pass $I_E \approx (I_{C1} + I_{C2})$, and R_4 must pass ($I_Z + I_E$).

Q_1 bias resistor R_1 is determined from Eq. 5-17, R_6 is selected equal to R_1, and R_5 is calculated in terms of R_6 to give the desired closed-loop gain. The impedance of C_2 is made equal to R_5 at the desired lower cutoff frequency (f_1), so that C_2 sets f_1. Capacitors C_1 and C_4 are determined in the usual way for capacitors that are not to affect f_1. An additional capacitor (C_6) might be included to set the upper cutoff frequency (f_2); ($X_{C6} = R_6$ at f_2).

Example 18-11

The amplifier circuit in Fig. 18-33 is to deliver 6 W to a 16 Ω load. Determine the required supply voltage and specify the output transistors.

Solution

from Eq. 18-8, $\quad V_p = \sqrt{(2\, R_L P_o)} = \sqrt{(2 \times 16\ \Omega \times 6\ W)}$

$$= 13.9\ V$$

$$V_{R14} = V_{R15} = 0.1\, V_p$$
$$\approx 1.4\ V$$

$$R_{14} = R_{16} = 0.1\, R_L$$
$$= 1.6\ \Omega\ \text{(use 1.5 Ω standard value)}$$

select, $\quad V_{CE3(min)} = V_{CE4(min)} = 1\ V$

and, $\quad V_{R9} = V_{R10} = 3\ V$

refer to Fig. 18-34,

$$V_{CC} = \pm(V_p + V_{R14} + V_{BE5} + V_{CE3(min)} + V_{R9})$$
$$= \pm(13.9 + 1.4\ V + 0.7\ V + 1\ V + 3\ V)$$
$$= \pm 20\ V$$

$$I_p = \frac{V_p}{R_L} = \frac{13.9\ V}{16\ \Omega}$$
$$= 869\ mA$$

dc power input from each supply line,

Eq. 18-17, $\quad P_{i(dc)} = [V_{CC} - (V_{EE})] \times 0.35\, I_p$
$$= [20\ V + 20\ V] \times 0.35 \times 869\ mA$$
$$\approx 12\ W$$

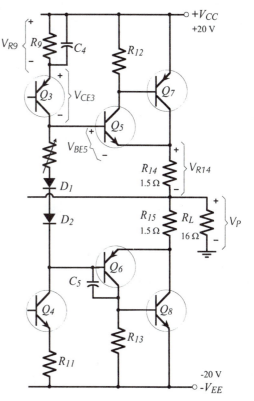

Figure 18-34
Output stage voltage drops for the circuit in Fig. 18-33.

Output transistor specification;

from Eq. 18-12, $P_T = 0.5 \, (P_{i(dc)} - P_o) = 0.5 \, (12 \text{ W} - 6 \text{ W})$

$$= 3 \text{ W}$$

$$V_{CE(max)} = 2 \, V_{CC} = 2 \times 20 \text{ V}$$
$$= 40 \text{ V}$$

$$I_{C(max)} \approx 1.1 \, I_p = 1.1 \times 869 \text{ mA}$$
$$\approx 956 \text{ mA}$$

Example 18-12

Determine suitable resistor values for the output and intermediate stages of the circuit in Example 18-11. Assume that Q_7 and Q_8 have $h_{FE} = 20$ and $I_{CBO(max)} = 50 \, \mu A$. Also, assume that Q_5 and Q_6 have $h_{FE} = 70$.

Solution
Refer to Fig. 18-35,

$$R_{12} = R_{13} = \frac{0.01 \text{ V}}{I_{CBO}} = \frac{0.01 \text{ V}}{50 \, \mu A}$$

$$= 200 \, \Omega \text{ (use 220 } \Omega \text{ standard value)}$$

$$I_{B5(peak)} = \frac{I_p}{h_{FE7} \times h_{FE5}} = \frac{869 \text{ mA}}{20 \times 70}$$

$$\approx 0.62 \text{ mA}$$

$$I_{C3} > I_{B5(peak)}$$

select $I_{C3} = 2 \text{ mA}$

$$R_9 = R_{11} = \frac{V_{R9}}{I_{C3}} = \frac{3 \text{ V}}{2 \text{ mA}}$$

$$= 1.5 \text{ k}\Omega \text{ (standard value)}$$

$$I_{Q78} \approx Q_7, \; Q_8 \text{ quiescent current}$$

select $I_{Q78} \approx 0.1 \, I_p = 0.1 \times 869 \text{ mA}$
$$\approx 86.9 \text{ mA}$$

$$V_{R14(dc)} = V_{R15(dc)} = I_{Q78} \times R_{15}$$
$$= 86.9 \text{ mA} \times 1.5 \, \Omega$$
$$\approx 0.13 \text{ V}$$

$$V_{R10(max)} = (V_{R14(dc)} + V_{R15(dc)}) + 50\%$$
$$= (0.13 \text{ V} + 0.13 \text{ V}) + 50\%$$
$$= 0.39 \text{ V}$$

$$R_{10} = \frac{V_{R10}}{I_{C3}} = \frac{0.39 \text{ V}}{2 \text{ mA}}$$

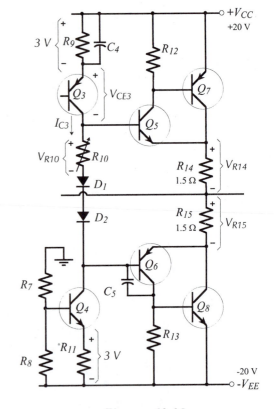

Figure 18-35
Intermediate and output stage dc voltage drops for the circuit in Fig. 18-33.

$$= 195 \ \Omega \ \text{(use 200 } \Omega \text{ variable resistor)}$$

select $R_8 = 10 \ R_{11} = 15 \ \text{k}\Omega \ \text{(standard value)}$

$$I_{R8} = \frac{V_{R11} + V_{BE}}{R_8} = \frac{3 \ \text{V} + 0.7 \ \text{V}}{15 \ \text{k}\Omega}$$

$$= 247 \ \mu\text{A}$$

$$R_7 = \frac{V_{EE} - (V_{R11} + V_{BE})}{I_{R8}} = \frac{20 \ \text{V} - (3 \ \text{V} + 0.7 \ \text{V})}{247 \ \mu\text{A}}$$

$$= 66 \ \text{k}\Omega \ \text{(use 56 k}\Omega + 10 \ \text{k}\Omega \text{)}$$

Practise   Problems

18-6.1   Design the input stage and feedback network for the circuit in Examples 18-11 and 18-12. The signal amplitude is ±0.5 V, and the lower cutoff frequency is to be 30 Hz. Assume that transistors Q_1 through Q_4 have $h_{FE(min)} = 60$.

18-7 Complementary MOSFET Common-Source Power Amplifier

Advantages of MOSFETs

Power *MOSFETs* (described in Section 9-5) have several advantages over power *BJTs* for large signal amplifier applications. One of the most important differences is that *MOSFET* transfer characteristics (I_D/V_{GS}) are more linear than I_C/V_{BE} characteristics for *BJTs*. This helps to minimize distortion in the output waveform. Thermal runaway does not occur with power *MOSFETs*, so the emitter resistors in the *BJT* output stage (R_{14} and R_{15} in Fig. 18-33) are not needed in a *MOSFET* amplifier. Thus, the wasted power dissipation in the emitter resistors is eliminated.

Power *MOSFETs* can be operated in parallel to reduce the total channel resistance and increase the output current level. Unlike *BJTs* operated in parallel, there is no need for resistors to equalize current distribution between parallel-connected *MOSFETs*. For Class-*AB* operation, the *MOSFET* gate-source should be biased to the *threshold voltage* (V_{TH}) for the device, to ensure that it is conducting at a low level when no signal is present.

Power Amplifier with MOSFET Output Stage

The four output transistors in the amplifier circuit in Fig. 18-33 could be replaced with two power *MOSFETs* operating as source followers. However, the *FET* gate-source voltage must be included when calculating the supply voltage, so that a larger supply voltage would be required than with a *BJT* amplifier. In the power amplifier in Fig. 18-36 the complementary *MOSFETs* output devices operate as common source-amplifiers. As will be seen, this permits the peak output voltage to approach the supply voltage level.

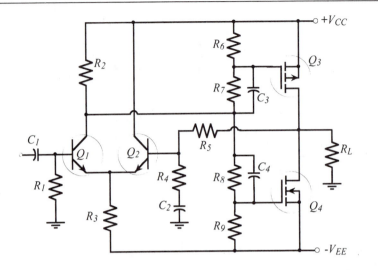

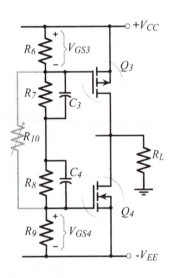

Figure 18-36
Class-AB power amplifier using a MOSFET output stage. Power MOSFETs Q_3 and Q_4 operate as common source amplifiers.

Figure 18-37
MOSFET gate-source bias voltage is provided by resistors R_6 through R_9.

As in the case of the *BJT* circuit in Fig. 18-33, the differential amplifier input stage in Fig. 18-36 allows the use of *dc* negative feedback (via R_5) to stabilize the *dc* output voltage level, and *ac* negative feedback (via R_4 and R_5) to set the closed-loop voltage gain. Output transistors Q_3 and Q_4 are complementary MOSFETs and both are operated as common-source amplifiers. Resistors R_6 through R_9 provide bias voltage to set the gate-source voltage of Q_3 and Q_4 to the threshold level (V_{TH}) for the MOSFETs, (see Fig. 18-37). Resistors R_{10} can be included, as shown, to facilitate bias voltage adjustment. Capacitors C_3 and C_4 short-circuit R_7 and R_8 at signal frequencies, so that all of the *ac* voltage from the first stage is applied to the MOSFET gate terminals.

A positive-going voltage at Q_1 collector increases V_{GS4} and decreases V_{GS3}. Thus, Q_4 drain current is increased and Q_3 is turned *off*, (see Fig. 18-38). I_{D4} flows through R_L producing a negative-going load voltage. When the voltage at Q_{1C} is negative-going, V_{GS4} is decreased and V_{GS3} is increased. This causes I_{D3} to increased and Q_4 to be turned *off*. Load current now flows via Q_3 to produce a positive-going load voltage.

MOSFET Power Amplifier Design
In a Class-*AB* MOSFET power amplifier, the *FET* gate-source voltages should be biased to the minimum specified threshold voltage for the devices. The peak output voltage and current are calculated in the usual way, and the minimum supply voltage is,

$$V_{CC} = \pm[V_P + I_{DS}\,R_{D(on)}] \qquad (18\text{-}21)$$

where $R_{D(on)}$ is the *FET* channel resistance, (see Section 9-3).

The required gate-source voltage swing (ΔV_{GS}) is determined from I_P/g_{FS}. The input stage must provide for $\pm\Delta V_{GS}$ at Q_1 collector (Fig. 18-38). Power dissipation in Q_3 and Q_4 is determined in the same way as for a *BJT* stage. The selected MOSFETs must survive the total supply voltage and pass a drain current approximately equal to $1.1\,I_P$. Capacitor values are determined in the usual way.

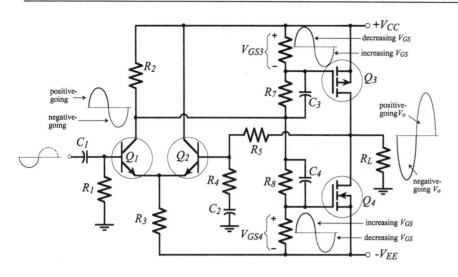

Figure 18-38
A positive-going voltage at Q_1 collector is applied to the gate-source terminals of Q_4 to produces a negative-going load voltage.

Example 18-13

The amplifier in Fig. 18-36 is to deliver 2.5 W to a 20 Ω load. The output MOSFETs have g_{FS} = 250 mA/V, V_{TH} = 1 V, and $R_{D(on)}$ = 4 Ω. Calculate the required supply voltage, and determine suitable dc voltage drops across R_2, R_3, and R_6 through R_9.

Solution

from Eq. 18-8, $V_p = \sqrt{(2\,R_L\,P_o)} = \sqrt{(2 \times 20\,\Omega \times 2.5\,W)}$
$$= 10\,V$$

$$I_P = \frac{V_P}{R_L} = \frac{10\,V}{20\,\Omega}$$
$$= 500\,mA$$

Eq. 18-21, $V_{CC} = \pm(V_P + I_P\,R_{D(on)}) = \pm[10\,V + (500\,mA \times 4\,\Omega)]$
$$= \pm 12\,V$$

$$V_{R6} = V_{R9} = V_{TH} = 1\,V$$

$$\Delta V_{R6} = \Delta V_{R9} = \frac{I_P}{g_{FS}} = \frac{500\,mA}{250\,mA/V}$$
$$= 2\,V$$

$$V_{R2(min)} = V_{CE1(min)} = \Delta V_{R6} + 1\,V = 2\,V + 1\,V$$
$$= 3\,V$$

select, $V_{CE1(dc)} = 3\,V$

then, $V_{R2(dc)} \approx V_{CC} - V_{CE1(dc)} = 12\,V - 3\,V$
$$= 9\,V$$

$$V_{R3} = V_{EE} - V_{BE} = 12\,V - 0.7\,V$$
$$= 11.3\,V$$

$$V_{R7} = V_{R2} - V_{R6} = 9\,V - 1\,V$$
$$= 8\,V$$

$$V_{R8} = [V_{CC} - (-V_{EE})] - V_{R6} - V_{R7} - V_{R9}$$
$$= 12\,V + 12\,V - 1\,V - 8\,V - 1\,V$$
$$= 14\,V$$

Example 18-14

Determine resistor values for the *MOSFET* amplifier circuit in Example 18-13. The input signal is to be ±800 mV.

Solution

$$R_6 = R_9 < 1\,M\Omega$$

select,
$$R_6 = R_9 = 100\,k\Omega$$

$$I_6 = \frac{V_{TH}}{R_6} = \frac{1\,V}{100\,k\Omega}$$
$$= 10\,\mu A$$

$$R_7 = \frac{V_{R7}}{I_6} = \frac{8\,V}{10\,\mu A}$$
$$= 800\,k\Omega \text{ (use 820 k}\Omega \text{ standard value)}$$

$$R_8 = \frac{V_{R8}}{I_6} = \frac{14\,V}{10\,\mu A}$$
$$= 1.4\,M\Omega \text{ (use 1.5 M}\Omega \text{ standard value)}$$

select,
$$I_{C1} = I_{C2} = 1\,mA$$

$$R_2 = \frac{V_{R2}}{I_{C1}} = \frac{9\,V}{1\,mA}$$
$$= 9\,k\Omega \text{ (use 8.2 k}\Omega \text{ standard value)}$$

$$R_3 = \frac{V_{R3}}{I_{C1} + I_{C2}} = \frac{11.3\,V}{1\,mA + 1\,mA}$$
$$= 5.65\,k\Omega \text{ (use 5.6 k}\Omega \text{ standard value)}$$

Select,
$$R_1 = 4.7\,k\Omega \text{ (from Eq. 5-17)}$$

$$R_5 = R_1 = 4.7\,k\Omega$$

$$A_{CL} = \frac{V_{p(out)}}{V_{p(in)}} = \frac{10\,V}{800\,mV}$$
$$= 12.5$$

$$R_4 = \frac{R_5}{A_{CL} - 1} = \frac{4.7\,k\Omega}{12.5 - 1}$$
$$= 408\,\Omega \text{ (use 390 }\Omega \text{ standard value)}$$

Practise   Problems

18-7.1   Calculate suitable capacitor values for the circuit in Examples 18-13 and 18-14 if the lower cutoff frequency is to be 20 Hz.

18-7.2   A *MOSFET* Class-*AB* amplifier as in Fig. 18-36 is to deliver 1 W to a 100 Ω load. The output devices have g_{FS} = 100 mS, V_{TH} = 1.3 V, and $R_{D(on)}$ = 6 Ω. Determine the supply voltage, and the *dc* voltage drops for the resistors.

18-7.3   Determine resistor and capacitor values for the circuit in Problem 18-7.2. The signal amplitude is ±0.4 V and the lower cutoff frequency is to be 40 Hz.

18-8   BJT Power Amplifier with Op-Amp Driver

Circuit   Operation

The Class-*AB* power shown in Fig. 18-39 uses an operational amplifier (A_1) for the input stage. Resistors R_4 and R_5 together with the two diodes provide bias for the complementary emitter-follower *BJT* output stage. There is 100% *dc* negative feedback via R_3 to keep the *dc* output at the same level as the op-amp noninverting input, which is grounded via R_1. Overall *ac* negative feedback via R_2 and R_3 controls the amplifier *ac* voltage gain.

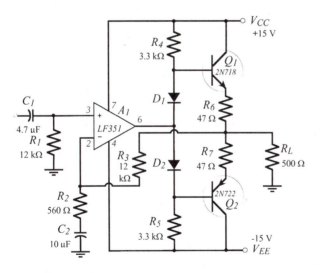

Figure 18-39
Complementary-symmetry Class-AB power amplifier using an operational amplifier and overall negative feedback.

No amplification is produced by the intermediate (output biasing) stage. Instead resistors R_4 and R_5 provide *active pull up* for transistors Q_1 and Q_2. When the op-amp output at the junction of D_1 and D_2 is increased in a positive direction, A_1 supplies current through D_2 and R_5. So, the voltage drop across R_4 is reduced, allowing it to pull the Q_1 base up to the required level while supplying increased base current to Q_1. This is illustrated by the example voltage levels shown in Fig. 18-40(a). Note that Q_2 is biased *off* when the output voltage is at its positive peak.

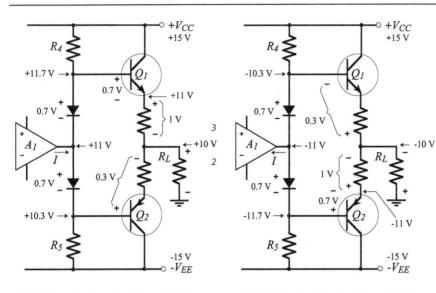

(a) Voltage levels when V_o = +10 V

(b) Voltage levels when V_o = -10 V

Figure   18-40
Output stage voltage levels for peak output voltages of ±10 V.

Figure 18-40(b) illustrates the situation when the op-amp output moves in a negative direction. A_1 pulls current through R_4 and D_1, leaving R_5 to pull the base of Q_2 down to the required voltage level while supplying the increased base current. Transistor Q_1 is biased *off* at this time, as indicated by the example voltage levels.

The circuit in Fig. 18-39 has no provision for adjusting the bias current in the output transistors. However, the diode voltage drops do bias Q_1 and Q_2 at least into a low-current *on* state. Although this might not seem enough to completely eliminate cross-over distortion, it should be recalled (from Eq. 13-28) that overall negative feedback (*NFB*) reduces distortion by a factor of $(1+ A_vB)$, where A_v is the circuit open-loop gain and B is the feedback factor. Thus, the high open-loop gain of the op-amp severely attenuates the cross-over distortion that would be present without *NFB*.

Use of Bootstrapping Capacitors
The resistance of (equal resistors) R_4 and R_5 (in Fig. 18-39) is limited by the need to supply base current to the output transistors. The calculation of R_C in Ex. 18-9 shows the process for determining R_4 and R_5. Also, there is a need for minimum voltage drop across R_4 and R_5 to produce the base current. Here again, this is shown in Ex. 18-9 where $V_{RC(min)}$ is the minimum voltage drop across R_C. This minimum resistor voltage requirement keeps the amplifier maximum peak output voltage well below the supply voltage level, and thus limits the amplifier efficiency.

The situation can be substantially improved by the use of the bootstrapping capacitors (C_3 and C_4) shown in Fig. 18-41. Resistors R_4 and R_5 are divided into two equal-value resistors (R_8 R_9 and R_{10} R_{11}), as illustrated, and the capacitors couple the output voltage back to the junctions of these components.

Consider the example supply voltage and *dc* bias levels shown in Fig. 18-42(a), where the Q_1 emitter resistor is left out for simplicity.

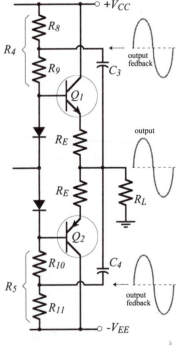

Figure   18-41
Bootstrapping capacitors used with a complementary emitter follower output stage can drive the output transistors close to saturation.

The output of A_1 is at 0 V, Q_1 base is at +0.7 V, and the load voltage is 0 V. The supply voltage is +15 V, the voltage at the junction of R_8 and R_9 is +7.5 V, and the voltage across C_3 is 7.5 V. Note that the voltage drop across R_9 is 6.8 V.

The new voltage levels that occur when the op-amp output increases by 3 V are shown in Fig. 18-42(b). Q_1 base is at +3.7 V, and the load voltage (V_o) is +3 V. Because C_3 is a large capacitor, its terminal voltage remains substantially constant at 7.5 V, so the junction of R_8 and R_9 is pushed up to;

$$V_o + V_{C3} = 3 \text{ V} + 7.5 \text{ V} = 10.5 \text{ V}$$

With 10.5 V at one end of R_9 and 3.7 V at the other end, the voltage across R_9 is 6.8 V. This is the same level of V_{R9} that occurs when the op-amp output is zero. Thus, C_3 keeps V_{R9} constant. Recall that, without the bootstrapping capacitor, V_{R9} decreases when the op-amp output rises.

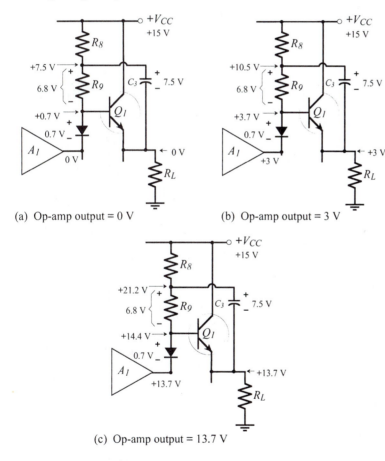

(a) Op-amp output = 0 V

(b) Op-amp output = 3 V

(c) Op-amp output = 13.7 V

Figure   18-42
Output stage voltage levels for various levels of op-amp output voltage. The voltage   on the bootstrapping capacitor remains constant.

Now consider the voltage levels shown in Fig. 18-42(c), where the op-amp output is +13.7 V. Q_1 base voltage is 14.4 V, V_o = +13.7, and the voltage at the $R_8 R_9$ junction is,

$$V_o + V_{C3} = 13.7 \text{ V} + 7.5 \text{ V} = 21.2 \text{ V}$$

Once again, V_{R9} remains constant, however, note that the bootstrapping capacitor has actually driven the $R_8 R_9$ junction to a level higher than the supply voltage. This allows the output transistors to be driven into saturation, and the voltage drop across $(R_8 + R_9)$ is no longer involved in the supply voltage calculation. In this case, the required supply voltage is,

$$V_{CC} = \pm[V_P + (I_P R_E) + V_{CE3(sat)}] \qquad \text{(18-22)}$$

The peak output voltage can also be limited by the output voltage range of the op-amp. For most op-amps the output voltage range is 1 V to 1.5 V less than the positive and negative supply levels. However, *rail-to-rail* op-amps are available with an output that ranges from $+V_{CC}$ to $-V_{EE}$.

Example 18-15

Analyze the amplifier circuit in Fig. 18-43 to determine the bootstrap capacitor terminal voltage (V_{C3}), peak output voltage (V_p), and the peak output power (P_o). The transistors have $V_{CE(sat)} = 1.5$ V.

Solution

When $V_o = 0$ V:

$$I_4 = \frac{V_{CC} - V_{D1}}{R_8 + R_9} = \frac{17\ V - 0.7\ V}{1.5\ k\Omega + 1.5\ k\Omega}$$

$$= 5.4\ mA$$

$$V_{C3} = V_{CC} - (I_4 R_8)$$
$$= 17\ V - (5.4\ mA \times 1.5\ k\Omega)$$
$$= 8.9\ V$$

$$V_P + V_{R6} = V_{CC} - V_{CE1(sat)} = 17\ V - 1.5\ V$$
$$= 15.5\ V$$

$$I_P = \frac{V_P + V_{R6}}{R_L + R_6} = \frac{15.5\ V}{100\ \Omega + 8.2\ \Omega}$$

$$= 143\ mA$$

$$V_P = I_P R_L = 143\ mA \times 100\ \Omega$$
$$= 14.3\ V$$

$$P_o = \frac{V_P^2}{2\ R_L} = \frac{(14.3\ V)^2}{2 \times 100\ \Omega}$$

$$= 1\ W$$

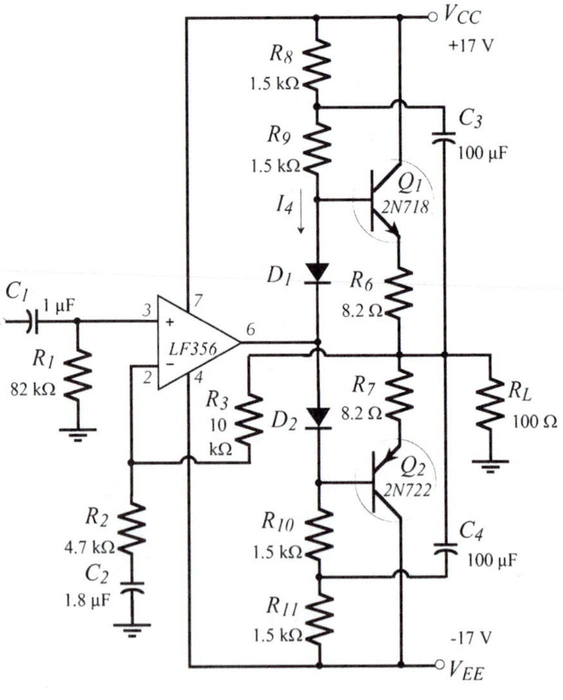

Figure 18-43
Power amplifier circuit for Examples 18-15.

Design Procedure

The amplifier in Fig. 18-44 uses four diodes to forward bias the base-emitter junctions of the Darlington output transistors. Otherwise, the circuit is exactly the same as in Fig. 18-43.

As always, the peak output voltage and current are calculated from the specified output power and load resistance. The supply voltage is determined using Eq. 18-22, and the emitter resistors for the output stage are typically selected as $0.1\ R_L$. The bias network current (I_4) should be larger than the peak base current for Q_1 and Q_2. The resistance of R_4 (which equals $R_8 + R_9$) is calculated from I_4 and the circuit dc voltage drops. R_8 should typically be selected as $0.5\ R_4$, and then R_9, R_{10}, and R_{11} are all equal to R_8.

R_1 and R_3 are equal-value resistors that bias the op-amp input terminals. R_2 is calculated from R_3 to give the required voltage gain. C_2 is selected to have its impedance equal to R_2 at the desired lower cutoff frequency (f_1). The bootstrapping capacitors are calculated in terms of the resistance in series with them; $R_8\|R_9$. The op-amp must have a suitable full power bandwidth (see Section 15-3) to produce the peak output voltage at the desired upper cutoff frequency for the amplifier.

Example 18-16

The circuit in Fig. 18-44 uses a *BIFET* op-amp. $R_L = 8\ \Omega$, P_o is to be 6 W, and $v_s = \pm 0.1$ V. Q_1 and Q_2 are Darlingtons with $h_{FE(min)} = 1000$ and $V_{CE(sat)} = 2$ V. Determine a suitable supply voltage and resistor values. Also, calculate the minimum op-amp slew rate to give $f_2 = 50$ kHz.

Solution

from Eq. 18-8, $\quad V_p = \sqrt{(2\ R_L\ P_o)} = \sqrt{(2 \times 8\ \Omega \times 6\ W)}$
$$= 9.8\ V$$

$$I_P = \frac{V_P}{R_L} = \frac{9.8\ V}{8\ \Omega}$$
$$= 1.2\ A$$

select, $\qquad R_6 = R_7 \approx 0.1\ R_L = 0.1 \times 8\ \Omega$
$$\approx 0.8\ \Omega$$

Eq. 18-22, $\quad V_{CC} = \pm[V_P + I_P\ R_6 + V_{CE3(sat)})$
$$= \pm[9.8\ V + (1.2\ A \times 0.8\ \Omega) + 2\ V]$$
$$\approx \pm 12.8\ V\ (use\ \pm 13\ V)$$

$$I_{B1(peak)} = \frac{I_p}{h_{FE1}} = \frac{1.2\ A}{1000}$$
$$= 1.2\ mA$$

$$I_4 > I_{B3(peak)}$$

select, $\qquad I_4 = 2\ mA$

$$R_4 = \frac{V_{CC} - V_{D1} - V_{D2}}{I_4} = \frac{13\ V - 0.7\ V - 0.7\ V}{2\ mA}$$
$$= 5.8\ k\Omega$$

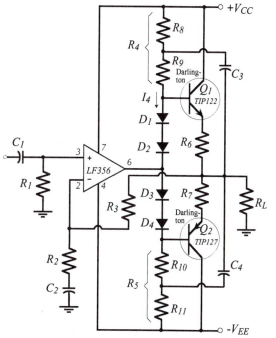

Figure 18-44
Power amplifier circuit for Examples 18-16 and 18-17.

$$R_8 = R_9 = R_{10} = R_{11} = 0.5\,R_4$$
$$= 0.5 \times 5.8\ \text{k}\Omega$$
$$= 2.9\ \text{k}\Omega \ (\text{use } 2.7\ \text{k}\Omega \text{ standard value})$$

$$A_{CL} = \frac{V_p}{V_{p(in)}} = \frac{9\ \text{V}}{0.1\ \text{V}}$$
$$= 90$$

select, $R_1 = R_3 = 100\ \text{k}\Omega$ (see *BIFET* in Section 14-2)

$$R_2 = \frac{R_3}{A_{CL} - 1} = \frac{100\ \text{k}\Omega}{90 - 1}$$
$$= 1.12\ \text{k}\Omega \ (\text{use } 1\ \text{k}\Omega)$$

Eq. 15-3, $SR = 2\,\pi\,f_2\,V_p = 2\,\pi \times 50\ \text{kHz} \times 9\ \text{V}$
$$= 2.8\ \text{V}/\mu s$$

Example 18-17

Calculate capacitor values for the circuit in Ex. 18-16. The lower cutoff frequency is to be 50 Hz.

Solution

$$X_{C1} = 0.1\,R_1 \ \text{at}\ f_1$$

$$C_1 = \frac{1}{2\,\pi\,f_1 \times 0.1\,R_1} = \frac{1}{2\,\pi \times 50\ \text{Hz} \times 0.1 \times 100\ \text{k}\Omega}$$
$$= 0.18\ \mu F \ (\text{use } 0.33\ \mu F)$$

$$X_{C2} = R_2 \ \text{at}\ f_1$$

$$C_2 = \frac{1}{2\,\pi\,f_1\,R_2} = \frac{1}{2\,\pi \times 50\ \text{Hz} \times 1\ \text{k}\Omega}$$
$$= 3.18\ \mu F \ (\text{use } 3.3\ \mu F)$$

$$X_{C3} = X_{C4} = 0.1(R_8 \| R_9) \ \text{at}\ f_1 = 0.1\,(2.7\ \text{k}\Omega \| 2.7\ \text{k}\Omega)$$
$$= 135\ \Omega$$

$$C_3 = C_4 = \frac{1}{2\,\pi\,f_1\,X_{C3}} = \frac{1}{2\,\pi \times 50\ \text{Hz} \times 135\ \Omega}$$
$$= 23.6\ \mu F \ (\text{use } 25\ \mu F)$$

Practise Problems

18-8.1   Calculate $V_{p(out)}$ and P_o for the circuit in Fig. 18-39 when $V_{in} = \pm0.54$ V. Also, determine f_2 when a *741* op-amp is used.

18-8.2   Calculate the supply voltage for an amplifier as in Fig. 18-44 to deliver 3 W to a 12 Ω load. Assume $h_{FE(min)} = 1500$ and $V_{CE(sat)} = 2$ V for Q_1 and Q_2. Also, determine the op-amp minimum slew rate to give $f_2 = 65$ kHz.

18-9 MOSFET Power Amplifier with OP-Amp Driver Stage

Basic Circuit Operation

The Class-*AB* power amplifier circuit in Fig. 18-45(a) consists of an operational amplifier (A_1), two *MOSFETs* (Q_3 and Q_4), and several resistors. The op-amp together with resistors R_4, R_5, and R_6, and capacitor C_2 constitutes a non-inverting amplifier. The two *MOSFETs* are a complementary common-source output stage, like the output stage for the circuit discussed in Section 18-7.

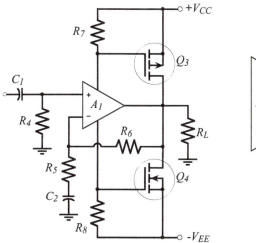

(a) Basic circuit of common-source amplifier

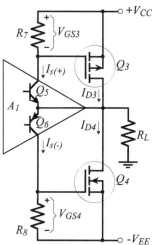

(b) Op-amp output stage controlls V_{GS3} and V_{GS34}

Figure 18-45
Complementary common-source power amplifier using an op-amp driver stage and a MOSFET output stage.

The gate-source bias voltages for Q_3 and Q_4 are provided by the voltage drops across resistors R_7 and R_8, which are in series with the op-amp supply terminals. So, the op-amp supply currents ($I_{S(+)}$ and $I_{S(-)}$) determine the levels of V_{GS3} and V_{GS4}:

$$V_{GS3} = I_{S(+)} \times R_7 \qquad\qquad \text{(18-23)}$$

and, $\qquad\quad V_{GS4} = I_{S(-)} \times R_8 \qquad\qquad \text{(18-24)}$

Suitable gate-source *threshold voltages* ($V_{GS(th)}$) to bias the output transistors for Class-*AB* operation, and typical op-amp supply currents (I_S) can be determined from the device data sheets.

Figure 18-45(b) shows that the op-amp supply currents are largely the collector currents in the op-amp output stage *BJTs* (Q_5 and Q_6). Thus, R_7 and R_8 are collector resistors for Q_5 and Q_6. When the base voltage for Q_5 and Q_6 is increased in a positive direction, Q_5 collector current ($I_{S(+)}$) increases, causing V_{GS3} to increase, and thus increasing the *MOSFET* drain current I_{D3}. At the same time, the Q_6 collector current ($I_{S(-)}$) decreases, reducing V_{GS4} and drain current I_{D4}. The result I_{D3} flows through R_L producing a positive output voltage swing (+V_o). Note that the op-amp output voltage is also positive at this time.

When the base voltage at Q_5 and Q_6 is negative-going, $I_{S(-)}$ increases, causing an increase in V_{GS4} and an increased level of I_{D4}. During this time, $I_{S(+)}$ decreases, reducing V_{GS3} and I_{D3}. Thus, I_{D4} flows through R_L producing a negative output voltage swing ($-V_o$). The op-amp output voltage is also going negative during the time that Q_4 is creating the negative output voltage across R_L.

It is seen that *BJTs* Q_5 and Q_6 operate as common-emitter amplifier stages, and that *MOSFETs* Q_3 and Q_4 function as common-source circuits. Both stages produce voltage gain which should be multiplied with the op-amp open-loop gain to determine the total open-loop gain for the circuit. Using typical quantities, the overall open-loop gain can be shown to be around 4×10^6.

Returning to Fig. 18-45(a), the complete (basic) circuit operates as a non-inverting amplifier with a closed-loop voltage gain;

$$A_{CL} = \frac{R_5 + R_6}{R_5} \qquad \textbf{(18-25)}$$

As explained, the power *MOSFETs* produce the high output current required by the amplifier load. An op-amp operating alone could not supply the load current.

Example 18-18
Determine the *MOSFET* gate-source bias voltages for the complementary common-source power amplifier in Fig. 18-46. Also, calculate the peak output voltage, peak output current, and output power if the ac input is ±100 mV.

Solution
From the *LF351* op-amp data sheet:

supply current, I_S = 1.8 mA to 3.4 mA

$$V_{GS3} = V_{GS4} = I_S \times R_7$$

$$V_{GS(min)} = I_{S(min)} \times R_7 = 1.8 \text{ mA} \times 820 \ \Omega$$
$$\approx 1.5 \text{ V}$$

$$V_{GS(max)} = I_{S(max)} \times R_7 = 3.4 \text{ mA} \times 820 \ \Omega$$
$$\approx 2.9 \text{ V}$$

Eq. 18-25, $$A_{CL} = \frac{R_5 + R_6}{R_5} = \frac{390 \ \Omega + 18 \text{ k}\Omega}{390 \ \Omega}$$
$$\approx 47.2$$

$$V_p = A_{CL} \times V_i = 47.2 \times 100 \text{ mV}$$
$$= 4.72 \text{ V}$$

$$I_p = \frac{V_p}{R_L} = \frac{4.72 \text{ V}}{10 \ \Omega}$$
$$= 472 \text{ mA}$$

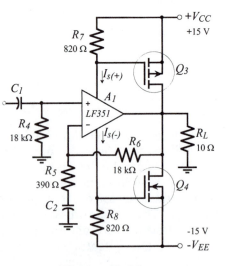

Figure 18-46
Power amplifier circuit for Example 18-18.

$$P_o = \frac{V_p I_p}{2} = \frac{4.72 \text{ V} \times 472 \text{ mA}}{2}$$

$$= 1.11 \text{ W}$$

Bias Control

As previously discussed, the bias current in the output transistors of a power amplifier should be adjustable. Control over the bias current flowing in transistor Q_3 in Fig. 18-47(a) might be achieved by using a variable resistor for R_7, as illustrated. This allows I_{D3} to be adjusted by varying V_{GS3}. When I_{D3} is increased, and the additional drain current flows through R_L, the *dc* feedback (via R_6 in Fig. 18-46) keeps the V_o equal to zero. The feedback produces the necessary change in V_{GS4} to make I_{D4} closely follow I_{D3}. Thus, adjustment of resistor R_7 controls the level of I_{D4} as well as I_{D3}.

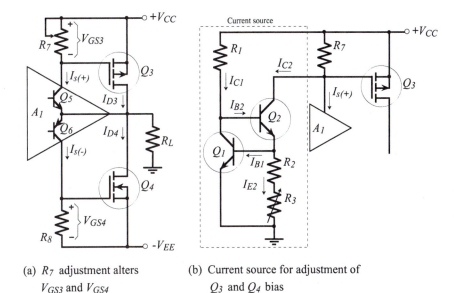

(a) R_7 adjustment alters V_{GS3} and V_{GS4}

(b) Current source for adjustment of Q_3 and Q_4 bias

Figure 18-47
The gate-source bias voltage for the two output MOSFETs can be adjusted by making R_7 adjustable, or by the use of a current source.

A disadvantage of using R_7 to adjust the bias current is that the voltage gains produced by Q_5 and Q_6 become unequal when R_7 and R_8 have different resistance values. This can be overcome by the negative feedback, however, the bias resistors can be kept equal by using the variable current source shown in Fig. 18-47(b).

In the variable current source, transistor Q_1 is biased from the emitter of Q_2, and the Q_2 base is connected to the collector of Q_1. The collector-emitter voltage of Q_1 is,

$$V_{CE1} = V_{BE1} + V_{BE2} = 2V_{BE}$$

Assuming that $I_{B2} \ll I_{C1}$, the Q_1 collector current is,

$$I_{C1} = \frac{V_{CC} - 2V_{BE}}{R_1} \qquad \text{(18-26)}$$

With $I_{B1} \ll I_{E2}$, the Q_2 collector current is approximately,

$$I_{C2} = \frac{V_{BE}}{R_2 + R_3} \qquad \text{(18-27)}$$

Variable resistor R_3 controls the level of I_{C2}. Also,

$$V_{GS3} = (I_{S(+)} + I_{C2})\, R_7 \qquad \text{(18-28)}$$

As explained, the *dc* negative feedback causes V_{GS4} to always be equal to V_{GS3}. So, R_3 controls the *MOSFET* gate-source voltages, and thus controls the quiescent drain current. The *ac* load offered to the collector of Q_5 [see Fig. 18-45(b)] is R_7, regardless of the Q_2 current. Consequently, the Q_5 and Q_6 stage gains are equal.

Example 18-19

Calculate the $V_{GS3(max)}$ and $V_{GS3(min)}$ obtainable by adjusting R_3 for the current source circuit in Fig. 18-48.

Solution

Eq. 18-27, $\qquad I_{C2} = \dfrac{V_{BE}}{R_2 + R_3}$

$$I_{C2(max)} = \frac{V_{BE}}{R_2} = \frac{0.7\ \text{V}}{560\ \Omega}$$

$$= 1.25\ \text{mA}$$

$$I_{C2(min)} = \frac{V_{BE}}{R_2 + R_3} = \frac{0.7\ \text{V}}{560\ \Omega + 1\ \text{k}\Omega}$$

$$= 449\ \mu\text{A}$$

Eq. 18-28, $\qquad V_{GS3} = (I_{S(+)} + I_{C2})\, R_7$

$$V_{GS3(min)} = (2\ \text{mA} + 449\ \mu\text{A}) \times 820\ \Omega$$
$$= 2\ \text{V}$$

$$V_{GS3(max)} = (2\ \text{mA} + 1.25\ \text{mA}) \times 820\ \Omega$$
$$\approx 2.7\ \text{V}$$

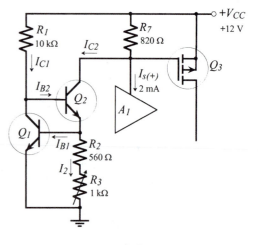

Figure 18-48
Current source circuit for Example 18-19.

Output Voltage Swing

One problem with the basic circuit in Fig. 18-45 is that the output voltage swing is limited by the voltage levels at the op-amp supply terminals. As illustrated in Fig. 18-49(a), the positive supply voltage to the op-amp is $(V_{CC} - V_{R7})$, and the negative supply voltage is $-(V_{EE} - V_{R8})$. Also, recall that V_{R7} must be increased by ΔV_{GS3} in order to drive the output in a positive direction, and V_{R8} must be increased by ΔV_{GS4} to drive V_o negative.

$$\Delta V_{GS} = \frac{I_p}{g_{fs}} \qquad \text{(18-29)}$$

In Eq. 18-29, I_p is the peak output voltage delivered to R_L, and g_{fs} is the *MOSFET* forward transconductance. Thus, the minimum supply voltage levels at the op-amp terminals are,

$$V_S = \pm(V_{CC} - V_{R7} - \Delta V_{GS}) \qquad\qquad (18\text{-}30)$$

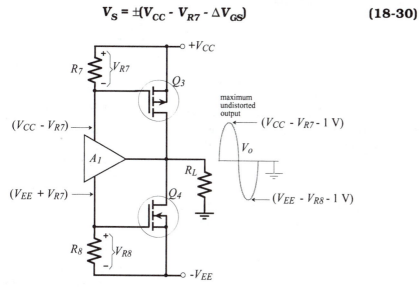

(a) The output voltage swing is limited to approximately $\pm(V_{CC} - V_{R7} - 1\text{ V})$

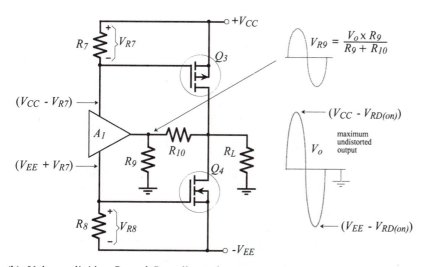

(b) Voltage divider R_9 and R_{10} allows the output swing to go to $\pm(V_{CC} - V_{RD(on)})$

Figure 18-49
The common source amplifier circuit can be modified to produce an output swing larger than the op-amp maximum output.

The op-amp peak output voltage is normally limited to approximately 1 V below the voltages at the supply terminals, (although, as previously noted, *rail-to-rail* op-amp are available). Consequently, the output voltage swing from the circuit is likely to be less than $\pm(V_{CC} - 4\text{ V})$, [see fig. 18-49(a)]. For greatest efficiency, it is necessary to drive the output as closely as possible to $\pm V_{CC}$. This can be achieved by the circuit modification shown in Fig. 18-49(b).

Resistors R_9 and R_{10} in Fig. 18-49(b) divide V_o, so that the op-amp output can be substantially lower than the peak output voltage developed across R_L. This means that the voltage levels at the op-amp supply terminals no longer limits the amplifier output voltage swing. Now, the largest peak output voltage that can be achieved is limited only by the supply voltage and $R_{D(on)}$.

$$V_p = \frac{V_{CC} \times R_L}{R_{D(on)} + R_L}$$ (18-31)

R_9 and R_{10} also provide negative feedback that controls the gain of the stage made up of the op-amp output *BJTs* and the common-source *MOSFETs*. This is illustrated in Fig. 18-50. A further function of R_9 and R_{10} is that they can be selected to limit the op-amp output current in the event of R_L becoming short-circuited.

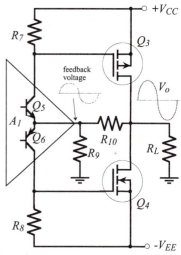

Example 18-20
The circuit in Fig. 18-51 has *MOSFETs* with g_{fs} = 2.5 S, and $R_{D(on)}$ = 0.5 Ω. Determine the maximum peak output voltage, the minimum supply voltage at op-amp terminals, and the op-amp peak output voltage when the circuit is producing maximum output power.

Figure 18-50
Negative feedback to the emitters of Q_5 and Q_6 controls the gain of the $(Q_5\text{-}Q_6) \text{—} (Q_3\text{-}Q_4)$ stage.

Solution
Eq. 18-31, $V_p = \dfrac{V_{CC} \times R_L}{R_{D(on)} + R_L} = \dfrac{12 \text{ V} \times 10 \text{ Ω}}{0.5 \text{ Ω} + 10 \text{ Ω}}$

$\qquad\qquad = 11.43 \text{ V}$

$\qquad I_p = \dfrac{V_p}{R_L} = \dfrac{11.43 \text{ V}}{10 \text{ Ω}}$

$\qquad\qquad = 1.14 \text{ A}$

Eq. 18-29, $\Delta V_{GS} = \dfrac{I_p}{g_{fs}} = \dfrac{1.14 \text{ A}}{2.5 \text{ S}}$

$\qquad\qquad = 0.46 \text{ V}$

$\qquad V_{R7(dc)} = I_S \times R_7 = 2 \text{ mA} \times 820 \text{ Ω}$

$\qquad\qquad = 1.64 \text{ V}$

Eq. 18-30, $V_{S(min)} = \pm(V_{CC} - V_{R7(dc)} - \Delta V_{GS}) = \pm(12 \text{ V} - 1.64 \text{ V} - 0.46 \text{ V})$

$\qquad\qquad = \pm 9.9 \text{ V}$

The op-amp peak output voltage is,

$$V_{R9} = \frac{V_p \times R_9}{R_9 + R_{10}} = \frac{11.43 \text{ V} \times 1 \text{ kΩ}}{1 \text{ kΩ} + 1 \text{ kΩ}}$$

$\qquad\qquad = 5.72 \text{ V}$

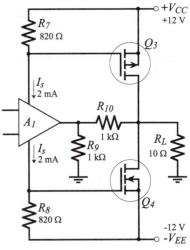

Figure 18-51
Circuit for Example 18-20.

Complete Amplifier Circuit
The complete circuit of the common-source power amplifier is shown in Fig. 18-52. Note the inclusion of resistors R_{11} and R_{12},

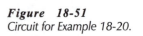

and capacitors C_3 and C_4. Resistors R_{11} and R_{12} are typically 100 Ω. They have no effect on the circuit *dc* conditions, but they help to reduce the possibility of oscillations in the output stage. The additional stage of voltage gain constituted by the *MOSFETs* and the op-amp (common-emitter) output transistors increases the possibility of circuit instability. Capacitor C_3 helps to ensure frequency stability by acting with resistor R_9 to introduce a phase lead in the output stage feedback loop, (see Section 15-2). The phase lead cancels some of the phase lag in the overall circuit.

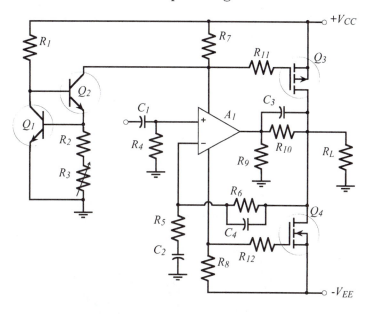

Figure 18-52
Complete circuit of Class-AB common-source power amplifier.

The additional stage of amplification extends the high cut-off frequency of the amplifier above the cut-off frequency of the op-amp operating alone. If the op-amp (full power) upper cutoff frequency for an overall voltage gain of 20 (or 26 dB) is 200 kHz, and the additional stage has a gain of 2 (or 6 dB), the circuit cut-off frequency is 400 kHz. For audio applications, it is normal to include capacitor C_4 (see Fig. 18-52) which is usually selected to set the amplifier upper cutoff frequency around 50 kHz, or lower.

Example 18-21
Analyse the circuit in Fig. 18-53 to determine the op-amp minimum supply voltage $(V_{S(dc)(min)})$, and the *MOSFET* maximum gate-source voltage $(V_{GS(max)})$. The op-amp supply current is 0.5 mA.

Solution

$$V_{C1} = 2\,V_{BE} = 2 \times 0.7 \text{ V}$$
$$= 1.4 \text{ V}$$

From Eq. 18-27, $\quad I_{C2(max)} = \dfrac{V_{BE}}{R_2} = \dfrac{0.7 \text{ V}}{470 \ \Omega}$

$$\approx 1.5 \text{ mA}$$

$$I_{C2(min)} = \frac{V_{BE}}{R_2 + R_3} = \frac{0.7 \text{ V}}{470 \ \Omega + 1 \text{ k}\Omega}$$

$$= 476 \ \mu\text{A}$$

Eq. 18-28, $V_{GS(max)} = [I_{S(max)} + I_{C2(max)}] \ R_7 = (0.5 \text{ mA} + 1.5 \text{ mA}) \times 1.5 \text{ k}\Omega$

$$= 3 \text{ V}$$

$$V_{S(dc)(min)} = \pm(V_{CC} - V_{R7}) = \pm(15 \text{ V} - 3 \text{ V})$$

$$= \pm 12 \text{ V}$$

Example 18-22

Analyse the circuit in Fig. 18-53 to determine: $P_{o(max)}$, A_{CL}, f_1, f_2. The op-amp supply current is 0.5 mA, and the MOSFETs have $R_{D(on)} = 0.3 \ \Omega$.

Solution

Power output:

Eq. 18-31, $V_p = \dfrac{V_{CC} \times R_L}{R_{D(on)} + R_L}$

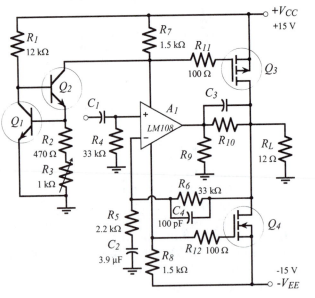

$$= \frac{15 \text{ V} \times 15 \ \Omega}{0.3 \ \Omega + 15 \ \Omega}$$

$$= 14.7 \text{ V}$$

$$I_p = \frac{V_p}{R_L} = \frac{14.7 \text{ V}}{15 \ \Omega}$$

$$= 980 \text{ mA}$$

$$P_{o(max)} = \frac{V_p I_p}{2} = \frac{14.7 \text{ V} \times 980 \text{ mA}}{2}$$

$$= 7.2 \text{ W}$$

Voltage gain:

$$A_v = \frac{R_5 + R_6}{R_5} = \frac{2.2 \text{ k}\Omega + 33 \text{ k}\Omega}{2.2 \text{ k}\Omega}$$

$$= 16$$

Figure 18-53
Common-source power amplifier circuit for Examples 18-21 and 18-22.

Cut-off frequencies:

$$f_1 = \frac{1}{2 \pi C_2 R_5} = \frac{1}{2 \pi \times 3.9 \ \mu\text{F} \times 2.2 \text{ k}\Omega}$$

$$= 18.5 \text{ Hz}$$

$$f_2 = \frac{1}{2 \pi C_4 R_6} = \frac{1}{2 \pi \times 100 \text{ pF} \times 33 \text{ k}\Omega}$$

$$= 48.2 \text{ kHz}$$

Practise Problems

18-9.1 A complementary common-source power amplifier is to deliver 5 W to a 12 Ω load. The available *MOSFETs* have $R_{D(on)} = 0.6$ Ω, $V_{TH} = 1.2$ V, and $g_{fs} = 3$ S. The op-amp to be used has 1 mA supply currents and maximum output of 20 mA. Design the output stage of the circuit as shown in Fig. 18-51.

18-9.2 Design a *BJT* current source bias control circuit for the amplifier in Problem 18-9.1 to adjust the V_{GS} of Q_3 and Q_4 by ±20%.

18-9.3 The amplifier in Problem 18-9.1 is to have $f_1 = 20$ Hz and $f_2 = 40$ kHz. If the ac input is ±600 mV, determine suitable values for: R_4, R_5, R_6, C_1, C_2, C_4, (see Fig. 18-52). Use a *BIFET* op-amp.

18-10 Integrated Circuit Power Amplifiers

IC Power Amplifier Driver

The *LM391* integrated circuit audio power driver contains amplification and driver stages for controlling an externally-connected Class-AB output stage delivering 10 W to 100 W. The voltage gain and bandwidth are set by additional components. Internal circuitry is included for overload and thermal protection, and for protection of the (externally-connected) amplifier output transistors. The circuit is designed for very low distortion, so that it can be used for high-fidelity amplifiers. Figure 18-54 illustrates the use of the device in an audio amplifier.

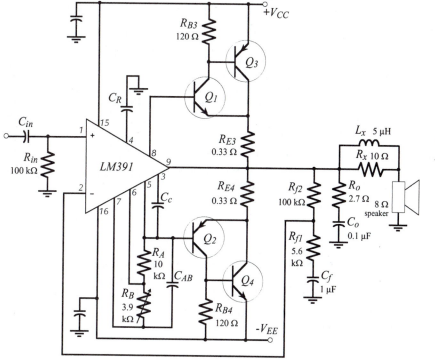

Figure 18-54
Audio power amplifier using an LM391 IC amplifier driver.

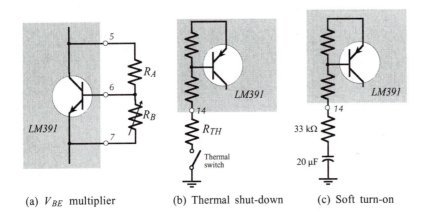

(a) V_{BE} multiplier   (b) Thermal shut-down   (c) Soft turn-on

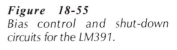

Figure 18-55
Bias control and shut-down circuits for the LM391.

The output stage in Fig. 18-54 is seen to be a complementary emitter follower with the low power and high power transistor pairs connected in quasi-complementary form. The *IC* output at terminal 9 is connected to the amplifier output, and the output stage transistors are controlled from *current source* terminal 8 and *current sink* terminal 5. The circuit uses a plus-minus supply, and the noninverting input terminal of the *IC* is biased to ground. The inverting input terminal receives feedback from the output, so that the complete circuit operates as a non-inverting amplifier.

Resistors R_A and R_B are connected to an internal transistor (via terminals 5, 6, and 7) to constitute a V_{BE} multiplier for controlling the bias voltage to the amplifier output stage, [see Fig. 18-55(a)]. Capacitor C_{AB} by-passes the V_{BE} multiplier circuit to improve the amplifier high frequency response. Capacitor C_R helps to reject power supply ripple, and C_C is a compensation capacitor for frequency stability. Components R_o, C_o, L_x and R_X are included for load compensation, (see Section 18-5).

The *LM391* has an internal transistor which can shut the circuit down when turned *on* by a thermal switch, [Fig. 18-55(b)]. This allows the device to be protected from overheating that might occur with an excessive load current demand. This same transistor can be employed for *soft turn-on* of the circuit, [Fig. 18-55(c)]. If the amplifier supply voltage is switched *on* at the instant that a peak input signal is applied, a high level output is passed to the speaker, causing a sharp unpleasant noise. Soft turn-*on* causes the output to increase slowly, thus eliminating the speaker noise. The circuit in Fig. 18-55(c) holds the amplifier in a shut-down condition until the capacitor charges.

Overload protection transistors are included in the *LM391*, as shown in Fig. 18-56. These transistors turn *on* when excessive voltage drops occur across the emitter resistors (R_{E3} and R_{E4}) in the output stage, (see Section 18-5).

The output stage components in Fig. 18-54 are selected in the same way as for other direct-coupled Class-*AB* amplifiers. The minimum levels of supply voltage are calculated by adding 5 V to the peak output voltage.

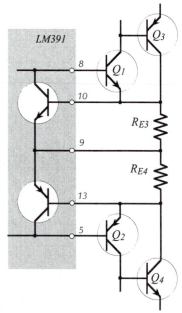

Figure 18-56
Overload protection for the LM391.

$$V_{CC} = \pm(V_P + 5 \text{ V}) \tag{18-32}$$

The input resistance at terminal 1 of the *LM391* is extremely high, so the circuit input resistance is set by resistor R_{in}, which is typically selected as 100 kΩ. Feedback resistors R_{f1} and R_{f2} set the amplifier closed-loop voltage gain. The feedback network components are determined in exactly the same way as for other feedback amplifiers. R_{f2} is made equal to R_{in} to minimize output offset, and R_{f1} is calculated from R_{f2} to give the desired voltage gain. Capacitor C_f is determined in terms of R_{f1} to set low cut-off frequency.

Example 18-23

Determine the maximum output power, the voltage gain, and the low cutoff frequency for the circuit shown in Fig. 18-54.

Solution

from Eq. 18-32, $\quad V_p = V_{CC} - 5 \text{ V} = 23 \text{ V} - 5 \text{ V}$

$$= 18 \text{ V}$$

$$P_o = \frac{V_p^2}{2 R_L} = \frac{(18 \text{ V})^2}{2 \times 8 \text{ Ω}}$$

$$\approx 20 \text{ W}$$

$$A_{CL} = \frac{R_{f1} + R_{f2}}{R_{f1}} = \frac{100 \text{ kΩ} + 5.6 \text{ kΩ}}{5.6 \text{ kΩ}}$$

$$= 18.9$$

$$f_1 = \frac{1}{2 \pi C_f R_{f1}} = \frac{1}{2 \times \pi \times 1 \text{ μF} \times 5.6 \text{ kΩ}}$$

$$= 28 \text{ Hz}$$

---

250 mW IC Power Amplifier

The *LM386* is a complete power amplifier circuit capable of delivering 250 mW to an 8 Ω load without any additional components. The supply voltage range is 5 V to 18 V, and the (inverting and non-inverting) input terminals are biased to ground (or to a negative supply) via internal 50 kΩ resistors. The output is automatically centered at half the supply voltage. Feedback resistors are also provided internally to set the voltage gain at 20.

The pin connections for the *LM386* are shown in Fig. 18-57(a), and the circuit connections for functioning as an amplifier with a gain of 20 is illustrated in part (b). Figure 18-57(c) shows how a capacitor and resistor can be connected at pins 1 and 8 to achieve a larger voltage gain. With the 10 μF capacitor alone, a maximum gain of 200 is obtained. The resistor in series with the capacitor allows the voltage gain to be set anywhere between 20 and 200.

(a) *LM386* pin connections

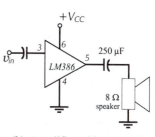

(b) Amplifier with $A_{CL} = 20$

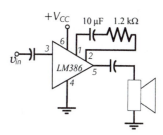

(c) Amplifier with $A_{CL} = 50$

Figure 18-57
The LM386 IC power amplifier can be connected to have a closed-loop gain from 20 to 200.

Bridge-Tied Load Amplifier

All of the power amplifiers already discussed have been *single-ended* (*SE*); meaning that they provide power to a load that has one terminal grounded and the other terminal connected to the amplifier output. These amplifiers either use a plus-minus supply with directly-coupled loads, or have a capacitor-coupled load and a single-polarity supply. A *bridge-tied load* (*BTL*) amplifier uses a single-polarity supply and a direct-coupled load.

Figure 18-58(a) shows the basic circuit of a *BTL* amplifier. The two op-amps are connected to function as inverting amplifiers, but note from the resistor values that A_1 has a voltage gain of 10 and that A_2 has a gain of 1. Each amplifier has a single-polarity supply (V_{CC}), and a voltage divider (R_5 and R_6) provides a bias voltage of $0.5\ V_{CC}$ to the op-amp noninverting input terminals. The load resistor (R_L) is connected from the output of A_1 to the output of A_2. This is the *bridge-tied load* configuration.

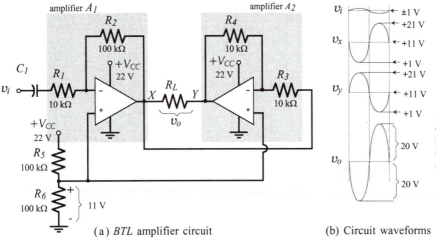

(a) *BTL* amplifier circuit   (b) Circuit waveforms

Figure   18-58
A bridge-tied load (BTL) amplifier uses two op-amp circuits with their outputs connected to opposite ends of the load.

When no *ac* signal is applied, the *dc* voltage level at the load terminals (*X* and *Y*) is $0.5\ V_{CC}$; in this case, 11 V for a 22 V supply. As illustrated by the waveforms in Fig. 18-58(b), a +1 V *ac* input to A_1 produces a -10 V change at load terminal *X*. This (-10 V) is also applied to the input of A_2, resulting in a +10 V change at load terminal *Y*. Thus, a peak of 20 V is developed across the load, negative at *X* and positive at *Y*. When the *ac* input goes to -1 V, a 20 V peak load voltage is again produced, but with the load polarity reversed. So, although a single-polarity +22 V supply is used, the output is 40 V peak-to-peak, and no load coupling capacitor is required. A (similar performance) single-ended amplifier producing a 40 V peak-to-peak output would require either a ±22 V supply for a direct-coupled load, or a +44 V supply for a capacitor-coupled load.

Figure 18-59 shows the pin connections and typical application of a *TPA4861*, 1 W integrated circuit *BTL* audio power amplifier. The load is connected across the two output terminals (*5* and *8*), external resistors R_1 and R_F set the voltage gain, and the signal is coupled to R_1 via C_1. The supply voltage (V_{DD}) is internally divided to bias the op-amp noninverting terminals to 0.5 V_{DD}. The bias point is externally accessible so that it can be bypassed to ground (via C_B) for soft start-up and to minimize noise.

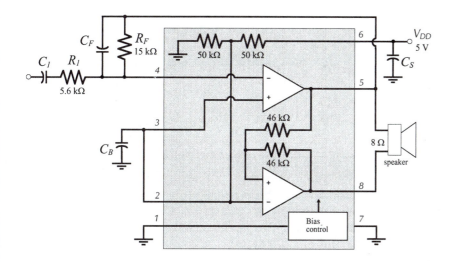

Figure 18-59
*TPA4861 bridge-tied load IC
amplifier.*

A *TPA4861* using a +5 V supply can dissipate 1 W in an 8 Ω load. The overall voltage gain for the (*BTL*) amplifier is twice the gain of the inverting amplifier stage,

$$A_{CL} = \frac{2\,R_F}{R_1} \qquad\qquad \textbf{(18-33)}$$

Because the signal is applied to an inverting amplifier, the input resistance is set by resistor R_1. The *IC* manufacturer recommends that R_1 should be selected in the range of 5 kΩ to 20 kΩ. Also, if R_F exceeds 50 kΩ, a small capacitor (C_F = 5 pF) should be connected in parallel with it for *ac* stability.

The input capacitor (C_1) sets the circuit low cutoff frequency. So,

$$X_{C1} = R_1 \text{ at } f_1 \qquad\qquad \textbf{(18-34)}$$

The internal voltage divider resistance (50 kΩ‖50 kΩ) is connected in series with the bypass capacitor (C_B). The impedance of C_B should typically be one tenth of the series resistance;

$$X_{CB} = 2.5 \text{ k}\Omega \text{ at } f_1 \qquad\qquad \textbf{(18-35)}$$

Example 18-24
Analyse the circuit in Fig. 18-59 to determine the load power dissipation when a ±0.5 V signal is applied at the input.

Solution

Eq. 18-33,
$$A_{CL} = \frac{2R_F}{R_1} = \frac{2 \times 15\ k\Omega}{5.6\ k\Omega}$$

$$\approx 5.4$$

$$V_o = A_{CL} \times v_s = \pm 5.4 \times 0.5\ V$$
$$= \pm 2.7\ V$$

$$P_o = \frac{V_p^2}{2R_L} = \frac{(2.7\ V)^2}{2 \times 8\ \Omega}$$

$$\approx 0.46\ W$$

7 W IC Power Amplifier

The *LM383* can deliver 7 W to a 4 Ω load. No additional output transistors are required because the amplifier can produce a 3.5 A peak output current. Overload protection circuitry is included, and internal bias is provided for the input terminals. The single-polarity supply voltage ranges from 5 V to 20 V. Amplifier voltage gain can be programmed by means of external components. The circuit bandwidth is 30 kHz at a gain of 40 dB.

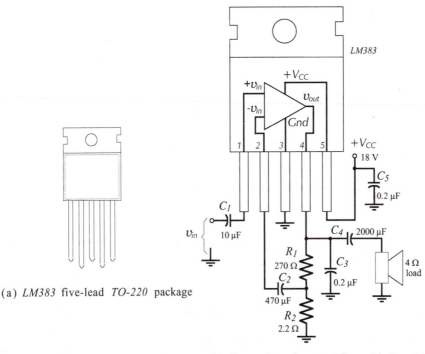

(a) *LM383* five-lead *TO-220* package

(b) Connection for amplifier with P_o = 7 W

Figure 18-60
LM383 IC power amplifier connected to dissipate 7 W in a 4 Ω load.

Figure 18-60 shows an *LM383* (in a 5 pin *TO220* package) connected to function as a (non-inverting) audio amplifier. Capacitors C_1 and C_4 are for coupling the signal and load. Resistors R_1 and R_2 are feedback components that set the circuit voltage gain, and capacitor C_2 couples the feedback voltage to the inverting input terminal. Other components are for circuit stability. Note that the resistance of R_2 is 2.2 Ω. This is because the inverting input terminal is connected to a transistor emitter terminal (internally) that has an input resistance around 20 Ω. Capacitor C_2 must be very large to couple the feedback voltage to the (low resistance) inverting input. The circuit functions as a noninverting amplifier.

68 W IC Power Amplifier

Figure 18-61(a) shows a power amplifier circuit using an *LM3886* IC audio amplifier. The *LM3886* can deliver 68 W to a 4 Ω load using a ±28 V supply. Alternatively, it can be used to dissipate 38 W in an 8 Ω load, again using a ±28 V supply. The circuit operates as a noninverting amplifier with the closed-loop gain set by resistors R_3 and R_4, and the low cutoff frequency set by capacitor C_1. Potentiometer R_1 allows the signal amplitude to be adjusted. Switch S_1 is a *mute* control.

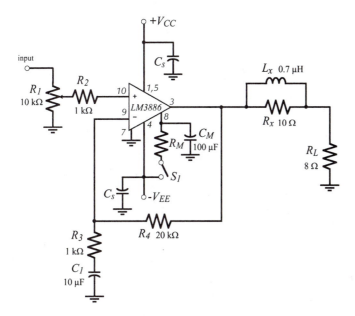

Figure   18-61
LM3886, 68 W audio power amplifier.

Practise   Problems

18-10.1   A power amplifier using an *LM391* driver (as in Fig. 18-54) has a ±20 V supply, a 50 Ω load, and power Darlington output *BJTs* with $h_{fe} = 600$. Calculate the maximum output power, and the peak output current from the integrated circuit.

18-10.2   Calculate the efficiency of the circuit in Fig. 18-59 when delivering 1 W to the speaker. The quiescent current for the *TPA4861* is specified as 2.5 mA.

18-10.3   Determine the signal amplitude for the circuit in Fig. 18-60 to dissipate 7 W in the load. Also, calculate the low cutoff frequency.

Chapter-18   Review   Questions

Section   18-1

18-1   Sketch the circuit of a transformer-coupled Class-*A* amplifier. Briefly explain the operation of the circuit.

18-2   Derive the equation for the *ac* load reflected into the primary of the transformer in a Class-*A* amplifier. Also, write the equations for *dc* and *ac* load resistances.

18-3   Sketch approximate I_C/V_{CE} transistor characteristics and an *ac* load line for a Class-*A* amplifier. Draw waveforms to show the transistor output voltage change with change in base current. Briefly explain.

18-4   Write equations for a Class-*A* transformer-coupled amplifier for: *dc* supply power, *ac* power to the transformer primary, and circuit efficiency. Show that the maximum theoretical efficiency of a Class-*A* amplifier is 50%.

Section   18-2

18-5   Sketch the circuit of a transformer-coupled Class-*B* power amplifier output stage. Explain the circuit operation.

18-6   Sketch approximate composite characteristics for a Class-*B* output stage and the *ac* load line. Draw waveforms to show the transistor output voltage change with change in base current. Briefly explain.

18-7   Sketch the circuit of a Class-*AB* transformer-coupled output stage, and explain the difference between Class-*B* and Class-*AB* amplifiers. Also, explain the difference in the performance of the two circuits.

18-8   Draw the complete circuit of a Class-*AB* transformer-coupled amplifier with a Class-*A* driver stage. Explain the operation of the circuit.

18-9   Sketch approximate composite characteristics for a Class-*AB* amplifier and the *ac* load line. Briefly explain.

18-10 Write equations for a Class-*B* transformer-coupled amplifier for: *dc* input power to the output stage, *ac* power delivered to the transformer primary, and circuit efficiency. Show that the maximum theoretical efficiency of a Class-*B* amplifier is 78.6%

Section   18-3

18-11 For a transformer-coupled Class-*B* power amplifier, write equations for the resistance seen when *looking-into* one-half of the center-tapped transformer primary, and the resistance seen when *looking-into* the whole winding of the center-tapped transformer primary.

18-12 For a transformer-coupled Class-*B* power amplifier, write an equation for the power delivered to the transformer primary in terms of peak primary voltage and reflected load. Also, write an equation for the peak current in terms of primary power and voltage.

18-13 Explain the *safe operating area (SOA)* for the output transistors in a power amplifier.

Section 18-4

18-14 Sketch the circuit of a complementary emitter follower, and explain its operation.

18-15 Sketch the basic circuit of a capacitor-coupled Class-*AB* complementary symmetry amplifier. Explain the *dc* biasing and *ac* operation of the circuit.

18-16 For the amplifier in Question 18-15, write equations for the average supply current to the output stage, the *dc* supply power, and the transistor power dissipation.

18-17 Sketch a direct-coupled Class-*AB* complementary symmetry amplifier. Explain the circuit operation.

Section 18-5

18-18 Sketch the circuit of a Darlington-connected complementary emitter follower output stage for a Class-*AB* power amplifier. Explain the operation of the circuit, and discuss its advantages.

18-19 Sketch the circuit of a quasi-complementary emitter follower output stage. Explain the circuit operation, and discuss its advantages.

18-20 Show how the maximum current can be limited in the output transistors of direct-coupled and capacitor-coupled Class-*AB* power amplifiers. Explain.

18-21 Show how a power amplifier can be protected from the effects of power supply ripple and transients. Explain.

Section 18-6

18-22 Sketch the circuit of a power amplifier with a single *BJT* input stage, an output driver stage, a complementary symmetry Darlington output with a capacitor-coupled load, and overall negative feedback.

18-23 Explain the biasing arrangement for the two amplification stages of the circuit in Question 18-22, and write an equation for the *ac* voltage gain.

18-24 Sketch the circuit of a direct-coupled power amplifier that uses a *npn* transistor differential input stage, a Class-*A* intermediate stage with a constant-current load, and a quasi-complementary emitter follower output stage.

18-25 Explain the *dc* and *ac* operation of the circuit in Question 18-24, and show how a Zener diode may be used to minimize the effect of ripple voltage on the negative supply line.

Section 18-7

18-26   Compare power *MOSFETs* to power *BJTs*.

18-27   Draw the circuit of a direct-coupled power amplifier with a differential amplifier *BJT* input stage, two *n*-channel *MOSFETs* in the output stage, and overall negative feedback.

18-28   Explain the *dc* and *ac* operation of the circuit in Question 18-27.

18-29   Modify the circuit in Question 18-27 to use complementary *MOSFETs* in the output stage.

Section 18-8

18-30   Draw the circuit of a power amplifier with a direct-coupled complementary symmetry *BJT* output stage, an op-amp driver stage, and overall negative feedback.

18-31   Explain the *dc* and *ac* operation of the circuit in Question 18-30.

18-32   Show how the circuit in Question 18-30 can be modified to use bootstrapping capacitors. Explain the function and advantage of the bootstrap capacitors.

Section 18-9

18-33   Draw the basic circuit of a complementary *MOSFET* common-source power amplifier that uses an op-amp driver with the op-amp supply currents controlling the *MOSFETs*.

18-34   Explain the *dc* and *ac* operation of the circuit in Question 18-33.

18-35   For the circuit in Question 18-33, show how the *MOSFET* bias currents can be controlled by a single variable resistor. Explain.

18-36   Show how a current source can be used to control the *MOSFET* bias currents in the circuit in Question 18-33. Draw the current source circuit and explain its operation.

18-37   Discuss the output voltage swing that can be achieved with the circuit in Question 18-33. Show how the output voltage swing can be made greater than the supply voltage levels at the op-amp terminals.

Section 18-10

18-38   Refer to the *IC* power amplifier circuit in Fig. 18-54. Explain the function of every component in the circuit.

18-39   Draw a circuit diagram to show how the current in the output transistors of a power amplifier can be limited to a desired maximum level. Explain the circuit operation.

18-40   Sketch the circuit of a bridge-tied load amplifier. Explain the circuit operation and discuss its advantages.

18-41   Explain the function of every component in the *IC* power amplifier circuit in Fig. 18-60.

Chapter-18 Problems

Section 18-1

18-1   A Class-*A* transformer-coupled amplifier, as in Fig. 18-1, has: V_{CC} = 20 V, R_1 = 3.9 kΩ, R_2 = 1 kΩ, R_E = 68 Ω, and R_L = 23 Ω. The transformer has: R_{PY} = 32 Ω, N_1 = 80, and N_2 = 20. Plot the *dc* load line and *ac* load line for this circuit on blank characteristics with vertical axis I_C = (0 to 100 mA) and horizontal axis V_{CE} = (0 to 40 V).

18-2.   Assuming an 85% transformer efficiency, calculate the maximum efficiency for the circuit in Problem 18-1.

18-3   A Class-*A* amplifier (as in Fig. 18-1) has the following components: R_1 = 68 kΩ, R_2 = 22 kΩ, R_E = 2.2 kΩ. The supply is V_{CC} = 25 V, and the transformer has: R_L = 5 kΩ, r_L' = 8 kΩ, and R_{PY} = 33 Ω. Plot the *dc* load line and *ac* load line for this circuit on blank characteristics with vertical axis I_C = (0 to 5 mA) and horizontal axis V_{CE} = (0 to 40 V).

18-4   Calculate the maximum peak load voltage for the circuit in Problem 18-3. Assume a 100% transformer efficiency.

Section 18-2

18-5   A class B transformer-coupled output stage, as in Fig. 18-8, has a load resistance R_L = 23 Ω and a supply voltage V_{CC} = 40 V. The transformer has N_1 = 80, N_2 = 20, and a total primary winding resistance R_{PY} = 64 Ω. Using blank characteristics with I_C = (0 to 100 mA) and V_{CE} = (0 to 40 V), plot the complete *ac* load line.

18-6   Determine the maximum output voltage and power for the circuit in Problem 18-5 if the transistors have $V_{CE(sat)}$ ≈ 0 V.

18-7   A Class-*B* amplifier uses a transformer with 80% efficiency and with N_p/N_s = 5, where N_p is the total number of primary turns on the center-tapped primary. The supply voltage is 45 V, and the load resistance is 8 Ω. Determine the maximum output voltage and power. Assume the transistors have $V_{CE(sat)}$ ≈ 0 V.

18-8   Using blank composite characteristics with I_C = (0 to 1 A) and V_{CE} = (0 to 45 V), plot the complete *ac* load line for the circuit in Problem 18-7. Assume R_{py} << r_L.

18-9   A Class-*AB* output stage (as in Fig. 18-11) has: V_{CC} = 30 V, R_4 = 6.8 kΩ, R_5 = 220 Ω, R_6 = R_7 = 22 Ω. The output transformer has: $R_L = 24\ \Omega$ and $r_L' = 800\ \Omega$. Assuming a 75% transformer efficiency, calculate the power delivered to the load. Assume Q_2 and Q_3 have $V_{CE(sat)} \approx 0.5$ V.

18-10   Prepare suitable blank composite characteristics and draw the *ac* load line for the circuit in Problem 18-9.

Section 18-3

18-11   Specify the maximum transistor voltage, current, and power dissipation for the circuit described in Problem 18-7.

18-12   A Class-*A* transformer-coupled amplifier with a 24 V supply is to deliver 1.25 W to a 50 Ω load. Assuming a transformer efficiency of 80%, specify the transformer and transistor.

18-13   Plot *dc* and *ac* load lines (on blank characteristics) for the circuit in Problem 18-12.

18-14   A Class-*B* amplifier is to supply 8 W to a 12 Ω load. The supply is $V_{CC} = 25$ V. Specify the output transformer and transistors. Assume a transformer efficiency of 75%.

18-15   A Class-*AB* transformer-coupled power amplifier (as in Fig. 18-13) is to deliver 0.5 W to a 4 Ω load. The output transformer has $r_L' = 312\ \Omega$ when $R_L = 4\ \Omega$, and has an efficiency of 75%. Calculate a suitable supply voltage and specify the output transistors.

Section 18-4

18-16   A capacitor-coupled power amplifier as in Fig. 18-18 is to deliver 0.6 W to a 250 Ω load. Specify the supply voltage and the output transistors.

18-17   Determine suitable resistor values for the circuit in Problem 18-16. Assume that the output transistors have $h_{FE(min)} = 40$, $h_{ie} = 1$ kΩ, and $h_{ib} = 18\ \Omega$.

18-18   Calculate suitable capacitance values for the circuit in Problems 18-16 and 18-17 for a 50 Hz lower cutoff frequency.

18-19   A direct-coupled amplifier as in Fig. 18-23 is to deliver 2 W to a 20 Ω load. Specify the supply voltage and the output transistors. Assume that the output transistors have $h_{FE(min)} = 200$.

18-20   Determine resistor values for the circuit in Problem 18-19.

Section 18-5

18-21   A power amplifier is required to deliver 5 W to a 20 Ω load. The output stage is to use Darlington-connected *BJTs* and is to be direct-coupled, as in Fig. 18-24. Determine a suitable supply voltage, and specify all the output stage transistors. Assume $h_{FE2} = h_{FE3} = 20$ and $h_{FE4} = h_{FE5} = 100$.

18-22   Determine suitable resistor values for the amplifier output stage in Problem 18-21.

18-23 A direct-coupled power amplifier using power Darlington BJTs with h_{fe} = 2000 and V_{BE} = 1.5 V is to dissipate 2 W in a 16 Ω load. Determine the required supply voltage and resistor values for the output stage, (see Fig. 18-25).

18-24 Modify the circuit in Problem 18-23 to include current limiting as in Fig. 18-27. The maximum current is to be limited to 20% above the calculated peak level.

18-25 Design a V_{BE} multiplier to replace the diode biasing stage for the output transistors in the circuit for Problem 18-23. Make V_B adjustable by ±20%.

18-26 The driver stage collector resistor (R_C) in the circuit designed for Problem 18-23 is to be replaced by a constant current circuit, as in Fig. 18-30. Design the constant current circuit.

18-27 The amplifier in Problems 18-16 and 18-17 is to be modified to use power supply decoupling as in Fig. 18-29(a). The ripple frequency is 120 Hz. Determine suitable values for R_{15} and C_D, and the new supply voltage level.

Section 18-6

18-28 A direct-coupled amplifier circuit as in Fig. 18-33 is to deliver 7 W to a 22 Ω load. Determine the supply voltage and specify transistors Q_5 through Q_8. Assume h_{FE5} = h_{FE6} = 90 and h_{FE7} = h_{FE8} = 15.

18-29 Determine suitable values for resistors R_9 through R_{15} for the circuit in Problem 18-30. Also, specify the diodes and transistors Q_3 and Q_4. Assume I_{CBO} = 10 µA for Q_7 and Q_8.

18-30 The input voltage to the circuit in Problems 18-28 and 18-29 is ±0.5 V. Determine suitable values for resistors R_1 through R_8.

18-31 The circuit in Problems 18-29 through 18-30 is to have a frequency range from 20 Hz to 50 kHz. Determine suitable capacitances for C_1, C_2, and C_6.

18-32 A direct-coupled amplifier as in Fig. 18-33 has V_{CC} = ±20 V, V_{R9} = V_{R11} = 3 V, and R_L = 16 Ω. Calculate the maximum power delivered to the load.

18-33 Specify Q_7 and Q_8 for the circuit in Problem 18-32.

18-34 Calculate the approximate efficiency for the circuit in Problem 18-32.

18-35 A direct-coupled amplifier, as in Fig. 18-33, uses *2N3904* and *2N3906 BJTs* for Q_5 and Q_6, Q_8 is a *2N3055* and Q_7 is complementary to Q_8. The supply voltage is ±25 V and the load resistance is 20 Ω. Determine the maximum output power and calculate the maximum collector current for Q_7 and Q_8, and the maximum base current for Q_5 and Q_6.

Section 18-7

18-36   A *MOSFET* power amplifier circuit as in Fig. 18-36 has V_{CC} = ±30 V and R_L = 50 Ω. Transistors Q_3 and Q_4 have $R_{D(on)}$ = 4 Ω. Determine the maximum power delivered to the load, and the power dissipated in each output transistor.

18-37   A *MOSFET* amplifier circuit as in Fig. 18-36 has a 25 Ω load resistance and uses output transistors with $R_{D(on)}$ = 3 Ω. Calculate the required supply voltage to dissipate 8 W in the load. Also, calculate the power dissipation in the output transistors.

18-38   The MOSFETs in Problem 18-37 have a threshold voltage of V_{TH} = 1.5 V and a transconductance of g_{fs} = 300 mA/V. Determine suitable *dc* voltage drops across resistors R_2 and R_3, and R_6 through R_9.

18-39   Calculate resistor and capacitor values for the circuit in Problems 18-37 and 18-38 if A_{CL} = 15 and f_1 = 80 Hz.

18-40   Calculate the approximate efficiency for the circuit in Problems 18-37 through 18-39.

18-41   Calculate the supply voltage for an amplifier circuit as in Fig. 18-36 to deliver 10 W to a 16 Ω load. The output *MOSFETs* have $R_{D(on)}$ = 1 Ω, V_{TH} = 2 V, and g_m = 2 S.

18-42   Determine suitable *dc* voltage levels for the circuit in Problem 18-41.

18-43   The amplifier in Problems 18-41 and 18-42 has v_{in} = ±0.7 V, and its low cutoff frequency is to be 40 Hz. Calculate all resistor and capacitor values.

Section 18-8

18-44   A direct-coupled Class-*AB* power amplifier using a complementary emitter follower output stage and a operational amplifier driver (as in Fig. 18-39) is to deliver 2.4 W to a 30 Ω load. Calculate the required supply voltage and specify the output transistors in terms of: $V_{CE(max)}$, $I_{C(max)}$, and power dissipation.

18-45   Determine suitable *dc* voltage and current levels for the circuit in Problem 18-44, and calculate all resistor values if v_{in} = ±0.5 V.

18-46   The circuit in Problems 18-44 and 18-45 is to have a frequency range from 30 Hz to 30 kHz. Calculate the capacitor values, and determine the minimum slew rate for the op-amp.

18-47   Modify the circuit in Problems 18-44 through 18-46 to use bootstrapping capacitors, as in Fig. 18-41. Determine suitable values for the bootstrapping capacitors, and calculate the new maximum peak output voltage that can be produced by the modified circuit.

Section 18-9

18-48   The common-source power amplifier circuit in Fig. 18-52 has the following components: $R_7 = R_8 = 680 \ \Omega$, $R_9 = R_{10} = 1.2 \ k\Omega$, $R_4 = R_6 = 18 \ k\Omega$, $R_5 = 820 \ \Omega$, $R_L = 32 \ \Omega$. The supply voltage is ±30 V, the op-amp supply current is 1 mA, the Q_2 collector current is 1 mA, and the *MOSFETs* have $R_{D(on)}$ = 1.5 Ω, and g_m = 1.2 S. Determine the gate-source bias voltage for Q_3 and Q_4, and the maximum output power.

18-49   Calculate the op-amp supply terminal voltages, and the op-amp peak output voltage for the circuit in Problem 18-48. Also, determine the required signal voltage to give maximum output.

18-50   Design the bias control circuit for the circuit in Problem 18-48 to give I_{C2} = 1 mA ±50%.

18-51   A direct-coupled Class-*AB* common-source power amplifier, as in Fig. 18-52 without the current source, is to deliver 8 W to a 12 Ω load. Calculate the required supply voltage if the output transistors have $R_{D(on)}$ = 0.95 Ω, V_{th} = 2 V, and g_m = 0.9 S.

18-52   The circuit in Problem 18-51 has v_{in} = ±0.9 V, and an op-amp with I_s = 1.3 mA and $I_{o(max)}$ = 15 mA. Determine suitable resistor values.

18-53   The circuit in Problems 18-51 and 18-52 is to have a frequency range from 40 Hz to 45 kHz. Calculate the capacitor values, and determine the minimum slew rate for the op-amp.

Section 18-10

18-54   A direct-coupled Class-*AB* audio power amplifier is to be designed to dissipate 5 W in a 16 Ω load. The circuit is to use quasi-complementary connected output transistors, and a *LM391* driver, as in Fig. 18-54. Calculate the required supply voltage, and specify the *BJTs*. Assume h_{FE} = 20 for Q_3 and Q_4.

18-55   Select suitable resistances for R_{E3} and R_{E4} in the circuit for Problem 18-54 to limit the output current to 20% above the required peak level, as illustrated in Fig. 18-56.

18-56   A *TPA4861 BTL* amplifier is to be used to dissipate 1 W in a 15 Ω load. The input voltage is ±0.7 V, and the low cutoff frequency is to be 25 Hz. Calculate the required supply voltage and suitable resistances for R_1 and R_F (in Fig. 18-59). Also, determine suitable capacitor values.

Practise Problem Answers

18-1.1   [Q point: I_C = 3.1 mA, V_{CE} = 10.2 V], [point A: I_C = 0, V_{CE} = 15 V], [point B: I_C = 0, V_{CE} = 21.1 V]

18-1.2   25.6%

18-2.1   37.7%

18-2.2   [Q point: I_C = 4.8 mA, V_{CE} = 26.7 V], [point A and A': I_C = 24.8 mA, V_{CE} = 25.6 V], [point B and B': I_C = 46.7 mA, V_{CE} = 2.55 V]

18-3.1   10 V, (20 V, 26 mA, 67.5 mW)

18-3.2   1.09 W, 50 V, 640 mA

18-4.1   21 V, (21 V, 90 mA, 200 mW)

18-4.2   1.8 kΩ, 50 Ω variable

18-5.1   ±16 V, 2.7 kΩ, 100 Ω

18-5.2   15 kΩ, 2.2 kΩ, 2 kΩ

18-5.3   10 kΩ, (56 kΩ + 1.5 kΩ), 1 kΩ, 500 Ω variable

18-6.1   4.7 kΩ, 3.3 kΩ, 4.7 kΩ, 1.2 kΩ, 180 Ω, 4.7 kΩ, 15 μF, 30 μF

18-7.1   18 μF, 22 μF, 0.82 μF

18-7.2   ±15 V, 12 V, 14.3 V, 1.3 V, 10.7 V, 17.7 V, 1.3 V

18-7.3   4.7 kΩ, 12 kΩ, 6.8 kΩ, 120 Ω, 4.7 kΩ, 100 kΩ, 820 kΩ, (1.2 MΩ + 150 kΩ), 100 kΩ, 10 μF, 33 μF, 0.47 μF, 0.47 μF

18-8.1   12.1 V, 146 mW, 6.6 kHz

18-8.2   ±12 V, 3.5 V/μs

18-9.1   1.2 kΩ, 1.2 kΩ, 1.2 kΩ, 1.2 kΩ

18-9.2   5.6 kΩ, 390 Ω, 1 kΩ variable, 560 Ω, 560 Ω

18-9.3   100 kΩ, 5.6 kΩ, 100 kΩ, 0.82 μF, 1.5 μF, 39 pF

18-10.1   2.25 W, 0.5 mA

18-10.2   62.5%

18-10.3   ±60 mV, 19.9 Hz

Chapter *19*
Thyristors

Chapter Contents

Objectives

You will be able to:

1 *Sketch the basic construction of an SCR and explain its operation. Draw typical SCR characteristics, and define the device parameters.*

2 *Sketch and explain the following SCR circuits: 90° phase control, 180° phase control, zero-point triggering, crowbar, heater control.*

3 *Design and analyze the above types of SCR circuits.*

4 *Sketch the basic construction of a TRIAC and explain its operation. Draw typical TRIAC characteristics, and define the device parameters.*

5 *Discuss TRIAC Quadrant I to IV triggering.*

6 *Sketch and explain TRIAC phase control and zero-point triggering circuits.*

7 *Design and analyze the above types of TRIAC circuits.*

8 *Sketch characteristics and graphic symbols for the following devices: DIAC, SUS, SBS, GTO, SIDAC. Explain the operation and applications for each device.*

9 *Sketch the basic construction of a UJT and explain its operation. Draw typical UJT characteristics, and define the device parameters.*

10 *Sketch the basic construction of a PUT and explain its operation. Draw typical PUT characteristics, and define the device parameters.*

11 *Sketch and explain relaxation oscillators and thyristor control circuit using UJTs and PUTs.*

12 *Design and analyze the above types of UJT and PUT circuits.*

Introduction

The *silicon-controlled rectifier* (*SCR*) can be thought of as an ordinary rectifier with a control element. The current flowing into the control element, which is termed the *gate*, determines the anode-to-cathode voltage at which the device commences to conduct. The *SCR* is widely applied as an *ac* power control device. The gate bias may keep the device *off*, or it may permit conduction to commence at any desired point in the forward half-cycle of a sinusoidal input. Many other devices, such as the *DIAC* and the *TRIAC*, are based on the *SCR* principle. Collectively, *SCR*-type devices are known as thyristors. This term is derived from *thyratron* and transistor, the thyratron being a gas-filled electron tube that behaves like an *SCR*.

The *unijunction transistor* (*UJT*) is a three-terminal device quite different from bipolar and field effect transistors. The device input, called the *emitter*, has a resistance that rapidly decreases when the input voltage reaches a certain level. This effect is termed a *negative resistance* and it makes the *UJT* useful in timing and oscillator circuits. The *programmable unijunction transistor*(*PUT*) is an *SCR*-type device that behaves like a *UJT*.

19-1 Silicon Controlled Rectifier (SCR)

SCR Operation

The *silicon-controlled rectifier* (*SCR*) consists of four layers of semiconductor material, alternately *p*-type and *n*-type as illustrated in Fig. 19-1(a). Because of its construction, the *SCR* is sometimes referred to as a *four-layer* diode, or a *pnpn* device. The layers are designated p_1, n_1, p_2, and n_2, as shown. There are three junctions; J_1, J_2, and J_3, and three terminals; *anode* (*A*), *cathode* (*K*), and *gate* (*G*). Figure 19-1(b) shows the *SCR* circuit symbol.

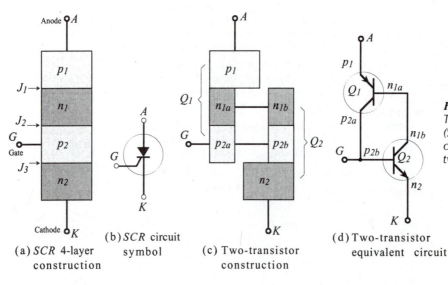

(a) *SCR* 4-layer construction

(b) *SCR* circuit symbol

(c) Two-transistor construction

(d) Two-transistor equivalent circuit

Figure 19-1
The silicon controlled rectifier (SCR) is a four-layer device that can be explained in terms of a two-transistor equivalent circuit.

To understand *SCR* operation, it is necessary to imagine layers n_1 and p_2 split into n_{1a}, n_{1b}, p_{2a} and p_{2b} as shown in Fig. 19-1(c). Since n_{1a} is connected to n_{1b}, and p_{2a} is connected to p_{2b}, nothing is really changed. However, it is now possible to think of p_1, n_{1a}, p_{2a} as a *pnp* transistor, and n_{1b}, p_{2b}, n_2 as an *npn* transistor. Replacing the transistor block representations in Fig. 19-1(c) with the *pnp* and *npn BJT* circuit symbols gives the *two-transistor equivalent circuit* in Fig. 19-1(d). It is seen that the Q_1 collector is connected to the Q_2 base, and the Q_2 collector is commoned with the Q_1 base. The Q_1 emitter is the *SCR* anode terminal, the Q_2 emitter is the cathode, and the junction of the Q_1 collector and the Q_2 base is the *SCR* gate terminal.

To forward bias an *SCR*, a voltage (V_{AK}) is applied positive on the anode (A), negative on the cathode (K), as shown in Fig. 19-2(a). If the gate (G) is left unconnected only small leakage currents (I_{CO}) flow, and both transistors remain *off*. Reference to Fig. 19-1(a) shows that the leakage currents are the result of junction J_2 being reverse biased when A is positive and K is negative.

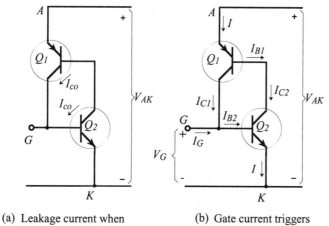

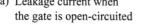

(a) Leakage current when the gate is open-circuited

(b) Gate current triggers the *SCR* on

Figure 19-2
When the SCR gate current is zero, the device normally remains off. The flow of gate current (I_G) triggers the SCR on.

When a negative gate-cathode voltage ($-V_G$) is applied, the Q_2 base-emitter junction is reverse biased, and only small leakage currents continue to flow, so both Q_1 and Q_2 remain *off*. A positive gate-cathode voltage forward biases the Q_2 base-emitter junction, causing a gate current ($I_G = I_{B2}$) to flow, and producing a Q_2 collector current (I_{C2}), [see Fig. 19-2(b)]. Because I_{C2} is the same as I_{B1}, Q_1 also switches *on* and I_{C1} flows providing base current I_{B2}. Each collector current provides much more base current than needed by the transistors, and even when I_G is switched *off*, the transistors remain *on*, conducting heavily with only a small anode-to-cathode voltage drop. The ability of the *SCR* to remain *on* when the triggering current is removed is referred to as *latching*.

To switch the *SCR on*, only a brief pulse of gate current is required. Once switched *on*, the gate has no further control and the device remains *on* until V_{AK} is reduced to near zero.

Consider Fig. 19-1(a) again. With a forward (anode-to-cathode) bias, junctions J_1 and J_3 are forward biased, while J_2 is reverse biased. When V_{AK} is made large enough, J_2 will break down and the resultant current flow across the junction constitutes collector current in each transistor. Each collector current flows into the base of the other transistor causing both transistors to switch *on*. Thus, the *SCR* can be triggered *on* with the gate open-circuited.

SCR Characteristics and Parameters

Figure 19-3(a) shows an *SCR* with a reverse bias anode-to-cathode voltage $(-V_{AK})$, (negative on *A*, positive on *K*). Note that the gate terminal is open-circuited. Figure 19-3(b) shows that the reverse bias voltage causes junction J_2 to be forward biased and J_1 and J_3 to be reverse biased. When $-V_{AK}$ is small, a *reverse leakage current* (I_{RX}) flows. This is plotted as the reverse characteristic $(-V_{AK}$ versus $I_R)$ on Fig. 19-3(c). I_{RX} is typically around 100 μA, and is sometimes referred to as the *reverse blocking current*.

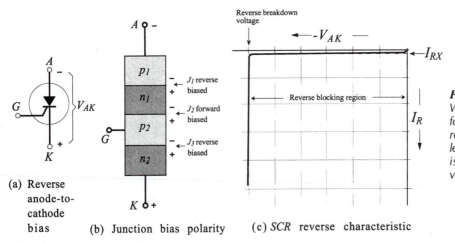

(a) Reverse anode-to-cathode bias

(b) Junction bias polarity

(c) *SCR* reverse characteristic

Figure 19-3
When V_{AK} is negative, J_2 is forward biased, and J_1 and J_3 are reverse biased. A small reverse leakage current flows while $-V_{AK}$ is less than the breakdown voltage.

When the level of $-V_{AK}$ is increased, I_{RX} remains approximately constant until the reverse breakdown voltage is reached. At this point the reverse-biased junctions $(J_1$ and $J_3)$ break down and the reverse current (I_R) increases very rapidly. If I_R is not limited (by additional circuit components) the device will be destroyed by excessive current flow. The region of the reverse characteristics before breakdown is termed the *reverse blocking region*.

An *SCR* with a forward bias anode-to-cathode voltage (positive on *A*, negative on *K*) is shown in Fig. 19-4(a). Here again, the gate terminal is open-circuited. As illustrated in Fig. 19-4(b), $+V_{AK}$ forward biases J_1 and J_3 and reverse biases J_2. With low levels of $+V_{AK}$, a small *forward leakage current* (I_{FX}) flows. This is actually the reverse leakage current at junction (J_2), and so (like I_{RX}), it is typically around 100 μA. Also like I_{RX}, I_{FX} remains substantially constant until $+V_{AK}$ is made large enough to cause (reverse biased) J_2 to break down. The applied voltage at this point is termed the forward breakover voltage $(V_{F(BO)})$. This is illustrated by the *forward characteristics* (I_F versus $+V_{AK}$) in Fig. 19-4(c). When $V_{F(BO)}$ is reached, the component transistors (Q_1 and Q_2) are immediately switched *on* into saturation as already explained, and the anode-to-cathode voltage falls rapidly to the *forward conduction voltage* V_F. The device is now into the *forward conduction region*, and I_F must be limited to protect the *SCR* from excessive current levels.

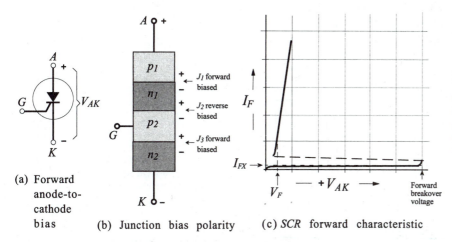

(a) Forward anode-to-cathode bias

(b) Junction bias polarity

(c) *SCR* forward characteristic

Figure 19-4
When V_{AK} is Positive, J_2 is reverse biased, and J_1 and J_3 are forward biased. A small forward leakage current flows while $+V_{AK}$ is less than the forward breakover voltage.

So far, the *SCR* forward characteristics have been discussed only for the case of $I_G = 0$. Now consider the effect of I_G levels greater than zero, [Fig. 19-5(a)]. As already shown, when $+V_{AK}$ is less than $V_{F(BO)}$ and I_G is zero, a small leakage current flows. This current is too small to have any effect on the level of $+V_{AK}$ that causes *SCR* switch *on*. When I_G is made just slightly larger than the junction leakage currents, it still has a negligible effect on the level of $+V_{AK}$ for switch-*on*. Now consider the opposite extreme. When I_G is made larger than the minimum base current required to switch Q_2 *on*, the *SCR* switches-*on* when $+V_{AK}$ forward biases the base-emitter junctions of Q_1 and Q_2, [Fig. 19-5(b) and Fig. 19-6].

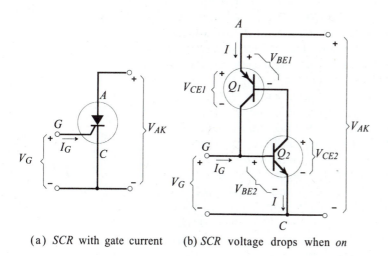

(a) *SCR* with gate current

(b) *SCR* voltage drops when *on*

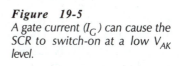

Figure 19-5
A gate current (I_G) can cause the *SCR* to switch-on at a low V_{AK} level.

The complete forward characteristics for an *SCR* are shown in Fig. 19-6. Note that when $I_G = I_{G4}$ switch-*on* occurs with $+V_{AK}$ at a relatively low level (V_4). Gate currents between I_{GO} and I_{G4} permit device switch-*on* at voltages greater than V_4 and less than $V_{F(BO)}$. The region of the forward characteristics before switch-*on* occurs is known as the *forward blocking region*, and the region after switch-*on* is termed the *forward conduction region*, as illustrated. In the forward conduction region, the *SCR* behaves as a forward-biased rectifier. The forward (anode-to-cathode) voltage (V_F) when the device is *on* is typically 1.7 V.

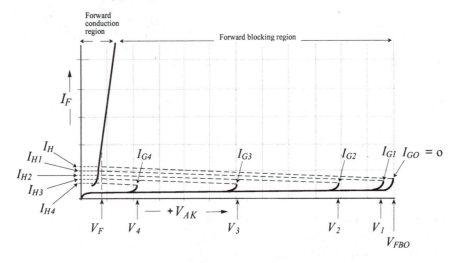

Figure 19-6
Forward characteristics for an SCR. Higher levels of gate current (I_G) cause the SCR to conduct at lower anode-to-cathode voltages ($+V_{AK}$).

To switch an *SCR off*, the forward current (I_F) must be reduced below the *holding current* (I_H), (see Fig. 19-6). The holding current is the minimum level of I_F that maintains *SCR* conduction. If a gate current greater than zero is maintained while the *SCR* is *on*, lower levels of holding current (I_{H1}, I_{H2}, etc.,) are possible.

SCR Specification

As in the case of most electronic devices, the *SCR* maximum voltage and current are important for any given application. The forward breakover voltage and reverse breakdown voltage have already been discussed. The maximum forward voltage that may be applied without causing the *SCR* to conduct is termed the *forward blocking voltage* (V_{DRM}). Similarly, the maximum reverse voltage that may be applied is the *reverse blocking voltage* (V_{RRM}).

The maximum *SCR* current is variously specified as: the *average current* ($I_{T(AV)}$), the *rms current* ($I_{T(RMS)}$), and the *peak non-repetitive surge current* (I_{TSM}). The first two of these need no explanation. The third is a relatively large current that can normally be permitted to flow for a maximum of a half-cycle of a 60 Hz sine wave. The *circuit fusing rating* (I^2t) is another parameter that defines the maximum nonrepetitive forward current. This can be used to calculate the maximum time duration for a given forward current surge. In many circuit applications the *SCR* current is limited by a series-connected load, so there is usually no need to consider surge current levels, except in the case of capacitive loads.

Some of the range of available *SCRs* is illustrated by the partial specifications and packages shown in Fig. 19-7. With 800 mA rms current and 30 V forward and reverse blocking voltage, the *2N5060* is a relatively low-current, low-voltage, device. This is packaged in the typical plastic *TO-92* transistor-type enclosure. Note that the *peak reverse gate voltage* (V_{GRM}) is 5 V. The *2N6396 SCR* is capable of handling a maximum *rms* current of 12 A, and has forward and reverse blocking voltage of 200 V. The package is a *TO-220* plastic enclosure with a metal tab for mounting on a heat sink. For the *C35N*, the peak forward and reverse voltage is 960 V, and maximum *rms* current is 35 A. The device package is designed for bolt-mounting to a heat sink.

	2N5060	*2N6396*	*C35N*
Peak forward & reverse voltage (V_{DRM} & V_{RRM})	30 V	200 V	960 V
Maximum rms current ($I_{T(RMS)}$)	0.8 A	12 A	35 A
Forward *on* voltage (V_{TM})	1.7 V	1.7 V	2 V
Holding current (I_H)	5 mA	6 mA	100 mA
Gate trigger current (I_{GT})	200 μA	12 mA	6 mA
Gate trigger voltage (V_{GT})	0.8 V	0.9 V	3 V
Gate reverse voltage (V_{GRM})	5 V	5 V	5 V

Figure 19-7
Partial specifications and packages for three SCRs for different voltage and current levels.

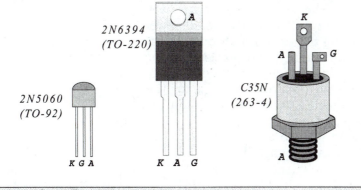

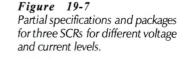

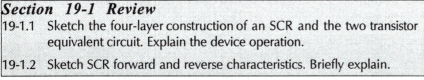

Section 19-1 Review

19-1.1   Sketch the four-layer construction of an SCR and the two transistor equivalent circuit. Explain the device operation.

19-1.2   Sketch SCR forward and reverse characteristics. Briefly explain.

19-2   SCR Control Circuits

Pulse Control

The simplest of *SCR* control circuits is shown in Fig. 19-8(a). If *SCR₁* was an ordinary rectifier, the *ac* supply voltage would be half-wave rectified and only the positive half-cycles would appear across the load (R_L). The same would be true if the SCR gate had a continuous bias voltage to keep it *on* when the anode-cathode voltage goes positive. A trigger pulse applied to the gate can switch the device *on* at any time during the positive half-cycle of the supply voltage. The SCR continues to conduct during the rest of the positive half-cycle, and then it switches *off* when the instantaneous level of the supply approaches zero. The resultant load waveform is a portion of the positive half-cycle commencing at the instant that the SCR is triggered [Fig. 19-8(b)]. Resistor R_G holds the gate-cathode voltage at zero when no trigger input is present.

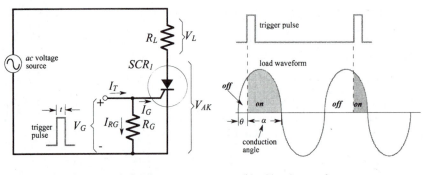

(a) *SCR* pulse control circuit   (b) Circuit waveforms

Figure 19-8
An SCR can be triggered on by a pulse applied to the gate. Once triggered, the device remains on until the load current falls below the holding current.

Load waveforms that result from the SCR being switched *on* at different points in the positive half-cycle of the supply voltage are shown in Fig. 19-9. It is seen that the average load current is controlled by the SCR conduction angle. Thus, the load power dissipation can be varied by adjusting the SCR switch-*on* point. It should be noted that the SCR cannot be triggered precisely at the 0° point in the waveform, because the anode-to-cathode voltage must be at lease equal to the forward *on* voltage (V_{TM}) for the device. Also, the SCR will switch *off* before the 90° point when the load current falls below the holding current.

The instantaneous level of the load voltage is the instantaneous supply voltage (e_s) minus the SCR forward voltage (V_{TM});

$$V_L = e_s - V_{TM} \qquad \textbf{(19-1)}$$

The load current can be calculated from V_L and R_L, and the instantaneous supply voltage ($e_{s(o)}$) that causes the SCR to switch *off* can be determined from V_{TM}, R_L, and the holding current;

$$e_{s(o)} = V_{TM} + (I_H \times R_L) \qquad \textbf{(19-2)}$$

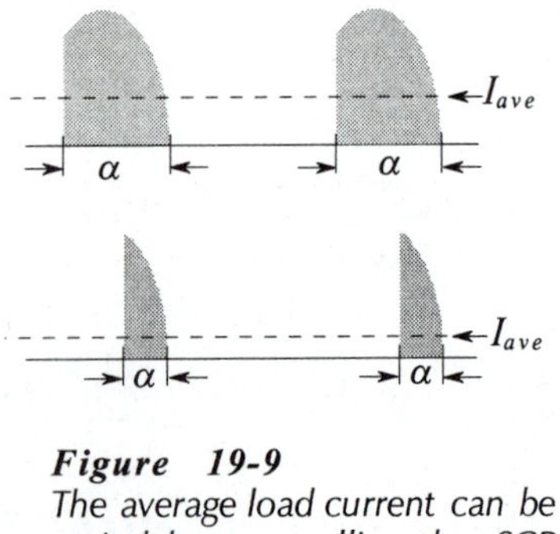

For any given application, the selected SCR must have forward and reverse blocking voltages greater that the peak supply voltage. Its specified maximum *rms* current must also be greater than the *rms* load current. When designing the circuit, the gate current used should be at least three times the specified I_G for the device.

Figure 19-9
The average load current can be varied by controlling the SCR conduction angle.

Note that the required triggering current (I_T) for the circuit in Fig. 19-8 is the sum of I_G and the resistor current I_{RG}, as illustrated.

Example 19-1
Select a suitable *SCR* for the circuit in Fig. 19-8(a), if the *rms* supply voltage is 24 V and the load resistance is 25 Ω. Also, calculate the instantaneous supply voltage that causes the *SCR* to switch *off*.

Solution
peak supply voltage,

$$V_{s(pk)} = 1.414 \times V_s = 1.414 \times 25 \text{ V}$$
$$= 33.9 \text{ V}$$

SCR forward and reverse blocking voltage,

$$V_{DRM} \& V_{RRM} > 33.9 \text{ V}$$

Referring to the partial specification for the *2N5060* to *2N5064* range of *SCRs* in Fig. 19-10, it is found that the *2N5060* has $V_{DRM} = 30$ V, and the *2N5061* has $V_{DRM} = 60$ V. So, the *2N5060* would *not* be suitable, while the *2N5061* would seem to be a suitable device.

$$I_{L(pk)} = \frac{V_{s(pk)} - V_{TM}}{R_L} = \frac{33.9 \text{ V} - 1.7 \text{ V}}{25 \text{ Ω}}$$
$$= 1.29 \text{ A}$$

For a half-wave rectified sinusoidal waveform,

$$I_{L(rms)} = 0.5 \ I_{L(pk)} = 0.5 \times 1.29 \text{ A}$$
$$= 0.64 \text{ A}$$

The *2N5061* has $I_{T(rms)} = 0.8$ A.
So, the *2N5061* is a suitable *SCR*.

Switch *off* voltage;

From Eq. 19-2, $\quad e_{s(o)} = V_{TM} + (I_H \times R_L)$
$$= 1.7 \text{ V} + (5 \text{ mA} \times 25 \text{ Ω})$$
$$\approx 1.8 \text{ V}$$

2N5060 to 2N5064

Peak forward	2N5060	30 V
& reverse voltage	2N5061	60 V
(V_{DRM} & V_{RRM})	2N5062	100 V
	2N5063	150 V
	2N5064	200 V
Maximum rms current ($I_{T(RMS)}$)		0.8 A
Forward *on* voltage (V_{TM})		1.7 V
Holding current (I_H)		5 mA
Gate trigger current (I_G)		200 µA

Figure 19-10
Partial specification for 2N5060 to 2N5064 SCRs.

90° Phase Control
In the *90° phase-control circuit* shown in Fig. 19-11, the gate triggering voltage is derived from the *ac* supply via resistors R_1, R_2, and R_3. When the moving contact is set to the top of R_2, the *SCR* can be triggered *on* almost immediately at the commencement of the positive half-cycle of the input. When the moving contact is set to the bottom of R_2, the *SCR* might not switch *on* until the peak of the positive half-cycle. Between these two extremes, the device can be switched *on* somewhere between the zero level and the peak of the positive half-cycle, (between *0°* and *90°*). If the triggering

voltage (V_T) is not large enough to trigger the *SCR* at *90°*, then the device will not trigger *on* at all, because V_T is greatest at the supply voltage peak and falls *off* past the peak.

Diode D_1 in Fig. 19-11 is included in the circuit to protect the *SCR* gate from the negative voltage that would otherwise be applied to it during the negative half-cycle of the *ac* supply.

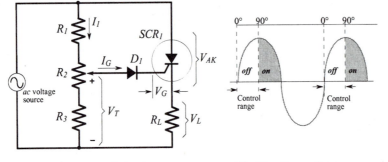

(a) 90° phase control circuit (b) Circuit waveforms

Figure 19-11
SCR 90° phase control circuit. The SCR can be triggered on anywhere between 0° and 90°.

The load for an *SCR* phase control circuit could be a permanent magnet motor, so that the circuit controls the motor speed. Alternatively, the load might be a heater or a light, and in this case the circuit controls the heater temperature or the light intensity.

The voltage divider (R_1 R_2 R_3) in Fig. 19-11 is designed in the usual way for the required range of adjustment of V_T. The voltage divider current (I_1) is selected much larger than the *SCR* gate current. The instantaneous triggering voltage at switch-*on* is,

$$V_T = V_{D1} + V_G \qquad\qquad (19\text{-}3)$$

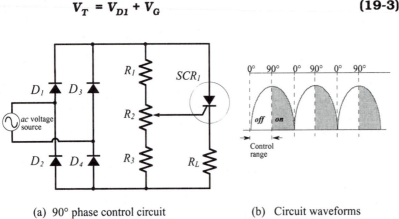

(a) 90° phase control circuit
with full wave rectified supply

(b) Circuit waveforms

Figure 19-12
SCR 90° phase control circuit with a full wave rectified supply.

Figure 19-12 shows a 90° phase control circuit with its *ac* voltage source full-wave rectified. This gives a larger maximum power dissipation in the load than a non-rectified source. Also, diode D_1 in Fig. 19-11 is not required in Fig. 19-12 because the *SCR* gate does not become reverse biased.

In the circuit in Fig. 19-13(a) the two *SCRs* are connected in inverse-parallel and they operate independently as 90° phase control circuits. SCR_1 controls the load current during the positive half-cycle of the supply voltage, and SCR_2 controls the current during the negative half-cycle. The triggering voltage for each *SCR* is set by the voltage divider network R_1 through R_4 and adjusted by variable resistor R_3. Diodes D_1 and D_2 protect the gate terminals of each *SCR* from excessive reverse voltage.

During the supply voltage positive half-cycle, D_2 is forward biased and current flows through R_2, R_3, and R_4. The voltage drop across R_4, triggers SCR_1 at the desired point in the positive half cycle. When triggered, the *SCR* forward voltage switches to a low level, and remains there until the instantaneous supply voltage level approaches zero. During the supply negative half-cycle, D_1 is forward biased to produce current flow through R_1, R_2, and R_3. With R_1 equal to R_4, the voltage drop across R_1 triggers SCR_2 at the same point in the negative half-cycle as SCR_1 in the positive half-cycle. The resultant 90° full-wave phase controlled load waveform is shown in Fig. 19-13(b).

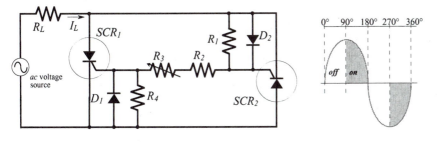
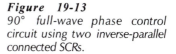

(a) 90° full-wave phase control circuit   (b) load current waveform

Figure 19-13
90° full-wave phase control circuit using two inverse-parallel connected SCRs.

Example 19-2

The *SCR* in Fig. 19-14 is to be triggered *on* between 5° and 90° during the positive half-cycle of the 30 V supply. The gate triggering current and voltage are 200 μA and 0.8 V. Determine suitable resistance values for R_1, R_2, and R_3.

Solution

Peak supply voltage,

$$V_{s(pk)} = 1.414 \times V_s = 1.414 \times 30\text{ V}$$
$$= 42.4\text{ V}$$

at 5°,

$$e_s = V_{s(pk)} \sin 5° = 42.4\text{ V} \sin 5°$$
$$\approx 3.7\text{ V}$$

at 90°,

$$e_s = V_{s(pk)} = 42.4\text{ V}$$

Eq. 19-3,

$$V_T = V_{D1} + V_G = 0.7\text{ V} + 0.8\text{ V}$$
$$= 1.5\text{ V}$$

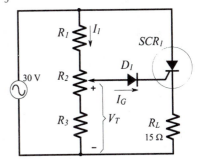

Figure 19-14
SCR 90° phase control circuit for Example 19-2.

To trigger at $e_s = 3.7$ V, the R_2 moving contact is at the top.

so, $V_{R2} + V_{R3} = V_T = 1.5$ V

and, $V_{R1} = e_s - V_T = 3.7$ V - 1.5 V
$$= 2.2 \text{ V}$$

$$I_{1(min)} >> (I_G = 200 \text{ } \mu A)$$

select $I_{1(min)} = 1$ mA

$$R_1 = \frac{V_{R1}}{I_1} = \frac{2.2 \text{ V}}{1 \text{ mA}}$$
$$= 2.2 \text{ k}\Omega \text{ (standard value)}$$

$$R_2 + R_3 = \frac{V_T}{I_1} = \frac{1.5 \text{ V}}{1 \text{ mA}}$$
$$= 1.5 \text{ k}\Omega$$

To trigger at $e_s = 42.4$ V, the R_2 moving contact is at the bottom.

so, $V_{R3} = V_T = 1.5$ V

and, $I_1 = \frac{e_s}{R_1 + R_2 + R_3} = \frac{42.4 \text{ V}}{2.2 \text{ k}\Omega + 1.5 \text{ k}\Omega}$
$$\approx 11.5 \text{ mA}$$

$$R_3 = \frac{V_T}{I_1} = \frac{1.5 \text{ V}}{11.5 \text{ mA}}$$
$$= 130 \text{ }\Omega \text{ (use 120 }\Omega \text{ standard value)}$$

$$R_2 = (R_2 + R_3) - R_3 = 1.5 \text{ k}\Omega - 120 \text{ }\Omega$$
$$= 1.38 \text{ k}\Omega \text{ (use 1.5 k}\Omega \text{ standard value potentiometer)}$$

180° Phase Control

In the circuit shown in Fig. 19-15, resistor R_1 and capacitor C_1 determine the point in the supply voltage cycle where the SCR switches on. During the negative half-cycle of the supply, C_1 is charged via diode D_1 to the negative peak of the supply voltage. When the negative peak is passed, D_1 is reverse biased because its anode (connected to C_1) is more negative than its cathode. With D_1 reversed, C_1 commences to discharge via R_1. While C_1 voltage remains negatively, D_2 is reverse biased and the gate voltage cannot go positive to trigger the SCR on. Depending on the values of C_1 and R_1, the capacitor might be completely discharged at the beginning of the positive half-cycle of the supply; allowing SCR_1 to switch on. Alternatively, C_1 might retain some negative charge past the end of positive half-cycle; keeping SCR_1 off. Resistor R_2 is included in the circuit to restrict the level of the gate current.

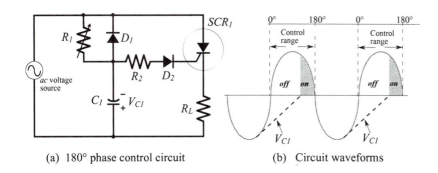

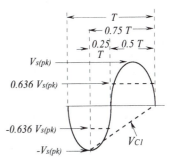

(a) 180° phase control circuit

(b) Circuit waveforms

Figure 19-15
SCR 180° phase control circuit. R_1 adjustment allows the SCR triggering point to be set anywhere between 0° and 180° in the positive half-cycle of the ac supply voltage.

Design of the 180° phase control circuit can commence with selection of a capacitor much larger than stray capacitance. A maximum resistance for R_1 should then be calculated to discharge the capacitor voltage to zero during the time from the negative peak of the supply voltage to the 180° point in the positive half-cycle. The capacitor voltage does not decrease linearly as Fig. 19-16 implies. However, the maximum resistance for R_1 can be most easily calculated by assuming a linear discharge. The average value of the discharging voltage (E) is first determined. Figure 19-16 shows that E is -0.636 $V_{s(pk)}$ for 0.25 T, and +0.636 $V_{s(pk)}$ for 0.5 T, which averages out to approximately 0.2 $V_{s(pk)}$ for the total discharge time of 0.75 T. Now the equation for discharge of a capacitor to zero volts via a resistor may be applied.

$$t = R\,C\ln\left[(E - E_o)/E\right]$$

Substituting the appropriate quantities into the equation gives,

$$R_1 \approx \frac{0.75\,T}{C_1\ln 6} \qquad\qquad \textbf{(19-4)}$$

Figure 19-16
Discharge times and voltages for C_1 in the circuit in Fig. 19-15.

Practise Problems

19-2.1 The 90° phase control circuit in Fig. 19-11 has a 115 V, 60 Hz supply, and R_L = 50 Ω. Specify the required *SCR*, and calculate suitable resistor values for switch *on* between 7° and 90°.

19-2.2 The 180° phase control circuit in Fig. 19-15 has a 50 V, 60 Hz supply, and the *SCR* has V_G = 0.5 V and I_G = 100 µA. Determine suitable values for R_1 and C_1. Also, calculate a resistance for R_2 to limit the gate current to a maximum of 50 mA.

19-3   More SCR Applications

SCR Circuit Stability

An *SCR* circuit is stable when it operates correctly; switching *on* and *off* only at the desired instants. Unwanted triggering (also called *false triggering*) can be produced by noise voltages at the gate, transient voltages at the anode terminal, or by very fast voltages changes at the anode (termed *dv/dt triggering*).

Obviously, gate noise voltages might be large enough to forward bias the gate-cathode junction and cause false triggering. Anode voltage transients (produced by other devices connected to the same *ac* supply) could exceed the *SCR* breakover voltage, and thus trigger it into conduction. The *dv/dt* effect occurs when the anode voltage changes instantaneously, such as when the supply is switched *on* at its peak voltage level. The *SCR* capacitance is charged very quickly, and the charging current is sufficient to trigger the device.

Gate noise problems can be minimized by keeping the gate connecting leads short, and by the use of a gate bias resistor [R_G in Fig. 19-17(a)]. This should be connected as close as possible to the *SCR* gate-cathode terminals, because connecting conductors between R_G and the device could pick up noise that might cause triggering. Biasing the gate negative with respect to the cathode can also be effective in combating noise. Capacitor C_1 in Fig. 19-17(b) can be used to short circuit gate noise voltages. C_1 also operates in conjunction with the anode-gate capacitance as a voltage divider that reduces the possibility of *dv/dt* triggering. C_1 is usually in the 0.01 μF to 0.1 μF range, and like R_G, it should be connected close to the *SCR* terminals.

An *RC snubber circuit* can be used to prevent triggering by anode terminal transients, [Fig. 19-17(c)]. A snubber is usually necessary for inductive loads, and might also be required for resistive loads. With an *ac* supply, there is a phase difference between an inductive load current and the supply voltage, and this can cause loss of *SCR* control. Also, the current through an inductor with a *dc* supply will not go to zero immediately when the *SCR* switches *off*. A snubber circuit is necessary in both cases.

Zero-Point Triggering

When an *SCR* is switched *on* while the instantaneous level of the supply voltage is greater than zero, surge currents occur that generate *electromagnetic interference* (*EMI*). The *EMI* can interfere with other nearby circuits and equipment, and the switching transients can affect control of the *SCR*. Circuits can be designed to trigger an *SCR on* at the instant the *ac* supply is crossing the zero voltage point from the negative half-cycle to the positive half-cycle. This is called *zero-point triggering*, and it effectively eliminates the *EMI* and the switching transients.

The zero-point triggering circuit in Fig. 19-18(a) shows two inverse-parallel connected *SCRs* that each have *RC* triggering circuits; C_1 and R_1 for SCR_1, and C_2 and R_2 for SCR_2. SCR_1 is held *off* while switch S_1 is closed, and because capacitor C_2 is uncharged SCR_2 remains *off*. With S_1 open, positive triggering current (I_{G1}) begins to flow when the supply voltage commences to go positive. As illustrated, I_{G1} flows via C_1 and R_1 to the gate of SCR_1 triggering it into conduction at the zero-crossing point. SCR_1 provides a path for (*positive*) load current (i_{L+}).

With SCR_1 *on*, capacitor C_2 charges (with the polarity shown) almost to the peak of the supply voltage. When the supply voltage

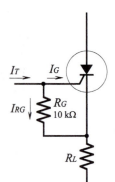

(a) Gate resistor

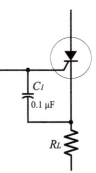

(b) Gate capacitor

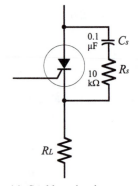

(c) Snubber circuit

Figure   19-17
Unwanted gate noise triggering can be prevented by R_G or C_1 at the gate-cathode terminals. The use of a snubber circuit prevents triggering by transients at the anode terminal.

crosses zero from the positive half-cycle to the negative half-cycle SCR_1 switches *off*. Also, D_1 becomes reverse biased, and the charge on C_1 provides triggering current (I_{G2}) to SCR_2. Thus, SCR_2 is switched *on* at the start of the supply negative half-cycle, providing a path for *(negative)* load current (i_L).

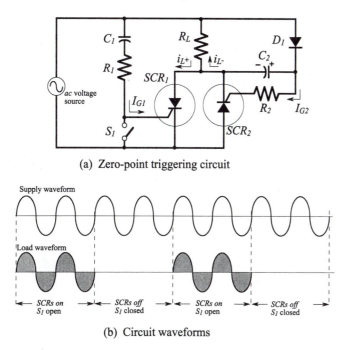

(a) Zero-point triggering circuit

Figure 19-18
In an SCR zero-point triggering circuit the devices are switched on only when the supply waveform crosses the zero-voltage point.

(b) Circuit waveforms

Both *SCRs* continue to switch *on* and *off* at the zero-crossing points while S_1 remains open, and both stay *off* when S_1 is closed. SCR_2 cannot switch *on* unless SCR_1 has first been *on*, and because of this the arrangement is sometimes termed a *master-slave* circuit; SCR_1 being the *master* and SCR_2 the *slave*. The waveforms in Fig. 19-18(b) show that power is supplied to the load for several cycles of the supply while S_1 is open, and no load power dissipation occurs for several cycles while S_1 remains closed. The switch might be controlled by a temperature sensor or other device.

Crowbar Circuit

A *crowbar circuit* (also known as an *overvoltage protection circuit*) is illustrated in Fig. 19-19. This circuit protects a sensitive load against an excessive *dc* supply voltage. When the supply (V_S) is at its normal voltage level, it is too low to cause the Zener diode (D_1) to conduct. Consequently, there is no current through the gate bias resistor (R_1), and no voltage drop across R_1. The gate voltage (V_G) remains equal to zero, and the *SCR* remains *off*. When the supply voltage exceeds V_Z, D_1 conducts, and the resultant voltage drop across R_1 triggers the *SCR* into conduction. The voltage across the load is now reduced to the *SCR* forward voltage drop. The voltage across V_Z and R_1 is also reduced to the *SCR* forward voltage, and the *dc* voltage source is short-circuited by the *SCR*.

The voltage source must have a current limiting circuit to protect the source and to minimize *SCR* power dissipation. The supply must be switched *off* for the SCR to cease conducting.

Example 19-3

The *dc* voltage source in the *SCR* crowbar circuit in Fig. 19-19 has $V_s = 5$ V and $I_{L(max)} = 300$ mA. The load voltage is not to exceed 7 V. Select suitable components for D_1 and R_1, and specify the SCR. Assume that $V_G = 0.8$ V.

Solution

$$V_z = V_{L(max)} - V_G = 7 \text{ V} - 0.8 \text{ V}$$
$$= 6.2 \text{ V}$$

For D_1, select a *1N753* with $V_z = 6.2$ V

Select $I_{z(min)} = 1$ mA

$$R_1 = \frac{V_G}{I_z} = \frac{0.8 \text{ V}}{1 \text{ mA}}$$

$$= 800 \ \Omega \quad \text{(use 820 } \Omega \text{ standard value)}$$

SCR specification:

$$V_{DRM} > 7 \text{ V}, \ I_{T(AV)} > 300 \text{ mA}$$

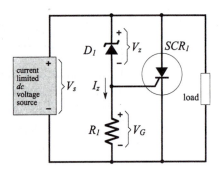

Figure 19-19
An SCR crowbar circuit (or overvoltage protection circuit) short-circuits the load when the supply voltage exceeds a predetermined level.

Heater Control Circuit

The circuit in Fig. 19-20 uses a temperature-sensitive control element (R_2). The resistance of R_2 decreases when the temperature increases, and increases when the temperature falls. Diode D_1 keeps capacitor C_1 charged to the supply voltage peak, and C_1 together with resistor R_1 behaves as a constant current source for R_2. When R_2 is raised to the desired temperature, V_G drops to a level that keeps the *SCR* from triggering. When the temperature drops, the resistance of R_2 increases, causing V_G to increase to the *SCR* triggering level. The result is that the load power is turned *off* when the desired temperature is reached, and turned *on* again when the temperature falls to a predetermined level. Rectifier D_2 might be included, as illustrated, to pass the negative half-cycle of the supply waveform to the load.

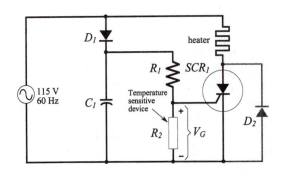

Figure 19-20
SCR heater control circuit. The SCR is triggered on when the temperature is below a specified level, and held off when the temperature is satisfactory.

19-4 TRIAC and DIAC

TRIAC Operation and Characteristics

The basic construction, equivalent circuit, and graphic symbol for a *TRIAC* are shown in Fig. 19-21. The *TRIAC* behaves as two inverse-parallel connected *SCRs* with a single gate terminal. Sections n_1, p_2, n_3, and p_3, in Fig. 19-21(a) form one *SCR* that can be represented by transistors Q_1 and Q_2 in Fig. 19-21(b). Similarly, p_1, n_2, p_2, and n_4, form another *SCR* with the transistor equivalent circuit Q_3 and Q_4. Layer p_2, common to the two *SCRs*, functions as a gate for both sections of the device. The two outer terminals cannot be identified as anode and cathode; instead they are designated *main terminal 1 (MT1)* and *main terminal 2 (MT2)*, as illustrated. The *TRIAC* circuit symbol is composed of two inverse-parallel connected *SCR* symbols, [Fig. 19-21(c)].

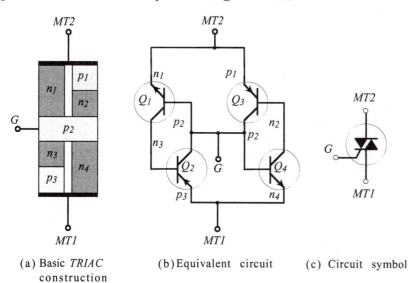

(a) Basic *TRIAC* construction   (b) Equivalent circuit   (c) Circuit symbol

Figure 19-21
Basic construction, equivalent circuit, and graphic symbol for a TRIAC.

When *MT2* is positive with respect to *MT1*, transistors Q_3 and Q_4 can be triggered *on* [Fig. 19-21(b)]. In this case current flow is from *MT2* to *MT1*. When *MT1* is positive with respect to *MT2*, Q_1 and Q_2 can be switched *on*. Now current flow is from *MT1* to *MT2*. It is seen that the *TRIAC* can be made to conduct in either direction. Regardless of the *MT2/MT1* voltage polarity, the characteristics for the *TRIAC* are those of a forward-biased *SCR*. This is illustrated by the typical *TRIAC* characteristics shown in Fig. 19-22.

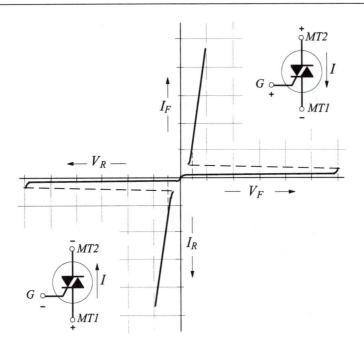

Figure 19-22
TRIAC characteristics. These are similar to the characteristics of two inverse-parallel connected SCRs.

TRIAC Triggering

The characteristics and circuit symbol in Fig. 19-22 show that when *MT2* is positive with respect to *MT1*, the *TRIAC* can be triggered *on* by application of a positive gate voltage. Similarly, when *MT2* is negative with respect to *MT1*, a negative gate voltage triggers the device into conduction. However, a negative gate voltage can also trigger the *TRIAC* when *MT2* is positive, and a positive gate voltage can trigger the device when *MT2* is negative.

Figure 19-23 shows the triggering conditions for a *2N6346*, 8 A, 200 V *TRIAC*. The voltage polarity for *MT2* is identified as *MT2*(+) or *MT2*(-), and the gate polarity is listed as *G*(+) or *G*(-). From the first line of the specifications, it is seen that with *MT2* positive the device gate triggering voltage is +0.9 V minimum and +2 V maximum. From the second line, still with *MT2* positive, triggering can be produced by a negative gate voltage; -0.9 V to -2.5 V. The third line shows *MT2* negative and the gate trigger voltage as -1.1 V to -2 V. Also, with *MT2* negative (fourth line), triggering can be effected by a positive gate voltage; +1.4 V to +2.5 V.

2N6346 TRIAC		
V_{GT}	Min	Max
MT2 (+), G (+)	0.9 V	2 V
MT2 (+), G (-)	0.9 V	2.5 V
MT2 (-), G (-)	1.1 V	2 V
MT2 (-), G (+)	1.4 V	2.5 V

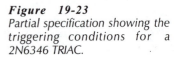

Figure 19-23
Partial specification showing the triggering conditions for a 2N6346 TRIAC.

The *TRIAC* triggering conditions are further illustrated by the diagram in Fig. 19-24. The vertical line identifies *MT2* as positive or negative, and the horizontal line shows the gate voltage as positive or negative. The *TRIAC* is defined as operating in one of the four quadrants: *I, II, III,* or *IV.* In *quadrant I*, *MT2* is positive, the gate voltage is positive, and current flow is from *MT2* to *MT1*, as shown. When *MT2* is positive and the device is triggered by a negative gate voltage, the *TRIAC* is operating in *quadrant II.* In this case, current flow is still from *MT2* to *MT1. Quadrant III* operation occurs when *MT2* is negative and the gate voltage is negative. Current flow is now from *MT1* to *MT2.* In *quadrant IV*, *MT2* is again negative, the gate voltage is positive, and current flow is from *MT1* to *MT2.*

Normally, a *TRIAC* is operated in either *quadrant I* or *quadrant III.* When this is the desired condition, it might be necessary to design the circuit to avoid *quadrant II* or *quadrant IV* triggering.

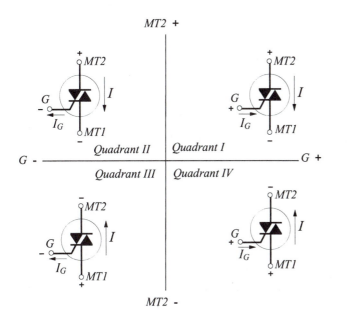

Figure 19-24
Quadrant diagram illustrating the TRIAC four-quadrant operating conditions.

DIAC

A *DIAC* is basically a low-current *TRIAC* without a gate terminal. Switch-*on* is effected by raising the applied voltage to the breakover voltage. Two different *DIAC* symbols in general use are shown in Fig. 19-25(a), and typical *DIAC* characteristics are illustrated in Fig. 19-25(b). Note that the terminals are identified as *anode 2* (A_2) and *anode 1* (A_1). Figure 19-26 shows partial specifications for two *DIACs.* The *HS-10* has a switching voltage that ranges from a minimum of 8 V to a maximum of 12 V. Switching current is a maximum of 400 μA. The *HS-60* switching voltage is 56 V to 70 V, and maximum switching current is 50 μA. Both devices have a 250 mW power dissipation, and each is contained in a cylindrical low-current diode-type package. *DIACs* are most often applied in triggering circuit for *SCRs* and *TRIACs.*

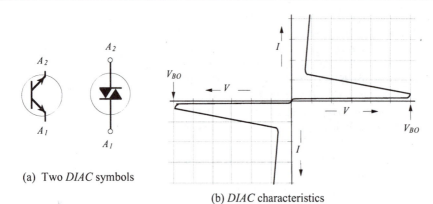

(a) Two *DIAC* symbols

(b) *DIAC* characteristics

Figure 19-25
The DIAC is basically a low-current TRIAC without a gate terminal.

DIACs

	V_S		$I_{S(max)}$	P_D
	Min	Max		
HS-10	8 V	12 V	400 μA	250 mW
HS-60	56 V	70 V	50 μA	250 mW

Figure 19-26
Partial specifications for two DIACs.

Section 19-4 Review

19-4.1 Sketch the construction and transistor equivalent circuit of a *TRIAC*. Explain the device operation.

19-4.2 Sketch *TRIAC* characteristics. Briefly explain.

19-5 TRIAC Control Circuits

TRIAC Phase Control Circuit

A *TRIAC* circuit that allows approximately 180° of phase control is shown in Fig. 19-27(a). The waveforms in Fig. 19-27(b) illustrate the circuit operation. With the *TRIAC* (Q_1) *off* at the beginning of the supply voltage positive half-cycle, capacitor C_1 is charged positively via resistors R_1 and R_2, as shown. When V_{C1} reaches the DIAC switching voltage plus the Q_1 gate triggering voltage, D_1 conducts producing gate current to trigger Q_1 *on*. C_1 discharges until the discharge current falls below the D_1 holding current level. The *TRIAC* switches *off* at the end of the supply positive half-cycle, and then the process is repeated during the supply negative half-cycle. The rate of charge of C_1 is set by variable resistor R_1, so that the Q_1 conduction angle is controlled by adjustment of R_1.

Example 19-4

Estimate the smallest conduction angle for Q_1 for the circuit in Fig. 19-27(a). The supply is 115 V, 60 Hz, and the components are: $R_1 = 25$ kΩ, $R_2 = 2.7$ kΩ, $C_1 = 3$ μF. The D_1 breakover voltage is 8 V, and $V_G = 0.8$ V for Q_1.

Solution

At Q_1 switch-*on*,
$$V_{C1} = V_{D1} + V_G = 8\,V + 0.8\,V$$
$$= 8.8\,V$$

Assume the average charging voltage is,
$$E = 0.636 \times V_{ac(pk)} = 0.636 \times 1.414 \times 115\,V$$
$$\approx 103\,V$$

Average charging current,
$$I_C \approx \frac{E}{R_1 + R_2} = \frac{103\,V}{25\,k\Omega + 2.7\,k\Omega}$$
$$\approx 3.7\,mA$$

Charging time,
$$t \approx \frac{C_1\,V_{C1}}{I_C} = \frac{3\,\mu F \times 8.8\,V}{3.7\,mA}$$
$$\approx 7.1\,ms$$

$$T = \frac{1}{f} = \frac{1}{60\,Hz} = 16.7\,ms$$

Q_1 switch-*on* point,
$$\varnothing \approx \frac{t \times 360°}{T} = \frac{7.1\,ms \times 360°}{16.7\,ms}$$
$$= 153°$$

Conduction angle,
$$\alpha = 180° - \varnothing = 180° - 153°$$
$$= 27°$$

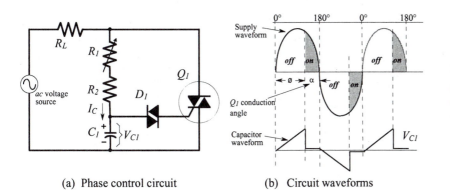

(a) Phase control circuit   (b) Circuit waveforms

Figure 19-27
TRIAC phase control circuit and circuit waveforms. Q_1 is switched on when the capacitor charges to D_1 breakover voltage.

TRIAC Zero-Point Switching Circuit

The *TRIAC zero-point switching circuit* in Fig. 19-28(a) produces a load waveform similar to that for the *SCR* zero-point circuit in Fig. 19-18. The load power dissipation is controlled by switching the *TRIAC on* for several cycles of the supply voltage and *off* for several cycles, with switch-*on* occurring only at the negative-to-positive zero crossing point of the supply waveform, and switch-*off* taking place at the positive-to-negative zero point. Q_1 is a low-current *SCR* that controls the switching point of Q_2.

With switch S_1 closed, Q_1 is *on*, and the Q_1 forward voltage drop is below the level required for triggering Q_2, ($V_{G2} + V_{D1} + V_{D2}$), [see Fig. 19-28(b)]. Thus, no gate current flows to Q_2, and no conduction occurs. Q_1 switches *off* when S_1 is opened, so that I_G flows to Q_2 gate via C_1, R_2, D_1, and D_2 to trigger Q_2 into conduction, [Fig. 19-28(c)]. With Q_2 conducting, capacitor C_2 is charged via D_3 almost to the positive peak of the load voltage, [Fig. 19-28(d)]. The *TRIAC* switches *off* at the end of the positive half-cycle. Then, the charge on C_2 (applied to the gate via D_2) triggers Q_2 on again just after the zero-crossing point into the negative half-cycle. (It should be noted this is *quadrant IV* triggering.)

The initial Q_2 switch-*on* occurs only at the beginning of the positive half-cycle of the supply voltage. If S_1 is opened during the supply positive half-cycle, Q_1 continues to conduct until the end of the half-cycle, thus keeping Q_2 *off*. With Q_2 *off*, C_2 remains uncharged, and so it cannot trigger Q_2 *on* during the supply negative half-cycle. Q_2 triggering now occurs at the beginning of the next positive cycle.

If S_1 is opened during the supply negative half-cycle, Q_2 cannot be triggered into conduction, again because of the lack of charge on C_2. It is seen that Q_2 conduction can commence only at the beginning of the positive half-cycle of the supply voltage. Also, once triggered, Q_2 conduction continues until the end of the cycle.

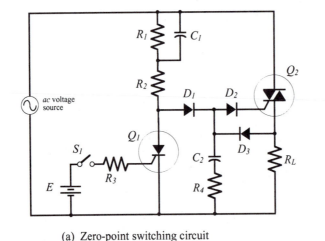

(a) Zero-point switching circuit

Figure 19-28
Zero-point switching circuit for a TRIAC. While Q_1 is on, Q_2 cannot switch on. With Q_1 off, Q_2 commences to conduct at the beginning of the supply positive half-cycle.

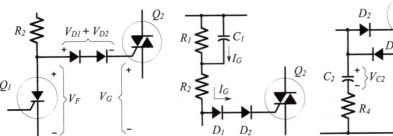

(b) With Q_1 on Q_2 is held *off*   (c) I_{G2} flows when Q_1 is off   (d) V_{C2} triggers Q_2

To design the circuit in Fig. 19-28, the *TRIAC* is first selected to pass the required load current and to survive the peak supply voltage. Resistor R_2 is a low-resistance component calculated to limit the peak surge current to the Q_2 gate in the event that the peak supply voltage is applied to the circuit without Q_1 being *on*. Capacitor C_1 has to supply triggering current (I_G) to Q_2 at the zero crossing point of the supply waveform when Q_1 is *off*. Usually I_{G2} is selected around three times the specified $I_{G(max)}$ for Q_2, and C_1 is then calculated from the simple equation for capacitor charge; $C = (I \times t)/\Delta V$. In this case $\Delta V/t$ can be replaced by the rate-of-change of the supply voltage at the zero crossing point, which is $(2\pi f V_p)$. So, the C_1 equation is,

$$C_1 = \frac{I_{G2}}{2\pi f V_p} \qquad\qquad \textbf{(19-5)}$$

Resistor R_1 can now be determined by using the selected gate current for Q_2 (I_{G2}) as the peak anode current for Q_1; $R_1 = V_p/I_{G2}$. The Q_1 gate resistor (R_3) is calculated from the Q_1 triggering current and the voltage of the *dc* source; $R_1 = (E - V_{G1})/I_{G1}$.

The Q_2 gate current is again used in the calculation of R_4 and C_2. To trigger Q_2 at the start of the supply negative half-cycle, I_{G2} must flow from C_2 into the Q_2 gate, so $R_4 \approx V_p/I_{G2}$. A suitable capacitance for C_2 is now calculated by again using the simple capacitance equation $C_2 = (I_{G2} \times t)/\Delta V$. In this case, time t is selected much larger than the Q_2 turn-on time, and ΔV is approximately $0.1\ V_p$.

SCR Q_1 must pass the selected anode current (I_{G2}) and survive the peak supply voltage. The diodes must each survive the peak supply voltage and pass the Q_2 triggering current.

IC Zero Voltage Switch

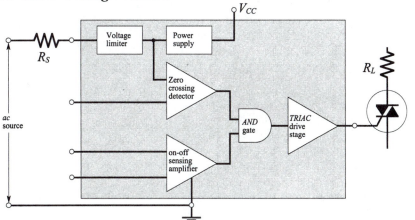

Figure 19-29
Functional block diagram for an integrated circuit zero voltage switch, or TRIAC driver.

The functional block diagram for a typical Integrated circuit *TRIAC* driver, known as a *zero voltage switch*, is shown in Fig. 19-29. The device contains a *voltage limiter* and a *dc power supply*, so that it operates directly from the *ac* supply to the load to be controlled. There is also a *zero crossing detector* (see Section 14-9) that provides an output pulse each time the supply waveform crosses the zero level. The zero crossing detector output is fed to an *AND gate* (see Section 3-11), and the *AND* gate output goes to the *TRIAC drive* stage that produces the current pulse to the *TRIAC* gate. An *on off sensing amplifier* is used to sense the voltage level from an externally-connected transducer; for example, a temperature sensing transducer would be used if the load is a heater. When the temperature drops to a predetermined level, the on-off sensing amplifier provides an input to the *AND* gate. The gate triggering pulse from the *TRIAC* drive stage occurs at the supply zero-crossing points only when the temperature is below the desired level. The circuit load waveforms are similar to those shown in Fig. 19-18.

Practise Problems

19-5.1 Determine suitable components for the *TRIAC* zero-point switching circuit in Fig. 19-28, given the following: (for Q_1; $I_G = 200\,\mu A$, $V_F = 2\,V$), (for Q_2; $I_G = 30\,mA$, $I_{GM} = 1\,A$, $t_{on} = 100\,\mu s$), $E = 6\,V$, ac source = 115 V, 60 Hz.

19-6 SUS, SBS, GTO, and SIDAC

SUS

The *silicon unilateral switch* (*SUS*), also known as a *four layer diode* and as a *Schokley diode*, can be treated as a low-current *SCR* without a gate terminal. Figure 19-30 shows the *SUS* circuit symbol and typical forward characteristics. The device triggers into conduction when a *forward switching voltage* (V_S) is applied. At this point a minimum *switching current* (I_S) must flow. Also, the voltage falls to a *forward conduction voltage* (V_F) at switch *on*, and conduction continues until the current level falls below the *holding current* (I_H). The 2N4988 *SUS* has V_S ranging from 7.5 V to 9 V, $I_S = 150\,\mu A$, and $I_H = 0.5\,mA$. *SUS* reverse characteristics are similar to *SCR* reverse characteristics; a very small reverse current flows until the reverse breakdown voltage is reached.

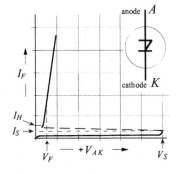

Figure 19-30
SUS circuit symbol and forward characteristics.

SBS

It might be convenient to think of a *silicon bilateral switch* (*SBS*) as an *SUS* with a gate terminal, or as a low-current *TRIAC*. However, the *SBS* is not simply another four-layer device. Silicon bilateral switches are actually integrated circuits constructed of matched transistors, diodes, and resistors. This produces better parameter stability than is possible with four-layer devices. The *SBS* equivalent circuit in Fig. 19-31(a) is similar to the *TRIAC* equivalent circuit with the addition of resistors R_1 and R_2 and Zener diodes D_1 and D_2. The device circuit symbol in Fig. 19-31(b) is seen to be composed of inverse-parallel connected *SUS* symbols with a gate terminal added. Note that the terminals are identified as *anode 1* (A_1), *anode 2* (A_2), and *gate* (*G*). The typical *SBS* characteristics shown in Fig. 19-31(c) are essentially the same shape as *TRIAC* characteristics.

Returning to the equivalent circuit, the *SBS* switches *on* when a positive A_1A_2 voltage is large enough to cause D_2 to break down. This produces base current in Q_1, resulting in Q_1 collector current that switches Q_2 *on*. Similarly, a negative A_1A_2 voltage causes D_1 to break down, producing base current in Q_4 that turns Q_4 and Q_3 *on*. The switching voltage is the sum of the Zener diode voltage and the transistor base-emitter voltage, ($V_S = V_Z + V_{BE}$). The Zener diode has a positive *temperature coefficient* (*TC*) and the transistor base-emitter voltage has a negative *TC*. This results in a very small *TC* for the *SBS* switching voltage.

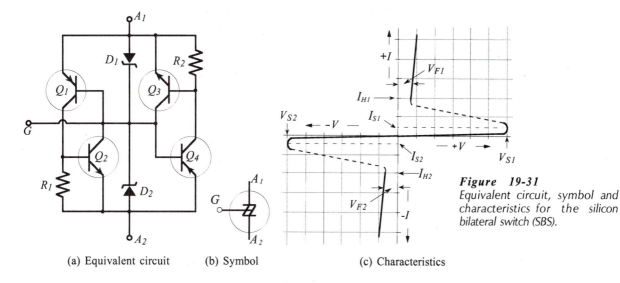

(a) Equivalent circuit   (b) Symbol   (c) Characteristics

Figure 19-31
Equivalent circuit, symbol and characteristics for the silicon bilateral switch (SBS).

The partial specification for a *MBS4991 SBS* in Fig. 19-32 shows a switching voltage that ranges from 6 V to 10 V. Note also that the *switching voltage differential* (the difference between the switching voltages in opposite directions), ($V_{S1} - V_{S2}$), is 0.5 V maximum. The maximum switching current is a 500 µA, and the *switching current differential* ($I_{S1} - I_{S2}$) is 100 µA.

MBS4991 SBS			
	Min	Typ	Max
Switching voltage (V_S)	6 V	8 V	10 V
Switching current (I_S)		175 µA	500 µA
Switching voltage differential		0.3 V	0.5 V
Switching current differential			100 µA
Gate trigger current (I_{GF})			100 µA
Forward on voltage (V_F)		1.4 V	1.7 V

Figure 19-32
Partial specification for the MBS4991 silicon bilateral switch.

SBS devices are frequently used with the gate open-circuited, so that they simply breakdown to the forward voltage drop when the applied voltage increases to the switching voltage level. The switching voltage can be reduced by connecting Zener diodes with V_Z lower than 6.8 V between the gate and the anodes, as shown in Fig. 19-33(a). The new switching voltage is approximately (V_Z + 0.7 V). The switching voltage can also be modified by the use of external resistors, [Fig. 19-33(b)]. Taking the gate current into account, it can be shown that the two 22 kΩ resistors reduce V_S to approximately 3.6 V.

The use of an SBS in a TRIAC phase control circuit is illustrated in Fig. 19-34. This is essentially the same as the circuit using a DIAC in Fig. 19-27. The SBS turns on and triggers the TRIAC when the capacitor voltage equals the SBS switching voltage plus the TRIAC gate triggering voltage.

For an SBS to switch on, the total resistance in series with it must have a maximum value that allows the switching current to flow. If the resistance is so large that it restricts the current to a level below the SBS switching current, the device will not switch on. Also, the series resistance must not be so small that it allows the holding current to flow when the SBS is supposed to switch off. These restrictions also apply to SCRs, TRIACS, DIACS, and other similar switching devices. Switch-off is usually no problem in thyristor circuits with ac supplies, because the devices normally switch off when the instantaneous supply voltage reduces to zero. With dc supplies, more care must be taken with resistor sizes.

Figure 19-35 shows a simple circuit that requires careful design to ensure that the SBS switches on and off as required. The circuit is a *relaxation oscillator* that produces an exponential output waveform, as illustrated. Capacitor C_1 is charged via resistor R_1 from the dc supply voltage (E). When the capacitor voltage (V_C) reaches the SBS switching voltage (V_S), D_1 switches on and rapidly discharges the capacitor to the D_1 forward voltage (V_F). Then D_1 switches off, and the capacitor commences to charge again. The SBS will not switch off if the D_1 holding current (I_H) continues to flow through R_1 when V_C equals V_F. SBS switch-on will normally occur when V_C equals V_S regardless of the R_1 resistance, because the capacitor discharge should provide the switching current (I_S). However, it is best to select R_1 small enough to allow I_S to flow at D_1 switch on.

The approximate oscillation frequency can be determined from the capacitor charging time (t), and the equation for t is derived from the RC charging equation.

$$t = CR \ln[\frac{E - V_F}{E - V_S}] \tag{19-6}$$

Example 19-5

The SBS in the circuit in Fig. 19-35 has the following parameters: V_S = 10 V, V_F = 1.7 V, I_S = 500 µA, I_H = 1.5 mA. Calculate the maximum and minimum

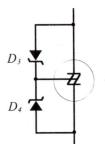

(a) V_S modification by external Zeners

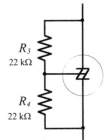

(b) V_S modification by external resistors

Figure 19-33
The switching voltage for an SBS can be modified by externally-connected Zener diodes or resistors.

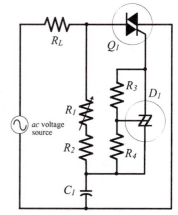

Figure 19-34
Use of an SBS in a TRIAC phase control circuit. The TRIAC is triggered by the current surge when the SBS switches on.

resistances for R_1 for correct circuit operation when $E = 30$ V. Also, determine the capacitor charging time when $R_1 = 27$ kΩ and $C_1 = 0.5$ μF.

Solution

$$R_{1(max)} = \frac{E - V_S}{I_S} = \frac{30\text{ V} - 10\text{ V}}{500\,\mu A}$$

$$= 40\text{ k}\Omega$$

$$R_{1(min)} = \frac{E - V_F}{I_H} = \frac{30\text{ V} - 1.7\text{ V}}{1.5\text{ mA}}$$

$$= 18.9\text{ k}\Omega$$

Eq. 19-6,
$$t = CR\,ln[\frac{E - V_F}{E - V_S}] = 0.5\,\mu F \times 27\text{ k}\Omega\ \ ln[\frac{30\text{ V} - 1.7\text{ V}}{30\text{ V} - 10\text{ V}}]$$

$$= 4.7\text{ ms}$$

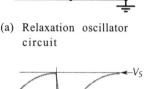

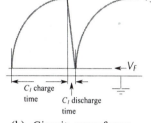

(a) Relaxation oscillator circuit

(b) Circuit waveforms

Figure 19-35
SBS relaxation oscillator. C_1 charges via R_1 to the SBS switching voltage. D_1 switches on at that point and rapidly discharges C_1.

GTO

When an *SCR* is triggered into conduction by application of a gate current, the gate looses control and the device continues to conduct until the forward current falls below the holding current. A *gate turn-off* *(GTO)* device is essentially an *SCR* designed to be switched *on* and *off* by an applied gate signal. The circuit symbol for a *GTO* is shown in Fig. 19-36(a), and the two-transistor equivalent circuit for the device is illustrated in Fig. 19-36(b) and (c). Note that at switch-*on*, the gate current has just got to be large enough to supply base current to transistor Q_2. However, at switch-*off*, the Q_1 collector current has to be diverted through the gate terminal in order to turn Q_2 off. Consequently, for device turn-*off* relative large levels of gate current are involved; approaching half the *GTO* forward current.

Figure 19-36
The gate turn-off device GTO is effectively an SCR that can be switched off by a voltage applied to the gate.

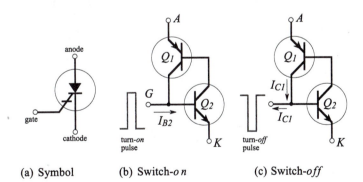

(a) Symbol   (b) Switch-*on*   (c) Switch-*off*

SIDAC

The *SIDAC* is a two-terminal thyristor designed mainly for use in over-voltage protection situations. As a bilateral device with no gate terminal, it simply breaks down to its forward voltage drop when the applied terminal voltage (of either polarity) rises to the breakover voltage level. Like other thyristors, there is a minimum current that must flow to latch the *SIDAC* into an *on* state. Also, when switched *on* conduction continues until the current falls below a holding current level.

The circuit symbol and typical characteristics for a *SIDAC* are shown in Fig. 19-37. Available devices have breakover voltages ranging from 110 V to 280 V. Typically *on* state voltages is 1.1 V, *rms* current is 1 A, and holding current is 100 mA.

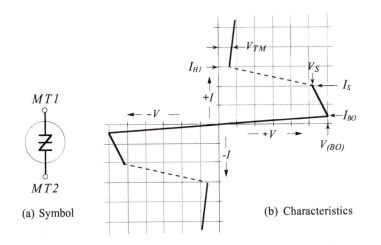

Figure 19-37
Circuit symbol and characteristics for a SIDAC.

Figure 19-38 shows a *SIDAC* used to protect a *dc* power supply from *ac* line transients. Normally, the *SIDAC* will behave as an open-circuit. A voltage transient on the *ac* line will cause it to break down to its forward voltage level, so that it essentially short-circuits the transformer output. This will cause a fuse to blow or a circuit breaker to trip, thus interrupting the *ac* supply.

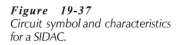

Figure 19-38
SIDAC used for protecting a dc power supply against transients on the supply line.

Practise Problems
19-6.1 The *SBS* in the circuit in Fig. 19-34 has $V_S \approx 4$ V, $I_S = 500\,\mu$A, and $I_{F(max)} = 200$ mA. The ac supply is 115 V, and $R_3 = R_4 = 22$ kΩ. Determine suitable resistances for R_1 and R_2.

19-7 UJT and PUT

UJT Operation
The *Unijunction transistor* (*UJT*) consists of a bar of lightly-doped *n*-type silicon with a block of *p*-type material on one side, [see Fig. 19-39(a)]. The end terminals of the bar are identified as *Base 1* (B_1) and *Base 2* (B_2), and the *p*-type block is named the *emitter* (*E*).

Figure 19-39(b) shows the *UJT* equivalent circuit. The resistance of the *n*-type silicon bar is represented as two resistors, r_{B1} from B_1

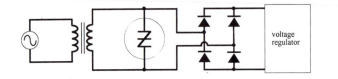

to *point C*, and r_{B2} from B_2 to C, as illustrated. The sum of r_{B1} and and r_{B2} is identified as R_{BB}. The p-type emitter forms a *pn*-junction with the n-type silicon bar, and this junction is shown as a diode (D_1) in the equivalent circuit.

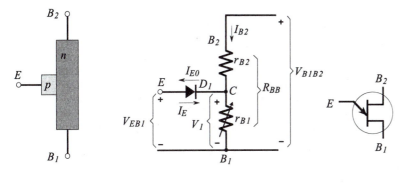

(a) Basic construction   (b) Equivalent circuit   (c) Circuit symbol

Figure 19-39
A unijunction transistor (UJT) is made up of a p-type emitter joined to a bar of n-type semiconductor.

With a voltage V_{B1B2} applied as illustrated, the voltage at the junction r_{B1} and r_{B2} is,

$$V_1 = V_{B1B2} \times \frac{r_{B1}}{R_{BB}}$$

Note that V_1 is also the voltage at the cathode of the diode; *point C* in the equivalent circuit.

With the emitter terminal open-circuited, the resistor current is,

$$I_{B2} = \frac{V_{B1B2}}{R_{BB}} \qquad \textbf{(19-7)}$$

If the emitter terminal is grounded, the pn-junction is *reverse* biased and a small *emitter reverse current* (I_{E0}) flows.

Now consider what happens when the emitter voltage (V_{EB1}) is slowly increased from zero. When V_{EB1} equals V_1 the emitter current is zero. (With equal voltage levels on each side of the diode, neither reverse nor forward current flows.) A further increase in V_{EB1} forward biases the *pn*-junction and causes a forward current (I_E) to flow from the p-type emitter into the n-type silicon bar. When this occurs, charge carriers are injected into the r_{B1} region. The resistance of the semiconductor material is dependent on doping, so the additional charge carriers cause the resistance of the r_{B1} region to rapidly decrease. The decrease in resistance reduces the voltage drop across r_{B1}, and so the *pn*-junction is more heavily forward biased. This in turn results in a greater emitter current, and more charge carriers that further reduce the resistance of the r_{B1} region. (The process is termed *regenerative*.) The input voltage is *pulled down*, and the emitter current (I_E) is increased to a limit determined by the V_{EB1} source resistance. The device remains in this *on* condition until the emitter input is open-circuited, or until I_E is reduced to a very low level.

The circuit symbol for a *UJT* is shown in Fig. 19-39(c). As always, the arrowhead points in the conventional current direction for a forward-biased junction. In this case it points from the *p*-type emitter to the *n*-type bar.

UJT Characteristics

A plot of emitter voltage V_{EB1} versus emitter current I_E gives the *UJT* emitter characteristics. Refer to the *UJT* terminal voltages and currents identified in Fig. 19-40(a) and to the equivalent circuit in Fig. 19-39(b). Note that when $V_{B1B2} = 0$, $I_{B2} = 0$ and $V_1 = 0$. If V_{EB1} is now increased from zero, the resultant plot of V_{EB1} and I_E is simply the characteristic of a forward-biased diode with some series resistance. This is the characteristic for $I_{B2} = 0$ in Fig. 19-40(b).

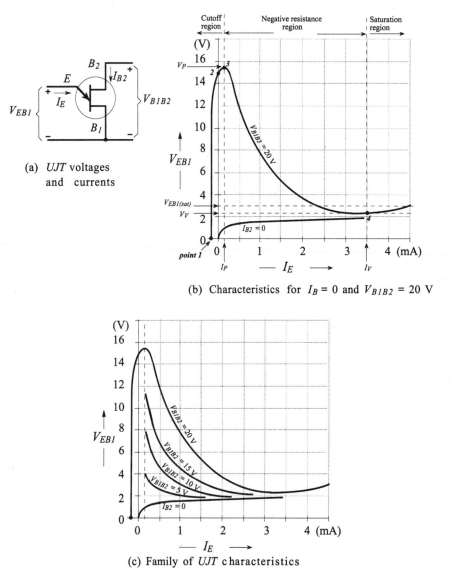

(a) *UJT* voltages and currents

(b) Characteristics for $I_B = 0$ and $V_{B1B2} = 20$ V

(c) Family of *UJT* characteristics

Figure 19-40
The UJT characteristics show that the device triggers on at various levels of emitter voltage V_{EB1}, depending upon the level of supply voltage V_{B1B2}.

When V_{B1B2} is 20 V the level of V_1 [Fig. 19-39(b)] might be around 15 V, depending on the resistances of r_{B1} and r_{B2}. With V_{B1B2} = 20 V and V_E = 0, the emitter junction is reverse biased and the emitter reverse current I_{E0} flows, as shown at *point 1* on the V_{B1B2} = 20 V characteristic in Fig. 19-40(b). Increasing the level of V_{EB1} until it equals V_1 gives I_E = 0; *point 2* on the characteristic. Further increase in V_{EB1} forward biases the emitter junction, and this gives the *peak point* on the characteristic (*point 3*). At the peak point, V_{EB1} is identified as the *peak voltage* (V_p) and I_E is termed the *peak current* (I_p).

Up until the peak point the *UJT* is said to be operating in the *cutoff region* of its characteristics. When V_{EB1} arrives at the peak voltage, charges carriers are injected from the emitter to decrease the resistance of r_{B1}, as already explained. The device enters the *negative resistance region*, r_{B1} falls rapidly to a *saturation resistance* (r_S), and V_{EB1} falls to the *valley voltage* (V_V), [*point 4* on the characteristic in Fig. 19-40(b)]. I_E also increases to the *valley current* (I_V) at this time. Further increase in I_E causes the device to enter the *saturation region* where V_E equals the sum of V_D and $I_E r_S$.

Starting with V_{B1B2} lower than 20 V gives a lower peak point voltage and a different characteristic. Thus, using various levels of V_{B1B2}, a family of V_{EB1}/I_E characteristics can be plotted for a given *UJT*, as shown in Fig. 19-40(c).

UJT Packages

Two typical *UJT* packages with the terminal identified are shown in Fig 19-41. These are similar to low-power *BJT* packages.

UJT Parameters

Interbase Resistance (R_{BB}): This is the sum of r_{B1} and r_{B2} when I_E is zero. Consider Fig. 19-42 that shows a portion of the manufacturer's data sheet for *2N4949 UJT*. R_{BB} is specified as 7 kΩ typical, 4 kΩ minimum, and 12 kΩ maximum. The value of R_{BB}, together with the maximum power dissipation P_D, determine the maximum value of V_{B1B2} that may be used. With I_E = 0,

$$V_{B1B2(max)} = \sqrt{(R_{BB} P_D)} \qquad (19\text{-}8)$$

Like all other devices, the P_D of the *UJT* must be derated for increased temperature levels.

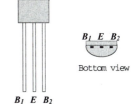

(a) Resin-encapsuled *UJT*

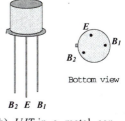

(b) *UJT* in a metal can

Figure 19-41
UJT packages and terminals.

2N4949 UJT

	Min	Typ	Max
Interbase resistance (R_{BB})	4 kΩ	7 kΩ	12 kΩ
Intrinsic standoff ratio (η)	0.74		0.86
Emitter saturation voltage ($V_{EB1(sat)}$)		2.5 V	3 V
Peak point current (I_P)		0.6 μA	1 μA
Valley point current (I_V)	2 mA	4 mA	

Figure 19-42
Partial specification for a 2N4949 UJT.

Example 19-6

A *UJT* has $R_{BB(min)}$ = 4 kΩ, P_D = 360 mW at 25 °C, and a power derating factor D = 2.4 mW/°C. Calculate the maximum V_{B1B2} that should be used at a temperature of 100 °C.

Solution

Eq. 8-20,

$$P_{D(100°)} = P_{D(25°)} - [D\,(T_2 - 25°)]$$
$$= 360\text{ mA} - [2.4\text{ mW/°C }(100° - 25°)]$$
$$= 180\text{ mW}$$

Eq. 19-8,

$$V_{B1B2(max)} = \sqrt{(R_{BB}\,P_D)} = \sqrt{(4\text{ kΩ} \times 180\text{ mW})}$$
$$= 26.8\text{ V}$$

Intrinsic Standoff Ratio (η): The *intrinsic standoff ratio* is simply the ratio of r_{B1} to R_{BB}. The peak point voltage is determined from η, the supply voltage, and the diode voltage drop;

$$\mathbf{V_p = \ V_D + \eta V_{B1B2}} \qquad\qquad \textbf{(19-9)}$$

Emitter Saturation Voltage ($V_{EB1(sat)}$). The emitter voltage when the *UJT* is operating in the saturation region of its characteristics; the minimum V_{EB1} level. Because it is affected by the emitter current and the supply voltage, $V_{EB1(sat)}$ is specified for given I_E and V_{B1B2} levels.

Example 19-7

Determine the maximum and minimum triggering voltages for a *2N4949 UJT* with V_{B1B2} = 25 V.

Solution

From Fig. 19-42, η = 0.74 minimum, 0.86 maximum

Eq. 19-9,

$$V_{p(max)} = \ V_D + (\eta_{max}\,V_{B1B2}) = 0.7\text{ V} + (0.86 \times 25\text{ V})$$
$$= 22.2\text{ V}$$

$$V_{p(min)} = \ V_D + (\eta_{min}\,V_{B1B2}) = 0.7\text{ V} + (0.74 \times 25\text{ V})$$
$$= 19.2\text{ V}$$

Peak Point Emitter Current (I_p): I_P is important as a lower limit to the emitter current. If the emitter voltage source resistance is so high that I_E is not greater than I_P the *UJT* will simply not trigger *on*. The maximum emitter voltage source resistance is,

$$R_{E(max)} = \frac{V_{B1B2} - V_P}{I_P} \qquad\qquad \textbf{(19-10)}$$

Valley Point Current (I_V): I_V is important in some circuits as an upper limit to the emitter current. If the emitter voltage source resistance is so low that I_E is equal to or greater than I_V the *UJT* will remain *on* once it is triggered; it will not switch *off*. So, the minimum emitter voltage source resistance is,

$$R_{E(min)} = \frac{V_{B1B2} - V_{EB1(sat)}}{I_V} \qquad \textbf{(19-11)}$$

UJT Relaxation Oscillator

The relaxation oscillator circuit in Fig. 19-43(a) consists of a *UJT* and a capacitor (C_1) charged via resistance R_E. When the capacitor voltage (V_C) reaches V_P the *UJT* fires and rapidly discharges C_1 to $V_{EB1(sat)}$. The device then cuts *off* and the capacitor commences charging again. The cycle is repeated continually, generating a sawtooth waveform across C_1, as illustrated in Fig. 19-43(b). The time (t) for the capacitor to charge from $V_{EB1(sat)}$ to V_P may be calculated, and the frequency of the sawtooth determined approximately as $1/t$. The discharge time (t_D) is difficult to calculate because the *UJT* is in its negative resistance region and its resistance is changing. However, t_D is much less than t, and so it can normally be neglected. Rewriting Eq. 19-6,

$$t = CR \; ln[\frac{V_{BB} - V_{EB1}}{V_{BB} - V_P}] \qquad \textbf{(19-12)}$$

Resistor R_3 in the circuit in Fig. 19-43 is included to produce a spike waveform output, as illustrated. When the *UJT* fires, the current surge through terminal B_2 produces the negative-going voltage spike across R_3. A resistor could also be included in series with terminal B_1 to produce positive-going spikes. Both resistor values should be much lower than the R_{BB} for the *UJT*.

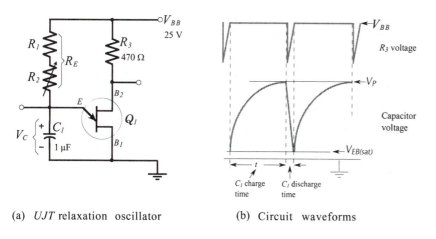

(a) *UJT* relaxation oscillator   (b) Circuit waveforms

Figure 19-43
UJT relaxation oscillator circuit and waveforms. C_1 charges to V_p when the UJT fires, then C_1 is discharged to $V_{BE(sat)}$.

Example 19-8

Calculate the $R_{E(max)}$ and $R_{E(min)}$ for the relaxation oscillator circuit in Fig. 19-43. The *UJT* is a 2N4949 with $I_p = 0.6\ \mu A$, $I_V = 2$ mA, and $V_{EB1(sat)} = 2.5$ V. Also, determine the approximate maximum oscillating frequency for the circuit when $R_E = 18$ kΩ, $C_1 = 1$ μF, and $V_P = 20$ V.

Solution

From Example 19-7,

$$V_{p(min)} = 19.2 \text{ V, and } V_{p(max)} = 22.2 \text{ V}$$

Eq. 19-10, $$R_{E(max)} = \frac{V_{B1B2} - V_{p(max)}}{I_p} = \frac{25 \text{ V} - 22.2 \text{ V}}{0.6 \text{ }\mu\text{A}}$$

$$= 4.7 \text{ M}\Omega$$

Eq. 19-11, $$R_{E(min)} = \frac{V_{B1B2} - V_{EB1(sat)}}{I_v} = \frac{25 \text{ V} - 2.5 \text{ V}}{2 \text{ mA}}$$

$$= 11.25 \text{ k}\Omega$$

Eq. 19-12, $$t = CR \ln[\frac{V_{B1B2} - V_{EB1(sat)}}{V_{B1B2} - V_p}]$$

$$= 1 \text{ }\mu\text{F} \times 18 \text{ k}\Omega \text{ } \ln[\frac{25 \text{ V} - 2.5 \text{ V}}{25 \text{ V} - 20 \text{ V}}]$$

$$= 27 \text{ ms}$$

Eq. 19-12, $$f = \frac{1}{t} = \frac{1}{27 \text{ ms}}$$

$$= 37 \text{ Hz}$$

UJT Control of an SCR

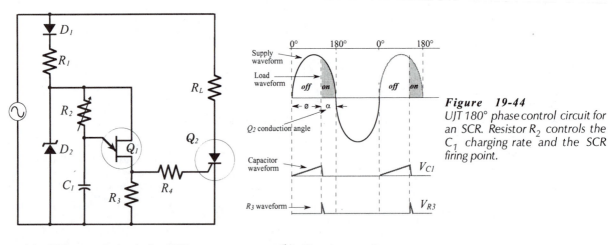

(a) *UJT* control circuit for *SCR*   (b) Circuit waveforms

Figure 19-44
UJT 180° phase control circuit for an SCR. Resistor R_2 controls the C_1 charging rate and the SCR firing point.

Unijunction transistors are frequently employed in *SCR* and *TRIAC* control circuits. In the typical circuit shown in Fig. 19-44(a) diode D_1, resistor R_1, and Zener diode D_2 provide a low-voltage *dc* supply to the *UJT* circuit derived from the positive half-cycle of the *ac* supply voltage. D_1 also isolates the *UJT* circuit during the supply negative half-cycle. Capacitor C_1 is charged via resistor R_2 to the *UJT* firing voltage, and the *SCR* is triggered by the voltage drop across R_3. By adjusting R_2 the charging rate of C_1 and the *UJT* firing time can be selected. The waveforms in Fig. 19-44(b) show that 180° of *SCR* phase control is possible.

Practise Problems

19-7.1 A 2N4870 UJT has the following parameters: P_D = 300 mW at 25°C, D = 3 mW/°C, η = 0.56 to 0.75, R_{BB} = 4 kΩ to 9.1 kΩ, $V_{EB1(sat)}$ = 2.5 V, I_P = 1 μA to 5 μA, I_V = 2 mA to 5 mA. Determine the maximum V_{B1B2} that may be used at 75°C.

19-7.2 Calculate the $V_{P(max)}$ and $V_{P(min)}$ for a 2N4870 when V_{BB} = 30 V.

19-7.3 A 2N4870 is used in the circuit in Fig. 19-44. If D_2 has V_Z = 30 V, determine the maximum and minimum resistance values for R_2.

19-8 Programmable Unijunction Transistor (PUT)

PUT Operation

The *programmable unijunction transistor (PUT)* is actually an *SCR*-type device used to simulate a *UJT*. The interbase resistance (R_{BB}) and the intrinsic standoff ratio (η) can be programmed to any desired values by selecting two resistors. This means that the device firing voltage (the peak voltage V_P) can also be programmed.

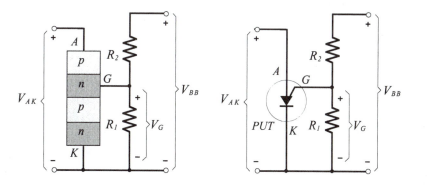

(a) *PUT* four-layer construction   (b) *PUT* circuit

Figure 19-45
The programmable unijunction transistor (PUT) is an SCR-type device that can be connected to function like a UJT.

Consider Fig. 19-45(a) which shows a four-layer device with its gate connected to the junction of resistors R_1 and R_2. Note that the gate terminal is close to the anode of the device, instead of the cathode as for an *SCR*. The anode-gate junction becomes forward biased when the anode is positive with respect to the gate. When this occurs, the device is triggered *on*. The anode-to-cathode voltage then drops to a low level, and the *PUT* conducts heavily until the current becomes too low to sustain conduction. To simulate the *UJT* performance, the anode of the device acts as the

UJT emitter, and R_1 and R_2 operate as r_{B1} and r_{B2}, respectively. Parameters R_{BB}, η, and V_P are programmed by selection of R_1 and R_2. The four-layer block diagram is replaced with the *PUT* graphic symbol in Fig. 19-45(b). Note that this is the same as the *SCR* symbol except that the gate terminal is at the anode.

PUT Characteristics

The typical *PUT* characteristic (V_{AK} plotted versus I_A) shown in Fig. 19-46 are seen to be very similar to *UJT* characteristics. A small gate reverse current (I_{AGO}) flows while the anode-gate junction is reverse biased. At this point the *PUT* is in the cutoff region of the characteristics. When the anode voltage is raised sufficiently above the gate voltage (V_G in Fig. 19-45), the *PUT* is triggered into the negative resistance region of its characteristics, and the anode-cathode voltage falls rapidly to the valley voltage (V_V). Further increase in I_A causes the device to operate in its saturation region.

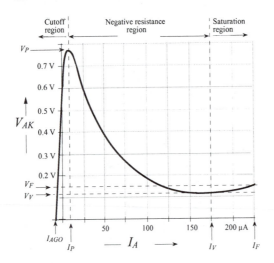

Figure 19-46
The typical V_{AK}/I_A characteristics for a PUT are similar to UJT characteristics.

PUT Parameters

The intrinsic standoff ratio for the *PUT* is,

$$\eta = \frac{R_1}{R_1 + R_2} \qquad \textbf{(19-13)}$$

The gate voltage is simply,

$$V_G = \eta\,V_{BB} \qquad \textbf{(19-14)}$$

and the peak voltage is,

$$V_P = V_D + \eta\,V_{BB} \qquad \textbf{(19-15)}$$

where V_D is the anode-gate junction voltage, typically 0.7 V.

The gate source resistance (R_G) is an important quantity because it affects the peak current and valley current for the *PUT*. R_G is the resistance at the junction of voltage divider R_1 and R_2, (Fig. 19-45).

$$R_G = R_1 \| R_2 \qquad\qquad \text{(19-16)}$$

Refer to the partial specification for a *2N6027 PUT* in Fig. 19-47, and note the typical quantities. With $R_G = 1$ MΩ, $I_P = 1.25$ µA and $I_V = 18$ µA; with $R_G = 10$ kΩ, $I_P = 4$ µA and $I_V = 150$ µA.

2N6027 PUT

		Typ	Max
Peak current (I_P)	($R_G = 1$ MΩ)	1.25 µA	2 µA
	($R_G = 10$ kΩ)	4 µA	5 µA
Valley current (I_V)	($R_G = 1$ MΩ)	18 µA	50 µA
	($R_G = 10$ kΩ)	150 µA	—
Forward voltage (V_F)	($I_F = 50$ mA)	0.8 V	1.5 V

Figure 19-47
Partial specification for a programmable unijunction transistor.

PUT Applications

A *PUT* can be applied in any circuit where a *UJT* might be used. Figure 19-48 shows a *PUT* relaxation oscillator used to control an *SCR*. This circuit operates in essentially the same way as the *UJT* circuit in Fig. 19-44. It should be noted that there are upper and lower limits to the resistance that can be connected in series with the *PUT* anode for correct operation of the device. This is similar to the $R_{E(min)}$ and $R_{E(max)}$ requirement for the *UJT*.

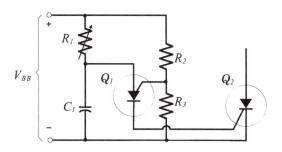

Figure 19-48
SCR phase control circuit using a PUT relaxation oscillator.

A battery charger circuit using a *PUT* (Q_1) and an *SCR* (Q_2) is shown in Fig. 19-49. The *ac* supply voltage is full-wave rectified and applied via current-limiting resistor R_5 to the anode of the *SCR*. The *SCR* is triggered into conduction by the *PUT* output coupled via transformer T_1. The *PUT* gate voltage (V_G) is set by the voltage divider (R_3, R_4, and R_5). While V_G is lower than the Zener diode voltage (V_Z), capacitor C_1 is charged via R_1 to the *PUT* peak voltage. At this point the *PUT* fires and triggers the *SCR on*.

As the battery charges, its voltage (E_B) increases, and thus V_G also increases. The increased V_G level raises the V_P of the *PUT* and causes C_1 to take a longer time charge. Consequently, the *SCR* is held *off* for a longer portion of the *ac* supply half-cycle. This means that the average charging current is gradually reduced as the battery approaches full charge. When E_B is fully charged, V_G is raised to the V_Z level, so that D_6 conducts and stops C_1 voltage increase before the *PUT* fires. Thus, the *SCR* remains *off*, and battery charging stops. The circuit will not operate if the battery is connected with the wrong polarity.

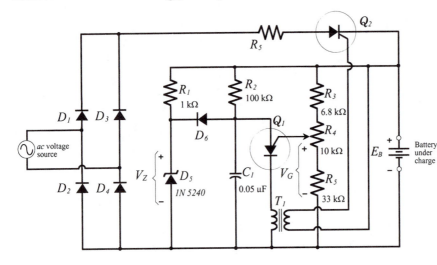

Figure 19-49
SCR battery charger using a PUT control circuit.

Example 19-9

Calculate $V_{P(max)}$ and $V_{P(min)}$ for the *PUT* in the circuit in Fig. 19-49 when $E_B =$ 12 V. Also, determine the gate bias resistance R_G, and calculate the maximum and minimum resistances for R_2 if the *PUT* is a 2N6027.

Solution

Eq. 19-13, $\eta_{(max)} = \dfrac{R_4 + R_5}{R_3 + R_4 + R_5} = \dfrac{10\ k\Omega + 33\ k\Omega}{6.8\ k\Omega + 10\ k\Omega + 33\ k\Omega}$

$= 0.86$

$\eta_{(min)} = \dfrac{R_5}{R_3 + R_4 + R_5} = \dfrac{33\ k\Omega}{6.8\ k\Omega + 10\ k\Omega + 33\ k\Omega}$

$= 0.66$

Eq. 19-15, $V_{P(max)} = V_D + (\eta_{(max)}\ V_{BB}) = 0.7\ V + (0.86 \times 12\ V)$
$= 11\ V$

$V_{P(min)} = V_D + (\eta_{(max)}\ V_{BB}) = 0.7\ V + (0.66 \times 12\ V)$
$= 8.6\ V$

Eq. 19-16, $R_G = (R_3 + 0.5\ R_4)\|(R_5 + 0.5\ R_4)$
$= (6.8\ k\Omega + 5\ \Omega)\|(33\ k\Omega + 5\ k\Omega)$
$= 9\ k\Omega$

$R_{2(max)} = \dfrac{E - V_{P(max)}}{I_P} = \dfrac{12\ V - 11\ V}{4\ \mu A}$

$= 250\ k\Omega$

$$R_{2(min)} = \frac{E - V_F}{I_V} = \frac{12\,V - 0.8\,V}{150\,\mu A}$$

$$= 74\,k\Omega$$

Practise Problems

19-8.1   The circuit in Fig. 19-48 has: $V_{BB} = 15$ V, $R_2 = 12$ kΩ, and $R_3 = 18$ kΩ. The *PUT* has $I_P = 10\,\mu A$, $I_V = 100\,\mu A$, and $V_F = 1$ V. Calculate R_G, η, V_P, and the maximum and minimum resistances for R_1.

19-8.2   Determine the voltage that the battery will be charged to in Fig. 19-49 when the moving contact is at the middle point on R_4. Assume that the *PUT* will stop firing when V_{AG} is reduced to 0.5 V.

Chapter-19 Review Questions

Section 19-1

19-1   Sketch the construction of a silicon controlled rectifier. Also, sketch the two-transistor equivalent circuit and show how it is derived from the *SCR* construction. Label all terminals and explain how the device operates.

19-2   Sketch typical *SCR* forward and reverse characteristics. Identify all regions of the characteristics and all important current and voltage levels. Explain the shape of the characteristics in terms of the *SCR* two-transistor equivalent circuit.

19-3   List the most important *SCR* parameters and state typical quantities for low, medium, and high current devices.

Section 19-2

19-4   Draw the circuit diagram to show how an *SCR* can be triggered by application of a pulse to the gate terminal. Sketch the circuit waveforms and explain its operation.

19-5   Sketch a 90° phase control circuit for an *SCR*. Draw the load waveform and explain the operation of the circuit. Also, show the circuit and load waveforms when the *ac* supply is full-wave rectified.

19-6   Draw the diagram for a 90° phase control circuit using two *SCRs* for full-wave phase control. Draw the load waveforms and briefly explain the circuit operation.

19-7   Sketch a 180° phase control for an *SCR*. Draw the load waveform and explain the circuit operation.

Section 19-3

19-8   Briefly discuss *SCR* circuit stability, and draw diagrams to show methods that can be used to improve stability.

19-9   Draw the diagram for an *SCR* zero-point triggering circuit. Explain the circuit operation and advantages, and draw the load waveform.

19-10   Sketch a circuit that uses an *SCR* to protect a load from excessive *dc* supply voltage. Briefly explain.

19-11   Draw a diagram for an *SCR* heater control circuit using a temperature-sensitive device. Explain the circuit operation.

Section 19-4

19-12   Draw sketches to show the construction, equivalent circuit, and characteristics of a *TRIAC*. Identify all important voltage and current levels on the characteristics and explain the operation of the device.

19-13   Using appropriate diagrams, explain the four quadrant operating conditions for a *TRIAC*.

19-14   Draw the typical characteristics for a *DIAC*. Explain the *DIAC* operation, and sketch the two circuit symbols used for the device.

Section 19-5

19-15   Draw the diagram for a *TRIAC* 180° phase control circuit. Draw all waveforms, and explain the circuit operation.

19-16   Draw the diagram for an *TRIAC* zero-point triggering circuit and carefully explain its operation

19-17   Sketch the functional block diagram for an *IC* zero voltage switch for *TRIAC* control. Discuss the components of the block diagram.

Section 19-6

19-18   Using appropriate diagrams, briefly explain a silicon unilateral switch (*SUS*). Draw the device circuit symbol.

19-19   Sketch the equivalent circuit and characteristics for a silicon bilateral switch (*SBS*). Explain how the device construction differs from other thyristors. Discuss the device operation, state typical parameters, and show how the switching voltage can be modified.

19-20   Sketch a relaxation oscillator circuit using an *SBS*. Draw the output waveform, and explain the circuit operation.

19-21   Draw the circuit symbol and equivalent circuit for a gate turnoff device (*GTO*) and discuss its operation.

19-22   Draw the circuit symbol and typical characteristics for a *SIDAC*. Discuss its operation and applications.

Section 19-7

19-23   Draw sketches to show the basic construction and equivalent circuit of a unijunction transistor (*UJT*). Briefly explain the device operation.

19-24   Sketch typical *UJT* V_{EB1}/I_E characteristics for $I_{B2} = 0$, V_{B1B2} = 20 V, and V_{B1B2} = 10 V. Identify each region and all important points on the characteristics, and explain the shape of the characteristics.

19-25   Define the following *UJT* parameters: intrinsic standoff ratio, interbase resistance, emitter saturation voltage, peak point current, valley point current.

19-26   Draw the circuit of a *UJT* relaxation oscillator with provision for frequency adjustment and spike waveform. Show all waveforms and explain the circuit operation.

19-27   Sketch a *UJT* circuit for controlling an *SCR*. Also, draw all waveforms, and briefly explain how the circuit operates.

Section 19-8

19-28   Draw the basic block diagram and basic circuit for a programmable unijunction transistor (*PUT*), and explain the device operation.

19-29   Sketch typical *PUT* characteristic, explain how the intrinsic stand-off ratio may be programmed, and identify the most important *PUT* parameters.

19-30   Draw a basic *PUT* circuit for controlling an *SCR*, and explain its operation.

19-31   Draw the circuit diagram of a battery charger using a *PUT* and an *SCR*. Explain the circuit operation.

Chapter-19   Problems

Section 19-1

19-1   Consult *2N6167* and *2N1595 SCR* specifications to determine the typical values for: V_{DRM}, I_T, V_{TM}, I_H, I_{GT}, V_{GT}.

Section 19-2

19-2   Select a suitable *SCR* from Fig. 19-10 for a circuit with a 115 V *ac* supply. Calculate the minimum load resistance that can be supplied, and determine the instantaneous voltage level when the *SCR* switches *off*.

19-3   A 33 Ω resistor is supplied from an *ac* source with a 60 V peak level. Current to the load is to be switched *on* and *off* by an *SCR*. Select a suitable device from the specifications in Appendix 1-17, and determine the instantaneous supply voltage at which the *SCR* switches off.

19-4   An *SCR* with a 115 V *ac* supply controls the current through a 150 Ω load resistor. A 90° phase-control circuit (as in Fig. 19-11) is employed to trigger the *SCR* between 12° and 90°. The gate trigger current is 50 µA and the trigger voltage is 0.5 V. Calculate suitable resistor values.

19-5   The circuit in Fig. 19-12 has a 50 V *ac* supply and $R_L = 20$ Ω. Determine suitable resistance values for R_1, R_2, and R_3 for the *SCR* to be triggered anywhere between 7.5° and 90°. The gate trigger current and voltage are 500 µA and 0.6 V.

19-6   The circuit in Problem 19-5 uses a *2N5170 SCR* (see Appendix 1-17). Calculate the instantaneous supply voltage level when the *SCR* switches *off*.

19-7   Design the circuit in Fig. 19-13 for 10° to 90° phase control and specify the *SCRs*. The *ac* supply is 115 V, $R_L = 12$ Ω, and the *SCRs* have $V_G = 0.6$ V and $I_G = 100$ µA.

19-8   A 180° phase circuit as in Fig. 19-15 has a 40 V, 400 Hz *ac* supply and $R_L = 22$ Ω. The *SCR* has $V_G = 0.5$ V, $I_G = 60$ µA, and $I_{GM} = 20$ mA. Determine suitable resistor and capacitor values, and specify the *SCRs* and the diodes.

Section 19-3

19-9   An *SCR* crowbar circuit (as in Fig. 19-19) is connected to a 12 V *dc* supply with a 200 mA current limiter. Design the crowbar circuit to protect the load from voltage levels greater than 13.5 V. Assume that $V_G = 0.7$ V for the *SCR*.

19-10   A zero-point triggering circuit (as in Fig. 19-18) is to control the power dissipation in a 12 Ω load resistor with a 115 V, 60 Hz *ac* supply. Assuming that the *SCRs* have $V_G = 0.5$ V and $I_{G(min)} = 10$ mA, determine suitable capacitor and resistor values.

19-11   An *SCR* heater control circuit (as in Fig. 19-20) is to switch *on* at 68°C and *off* at 71°C. The circuit is to operate from a 50 V, 60 Hz supply, and the available temperature-sensitive device has a resistance of 500 Ω at 68°C and 350 Ω at 71°C. The load resistance is $R_L = 2.5$ Ω and the *SCR* has $V_G = 0.6$ V. Determine suitable component values, and calculate the *SCR* gate voltage at 71°C.

19-12   Specify the *SCRs* required for the circuits in Problems 19-9, 10, and 11 in terms of maximum anode-cathode voltage and maximum anode current.

Section 19-4

19-13   Consult specifications for *2N6071* and *2N6343 TRIACs* to determine the maximum supply voltage, maximum *rms* current, and the typical quadrant *I* gate triggering voltage.

Section 19-5

19-14   A light dimmer uses a *TRIAC* 180° phase control circuit, as in Fig. 19-27. The *ac* supply is 220 V, 60 Hz, and the total load is 750 W. The *TRIAC* has triggering current $I_G = 200$ µA, maximum gate current $I_{GM} = 50$ mA, and $V_G = 0.7$ V. The *DIAC* has $V_S = 9.2$ V and $I_S = 400$ µA. Determine suitable component values.

19-15 The *TRIAC* control circuit in Fig. 19-27 has the following components: R_L = 100 Ω, R_1 = 10 kΩ, R_2 = 500 Ω, C_1 = 3.9 μF. The *DIAC* has V_S = 7 V and the *TRIAC* has V_G = 1 V. The *ac* supply is 60 V, 60 Hz. Determine the minimum conduction angle for the *TRIAC*.

19-16 Specify the *TRIACs* required for the circuits in Problems 19-14 and 15.

19-17 The *TRIAC* zero-point switching circuit in Fig. 19-28 uses an *SCR* with I_G = 100 μA and V_F = 1.5 V. The *TRIAC* has I_G = 5 mA, I_{GM} = 500 mA, and t_{on} = 50 μs. The control voltage is E = 5 V, the *ac* source is 60 V, 60 Hz, and the load is R_L = 15 Ω. Determine suitable component values.

Section 19-6

19-18 A relaxation oscillator (as in Fig. 19-35) uses an *SBS* with V_S = 8 V, V_F = 1 V, I_S = 300 μA, and I_H = 1 mA. The *dc* supply is E = 40 V. Calculate maximum and minimum R_1 values for correct operation of the circuit.

19-19 A relaxation oscillator (as in Fig. 19-35) has a 25 V supply, a 1 μF capacitor, and a 12 kΩ series resistor. The capacitor is to charge up to 15 V and then discharge to approximately 1 V. Specify the required *SBS* in terms of forward conduction voltage, switching voltage, switching current, and holding current.

19-20 The phase control circuit in Fig. 19-34 has the following components: R_L = 18 Ω, R_1 = 12 kΩ, R_2 = 470 Ω, C_1 = 10 μF. Resistors R_3 and R_4 are replaced with 3.3 V Zener diodes, and the *TRIAC* triggering voltage is V_G = 1 V. The *ac* supply is 115 V, 60 Hz. Determine the *TRIAC* minimum conduction angle.

Section 19-7

19-21 Determine the maximum power dissipation for a *2N2647* *UJT* at an ambient temperature of 70°C. Also, calculate the maximum level of V_{B1B2} that may be used at 70°C. Appendix 1-18 gives partial specification for the *2N2647*.

19-22 Calculate the minimum and maximum V_{EB1} triggering levels for a *2N2647* *UJT* when V_{B1B2} = 20 V.

19-23 A relaxation oscillator (as in Fig. 19-43) uses a *2N2647* *UJT* with V_{BB} = 25 V. Calculate the typical oscillation frequency if C_1 = 0.5 μF and R_E = 3.3 kΩ.

19-24 Calculate the maximum and minimum charging resistance values that can be used in the circuit of Problem 19-23.

19-25 The *UJT* phase control circuit in Fig. 19-44 has a 115 V, 60 Hz *ac* supply, and an *SCR* with V_G = 1 V and I_{GM} = 25 mA. Design the circuit to use a *2N2647* *UJT* and a Zener diode with $V_Z \approx$ 15 V.

Section 19-8

19-26 A *PUT* operating from a 25 V supply has $V_F = 1.5$ V and $I_G = 50$ μA. Determine values for R_1 and R_2 to program η to 0.75. Also, calculate V_P, V_V, R_{BB}, and R_G.

19-27 A *PUT* relaxation oscillator (as in Fig. 19-48) has a 20 V supply and a 0.68 μF capacitor. The *PUT* has $V_G = 1$ V and $I_G = 100$ μA. The peak capacitor voltage is to be 5 V and the oscillating frequency is to be 300 Hz. Determine suitable resistor values.

19-28 The *UJT* in Problem 19-23 is to be replaced with a *PUT*. Determine suitable resistance values for the gate bias voltage divider.

19-29 A *PUT* has a forward voltage of $V_F = 0.9$ V and $I_G = 200$ μA. The device is to be programmed to switch on at $V_G = 15$ V when operating from a 24 V supply. Determine values for R_1 and R_2, and calculate V_P, V_V, R_{BB}, and R_G.

Practise Problem Answers

19-2.1   163 V, 1.6 A, 18 kΩ, 1.5 kΩ, 180 Ω
19-2.2   10 kΩ, 0.82 μF, 1.5 kΩ
19-3.1   (68 kΩ + 12 kΩ), 4 μF
19-5.1   1 kΩ, 180 Ω, 5.6 kΩ, 1 kΩ, 2.5 μF, 1 μF
19-6.1   500 kΩ, 820 Ω
19-7.1   24.5 V
19-7.2   23.2 V, 17.5 V
19-7.3   1.36 MΩ, 6.25 kΩ
19-8.1   7.2 kΩ, 0.6, 9.7 V, 530 kΩ, 140 kΩ
19-8.2   13.4 V

Chapter *20*

Optoelectronic Devices

Chapter Contents

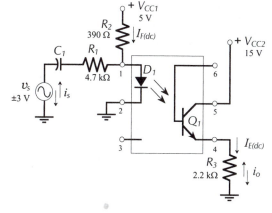

Objectives

You will be able to:

1 Define important illumination units, and calculate illumination intensity at a given distance from a source.

2 Explain the fabrication and operation of a light emitting diode (LED), discuss its parameters, and design LED circuits.

3 Explain the operation of liquid crystal displays (LCDs). Show how LCDs and LEDs are used in seven-segment numerical displays, and calculate power dissipations in each type of display.

4 Explain the construction and operation of a photoconductive cell. Sketch the device characteristics, and discuss its parameters.

5 Design photoconductive cell circuits to bias BJTs on or off, energize relays, trigger Schmitt circuits etc.

6 Explain the construction and operation of a photodiode. Sketch photodiode characteristics, and discuss its parameters.

7 Design photodiodes into circuits where they operate as a photo-conductive device, and as a photovoltaic device.

8 Explain solar cells, and design solar cell battery charger circuits.

9 Explain the operation of a photo-transistor. Sketch phototransistor characteristics, and discuss its parameters. Also, explain photo-darlingtons and photoFETs.

10 Design phototransistor circuit for energizing relays, triggering SCRs, etc.

11 Explain the construction and operation of optocouplers, and discuss optocoupler parameters.

12 Design optocoupler circuits for coupling pulse-type and linear signals between systems with different supply voltages.

Introduction

Optoelectronic devices emit light, modify light, have their resistance affected by light, or produce currents and voltages proportional to light intensity.

Light-emitting diodes (*LEDs*) produce light, and are typically used as indicating lamps and in numerical displays. Liquid crystal displays (*LCDs*) which modify light are also used as numerical displays. Photoconductive cells have a resistance that depends upon illumination intensity. They are used in circuits designed to produce an output change when the light level changes. The current and voltage levels in photodiodes and phototransistors are affected by illumination. These devices are also used in circuit that have their conditions altered by light level changes. Illumination is converted into electrical energy by means of solar cells, and this energy is often used to charge storage batteries. Optocouplers combine *LEDs* and phototransistors to provide a means of coupling between circuits that have different supply voltages, while maintaining a high level of electrical insolation.

20-1   Light Units

The total light energy output, or *luminous flux* (ϕ_s), from a source can be measured in *milliwatts* (mW) or in *lumens* (lm), where 1 lm = 1.496 mW. The *luminous intensity* (E_s) (also termed *illuminance*) of a light source is defined as the luminous flux density per unit solid angle (or cone) emitting from the source, [see Fig. 20-1(a)]. This is measured in *Candelas* (cd), where one candela is equal to one lumen per unit solid angle, (assuming a point source that emits light evenly in all directions).

$$E_s = \frac{\phi_s}{4\pi} \qquad\qquad \textbf{(20-1)}$$

The light intensity (E_A) on an area at a given distance from the source is determined from the surface area of a sphere surrounding the source, [Fig. 20-1(b)]. At a distance of r meters, the luminous flux is spread over a spherical area of $4\pi r^2$ square meters, so

$$E_A = \frac{\phi_s}{4\pi r^2} \qquad\qquad \textbf{(20-2)}$$

When the total flux is expressed in lumens, Eq. 20-2 gives the luminous intensity in *lumens per square meter* (lm/m²), also termed *lux* (*lx*). Comparing Eq. 20-2 to Eq. 20-1, it is seen that the luminous intensity per unit area at any distance r from a point source is determined by dividing the source intensity by r^2.

Luminous intensity can also be measured in *milliwatts per square centimeter* (mW/cm²), or *lumens per square foot* (lm/ft²), also known as a *foot candle* (*fc*), where 1 fc = 10.764 lx.

The light intensity of sunlight on the earth at noon on a clear day is approximately 107,640 lx, or 161 W/m². The light intensity from a 100 W lamp is approximately 4.8 x 10³ cd, (allowing for a 90% lamp efficiency). At a distance of 2 m, this is 1.2 x 10³ lx. An indicating lamp with a 3 mcd output can be clearly seen at a distance of several meters in normal room lighting conditions.

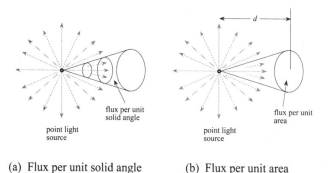

(a) Flux per unit solid angle   (b) Flux per unit area

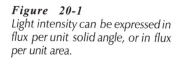

Figure 20-1
Light intensity can be expressed in flux per unit solid angle, or in flux per unit area.

Example 20-1
Calculate the light intensity 3 m from a lamp that emits 25 W of light energy. Also, determine the total luminous flux striking an area of 0.25 m² at 3 m from the lamp.

Solution

Eq. 20-2,
$$E_A = \frac{\phi_s}{4\pi r^2} = \frac{25\text{ W}}{4\pi \times (3\text{ m})^2}$$
$$= 0.221\text{ W/m}^2 = 221\text{ mW/m}^2$$

Total flux $= E_A \times$ area $= 221$ mW/m² x 0.25 m²
$$\approx 55\text{ mW}$$

Example 20-2
Determine the light intensity at a distance of 2 m from a 10 mcd source.

Solution

$$E_A = \frac{E_s\text{ (cd)}}{r^2} = \frac{10\text{ mcd}}{2^2}$$
$$= 2.5\text{ mlx}$$

Light energy is electromagnetic radiation; it is in the form of electromagnetic waves. So, it can be defined in terms of *frequency* or *wavelength,* as well as intensity. Wavelength, frequency, and velocity are related by the equation;

$$c = f\lambda \qquad\qquad \textbf{(20-3)}$$

where c = velocity = 3 x 10⁸ m/s for electromagnetic waves
       f = frequency in Hz
       λ = wavelength in meters

The wavelength of visible light ranges from violet at approximately 380 nm (*nanometers*) to red at 720 nm. From Eq. 20-3 , the frequency extremes are;

$$f_{\text{violet}} = \frac{c}{\lambda} = \frac{3 \times 10^8 \, \text{m/s}}{380 \, \text{nm}}$$

$$\approx 8 \times 10^{14} \, \text{Hz}$$

$$f_{\text{red}} = \frac{c}{\lambda} = \frac{3 \times 10^8 \, \text{m/s}}{720 \, \text{nm}}$$

$$\approx 4 \times 10^{14} \, \text{Hz}$$

Practise Problems

20-1.1 A lamp is required to produce a light intensity of 213 lx at a distance of 5 m. Calculate the total light energy output of the lamp in watts.

20-1.2 Calculate the frequency of yellow light with a 585 nm wavelength.

20-1.3 Determine the light intensity 3.3 m from an 8 mcd lamp, and the total luminous flux striking a 4 cm² area at that location.

20-2   Light Emitting Diode (LED)

LED Operation and Construction

Charge carrier recombination occurs at a forward biased *pn-*junction as electrons cross from the *n-*side and recombine with holes on the *p-*side. Free electrons have a higher energy level than holes, and some of this energy is dissipated in the form of heat and light when recombination takes place. If the semiconductor material is translucent, the light is emitted and the junction becomes a light source; that is, a *light-emitting diode* (*LED*).

A cross-sectional view of an *LED* junction is shown in Fig. 20-2(a). The semiconductor material is gallium arsenide (*GaAs*), gallium arsenide phosphide (*GaAsP*), or gallium phosphide (*GaP*). An *n-*type epitaxial layer is grown upon a substrate, and the *p-*region is created by diffusion. Charge carrier recombinations occur in the *p-*region, so the *p-*region is kept uppermost to allow the light to escape. The metal film anode connection is patterned to allow most of the light to be emitted. A gold film is applied to the bottom of the substrate to reflect as much light as possible toward the surface of the device and to provide a cathode connection. *LEDs* made from *GaAs* emit infrared (invisible) radiation. *GaAsP* material provides either red light or yellow light, while red or green emission can be produced by using *GaP*. Using various materials, *LEDs* can be manufactured to produce light of virtually any color.

Figure 20-2(b) shows the typical construction of a *LED*. The *pn-*junction is mounted on a cup-shaped reflector, as illustrated, wires are provided for anode and cathode connection, and the device is encapsulated in an epoxy lens. The lens can be clear or colored, and (when not energized) the lens color identifies the *LED* light color. The color of the light emitted by the energized *LED* is determined solely by the *pn*-junction material. Some *LEDs* have glass particles embedded in the epoxy lens to diffuse the emitted light and increase the viewing angle of the device. The *LED* circuit symbol is shown in Fig. 20-2(c). Note that the arrow directions indicate emitted light.

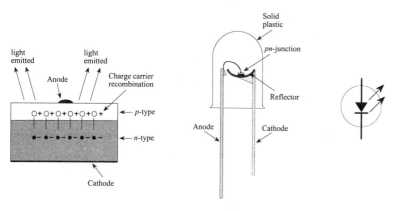

(a) *LED* junction   (b) Typical construction   (c) Graphic symbol

Figure 20-2
A light-emitting diode (LED) produces energy in the form of light when charge carrier recombination occurs at the pn-junction.

Characteristics and Parameters

LED characteristics are similar to those of other semiconductor diodes, except that (as shown in the partial specification in Fig. 20-3) the typical forward voltage drop is 1.6 V. Note also, that the reverse breakdown voltage can be as low as 3 V. In some circuits it is necessary to include a diode with a high reverse breakdown voltage in series with a *LED*. The forward current used with a *LED* is usually in the 10 mA to 20 mA range, but (depending on the particular device) the peak current can be as high as 90 mA. *LED* luminous intensity depends on the forward current level; it is usually specified at 20 mA. The peak wavelength of the light output is also normally listed on the specification.

Typical LED Specification			
	Min	Typ	Max
Luminous intensity (I_V at 20 mA)	4 mcd	8 mcd	
Forward voltage (V_F)	1.4 V	1.6 V	2.0 V
Reverse Breakdown voltage (V_{FBR})	3 V	10 V	
Peak forward current ($I_{F(max)}$)		90 mA	
Average forward current ($I_{F(av)}$)		20 mA	
Power dissipation (P_D)		100 mW	
Response speed (τ_S)		90 ns	
Peak wavelength (λ_p)		660 nm	

Figure 20-3
Partial specification for a typical light emitting diode.

LED Circuits

As explained, an *LED* is a semiconductor diode that emits light when a forward current is passed through the device. A single *LED* might simply be employed as a supply voltage *on/off* indicator, as illustrated in Fig. 20-4(a). A series-connected resistor (R_1) must be included to limit the current to the desired level. Figure 20-4(b) shows an *LED* connected at the output of a comparator to indicate a *high* output voltage. As well as the current-limiting resistor (R_1), an ordinary semiconductor diode (D_1) is connected in series with the *LED* to protect it from an excessive reverse voltage when the comparator output is negative.

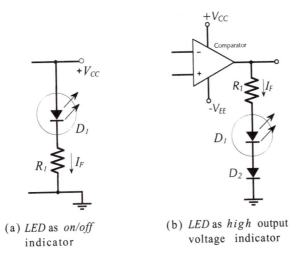

(a) *LED* as *on/off* indicator

(b) *LED* as *high* output voltage indicator

Figure 20-4
Light emitting diode indicator circuits.

LEDs are often controlled by a *BJT*, as illustrated in Fig. 20-5(a) and (b). Transistor Q_1 in Fig. 20-5(a) is switched into saturation by the input voltage (V_B). Resistor R_1 limits the transistor base current, and R_2 limits the *LED* current. In Fig. 20-5(b), the emitter resistor (R_1) limits the *LED* current to (V_B -V_{BE})/R_1.

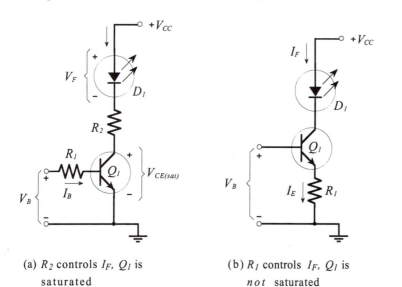

(a) R_2 controls I_F, Q_1 is saturated

(b) R_1 controls I_F, Q_1 is *not* saturated

Figure 20-5
BJT control circuits for light emitting diodes.

Example 20-3

The *LED* in Fig. 20-5(a) is to have a forward current of approximately 10 mA. The circuit voltages are $V_{CC} = 9$ V, $V_F = 1.6$ V, and $V_B = 7$ V; and Q_1 has $h_{FE(min)} = 100$. Calculate suitable resistance values for R_1 and R_2.

Solution

$$R_2 = \frac{V_{CC} - V_F - V_{CE(sat)}}{I_C} = \frac{9\text{ V} - 1.6\text{ V} - 0.2\text{ V}}{10\text{ mA}}$$

$$= 720\ \Omega\ \text{(use 680 }\Omega\text{ standard value)}$$

I_C becomes,

$$I_C = \frac{V_{CC} - V_F - V_{CE(sat)}}{R_2} = \frac{9\text{ V} - 1.6\text{ V} - 0.2\text{ V}}{680\ \Omega}$$

$$= 10.6\text{ mA}$$

$$I_B = \frac{I_C}{h_{FE(min)}} = \frac{10.6\text{ mA}}{100}$$

$$= 106\ \mu\text{A}$$

$$R_B = \frac{V_B - V_{BE}}{I_B} = \frac{7\text{ V} - 0.7\text{ V}}{106\ \mu\text{A}}$$

$$= 59\text{ k}\Omega\ \text{(use 56 k}\Omega\text{ standard value)}$$

Practise Problems

20-2.1 The *LED* in the circuit in Fig. 20-5(b) is to pass a 20 mA current. The circuit voltages are $V_{CC} = 15$ V, $V_F = 1.9$ V, and $V_B = 5$ V. Determine a suitable resistance for R_1, and calculate V_{CE} for Q_1.

20-2.2 Determine suitable resistances for the circuits in Fig. 20-4 to give $I_F = 15$ mA. The *LEDs* have $V_F = 1.8$ V. Also, $V_{CC} = 6$ V in Fig. 20-4(a), and in Fig. 20-4(b) the op-amp output is $V_o = \pm 9$ V.

20-3   Seven-Segment Displays

LED Seven-Segment Display

The arrangement of a *seven-segment LED* numerical display is shown in Fig. 20-6(a). The actual *LED* devices are very small, so, to enlarge the lighted surface, solid plastic *light pipes* are often employed, as shown. Any desired numeral from 0 to 9 can be indicated by passing current through the appropriate segments, [Fig 20-6(b)]. Part (c) in Fig. 20-6 shows three seven-segment displays together with a two-segment display referred to as a *half digit*. The whole display, termed a *three-and-a-half digit display*, can be used to indicate numerical values up to a maximum of 1999.

The *LEDs* in a seven-segment display may be connected in *common-anode* or in *common-cathode* configuration, [Fig. 20-6(d)]. When selecting a *LED* seven-segment display, it is important to

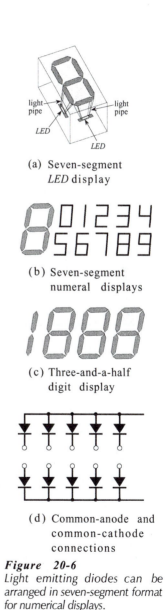

(a) Seven-segment
    LED display

(b) Seven-segment
    numeral displays

(c) Three-and-a-half
    digit display

(d) Common-anode and
    common-cathode
    connections

Figure 20-6
Light emitting diodes can be arranged in seven-segment format for numerical displays.

determine which of the two connecting arrangements is required.

The relatively large amounts of current consumed by *LED* seven-segment displays are their major disadvantage. Apart from this, *LEDs* have the advantage of long life and ruggedness.

Example 20-4

Calculate the total power supplied to a $3\frac{1}{2}$ digit *LED* display when it indicates 1999. A 5 V supply is used, and each *LED* has a 10 mA current.

Solution

total segments $N = [3 \times (\text{segments for } 8)] + [1 \times (\text{segments for } 1)]$

$$= (3 \times 7) + (1 \times 2) = 23$$

total current $I_T = N \times (\text{current per segment}) = 23 \times 10 \text{ mA}$

$$= 230 \text{ mA}$$

power $P = I_T \times V_{CC} = 230 \text{ mA} \times 5 \text{ V}$

$$= 1.15 \text{ W}$$

Liquid Crystal Cells

Liquid crystal material is a liquid that exhibits some of the properties of a solid. The molecules in ordinary liquids normally have random orientations. In liquid crystals the molecules are oriented in a definite crystal pattern.

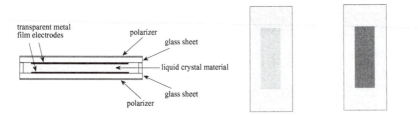

(a) Cross-section of a liquid crystal cell   (b) Unergized cell   (c) Energized cell

Figure 20-7
A liquid crystal cell consists of a layer of liquid crystal material sandwiched between glass sheets with transparent metal film electrodes and polarizers.

A liquid crystal cell consists of a very thin layer of liquid crystal material sandwiched between glass sheets, as illustrated in Fig. 20-7(a). The glass sheets have transparent metal film electrodes deposited on the inside surfaces. In the commonly used *twisted nematic* cell two thin polarizing optical filters are placed at the surface of each glass sheet. The liquid-crystal material employed twists the light passing through when the cell is not energized. This twisting allows the light to pass through the polarizing filters, so that the cell is semi-transparent. When energized, the liquid molecules are reoriented so that no twisting occurs, and no light can pass through. Thus, the energized cell can appear dark against a bright background. The cells can also be manufactured to appear bright against a dark background.

Seven-segment numerical (and other type) displays made from liquid crystal cells are referred to as *liquid crystal displays* (*LCDs*). Two types of *LCDs* are illustrated in Fig. 20-8. The *reflective*-type shown in Fig. 20-8(a) relies on reflected light. The cell is placed on a reflective surface, so that when not energized it is just as reflective as the surrounding material, consequently, it disappears. When energized, no light is reflected from the cell, and it appears dark against the bright background. The *transmittive* cell in Fig. 20-8(b) allows light to pass through from the back of the cell when not energized. When energized, the light is blocked, and here again the cell appears dark against a bright background. The *trans-reflective* cell is a combination of transmittive and reflective types.

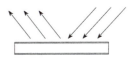

(a) Reflective cell

(b) Transmittive cell

Figure 20-8
Reflective-type liquid crystal cells reflect incident light. Transmittive-type pass light from behind.

LCD Seven-Segment Display

Because liquid-crystal cells are light reflectors or transmitters rather than light generators, they consume very small quantities of energy. The only energy required by the cell is that needed to activate the liquid crystal. The total current flow through four small seven-segment *LCDs* is typically about 20 µA. However, *LCDs* require an *ac* voltage supply, either in the form of a sine wave or a square wave. This is because a continuous direct current flow produces a plating of the cell electrodes that could damage the device. Repeatedly reversing the current avoids this problem.

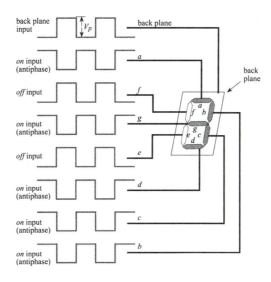

Figure 20-9
Liquid crystal display using a square wave supply. A segment is energized when its input is in antiphase with the back plane input.

A typical *LCD* supply is a 3 V to 8 V peak-to-peak square wave with a frequency of 60 Hz. Figure 20-9 illustrates the square wave drive method. The *back plane*, which is common to all of the cells, is supplied with a square wave, (with peak voltage V_p). Similar square wave applied to each of the other terminals are either in phase or in antiphase with the back plane square wave. Those cells with waveforms in phase with the back plane waveform (cells *e* and *f* in Figure 20-9) have no voltage developed across them. Both terminals of the segment are at the same potential, so they are not energized. The cells with square waves in antiphase with the back plane input have a square wave with peak voltage $2V_p$ developed across them, consequently, they are energized.

Unlike *LED* displays, which are usually quite small, *LCDs* can be fabricated in almost any convenient size. The major advantage of *LCDs* is their low power consumption. Perhaps the major disadvantage of the *LCD* is its decay time of 150 ms (or more). This is very slow compared to the rise and fall times of *LEDs*. In fact, the human eye can sometimes observe the fading out of *LCD* segments switching *off*.

Practise Problems

20-3.1 Calculate the total power supplied to a 3¹/₂ digit *LCD* display indicating 1999. The supply is a square wave with an 8 V peak level, and the current to each segment is 5 μA.

20-4   Photoconductive Cell

Cell Construction

Light striking the surface of a material can provide sufficient energy to cause electrons within the material to break away from their atoms. Thus, free electrons and holes (*charge carriers*) are created within the material, and consequently its resistance is reduced. This is known as the *photoconductive effect*.

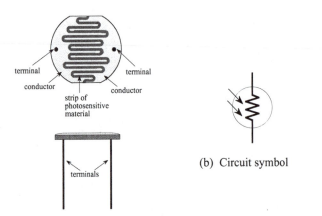

terminal   terminal

conductor   conductor

strip of
photosensitive
material

terminals

(b) Circuit symbol

(a) Photoconductive cell construction

Figure 20-10
A photoconductive cell consists of a strip of light-sensitive material situated between two conductors.

The construction of a typical photoconductive cell is illustrated in Fig. 20-10(a), and the graphic symbol is shown in Fig. 20-10(b). Light-sensitive material is arranged in the form of a long strip zigzagged across a disc-shaped base. The connecting terminals are fitted to the conducting material on each side of the strip; they are *not* at the ends of the strip. Thus, the light sensitive material is actually a short, wide strip between the two conductors. For added protection, a transparent plastic cover is usually included.

Cadmium sulfide (*CdS*) and cadmium selenide (*CdSe*) are the two materials normally used in photoconductive cell manufacture. Both respond rather slowly to changes in light intensity. For cadmium selenide, the response time (t_{res}) is around 10 ms, while for cadmium sulfide it may be as long as 100 ms. Temperature sensitivity is another important difference between the two materials. There is a large change in the resistance of a cadmium selenide cell with changes in ambient temperature, but the resistance of cadmium sulfide remains relatively stable. As with all other devices, care must be taken to ensure that the power dissipation is not excessive. The *spectral response* of a cadmium sulfide cell is similar to that of the human eye; it responds to visible light. For a cadmium selenide cell, the spectral response is at the longer wavelength end of the visible spectrum and extends into the infrared region.

Characteristics and Parameters

Typical illumination characteristic for a photoconductive cell are shown in Fig. 20-11. It is seen that, when the cell is not illuminated its resistance can be greater than 100 kΩ. This is known as the *dark resistance* of the cell. When the cell is illuminated, its resistance might fall to a few hundred ohms. Note that the scales on the illumination characteristic are logarithmic.

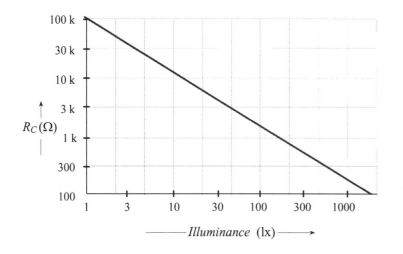

Figure 20-11
Typical photoconductive cell illumination characteristics. The resistance is usually very high when the cell is dark, and relatively low when illuminated.

Typical Photoconductive Cell Specification				
PD (mW)	Max ac (V)	Dark Resistance	λP (nm)	Resistance at 10 lx
200	180	100 kΩ	550	6 kΩ (min) 18 kΩ (max)

Figure 20-12
Partial specification for a typical photoconductive cell.

A typical photoconductive cell specification is shown in Fig. 20-12. As well as maximum voltage and power dissipation, the cell dark resistance and the resistance at a 10 lx illumination is listed. Note the wide range of cell resistance at a 10 lx. The light wavelength that gives peak response (λP) is also given on the specification. Cell *sensitivity* is sometimes used, and this is simply the cell current for a given voltage and given level of illumination.

Applications

Figure 20-13 shows a photoconductive cell used for relay control. When the cell is illuminated, its resistance is low and the relay current is at its maximum. Thus, the relay is energized. When the cell is dark, its high resistance keeps the current down to a level too low to energize the relay. Resistance R_1 is included to limit the relay current to the desired level when the cell resistance is low.

Example 20-5

A relay is to be controlled by a photoconductive cell as in Fig. 20-13. The cell has the characteristics shown in Fig. 20-11. The relay is to be supplied with 10 mA from a 30 V supply when the cell is illuminated with about 200 lm/m². Calculate the required series resistance and the level of the dark current. Assume that the coil resistance is much smaller than R_1 and R_C.

Solution

From the characteristics; at 200 lm/m², $R_C \approx 1$ kΩ

When the cell is illuminated,

$$I = \frac{E}{R_1 + R_C}$$

or,

$$R_1 = \frac{E}{I} - R_C = \frac{30 \text{ V}}{10 \text{ mA}} - 1 \text{ k}\Omega$$

$$= 2 \text{ k}\Omega \text{ (use 1.8 k}\Omega \text{ standard value)}$$

From the characteristics;

when dark, $R_C \approx 100$ kΩ

dark current, $$I = \frac{E}{R_1 + R_C} = \frac{30 \text{ V}}{1.8 \text{ k}\Omega + 100 \text{ k}\Omega}$$

$$\approx 0.3 \text{ mA}$$

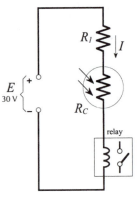

Figure 20-13
Circuit of a relay controlled by a photoconductive cell. The relay is energized when the cell is illuminated.

Photoconductive cells employed to switch transistors *on* and *off* are shown in Fig. 20-14. When cell in Fig. 20-14(a) is dark, the cell resistance (R_C) is high. Consequently, the transistor base is biased above its emitter voltage level, and Q_1 is turned *on*. When the cell is illuminated, its resistance is reduced, and the lower cell resistance in series with R_1 biases the transistor base below its emitter voltage level. Thus, Q_1 is turned *off* when the cell is illuminated.

In Fig. 20-14(b), Q_1 is biased *off* when the cell is dark, because R_C is high. When illuminated, the reduced cell resistance causes Q_1 to be biased *on*.

Example 20-6

The transistor in Fig. 20-14(a) is to be biased *on* when the photoconductive cell is dark, and *off* when it is illuminated. The supply is ±6 V, and the transistor base current is to be 200 μA when *on*. Design the circuit using the photoconductive cell characteristics in Fig. 20-11. Also, determine the minimum light level when the transistor is *off*.

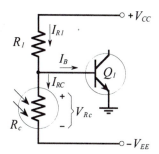

(a) Circuit to switch a *BJT off* when a cell is illuminated

Solution

when dark, $\qquad R_C \approx 100 \text{ k}\Omega$

when Q_1 is on, $\qquad V_{RC} = V_{EE} + V_{BE} = 6 \text{ V} + 0.7 \text{ V}$
$$= 6.7 \text{ V}$$

$$I_{RC} = \frac{V_{RC}}{R_C} = \frac{6.7 \text{ V}}{100 \text{ k}\Omega}$$
$$= 67 \text{ }\mu A$$

$$I_{R1} = I_{RC} + I_B = 67 \text{ }\mu A + 200 \text{ }\mu A$$
$$= 267 \text{ }\mu A$$

$$V_{R1} = V_{CC} - V_B = 6 \text{ V} - 0.7 \text{ V}$$
$$= 5.3 \text{ V}$$

$$R_1 = \frac{V_{R1}}{I_{R1}} = \frac{5.3 \text{ V}}{267 \text{ }\mu A}$$
$$\approx 20 \text{ k}\Omega \text{ (use 18 k}\Omega \text{ standard value)}$$

when Q_1 is off, $\qquad V_{R1} \geq 6 \text{ V}$

and, $\qquad I_{R1} = \frac{V_{R1}}{R_1} = \frac{6 \text{ V}}{18 \text{ k}\Omega}$
$$= 333 \text{ }\mu A$$

$$R_c = \frac{V_{RC}}{I_{R1}} = \frac{6 \text{ V}}{333 \text{ }\mu A}$$
$$= 18 \text{ k}\Omega$$

From the characteristics;

$$\text{when } R_C \approx 18 \text{ k}\Omega, \text{ illumination} \approx 7 \text{ lm/m}^2$$

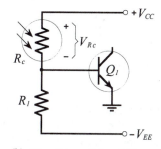

(b) Circuit to switch a *BJT on* when a cell is illuminated

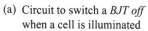

Figure 20-14
BJTs controlled by photo-conductive cells. The position of the cell in the circuit determines whether the transistor is switched on or off by an increase in illuminance.

Figure 20-15 shows a photoconductive cell used with an op-amp Schmitt trigger circuit (see Section 14-10). When the cell resistance is low (cell illuminated), the voltage across R_1 is higher than the upper trigger point (*UTP*) for the Schmitt. Consequently, the op-amp output is *low* (negative). The output switches to a *high* (positive) level when V_{R1} falls to the Schmitt circuit lower trigger point (*LTP*). This occurs when the cell illumination level falls, causing R_C to rise.

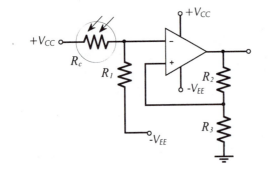

Figure 20-15
Op-amp Schmitt trigger circuit with a photo-conductive cell for light level detection.

The circuit in Fig. 20-16 uses a photoconductive cell to control the current level in an *LED*. The *LED* current is low when the ambient light level is low, because the cell resistance is high. The *LED* current is increased as a result of the decreased resistance of R_C when the ambient light level is high. This increased current gives greater *LED* brightness so that it can be easily seen.

Example 20-7

Design the circuit in Fig. 20-16 for a *LED* current of $I_F \approx 10$ mA at $R_C = 5$ kΩ (light level H_1), and for $I_F \approx 20$ mA at $R_C = 1$ kΩ (light level H_2).

Solution

at H_1,
$$V_{B1} = \frac{V_{CC} \times R_2}{R_{C1} + R_2} = \frac{V_{CC} \times R_2}{5 \text{ k}\Omega + R_2}$$

at H_2,
$$V_{B2} = \frac{V_{CC} \times R_2}{R_{C2} + R_2} = \frac{V_{CC} \times R_2}{1 \text{ k}\Omega + R_2} = 2 V_{B1}$$

so,
$$\frac{V_{CC} \times R_2}{1 \text{ k}\Omega + R_2} = \frac{2 V_{CC} \times R_2}{5 \text{ k}\Omega + R_2}$$

giving,
$$R_2 = 3 \text{ k}\Omega \text{ (use 2.2 k}\Omega + 2 \text{ k}\Omega \text{ variable)}$$

$$V_{B1} = \frac{V_{CC} \times R_2}{R_{C1} + R_2} = \frac{12 \text{ V} \times 3 \text{ k}\Omega}{5 \text{ k}\Omega + 3 \text{ k}\Omega}$$
$$= 4.5 \text{ V}$$

$$V_{B2} = 2 V_{B1} = 9 \text{ V}$$

For $I_C = 10$ mA,
$$R_3 \approx \frac{V_{B1} - V_{BE}}{I_C} = \frac{4.5 \text{ V} - 0.7 \text{ V}}{10 \text{ mA}}$$
$$= 380 \ \Omega \text{ (use 390 }\Omega \text{ standard value)}$$

at H_2,
$$I_C \approx \frac{V_{B2} - V_{BE}}{R_3} = \frac{9 \text{ V} - 0.7 \text{ V}}{390 \ \Omega}$$
$$= 21.3 \text{ mA}$$

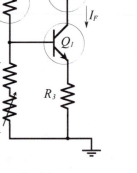

Figure 20-16
Photo-conductive cell circuit for controlling LED current level.

Figure 20-17
A photo-diode has a reverse-biased pn-junction designed to be light sensitive. A minority charge carrier current flows when the junction is illuminated.

20-5 Photodiode and Solar Cell

Photodiode Operation

When a *pn*-junction is reverse biased, a small reverse saturation current flows due to thermally generated holes and electrons being swept across the junction as minority charge carriers, (see Section 1-6). Increasing the junction temperature generates more hole-electron pairs, and so the minority carrier (reverse) current is increased. The same effect occurs if the junction is illuminated, (see Fig. 20-17). Hole-electron pairs are generated by the incident light energy, and minority charge carriers are swept across the junction to produce a reverse current flow. Increasing the junction illumination increases the number of charge carriers generated, and thus increases the level of reverse current. Diodes designed to be sensitive to illumination are known as *photodiodes*.

Characteristics

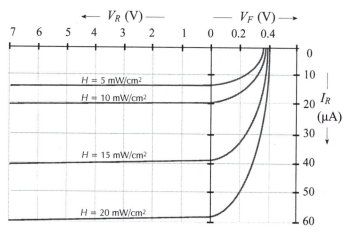

Figure 20-18
Typical photodiode characteristics. The reverse current remains substantially constant for each level of illumination.

Consider the typical photodiode illumination characteristics in Fig. 20-18. When the junction is dark, the *dark current* (I_D) would seem to be zero. Typically, I_D is around 2 nA. A 20 mW/cm² illumination level produces a reverse current of approximately 60 µA. Increasing the reverse voltage does not increase I_R significantly. So, each characteristic is approximately a horizontal line.

Figure 20-19 shows a simple photodiode circuit using a 2 V reverse bias. (Note the device circuit symbol.) Assuming that D_1 has the characteristics in Fig. 20-18, the current at a 5 mW/cm² illumination level is approximately 13 µA. At 20 mW/cm² the diode current is around 60 µA. The device resistance at each illumination level is readily calculated: (at a 5 mW/cm², R = 2 V/13 µA = 154 kΩ), (at a 20 mW/cm², R = 2 V/60 µA = 33 kΩ). The resistance changed by a factor of approximately 5 from the low to the high illumination level, showing that a photodiode can be employed as a photoconductive device.

When the reverse-bias voltage across a photodiode is removed, minority charge carriers continue to be swept across the junction while the diode is illuminated. With an external circuit connected across the diode terminals, the minority carriers flow back to their original sides. The electrons that crossed the junction from p to n will now flow out through the n-terminal and into the p-terminal. This means that the device is behaving as a voltage cell, with the n-side being the negative terminal and the p-side the positive terminal, as illustrated in Fig. 20-20. In fact, a voltage can be measured at the photodiode terminal, positive on the p-side and negative on the n-side. So, the photodiode is a photovoltaic device as well as a photoconductive device. The characteristics in Figure 20-18 show that, when illuminated, the photodiode actually has to be forward biased to reduce the reverse current to zero.

It should be noted that V_R and V_F have different scales on the photodiode characteristics shown in Fig. 20-18. A *dc* load line that crosses between the forward and reverse biased regions cannot be drawn on these characteristics. Equal scales must be used for each part of the characteristics to draw such a load line.

Specification

A partial specification for a typical photodiode is shown in Fig. 20-21. The *light current* (I_L) is listed as 10 µA at an illumination level of 5 mW/cm² when the reverse bias is 2 V. This is sometimes defined as a *short-circuit current* (I_{SC}). The *dark current* (I_D) is specified as 2 nA maximum when the reverse voltage is 20 V, and the *open-circuit terminal voltage* (V_{OC}) is given as 350 mV. Note that the typical response time (t_{res}) of 2 ns for a photodiode is very much superior to that for a photoconductive cell. The diode *sensitivity* (S) is the change in diode current produced by a given change in light intensity. The power dissipation, reverse breakdown voltage, and peak output wavelength are also listed.

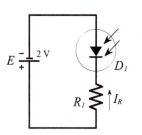

Figure 20-19
Photodiode circuit with a reverse bias voltage.

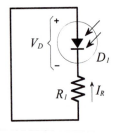

Figure 20-20
An illuminated photodiode without an external bias operates as a photovoltaic device.

Typical Photodiode Specification							
PD	V_{OC}	$BV_{R(max)}$	$I_{D(max)}$ (dark)	I_L [V_R = 2 V, H = 5 mW/cm²]	t_{res}	S	λP
100 mW	350 mV	100 V	2 nA	10 µA	2 ns	7 µA/ mW/cm²	900 nm

Figure 20-21
Partial specification for a low-current photodiode.

Construction

Figure 20-22(a) shows the cross-section of a diffused photo diode. It is seen that a thin heavily-doped *p*-type layer is situated at the top where it is exposed to incident light. The junction depletion region penetrates deeply into the lightly-doped *n*-type layer. This is in contact with a lower heavily-doped *n*-type layer which connects to a metal film contact. A ring-shaped contact is provided at the top of the *p*-type layer. Low-current photodiodes (also called *signal photodiodes*) are usually contained in a *TO*-type can with a lens at the top, [see Fig. 20-22(b)]. Clear plastic encapsulation is also used, [Fig. 20-22(c)].

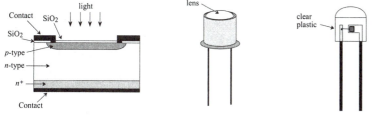

(a) Cross-section   (b) *TO* can-type package   (c) Clear plastic encapsulation

Figure 20-22
Photodiode cross-section and typical packages.

Photodiode Applications

Photodiodes can be used as photoconductive devices in the type of circuits discussed in Section 20-4. They can also be used in circuits where they function as photovoltaic devices. Figure 20-23 shows typical photodiode characteristics plotted in the first and second quadrants for convenience. When the device is operated with a reverse voltage, it functions as a photoconductive device. When operating without the reverse voltage, it operates as a photovoltaic device. In some circuits the photodiode can change between the photoconductive mode and the photovoltaic mode.

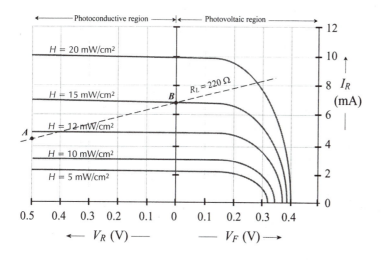

Figure 20-23
Photodiode characteristics have a photoconductive region and a photovoltaic region.

Example 20-8

The circuit in Fig. 20-24 uses a photodiode with the illumination characteristics in Fig. 20-23. Draw the *dc* load line, and determine the diode currents and voltages at light levels of 12, 15 and 20 mW/cm².

Solution

when $V_D = 0.5$ V, $V_{R1} = E - V_D = 1.5$ V - 0.5 V
$$= 1 \text{ V}$$

$$I_D = \frac{V_{R1}}{R_1} = \frac{1 \text{ V}}{220 \text{ }\Omega}$$

$$\approx 4.5 \text{ mA}$$

Plot *point A* on the characteristics at $I_D = 4.5$ mA and $V_D = 0.5$ V

when $V_D = 0$, $V_{R1} = E = 1.5$ V

so, $I_D = \dfrac{E}{R_1} = \dfrac{1.5 \text{ V}}{220 \text{ }\Omega}$

$$= 6.8 \text{ mA}$$

Plot *point B* on the characteristics at $V_D = 0$ and $I_D = 6.8$ mA
Draw the *dc* load line through points *A* and *B*.

From the load line;

      at 12 mW/cm², $I_D \approx$ -5 mA, $V_D \approx$ -0.4 V
      at 15 mW/cm², $I_D \approx$ -6.8 mA, $V_D \approx$ 0 V
      at 20 mW/cm², $I_D \approx$ -8 mA, $V_D \approx$ +0.28 V

Note the V_D polarity change at the highest illumination level.

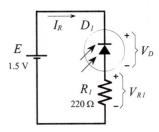

Figure 20-24
Photodiode with a reverse bias and a load resistor.

Solar Cells

The *solar cell*, or *solar energy converter*, is essentially a large photodiode designed to operate solely as a photovoltaic device and to give as much output power as possible. To provide maximum output current, solar cell surface areas are much larger than those of signal photodiodes, Typical solar cell output characteristics are illustrated in Fig. 20-25. Consider the characteristic for a 100 mW/cm² illumination level. If the cell is short-circuited, the output current (I_o) is 50 mA. Because the cell voltage (V_o) is zero at this point, the output power (P_o) is zero. Open-circuiting the cell gives $V_o \approx 0.55$ V, but $I_o = 0$. So, P_o is again zero. At the *knee* of the characteristic $V_o \approx 0.44$ V and $I_o \approx 45$ mA; giving $P_o \approx 20$ mW. Therefore, for maximum output power, the device must be operated on the knee of the characteristic. As in the case of all other devices, the power must be derated at high temperatures.

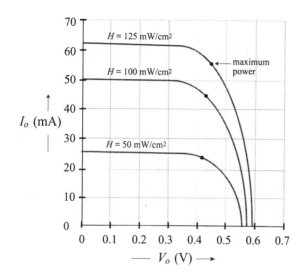

Figure 20-25
Typical solar cell characteristics. The device must be operated at the knee of its characteristic for maximum output power.

Figure 20-26 shows a group of series-parallel connected solar cells operating as a battery charger. Several cells must be series connected to produce the required output voltage, and several of these series-connected groups must be connected in parallel to provide the necessary output current.

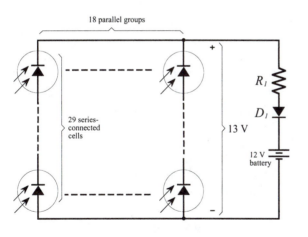

Figure 20-26
Series-parallel arrangement of solar cells connected to function as a solar battery charger.

Example 20-9

An earth satellite has 12 V batteries that supply a continuous current of 0.5 A. Solar cells with the characteristics shown in Fig. 20-25 are employed to keep the batteries charged. If the illumination from the sun for 12 hours in every 24 is 125 mW/cm², determine approximately the total number of cells required.

Solution

From Fig. 20-25, maximum output power at 125 mW/cm² is achieved when each cell is operated at approximately $V_o = 0.45$ V and $I_o = 57$ mA.

Allowing for the voltage drop across the rectifier, a maximum charging voltage of approximately $V_{CH} = 13$ V is required.

Number of series-connected cells,

$$N_s = \frac{V_{CH}}{V_o/\text{cell}} = \frac{13 \text{ V}}{0.45 \text{ V}}$$

$$\approx 29$$

The charge taken from the batteries over a 24-hour period is

$$Q = I_L \times t = 0.5 \text{ A} \times 24 \text{ hours}$$

$$= 12 \text{ Ah}$$

So, the charge delivered by the solar cells must be 12 Ah.

The solar cells deliver current only while they are illuminated; that is, for 12 hours in every 24. Therefore, the charging current from the solar cells is,

$$I_{CH} = \frac{Q}{t} = \frac{12 \text{ Ah}}{12 \text{ h}}$$

$$= 1 \text{ A}$$

Number of parallel-connected groups of cells,

$$N_P = \frac{I_{CH}}{I_o/\text{cell}} = \frac{1 \text{ A}}{57 \text{ mA}}$$

$$\approx 18$$

The total number of cells required,

$$N_T = N_P \times N_S = 18 \times 29$$

$$= 522$$

Practise Problems

20-5.1 A photodiode with the characteristics in Fig. 20-23 is connected in series with a 330 Ω resistor and a (reverse-biasing) 3 V battery. Draw the *dc* load line for the circuit, and determine the diode voltage at an illumination level of 20 mW/cm².

20-5.2 A solar voltage source is to be designed to produce a 3 V, 20 mA output from a 15 mW/cm² illumination level. Calculate the required number of cells if the available devices have the characteristics shown in Fig. 20-23.

20-6 Phototransistors

Phototransistor (BJT)

A *phototransistor* is similar to an ordinary *BJT*, except that its collector-base junction is constructed like a photodiode. Instead of a base current, the input to the transistor is in the form of illumination at the junction. Consider an ordinary *BJT* with its

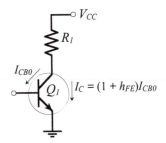

(a) *BJT* currents with open-circuited base

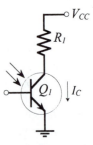

(b) Phototransistor circuit

Figure 20-27
In a phototransistor, the collector current depends upon the level of illumination at the CB junction.

base terminal open-circuited, [Fig. 20-27(a)]. The collector-base leakage current (I_{CBO}) acts as a base current, giving a collector current; $I_C = (h_{FE} + 1) I_{CBO}$. In the case of the photodiode, it was shown that the reverse saturation current is increased by the light energy at the junction. Similarly, in the phototransistor I_{CBO} is proportional to the collector-base illumination, [Fig. 20-27(b)]. This results in I_C also being proportional to the illumination level.

For a given amount of illumination on a very small area, the phototransistor provides a much larger output current than that available from a photodiode. Thus, the phototransistor is the most sensitive of the two devices. The phototransistor circuit symbol shows a base terminal, and this is often left unconnected, but is sometime used to provide stable bias conditions.

The cross-section in Fig. 20-28 illustrates the construction of a phototransistor. The emitter area is seen to be quite small, to allow incident illumination to pass to the collector-base junction. Phototransistor packages are similar to the photodiode packages in Fig. 20-22, except that three terminals are provided.

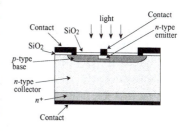

Figure 20-28
Phototransistor cross-section.

Characteristics and Specification

Typical phototransistor output characteristics are shown in Fig. 20-29. These are seen to be similar to *BJT* characteristics except that the base current levels are replaced with illumination levels. A *dc* load line can be drawn on the characteristics in the usual way.

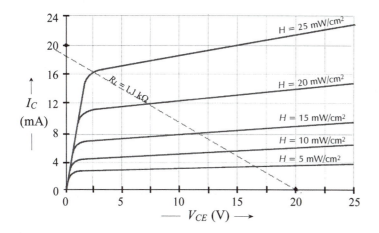

Figure 20-29
The I_C/V_{CE} characteristics for a phototransistor are similar to those of an ordinary BJT except that base current level is replaced by illumination level.

Typical Phototransistor Specification							
P_D	$V_{CE(max)}$	$I_{CEO(max)}$ (dark) [V_{CE} = 10 V]	$I_{C(min)}$ [V_{CE} = 5 V] [H = 5 mW/cm²]	t_r [I_{CE} = 1 mA]	t_f	S [H = 5 mW/cm²]	λP (nm)
200 mW	40 V	100 nA	3 mA	5 µs	8 µs	500 µA/mW/cm²	900 nm

Figure 20-30
Partial specification for a photo-transistor.

The partial specification for a phototransistor in Fig. 20-30 shows a 3 mA minimum current at 5 mW/cm², and a sensitivity of 500 µA/mW/cm². Comparing this to the photodiode specification

(Fig. 20-21) shows (as stated above) that a phototransistor is very much more sensitive than a photodiode. However, the rise and fall times (typically 5 μs and 8 μs) for a phototransistor are very much slower than the 2 ns response time for a photodiode.

Applications

Two phototransistor applications are shown in Fig. 20-31 and 20-32. The relay in Fig. 20-31 is energized when the incident light on the phototransistor is raised to a particular level. This occurs when the Q_1 emitter current produces sufficient voltage drop across R_2 to forward bias the BE junction of Q_2. The relay current falls again when Q_2 turns *off* as the light level decreases.

In Fig. 20-32, SCR_1 remains untriggered while the illumination keeps Q_1 in saturation. If the light fails, Q_1 turns *off* and V_{R3} triggers the SCR *on*. This kind of circuit can be used to switch *on* an emergency lighting system when the normal lighting fails.

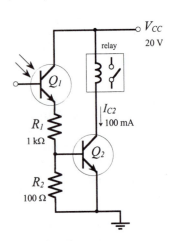

Figure 20-31
Use of a phototransistor to energize a relay when the illumination increases to a predetermined level.

Example 20-10

Transistor Q_1 in the circuit in Fig. 20-31 has the characteristics in Fig. 20-29, and Q_2 has $h_{FE} = 80$. Determine the light level required to energize the relay.

Solution

when Q_2 is on,

$$V_{R2} = V_{BE2} = 0.7 \text{ V}$$

$$I_{R2} = \frac{V_{R2}}{R_2} = \frac{0.7 \text{ V}}{100 \ \Omega}$$

$$= 7 \text{ mA}$$

$$I_{B2} = \frac{I_{C2}}{h_{FE}} = \frac{100 \text{ mA}}{80}$$

$$= 1.25 \text{ mA}$$

$$I_{E1} = I_{B2} + I_{R2} = 1.25 \text{ mA} + 7 \text{ mA}$$

$$= 8.25 \text{ mA}$$

The *dc* load resistance for Q_1 is,

$$R_L = R_1 + R_2 = 1 \text{ k}\Omega + 100 \ \Omega$$

$$= 1.1 \text{ k}\Omega.$$

Draw the *dc* load line on the characteristics for $R_L = 1.1$ kΩ

At $I_C = 8.25$ mA on the load line,

$$H \approx 15 \text{ mW/cm}^2$$

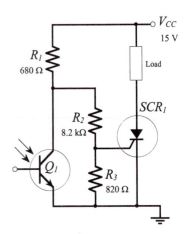

Figure 20-32
Phototransistor circuit for triggering an SCR at low light levels.

Photodarlington

The photodarlington (Fig. 20-33) consists of a phototransistor connected in Darlington arrangement with another transistor. This device is capable of producing much higher output currents than a phototransistor, and so it has a greater sensitivity to illumination levels than either a phototransistor or a photodiode. With the additional transistor involved, the photodarlington has a considerably longer switching time than a phototransistor.

PhotoFET

A *photoFET* is a *JFET* designed to have its gate-channel junction illuminated. The illumination controls the level of the device drain current. Consider the *n*-channel *JFET* and the *photoFET* in Fig. 20-34. The gate-source leakage current (I_{GSS}) is the reverse saturation current at a *pn*-junction. The voltage drop across R_G produced by I_{GSS} is normally too small to affect the *JFET* circuit. In the *photo-FET*, the junction reverse current (λI_G) is susceptible to light. Illumination on the junction generates additional minority charge carriers, thus increasing λI_G. This current flows through the bias resistance (R_G) and produces a voltage drop (V_{RG}), as illustrated. If the gate bias voltage (-V_G) is just sufficient to bias the device *off* when dark, then when the junction is illuminated, V_{RG} can raise the level of the gate voltage to bias the *photoFET on*.

The external bias voltage (-V_G) might be selected at a level that biases the device *on*, so that light level variations cause I_D to increase and decrease. In a *photoFET*, λI_G is termed the *gate current*, and the normal I_{GSS} at the junction is the *dark gate-leakage current*. The light-controlled drain current is designated λI_D.

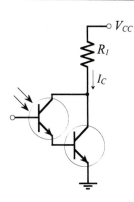

Figure 20-33
A photodarlington is made up of a phototransistor connected in Darlington with another BJT.

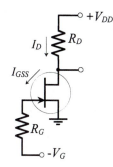

(a) *JFET* circuit

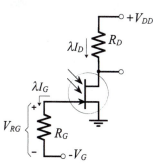

(b) *PhotoFET* circuit

Figure 20-34
In a photoFET, the gate-channel leakage current depends upon the illumination level. This current produces a voltage drop across the bias resistor to control the gate-source voltage.

Practise Problems

20-6.1 A phototransistor with the characteristics in Fig. 20-29 is connected in series with a resistor R_1 = 820 Ω and V_{CC} = 18 V. Determine V_{R1} at 5 mW/cm² and 25 mW/cm² illumination levels .

20-6.2 The *SCR* in Fig. 20-32 triggers at V_G = 0.8 V and I_G = 100 µA. Determine the approximate light level to trigger the *SCR* on if Q_1 has the characteristics in Fig. 20-29.

20-7 Optocouplers

Operation and Construction

An *optocoupler* (*optoelectronic coupler*) is essentially a photo-transistor and an *LED* combined in one package. Figure 20-35(a) and (b) shows the typical circuit and terminal arrangement for one such device contained in a *DIL* plastic package. When current flows in the *LED*, the emitted light is directed to the phototransistor, producing current flow in the transistor. The coupler may be operated as a switch, in which case both the *LED* and the phototransistor are normally *off*. A pulse of current through the *LED* causes the transistor to be switched *on* for the duration of the pulse. Linear signal coupling is also possible. Because the coupling is optical, there is a high degree of electrical isolation between the input and output terminals, and so the term *optoisolator* is sometimes used. The output (detector) stage has no effect on the input, and the electrical isolation allows a low-voltage *dc* source to control high voltage circuits.

The cross-section diagram in Fig. 20-35(c) illustrates the construction of an optocoupler. The emitter and detector are contained in a transparent insulating material that allows the passage of illumination while maintaining electrical isolation.

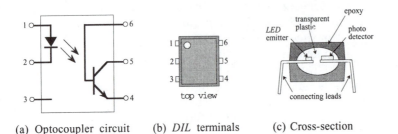

(a) Optocoupler circuit   (b) *DIL* terminals   (c) Cross-section

Figure 20-35
An optocoupler is composed of an LED and a phototransistor. The input and output are electrically isolated.

Specification

The partial specification for an optoelectronic coupler in Fig. 20-36 has three parts. The first part specifies the current and voltage conditions for the input (*LED*) stage. The second deals with the output (phototransistor) stage. The third part defines the coupling parameters. The transistor collector current is listed as 5 mA (typical) when its V_{CE} = 10 V and the *LED* has I_F = 10 mA. In this particular case, the ratio of output current to input current is 50%. This is known as the *current transfer ratio* (*CTR*), and for an optoelectronic coupler with a transistor output it can range from 10% to 150%.

Typical Optocoupler Specification

Input Stage

$I_{F(max)}$	$V_{F(max)}$ [I_F = 20 mA]	V_R
60 mA	1.5 V	3 V

Output Stage

$V_{CE(max)}$	$I_{C(max)}$	P_D	$V_{CE(sat)}$	I_{CEO} (dark)
30 V	150 mA	150 mW	0.2 V	50 nA

Coupled

$I_{C(out)}$ [I_F = 10 mA]	$t_{on(max)}$	$t_{off(max)}$	Isolation voltage
5 mA	2.5 µs	4 µs	7500 V

Figure 20-36
The partial specification for an optocoupler is made up of three parts: input, output, and coupling.

Applications

The circuit of an optocoupler in a *dc* or pulse-type coupling application is shown in Fig. 20-37. The diode current is switched *on* and *off* by the action of transistor Q_1 operating from a 24 V supply. Transistor Q_2 is turned *on* into saturation when D_1 is energized. The collector current of Q_2 provides the load (*sinking*) current and the current through resistor R_2. *Pull-up* resistor R_2 is necessary to ensure that the load terminal is held at the 5 V supply level when Q_2 is *off*.

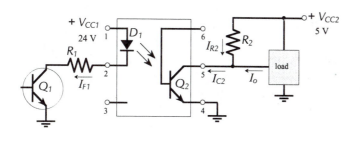

Figure 20-37
Optocoupler used for coupling a signal from a 24 V system to a 5 V system.

Example 20-11

The optocoupler in Fig. 20-37 is required to sink a 2 mA load current when Q_2 is in saturation and $I_{F1} = 10$ mA. The optocoupler has the specification in Fig. 20-36. Determine suitable resistor values.

Solution

From the specification; $I_{C2} = 5$ mA when $I_{FD1} = 10$ mA

$$I_{R2} = I_{C2} - I_o = 5 \text{ mA} - 2 \text{ mA}$$
$$= 3 \text{ mA}$$

$$R_2 = \frac{V_{CC2} - V_{CE(sat)}}{I_{R2}} = \frac{5 \text{ V} - 0.2 \text{ V}}{3 \text{ mA}}$$

$$= 1.6 \text{ k}\Omega \text{ (use 1.8 k}\Omega \text{ to ensure } Q_2 \text{ saturation)}$$

$$R_1 = \frac{V_{CC1} - V_{FD1} - V_{CE(sat)}}{I_{FD1}} = \frac{24 \text{ V} - 1.5 \text{ V} - 0.2 \text{ V}}{10 \text{ mA}}$$

$$= 2.23 \text{ k}\Omega \text{ (use 2.2 k}\Omega\text{)}$$

A linear application of an optocoupler is shown in Fig. 20-38. The 5 V supply provides a *dc* bias current to D_1 via R_2, and the *ac* signal coupled via C_1 and R_1 increases and decreases the diode current. Transistor Q_1 is biased into an *on* state by the direct current through D_1, and its emitter current is increased and decreased by the variation in light level produced by the alternating current in D_1. An output voltage is developed across R_3.

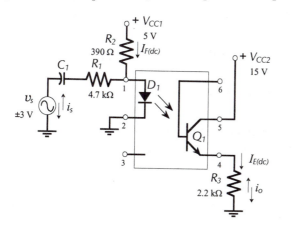

Figure 20-38
Linear signal coupling by means of an optocoupler.

Other Optocouplers

Other types of optocouplers involve different types of output stage. The three types illustrated in Fig. 20-39 are; (a) *Darlington-output* type, (b) *SCR*-output, and (c) *TRIAC*-output. In (a), the photo-darlington output stage provides much higher *CTR* than a *BJT* phototransistor output stage (typically 500%), but it also has a slower response time. The output stages in (b) and (c) are a *light-activated SCR* and a *light-activated TRIAC*, respectively. They are used with the kind of control circuits discussed in Chapter 19, where high electrical isolation between the triggering circuit and the control device is an additional requirement. *CTR* does not apply to *SCR* and *TRIAC* output stages; instead, the *LED* current needed to trigger the thyristor is of interest.

Optocoupler output stages are *not* designed for high load currents. Maximum current levels for Darlington outputs are around 150 mA, and 300 mA is typical for *SCR* and *TRIAC* outputs. When high load currents are to be switched, the optocoupler output stage is used as a trigger circuit for a high power device.

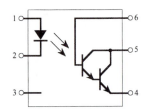

(a) Darlington output

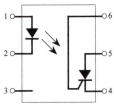

(b) *SCR* output

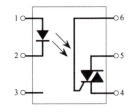

(c) *TRIAC* output

Figure 20-39
Optocouplers are available with various types of output stage.

Practise Problems

20-7.1 An optocoupler with the specification in Fig. 20-36 is to control a 10 mA relay with a 30 V supply. The input stage is connected via a resistor (R_1) to a 5 V supply. Calculate a suitable resistance for R_1.

20-7.2 Analyze the circuit in Fig. 20-38 to determine the *dc* bias current through D_1, and the *ac* signal current peaks. Also, calculate the maximum and minimum *dc* and *ac* output voltage levels. The optocoupler used has a diode with $V_F = 1.5$ V, and a *CTR* ranging from 20% to 70%.

Chapter-20 Review Questions

Section 20-1

20-1 State measurement units for luminous flux and luminous intensity. Using diagram, explain flux per solid angle.

20-2 Define: Candela, Lumen, and foot candle.

Section 20-2

20-3 Sketch diagrams to show the operation and construction of an *LED*. Briefly explain.

20-4 For an *LED*, state typical values of forward current, forward voltage, reverse breakdown voltage.

20-5 Draw circuit diagrams showing *LEDs* used to indicate (a) a *dc* supply voltage switched *on*, (b) a *high* output level from an op-amp. Explain each circuit.

20-6 The current level in an *LED* is to be controlled by use of a *BJT*. Sketch two possible circuits, and explain the operation of each.

20-7   An op-amp is to be used to control the current level in an *LED*. Draw a suitable circuit diagram and explain its operation.

Section 20-3

20-8   Sketch a seven-segment *LED* display. Explain common-anode and common-cathode connections. State total current requirements for an *LED* four numeral, seven-segment display.

20-9   Using illustrations, explain the operation of liquid-crystal cells. Discuss the difference between reflective-type and transmittive-type cells.

20-10   Sketch a seven-segment *LCD* and show the waveforms involved in controlling the cells. Explain.

Section 20-4

20-11   Sketch the typical construction and illumination characteristics for a photoconductive cell. Explain its operation.

20-12   Draw circuit diagrams to show how a photoconductive cell can be used for: (a) biasing a *pnp* transistor *off* when the cell is illuminated, (b) biasing an *npn* transistor *on* when the cell is illuminated. Explain how each circuit operates.

20-13   Draw circuit diagrams to show a photoconductive cell used for: (a) triggering an op-amp Schmitt trigger circuit, (b) energizing a relay when the cell is illuminated. Explain the operation of each circuit.

Section 20-5

20-14   Sketch the cross-section of a typical photodiode and explain its operation. Sketch typical photodiode characteristics and discuss their shape.

20-15   For photodiode, define; dark current, light current, and sensitivity. State typical values for each quantity.

20-16   Explain how a solar cell differs from a photodiode. Sketch typical solar cell characteristics, and discuss the best operating point on the characteristics.

20-17   Sketch the circuit diagram for an array of solar cells employed as a battery charger. Briefly explain.

Section 20-6

20-18   Sketch characteristics for a phototransistor, and explain how the device operates.

20-19   Draw a circuit diagram to show how a phototransistor can be used to energize a relay when the incident illumination is increased to a given level. Explain the circuit operation.

20-20 Modify the circuit drawn for Question 20-19 to have the relay energized until the illumination is increased to a given level. Explain.

20-21 Draw a circuit diagram for phototransistor control of an *SCR*; to have the *SCR* trigger *on* when the incident illumination falls to a low level. Explain how the circuit operates.

20-22 Sketch a circuit diagram for a photodarlington. Compare the performance of photodarlingtons to phototransistors.

20-23 Sketch a circuit diagram to show the operation of a photo*FET* circuit. Briefly explain the principle of the device.

Section 20-7

20-24 Draw the circuit diagram of an optocoupler with a *BJT* output stage. Also, sketch a cross-section to show the construction of an optocoupler. Explain the device operation.

20-25 Discuss the most important parameters of optocouplers.

20-26 Draw a circuit diagram to show how an optocoupler can use a pulse signal from a low-voltage source to control a circuit with a high-voltage supply, or vice versa. Explain how the circuit operates.

20-27 Draw a circuit diagram to show how an optocoupler can be used to pass a linear signal between two circuits with different supply voltages. Explain how the circuit operates.

20-28 Sketch circuit diagrams for optocouplers with Darlington, *SCR*, and *TRIAC* outputs. Briefly discuss each optocoupler.

Chapter-20 Problems

Section 20-1

20-1 A total luminous flux striking a 4 cm^2 photocell at 7 m from a lamp is to be 80 mlm. Determine the required energy output from the lamp in watts.

20-2. Calculate the total luminous flux striking the surface of a solar cell located 4.5 m from a lamp with a 509 W output. The surface area of the solar cell is 5 cm^2.

20-3 Calculate the frequency of the light output from red, yellow, and green *LEDs* with the following peak wavelengths: 635 nm, 583 nm, 565 nm.

Section 20-2

20-4 An *LED* with I_F = 20 mA current and V_F = 1.4 V is to indicate when a 25 V supply is switched *on*. Sketch a suitable circuit and make all necessary calculations.

20-5   Two series-connected *LEDs* are to be controlled by a *2N3903* transistor with a 12 V supply and V_B = 5 V. The diode current is to be approximately 15 mA. Design a suitable circuit.

20-6   An op-amp Schmitt trigger circuit with V_{CC} = ±15 V is to have the state of its output indicated by *LEDs*. A green *LED* is to indicate *high*, and a red *LED* is to indicate *low*. Design the circuit for 10 mA diode currents. Include reverse-voltage protection diodes in series with each *LED*.

20-7   The *BJT-LED* circuit in Fig. 20-5(a) has: V_B = 5 V, V_{CC} = 20 V, and $h_{FE(min)}$ = 40 for Q_1. Design the circuit to give a 20 mA *LED* current with V_F = 2 V.

Section 20-3

20-8   Calculate the maximum power used by a three-and-a-half digit seven-segment *LED* display with a 5 V supply and 10 mA *LED* currents. Also, determine the power dissipated in each *LED* series resistor, if the *LEDs* have V_F = 1.4 V.

20-9   Determine the maximum power consumed by a three-and-a-half digit seven-segment *LCD* display with a 15 V peak square-wave supply and 1 μA *LCD* segment currents.

Section 20-4

20-10   A *pnp BJT* is to be biased *on* when the level of illumination on a photoconductive cell is greater than 100 lx, and *off* when the cell is dark. A ±5 V supply is to be used, and the *BJT* collector current is to be 10 mA when *on*. Design a suitable circuit to use a *BJT* with h_{FE} = 50 and a photoconductive cell with the characteristics in Fig. 20-11.

20-11   An inverting Schmitt trigger circuit has V_{CC} = ±12 V and *UTP/LTP* = ±5 V. The Schmitt output is to switch positively when the illumination level exceeds 30 lx on a photoconductive cell with the characteristics in Fig. 20-11. Design the circuit, and estimate the light level that causes the output to switch negatively.

20-12   A photoconductive cell with the characteristics in Fig. 20-11 is connected in series with an 820 Ω resistor and a 12 V supply. Determine the illumination level when the circuit current is approximately 6.5 mA, and when it is 1.1 mA.

20-13   A photoconductive cell circuit for controlling the current in an *LED* (as in Fig. 20-16) has: V_{CC} = 9 V, R_2 = 3.3 kΩ, R_3 = 270 Ω. The photoconductive cell has a dark resistance of 100 kΩ, and R_C = 3 kΩ at 10 lx. Determine the *LED* current at light levels of 3 lx and 30 lx.

20-14   The circuit in Fig. 20-14(a) has V_{CC} = ±5 V, R_1 = 12 kΩ, and a photoconductive cell with the specification in Fig. 20-12. Calculate the transistor maximum and minimum base voltage at 10 lx.

Section 20-5

20-15   A photodiode with the illumination characteristics in Fig. 20-23 is connected in series with a resistance and a 1 V reverse bias supply. The diode is to produce a +0.2 V output when illuminated with 20 mW/cm². Calculate the required series resistance value, and determine the device voltage and current at a 15 mW/cm² illumination level.

20-16   A photodiode with the characteristics in Fig. 20-23 is connected in series with a 1.2 V reverse bias supply and a 100 Ω resistance. Determine the resistance offered by the photodiode at illumination levels of 15 mW/cm² and 20 mW/cm².

20-17   Two photodiodes that each have a 100 Ω series resistor are connected to a 0.5 V reverse bias supply. A voltmeter is connected to measure the voltage difference between the diode cathodes. Assuming that each photodiode has the characteristics illustrated in Fig. 20-23, determine the voltmeter reading when the illumination level is 10 mW/cm² on one diode and 15 mW/cm² on the other.

20-18   Six photodiodes with the characteristics in Fig. 20-23 are connected in series. Determine the maximum output current and voltage at illumination levels of 15 mW/cm² and 12 mW/cm².

20-19   A rural telephone system uses 6 V rechargeable batteries which supply an average current of 50 mA. The batteries are recharged from an array of solar cells, each with the characteristics in Fig. 20-25. The average level of sunshine is 50 mW/cm² for 10 hours of each 24 hour period. Calculate the number of solar cells required.

20-20   The roof of a house has an area of 200 m² and is covered with solar cells which are each 2 cm x 2 cm. If the cells have the output characteristics shown in Fig. 20-25, determine how they should be connected to provide an output voltage of approximately 120 V. Take the average daytime level of illumination as 100 mW/cm². If the sun shines for an average of 12 hours in every 24 hours, calculate the energy in kilowatt-hours generated by the solar cells each day.

Section 20-6

20-21   The phototransistor circuit in Fig. 20-27(b) has a 25 V supply, and the device has the output characteristics in Fig. 20-29. Determine the collector resistance required to give V_{CE} = 10 V when the illumination level is 20 mW/cm².

20-22   Estimate V_{CE} for the circuit in Problem 20-21 at a 5 mW/cm² illumination level. If the phototransistor has the specification in Fig. 20-30, calculate the V_{CE} variation produced by a ±0.5 mW/cm² illumination change.

20-23 A phototransistor with the characteristics in Fig. 20-29 is connected in series with a 600 Ω relay coil. The coil current is to be 8 mA when the illumination level is 15 mW/cm². Determine the required supply voltage. Also, estimate the coil current at 10 mW/cm².

20-24 A phototransistor circuit for controlling an *SCR* (as in Fig. 20-32) is to be designed. The *SCR* has triggering conditions of $V_G = 0.7$ V and $I_G = 50$ μA, and the phototransistor has the specification in Fig. 20-30. Calculate suitable resistor values if the *SCR* is to switch *on* when the light level drops to 5 mW/cm². The supply voltage is $V_{CC} = 12$ V.

Section 20-7

20-25 An optocoupler with the specification in Fig. 20-36 is to control a 12 mA load that has a 6 V supply. The input is a 10 V square wave connected via a resistor (R_1). Determine a suitable resistance for R_1.

20-26 A 25 V, 0.5 W lamp is to be switched *on* and *off* by an *BJT* circuit with $V_{CC} = 9$ V and $I_C = 6$ mA. Design a suitable optocoupler circuit and estimate the required *CTR*.

20-27 An optocoupler circuit has its input connected via a 820 Ω resistor (R_1) to a pulse source. Its output transistor has a 5 V collector supply and a 470 Ω emitter resistor (R_2). The gate-cathode terminals of an *SCR* are connected across R_2. The *SCR* requires $V_G = 1.1$ V and $I_G = 500$ μA for triggering. If the optocoupler has *CTR* = 40%, calculate the required amplitude of the pulse input to trigger the *SCR*.

20-28 A optocoupler linear circuit, as in Fig. 20-38, has: $V_{CC1} = 15$ V, $V_{CC2} = 25$ V, $v_s = \pm 0.1$ V, $R_1 = 100$ Ω, $R_2 = 1.2$ kΩ, $R_3 = 1.5$ kΩ. Calculate the *dc* and *ac* output voltages, and the overall voltage gain. The optocoupler has *CTR* = 30%.

20-29 An optocoupler switching circuit, as in Fig. 20-37, has: $V_{CC1} = 18$ V, $V_{CC2} = 3$ V, $R_1 = 1.8$ kΩ, $R_2 = 820$ Ω, and $I_o = 1$ mA. Analyze the circuit to determine I_{F1}, I_{C2}, and *CTR*.

Practise Problem Answers

20-1.1	100 W	20-6.2	13 mW/cm²
20-1.2	5.13 x 10¹⁴ Hz	20-7.1	150 Ω
20-1.3	0.73 mlx, 0.29 μlm	20-7.2	9 mA, ±638 μA, (3.96 V to 13.9 V),
20-2.1	220 Ω, 8.4 V		(±0.27 V to ±0.98 V)
20-2.2	270 Ω, 390 Ω		
20-3.1	1.84 mW		
20-4.1	422 μA, 968 μA, 1.2 mA		
20-4.2	33 kΩ + 2.2 kΩ		
20-4.3	30 lx, 15 lx		
20-5.1	+0.19 V		
20-5.2	48		
20-6.1	2.5 V, 13.9 V		

Chapter *21*
Miscellaneous Devices

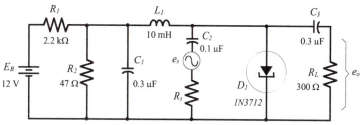

Chapter Contents

Objectives

You will be able to:

1 Explain the construction and operation of voltage variable capacitor diodes (VVCs).

2 Sketch typical VVC voltage/ capacitance characteristics, draw the equivalent circuit, and discuss typical VVC parameters.

3 Design and analyze resonance circuits using VVCs for frequency tuning.

4 Discuss the construction and operation of thermistors, sketch typical thermistor resistance/ temperature characteristics, and discuss typical thermistor parameters.

5 Calculate thermistor resistance at various temperature levels from the data sheet information.

6 Design and analyze circuit using thermistors for temperature level detecting.

7 Explain the construction and operation of tunnel diodes.

8 Sketch typical forward and reverse characteristics for a tunnel diode, explain their shape, and identify the important points and regions of the characteristics.

9 Draw tunnel diode piecewise linear characteristics from data sheet information.

10 Design and analyze tunnel diode parallel amplifier circuits.

Introduction

Three major devices are examined in this chapter: *voltage variable capacitance diodes (VVCs)*, *thermistors*, and *tunnel diodes*. VVCs are *pn*-junction devices designed to produce substantial junction capacitance change when the reverse bias voltage is adjusted. They can be applied to tune resonant circuits over a range of frequencies. The resistance of a thermistor changes significantly with change in temperature, so its major application is control of circuits that must respond to temperature change. The tunnel diode is a two-terminal negative-resistance device that can be employed as an oscillator, an amplifier, or a switch.

21-1   Voltage Variable Capacitors

VVC Operation

Voltage-variable capacitor diodes (VVCs) are also known as *varicaps*, *varactors*, and as *tuning diodes*. Basically, a VVC is a reverse biased diode, and its capacitance is the junction capacitance. Recall that the width of the depletion region at a *pn*-junction depends upon the reverse bias voltage, (Fig. 21-1). A large reverse bias produces a wide depletion region, and a small reverse bias gives a narrow depletion region. The depletion region acts as a dielectric between two conducting plates, so the junction behaves as a capacitor. The depletion layer capacitance (C_{pn}) is proportional to the junction area and inversely proportional to the width of the depletion region. Because the depletion region width is proportional to the reverse bias voltage, C_{pn} is inversely proportional to the reverse bias voltage. This is not a direct proportionality; instead C_{pn} is proportional to $1/V^n$, where V is the reverse bias voltage, and n depends upon doping density.

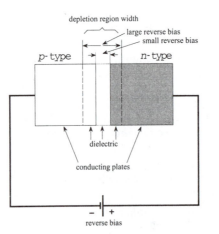

Figure 21-2 shows the doping profiles for two types of VVC classified as *abrupt junction* and *hyperabrupt junction* devices. In the abrupt junction VVC, the semiconductor material is uniformly doped, and it changes abruptly from *p*-type to *n*-type at the junction. The hyperabrupt junction device has the doping density increased close to the junction. This increasing density produces a narrower depletion region, and so it results in a larger junction capacitance. It also causes the depletion region width to be more sensitive to bias voltage variations, thus it produces the largest capacitance change for a given voltage variation. VVCs are packaged just like ordinary low-current diodes.

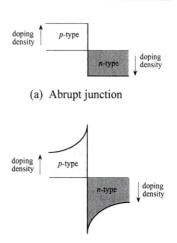

(a) Abrupt junction

(b) Hyperabrupt junction

Figure 21-2
Doping profiles for abrupt junct-ion and hyperabrupt junction VVCs.

Equivalent Circuit

The complete equivalent circuit for a VVC is shown in Fig. 21-3(a), and a simplified version is given in Fig. 21-3(b). In the complete circuit, the junction capacitance (C_J) is shunted by the junction reverse leakage resistance (R_J). The resistance of the semiconductor material is represented by R_S, the terminal inductance is L_S, and the capacitance of the terminals (or the device package) is C_C. Because L_S is normally very small and R_J is very large, the equival-ent circuit can be simplified [Fig 21-3(b)] to R_S in series with C_T, where C_T is the sum of the junction and terminal capacitances, $(C_T = C_J + C_C)$. The Q-factor for a VVC can be as high as 600 at a 50 MHz frequency. However, Q-factor varies with bias voltage and frequency, so it is used only as a figure of merit for comparing the performance of different VVCs.

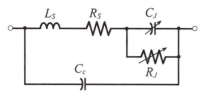

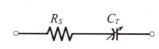

(a) Complete equivalent circuit   (b) Simplified equivalent circuit

Figure 21-3
The complete equivalent circuit of a VVC has five components. The simplified circuit is made up of the semiconductor resistance R_S and the total (junction + terminal) capacitance C_T.

Typical VVC Specification						
C_T $V_R = 1$ V, $f = 1$ MHz		C_1 / C_{10} $f = 1$ MHz	Q $V_R = 1$ V, $f = 1$ MHz	$V_{R(max)}$	$I_{R(max)}$	$I_{F(max)}$
min	max					
400 pF	600 pF	14	200	15 V	100 nA	200 mA

Figure 21-4
Partial specification for a voltage-variable capacitor diode (VVC).

Specification and Characteristics

A wide selection of nominal VVC capacitances is available, ranging approximately from 6 pF to 700 pF. The *capacitance tuning ratio* (*TR*) is the ratio of C_T at a small reverse voltage to C_T at a large reverse voltage. In the partial specification for a VVC shown in Fig. 21-4, the tuning ratio is listed as C_1/C_{10}. This is the ratio of the device capacitance at 1 V reverse bias to that at a 10 V reverse bias. Using the 400 pF minimum capacitance (C_T) listed for a 1 V bias, the capacitance is changed to 400 pF/14 when the bias is 10 V. The specification also lists the Q-factor, as well as maximum reverse voltage, reverse leakage current, and the maximum forward current that can be passed when the device is forward biased.

A typical graph of capacitance (C_T) versus reverse bias voltage (V_R) for a hyperabrupt junction *VVC* is reproduced in Fig. 21-5 together with the *VVC* circuit symbol. It is seen that C_T varies (approximately) from 500 pF to 25 pF when V_R is changed from 1 V to 10 V. It should be noted from the specification in Fig. 21-4 that the nominal capacitance has a large tolerance (400 pF to 600 pF), and this must be taken into account when using the C_T/V_R graphs.

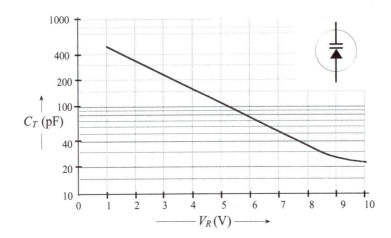

Figure 21-5
Capacitance/voltage characteristics for a hyperabrupt junction VVC.

Applications

The major application of *VVCs* is as tuning capacitors to adjust the frequency of resonance circuits. An example of this is the circuit shown in Fig. 21-6, which is an amplifier with a tuned circuit load. The amplifier produces an output at the resonance frequency of the tuned circuit. The *VVC* provides the capacitance (C_T) of the resonant circuit, and this can be altered by adjusting the diode (reverse) bias voltage (V_D). So, the resonance frequency of the circuit can be varied. C_1 is a coupling capacitor with a capacitance much larger than that of the *VVC*, and R_2 limits the *VVC* forward current in the event that it becomes forward biased.

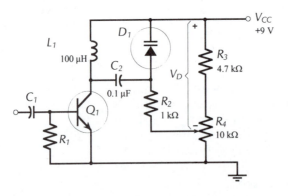

Figure 21-6
Amplifier stage with an LC tank circuit load. The resonance frequency of the LC circuit can be varied by adjusting the VVC reverse bias voltage.

Example 21-1

Determine the maximum and minimum resonance frequency for the circuit in Fig. 21-6. Assume that D_1 has the C_T/V_R characteristic in Fig. 21-5.

Solution

$$V_{D(min)} = \frac{V_{CC} \times R_3}{R_3 + R_4 + R_5} = \frac{9\text{ V} \times 4.7\text{ k}\Omega}{4.7\text{ k}\Omega + 5\text{ k}\Omega + 4.7\text{ k}\Omega}$$

$$= 2.9\text{ V}$$

$$V_{D(max)} = \frac{V_{CC}\,(R_3 + R_4)}{R_3 + R_4 + R_5} = \frac{9\text{ v} \times (4.7\text{ k}\Omega + 5\text{ k}\Omega)}{4.7\text{ k}\Omega + 5\text{ k}\Omega + 4.7\text{ k}\Omega}$$

$$\approx 6.1\text{ V}$$

From Fig. 21-5, at $V_D = 2.9$ V, $C_T \approx 250$ pF

$$f_{(min)} = \frac{1}{2\,\pi\,\sqrt{(L\,C_T)}} = \frac{1}{2\,\pi\,\sqrt{(100\,\mu H \times 250\text{ pF})}}$$

$$\approx 1\text{ MHz}$$

From Fig. 21-5, at $V_D = 6.1$ V, $C_T \approx 70$ pF

$$f_{(min)} = \frac{1}{2\,\pi\,\sqrt{(L\,C_T)}} = \frac{1}{2\,\pi\,\sqrt{(100\,\mu H \times 70\text{ pF})}}$$

$$\approx 1.9\text{ MHz}$$

Practise Problems

21-1.1 A tuned amplifier circuit as in Fig. 21-6 is to have a resonance frequency adjustable from 1.5 MHz to 2.5 MHz. A 12 V supply is used, the inductor (L_1) is 80 μH, and the VVC (D_1) has the characteristics in Fig. 21-5 and the specification in Fig 21-4. Determine suitable resistance values for R_3, R_4, and R_5.

21-2 Thermistors

Thermistor Operation

The word *thermistor* is a combination of thermal and resistor. A thermistor is a resistor with definite thermal characteristics. Most thermistors have a negative temperature coefficient (*NTC*), but positive temperature coefficient (*PTC*) devices are also available. Thermistors are widely applied for measurement and control of temperature, liquid level, gas flow, etc.

Silicon and germanium are not normally used for thermistor manufacture, because larger and more predictable temperature coefficients are available with metallic oxides. Various mixtures of manganese, nickel, cobalt, copper, iron, and uranium are pressed into desired shapes and sintered (or baked) at high temperature to form thermistors. Electrical connections are made either by including fine wires during the shaping process, or by silvering the surfaces after sintering, [see Fig. 21-7(a)]. Thermistors are made in the shape of beads, probes, discs, washers, etc. [Fig. 21-7(b)]. Beads may be glass-coated or enclosed in evacuated or gas-filled glass envelopes for protection against corrosion.

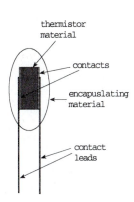

(a) Thermistor construction

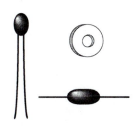

(b) Some thermistor shapes

Figure 21-7
Thermistor are resistors that are very sensitive to temperature.

Characteristics and Specifications

The typical thermistor resistance/temperature characteristic in Fig. 21-8 shows that the device resistance (R) decreases substantially when its temperature is raised. At 0°C, $R \approx 1.5$ kΩ; and at 60°C, $R \approx 70$ Ω. Current flow through a thermistor causes power dissipation that can raise its temperature and change its resistance. This could introduce errors in the thermistor application, so device currents are normally kept to a minimum.

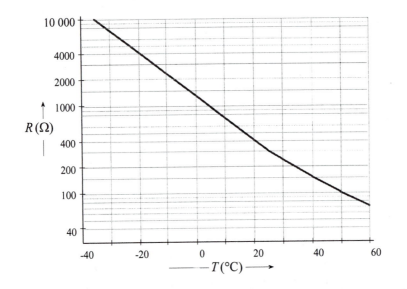

Figure 21-8
Typical resistance/temperature characteristics for a negative temperature coefficient (NTC) thermistor.

Figure 21-9 shows partial specifications for two thermistors with widely differing resistance values. Both devices have the resistance specified at 25°C as the *zero power resistance*. This, of course, means that there must be zero power dissipation in the thermistor to give this resistance value. The *dissipation constant* is the device power dissipation that can raise its temperature through 1°C. The dissipation constant in both cases is specified as *1 mW/°C in still air*, and *8 mW/°C in moving liquid*. Thus, a thermistor located in still air conditions could have its temperature increased by 1°C if it has 1 mW of power dissipation.

Typical Thermistor Specifications

Thermistor	Zero power resistance at 25°C	Resistance ratio 25°C/125°C	β (0 to 50°C)	Maximum working temperature	Dissipation constant
44002A	300 Ω	15.15	3118	100°C	1 mW/°C in still air
44008	30 kΩ	29.15	3810	150°C	8 mW/°C in moving liquid

Figure 21-9
Partial specifications for two thermistors, one with a 300 Ω 25°C resistance, and the other with a 30 kΩ 25°C resistance.

An indication of how much the thermistor resistance changes is given by the *resistance ratio at 25/125°C*. Clearly, with this ratio specified as 15.15, the resistance at 25°C is divided by 15.15 to determine the resistance at 125°C. Note that both devices have maximum working temperatures listed. The resistance change with temperature is also defined by the constant *Beta* (β), this time for the range 0°C to 50°C. This constant is used in an equation that relates resistance values at different temperatures:

$$\ln \frac{R_1}{R_2} = \beta \left(\frac{1}{T_1} - \frac{1}{T_2} \right) \qquad \text{(21-1)}$$

In Eq. 21-1, R_1 is the resistance at temperature T_1, and R_2 is the resistance T_2. It is important to note that T_1 and T_2 are *absolute* (or *Kelvin*) temperature values, (°C + 273) K.

Example 21-2

Calculate the resistance of the 300 Ω thermistor specified in Fig. 21-9 at temperatures of 20°C and 30°C.

Solution

For $T = 20°C$: $T_1 = 25°C + 273 = 298$ K

and $T_2 = 20°C + 273 = 293$ K

from Eq, 21-1, $R_2 = \dfrac{R_1}{\varepsilon^{\beta(1/T_1 - 1/T_2)}} = \dfrac{300\ \Omega}{\varepsilon^{3118(1/298 - 1/293)}}$

$= 358\ \Omega$

For $T = 30°C$: $T_1 = 25°C + 273 = 298$ K

and $T_2 = 30°C + 273 = 303$ K

from Eq, 21-1, $R_2 = \dfrac{R_1}{\varepsilon^{\beta(1/T_1 - 1/T_2)}} = \dfrac{300\ \Omega}{\varepsilon^{3118(1/298 - 1/303)}}$

$= 252\ \Omega$

Applications

Figure 21-10 shows a thermistor connected as a feedback resistor in an inverting amplifier circuit. (Note the device circuit symbol.) In this case, the thermistor is supplied with a constant current determined by R_1 and V_i. The output voltage is directly proportional to the thermistor resistance, and so V_o varies with temperature change.

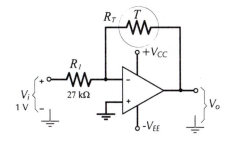

Figure 21-10
Use of an inverting amplifier to produce a constant current through a thermistor.

The circuit in Fig. 21-11 illustrates how a thermistor can be used for triggering a Schmitt circuit at a predetermined temperature. This could be air temperature, or the temperature of a liquid, or perhaps the temperature of some type of heating appliance. When the thermistor resistance (R_T) is increased to by the device temperature decrease, the Schmitt input voltage is raised to the upper trigger point, causing the output to switch negatively.

Example 21-3

Calculate V_i for the Schmitt circuit in Fig. 21-11 at 25°C and at 28°C if the thermistor is the 300 Ω device specified in Fig. 21-9.

Solution

at 25°C, $R_T = 300$ Ω,

$$V_i = \frac{V_{CC} \times R_T}{R_1 + R_T} = \frac{5\text{ V} \times 300\ \Omega}{47\text{ k}\Omega + 300\ \Omega}$$

$$= 31.7\text{ mV}$$

For $T = 28$°C: $T_1 = 25°C + 273 = 298$ K

and $T_2 = 28°C + 273 = 301$ K

from Eq. 21-1, $R_2 = \dfrac{R_1}{\varepsilon^{\beta(1/T_1 - 1/T_2)}} = \dfrac{300\ \Omega}{\varepsilon^{3118(1/298 - 1/301)}}$

$$= 270\ \Omega$$

$$V_i = \frac{V_{CC} \times R_T}{R_1 + R_T} = \frac{5\text{ V} \times 270\ \Omega}{47\text{ k}\Omega + 270\ \Omega}$$

$$= 28.6\text{ mV}$$

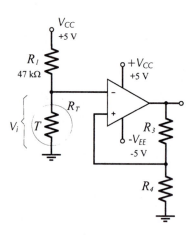

Figure 21-11
Schmitt trigger circuit using a thermistor input stage for temperature level detection.

Practise Problems

21-2.1 Calculate the output voltage from the circuit in Fig. 21-10 at 25°C and 28°C, if the 30 kΩ thermistor specified in Fig. 21-9 is used.

21-2.2 If the Schmitt circuit in Fig. 21-11 has *UTP* = 1 V, calculate a suitable resistance value for R_1 for the circuit to trigger at 18°C. The thermistor used is the 300 Ω device specified in Fig. 21-9.

21-3 Tunnel Diodes

Tunnel Diode Operation

A *tunnel diode* (sometimes called an *Esaki diode* after its inventor, Leo Esaki) is a two-terminal *negative resistance* device that can be employed as an amplifier, an oscillator, or a switch. Recall from Ch. 1 that the width of the depletion region at a *pn*-junction depends upon the doping density of the semi-conductor material. Lightly doped material has a wide depletion region, while heavily doped material has a narrow region. A tunnel diode uses very heavily doped semiconductor material, so the depletion region is extremely narrow. This is illustrated in Fig. 21-12 along with three frequently-used tunnel diode circuit symbols.

The depletion region is an insulator because it lacks charge carriers, and usually charge carriers can cross it only when the external bias is large enough to overcome the barrier potential. However, because the depletion region in a tunnel diode is so narrow, it does not constitute a large barrier to electron flow. Consequently, a small forward or reverse bias (not large enough to overcome the barrier potential) can give charge carriers sufficient energy to cross the depletion region. When this occurs, the charge carriers are said to be *tunnelling* through the barrier.

When a tunnel diode junction is reverse biased, (negative on the *p*-side, positive on the *n*-side), substantial current flow occurs due to the tunnelling effect, (electrons moving from the *p*-side to the *n*-side). Increasing levels of reverse bias voltage produce more tunnelling and a greater reverse current. So, as shown in Fig. 21-13, the reverse characteristic of a tunnel diode is linear, just like that of a resistor.

A forward biased tunnel diode initially behaves like a reverse biased device. Electron tunnelling occurs from the *n*-side to the *p*-side, and the forward current (I_F) continues to increase with increasing levels of forward voltage (E_F). Eventually, a peak level of tunnelling is reached, and then further increase in E_F actually causes I_F to decrease. (See the forward characteristic in Fig. 21-13.) The decrease in I_F with increasing E_F continues until the normal process of current flow across a forward biased junction begins to take over when the bias voltage becomes large enough to overcome the barrier potential. I_F now commences to increase with increasing levels of E_F, so that the final portion of the tunnel diode forward characteristics is similar to that for an ordinary *pn*-junction. The shape of the tunnel diode characteristics can be explained in terms of energy band diagrams for the semiconductor material.

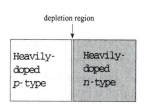

depletion region

Heavily-doped p-type	Heavily-doped n-type

(a) A heavily-doped *pn*-junction has a very narow depeltion region

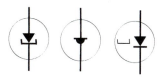

(b) Tunnel diode circuit symbols

Figure 21-12
A tunnel diode has a heavily-doped pn-junction which results in a very narrow depletion region.

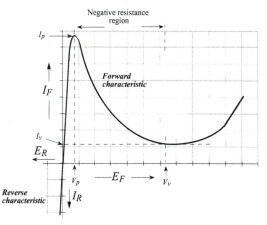

Figure 21-13
Tunnel diode characteristics. The current increases to a peak level (I_P) as the forward bias is increased, then falls off to a valley current (I_V) with increasing bias voltage.

Characteristics and Parameters

Consider the typical tunnel diode forward characteristics shown in Fig. 21-14. The *peak current* (I_p) and *valley current* (I_v) are easily identified on the forward characteristic as the maximum and minimum levels of I_F prior to the junction being completely forward biased. The peak voltage (V_p) is the level of forward bias voltage (E_F) corresponding to I_p, and the *valley voltage* (V_v) is the E_F level at I_v. V_F is the forward voltage drop when the device is completely forward biased. The dashed line at the bottom of the forward characteristic shows the characteristic for an ordinary forward biased diode. It is seen that this joins the tunnel diode characteristic as V_F is approached.

When a voltage is applied to a resistance, the current normally increases as the applied voltage is increased. Between I_p and I_v on the tunnel diode characteristic, I_F actually decreases as E_F is increased. So, this region of the characteristic is named the *negative resistance region*, and the *negative resistance* (R_D) of the tunnel diode is its most important property.

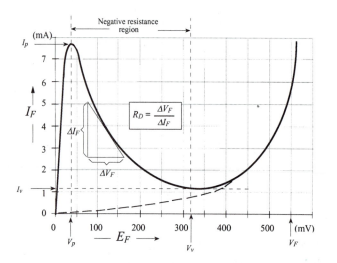

Figure 21-14
Typical forward characteristic for a tunnel diode. Note that the negative resistance region exists between forward bias voltages levels of approximately 50 mV and 325 mV.

The negative resistance value can be determined as the reciprocal of the slope of the characteristic in the negative resistance region. From Fig. 21-14, the negative resistance is $R_D = \Delta V_F / \Delta I_F$, and the *negative conductance* is $G_D = \Delta I_F / \Delta V_F$. If R_D is measured at different points on the negative resistance portion of the characteristic, slightly different values will be obtained at each point because the slope is not constant. Therefore, R_D is usually specified at the center of the negative resistance region. Figure 21-15 lists typical tunnel diode parameters.

Typical Tunnel Diode Parameters					
I_p (mA)	V_p (mV)	I_v (mA)	V_v (mV)	V_F (V)	R_D (Ω)
1 to 100	50 to 200	0.1 to 5	350 to 500	0.5 to 1	-10 to -200

Figure 21-15
Tunnel diode specification data showing the range of parameters.

It is shown in Ch. 2 that a straight-line approximation of diode characteristics can sometimes be conveniently employed. For a tunnel diode, the *piecewise linear characteristics* can usually be constructed from data provided by the device manufacturer.

Example 21-4

Construct the piecewise linear characteristics and determine R_D for a *1N3712* tunnel diode from the following data: $I_p = 1$ mA, $I_v = 0.12$ mA, $V_p = 65$ mV, $V_v = 350$ mV, and $V_F = 500$ mV at $I_F = I_p$.

Solution

Refer to Fig. 21-16.

Plot *point 1* at, $I_p = 1$ mA and $V_p = 65$ mV

Plot *point 2* at, $I_v = 0.12$ mA and $V_v = 350$ mV

Draw the first portion of the characteristic from the zero point to *point 1*. Draw the negative resistance portion between *points 1* and *2*.

Plot *point 3* at, $I_F = I_p$ and $V_F = 500$ mV

Draw the final portion of the characteristics at the same slope as the line between *point 0* and *point 1*.

Draw the horizontal part of the characteristic from *point 2* to the final portion.

$$R_D = \frac{\Delta E_F}{\Delta I_F} = \frac{350 \text{ mV} - 65 \text{ mV}}{-(1 \text{ mA} - 0.12 \text{ mA})}$$

$$= -324 \ \Omega$$

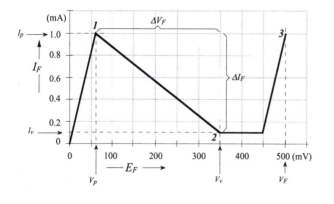

Figure 21-16
Piecewise linear characteristics for a tunnel diode, drawn from information provided on the device specification.

Parallel Amplifier

For operation as an amplifier, a tunnel diode must be biased to the center of its negative resistance region. Figure 21-17(a) shows the basic circuit of a tunnel diode *parallel amplifier*. Load resistor R_L is connected in parallel with diode D_1 and supplied with current from voltage source E_B and signal source e_s. Figure 21-17(b) uses the tunnel diode piecewise linear characteristics to show the *dc* conditions of the diode when the signal voltage is zero ($e_s = 0$), and when $e_s = \pm 100$ mV. Operation of the circuit is explained by the analysis in Ex. 21-5, which also demonstrates that a parallel amplifier has current gain but no voltage gain.

Example 21-5

Assuming that E_B and e_s have zero source resistance, calculate the current gain and voltage gain for the tunnel diode parallel amplifier in Fig. 21-17(a). The device piecewise linear characteristics are given in Fig. 21-17(b).

Solution

When $e_s = 0$;

$$E_{DQ} = E_B = 200 \text{ mV} \quad [\text{point } Q \text{ on Fig. 21-17(b)}]$$

at the Q point, $I_{DQ} = 2$ mA

also, $E_{RLQ} = E_B = 200$ mV

$$I_{RLQ} = \frac{E_{RL}}{R_L} = \frac{200 \text{ mV}}{80 \text{ }\Omega}$$

$$= 2.5 \text{ mA}$$

$$I_{BQ} = I_{RL} + I_D = 2.5 \text{ mA} + 2 \text{ mA}$$

$$= 4.5 \text{ mA}$$

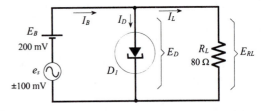

(a) Basic parallel amplifier circuit

When $e_s = +100$ mV;

$$E_B + e_s = 200 \text{ mV} + 100 \text{ mV} = 300 \text{ mV}$$

$$E_D = E_{RL(A)} = 300 \text{ mV} \quad [\text{point } A \text{ on Fig. 21-17(b)}]$$

and, $I_{D(A)} = 1$ mA

also, $I_{RL(A)} = \dfrac{E_{RL(A)}}{R_L} = \dfrac{300 \text{ mV}}{80 \text{ }\Omega}$

$$= 3.75 \text{ mA}$$

$$I_{B(A)} = I_{RL(A)} + I_{D(A)} = 3.75 \text{ mA} + 1 \text{ mA}$$

$$= 4.75 \text{ mA}$$

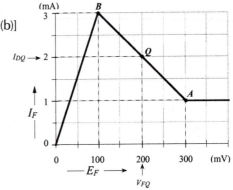

(b) Circuit current and voltage levels

When $e_s = -100$ mV;

$$E_B + e_s = 200 \text{ mV} - 100 \text{ mV} = 100 \text{ mV}$$

$$E_D = E_{RL(B)} = 100 \text{ mV} \quad [\text{point } B \text{ on Fig. 21-17(b)}]$$

and, $I_{D(B)} = 3$ mA

also, $I_{RL(B)} = \dfrac{E_{RL(B)}}{R_L} = \dfrac{100 \text{ mV}}{80 \text{ }\Omega}$

$$= 1.25 \text{ mA}$$

Figure 21-17
A basic tunnel diode parallel amplifier has a load resistance in parallel with the diode, and the (series-connected) bias and signal sources applied directly to the diode and load.

$$I_{B(B)} = I_{RL(B)} + I_{D(B)} = 1.25 \text{ mA} + 3 \text{ mA}$$
$$= 4.25 \text{ mA}$$

total load current change,

$$\Delta I_{RL} = I_{RL(A)} - I_{RL(B)} = 3.75 \text{ mA} - 1.25 \text{ mA}$$
$$= 2.5 \text{ mA}$$

total signal current change,

$$\Delta I_B = I_{B(A)} - I_{B(B)} = 4.75 \text{ mA} - 4.25 \text{ mA}$$
$$= 0.5 \text{ mA}$$

current gain, $A_i = \dfrac{\Delta I_{RL}}{\Delta I_B} = \dfrac{2.5 \text{ mA}}{0.5 \text{ mA}}$

$$= 5$$

voltage gain, $A_v = \dfrac{\Delta E_{RL}}{e_s} = \dfrac{\pm 100 \text{ mV}}{\pm 100 \text{ mV}}$

$$= 1$$

The current gain equation for a tunnel diode parallel amplifier can be shown to be,

$$A_i = \frac{R_D}{R_D - R_L} \tag{21-2}$$

Note that R_D is already taken as negative in Eq. 21-2, so that only the absolute value should be used in calculating A_i. For $R_D = 100$ Ω and $R_L = 80$ Ω, as in Ex. 21-5,

$$A_i = \frac{100}{100 \ \Omega - 80 \ \Omega} = 5$$

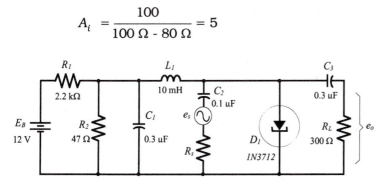

Figure 21-18
In a practical parallel amplifier circuit the load and signal source are capacitor-coupled to the tunnel diode.

From Eq. 21-2, it is seen that (when $R_L \ll R_D$, $A_i \approx 1$), (when $R_L \gg R_D$, $A_i < 1$), and (when $R_L = R_D$, $A_i = \infty$). A current gain of infinity means that the circuit is likely to oscillate. For maximum stable current gain, R_L should be selected just slightly less than R_D.

Figure 21-18 shows the circuit of a practical tunnel diode parallel amplifier. The signal voltage e_s and load resistor R_L are capacitor-coupled to the diode, while *dc* bias is provided by source voltage E_B and voltage divider R_1 and R_2. Inductor L_1 and capacitor C_1 isolate the bias supply from *ac* signals.

A tunnel diode *series amplifier* can be constructed. In this case the device is connected in series with the load, and voltage amplification is obtained instead of current amplification. Oscillators and switching circuits can also be constructed using tunnel diodes.

Practise Problems

21-3.1 Draw the *dc* and *ac* equivalent circuits for the tunnel diode parallel amplifier in Fig. 21-18. Also, draw the *dc* load line on the device piecewise linear characteristics in Fig. 21-16. Determine the bias conditions and calculate the current gain. The inductor has a 35 Ω winding resistance.

Chapter-21 Review Questions

Section 21-1

21-1 Using illustrations, explain the operation of a *VVC* diode. Sketch the doping profile at abrupt and hyperabrupt junctions, and explain the difference between the two.

21-2 Sketch the equivalent circuit for a *VVC*. Explain the origin of each component and show how the circuit may be simplified.

21-3 List the most important *VVC* parameters and state typical parameter values.

21-4 Sketch a circuit to show a typical *VVC* application. Briefly explain.

Section 21-2

21-5 Sketch typical resistance/temperature characteristics for a thermistor, and discuss the thermistor operation.

21-6 List the most important parameters for a thermistor, and state typical parameter values.

21-7 Sketch a circuit diagram to show how a thermistor can be used to control a Schmitt trigger circuit. Explain the circuit operation.

21-8 Draw a diagram to show how a thermistor might be used to compensate for V_{BE} variations (due to temperature change) in an emitter current biased *BJT* circuit.

Section 21-3

21-9 Discuss the difference between a tunnel diode and an ordinary *pn*-junction diode. Explain what is meant by *tunnelling*.

21-10 Sketch typical forward and reverse characteristics for a tunnel diode. Discuss the shape of the characteristics, and identify the regions and important points on the characteristics.

21-11 List the most important parameters for a tunnel diode, and state typical parameter values.

21-12 Sketch the basic circuit of a tunnel diode parallel amplifier, explain its operation, and write the equation for amplifier current gain.

21-13 Sketch a practical tunnel diode parallel amplifier circuit, and discuss the function of each component.

Chapter-21 Problems

Section 21-1

21-1   A tuner amplifier circuit similar to Fig. 21-6 has $V_{CC} = 15$ V and the following component values: $L_1 = 80$ µH, $R_3 = 1$ kΩ, $R_4 = 10$ kΩ, and $R_5 = 4.7$ kΩ. Assuming that D_1 has the C_T/V_R characteristic in Fig. 21-5, determine the maximum and minimum resonance frequency for the circuit.

21-2   If the *VVC* in the circuit in Problem 21-1 is replaced with the *VVC* specified in Fig. 21-4. Calculate the highest and lowest possible resonance frequency for the circuit.

21-3   A tuned amplifier circuit as in Fig. 21-6 is to have its resonance frequency adjustable from 0.8 MHz to 1.2 MHz. Determine suitable resistance values for R_3, R_4, and R_5 if $V_{CC} = 18$ V, $L_1 = 100$ µH, and the *VVC* characteristics are those in Fig. 21-5.

21-4   Determine the bias voltage for the *VVC* in Problem 21-3 to give a 1 MHz resonance frequency.

21-5   The *VVC* in the circuit in Problem 21-1 is replaced with another one that gives a resonance frequency ranging from 900 kHz to 3.5 MHz. Specify the new *VVC* in terms of its capacitance and tuning ratio from 1 V to 10 V.

Section 21-2

21-6   A thermistor with a 1 kΩ resistance at 25°C has β specified as 3395. Calculate the thermistor resistance at 5°C and at 35°C temperatures.

21-7   Calculate the temperature of the 30 kΩ thermistor specified in Fig. 21-9 when its resistance is measured as 24.5 kΩ.

21-8   A thermistor circuit as in Fig. 21-10 has $V_i = -1$ V and $R_1 = 22$ kΩ. Calculate the output voltage at 25°C and 35°C if the thermistor is the 30 kΩ device specified in Fig. 21-9.

21-9   The 300 Ω thermistor specified in Fig. 21-9 is connected in series with a 1.5 kΩ resistor (R_1) and a 12 V supply.

Determine the voltage drop across R_1 at temperatures of 22°C, 28°C, and 31°C.

Section 21-3

21-10 A tunnel diode is specified as having $I_p = 6$ mA, $V_p = 50$ mV, $I_v = 0.5$ mA, $V_v = 400$ mV, and $V_F = 550$ mV at $I_F = I_p$. Construct the piecewise linear characteristics for the device, and determine its negative resistance value.

21-11 Construct the piecewise linear characteristics for a 1N3715 from the following data: $I_p = 2.2$ mA, $I_v = 0.21$ mA, $V_p = 65$ mV, $V_v = 355$ mV, and $V_F = 510$ mV at $I_F = I_p$. Also, determine R_D for the device.

21-12 A parallel amplifier uses the tunnel diode specified in Problem 21-10 and a load resistance of 47 Ω. Calculate the circuit current gain.

21-13 A 1N3715 is to be connected as a parallel amplifier. Using the piecewise linear characteristics drawn for Problem 21-11, draw an appropriate dc load line and determine suitable values for R_L, E_B, and e_s. Also, calculate the current gain.

21-14 A practical tunnel diode parallel amplifier circuit as in Fig. 21-18 has the following components: $E_B = 5$ V, $R_1 = 220$ Ω, $R_2 = 12$ Ω, $C_1 = 0.5$ µF, $R_w = 0.5$ Ω, $L_1 = 20$ mH, $C_2 = 0.2$ µF, $C_3 = 0.5$ µF and $R_L = 75$ Ω. The tunnel diode used has $I_p = 5$ mA, $V_p = 50$ mV, $I_v = 1$ mA, and $V_v = 400$ mV. Construct the piecewise linear characteristics, draw the dc load line, and calculate the circuit current gain.

Practise Problem Answers

21-1.1   39 kΩ, 35 kΩ, 47 kΩ
21-2.1   -1.1 V, -0.98 V
21-2.2   1.5 kΩ
21-3.1   0.57 mA, 204 mV, 13.5

Appendices

Appendix 1-1
1N914 through 1N917 Switching Diodes

absolute maximum ratings at 25°C ambient temperature (unless otherwise noted)

		1N914	1N914A	1N914B	1N915	1N916	1N916A	1N916B	1N917	Unit
V_R	Reverse Voltage at — 65 to + 150°C	75	75	75	50	75	75	75	30	v
I_o	Average Rectified Fwd. Current	75	75	75	75	75	75	75	50	ma
I_o	Average Rectified Fwd. Current at + 150°C	10	10	10	10	10	10	10	10	ma
i_f	Recurrent Peak Fwd. Current	225	225	225	225	225	225	225	150	ma
$i_{f(surge)}$	Surge Current, 1 sec	500	500	500	500	500	500	500	300	ma
P	Power Dissipation	250	250	250	250	250	250	250	250	mw
T_A	Operating Temperature Range	— 65 to + 175								°C
T_{stg}	Storage Temperature Range	200								°C

maximum electrical characteristics at 25°C ambient temperature (unless otherwise noted)

		1N914	1N914A	1N914B	1N915	1N916	1N916A	1N916B	1N917	Unit
BV_R	Min Breakdown Voltage at 100 μa	100	100	100	65	100	100	100	40	v
I_R	Reverse Current at V_R	5	5	5	5	5	5	5		μa
I_R	Reverse Current at — 20 v	0.025	0.025	0.025		0.025	0.025	0.025		μa
I_R	Reverse Current at — 20 v at 100°C	3	3	3	5	3	3	3	25	μa
I_R	Reverse Current at — 20 v at + 150°C	50	50	50		50	50	50		μa
I_R	Reverse Current at — 10 v				0.025				0.05	μa
I_R	Reverse Current at — 10 v at 125°C									μa
I_F	Min Fwd Current at $V_F = 1$ v	10	20	100	50	10	20	30	10	ma
V_F	at 250 μa								0.64	v
V_F	at 1.5 ma								0.74	v
V_F	at 3.5 ma								0.83	v
V_F	at 5 ma			0.72	0.73			0.73		v
V_F	Min at 5 ma				0.60					v
C	Capacitance at $V_R = 0$	4	4	4	4	2	2	2	2.5	pf

operating characteristics at 25°C ambient temperature (unless otherwise noted)

		1N914	1N914A	1N914B	1N915	1N916	1N916A	1N916B	1N917	Unit
t_{rr}	Max Reverse Recovery Time	**4 °8	**4 °8	**4 °8	°10	**4 °8	**4 °8	**4 °8	°3	nsec nsec
V_f	Fwd Recovery Voltage (50 ma Peak Sq. wave, 0.1 μsec pulse width, 10 nsec rise time, 5 kc to 100 kc rep. rate)	2.5	2.5	2.5	2.5	2.5	2.5	2.5	2.5	v

* Trademark of Texas Instruments
° Lumatron (10 ma I_F 10 ma I_R, recover to 1 ma)
** EG&G (10 ma I_F, 6v V_R, recover to 1 ma)

* Courtesy of Texas Instruments, Incorporated.

Appendix 1-2
1N4001 through *1N4007* **Rectifier Diodes**

CASE 59

CATHODE

Low-current, passivated silicon rectifiers in subminiature void-free, flame-proof silicone polymer case. Designed to operate under military environmental conditions.

MAXIMUM RATINGS (At 60 cps Sinusoidal, Input, Resistive or Inductive Load)

Rating	Symbol	1N4001	1N4002	1N4003	1N4004	1N4005	1N4006	1N4007	Unit
Peak Repetitive Reverse Voltage DC Blocking Voltage	$V_{RM(rep)}$ V_R	50	100	200	400	600	800	1000	Volts
RMS Reverse Voltage	V_r	35	70	140	280	420	560	700	Volts
Average Half-Wave Rectified Forward Current (75°C Ambient) (100°C Ambient)	I_O	1000 750	1000 750	1000 750	1000 750	1000 750	1000 750	1000 750	mA mA
Peak Surge Current 25°C (1/2 Cycle Surge, 60 cps) Peak Repetitive Forward Current	$I_{FM(surge)}$ $I_{FM(rep)}$	30 10	30 10	30 10	30 10	30 10	30 10	30 10	Amps Amps
Operating and Storage Temperature Range	T_J, T_{stg}	-65 to + 175							°C

ELECTRICAL CHARACTERISTICS

Characteristic	Symbol	Rating	Unit
Maximum Forward Voltage Drop (1 Amp Continuous DC, 25°C)	V_F	1.1	Volts
Maximum Full-Cycle Average Forward Voltage Drop (Rated Current @ 25°C)	$V_{F(AV)}$	0.8	Volts
Maximum Reverse Current @ Rated DC Voltage (25°C) (100°C)	I_R	0.01 0.05	mA
Maximum Full-Cycle Average Reverse Current (Max Rated PIV and Current, as Half-Wave Rectifier, Resistive Load, 100°C)	$I_{R(AV)}$	0.03	mA

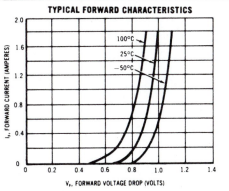

TYPICAL FORWARD CHARACTERISTICS

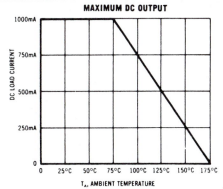

MAXIMUM DC OUTPUT

* Courtesy of Motorola, Inc.

Appendix 1-3
1N5391 through 1N5399 Low-power Rectifiers

*MAXIMUM RATINGS

Rating	Symbol	1N5391	1N5392	1N5393	1N5395	1N5397	1N5398	1N5399	Unit
Peak Repetitive Reverse Voltage Working Peak Reverse Voltage DC Blocking Voltage	V_{RRM} V_{RWM} V_R	50	100	200	400	600	800	1000	Volts
Nonrepetitive Peak Reverse Voltage (Halfwave, Single Phase, 60 Hz)	V_{RSM}	100	200	300	525	800	1000	1200	Volts
RMS Reverse Voltage	$V_{R(RMS)}$	35	70	140	280	420	560	700	Volts
Average Rectified Forward Current (Single Phase, Resistive Load, 60 Hz, T_L = 70°C, 1/2'' From Body)	I_O			←	1.5	→			Amp
Nonrepetitive Peak Surge Current (Surge Applied at Rated Load Conditions, See Figure 2)	I_{FSM}		←	50 (for 1 cycle)		→			Amp
Storage Temperature Range	T_{stg}		←	−65 to +175		→			°C
Operating Temperature Range	T_L		←	−65 to +170		→			°C
DC Blocking Voltage Temperature	T_L		←	150		→			°C

*ELECTRICAL CHARACTERISTICS

Characteristic and Conditions	Symbol	Typ	Max	Unit
Maximum Instantaneous Forward Voltage Drop (i_F = 4.7 Amp Peak, T_L = 170°C, 1/2 Inch Leads)	v_F	−	1.4	Volts
Maximum Reverse Current (Rated dc Voltage) (T_L = 150°C)	I_R	250	300	µA
Maximum Full-Cycle Average Reverse Current (1) (I_O = 1.5 Amp, T_L = 70°C, 1/2 Inch Leads)	$I_{R(AV)}$	−	300	µA

* Courtesy of Motorola, Inc.

Appendix 1-4
Zener Diodes: 1N745 thru 1N759, 1N957A
thru 1N986A, 1N4370 thru 1N4372

MAXIMUM RATINGS

Rating	Symbol	Value	Unit
DC Power Dissipation @ T_L ≤ 50°C, Lead Length = 3/8'' *JEDEC Registration *Derate above T_L = 50°C Motorola Device Ratings Derate above T_L = 50°C	P_D	 400 3.2 500 3.33	 mW mW/°C mW mW/°C
Operating and Storage Junction Temperature Range *JEDEC Registration Motorola Device Ratings	T_J, T_{stg}	 −65 to +175 −65 to +200	°C

*Indicates JEDEC Registered Data.

ELECTRICAL CHARACTERISTICS ($T_A = 25^\circ C$, $V_F = 1.5$ V max at 200 mA for all types)

Type Number (Note 1)	Nominal Zener Voltage V_Z @ I_{ZT} (Note 2) Volts	Test Current I_{ZT} mA	Maximum Zener Impedance Z_{ZT} @ I_{ZT} (Note 3) Ohms	*Maximum DC Zener Current I_{ZM} (Note 4) mA		Maximum Reverse Leakage Current $T_A = 25^\circ C$ I_R @ $V_R = 1$ V µA	$T_A = 150^\circ C$ I_R @ $V_R = 1$ V µA
1N4370	2.4	20	30	150	190	100	200
1N4371	2.7	20	30	135	165	75	150
1N4372	3.0	20	29	120	150	50	100
1N746	3.3	20	28	110	135	10	30
1N747	3.6	20	24	100	125	10	30
1N748	3.9	20	23	95	115	10	30
1N749	4.3	20	22	85	105	2	30
1N750	4.7	20	19	75	95	2	30
1N751	5.1	20	17	70	85	1	20
1N752	5.6	20	11	65	80	1	20
1N753	6.2	20	7	60	70	0.1	20
1N754	6.8	20	5	55	65	0.1	20
1N755	7.5	20	6	50	60	0.1	20
1N756	8.2	20	8	45	55	0.1	20
1N757	9.1	20	10	40	50	0.1	20
1N758	10	20	17	35	45	0.1	20
1N759	12	20	30	30	35	0.1	20

Type Number (Note 1)	Nominal Zener Voltage V_Z (Note 2) Volts	Test Current I_{ZT} mA	Maximum Zener Impedance (Note 3) Z_{ZT} @ I_{ZT} Ohms	Z_{ZK} @ I_{ZK} Ohms	I_{ZK} mA	*Maximum DC Zener Current I_{ZM} (Note 4) mA		Maximum Reverse Current I_R Maximum µA	Test Voltage Vdc 5% V_R	10%
1N957A	6.8	18.5	4.5	700	1.0	47	61	150	5.2	4.9
1N958A	7.5	16.5	5.5	700	0.5	42	55	75	5.7	5.4
1N959A	8.2	15	6.5	700	0.5	38	50	50	6.2	5.9
1N960A	9.1	14	7.5	700	0.5	35	45	25	6.9	6.6
1N961A	10	12.5	8.5	700	0.25	32	41	10	7.6	7.2
1N962A	11	11.5	9.5	700	0.25	28	37	5	8.4	8.0
1N963A	12	10.5	11.5	700	0.25	26	34	5	9.1	8.6
1N964A	13	9.5	13	700	0.25	24	32	5	9.9	9.4
1N965A	15	8.5	16	700	0.25	21	27	5	11.4	10.8
1N966A	16	7.8	17	700	0.25	19	37	5	12.2	11.5
1N967A	18	7.0	21	750	0.25	17	23	5	13.7	13.0
1N968A	20	6.2	25	750	0.25	15	20	5	15.2	14.4
1N969A	22	5.6	29	750	0.25	14	18	5	16.7	15.8
1N970A	24	5.2	33	750	0.25	13	17	5	18.2	17.3
1N971A	27	4.6	41	750	0.25	11	15	5	20.6	19.4
1N972A	30	4.2	49	1000	0.25	10	13	5	22.8	21.6
1N973A	33	3.8	58	1000	0.25	9.2	12	5	25.1	23.8
1N974A	36	3.4	70	1000	0.25	8.5	11	5	27.4	25.9
1N975A	39	3.2	80	1000	0.25	7.8	10	5	29.7	28.1
1N976A	43	3.0	93	1500	0.25	7.0	9.6	5	32.7	31.0
1N977A	47	2.7	105	1500	0.25	6.4	8.8	5	35.8	33.8
1N978A	51	2.5	125	1500	0.25	5.9	8.1	5	38.8	36.7
1N979A	56	2.2	150	2000	0.25	5.4	7.4	5	42.6	40.3
1N980A	62	2.0	185	2000	0.25	4.9	6.7	5	47.1	44.6
1N981A	68	1.8	230	2000	0.25	4.5	6.1	5	51.7	49.0
1N982A	75	1.7	270	2000	0.25	1.0	5.5	5	56.0	54.0
1N983A	82	1.5	330	3000	0.25	3.7	5.0	5	62.2	59.0
1N984A	91	1.4	400	3000	0.25	3.3	4.5	5	69.2	65.5
1N985A	100	1.3	500	3000	0.25	3.0	4.5	5	76	72
1N986A	110	1.1	750	4000	0.25	2.7	4.1	5	83.6	79.2

NOTE 1. TOLERANCE AND VOLTAGE DESIGNATION

Tolerance Designation

The type numbers shown have tolerance designations as follows:

1N4370 series: ±10%, suffix A for ±5% units,
                    C for ±2%, D for ±1%.
1N746 series: ±10%, suffix A for ±5% units,
                 C for ±2%, D for ±1%.
1N957 series: ±10%, suffix A for ±10% units,
                 C for ±2%, D for ±1%,
                 suffix B for ±5% units,
                 C for ±2%, D for ±1%.

* Courtesy of Motorola, Inc.

Appendix 1-5
2N3903 and 2N3904 NPN BJTs

MAXIMUM RATINGS

Rating	Symbol	Value	Unit
Collector-Emitter Voltage	V_{CEO}	40	Vdc
Collector-Base Voltge	V_{CBO}	60	Vdc
Emitter-Base Voltage	V_{EBO}	6.0	Vdc
Collector Current — Continuous	I_C	200	mAdc
Total Device Dissipation @ T_A = 25°C Derate above 25°C	P_D	625 5.0	mW mW/°C
*Total Device Dissipation @ T_C = 25°C Derate above 25°C	P_D	1.5 12	Watts mW/°C
Operating and Storage Junction Temperature Range	T_J, T_{stg}	− 55 to + 150	°C

*THERMAL CHARACTERISTICS

Characteristic	Symbol	Max	Unit
Thermal Resistance, Junction to Case	$R_{\theta JC}$	83.3	°C/W
Thermal Resistance, Junction to Ambient	$R_{\theta JA}$	200	°C/W

*Indicates Data in addition to JEDEC Requirements.

2N3903
2N3904

CASE 29-04, STYLE 1
TO-92 (TO-226AA)

3 Collector

2 Base

1 Emitter

1
2
3

GENERAL PURPOSE TRANSISTORS

NPN SILICON

ELECTRICAL CHARACTERISTICS (T_A = 25°C unless otherwise noted.)

Characteristic		Symbol	Min	Max	Unit
OFF CHARACTERISTICS					
Collector-Emitter Breakdown Voltage(1) (I_C = 1.0 mAdc, I_B = 0)		$V_{(BR)CEO}$	40	—	Vdc
Collector-Base Breakdown Voltage (I_C = 10 μAdc, I_E = 0)		$V_{(BR)CBO}$	60	—	Vdc
Emitter-Base Breakdown Voltage (I_E = 10 μAdc, I_C = 0)		$V_{(BR)EBO}$	6.0	—	Vdc
Base Cutoff Current (V_{CE} = 30 Vdc, V_{EB} = 3.0 Vdc)		I_{BL}	—	50	nAdc
Collector Cutoff Current (V_{CE} = 30 Vdc, V_{EB} = 3.0 Vdc)		I_{CEX}	—	50	nAdc
ON CHARACTERISTICS					
DC Current Gain(1)		h_{FE}			—
(I_C = 0.1 mAdc, V_{CE} = 1.0 Vdc)	2N3903 2N3904		20 40	— —	
(I_C = 1.0 mAdc, V_{CE} = 1.0 Vdc)	2N3903 2N3904		35 70	— —	
(I_C = 10 mAdc, V_{CE} = 1.0 Vdc)	2N3903 2N3904		50 100	150 300	
(I_C = 50 mAdc, V_{CE} = 1.0 Vdc)	2N3903 2N3904		30 60	— —	
(I_C = 100 mAdc, V_{CE} = 1.0 Vdc)	2N3903 2N3904		15 30	— —	
Collector-Emitter Saturation Voltage(1)		$V_{CE(sat)}$			Vdc
(I_C = 10 mAdc, I_B = 1.0 mAdc)			—	0.2	
(I_C = 50 mAdc, I_B = 5.0 mAdc)			—	0.3	
Base-Emitter Saturation Voltage(1)		$V_{BE(sat)}$			Vdc
(I_C = 10 mAdc, I_B = 1.0 mAdc)			0.65	0.85	
(I_C = 50 mAdc, I_B = 5.0 mAdc)			—	0.95	
SMALL-SIGNAL CHARACTERISTICS					
Current-Gain — Bandwidth Product (I_C = 10 mAdc, V_{CE} = 20 Vdc, f = 100 MHz)	2N3903 2N3904	f_T	250 300	— —	MHz

ELECTRICAL CHARACTERISTICS (continued) (T_A = 25°C unless otherwise noted.)

Characteristic		Symbol	Min	Max	Unit
Output Capacitance (V_{CB} = 5.0 Vdc, I_E = 0, f = 1.0 MHz)		C_{obo}	—	4.0	pF
Input Capacitance (V_{BE} = 0.5 Vdc, I_C = 0, f = 1.0 MHz)		C_{ibo}	—	8.0	pF
Input Impedance (I_C = 1.0 mAdc, V_{CE} = 10 Vdc, f = 1.0 kHz)	2N3903 2N3904	h_{ie}	1.0 1.0	8.0 10	k ohms
Voltage Feedback Ratio (I_C = 1.0 mAdc, V_{CE} = 10 Vdc, f = 1.0 kHz)	2N3903 2N3904	h_{re}	0.1 0.5	5.0 8.0	X 10^{-4}
Small-Signal Current Gain (I_C = 1.0 mAdc, V_{CE} = 10 Vdc, f = 1.0 kHz)	2N3903 2N3904	h_{fe}	50 100	200 400	—
Output Admittance (I_C = 1.0 mAdc, V_{CE} = 10 Vdc, f = 1.0 kHz)		h_{oe}	1.0	40	μmhos
Noise Figure (I_C = 100 μAdc, V_{CE} = 5.0 Vdc, R_S = 1.0 k ohms, f = 1.0 kHz)	2N3903 2N3904	NF	— —	6.0 5.0	dB

SWITCHING CHARACTERISTICS

			Symbol	Min	Max	Unit
Delay Time	(V_{CC} = 3.0 Vdc, V_{BE} = 0.5 Vdc, I_C = 10 mAdc, I_{B1} = 1.0 mAdc)		t_d	—	35	ns
Rise Time			t_r	—	35	ns
Storage Time	(V_{CC} = 3.0 Vdc, I_C = 10 mAdc, I_{B1} = I_{B2} = 1.0 mAdc)	2N3903 2N3904	t_s	— —	175 200	ns
Fall Time			t_f	—	50	ns

(1) Pulse Test: Pulse Width ≤ 300 μs, Duty Cycle ≤ 2.0%.

* Courtesy of Motorola, Inc.

Appendix 1-6
2N3905 and 2N3906 PNP BJTs

MAXIMUM RATINGS

Rating	Symbol	Value	Unit
Collector-Emitter Voltage	V_{CEO}	−40	Vdc
Collector-Base Voltage	V_{CBO}	−40	Vdc
Emitter-Base Voltage	V_{EBO}	−5.0	Vdc
Collector Current — Continuous	I_C	−200	mAdc
Total Device Dissipation @ T_A = 25°C Derate above 25°C	P_D	625 5.0	mW mW/°C
Total Power Dissipation @ T_A = 60°C	P_D	250	mW
Total Divice Dissipation @ T_C = 25°C Derate above 25°C	P_D	1.5 12	Watts mW/°C
Operating and Storage Junction Temperature Range	T_J, T_{stg}	−55 to +150	°C

*THERMAL CHARACTERISTICS

Characteristic	Symbol	Max	Unit
Thermal Resistance, Junction to Ambient	$R_{\theta JA}$	200	°C/W
Thermal Resistance, Junction to Case	$R_{\theta JC}$	83.3	°C/W

2N3905
2N3906★

CASE 29-04, STYLE 1
TO-92 (TO-226AA)

3 Collector

2 Base

1 Emitter

GENERAL PURPOSE TRANSISTORS

PNP SILICON

★This is a Motorola
designated preferred device.

ELECTRICAL CHARACTERISTICS (T_A = 25°C unless otherwise noted.)

Characteristic		Symbol	Min	Max	Unit
OFF CHARACTERISTICS					
Collector-Emitter Breakdown Voltage (1) (I_C = −1.0 mAdc, I_B = 0)		$V_{(BR)CEO}$	−40	—	Vdc
Collector-Base Breakdown Voltage (I_C = −10 μAdc, I_E = 0)		$V_{(BR)CBO}$	−40	—	Vdc
Emitter-Base Breakdown Voltage (I_E = −10 μAdc, I_C = 0)		$V_{(BR)EBO}$	−5.0	—	Vdc
Base Cutoff Current (V_{CE} = −30 Vdc, V_{EB} = −3.0 Vdc)		I_{BL}	—	−50	nAdc
Collector Cutoff Current (V_{CE} = −30 Vdc, V_{EB} = −3.0 Vdc)		I_{CEX}	—	−50	nAdc
ON CHARACTERISTICS(1)					
DC Current Gain (I_C = −0.1 mAdc, V_{CE} = −1.0 Vdc)	2N3905 2N3906	h_{FE}	30 60	— —	—
(I_C = −1.0 mAdc, V_{CE} = −1.0 Vdc)	2N3905 2N3906		40 80	— —	
(I_C = −10 mAdc, V_{CE} = −1.0 Vdc)	2N3905 2N3906		50 100	150 300	
(I_C = −50 mAdc, V_{CE} = −1.0 Vdc)	2N3905 2N3506		30 60	— —	
(I_C = −100 mAdc, V_{CE} = −1.0 Vdc)	2N3905 2N3906		15 30	— —	
Collector-Emitter Saturation Voltage (I_C = −10 mAdc, I_B = −1.0 mAdc) (I_C = −50 mAdc, I_B = −5.0 mAdc)		$V_{CE(sat)}$	— —	−0.25 −0.4	Vdc
Base-Emitter Saturation Voltage (I_C = −10 mAdc, I_B = −1.0 mAdc) (I_C = −50 mAdc, I_B = −5.0 mAdc)		$V_{BE(sat)}$	−0.65 —	−0.85 −0.95	Vdc
SMALL-SIGNAL CHARACTERISTICS					
Current-Gain — Bandwidth Product (I_C = −10 mAdc, V_{CE} = −20 Vdc, f = 100 MHz)	2N3905 2N3906	f_T	200 250	— —	MHz
Output Capacitance (V_{CB} = −5.0 Vdc, I_E = 0, f = 1.0 MHz)		C_{obo}	—	4.5	pF

ELECTRICAL CHARACTERISTICS (continued) (T_A = 25°C unless otherwise noted.)

Characteristic		Symbol	Min	Max	Unit
Input Capacitance (V_{EB} = −0.5 Vdc, I_C = 0, f = 1.0 MHz)		C_{ibo}	—	10.0	pF
Input Impedance (I_C = −1.0 mAdc, V_{CE} = −10 Vdc, f = 1.0 kHz)	2N3905 2N3906	h_{ie}	0.5 2.0	8.0 12	k ohms
Voltage Feedback Ratio (I_C = −1.0 mAdc, V_{CE} = −10 Vdc, f = 1.0 kHz	2N3905 2N3906	h_{re}	0.1 0.1	5.0 10	X 10^{-4}
Small-Signal Current Gain (I_C = −1.0 mAdc, V_{CE} = −10 Vdc, f = 1.0 kHz)	2N3905 2N3906	h_{fe}	50 100	200 400	—
Output Admittance (I_C = −1.0 mAdc, V_{CE} = −10 Vdc, f = 1.0 kHz)	2N3905 2N3906	h_{oe}	1.0 3.0	40 60	μmhos
Noise Figure (I_C = −100 μAdc, V_{CE} = −5.0 Vdc, R_S = 1.0 k ohm, f = 1.0 kHz)	2N3905 2N3906	NF	— —	5.0 4.0	dB

SWITCHING CHARACTERISTICS

			Symbol	Min	Max	Unit
Delay Time	(V_{CC} = −3.0 Vdc, V_{BE} = −0.5 Vdc		t_d	—	35	ns
Rise Time	I_C = −10 mAdc, I_{B1} = −1.0 mAdc)		t_r	—	35	ns
Storage Time		2N3905 2N3906	t_s	— —	200 225	ns
Fall Time	(V_{CC} = −3.0 Vdc, I_C = −10 mAdc, I_{B1} = I_{B2} = −1.0 mAdc)	2N3905 2N3906	t_f	— —	60 75	ns

(1) Pulse Width ≤ 300 μs, Duty Cycle ≤ 2.0%.

* Courtesy of Motorola, Inc.

Appendix 1-7
2N3251 PNP BJT

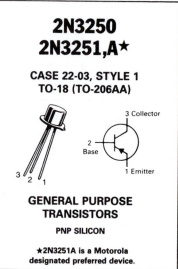

2N3250
2N3251,A★

CASE 22-03, STYLE 1
TO-18 (TO-206AA)

3 Collector
2 Base
1 Emitter

GENERAL PURPOSE TRANSISTORS

PNP SILICON

★2N3251A is a Motorola designated preferred device.

MAXIMUM RATINGS

Rating	Symbol	2N3250 2N3251	2N3251A	Unit
Collector-Emitter Voltage	V_{CEO}	−40	−60	Vdc
Collector-Base Voltage	V_{CBO}	−50	−60	Vdc
Emitter-Base Voltage	V_{EBO}	−5.0		Vdc
Collector Current	I_C	−200		mAdc
Total Device Dissipation @ T_A = 25°C Derate above 25°C	P_D	0.36 2.06		Watt mW/°C
Total Device Dissipation @ T_C = 25°C Derate above 25°C	P_D	1.2 6.9		Watts mW/°C
Operating and Storage Temperature Temperature Range	T_J, T_{stg}	−65 to +200		°C

THERMAL CHARACTERISTICS

Characteristic	Symbol	Max	Unit
Thermal Resistance, Junction to Ambient	$R_{\theta JA}$	486	°C/W
Thermal Resistance, Junction to Case	$R_{\theta JC}$	146	°C/W

ELECTRICAL CHARACTERISTICS (T_A = 25°C unless otherwise noted.)

Characteristic		Symbol	Min	Max	Unit
OFF CHARACTERISTICS					
Collector-Emitter Breakdown Voltage(1) (I_C = −10 mAdc)	2N3250, 2N3251 2N3251A	$V_{(BR)CEO}$	−40 −60	—	Vdc
Collector-Base Breakdown Voltage (I_C = −10 μAdc)	2N3250, 2N3251 2N3251A	$V_{(BR)CBO}$	−50 −60	—	Vdc
Emitter-Base Breakdown Voltage I_E = −10 μAdc)		$V_{(BR)EBO}$	−5.0	—	Vdc
Collector Cutoff Current (V_{CE} = −40 Vdc, V_{EB} = −3.0 Vdc)		I_{CEX}	—	−20	nA
Base Cutoff Current (V_{CE} = −40 Vdc, V_{EB} = −3.0 Vdc)		I_{BL}	—	−50	nAdc
ON CHARACTERISTICS					
DC Forward Current Transfer Ratio (I_C = −0.1 mAdc, V_{CE} = −10 Vdc)	2N3250 2N3251, 2N3251A	h_{FE}	40 80	— —	—
(I_C = −1.0 mAdc, V_{CE} = −1.0 Vdc)	2N3250 2N3251, 2N3251A		45 90	— —	
(I_C = −10 mAdc, V_{CE} = −1.0 Vdc)(1)	2N3250 2N3251, 2N3251A		50 100	150 300	
(I_C = −50 mAdc, V_{CE} = −1.0 Vdc)(1)	2N3250 2N3251, 2N3251A		15 30	— —	
Collector-Emitter Saturation Voltage (1) (I_C = −10 mAdc, I_B = −1.0 mAdc) (I_C = −50 mAdc, I_B = −5.0 mAdc		$V_{CE(sat)}$	— —	−0.25 −0.5	Vdc
Base-Emitter Saturation Voltage (1) (I_C = −10 mAdc, I_B = −1.0 mAdc) (I_C = −50 mAdc, I_B = −5.0 mAdc)		$V_{BE(sat)}$	−0.6 —	−0.9 −1.2	Vdc
SMALL-SIGNAL CHARACTERISTICS					
Current-Gain — Bandwidth Product (I_C = −10 mAdc, V_{CE} = −20 Vdc, f = 100 MHz)	2N3250 2N3251, 2N3251A	f_T	250 300	— —	MHz
Output Capacitance (V_{CB} = −10 Vdc, I_E = 0, f = 1.0 MHz)		C_{obo}	—	6.0	pF
Input Capacitance (V_{EB} = −1.0 Vdc, I_C = 0, f = 1.0 MHz)		C_{ibo}	—	8.0	pF

ELECTRICAL CHARACTERISTICS (continued) (T_A = 25°C unless otherwise noted.)

Characteristic		Symbol	Min	Max	Unit
Input Impedance (I_C = 1.0 mA, V_{CE} = 10 V, f = 1.0 kHz)	2N3250, 2N3250A 2N3251, 2N3251A	h_{ie}	1.0 2.0	6.0 12	kohms
Voltage Feedback Ratio (I_C = 1.0 mA, V_{CE} = 10 V, f = 1.0 kHz)	2N3250, 2N3250A 2N3251, 2N3251A	h_{re}	— —	10 20	X 10^{-4}
Small-Signal Current Gain (I_C = 1.0 mA, V_{CE} = 10 V, f = 1.0 kHz)	2N3250, 2N3250A 2N3251, 2N3251A	h_{fe}	50 100	200 400	—
Output Admittance (I_C = 1.0 mA, V_{CE} = 10 V, f = 1.0 kHz)	2N3250, 2N3250A 2N3251, 2N3251A	h_{oe}	4.0 10	40 60	μmhos
Collector Base Time Constant (I_C = 10 mA, V_{CE} = 20 V, f = 31.8 MHz)		$rb'C_C$	—	250	ps
Noise Figure (I_C = 100 μA, V_{CE} = 5.0 V, R_S = 1.0 k Ω, f = 100 Hz)		NF	—	6.0	dB

SWITCHING CHARACTERISTICS

Characteristic			Symbol	Max	Unit
Delay Time	(V_{CC} = 3.0 Vdc, V_{BE} = 0.5 Vdc		t_d	35	ns
Rise Time	I_C = 10 mAdc, I_{B1} = 1.0 mA)		t_r	35	ns
Storage Time	(I_C = 10 mAdc, I_{B1} = I_{B2} = 1.0 mAdc V_{CC} = 3.0 V)	2N3250, 2N3250A 2N3251, 2N3251A	t_s	175 200	ns
Fall Time			t_f	50	ns

(1) Pulse Test: PW = 300 μs, Duty Cycle = 2.0%.

* Courtesy of Motorola, Inc.

Appendix 1-8
2N3055 and *MJ2955* Complementary Power *BJTs*

ELECTRICAL CHARACTERISTICS (T_C = 25°C unless otherwise noted)

Characteristic	Symbol	Min	Max	Unit
*OFF CHARACTISTICS				
Collector-Emitter Sustaining Voltage (1) (I_C = 200 mAdc, I_B = 0)	$V_{CEO(sus)}$	60	—	Vdc
Collector-Emitter Sustaining Voltage (1) (I_C = 200 mAdc, R_{BE} = 100 Ohms)	$V_{CER(sus)}$	70	—	Vdc
Collector Cutoff Current (V_{CE} = 30 Vdc, I_B = 0)	I_{CEO}	—	0.7	mAdc
Collector Cutoff Current (V_{CE} = 100 Vdc, $V_{BE(off)}$ = 1.5 Vdc) (V_{CE} = 100 Vdc, $V_{BE(off)}$ = 1.5 Vdc, T_C = 150°C)	I_{CEX}	— —	1.0 5.0	mAdc
Emitter Cutoff Current (V_{BE} = 7.0 Vdc, I_C = 0)	I_{EBO}	—	5.0	mAdc
*ON CHARACTERISTICS (1)				
DC Current Gain (I_C = 4.0 Adc, V_{CE} = 4.0 Vdc) (I_C = 10 Adc, V_{CE} = 4.0 Vdc)	h_{FE}	20 5.0	70 —	—
Collector-Emitter Saturation Voltage (I_C = 4.0 Adc, I_B = 400 mAdc) (I_C = 10 Adc, I_B = 3.3 Adc)	$V_{CE(sat)}$	—	1.1 3.0	Vdc
Base-Emitter On Voltage (I_C = 4.0 Adc, V_{CE} = 4.0 Vdc)	$V_{BE(on)}$	—	1.5	Vdc
SECOND BREAKDOWN				
Second Breakdown Collector Current with Base Forward Biased (V_{CE} = 40 Vdc, t = 1.0 s; Nonrepetitive)	$I_{s/b}$	2.87	—	Adc
DYNAMIC CHARACTERISTICS				
Current Gain — Bandwidth Product (I_C = 0.5 Adc, V_{CE} = 10 Vdc, f = 1.0 MHz)	f_T	2.5	—	MHz
*Small-Signal Current Gain (I_C = 1.0 Adc, V_{CE} = 4.0 Vdc, f = 1.0 kHz)	h_{fe}	15	120	—
*Small-Signal Current Gain Cutoff Frequency (V_{CE} = 4.0 Vdc, I_C = 1.0 Adc, f = 1.0 kHz)	f_{hfe}	10	—	kHz

* Indicates Within JEDEC Registration. (2N3055)
(1) Pulse Test: Pulse Width ≤ 300 µs, Duty Cycle ≤ 2.0%.

MAXIMUM RATINGS

Rating	Symbol	Value	Unit
Collector-Emitter Voltage	V_{CEO}	60	Vdc
Collector-Emitter Voltage	V_{CER}	70	Vdc
Collector-Base Voltage	V_{CB}	100	Vdc
Emitter-Base Voltage	V_{EB}	7	Vdc
Collector Current — Continuous	I_C	15	Adc
Base Current	I_B	7	Adc
Total Power Dissipation @ T_C = 25°C Derate above 25°C	P_D	115 0.657	Watts W/°C
Operating and Storage Junction Temperature Range	T_J, T_{stg}	-65 to +200	°C

THERMAL CHARACTERISTICS

Characteristic	Symbol	Max	Unit
Thermal Resistance, Junction to Case	$R_{\theta JC}$	1.52	°C/W

* Courtesy of Motorola, Inc.

Appendix 1-9
2N6121 and 2N6124 Complementary Power *BJTs*

COMPLEMENTARY SILICON PLASTIC POWER TRANSISTORS

. . . designed for use in power amplifier and switching circuits, — packaged in the compact TO-220AB outline. TO-66 leadform also available.

4 AMPERE
POWER TRANSISTORS
COMPLEMENTARY SILICON
45-80 VOLTS
40 WATTS

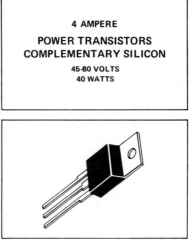

*MAXIMUM RATINGS

Rating	Symbol	2N6121 2N6124	2N6122 2N6125	2N6123	Unit
Collector-Emitter Voltage	V_{CEO}	45	60	80	Vdc
Collector-Base Voltage	V_{CB}	45	60	80	Vdc
Emitter-Base Voltage	V_{EB}	◄——— 5.0 ———►			Vdc
Collector Current	I_C	◄——— 4.0 ———►			Adc
Base Current	I_B	◄——— 1.0 ———►			Adc
Total Power Dissipation @ T_C = 25°C Derate above 25°C	P_D	◄——— 40 ———► ◄——— 320 ———►			Watts mW/°C
Operating and Storage Junction Temperature Range	T_J, T_{stg}	◄——— –65 to +150 ———►			°C

THERMAL CHARACTERISTICS

Characteristic	Symbol	Max	Unit
Thermal Resistance, Junction to Case	$R_{\theta JC}$	3.12	°C/W

*ELECTRICAL CHARACTERISTICS (T_C = 25°C unless otherwise noted)

Characteristic	Symbol	Min	Max	Unit
OFF CHARACTERISTICS				
Collector-Emitter Sustaining Voltage (1) (I_C = 0.1 Adc, I_B = 0) 2N6121, 2N6124 2N6122, 2N6125 2N6123	$V_{CEO(sus)}$	45 60 80	– – –	Vdc
Collector Cutoff Current (V_{CE} = 45 Vdc, I_B = 0) 2N6121, 2N6124 (V_{CE} = 60 Vdc, I_B = 0) 2N6122, 2N6125 (V_{CE} = 80 Vdc, I_B = 0) 2N6123	I_{CEO}	– – –	1.0 1.0 1.0	mAdc
Collector Cutoff Current (V_{CE} = 45 Vdc, $V_{EB(off)}$ = 1.5 Vdc) 2N6121, 2N6124 (V_{CE} = 60 Vdc, $V_{EB(off)}$ = 1.5 Vdc) 2N6122, 2N6125 (V_{CE} = 80 Vdc, $V_{EB(off)}$ = 1.5 Vdc) 2N6123 (V_{CE} = 45 Vdc, $V_{EB(off)}$ = 1.5 Vdc, 2N6121, 2N6124 T_C = 125°C) (V_{CE} = 60 Vdc, $V_{EB(off)}$ = 1.5 Vdc, 2N6122, 2N6125 T_C = 125°C) (V_{CE} = 80 Vdc, $V_{EB(off)}$ = 1.5 Vdc, 2N6123, 2N6126 T_C = 125°C)	I_{CEX}	– – – – – –	0.1 0.1 0.1 2.0 2.0 2.0	mAdc
Collector Cutoff Current (V_{CB} = 45 Vdc, I_E = 0) 2N6121, 2N6124 (V_{CB} = 60 Vdc, I_E = 0) 2N6122, 2N6125 (V_{CB} = 80 Vdc, I_E = 0) 2N6123	I_{CBO}	– – –	0.1 0.1 0.1	mAdc
Emitter Cutoff Current (V_{BE} = 5.0 Vdc, I_C = 0)	I_{EBO}	–	1.0	mAdc
ON CHARACTERISTICS				
DC Current Gain (1) (I_C = 1.5 Adc, V_{CE} = 2.0 Vdc) 2N6126, 2N6124 2N6122, 2N6125 2N6123 (I_C = 4.0 Adc, V_{CE} = 2.0 Vdc) 2N6121, 2N6124 2N6122, 2N6125 2N6123	h_{FE}	25 25 20 10 10 7.0	100 100 80 – – –	–
Collector-Emitter Saturation Voltage (1) (I_C = 1.5 Adc, I_B = 0.15 Adc) (I_C = 4.0 Adc, I_B = 1.0 Adc)	$V_{CE(sat)}$	– –	0.6 1.4	Vdc
Base-Emitter On Voltage (1) (I_C = 1.5 Adc, V_{CE} = 2.0 Vdc)	$V_{BE(on)}$	–	1.2	Vdc
DYNAMIC CHARACTERISTICS				
Small-Signal Current Gain (I_C = 0.1 Adc, V_{CE} = 2.0 Vdc, f = 1.0 kHz)	h_{fe}	25	–	–
Current-Gain-Bandwidth Product (I_C = 1.0 Adc, V_{CE} = 4.0 Vdc, f = 1.0 MHz)	f_T	2.5	–	MHz

(1)Pulse Test: Pulse Width ≤300 μs, Duty Cycle ≤2.0%.
*Indicates JEDEC Registered Data.

* Courtesy of Motorola, Inc.

Appendix 1-10
Transistor Heat Sinks

401

MODEL 401
NATURAL CONVECTION COOLER

Produces maximum semiconductor cooling per unit volume, often eliminating the need for costly redesigns required to accommodate bulkier, less efficient coolers. Unexcelled where space is critical. Additionally affords excellent forced convection characteristics, with a thermal resistance under 0.5°C/W at moderate air flows. Uses 3 Teflon mounting washers

403

MODEL 403
NATURAL CONVECTION COOLER

Affords heat dissipation characteristics which are difficult to equal with less "engineered" products. A thermal resistance of 1.8°C/W under medium power input permits adequate cooling in all but the most stringent applications. Under forced convection conditions, a thermal resistance of only 0.3°C/W is achieved with moderate air flows! Permits safe, reliable operation of the semiconductor under most power requirements where space is limited, and allowable temperature rise is restricted. Uses 4 Teflon mounting washers

413, 421

MODEL 421
NATURAL CONVECTION COOLER

Permits design engineers to free themselves from the question of how to maintain their power transistors at safe operating temperatures. Requiring no greater chassis base mounting area than that required by Model 403, this unit affords a still lower thermal resistance that is generally adequate for most high-power applications. Natural convection thermal resistance as low as 1.3°C/W; forced convection thermal resistance as low as 0.3°C/W. Uses 4 Teflon mounting washers

423

MODEL 423
NATURAL CONVECTION COOLER

Will provide semiconductor cooling never before possible under natural convection conditions, with thermal resistance as low as 0.8°C/W, and only 0.25°C/W with moderate air flows! Ideally suited to high-power transistors, rectifiers, and silicon-controlled rectifiers. Where installation area is not greatly restricted, and design conditions do not permit forced convection, Model 423 will achieve the lowest possible temperature rise per unit volume — and at the lowest possible cost.

MODEL 441
NATURAL CONVECTION HEAT SINKS

441

Model 441 is designed for mounting on a vertical surface (no special bracket required) to take better advantage of natural convection. Thermal resistance of .55°C/W at higher power levels is obtained. With moderate forced convection, thermal resistance of .18°C/W may be achieved. Electrical isolation may be achieved by the use of 4 teflon mounting washers (not included).

* Courtesy of EG & G, Wakefield Engineering.

Appendix 1-11
2N5457 FET

**2N5457
thru
2N5459★**

**CASE 29-04, STYLE 5
TO-92 (TO-226AA)**

1 Drain

3
Gate

2 Source

1
2
3

**JFETs
GENERAL PURPOSE**

N-CHANNEL — DEPLETION

★These are Motorola
designated preferred devices.

MAXIMUM RATINGS

Rating	Symbol	Value	Unit
Drain-Source Voltage	V_{DS}	25	Vdc
Drain-Gate Voltage	V_{DG}	25	Vdc
Reverse Gate-Source Voltage	V_{GSR}	−25	Vdc
Gate Current	I_G	10	mAdc
Total Device Dissipation @ T_A = 25°C Derate above 25°C	P_D	310 2.82	mW mW/°C
Junction Temperature Range	T_J	125	°C
Storage Channel Temperature Range	T_{stg}	−65 to +150	°C

ELECTRICAL CHARACTERISTICS (T_A = 25°C unless otherwise noted.)

Characteristic		Symbol	Min	Typ	Max	Unit
OFF CHARACTERISTICS						
Gate-Source Breakdown Voltage (I_G = −10 μAdc, V_{DS} = 0)		$V_{(BR)GSS}$	−25	—	—	Vdc
Gate Reverse Current (V_{GS} = −15 Vdc, V_{DS} = 0) (V_{GS} = −15 Vdc, V_{DS} = 0, T_A = 100°C)		I_{GSS}	 — —	 — —	 −1.0 −200	nAdc
Gate Source Cutoff Voltage (V_{DS} = 15 Vdc, I_D = 10 nAdc)	2N5457 2N5458 2N5459	$V_{GS(off)}$	−0.5 −1.0 −2.0	— — —	−6.0 −7.0 −8.0	Vdc
Gate Source Voltage (V_{DS} = 15 Vdc, I_D = 100 μAdc) (V_{DS} = 15 Vdc, I_D = 200 μAdc) (V_{DS} = 15 Vdc, I_D = 400 μAdc)	2N5457 2N5458 2N5459	V_{GS}	— — —	−2.5 −3.5 −4.5	— — —	Vdc
ON CHARACTERISTICS						
Zero-Gate-Voltage Drain Current* (V_{DS} = 15 Vdc, V_{GS} = 0)	2N5457 2N5458 2N5459	I_{DSS}	1.0 2.0 4.0	3.0 6.0 9.0	5.0 9.0 16	mAdc
SMALL-SIGNAL CHARACTERISTICS						
Forward Transfer Admittance Common Source* (V_{DS} = 15 Vdc, V_{GS} = 0, f = 1.0 kHz)	2N5457 2N5458 2N5459	$|y_{fs}|$	1000 1500 2000	— — —	5000 5500 6000	μmhos
Output Admittance Common Source* (V_{DS} = 15 Vdc, V_{GS} = 0, f = 1.0 kHz)		$|y_{os}|$	—	10	50	μmhos
Input Capacitance (V_{DS} = 15 Vdc, V_{GS} = 0, f = 1.0 MHz)		C_{iss}	—	4.5	7.0	pF
Reverse Transfer Capacitance (V_{DS} = 15 Vdc, V_{GS} = 0, f = 1.0 MHz)		C_{rss}	—	1.5	3.0	pF

*Pulse Test: Pulse Width ≤ 630 ms; Duty Cycle ≤ 10%.

* Courtesy of Motorola, Inc.

Appendix 1-12
2N4856 FET

N-CHANNEL JUNCTION FIELD-EFFECT TRANSISTORS

Depletion Mode symmetrical Field-Effect transistors designed for low-power switching and chopper applications.

- Low Drain-Source "ON" Resistance —
  $r_{ds(on)}$ = 25 Ohms (Max) @ f = 1.0 kHz — 2N4856,A, 2N4859,A

- Low Drain Cutoff Current —
  $I_{D(off)}$ = 250 pAdc (Max) @ V_{DS} = 15 Vdc

N-CHANNEL JUNCTION FIELD-EFFECT TRANSISTORS

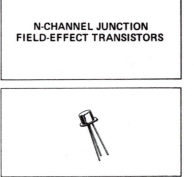

*MAXIMUM RATINGS

Rating	Symbol	2N4856,A 2N4857,A 2N4858,A	2N4859,A 2N4860,A 2N4861,A	Unit
Drain-Gate Voltage	V_{DG}	+40	+30	Vdc
Drain-Source Voltage	V_{DS}	+40	+30	Vdc
Reverse Gate-Source Voltage	V_{GSR}	–40	–30	Vdc
Forward Gate Current	I_{GF}	50		mAdc
Total Device Dissipation @ T_A = 25°C Derate above 25°C	P_D	360 2.4		mW mW/°C
Storage Temperature Range	T_{stg}	–65 to +200		°C

*Indicates JEDEC Registered Data.

*ELECTRICAL CHARACTERISTICS (T_A = 25°C unless otherwise noted)

Characteristic	Symbol	Min	Max	Unit
OFF CHARACTERISTICS				
Gate-Source Breakdown Voltage	$V_{(BR)GSS}$			Vdc
(I_G = 1.0 µAdc, V_{DS} = 0) 2N4856,A, 2N4857,A, 2N4858,A		−40	−	
2N4859,A, 2N4860,A, 2N4861,A		−30	−	
Gate-Source Cutoff Voltage	$V_{GS(off)}$			Vdc
(V_{DS} = 15 Vdc, I_D = 0.5 nAdc) 2N4856,A, 2N4859,A		−4.0	−10	
2N4857,A, 2N4860,A		−2.0	−6.0	
2N4858,A, 2N4861,A		−0.8	−4.0	
Gate Reverse Current	I_{GSS}			
(V_{GS} = −20 Vdc, V_{DS} = 0) 2N4856,A, 2N4857,A, 2N4858,A		−	0.25	nAdc
(V_{GS} = −15 Vdc, V_{DS} = 0) 2N4859,A, 2N4860,A, 2N4861,A		−	0.25	
(V_{GS} = −20 Vdc, V_{DS} = 0, T_A = 150°C) 2N4856,A, 2N4857,A, 2N4858,A		−	0.5	µAdc
(V_{GS} = −15 Vdc, V_{DS} = 0, T_A = 150°C) 2N4859,A, 2N4860,A, 2N4861,A		−	0.5	
Drain Cutoff Current	$I_{D(off)}$			
(V_{DS} = 15 Vdc, V_{GS} = −10 Vdc)		−	0.25	nAdc
(V_{DS} = 15 Vdc, V_{GS} = −10 Vdc, T_A = 150°C)		−	0.5	µAdc
ON CHARACTERISTICS				
Zero-Gate Voltage Drain Current (1) 2N4856,A, 2N4859,A	I_{DSS}	50	−	mAdc
(V_{DS} = 15 Vdc, V_{GS} = 0) 2N4857,A, 2N4860,A		20	100	
2N4858,A, 2N4861,A		8.0	80	
Drain-Source "ON" Voltage	$V_{DS(on)}$			Vdc
(I_D = 20 mAdc, V_{GS} = 0) 2N4856,A, 2N4859,A		−	0.75	
(I_D = 10 mAdc, V_{GS} = 0) 2N4857,A, 2N4860,A		−	0.5	
(I_D = 5.0 mAdc, V_{GS} = 0) 2N4858,A, 2N4861,A		−	0.5	
SMALL-SIGNAL CHARACTERISTICS				
Drain-Source "ON" Resistance	$r_{ds(on)}$			Ohms
(V_{GS} = 0, I_D = 0, f = 1.0 kHz) 2N4856,A, 2N4859,A		−	25	
2N4857,A, 2N4860,A		−	40	
2N4858,A, 2N4861,A		−	60	
Input Capacitance 2N4856 thru 2N4861	C_{iss}	−	18	pF
(V_{DS} = 0, V_{GS} = −10 Vdc, f = 1.0 MHz) 2N4856 A thru 2N4861 A			10	
Reverse Transfer Capacitance	C_{rss}			pF
(V_{DS} = 0, V_{GS} = −10 Vdc, f = 1.0 MHz) 2N4856 thru 2N4861		−	8.0	
2N4856 A, 2N4859 A		−	4.0	
2N4857 A, 2N4858 A, 2N4860 A, 2N4861 A		−	3.5	

SWITCHING CHARACTERISTICS (See Figure 1) (2)

			Symbol	Min	Max	Unit
Turn-On Delay Time	Conditions for 2N4856,A, 2N4859,A:	2N4856, 2N4859	$t_{d(on)}$	−	6.0	ns
		2N4856A, 2N4859A		−	5.0	
	(V_{DD} = 10 Vdc, $I_{D(on)}$ = 20 mAdc,	2N4857, 2N4860		−	6.0	
	$V_{GS(on)}$ = 0, $V_{GS(off)}$ = −10 Vdc)	2N4857A, 2N4860A		−	6.0	
		2N4858, 2N4861		−	10	
		2N4858A, 2N4861A		−	8.0	
Rise Time	Conditions for 2N4857,A, 2N4860,A:	2N4856,A, 2N4859,A	t_r	−	3.0	ns
		2N4857,A, 2N4860,A		−	4.0	
	(V_{DD} = 10 Vdc, $I_{D(on)}$ = 10 mAdc,	2N4858, 2N4861		−	10	
	$V_{GS(on)}$ = 0, $V_{GS(off)}$ = −6.0 Vdc)	2N4858A, 2N4861A		−	8.0	
Turn-Off Time		2N4856, 2N4859	t_{off}	−	25	ns
	Conditions for 2N4858,A, 2N4861,A:	2N4856A, 2N4859A		−	20	
		2N4857, 2N4860		−	50	
	(V_{DD} = 10 Vdc, $I_{D(on)}$ = 5.0 mAdc,	2N4857A, 2N4860A		−	40	
	$V_{GS(on)}$ = 0, $V_{GS(off)}$ = −4.0 Vdc)	2N4858, 2N4861		−	100	
		2N4858A; 2N4861A		−	80	

*Indicates JEDEC Registered Data.
(1) Pulse Test: Pulse Width = 100 ms, Duty Cycle ≤10%.
(2) The $I_{D(on)}$ values are nominal; exact values vary slightly with transistor parameters.

* Courtesy of Motorola, Inc.

Appendix 1-13
741 Operational Amplifier

GENERAL DESCRIPTION — The µA741 is a high performance monolithic Operational Amplifier constructed using the Fairchild Planar* epitaxial process. It is intended for a wide range of analog applications. High common mode voltage range and absence of "latch-up" tendencies make the µA741 ideal for use as a voltage follower. The high gain and wide range of operating voltage provides superior performance in integrator, summing amplifier, and general feedback applications.

- **NO FREQUENCY COMPENSATION REQUIRED**
- **SHORT CIRCUIT PROTECTION**
- **OFFSET VOLTAGE NULL CAPABILITY**
- **LARGE COMMON-MODE AND DIFFERENTIAL VOLTAGE RANGES**
- **LOW POWER CONSUMPTION**
- **NO LATCH UP**

ABSOLUTE MAXIMUM RATINGS

Supply Voltage	
Military (741)	±22 V
Commercial (741C)	±18 V
Internal Power Dissipation (Note 1)	
Metal Can	500 mW
DIP	670 mW
Mini DIP	310 mW
Flatpak	570 mW
Differential Input Voltage	±30 V
Input Voltage (Note 2)	±15 V
Storage Temperature Range	
Metal Can, DIP, and Flatpak	−65°C to +150°C
Mini DIP	−55°C to +125°C
Operating Temperature Range	
Military (741)	−55°C to +125°C
Commercial (741C)	0°C to +70°C
Lead Temperature (Soldering)	
Metal Can, DIP, and Flatpak (60 seconds)	300°C
Mini DIP (10 seconds)	260°C
Output Short Circuit Duration (Note 3)	Indefinite

CONNECTION DIAGRAMS

8-LEAD METAL CAN
(TOP VIEW)
PACKAGE OUTLINE 5B

Note: Pin 4 connected to case

ORDER INFORMATION

TYPE	PART NO.
741	741HM
741C	741HC

14-LEAD DIP
(TOP VIEW)
PACKAGE OUTLINE 6A

ORDER INFORMATION

TYPE	PART NO.
741	741DM
741C	741DC

10-LEAD FLATPAK
(TOP VIEW)
PACKAGE OUTLINE 3F

ORDER INFORMATION

TYPE	PART NO.
741	741FM

8-LEAD MINIDIP
(TOP VIEW)
PACKAGE OUTLINE 9T

ORDER INFORMATION

TYPE	PART NO.
741C	741TC

EQUIVALENT CIRCUIT

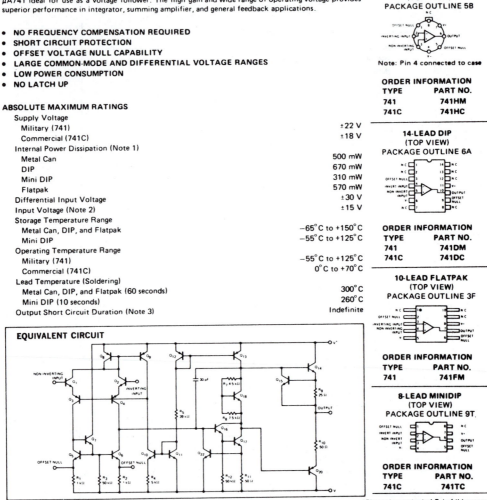

Notes on following pages.

*Planar is a patented Fairchild process.

ELECTRICAL CHARACTERISTICS ($V_S = \pm 15$ V, $T_A = 25°$C unless otherwise specified)

PARAMETERS (see definitions)		CONDITIONS	MIN.	TYP.	MAX.	UNITS
Input Offset Voltage		$R_S \leqslant 10$ kΩ		1.0	5.0	mV
Input Offset Current				20	200	nA
Input Bias Current				80	500	nA
Input Resistance			0.3	2.0		MΩ
Input Capacitance				1.4		pF
Offset Voltage Adjustment Range				±15		mV
Large Signal Voltage Gain		$R_L \geqslant 2$ kΩ, $V_{OUT} = \pm 10$ V	50,000	200,000		
Output Resistance				75		Ω
Output Short Circuit Current				25		mA
Supply Current				1.7	2.8	mA
Power Consumption				50	85	mW
Transient Response (Unity Gain)	Risetime	$V_{IN} = 20$ mV, $R_L = 2$ kΩ, $C_L \leqslant 100$ pF		0.3		μs
	Overshoot			5.0		%
Slew Rate		$R_L \geqslant 2$ kΩ		0.5		V/μs

The following specifications apply for $-55°$C $< T_A <$ $+125°$C:

			MIN.	TYP.	MAX.	UNITS
Input Offset Voltage		$R_S \leqslant 10$ kΩ		1.0	6.0	mV
Input Offset Current	$T_A = +125°$C			7.0	200	nA
	$T_A = -55°$C			85	500	nA
Input Bias Current	$T_A = +125°$C			0.03	0.5	μA
	$T_A = -55°$C			0.3	1.5	μA
Input Voltage Range			±12	±13		V
Common Mode Rejection Ratio		$R_S \leqslant 10$ kΩ	70	90		dB
Supply Voltage Rejection Ratio		$R_S \leqslant 10$ kΩ		30	150	μV/V
Large Signal Voltage Gain		$R_L \geqslant 2$ kΩ, $V_{OUT} = \pm 10$ V	25,000			
Output Voltage Swing	$R_L \geqslant 10$ kΩ		±12	±14		V
	$R_L \geqslant 2$ kΩ		±10	±13		V
Supply Current	$T_A = +125°$C			1.5	2.5	mA
	$T_A = -55°$C			2.0	3.3	mA
Power Consumption	$T_A = +125°$C			45	75	mW
	$T_A = -55°$C			60	100	mW

TYPICAL PERFORMANCE CURVES FOR 741

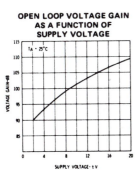

OPEN LOOP VOLTAGE GAIN
AS A FUNCTION OF
SUPPLY VOLTAGE

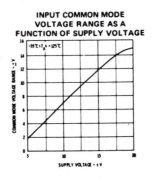

OUTPUT VOLTAGE SWING
AS A FUNCTION OF
SUPPLY VOLTAGE

INPUT COMMON MODE
VOLTAGE RANGE AS A
FUNCTION OF SUPPLY VOLTAGE

* Courtesy of Fairchild Camera and Instrument Corporation ©1982

Appendix 1-14
108 and 308 Operational Amplifiers

Absolute Maximum Ratings

	LM108/LM208	LM308
Supply Voltage	±20V	±18V
Power Dissipation (Note 1)	500 mW	500 mW
Differential Input Current (Note 2)	±10 mA	±10 mA
Input Voltage (Note 3)	±15V	±15V
Output Short-Circuit Duration	Indefinite	Indefinite
Operating Temperature Range (LM108)	−55°C to +125°C	0°C to +70°C
(LM208)	−25°C to +85°C	
Storage Temperature Range	−65°C to +150°C	−65°C to +150°C
Lead Temperature (Soldering, 10 seconds)	300°C	300°C

Open Loop Frequency Response

Electrical Characteristics (Note 4)

PARAMETER	CONDITIONS	LM108/LM208			LM308			UNITS
		MIN	TYP	MAX	MIN	TYP	MAX	
Input Offset Voltage	$T_A = 25°C$		0.7	2.0		2.0	7.5	mV
Input Offset Current	$T_A = 25°C$		0.05	0.2		0.2	1	nA
Input Bias Current	$T_A = 25°C$		0.8	2.0		1.5	7	nA
Input Resistance	$T_A = 25°C$	30	70		10	40		MΩ
Supply Current	$T_A = 25°C$		0.3	0.6		0.3	0.8	mA
Large Signal Voltage Gain	$T_A = 25°C$, $V_S = ±15V$, $V_{OUT} = ±10V$, $R_L \geq 10\,k\Omega$	50	300		25	300		V/mV
Input Offset Voltage				3.0			10	mV
Average Temperature Coefficient of Input Offset Voltage			3.0	15		6.0	30	µV/°C
Input Offset Current				0.4			1.5	nA
Average Temperature Coefficient of Input Offset Current			0.5	2.5		2.0	10	pA/°C
Input Bias Current				3.0			10	nA
Supply Current	$T_A = 125°C$		0.15	0.4				mA
Large Signal Voltage Gain	$V_S = ±15V$, $V_{OUT} = ±10V$ $R_L \geq 10\,k\Omega$	25			15			V/mV
Output Voltage Swing	$V_S = ±15V$, $R_L = 10\,k\Omega$	±13	±14		±13	±14		V
Input Voltage Range	$V_S = ±15V$	±13.5			±14			V
Common-Mode Rejection Ratio		85	100		80	100		dB
Supply Voltage Rejection Ratio		80	96		80	96		dB

Note 1: The maximum junction temperature of the LM108 is 150°C, for the LM208, 100°C and for the LM308, 85°C. For operating at elevated temperatures, devices in the TO-5 package must be derated based on a thermal resistance of 150°C/W, junction to ambient, or 45°C/W, junction to case. The thermal resistance of the dual-in-line package is 100°C/W, junction to ambient.

Note 2: The inputs are shunted with back-to-back diodes for overvoltage protection. Therefore, excessive current will flow if a differential input voltage in excess of 1V is applied between the inputs unless some limiting resistance is used.

Note 3: For supply voltages less than ±15V, the absolute maximum input voltage is equal to the supply voltage.

Note 4: These specifications apply for ±5V ≤ V_S ≤ ±20V and −55°C ≤ T_A ≤ 125°C, unless otherwise specified. With the LM208, however, all temperature specifications are limited to −25°C ≤ T_A ≤ 85°C, and for the LM308 they are limited to 0°C ≤ T_A ≤ 70°C.

* Reprinted with permission of National Semiconductor Corporation.
(Not authorized for use as critical components in a life support system.)

Appendix 1-15
353 BIFET **Operational Amplifier**

Absolute Maximum Ratings

Supply Voltage	±18V
Power Dissipation (Note 1)	500 mW
Operating Temperature Range	0°C to +70°C
T_j(MAX)	115°C
Differential Input Voltage	±30V
Input Voltage Range (Note 2)	±15V
Output Short Circuit Duration (Note 3)	Continuous
Storage Temperature Range	−65°C to +150°C
Lead Temperature (Soldering, 10 seconds)	300°C

Open Loop Frequency Response

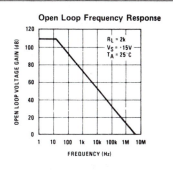

DC Electrical Characteristics (Note 4)

SYMBOL	PARAMETER	CONDITIONS	LF353A MIN	LF353A TYP	LF353A MAX	LF353B MIN	LF353B TYP	LF353B MAX	LF353 MIN	LF353 TYP	LF353 MAX	UNITS
V_{OS}	Input Offset Voltage	$R_S = 10\,k\Omega$, $T_A = 25°C$		1	2		3	5		5	10	mV
		Over Temperature			4			7			13	mV
$\Delta V_{OS}/\Delta T$	Average TC of Input Offset Voltage	$R_S = 10\,k\Omega$		10	20		10	30		10		µV/°C
I_{OS}	Input Offset Current	$T_j = 25°C$, (Notes 4, 5)		25	100		25	100		25	100	pA
		$T_j \le 70°C$			2			4			4	nA
I_B	Input Bias Current	$T_j = 25°C$, (Notes 4, 5)		50			50	200		50	200	pA
		$T_j \le 70°C$			4			8			8	nA
R_{IN}	Input Resistance	$T_j = 25°C$		10^{12}			10^{12}			10^{12}		Ω
A_{VOL}	Large Signal Voltage Gain	$V_S = ±15V$, $T_A = 25°C$, $V_O = ±10V$, $R_L = 2\,k\Omega$	50	100		50	100		25	100		V/mV
		Over Temperature	25			25			15			V/mV
V_O	Output Voltage Swing	$V_S = ±15V$, $R_L = 10\,k\Omega$	±12	±13.5		±12	±13.5		±12	±13.5		V
V_{CM}	Input Common-Mode Voltage Range	$V_S = ±15V$	±11	+15 −12		±11	+15 −12		±11	+15 −12		V
CMRR	Common-Mode Rejection Ratio	$R_S \le 10\,k\Omega$	80	100		80	100		70	100		dB
PSRR	Supply Voltage Rejection Ratio	(Note 6)	80	100		80	100		70	100		dB
I_S	Supply Current			3.6	5.6		3.6	5.6		3.6	6.5	mA

AC Electrical Characteristics (Note 4)

SYMBOL	PARAMETER	CONDITIONS	LF353A MIN	LF353A TYP	LF353A MAX	LF353B MIN	LF353B TYP	LF353B MAX	LF353 MIN	LF353 TYP	LF353 MAX	UNITS
	Amplifier to Amplifier Coupling	$T_A = 25°C$, $f = 1\,Hz - 20\,kHz$ (Input Referred)		−120			−120			−120		dB
SR	Slew Rate	$V_S = ±15V$, $T_A = 25°C$	10	13			13			13		V/µs
GBW	Gain-Bandwidth Product	$V_S = ±15V$, $T_A = 25°C$	3	4			4			4		MHz
e_n	Equivalent Input Noise Voltage	$T_A = 25°C$, $R_S = 100\Omega$, $f = 1000\,Hz$		16			16			16		nV/$\sqrt{Hz}$
i_n	Equivalent Input Noise Current	$T_j = 25°C$, $f = 1000\,Hz$		0.01			0.01			0.01		pA/$\sqrt{Hz}$

Note 1: For operating at elevated temperature, the device must be derated based on a thermal resistance of 160°C/W junction to ambient for the N package, and 150°C/W junction to ambient for the H package.

Note 2: Unless otherwise specified the absolute maximum negative input voltage is equal to the negative power supply voltage.

Note 3: The power dissipation limit, however, cannot be exceeded.

Note 4: These specifications apply for $V_S = ±15V$ and $0°C \le T_A \le +70°C$. V_{OS}, I_B and I_{OS} are measured at $V_{CM} = 0$.

Note 5: The input bias currents are junction leakage currents which approximately double for every 10°C increase in the junction temperature, T_j. Due to limited production test time, the input bias currents measured are correlated to junction temperature. In normal operation the junction temperature rises above the ambient temperature as a result of internal power dissipation, P_D. $T_j = T_A + \Theta_{jA} P_D$ where Θ_{jA} is the thermal resistance from junction to ambient. Use of a heat sink is recommended if input bias current is to be kept to a minimum.

Note 6: Supply voltage rejection ratio is measured for both supply magnitudes increasing or decreasing simultaneously in accordance with common practice.

Appendix 1-16
723 Voltage Regulator

	Min	Typ	Max
Continuous voltage $V_{CC}+$ to $V_{CC}-$			40 V
Input-output voltage differential			40 V
Input voltage range	9.5 V		37 V
Output voltage range	2 V		37 V
Output current			150 mA
Standby current		2.3 mA	3.5 mA
Reference voltage	6.95 V	7.15 V	7.35 V
Current from V_{ref}			15 mA
Maximum power dissipation at or below 25°C free air temperature:			
Metal can			1000 mW
DIP			12.5 W

Appendix 1-17
SCRs - Comparison of Typical Parameters

		Device		
Parameter		2N1597	2N5170	2N6400
V_{RRM}	Peak repetitive blocking voltage	200 V	400 V	50 V
$I_{T(rms)}$	RMS on-state current	1.6 A	20 A	16 A
V_{TM}	Peak on-state voltage	1.1 V	1.5 V	1.7 V
I_{GT}	Gate trigger current	2 mA	40 mA	5 mA
V_{GT}	Gate trigger voltage	0.7 V	1.5 V	0.7 V
I_H	Holding current	5 mA	50 mA	6 mA
t_{gt}	Turn-on time	0.8 μs	1 μs	1 μs
t_q	Turn-off time	10 μs	20 μs	15 μs

Appendix 1-18
UJTs - Comparison of Typical Parameters

		Device		
Parameter		2N4870	2N4949	2N2647
P_D	Maximum power dissipation	300 mW	360 mW	300 mW
	Derate linearly at	3 mW/°C	2.4 mW/°C	3 mW/°C
R_{BB}	Interbase resistance	6 kΩ	7 kΩ	7 kΩ
η	Intrinsic stand-off ratio	0.56 to 0.75	0.55 to 0.82	0.68 to 0.82
$V_{EB1(sat)}$	Emitter - B1 saturation voltage	2.5 V	2.5 V	3.5 V
I_p	Peak point current	1 μA	0.6 μA	1 μA
I_v	Valley point current current	5 mA	4 mA	10 mA
$I_{E(rms)}$	Maximum rms emitter current	50 mA	50 mA	50 mA
i_E	Peak emitter pulse current	1.5 A	1 A	2 A
V_{B1B2}	Maximum B1 - B2 voltage	35 V	35 V	35 V
V_{B2E}	Maximum emitter reverse voltage	30 V	30 V	30 V

Appendix 2-1
Typical Standard-Value Resistors

10% tolerance resistors

Ω	Ω	Ω	kΩ	kΩ	kΩ	MΩ	MΩ
-	10	100	1	10	100	1	10
-	12	120	1.2	12	120	1.2	12
-	15	150	1.5	15	150	1.5	15
-	18	180	1.8	18	180	1.8	18
-	22	220	2.2	22	220	2.2	22
2.7	27	270	2.7	27	270	2.7	-
3.3	33	330	3.3	33	330	3.3	-
3.9	39	390	3.9	39	390	3.9	-
4.7	47	470	4.7	47	470	4.7	-
5.6	56	560	5.6	56	560	5.6	-
6.8	68	680	6.8	68	680	6.8	-
8.2	82	820	8.2	82	820	8.2	-

5% tolerance resistors

Ω	Ω	Ω	kΩ	kΩ	kΩ	MΩ	MΩ
-	10	100	1	10	100	1	10
-	11	110	1.1	11	110	1.1	11
-	12	120	1.2	12	120	1.2	12
-	13	130	1.3	13	130	1.3	13
-	15	150	1.5	15	150	1.5	15
-	16	160	1.6	16	160	1.6	16
-	18	180	1.8	18	180	1.8	18
-	20	200	2	20	200	2	20
-	22	220	2.2	22	220	2.2	22
-	24	240	2.4	24	240	2.4	-
2.7	27	270	2.7	27	270	2.7	-
3	30	300	3	30	300	3	-
3.3	33	330	3.3	33	330	3.3	-
3.6	36	360	3.6	36	360	3.6	-
3.9	39	390	3.9	39	390	3.9	-
4.3	43	430	4.3	43	430	4.3	-
4.7	47	470	4.7	47	470	4.7	-
5.1	51	510	5.1	51	510	5.1	-
5.6	56	560	5.6	56	560	5.6	-
6.2	62	620	6.2	62	620	6.2	-
6.8	68	680	6.8	68	680	6.8	-
7.5	75	750	7.5	75	750	7.5	-
8.2	82	820	8.2	82	820	8.2	-
9.1	91	910	9.1	91	910	9.1	-

1% tolerance resistors (Basic values in Ω)

For values above 10 Ω multiply by 10, 100, etc.

1.00	6.04	11.0	20.0	36.5	66.5	121	221
1.10	6.19	11.3	20.5	37.4	68.1	124	226
1.21	6.34	11.5	21.0	38.3	69.8	127	232
1.30	6.49	11.8	21.5	39.2	71.5	130	237
1.50	6.65	12.1	22.1	40.2	73.2	133	243
1.62	6.81	12.4	22.6	41.2	75.0	137	249
1.82	6.98	12.7	23.2	42.2	76.8	140	255
2.00	7.15	13.0	23.7	43.2	78.7	143	261
2.21	7.32	13.3	24.3	44.2	80.6	147	267
2.43	7.50	13.7	24.9	45.3	82.5	150	274
2.67	7.68	14.0	25.5	46.4	84.5	154	280
3.01	7.87	14.3	26.1	47.5	86.6	158	287
3.32	8.06	14.7	26.7	48.7	88.7	162	294
3.57	8.25	15.0	27.4	49.9	90.9	165	301
3.92	8.45	15.4	28.0	51.1	93.1	169	309
4.32	8.66	15.8	28.7	52.3	95.3	174	316
4.75	8.87	16.2	29.4	53.6	97.6	178	324
4.99	9.09	16.5	30.1	54.9	100	182	332
5.11	9.31	16.9	30.9	56.2	102	187	340
5.23	9.53	17.4	31.6	57.6	105	191	348
5.36	9.76	17.8	32.4	59.0	107	196	356
5.49	10.0	18.2	33.2	60.4	110	200	365
5.62	10.2	18.7	34.0	61.9	113	205	374
5.76	10.5	19.1	34.8	63.4	115	210	383
5.90	10.7	19.6	35.7	64.9	118	215	392

Potentiometers

Ω	Ω	kΩ	kΩ	kΩ	MΩ
10	100	1	10	100	1
-	150	1.5	15	-	-
20	200	2	20	200	2
-	250	2.5	25	250	2.5
-	350	3.5	35	-	-
50	500	5	50	500	-
-	750	7.5	75	750	-

Appendix 2-2
Typical Standard-Value Capacitors

pF	pF	pF	pF	µF	µF	µF	µF	µF	µF	µF
5	50	500	5000	-	0.05	0.5	5	50	500	5000
-	51	510	5100	-	-	-	-	-	-	-
-	56	560	5600	-	0.056	0.56	5.6	56	-	5600
-	-	-	6000	-	0.06	-	6	-	-	6000
-	62	620	6200	-	-	-	-	-	-	-
-	68	680	6800	-	0.068	0.68	6.8	68	680	6800
-	75	750	7500	-	-	-	-	75	-	-
-	-	-	8000	-	-	-	8	80	-	-
-	82	820	8200	-	0.082	0.82	8.2	82	-	-
-	91	910	9100	-	-	-	-	-	-	-
10	100	1000	-	0.01	0.1	1	10	100	1000	10000
-	110	1100	-	-	-	-	-	-	-	-
12	120	1200	-	0.012	0.12	1.2	-	-	-	-
-	130	1300	-	-	-	-	-	-	-	-
15	150	1500	-	0.015	0.15	1.5	15	150	1500	15000
-	160	1600	-	-	-	-	-	-	-	-
18	180	1800	-	0.018	0.18	1.8	18	180	-	-
20	200	2000	-	0.02	0.2	2	20	200	2000	-
22	220	2200	-	-	0.22	2.2	22	220	2200	22000
24	240	2400	-	-	-	-	-	240	-	-
-	250	2500	-	-	0.25	-	25	250	2500	-
27	270	2700	-	0.027	0.27	2.7	27	270	-	-
30	300	3000	-	0.03	0.3	3	30	300	3000	-
33	330	3300	-	0.033	0.33	3.3	33	330	3300	-
36	360	3600	-	-	-	-	-	-	-	-
39	390	3900	-	0.039	0.39	3.9	39	-	-	-
-	-	4000	-	0.04	-	4	-	400	-	-
43	430	4300	-	-	-	-	-	-	-	-
47	470	4700	-	0.047	0.47	4.7	47	470	4700	-

Appendix 3
Answers for Odd-Numbered Problems

2-1	$4.25\,\Omega$, $1.5 \times 10^9\,\Omega$
2-3	$13\,\Omega$, $1.7\,\Omega$
2-5	6 V
2.7	Pt. A (0 mA, 0.7 V), Pt. B (300 mA, 0.85 V)
2-9	170 mA, 260 mA
2-11	9.1 V
2-13	625 mA
2-15	1.7 A, 1.25 A
2-17	26.7 mA, 28 mA
2-19	20 ns
2-21	80 ns
2-23	800 V, 1.5 A
2-25	2 V/cm, $200\,\Omega$, 0.5 V/cm
2-27	44 mA, 33.4 mA
3-1	34.7 V, 57.8 mA, 35.4 V
3-3	536 mW
3-5	34 V, 56.6 mA, 14 mW, 35.4 V
3-7	$3300\,\mu F$
3-9	50.4 V, 200 mA, 4.38 A, 1N4002, $0.84\,\Omega$
3-11	34.4 V, 300 mA, 3.68 A, 1N4001, $0.57\,\Omega$
3-13	$1500\,\mu F$
3-15	25.9 V, 200 mA, 2.2 A, 1N4001, $0.86\,\Omega$
3-17	21.2 V, 300 mA, 1.54 A, 1N4401, $0.7\,\Omega$
3-19	0.4 V, 0.5 V, 3.3%, 4.2%
3-21	10%, 2.08%
3-23	$680\,\Omega$, 0.27 W, ±3.7 mA
3-25	129 mV, 1.16 V, 5.85×10^{-2}
3-27	4.59 V, 5.61 V, $92.2\,\Omega$
3-29	$10\,k\Omega$, 12 V, 1 mA
3-31	23 mA, 7 V
3-33	$2.7\,k\Omega$, 6 V, 1.96 mA
3-35	±3.3 V, $6.8\,k\Omega$
3-37	1N756, $1\,k\Omega$
3-39	1N751, $820\,\Omega$
3-41	132 mV
3-43	$2.7\,\mu F$, $5.6\,k\Omega$
3-45	+3.6 V, -12.4 V, 4.4 V
3-47	1N759, $1.5\,\mu F$, $1,5\,k\Omega$
3-49	$1.5\,\mu F$, $3\,\mu F$
3-51	$1.2\,k\Omega$
4-1	1.62 mA, 1.67 mA, 32.3
4-3	1.98 mA, 2 mA
4-5	1 mA
4-7	120
4-9	68
4-13	4 mA
4-15	$40\,\mu A$
4-17	7.7 mA, 3.7 mA, 123
4-19	9.3 V

4-21	200
4-23	133, $100\,\mu A$
4-25	$7.46\,\Omega$, -0.993, 746 mS
4-27	$20\,\mu S$, 125
4.29	$333\,\Omega$
4-31	0.992, $2.48\,M\Omega$
4-33	0.1 V/cm, $20\,k\Omega$, 50 mV/cm
5-1	(0.97 mA, 7.7 V), (9.4 mA, 4.5 V)
5-3	±7.3 V, ±4.5 V
5-5	2.8 V, 2.35 V, 12.8 V
5-7	5.95 mA, 4.3 V
5-9	25 V, 4 V
5-11	5.45 mA, 5.1 V
5-13	33.2, 3.86 V
5-15	4.7 mA, 6.5 V
5-17	10.6 V
5-19	5.93 V
5-21	(1.5 V, 10.72 V), (4.46 V, 8.7 V), (6.98 V, 7.66 V)
5-23	$V_{CE} = 0.45$ V
5-25	$4.7\,k\Omega$
5-27	10.7 V, (≈ 0.2 V)
5-29	$270\,k\Omega$, $2.2\,k\Omega$
5-31	$270\,k\Omega$, $4.7\,k\Omega$
5-33	$220\,k\Omega$, $56\,k\Omega$, $15\,k\Omega$, $5.6\,k\Omega$
5-35	$100\,k\Omega$, $33\,k\Omega$, $2.7\,k\Omega$
5-37	$7.8\,\sqrt{}$, 1.4 mA
5-39	$680\,\Omega$, $1.2\,k\Omega$, $1.8\,k\Omega$
5-41	$2.7\,k\Omega$, $6.8\,k\Omega$
5-43	$10\,k\Omega$, $5.6\,k\Omega$, $1.2\,k\Omega$, $470\,\Omega$
5-45	$39.5\,\mu A$, $12.3\,\mu A$, $2.6\,\mu A$
5-47	$16\,\mu A$, $33\,\mu A$, $50\,\mu A$
5-49	9.9
5-51	$1.8\,k\Omega$, $10\,k\Omega$
5-53	$5.6\,k\Omega$, $22\,k\Omega$
6-1	7.92 V, 6 V
6-3	point A ($I_C = 0$, $V_{CE} = 18$ V);
	point B ($I_C = 1.6$ mA, $V_{CE} = 0$)
	point Q ($I_C = 0.9$ mA, $V_{CE} = 7.8$ V);
	point C ($I_C = 0$, $V_{CE} = 12.1$ V)
	$\Delta V_C = \pm4.3$ V
6-5	point A ($I_C = 0$, $V_{CE} = 15$ V);
	point B ($I_C = 1.39$ mA, $V_{CE} = 0$)
	point Q ($I_C = 1$ mA, $V_{CE} = 4$ V);
	point C ($I_C = 0$, $V_{CE} = 12.1$ V)
6-7	$1\,k\Omega$, $62.5\,\mu S$, 135
6-9	123
6-11	$50\,\mu S$
6-13	$15.4\,\Omega$
6-15	$893\,\Omega$, $5.54\,k\Omega$, -425

6-17   1.29 kΩ, 3.28 kΩ, -98.3
6-19   1.48 kΩ, 3.29 kΩ, -213
6-21   8.2 kΩ, -0.88
6-23   13.1 kΩ, 18.5 Ω
6-25   62.9 kΩ, 55 Ω
6-27   11.6 Ω, 5.5 kΩ, 425
6-29   31.4 Ω, 1.8 kΩ, 54.9
6-31   7.9 Ω, 3.28 kΩ, 380
6-33   (12.9 kΩ to 13.3 kΩ), (21 Ω to 42.3 Ω)

8-1   60 V, 4 A, 5 V, 25, 100
8-3   -1.25 dB
8-5   1.26 V
8-7   141 pF
8-9   100 kΩ
8-11   430 μs
8-13   2.79 pF
8-15   1038 pF, 31.94 nF, 938 pF
8-17   19.13 μs
8-19   417
8-21   31 V
8-23   78.4°C
8-25   8.1 kΩ
8-27   44.8°C, 58.5°C
8-29   4.6°C/W

9-1   5 mA, 5 V
9-3   9 mA, 8 V
9-5   2 mS
9-7   40 V, 360 mW, -6 V, 100 mA, 40 Ω
9-9   540 μS, 1150 μS
9-11   20 mS, 30 mS
9-13   20 mS, 30 mS
9-15   25 V, 310 mW, 6 V, 5 mA
9-17   ±0.5 V, ±0.1 V, 5, 1
9-19   3.3 kΩ
9-21   4 V, 2.5 V, 12.5 W
9-23   5.7 V, 1.5 V, 7.5 W
10-1   -1.5 V
10-3   4 kΩ
10-5   2.3 mA, 8.1 V
10-7   7.8 V, 19.7 V
10-9   11.9 V, 19.1 V
10-11   7.5 V, 18.4 V
10-13   17.3 V
10-15   6.5 V, 9.2 V
10-17   9.6 V, 14.2 V
10-19   -0.9 V, 1 MΩ, 3.3 kΩ
10-21   1 MΩ, 5.6 kΩ, 1 kΩ
10-23   3.9 MΩ, 1 MΩ, 3.3 kΩ, 3.3 kΩ
10-25   (6.8 MΩ + 1.8 MΩ), 1 MΩ, 3.9 kΩ, 3.9 kΩ
10-27   1 MΩ, 3.9 kΩ, 10 kΩ
10-29   1 MΩ, 4.7 kΩ, 10 kΩ
10-31   1.8 MΩ, 1 MΩ, 3.3 kΩ, 3.3 kΩ
10-33   6.8 MΩ, 1 MΩ, 3.9 kΩ, 3.9 kΩ
10-35   (82 kΩ + 1.5 kΩ), 27 kΩ, 3.9 kΩ, 2.7 kΩ

10-37   5.2 V
10-39   8.9 V
10-41   (5.6 MΩ + 2.2 MΩ), 1 MΩ, 1.8 kΩ
10-43   3.9 MΩ, 1 MΩ, 2.2 kΩ, 2.2 kΩ
10-45   1 MΩ, 1 Ω, 5.5 V
10-47   1 MΩ, 10 kΩ, -5 V
10-49   0.6 Ω, 5.7 V

11-1 $R_{L(dc)}$ = 5.4 kΩ, $R_{L(ac)}$ = 2.7 kΩ
11-3 $R_{L(dc)}$ = 3.6 kΩ, $R_{L(ac)}$ = 1.6 kΩ
11-5   880 kΩ, 3.7 kΩ, -12.5
11-7   470 kΩ, 4.5 kΩ, -19.3
11-9   9.1 kΩ, -36
11-11   470 kΩ, 4.5 kΩ, -1.8
11-13   629 kΩ, 161 Ω, 0.96
11-15   820 kΩ, 241 Ω, 0.96
11-17   685 kΩ, 318 Ω, 0.95
11-19   160 Ω, 3.7 kΩ, 21.2
11-21   266 Ω, 3.7 kΩ, 12.5
11-23   241 Ω, 6.4 kΩ, 14.1
11-25   1.3 V
11-27   693 mV, 217 mV

12-1   100 kΩ, 56 kΩ, 6.8 kΩ, 5.6 kΩ
12-3   620, 493 Ω, 5.3 kΩ
12-5   15 μF, 200 μF, 0.33 μF, 430 pF
12-7   56 kΩ, 39 kΩ, 3.9 kΩ, 3.9 kΩ
12-9   360, 958 Ω, 3.4 kΩ
12-11   0.039 μF, 15 μF, 0.39 μF, 510 pF
12-13   3.3 MΩ, 1 MΩ, 6.8 kΩ, 6.8 kΩ
12-15   24.3, 767 kΩ, 6.8 kΩ
12-17   0.027 μF, 15 μF, 0.39 μF

12-19   (56 kΩ + 1.5 kΩ), 47 kΩ, 3.9 kΩ, 4.7 kΩ,
     (56 kΩ + 1.5 kΩ), 47 kΩ, 3.9 kΩ, 4.7 kΩ
12-21   15 701, 491 Ω, 3.9 kΩ
12-23   5.6 μF, 100 μF, 3 μF, 100 μF, 0.2 μF
12-25   100 kΩ, 47 kΩ, 8.2 kΩ, 4.7 kΩ,6.8 kΩ, (8.2 kΩ +
     560 Ω)
12-27   29 343, 492 Ω, 6.8 kΩ
12-29   20 μF, 500 μF, 240 μF, 0.33 μF
12-31   33 kΩ, 56 kΩ, 3.3 kΩ, 4.7 kΩ, 6.8 kΩ,
     3.9 kΩ, 25 μF, 330 μF, 330 μF, 0.25 μF
12-33   -389
13-35   -256
12-37   36 464
12-39   8.2 kΩ, 220 kΩ, 220 kΩ, 3.3 kΩ, 22 μF, 1.2 μF,
     75 μF
12-41   3.3 kΩ, 120 kΩ, 220 kΩ, 3.3 kΩ, 15 μF, 1 μF, 15
     μF
12-43   6.8 MΩ, 820 kΩ, 3.9 kΩ, 3.9 kΩ,
     (120 kΩ + 15 kΩ), 39 kΩ, 12 kΩ, 4.7 kΩ
12-45   3818, 732 kΩ, 12 kΩ
12-47   0.068 μF, 15 μF, 150 μF, 0.18 μF
12-49   1 MΩ, 6.8 kΩ, 2.2 kΩ, 220 kΩ, 68 kΩ, 18 kΩ, 6.8
     kΩ, 0.02 μF, 25 μF, 22 μF, 330 μF, 0.1 μF

12-51 1 MΩ, 6.8 kΩ, 2.2 kΩ, 12 kΩ, 3.3 kΩ,
0.02 μF, 25 μF, 330 μF, 0.1 μF

12-53 0.82 μF, 0.15 μF

12-55 (33 kΩ + 3.3 kΩ), 22 kΩ, 6.8 kΩ, 2.2 kΩ, 6.8
kΩ,(33 kΩ + 3.3 kΩ), 22 kΩ, 2.2 μF, 0.3 μF

12-57 3.9 kΩ, 27 kΩ, 2.7 kΩ, 2.7 kΩ, 22 μF, 240 μF,
0.82 μF

12-59 (56 kΩ + 3.3 kΩ), 18 kΩ, 33 kΩ, 4.7 kΩ,
3.3 kΩ, 20 μF, 20 μF, 200 μF, 0.39 μF

12-61 15.1 kΩ

12-63 1.91 kΩ

13-1 124.9, 123.9

13-3 44.3 Ω, 629 Ω

13-5 470 Ω, (22 kΩ + 1 kΩ), 2.7 kΩ, 4.7 μF

13-7 470 Ω, (82 kΩ + 2.2 kΩ), 3.9 μF, 2.2 μF

13-9 15 μF, 220 Ω, (15 kΩ + 1.2 kΩ), 15 μF

13-11 27 kΩ, 220 Ω, 30 μF, 36 μF

13-13 18 kΩ, 220 Ω, 8 μF, 6 μF, 180 pF

13-15 2.7 kΩ, 150 Ω, 15 μF

13-17 4.7 kΩ, 10 kΩ, 5.6 kΩ, 10 kΩ, 820 kΩ, 4.7 kΩ,
4.7 kΩ, 12 kΩ

13-19 10 kΩ, 6.8 kΩ, 3.9 kΩ, 6.8 kΩ, (680 kΩ + 68 kΩ),
10 kΩ, 2.7 kΩ, 8.2 kΩ

13-21 100 kΩ, 47 kΩ, 10 kΩ, 180 Ω, 4.7 kΩ, 2.7 μF, 18
μF, 0.2 μF

13-23 (R_{E1} = 270 Ω), 2.2 μF, 10 μF, 0.39 μF

13-25 (R_4 = 82 Ω), (R_9 = 150 Ω), 5 μF, 75 μF, 2 μF, 33
μF, 0.47 μF

13-27 1535, 6.5 kΩ, 6.3 kΩ

13-29 8.2 kΩ, 270 Ω, 270 kΩ, 6.8 kΩ, 270 Ω, 4.7 kΩ, 1
μF, 10 μF, 0.27 μF

13-31 18 kΩ, 12 kΩ, 390 kΩ, 15 kΩ, 560 Ω, 4.7 kΩ,
0.68 μF, 1.5 μF

13-33 12 kΩ, 8.2 kΩ, 330 kΩ, 4.7 kΩ, 270 Ω, 2.7 kΩ, 8
μF, 2.7 μF

13-35 0.006%

13-37 22.5 MHz

14-1 2 MΩ, 80 nA, 75 Ω, 25 mA, 1.7 mA,
50 mW

14-3 3.6 mA, 100 dB, ±13.5 V, 1 mV

14-5 180 kΩ, 180 kΩ, 82 kΩ

14-7 1 MΩ, 1 MΩ, 470 kΩ, 8.75 V, 8.75 V, 8.985 V

14-9 10^{11}Ω, 0.5 x 10^{-3} Ω

14-11 68 kΩ, 0.2 μF, 1.5 μF

14-13 1 kΩ, 120 kΩ, 1 kΩ

14-15 15 kΩ, 1 MΩ, 15 kΩ

14-17 5.9 x 10^9 Ω, 1.5 kΩ

14-19 22 kΩ, 1 MΩ, 22 kΩ

14-21 820 Ω, (150 kΩ + 15 kΩ), 820 Ω, 820 Ω

14-23 (47 kΩ + 3.3 kΩ), 1 MΩ, 47 kΩ

14-25 680 Ω, 33 kΩ, 680 Ω, 6 μF, 3.3 μF

14-27 1.2 kΩ, (56 kΩ + 3.9 kΩ), 120 kΩ, 120 kΩ,
1.8 μF, 4.7 μF

14-29 60 μA, 80 μA, 140 μA, -1.4 V

14-31 -0.528 V

14-33 5.6 mV, -5.92 mV, -0.11 mV

14-35 50 μA, 50 μA, 50 μA, 22.2 μA, 22.2 μA, 88.9 μA,
88.9 μA, +1.8 V, +450 mV,
+150 mV, -1.2 V, 2.4 V, -0.6 V, 3 V

14-37 (270 kΩ + 22 kΩ), 39 kΩ

14-39 39 kΩ, 330 kΩ

14-41 4.7 kΩ, (39 kΩ + 6.8 kΩ),
(82 kΩ + 10 kΩ)

15-1 θ_L = -370° (unstable)

15-3 θ_m = 45° (stable)

15-5 44°, 1.1°

15-7 200 pF

15-9 800 kHz, 20 kHz

15-11 80 kHz, 3.4 MHz

15-13 20 kHz, 5.6 kHz

15-15 31.8 kHz, 159 kHz

15-17 39.8 kHz, 3.97 V (peak)

15-19 0.4 pF, 15.4 pF

15-21 8.2 pF

15-23 4.4 pF, 0.4 pF

15-25 265 pF

15-27 147 pF

15-29 265 pF

16-1 6.8 kΩ, 220 kΩ, 6.8 kΩ, 3300 pF, 6.8 kΩ

16-3 39 kΩ, 22 kΩ, 1.2 kΩ, 2.2 kΩ, 2.2 kΩ,
2.2 kΩ, 3300 pF, 25 μF

16-5 0.5 μF, 0.068 μF

16-7 8.2 kΩ, 82 kΩ, 8.2 kΩ, 3900 pF, 470 pF

16-9 0.06 μF, 5600 pF, 0.082 μF, 220 kΩ,
82 kΩ, 12 kΩ, 8.2 kΩ, 0.68 μF

16-11 17.8 kHz

16-13 1300 pF, 8.2 kΩ, 1300 pF, 8.2 kΩ, 18 kΩ, 8.2 kΩ,
45 kHz, 1.2 V/μs

16-15 10.5 kHz

16-17 220 Ω, 1.5 kΩ, 5.6 kΩ

16-19 0.015 μF, 3.3 kΩ, 0.015 μF, 3.3 kΩ, 3.3 kΩ, 5.6
kΩ, 1.5 kΩ

16-21 ±12 V, 1.35 kHz

16-23 (1.5 kΩ + 15 kΩ pot.), 1 μF, (15 kΩ + 1.5 kΩ),
470 Ω

16-25 10 kΩ, 0.27 μF, 1 kΩ, (10 kΩ + 1 kΩ)

16-27 ±1.95 V, 282 Hz

16-29 crystal in series with R_1 = 270 Ω, 0.24 mW

16-31 1.5 kΩ, 2.2 kΩ, 4.7 kΩ, 2.2 kΩ, 330 pF, 330 pF,
0.85 mW

17-1 220 Ω, *IN758*, 8.2 kΩ

17-3 36 mA, 19.6 mA, 0.77 mA, 21.4 mA

17-5 12.5 mV, 10 mV, 0.07%, 0.06%

17-7 680 Ω, 470 Ω, *IN758*, 5.6 kΩ, 10 kΩ

17-9 0.3%, 0.15%, 33 dB

17-11   1.2 kΩ, 8.2 kΩ, 2.5 kΩ

17-13   2.7 kΩ, 15 kΩ

17-15   3.3 kΩ, *1N758*, 470 Ω, 3.9 kΩ, 8.2 kΩ, 2.5 kΩ, 15 kΩ

17-17   470 Ω, (2 x *1N758*)

17-19   *1N751*, 3.9 kΩ, 1.8 kΩ, 23 V

17-21   2.27 Ω, 4.4 W

17-23   1 A, 313 mA

17-25   220 Ω, 2.2 kΩ, 4.7 kΩ, 2.5 kΩ, 12 kΩ, *1N754*

17-27 0.75×10^{-3}%, 0.4×10^{-3}%

17-29   25 V, 6.8 kΩ, 4.7 kΩ, 2.5 kΩ, 37.5 mA

17-31   -17 V, 1 kΩ, 3.9 kΩ, 400 mW

17-33   61.5%, 82.5%

17-35   330 μH, 15 μF

17-37   1.2 kΩ, (12 kΩ + 1.8 kΩ), 0.55 Ω, 620 pF

17-39   1.2 kΩ, (10 kΩ + 820 Ω), 0.39 Ω, 510 pF

18-1   Point *A*: (0 mA, 20 V), Point *Q*: (49.7 mA, 15 V), Point *B*: (49.7 mA and 20 V from *Q*)

18-3   Point *A*: (0 mA, 25 V), Point *Q*: (1.19 mA, 22.3 V), Point *B*: (1.19 mA and 9.6 V from *Q*)

18-5   Point *Q*: (0 mA, 40 V), Point *B*: (92.6 mA, 0 V)

18-7   16.1 V, 16.2 V

18-9   386 mW

18-11   90 V, 900 mA, 2.8 W

18-13   Point *Q*: (183 mA, 19 V), Point *A*: (0 mA, 24 V), Point *B*: (19.03 V and 183 mA from *Q*)

18-15   12 V, 24 V, 144 mA, 215 mW

18-17   23 kΩ, 1.8 kΩ, 1.5 kΩ, 25 Ω, 180 Ω

18-19   ±15 V, 30 V, 492 mA, 675 mW

18-21   ±21 V, Q_2 Q_3: (2.7 W, 42 V, 778 mA), Q_4 Q_5: (135 mW, 42 V, 39 mA)

18-23   ±15 V, 3.3 kΩ, 50 Ω, 680 Ω, 1.5 Ω, 1.5 Ω

18-25   5.6 kΩ, 1.2 kΩ, 1 kΩ

18-27   330 Ω, 400 μF, 53 V

18-29   1.8 kΩ, 250 Ω, 1.8 kΩ, 1 kΩ, 1 kΩ, 2.2 Ω, 2.2 Ω

18-31   18 μF, 68 μF, 680 pF

18-33   1.72 W, 30 V, 644 mA

18-35   0.85 mA, 1.01 A

18-37   ±23 V, 2.4 W

18-39   4.7 kΩ, 18 kΩ, 10 kΩ, 330 Ω, 4.7 kΩ, 100 kΩ, 1.2 MΩ, 1.8 MΩ, 100 kΩ, 4.7 μF, 6 μF, 0.2 μF, 0.2 μF

18-41   ±19 V

18-43   4.7 kΩ, 15 kΩ, 8.2 kΩ, 180 Ω, 4.7 kΩ, 100 kΩ, 680 kΩ, 1 MΩ, 100 kΩ, 8.2 μF, 22 μF, 0.47 μF, 0.47 μF

18-45   15 kΩ, (560 Ω + 82 Ω), 15 kΩ, 560 Ω, 560 Ω

18-47   ($R_8 = R_9 = R_{10} = R_{11} = 2.2$ kΩ), 5 μF, 17 V

18-49   ±28 V, 14.3 V, ±1.25 V

18-51   ±15 V

18-53   2.7 μF, 4 μF, 1600 pF, 240 pF

18-55   0.53 Ω

19-1

	2N1595	2N6167
V_{DRM}	50 V	100 V
$I_{T(rms)}$	1.6 A	13 A
V_{TM}	1.1 V	1.5 V
I_H	5 mA	3.5 mA
I_{GT}	2 mA	2.1 mA
V_{GT}	0.7 V	0.63 V

19-3   2N1597, 1.27 V

19-5   1.5 kΩ, 100 Ω, 12 Ω

19-7   560 Ω, 22 kΩ, 200 kΩ

19-9   12.8 V, 680 Ω

19-11   56 kΩ, 6 μF, 0.43 V

19-13

	2N6071	2N6343
V_{DRM}	200 V	400 V
$I_{T(rms)}$	4 A	8 A
V_G	1.4 V	0.9 V

19-15   49°

19-17   3.3 kΩ, 180 Ω, 10 kΩ, 3.3 kΩ, 0.82 μF, 0.15 μF

19-19   1 V, 15 V, 830 μA, 2 mA

19-21   165 mW, 34 V

19-23   435 Hz

19-25   6.8 kΩ, 10 kΩ, 100 Ω, 330 Ω, 0.56 μF

19-27   25 kΩ, (12 kΩ + 2.2 kΩ), 3.9 kΩ

19-29   15 kΩ, (8.2 kΩ + 820 Ω), 15.7 V, 0.9 V, 24 kΩ, 5.6 kΩ

20-1   184 W

20-3 4.72×10^{14} Hz, 5.15×10^{14} Hz, 5.31×10^{14} Hz

20-5 $R_F = 270$ Ω

20-7   10 kΩ, 820 Ω

20-9   690 μW

20-11   4.7 kΩ, 10 lx

20-13   2 mA, 26 mA

20-15   120 Ω, -0.12 V, -7 mA

20-17   0.38 V

20-19   102

20-21   1.2 kΩ

20-23   13 V, 5.5 mA

20-25   330 Ω

20-27   5.7 V

20-29   9 mA, 4.4 mA, 49%

21-1   3.6 MHz, 796 kHz

21-3   15 kΩ, 20 kΩ, 150 kΩ

21-5   390 pF, 15

21-7   30°C

21-9   9.8 V, 10.2 V, 10.3 V

21-11   146 Ω

21-13   120 Ω, 225 mV, ±145 mV, 5.6

Index